Advances in Intelligent Systems and Computing

Volume 1172

The series "Advances in Intelligent Systems and Computing" contains publications on theory, applications, and design methods of Intelligent Systems and Intelligent Computing. Virtually all disciplines such as engineering, natural sciences, computer and information science, ICT, economics, business, e-commerce, environment, healthcare, life science are covered. The list of topics spans all the areas of modern intelligent systems and computing such as: computational intelligence, soft computing including neural networks, fuzzy systems, evolutionary computing and the fusion of these paradigms, social intelligence, ambient intelligence, computational neuroscience, artificial life, virtual worlds and society, cognitive science and systems, Perception and Vision, DNA and immune based systems, self-organizing and adaptive systems, e-Learning and teaching, human-centered and human-centric computing, recommender systems, intelligent control, robotics and mechatronics including human-machine teaming, knowledge-based paradigms, learning paradigms, machine ethics, intelligent data analysis, knowledge management, intelligent agents, intelligent decision making and support, intelligent network security, trust management, interactive entertainment, Web intelligence and multimedia.

The publications within "Advances in Intelligent Systems and Computing" are primarily proceedings of important conferences, symposia and congresses. They cover significant recent developments in the field, both of a foundational and applicable character. An important characteristic feature of the series is the short publication time and world-wide distribution. This permits a rapid and broad dissemination of research results.

**** Indexing: The books of this series are submitted to ISI Proceedings, EI-Compendex, DBLP, SCOPUS, Google Scholar and Springerlink ****

More information about this series at http://www.springer.com/series/11156

Subhransu Sekhar Dash · Swagatam Das ·
Bijaya Ketan Panigrahi
Editors

Intelligent Computing and Applications

Proceedings of ICICA 2019

Springer

Editors
Subhransu Sekhar Dash
Department of Electrical Engineering
Government College of Engineering
Keonjhar, Odisha, India

Swagatam Das
Electronics and Communication
Sciences Unit
Indian Statistical Institute
Kolkata, West Bengal, India

Bijaya Ketan Panigrahi
Indian Institute of Technology Delhi
New Delhi, Delhi, India

ISSN 2194-5357 ISSN 2194-5365 (electronic)
Advances in Intelligent Systems and Computing
ISBN 978-981-15-5565-7 ISBN 978-981-15-5566-4 (eBook)
https://doi.org/10.1007/978-981-15-5566-4

This Springer imprint is published by the registered company Springer Nature Singapore Pte Ltd.
The registered company address is: 152 Beach Road, #21-01/04 Gateway East, Singapore 189721, Singapore

Preface

This AISC volume contains the papers presented at the Fifth International Conference on Intelligent Computing and Applications (ICICA 2019) held during December 6–8, 2019, at, SRM Institute of Science and Technology, Delhi-NCR Campus, Modinagar, Ghaziabad, India.

ICICA 2019 is the Fifth International Conference aiming at bringing together the researchers from academia and industry to report and review the latest progresses in the cutting-edge research on various research areas of electronic circuits, power systems, renewable energy applications, image processing, computer vision and pattern recognition, machine learning, data mining and computational life sciences, management of data including big data and analytics, distributed and mobile systems including grid and cloud infrastructure, information security and privacy, VLSI, antenna, intelligent manufacturing, signal processing, intelligent computing, soft computing, web security, privacy and e-commerce, e-governance, optimization, communications, smart wireless and sensor networks, networking and information security, mobile computing and applications, industrial automation and MES, cloud computing, green IT and finally to create awareness about these domains to a wider audience of practitioners.

ICICA 2019 received 290 paper submissions including two foreign countries across the globe. All the papers were peer-reviewed by the experts in the area in India and abroad, and comments have been sent to the authors of accepted papers. Finally 75 were accepted for oral presentation in the conference. This corresponds to an acceptance rate of 32% and is intended to maintain the high standards of the conference proceedings. The papers included in this AISC volume cover a wide range of topics in intelligent computing and algorithms and their real-time applications in problems from diverse domains of science and engineering

The conference was inaugurated by Prof. Nattachote Rugthaicharoencheep, Senior IEEE Member, Rajamangala University of Technology, Thailand, on December 6th, 2019. The conference featured distinguished keynote speakers as follows: Dr. Ikechi Ukaegbu, Nazarbayev University, Republic of Kazakhstan; Prof. D. P. Kothari, FIEEE, Nagpur, India; Shri. Aninda Bose, Senior Editor, Springer New Delhi, India; Dr. Bijaya Ketan Panigrahi, IIT Delhi, India;

Dr. Swagatam Das, ISI, Kolkata, India; Dr. Subhransu Sekhar Dash, Professor and Head, GCE Keonjhar, Odisha, India.

We take this opportunity to thank the authors of the submitted papers for their hard work, adherence to the deadlines, and patience with the review process. The quality of a referred volume depends mainly on the expertise and dedication of the reviewers. We are indebted to the Technical Committee members, who produced excellent reviews in short time frames. First, we are indebted to the Hon'ble Dr. T. R. Paarivendhar, Member of Parliament (Lok Sabha), Founder-Chancellor, SRM Institute of Science and Technology; Shri. Ravi Pachamoothoo, Chairman, SRM Institute of Science and Technology; Dr. P. Sathyanarayanan, President, SRM Institute of Science and Technology; Dr. R. Shivakumar, Vice President, SRM Institute of Science and Technology; Dr. Sandeep Sancheti, Vice Chancellor, SRM Institute of Science and Technology for supporting our cause and encouraging us to organize the conference there. In particular, we would like to express our heartfelt thanks for providing us with the necessary financial support and infrastructural assistance to hold the conference. Our sincere thanks to Dr. D. K. Sharma, Professor and Dean; Dr. S. Viswanathan, Deputy Registrar; Dr. Navin Ahalawat, Professor and Dean (Campus Life), SRM Institute of Science and Technology, Delhi-NCR Campus, Modinagar, Ghaziabad, for their continuous support and guidance. We specially thank Dr. R. P. Mahapatra, Professor and Head (CSE), Convener of ICICA 2019 and Dr. Dambarudhar Seth, Co-converner, SRM Institute of Science and Technology, Delhi-NCR Campus, of this conference for their excellent support and arrangements, without them it is beyond imagination to conduct this conference. We thank the International Advisory Committee members for providing valuable guidelines and inspiration to overcome various difficulties in the process of organizing this conference. We would also like to thank the participants of this conference. The members of faculty and students of SRM Institute of Science and Technology, Delhi-NCR Campus, Modinagar, Ghaziabad, deserve special thanks because without their involvement, we would not have been able to face the challenges of our responsibilities. Finally, we thank all the volunteers who made great efforts in meeting the deadlines and arranging every detail to make sure that the conference could run smoothly. We hope the readers of these proceedings find the papers inspiring and enjoyable.

Keonjhar, India
Kolkata, India
New Delhi, India
December 2019

Subhransu Sekhar Dash
Swagatam Das
Bijaya Ketan Panigrahi

Contents

About the Editors

Prof. Subhransu Sekhar Dash is currently a Professor at the Department of Electrical Engineering, Government College of Engineering, Keonjhar, Odisha, India. Holding a Ph.D. from the College of Engineering, Guindy, Anna University, Chennai, India, he has more than 22 years of research and teaching experience. His research interests include power electronics and drives, modeling of FACTS controllers, power quality, power system stability, and smart grids. He is a Visiting Professor at Francois Rabelais University, POLYTECH, France; the Chief Editor of the International Journal of Advanced Electrical and Computer Engineering; and an Associate Editor of IJRER. He has published more than 200 research articles in peer-reviewed international journals and conference proceedings.

Prof. Swagatam Das received his B.E. Tel. E., M.E. Tel. E (Control Engineering specialization) and Ph.D. degrees from Jadavpur University, India, in 2003, 2005, and 2009, respectively. He is currently serving as an Associate Professor at the Electronics and Communication Sciences Unit of the Indian Statistical Institute, Kolkata, India. His research interests include evolutionary computing, pattern recognition, multi-agent systems, and wireless communication. Dr. Das has published one research monograph, one edited volume, and more than 200 research articles in peer-reviewed journals and international conference proceedings. He is the Founding Co-Editor-in-Chief of Swarm and Evolutionary Computation, an international journal from Elsevier. He has also served as or is serving as an Associate Editor of IEEE Trans. on Systems, Man, and Cybernetics: Systems, IEEE Computational Intelligence Magazine, IEEE Access, Neurocomputing (Elsevier), Engineering Applications of Artificial Intelligence (Elsevier), and Information Sciences (Elsevier). He is also an editorial board member for many journals. He has been associated with the international program committees and organizing committees of several regular international conferences including IEEE CEC, IEEE SSCI, SEAL, GECCO, and SEMCCO. He is the recipient of the 2012 Young Engineer Award from the Indian National Academy of Engineering (INAE).

Prof. Bijaya Ketan Panigrahi is a Professor at the Electrical Engineering Department, IIT Delhi, India. Prior to joining IIT Delhi in 2005, he has served as a faculty in Electrical Engineering Department, UCE Burla, Odisha, India, from 1992 to 2005. Dr. Panigrahi is a senior member of IEEE and Fellow of INAE, India. His research interest includes the application of soft computing and evolutionary computing techniques to power system planning, operation, and control. He has also worked in the fields of bio-medical signal processing and image processing. He has served as the editorial board member, associate editor, and special issue guest editor of different international journals. He is also associated with various international conferences in various capacities. Dr. Panigrahi has published more than 150 research papers in various international and national journals.

Performance Analysis of Smart Meters for Enabling a New Era for Power and Utilities with Securing Data Transmission and Distribution Using End-to-End Encryption (E2EE) in Smart Grid

M. Manimegalai and K. Sebasthirani

Abstract The most outstanding engineering achievement of the twenty-first century in power systems is smart grid. The SG is an emerging technology which is revolutionizing the typical electrical grid by incorporating Information Communication Technology (ICT); this can be the main enabler of smart grid. This ICT can bring increased connectivity with increased severe security vulnerabilities and challenges. Cyberterrorism can be a prime target in smart grid because of its complex nature. To eliminate such cybersecurity issues, the most reliable end-to-end security algorithm was proposed. In this paper, information transmitted from customer through smart meter to the distributor and from distributor to the customer via smart meter was secured by End-to-End Encryption (E2EE). Data to and from the user and distributor cannot be modified in between by third parties. Experimental setup has customized design of Smart Meter (SM) for Home Area Network (HAN) and the proposed design will monitor the smart meter data transmission online. Performance analysis will show the reliability, integrity, and availability maintained in the communication network.

Keywords Smart grid (SG) · Smart meter (SM) · Information and communication technology (ICT) · End-to-end encryption (E2EE) · Home area network (HAN)

M. Manimegalai (✉)
Department of Computer Science and Engineering, P.S.R. Rengasamy College of Engineering for Women, Sivakasi, India
e-mail: mm11.1990@gmail.com

K. Sebasthirani
Department of Electrical and Electronics Engineering, Sri Ramakrishna Engineering College, Coimbatore, India
e-mail: sebasthirani.kathalingam@srec.ac.in

S. S. Dash et al. (eds.), *Intelligent Computing and Applications*,
Advances in Intelligent Systems and Computing 1172,
https://doi.org/10.1007/978-981-15-5566-4_1

1 Introduction

By integrating electrical distribution system with communication system forms, modern power system which is called smart grid, where power and information flow will be bi-directional. Transformation from conventional power grid to the smart grid will increase reliability, performance, and manageability because of full-duplex communication. SG has advanced communication systems in order to improve the performance. With larger benefits of advanced communication system, some security vulnerabilities are also there in the ICT. Figure 1 shows the base architecture diagram of smart grid.

In Fig. 1, electricity generation and distribution flow are shown in the straight line. Directed lines will show the communication network flow. In the communication network [1], smart meter plays an important role, which collects information from each house and sends it to the distributor. Through Internet, the data from all the houses will be sent to the distributor side. The following diagram will show the flow of information from customer to the distributor (Fig. 2).

Security threats [2] are more in the communication system. Security attack mainly depends on three factors as shown in Fig. 3.

Risk can be defined as

$$\text{Risk} = \text{Assets} \times \text{Vulnerabilities} \times \text{Threats} \tag{1}$$

where assets are smart grid devices such as smart meters, substations, data, and network devices. Vulnerabilities will allow the attackers to reduce the reliability and integrity of the systems information. Threats are the causes of inside or outside of the smart grid systems. If the vulnerabilities in the system are smaller, risk in the system will be minimized. Particularly, the assets and the threats cannot be zero. Security policies in smart grid have three important objectives, which are confidentiality, integrity, and availability [3]. CIA triad has been shown in Fig. 4.

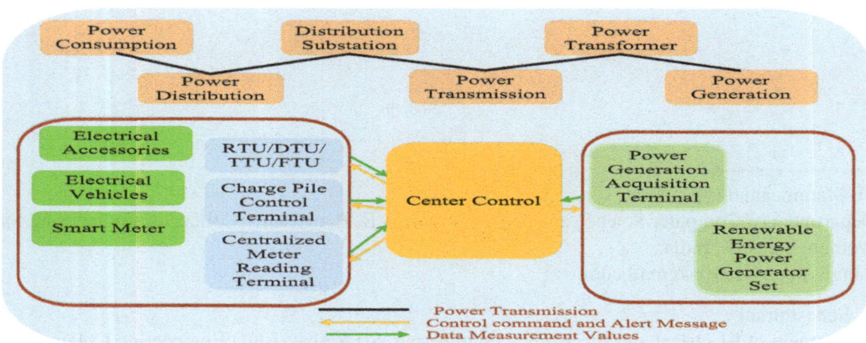

Fig. 1 Architecture model of smart grid

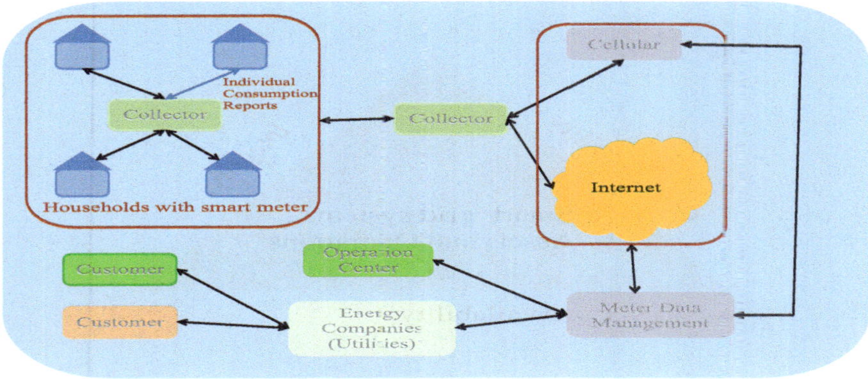

Fig. 2 Flow of information in the smart meter

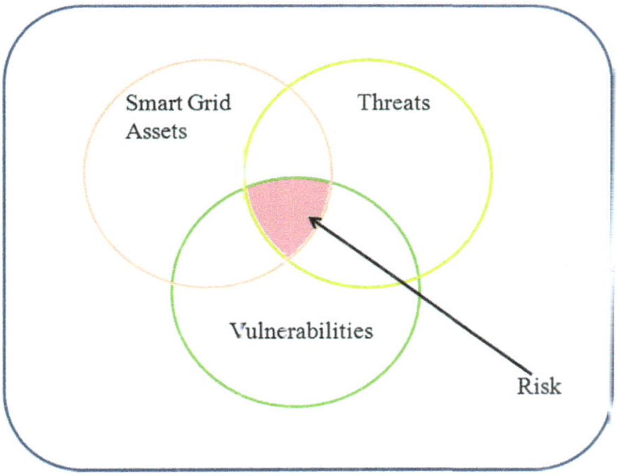

Fig. 3 Risk assessment in smart grid

There are different levels of attacks in the smart grid to cause security threat. These levels can be classified based on the networks,

- Home Area Network Attacks
- Neighborhood Area Network Attacks
- Wide area Network Attacks

Based on these network attacks, different levels of security protection will be given. This is shown in Fig. 5. Proposed method concentrates on the HAN and WAN level attacks. In the HAN level, smart meter was designed and data transmitted from customer household will be encrypted by E2E encryption via WAN [4]. So the security will be given in both HAN and WAN. Online database will be used to store

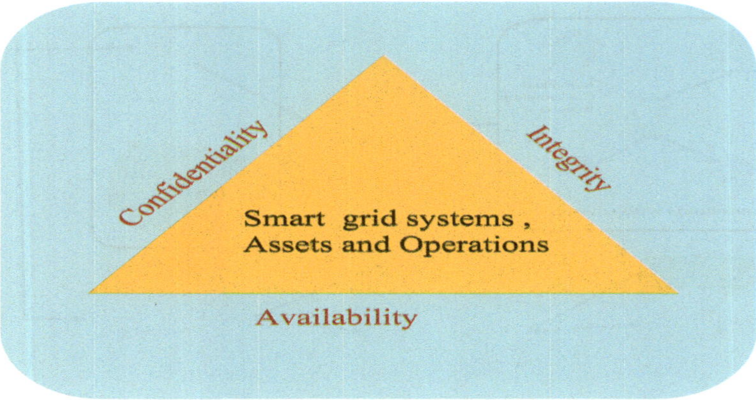

Fig. 4 CIA triad for smart grid systems

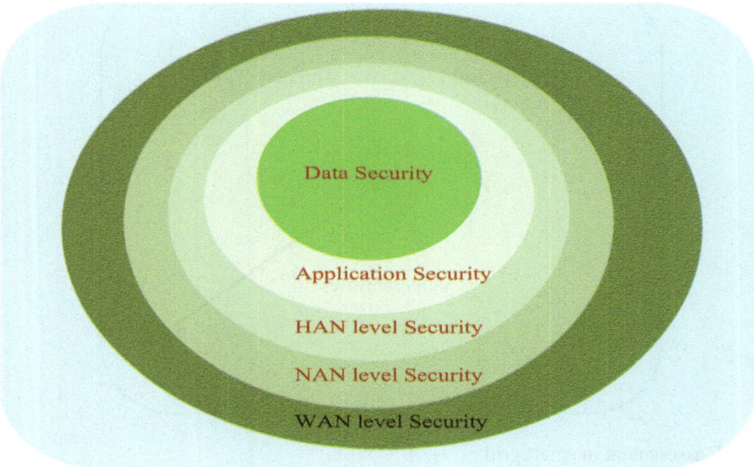

Fig. 5 Levels of security protection in smart grid

the customer information, and distributor also can make the security protection over this information. Distributors also do the E2E decryption to view the data in the WAN.

2 Cybersecurity Issues in Smart Meter

From household, the information like electricity usage of the single house, amount for the used electricity, etc. can be transmitted to the main server. The distributor also can

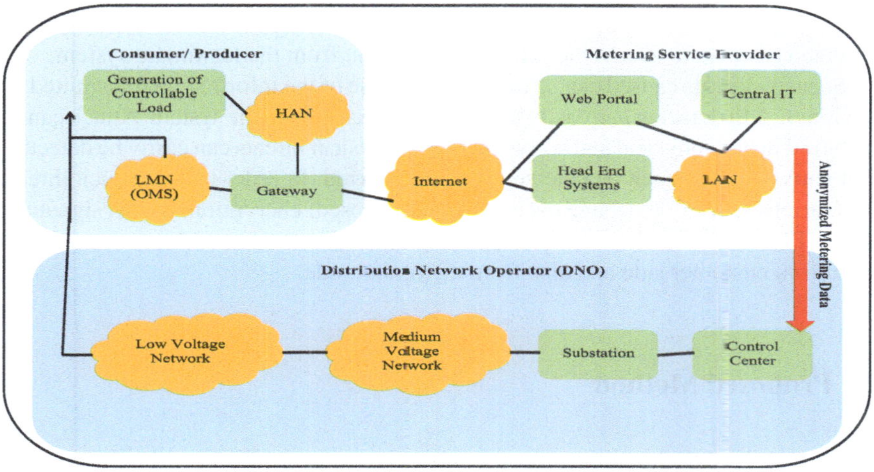

Fig. 6 Smart metering system

view these details of the customer and access control mechanisms applied. Security threats which affect the confidentiality, integrity, availability, and non-repudiation can occur in the smart meter from the customer and also from the distributor. The information flow from the customer to the central server is shown in Fig. 6.

Smart meter information from different households is sent through Internet using Wi-Fi, ZigBee, or HomePlug technologies. Smart meter communicates in the home area network and the neighborhood area network. Information flow from household to the center control is shown in Fig. 7.

When information flow from the HAN to NAN or WAN, attacker can masquerade as a legitimate meter data management, and he can change the data or view the confidential information. For example, the amount calculated for the used electricity can be changed (maximized or minimized) by the attacker [5]. Attacker can also

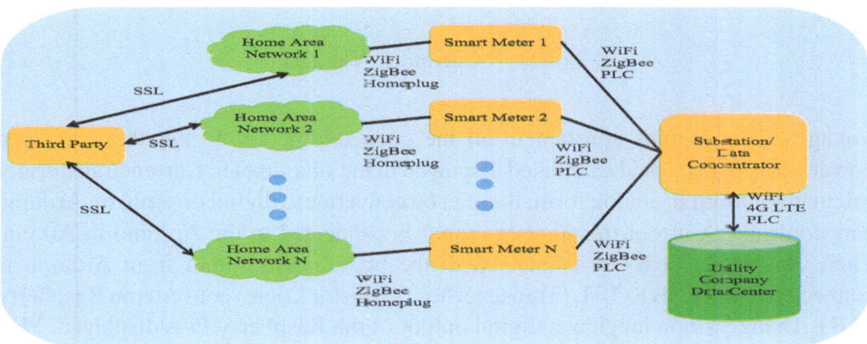

Fig. 7 Usage of information and communication technology in smart meter

change the commands to control the system, deny access to the distributor system, and meter unable to access the critical information from the legitimate system.

Security threats caused can be occurred because of the information transmitted on the Internet. So attacker can easily hack the center control of the system. Attack can be occurred in both physical and software forms. Physical threats can easily be detected. But software attack can be detected after it occurred. In order to avoid such threats [6], various security mechanisms have been proposed: encryption, digital signature, firewall, access control, and trusted party. Proposed method has concentrated on the E2E from customer side and also from distributor side.

3 Proposed Method

From each household, smart meter information sent via Wi-Fi technology to the distributor. Customer and the distributor can monitor the system through online. Customized smart meter was designed, and the details are sent to the distributor. Proposed system consists of the following functionalities:

- User can view the usage of electricity and amount need to pay at particular interval through online.
- Meter will show the electricity usage and amount at any time.
- Data transmitted from customer to distributor and distributor to the customer will be encrypted by E2E algorithm.
- Other side needs to do the E2E decryption algorithm to view the details.
- Two-way communication and ICT will be implemented via Raspberry Pi and Arduino controller.
- By monitoring systems online, user can control the usage of electricity based on their needs.

Simplified architecture diagram of the proposed system is shown in Fig. 8.

Smart meter reads the current value from the customer household by the current transformer using the formula,

$$\text{Power} = \text{Current} \times \text{Voltage} \tag{2}$$

Voltage value remains constant in all the connections 230 V. For the calculated power, amount to be paid calculated displayed in the smart meter. Current transformer calculates current in analog form; this can be converter to digital or serial by Arduino nanocontroller. Current transformer output is connected to the Arduino in A0 pin. Then, Raspberry Pi B+ is connected to the broadband. Output from Arduino is connected to the USB to TTL (Transistor-to-Transistor Logic) converter in Raspberry Pi B+. Using Python language, digital output of the Raspberry Pi is displayed. PHP language is used to store the output from Raspberry Pi [7] to the online database. From the online database, both the customer and the distributor can view the details.

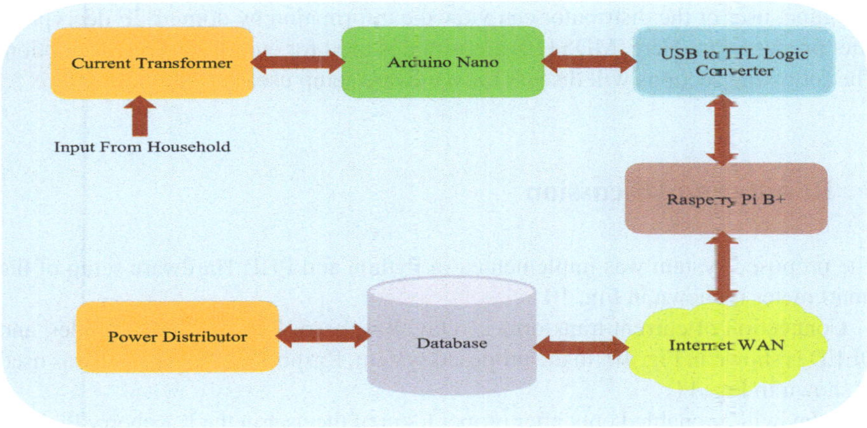

Fig. 8 Block diagram of the proposed system

4 Cybersecurity in Smart Meter

In cryptography [8] and network security, more encryption algorithms were proposed. Particularly, end-to-end encryption algorithm will provide better security in terms of confidentiality, integrity, and availability. In both, the end encryption is done when sending the information. Do the corresponding decryption to view the original information. Figure 9 shows the flow of information from customer to the distributor.

When the user login into the system, relay will be on. Power and amount values are calculated and after the E2E encryption [9]; values are stored in online database.

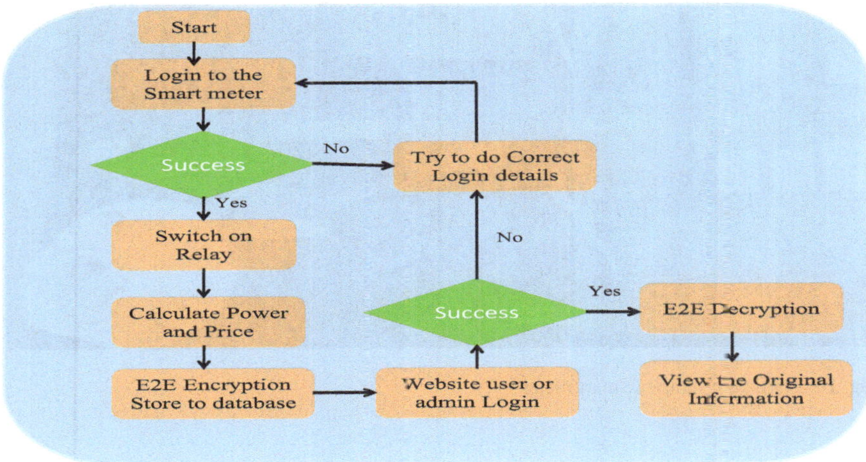

Fig. 9 Flow of E2E encryption and decryption

In online, user or the distributor can view the information by doing E2E decryption after proper login. Here, MD5 [10] algorithm is used for encryption and decryption. The following sections will discuss the hardware setup used in the smart meter.

5 Results and Discussion

The proposed system was implemented in Python and PHP. Hardware setup of the smart meter is shown in Fig. 10.

Connection of current transformer, relay, Raspberry Pi, Arduino controller, and OLED is shown in Fig. 10. In the proposed system, Raspberry Pi B+ model was used as shown in Fig. 11.

Relay will be enabled only after proper login of the user in the Raspberry Pi. User login page is shown in Fig. 12. User needs to give username, password, and service number as input to the system. This information will be given by distributor only. Figure 13 shows the details of entered page of user.

After relay on, power can be calculated. For each and every minute, power and price values are shown in the smart meter. This can be shown in Figs. 13 and 14.

Data are stored in online database which is unlimited. User and the distributor have to login into the system to view the details. Figure 16 shows the login page in website (Fig. 15).

After login, user can view the details stored in the database as shown in Fig. 17.

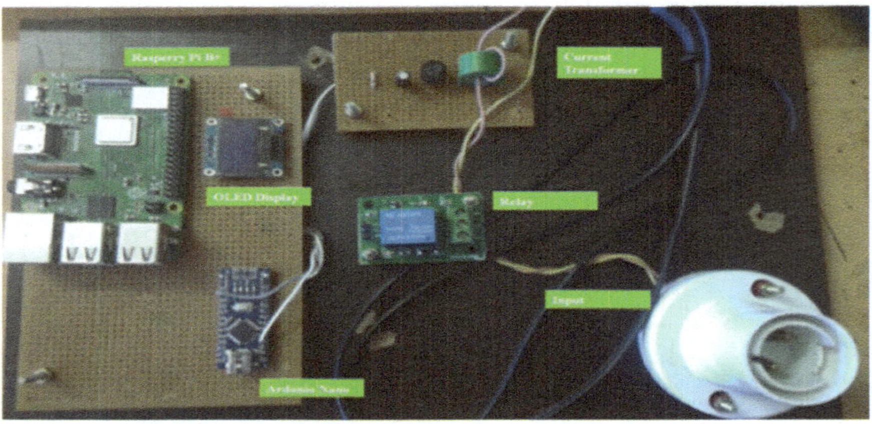

Fig. 10 Hardware setup of smart meter

Fig. 11 Rasperry Pi B+ model

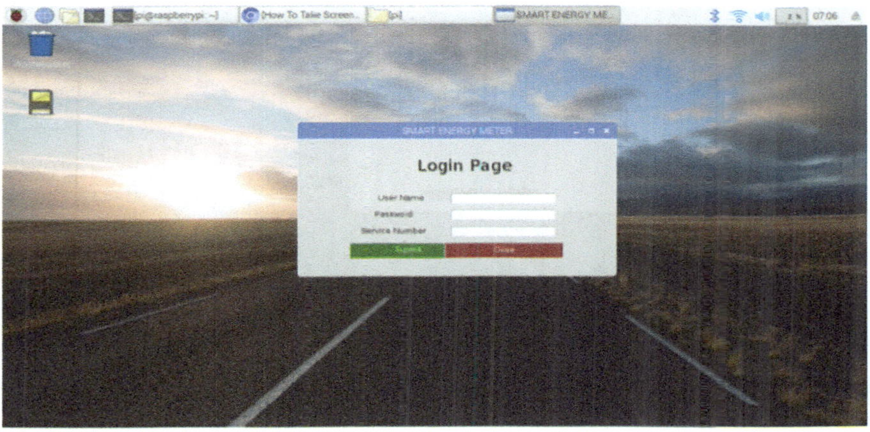

Fig. 12 User login page in Raspberry Pi

User can filter the information based on the date as shown in Fig. 18. Data are in the encrypted form only. Distributor also can view the details in encrypted form; after decryption, he can view the details.

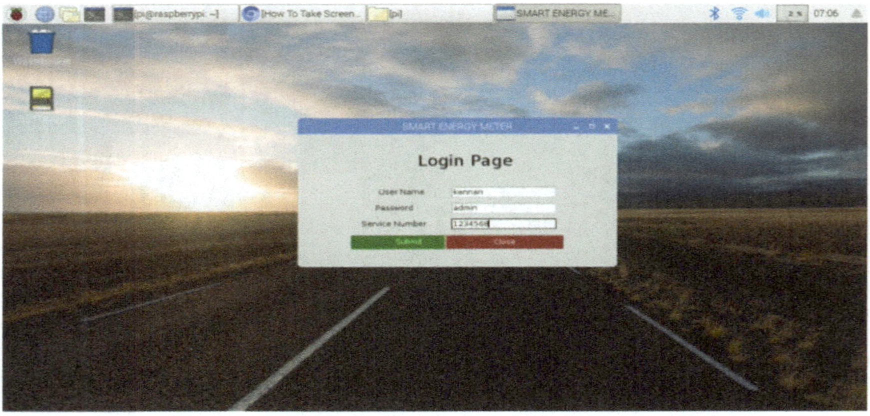

Fig. 13 Details of user entered in login page

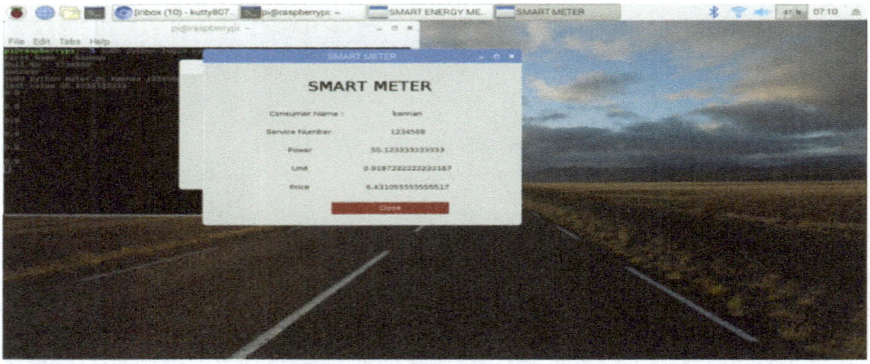

Fig. 14 Calculation of power and price in smart meter

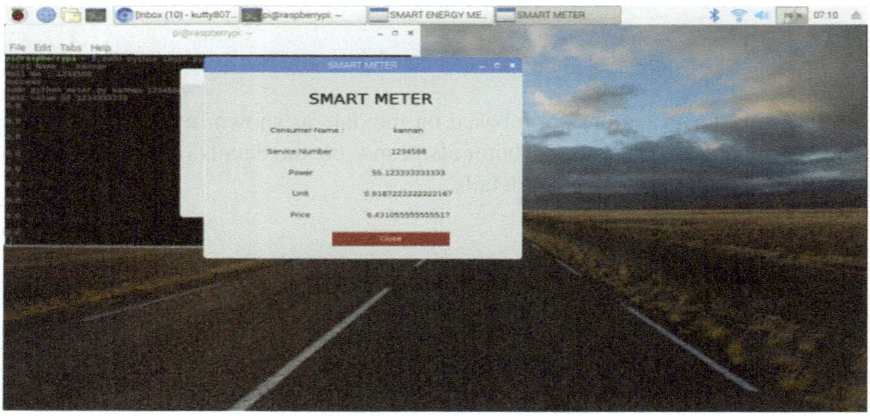

Fig. 15 Calculation of power and price

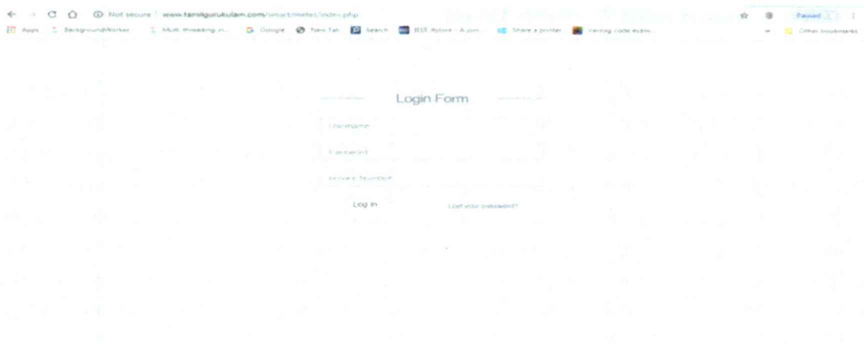

Fig. 16 User and admin login page

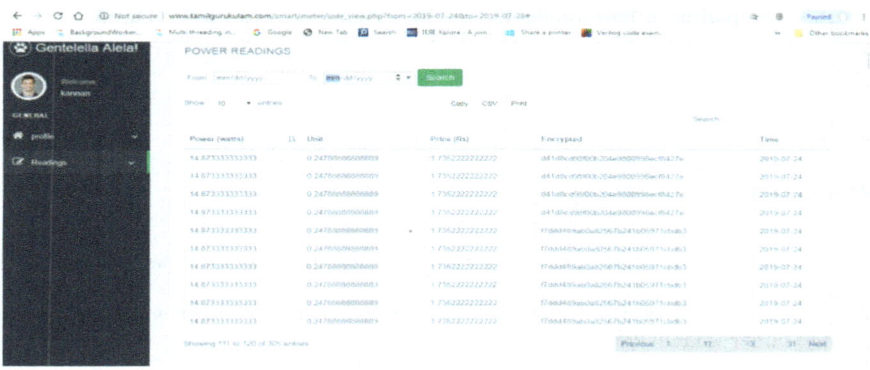

Fig. 17 Online database viewed by user

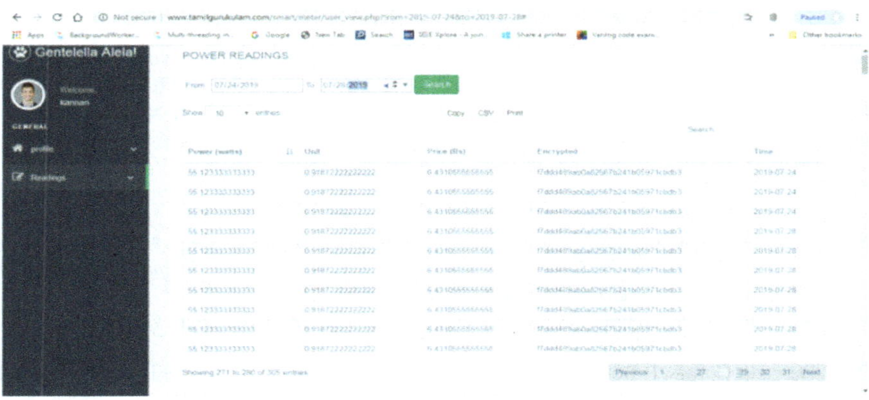

Fig. 18 Online database viewed by user after filtering applied

6 Conclusion and Future Work

Customized user-friendly smart meter was designed and Internet of Things (IoT) concepts have been implemented. Power and amount to be paid have been calculated based on the user needs. This can be seen digitally in the smart meter itself. Raspberry Pi B+ is used to transmit the data from the meter to the online database. Python language used to convert the data from meter to the database. MD5 algorithm was used to encrypt data from end to end. Service number was given as the additional authentication information to the user. Distributor no needs to enter the service number. Thus, information from customer to the distributor was sent through WAN with end-to-end encryption. Encryption was given in HAN and WAN level networks.

In future, smart meter can also include power quality improvement techniques like the elimination of harmonics, flicker, and voltage sag or swell. Through Wi-Fi and ZigBee technologies, power quality improvement monitoring can be done in dynamic mode.

References

1. NIST Special publication 1108, "NIST Framework and Roadmap for Smart Grid Interoperability Standards", Release 1.0, Jan 2010
2. C. Kaufman, R. Perlman, M. Speciner, *Network Security: Private Communication in a Public World* (Prentice Hall Press, 2002)
3. S. Clements, H. Krishnan, Cyber security considerations for the smart grid, in *2010 IEEE Power and Energy Society General Meeting* (2010), pp. 1–5
4. A. Giani, E. Bitar, M. Garcia, M. McQueen, P. Khargonekar, K. Poolla, Smart grid data integrity attacks: characterizations and countermeasures, in *2011 IEEE International Conference on Smart Grid Communications (SmartGridComm)* (2011), pp. 232–237
5. S. Ranjan, R. Swaminathan, M. Uysal, A. Nucci, E. Knightly, DDoS-shield: DDoS-resilient scheduling to counter application layer attacks. IEEE/ACM Trans. Netw. **17**(1), 26–39 (2009)
6. C. Bekara, Security issues and challenges for the IoT-based Smart Grid. Procedia Comput. Sci. **34**, 532–537 (2014), *The 9th International Conference on Future Networks and Communications (FNC'14)*
7. A. Metke, R. Ekl, Security technology for Smart Grid networks. IEEE Trans. Smart Grid **1**(1), 99–107 (2010)
8. Y. Wang, D. Ruan, D. Gu, J. Gao, D. Liu, J. Xu, F. Chen, F. Dai, J. Yang, Analysis of Smart Grid security standards, in *2011 IEEE International Conference on Computer Science and Automation Engineering (CSAE)*, vol. 4 (2011), pp. 697–701
9. P. McDaniel, S. McLaughlin, Security and privacy challenges in the Smart Grid. IEEE Secur. Priv. **7**(3), 75–77 (2009)
10. V. Delgado-Gomes, P. Borza, A biological approach for energy management in smart grids and hybrid energy storage systems, in *2014 International Conference on Optimization of Electrical and Electronic Equipment (OPTIM)* (2014), pp. 1082–1086

Energy Efficient Data Centre Selection Using Service Broker Policy

Sameena Naaz, Iffat Rehman Ansari, Insha Naz, and Ranjit Biswas

Abstract Over the past few years human beings have become completely dependent on computers and IT technologies, which in turn have led to the issue of energy and power consumption in IT industries. Since the energy cost as well as the electrical requirements of the industry has increased drastically throughout the world, therefore it has become necessary to shift our focus to Green computing, which refers to environmentally sustainable computing and aims to limit energy and power consumption and shrink costs thus maximizing the efficiency of the system. Green Computing provides the proficient use of computing power.

Keywords Green computing · Power consumption · Sustainable computing

1 Introduction

Owing to the escalating growth of internet users and falling price of computer hardware various fields like finance, medical, education and many more are becoming heavily dependent on the services offered by IT technologies [1]. All this has resulted in the production of huge amount of power and energy consumption which in turn not only produces great amount of CO_2 which is harming the environment badly also the increasing energy costs is taking toll on the IT industries. Green computing

S. Naaz (✉) · I. Naz · R. Biswas
Department of Computer Science and Engineering, School of Engineering Sciences and Technology, Jamia Hamdard, New Delhi, India
e-mail: snaaz@jamiahamdard.ac.in

I. Naz
e-mail: inshanaz_sch@jamiahamdard.ac.in

R. Biswas
e-mail: rbiswas@jamiahamdard.ac.in

I. R. Ansari
University Women's Polytechnic, Aligarh Muslim University, Aligarh, Uttar Pradesh, India
e-mail: iffat_rehman2002@yahoo.co.in

S. S. Dash et al. (eds.), *Intelligent Computing and Applications*,
Advances in Intelligent Systems and Computing 1172,
https://doi.org/10.1007/978-981-15-5566-4_2

tries to provide solution to these alarming issues. One of the vital segments of green computing is to focus on how to reduce the carbon footprint and save the energy. It covers the complete lifespan of the computing starting from designing to the disposal of e-waste in such a way that there is very less or no effect on the environment [2].

Green computing provides various approaches that help the IT technology firms to tackle crucial computing requirements in a more sustainable way so that the damage on environment and resources is either reduced or eliminated hence reducing the carbon emission and energy consumptions.

Rest of the paper is organized as follows: Sect. 2 discusses the reasons for the need of going green, Sect. 3 gives a review of the latest literature in this area. Section 4 is dedicated to current trends in green computing, Sect. 5 highlights the challenges or hurdles in the way of going green and finally Sect. 6 gives the conclusions drawn from this study.

2 Need for Greening

We need to go green to increase the energy efficiency and reduce the resource consumption so that we can achieve the goal of limiting the carbon footprint. The main aim of green computing is to provide a healthier computing environment. No doubt the use of computing services and IT has made our lives simpler and our tasks easier. But the fact that due to increase in usage, power consumption also increases can't be ignored, which in turn increases the production of greenhouse gasses (CO_2). Thus a data centre needs an efficient cooling system, but if the data centre doesn't have an appropriate and capable cooling system then there will be a loss of energy which will lead to the environmental degradation [3]. Some of the other issues that amplify the green IT movement are managing harmful e-waste, rising gasoline expenditures and growing real estate costs [4]. As a result, there is a dire need to achieve or go green to save our environment.

3 Literature Review

A basic transformation happening in the area of IT nowadays is cloud computing. The key element of cloud computing is virtualization which brings ease and efficiency. Virtualization increases the security of cloud computing, shielding the reliability of guest VM's and the cloud infrastructure components [5, 6].

Since data centres that host the Cloud applications consume a large amount of energy, which results in increased costs and carbon emission rates, therefore, we require green cloud computing that reduces functional costs and minimizes the environmental impact [7, 8].

Due to the increasing demand of cloud computing, a large number of companies are investing in constructing huge data centres to host Cloud services. These data centres consume huge amount of energy and are very complex in the infrastructure. In order to make data centre energy efficient various technologies like virtualization and consolidation are employed [9].

4 Current Trends in Green Computing

In order to address the impact of IT on the environment, we need to implement an integrated procedure that can solve the problems using different paths from the manufacture to disposal of the devices. Everything needs to be green like green use, green designing, green manufacturing and green disposal thereby making the whole lifecycle of IT greener [10].

Different trends in green computing given in Fig. 1 are discussed.

4.1 Virtualization

It refers to the creation of virtual version of computer resources, for example, creation of multiple logical versions of a single physical resource or hardware [11]. With virtualization, solo equipment can operate as multiple equipment's running individually. It is the best possible way to greening and conserving the space, resources and environment. Management software and virtualization software both are provided for

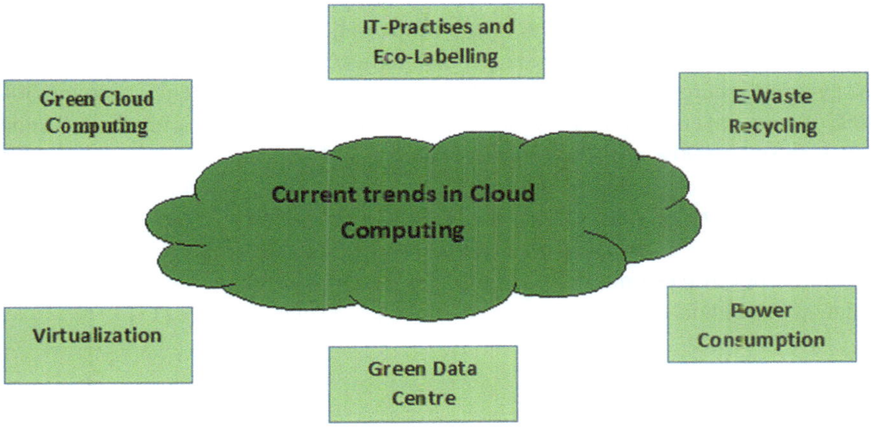

Fig. 1 Current trends in cloud computing

efficient working of a virtualized environment. Virtualization directs to Server consolidation. The security of the system is also enhanced. It permits complete consumption of resources, reduces the overall quantity of hardware used, switches off the idle machines to conserve energy and also the cost of space and rent is decreased [2]. Cooling expenses can also be curtailed by making the proficient use of available resources [12].

4.2 Green Data Centres

The term "Green Data centres" means designing an algorithm and infrastructure for data centres that provide efficient use of the resources that are available, help in reduction of cooling costs and tackle energy consumption issues [13]. Owing to the good environmental and financial impact, green data centres have gained substantial focus. As the usage of the internet is increasing day by day so is the power consumption in our data centres which results in higher energy costs and environmental degradation. So in order to overcome this issue, IT companies need to necessarily go green and build a sustainable data centres to save the energy, costs and environment. According to the US department of energy, an efficient data centre should focus on five basic areas while designing a data centre, cooling system, conditions of the environment, management of air, IT systems and electrical system, other things that need to be kept in mind is e-waste recycling [11, 14].

4.3 Power Consumption

Reducing the power consumption is one of the major issues these days so in order to tackle this IT industries have started making use of the devices that are efficient in saving the energy. According to the environment protection agency, "around 30–40% of personal computers are kept 'ON' after office hours and during the weekend and even around 90% of those computers are idle" [15]. There are different ways in which the power usage can be limited like switching off the system when it's not in use or sending the monitor in low power state. The use of LCD and LED can also be helpful. Hibernate and standby mode saves the 80–90% of the power so devices should be put in these modes when idle [16]. Combining the efficient coding and efficient algorithms can also provide great results in saving energy [17].

4.4 E-Waste Recycling

Since electronic devices and computer systems contain toxic metals and poisonous substances like lead and mercury that can be harmful to the environment so we need to find the proper ways for their disposal [16]. Reusing the discarded devices or old computers can save a lot of energy and also help in combating the harmful effect of e-waste on the environment [15]. Recycling can be done in many ways like taking the components of old computer systems and using them for repairing or upgradation purpose. Also the practice of changing the computers every 2–3 years needs to be changed for the benefit of cost and environment [17].

Growing use of technology nowadays has led to the creation of an enormous quantity of electronic wastes resulting in environmental degradation; therefore, the safety of environment and keeping the check on environmental pollution has become the chief concern of scientists all over the world. The most important concern related to e-wastes is that these are non-biodegradable and their dumping has led to the accretion of toxic material like lead, cadmium, etc. in the environment resulting in global warming and contamination of the soil and groundwater. Thereby, disturbing the plant and animal life which in turn has an effect on the entire living organisms yielding harsh health risks and disorders. Increasing global warming and mounting energy expenses has led the government as well as the private organizations to think and examine different ways to safeguard the environment worldwide [18].

4.5 IT-Practises and Eco-Labelling

For companies to create products to be given eco-label, different policies need to be introduced all over the world. Many organizations in the world support eco-labelling and provide certificates to IT products on the basis of several features like energy consumption, recycling, power consumption, etc. [12].

Eco-labelling basically originated because of the increasing environmental concerns worldwide. Labels like eco-friendly, recyclable, energy efficient, reusable, etc. attracted consumers who were finding ways to reduce the impact of hazardous material on the environment.

Eco-labels give the information regarding the existence or nonexistence of particular feature in any product. It enables the customers to get an insight about the environmental quality of items at the time of purchasing, allowing them to pick or buy products that are suitable from an environmental point of view. Therefore, minimizing the utilization of harmful substances that may be detrimental to the environment [19].

Companies have to make sure that they manufacture and design products in such a way that they can obtain the eco-label. Many organizations grant certificates to IT products after reviewing the features like energy and resource consumption, recycling and refurbishing, etc. thereby enabling the customers to make the environmentally suitable decision at the time of purchasing a product [20].

4.6 Green Cloud Computing

The research carried by Pike Research, "the wide-spread adoption of cloud computing could lead to a potential 38% reduction in worldwide data centre energy expenditures by 2020" [21]. Cloud computing has proved to be a vital and sound means for virtualization of data centres and servers so that they can be a resource as well as energy efficient. Due to the high consumption of power and energy in IT firms, which produces the harmful gases in the environment resulting in global climate changes. As a result, there is a dire need of cloud computing to go green [22, 23].

Cloud computing is one of the most important paradigm in modern world because of the fact it has dynamic, high-powered computing abilities, with access to intricate applications and data archiving, with no requirement of any extra computing resources. Since cloud computing offers reliability, scalability and high performance at fewer expenses thus cloud computing technologies have a diversity of application domains. By offering promising environmental protection, economic and technological advantages, cloud computing has revolutionized the modern computing. These technologies have the potential to improve. The various technologies of cloud computing like energy efficiency, reduced carbon footprints and e-waste can convert cloud computing into green cloud computing [24, 25] (Table 1).

Table 1 Summary

Approach	Observation
Virtualisation	Creates virtual version of computer resources like creating multiple logical versions of a single physical resource or hardware
Green data centre	To design an infrastructure for data centres which makes efficient use of the available resources thus helping in reduction of cooling costs
E-Waste Recycling	Recycling can be done in many ways like taking the components of old computer systems and using them for repairing or upgradation purpose thereby saving the cost and tackling the hazards related to e-wastes
IT-Practises and Eco-Labelling	Labels like eco-friendly, recyclable, energy efficient, reusable, etc. attracted consumers who were finding ways to reduce the impact of hazardous material on the environment
Green cloud computing	The various technologies of cloud computing like energy efficiency, reduced carbon footprints and e-waste can convert cloud computing into green cloud computing
Power consumption	There are different ways in which the power usage can be limited like switching off the system when it's not in use or sending the monitor in low power state

5 Proposed Technique for Energy Efficiency in Virtual Machines

This work is an extension of the work carried out in [26]. Here various load balancing algorithms have been implemented on Cloud Analyst to study the energy efficiency. The proposed algorithm has been simulated on Cloud Analyst, and its result has been compared with that of throttled and round-robin algorithm.

There are three parts in this load balancing algorithm.

(i) The energy consumption of each VM is calculated.
(ii) The most efficient VM is found
(iii) ID of the most efficient VM is returned.

Simulation package CloudSim [27] has been used for the experiments. A single data centre having 100 virtual machines (VM memory—1024 Mb, PM speed—100 MIPS, Data Centre OS—Linux) has been used.

6 Results and Discussions

CloudSim toolkit has been used for simulating the three algorithms. The parameters studied here are the overall response time, data centre processing time and the energy consumption.

Average Response Time and Energy Consumption for various algorithms are shown in Table 3.

Table 2 shows that the average processing time at data centre is much higher for round robin and throttled algorithm as compared to that of Service Broker-based VM load balancing policy. Response time for any user query in round-robin scheduling and throttled policy is much higher as compared to VM load balancing policy as can be seen from Table 3. The same results are also depicted in Figs. 2 and 3, respectively.

Table 2 Request processing time for different algorithms

Algorithm	Data centre	Request processing time (ms)
Round-robin	DC1	20.499
	DC2	56.296
Throttled	DC1	20.631
	DC2	55.859
Service broker	DC1	10.575
	DC2	28.659

Table 3 Average response time and energy consumption for various algorithms

Algorithm	Average response time (ms)	Energy consumption (mW)
Round robin	123.98	10.49
Throttled	123.85	8.33
Service broker	109.54	4.01

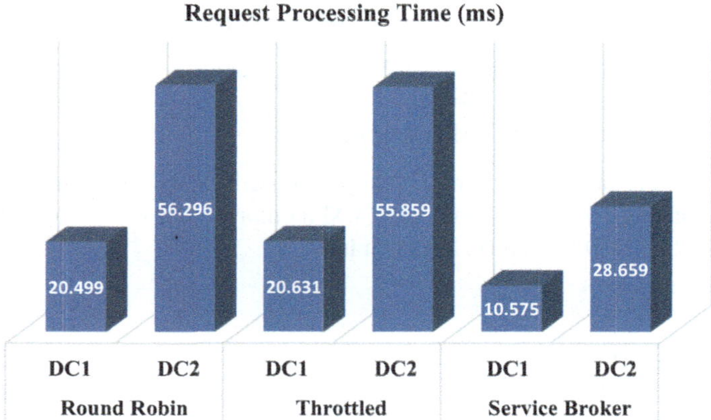

Fig. 2 Request processing time for different algorithm

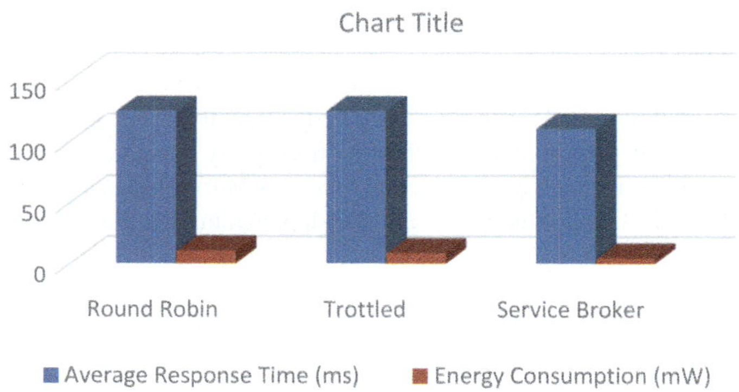

Fig. 3 Average response time and energy consumption for various algorithms

7 Conclusion

In this paper, an energy-efficient service broker-based policy for data centre has been implemented in CloudSim cloud computing environment. The proposed technique first calculates the expected response time of each virtual machine and then sends the identity of that machine to the data centre for allocation of any new request. The average request processing time and energy consumption are also calculated for the three algorithms, viz, round-robin, throttled and service broker based, and it has been observed that service broker-based algorithm performs the best in terms of all the three parameters.

References

1. Q. Li, *The Survey and Future Evolution of Green Computing*, pp. 0–3 (2011)
2. T.R. Soomro, M. Sarwar, Green computing from current to future trends, vol. 6 no. 3, pp. 326–329 (2012)
3. P. Malviya, S. Singh, A study about green computing. Int. J. Adv. Res. **3**(6), 790–794 (2013)
4. R. Sharma, Approaches for green computing. Int. J. Innov. Comput. Sci. Eng. **2**(3), 52–55 (2015)
5. Y. Xing, Y. Zhan, Virtualization and cloud computing, pp. 305–306 (2012)
6. F. Lombardi, R. Di, C. Nazionale, D. Informativi, P.A. Moro, Secure virtualization for cloud computing. J. Netw. Comput. Appl. **34**(4), 1113–1122 (2011)
7. A. Beloglazov, J. Abawajy, R. Buyya, Energy-aware resource allocation heuristics for efficient management of data centers for Cloud computing. Futur. Gener. Comput. Syst. **28**(5), 755–768 (2012)
8. S. Naaz, A. Alam, R. Biswas, Effect of different defuzzification methods in a fuzzy based load balancing application. Ijcsi **8**(5), 261–267 (2011)
9. S.K. Garg, C.S. Yeo, R. Buyya, Green cloud framework for improving carbon, pp. 491–502
10. S. Murugesan, Harnessing green IT: principles and practices—adopting a holistic approach to greening IT is our responsibility toward creating a more sustaining environment. Green Comput. 24–33 (2008)
11. R. Thomas, Approaches in green computing
12. D. Ranjith, G.S. Tamizharasi, B. Balamurugan, A survey on current trends to future trends in green computing, in *Proceedings of International on Conference Electronics, Communication and Aerospace Technologies ICECA 2017*, vol. 2017 (2017), pp. 632–637
13. X. Jin, F. Zhang, A.V. Vasilakos, Z. Liu, Green data centers: a survey, perspectives, and future directions (2016)
14. S. Naaz, A. Alam, R. Biswas, Load balancing algorithms for peer to peer and client server distributed environments. Int. J. Comput. Appl. **47**(8), 17–21 (2012)
15. I. AlMusbahi, O. Anderkairi, R.H. Nahhas, B. AlMuhammadi, M. Hemalatha, Survey on green computing: vision and challenges. Int. J. Comput. Appl. **167**(10), 4–6 (2017)
16. N. Mumbai, N. Mumbai, Green computing: an essential trend for secure, Apr 2013, pp. 19–20
17. S. Singh, Green computing strategies & challenges, in *Proceedings of 2015 International Conference on Green Computing Internet Things, ICGCIoT 2015*, no. 1, pp. 758–760 (2016)
18. R. Panda, E-waste management: a step towards, vol. 4, no. 5, pp. 417–424 (2013)
19. Green IT
20. K. Saikumar, Design of data center, vol. 2, no. 6, pp. 147–149 (2014)
21. K. Sourabh, S. M. Aqib, A. Elahi, Sustainable green computing: objectives and approaches, pp. 672–681
22. Y.S. Patel, N. Mehrotra, S. Soner, Green cloud computing: a review on green IT areas for cloud computing environment, pp. 327–332 (2015)
23. I. Naz, S. Naaz, R. Biswas, A parametric study of load balancing techniques in cloud environment
24. L. Radu, SS symmetry Green Cloud computing: a literature survey (2017)
25. S. Naaz, A. Alam, R. Biswas, Implementation of a new fuzzy based load balancing algorithm for hypercubes. Int. J. Comput. Sci. Inf. Secur. 270–274 (2010)

26. S. Naaz, I.R. Ansari, A. Khan, Load balancing of virtual machines using service broker algorithm. Int. J. Adv. Technol. Eng. Sci. **4**(9), 232–238 (2016)
27. R. Buyya, R. Ranjan, R.N. Calheiros, Modeling and simulation of scalable Cloud computing environments and the CloudSim toolkit: challenges and opportunities, in *2009 International Conference on High Performance Computing & Simulation* (2009), pp. 1–11

Co-ordinate Measurement of Roll-Cage Using Digital Image Processing

Ritwik Dhar, Parth Kansara, Sanket Shegade, Atharv Bagde, and Sunil Karamchandani

Abstract This paper proposes a unique and novel approach for co-ordinate measurement of an all-terrain vehicle. The manufactured roll-cage is replete with minute errors which are introduced during the process. The suspension points on the roll-cage have been considered as dataset for this algorithm, as they are an integral part in the performance of the vehicle and for maintaining suspension geometry. A feasible method using image processing techniques such as shade correction, adaptive binarization, and segmentation has been used to improve fault tolerance by analyzing and reducing the manufacturing errors with respect to the computer-aided design model. A MATLAB script has been developed for the proposed method with the help of image processing toolbox.

Keywords Co-ordinate measurement · Computer vision · Adaptive thresholding · Bounding box · Object detection · Edge detection · Binarization · Integral images · Shape detection

---

R. Dhar · A. Bagde · S. Karamchandani
Electronics & Telecommunication, Dwarkadas J. Sanghvi College of Engineering, Mumbai 400056, India
e-mail: ritwik1798@gmail.com

A. Bagde
e-mail: atharva17bagde98@gmail.com

S. Karamchandani
e-mail: skaramchandani@rediffmail.com

P. Kansara (✉)
Information Technology, Dwarkadas J. Sanghvi College of Engineering, Mumbai 400056, India
e-mail: parthskansara@gmail.com

S. Shegade
Mechanical Engineering, Dwarkadas J. Sanghvi College of Engineering, Mumbai 400056, India
e-mail: sanketshegade@gmail.com

S. S. Dash et al. (eds.), *Intelligent Computing and Applications*,
Advances in Intelligent Systems and Computing 1172,
https://doi.org/10.1007/978-981-15-5566-4_3

1 Introduction

Dimensional accuracy of fabricated components is an important factor, particularly in instances where they are to be in contact with each other. The need to fabricate within tolerances imposes accuracy in measurement. In addition, to ensure accuracy in any outcome, traceability must be maintained. Such precision in measurements is required in the manufacturing of all-terrain vehicles, to fulfill the desired suspension geometry. Failure in incorporating this precision results in ample deviation in the position of the in-board and out-board points of the control links. These control links decide the point of application of the reaction force from the ground to the vehicle body. Thus, co-ordinate measurement of these points in the spatial domain helps in quantifying the manufacturing anomalies by calculating the deviations from the computer-aided design model, which is synonymous to degradation in the overall stability of the vehicle. Due to the limitations of the commercial co-ordinate measurement machines, an unconventional yet reliable approach has been implemented with the help of images to process the distance of the chassis points of vehicle with respect to a reference point [3].

Wojnar [5] and Zhao et al. [6] applied Fourier Transform as a solution to reduce the noises in periodic signals. To remove noise, they can filter high frequencies and keep just the low frequencies, so the filters that eliminate noise are most often pointed as low-pass filters. Thresholding is one of the most common techniques for segregating objects from the background of the image, Slezak et al. [7]. According to Cheriet et al. [8], Thresholding can be used to differentiate text from the background, e.g., by filtering the areas above or below the threshold to a certain gray value in a greyscale image.

Corners in images have distinct characteristics that clearly distinguish them from the pixels around them. There are many algorithms for detecting corners that are accurate even after the image is geometrically distorted, Jeong and Moon [9]. Object detection algorithms often use the position of corners in the image, Harris corner detector is prevalent in this class. The most successful edge detector is the Canny edge detector which aims at the three basic criteria, i.e., good localization of edge points, low error rate, and avoidance of similar detector results from the same edge, Tang and Shen [10]. The algorithm's final phase consists of selecting edge points using a double threshold. According to Rong et al. [11] and Li et al. [12] using double threshold is an effective way to reduce the noise in the last stages of edge detection.

Our approach has been chosen on the basis of its ease of setup and also the optimization with the help of different algorithms involved has resulted in better outputs. Aldasoro et al. [1] described an algorithm for removing the image shading element by estimating the signal envelope. This technique was implemented for shade correction to normalize the light intensity throughout the image. Bradley et al. [2] presented an extension of Wellner's work but reduced the computational complexity by using integral images rather than the moving average throughout the pixels.

Followed by binarization we perform threshold-based segmentation to extract the required interest region from image with the help of shape detection. The distance

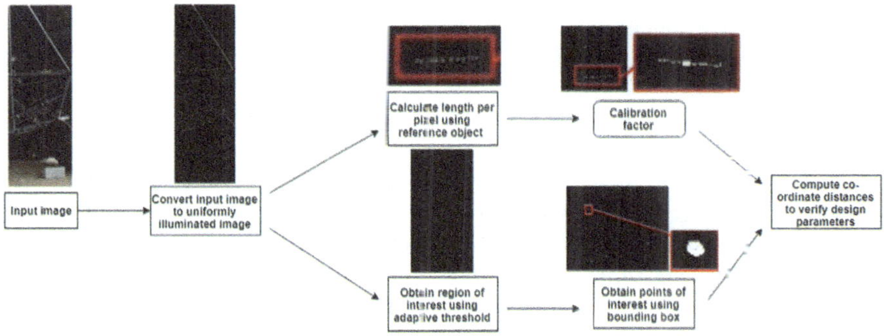

Fig. 1 Algorithm flow chart

from the region of interest, i.e., the suspension points in the front box used for analysis are calibrated in pixels with respect to reference object of pre-determined dimensions placed in the images to get the final results (Fig. 1).

2 Pre-processing

2.1 Shade Correction

It is not easy to obtain a digital image with uniform illumination over the entire pixels. The light due to reflectance and illuminance components in the image background is not generally uniform. The technique through which images have been captured as well as the relationship among various objects present throughout the field of vision and camera illumination often contributes to cases in which the image shows irregular patterns of shading throughout the image. The basic approach toward shade correction is using simple low-pass filters or morphological operators due to the ease of use and universal applicability. The above techniques are effective though fail in cases when the distortion is bigger than the background, basically it assumes only two cases with background being below or above the median pixel value than the region of interest. Aldasoro et al. [1] described an algorithm for removing the image shading element by estimating the signal envelope. The primary feature of using such a technique was that it did not make any assumptions about whether the objects were of lesser or greater intensity than the background and worked well with objects of all sizes.

It was presumed that the shaded image used, $I(x, y)$ was created by an additive shading element $S(x, y)$ that damaged an initially unbiased image $U(x, y)$.

$$I(x, y) = U(x, y) + S(x, y) \tag{1}$$

We can also say that the corrected image $\hat{U}(x, y)$, which was an estimation of $U(x, y)$, was obtained by

$$U(x, y) \cong \hat{U}(x, y) = I(x, y) - S(x, y) \tag{2}$$

The effects of noise are first minimized by low-pass filtering the input signal with a 3 × 3 Gaussian kernel. The envelope is then obtained by comparing the intensity of every pixel against the average of opposite eight-connectivity neighbors with increasing the distance of pairs in 45° directions from 0° to 135°. Two new envelope series $S_{\max}/S_{\min}$ were obtained by changing the existing intensity with the intensity of the pixel by the maximum/minimum value of the comparison of each neighboring pixel. The upper envelope $S_{\max}$ can be produced by replacing the maximum value of the averages with the pixel as given by

$$S^i_{\max}(x, y) = \max\left\{\begin{array}{c} \frac{I(x-d_i, y-d_i)+I(x+d_i, y+d_i)}{2} \\ \frac{I(x+d_i, y-d_i)+I(x-d_i, y+d_i)}{2} \\ \frac{I(x-d_i, y)+I(x+d_i, y)}{2} \\ \frac{I(x, y-d_i)+I(x, y+d_i)}{2}, I(x, y) \end{array}\right\} \tag{3}$$

The replacement with respect to the minimum value results in the lower envelope $S_{\min}$. The max/min values formed two stacks from which the maximum intensity projection corresponded to the current envelope estimation: Both $S_{i\max} = S_{i\min}$ surfaces have been filtered with a low-pass Gaussian filter of size equal to the range di to limit the measurement of the envelope to the intermediate orientation pixels. The method was replicated by expanding d_i and filter size, enabling the envelope to be adapted to all objects. To determine a stop criterion, magnitude and local derivatives were calculated.

$$\frac{\partial s^i_{\frac{\max}{\min}}}{\partial x}, \frac{\partial s^i_{\frac{\max}{\min}}}{\partial y} \tag{4}$$

In each iteration of d_i, $\mathrm{MG}^i_{\mathrm{tot}}$ was compared with the $i - 1$ gradient $\mathrm{MG}^{i-1}_{\mathrm{tot}}$, and when $\frac{\mathrm{MG}^i_{\mathrm{tot}} - \mathrm{MG}^{i-1}_{\mathrm{tot}}}{\mathrm{MG}^{i-1}_{\mathrm{tot}}} < 0.01$ the iterations were stopped. Finally, the more smoother surface, either $S^i_{\max}$ or $S^i_{\min}$ which had the lower $\mathrm{MG}^i_{\mathrm{tot}}$, was used as the shading S.

$$\mathrm{MG}^i = \left(\left(\frac{\partial S^i_{\frac{\max}{\min}}}{\partial x}\right)^2 + \left(\frac{\partial S^i_{\frac{\max}{\min}}}{\partial y}\right)^2\right)^{\frac{1}{2}} \qquad \mathrm{MG}^i_{\mathrm{tot}} = \sum_{x,y} \mathrm{MG}^i \tag{5}$$

Figure 2a, b here illustrates the image before and after the shading correcting algorithm has been applied.

Fig. 2 **a** Input image. **b** Output image from shade correction algorithm

3 Co-ordinate Measurement

The results from the shade correction algorithm are further used for two applications:

A. Calibration factor from reference object using edge detection
B. Binarization for co-ordinate measurement.

3.1 Calibration Factor from Reference Object Using Edge Detection

The output as shown in Fig. 2b has been used for computing the calibration factor. A plate with pre-determined measurements had been placed along the axis of the roll-cage and camera to produce no optical errors. According to Kainz et al. [4], the dimensions of an object in a 2D image can be estimated by taking into account its position from capturing element and the positional angles involved. Taking two cases of input image with the reference object and without the reference object gives the alignment of the object. In our case, the reference object is carefully placed in the same plane as the capturing element is helping in reducing the errors.

As shown in Fig. 3, the width is estimated using the following relations:

$$w = \alpha \times y + \beta \tag{6}$$

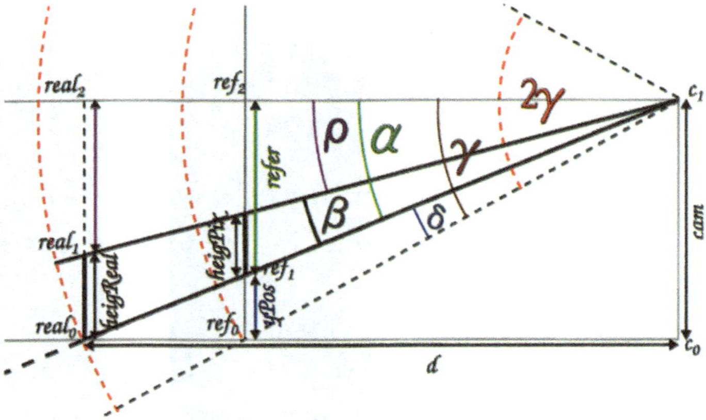

Fig. 3 Positional angles and measurements

$$\alpha = m_1 \times \frac{\text{cam}}{\text{real}} - b_1 \tag{7}$$

$$\beta = m_2 \times \frac{\text{cam}}{\text{real}} - b_2 \tag{8}$$

$$\text{Image width} = \text{cam} \times \frac{w + b_1 \times y + b_2}{m_1 \times y + m_2} \tag{9}$$

Here, m_1, b_1, m_2,b_2 are constants analyzed from previous tests performed, w is the object width in pixels, α and β give the x and y intercept values, cam gives the distance of camera from ground, and real is the actual object width.

The length of the edge is then obtained in pixels and the calibration factor is obtained as (Eq. 10)

$$\text{Calibration Factor} = \frac{\text{Image width}}{\text{real}} \tag{10}$$

3.2 Binarization For Obtaining Region of Interest & Distance Measurement

After the shade correction to produce uniform illumination, the image from Fig. 2b is binarized, i.e., all the pixels are given values of 0 or 1 with respect to a threshold that is to be determined for optimum results. The task of determining the threshold value is of utmost importance in this stage. While it is generally known that a single fixed threshold value will not yield optimum results in non-uniformly illuminated or damaged images, an adaptive threshold technique is used. Bradley et al. [2] presented

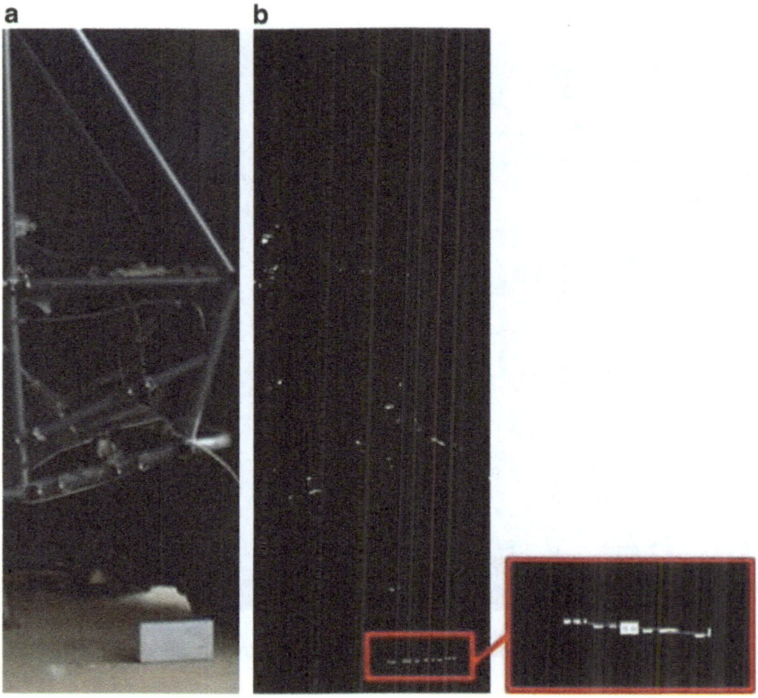

Fig. 4 **a** Reference object. **b** Dimensions

an extension of Wellner's work but reduced the computational complexity by using integral images rather than the moving average throughout the pixels. Integral images do not take into consideration the neighborhood size and help in fast summation of pixels. These are calculated using

$$I(x, y) = f(x, y) + I(x - 1, y) + I(x, y - 1) - I(x - 1, y - 1) \tag{11}$$

$$\sum_{x=x1}^{x2} \sum_{y=y1}^{y2} f(x, y) = I(x2, y2) - I(x2, y1 - 1) - I(x1 - 1, y2) + I(x1 - 1, y1 - 1) \tag{12}$$

The integral image using Eqs. 11 and 12 are compared with a s x s average integral image pixel to pixel. The value is set to black, if it is t percent less than average and white otherwise.

Figure 5a shows the results from binarization using adaptive thresholding from which we have obtained the region of interest (ROI), i.e., the suspension points that are to be taken into consideration for co-ordinate measurement. The white pixel arrays are the marked points for computation of distance as shown in Fig. 5b with

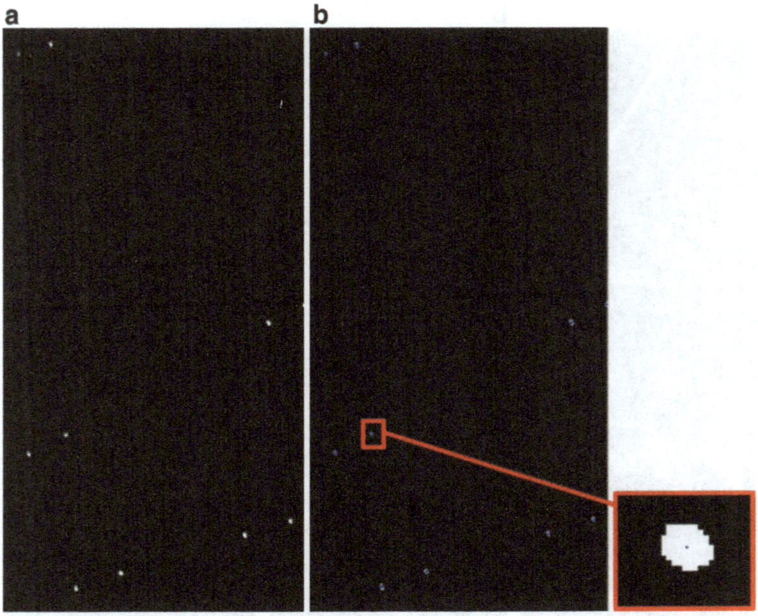

Fig. 5 **a** Region of Interest extraction. **b** Centroid detection using *regionprops*

centroid marked with blue points. *Regionprops* function has been used from image processing toolbox to measure the centroid of the white pixel arrays whose co-ordinates are saved in matrix *Centroid*. Any co-ordinates are chosen as the reference co-ordinate from Centroid matrix and the distances are computed between each point in pixel length.

The actual distance makes use of the calibration factor computed in Sect. 3.1:

$$\text{Actual distance (in cm)} = \text{Image width} \times \text{calibration factor} \tag{13}$$

4 Observation

The all-terrain vehicle was also appointed to industrial co-ordinate measurement for roll-cage. The readings in Table 1 given below gives the readings obtained from the CAD design, the industrial co-ordinate measurement, and the readings from output of MATLAB code for some of the points used to determine the accuracy of the algorithm and technique with industrial standards and ideal requirements. For ease of understanding, the points have been named as shown in Figure 6.

The tab points, i.e., A-B, E-F, C-D, G-H, I-J give appropriate results with comparatively larger deviations as seen in case of E-F, G-H, and I-J tending to >0.01%

Table 1 Comparison of algorithm results with CAD and industrial CMM

Points	CAD measurements (cm)	Industrial CMM (cm)	Algorithm results (cm)	Deviation w.r.t I.CMM (%)
A-B	2.4	2.88	2.90	0.0069
E-F	3.0	3.37	3.43	0.017
C-D	3.0	3.54	3.57	0.0084
G-H	3.6	3.78	4.15	0.089
I-J	3.6	3.89	4.14	0.060
A-C	28.9	31.40	31.41	0.0003
A-E	28.94	30.36	30.41	0.0016
A-G	39.62	42.25	42.27	0.0004
A-I	39.91	42.47	42.67	0.0046

Fig. 6 Labelling of suspension points

deviations from industrial CMM reports. The inter-tabular point distances taken into consideration, i.e., A-C, A-E, A-G, and A-I give better results from tab points and also overall deviations of <0.005% from the industrial CMM reports.

5 Conclusion

The paper introduces a fast and accurate low computation image processing technique that is used to detect manufacturing defects in all-terrain vehicles. The pre-processing takes into account external factors such as non-uniform illumination which is rectified using signal enveloping to removing shading element without any user intervention. The use of reference object for calibration makes the technique useful over wider range of setups. The reference object calculations consist of simple trigonometric rules which are computed using the image area and co-ordinates. The use of bounding box for finding centroid improves the accuracy of the technique reducing placement redundancies. A comparative study was performed on the roll-cage front box of an all-terrain vehicle with the technique and industrial co-ordinate measurement machine which showed results with lesser than 0.1%.

References

1. C.C. Reyes-Aldasoro, Retrospective shading correction algorithm based on signal envelope estimation. Electron. Lett. **45**(9), 454–456 (2009)
2. D. Bradley, G. Roth, Adaptive thresholding using the integral image. J. Graph. Tools **12**, 13–21 (2007). https://doi.org/10.1080/2151237X.2007.10129236
3. R. Bello, T. Suleiman, U.S. Tunga, Determination of object position and orientation using image processing technique. Res. Rev. J. Phys. **7**(2), 18–24 (2018)
4. O. Kainz, F. Jakab, M.W. Horečný, D. Cymbalák, Estimating the object size from static 2D image, in *2015 International Conference and Workshop on Computing and Communication (IEMCON)*, Vancouver, BC (2015), pp. 1–5
5. L. Wojnar, *Image Analysis: Applications in Materials Engineering* (CRC Press, Boca Raton, FL, 1999)
6. L. Zhao, J. Chen, J. Benesty, A multichannel widely linearwiener filter for binaural noise reduction in the short-time-fourier-transform domain, in *IEEE China Summit & International Conference on Signal and Information Processing (ChinaSIP)*, July 2014, pp. 227–231
7. D. Slezak, S.K. Pal, B.-H. Kang, J. Gu, H. Kuroda, T.-H. Kim, *Signal Processing, Image Processing and Pattern Recognition* (Springer Science & Business Media, Germany, 2009)
8. M. Cheriet, N. Kharma, C.-L. Liu, Ch. Suen, *Character Recognition Systems: A Guide for Students and Practitioners* (Wiley, Hoboken, NJ, 2007)
9. K. Jeong, H. Moon, Object detection using FAST corner detector based on smartphone platforms, in *First ACIS/JNU International Conference on Computers, Networks, Systems and Industrial Engineering (CNSI)*, May 2011, pp. 111–115
10. Z. Tang, D. Shen, Canny edge detection codec using VLib on Davinci series DSP, in *International Conference on Computer Science & Service System (CSSS)*, Aug 2012, pp. 221–224

11. W. Rong, Z. Li, W. Zhang, L. Sun, An improved Canny edge detection algorithm, in *IEEE International Conference on Mechatronics and Automation (ICMA)*, Aug 2014, pp. 557–582
12. D. Li, P. Qiao, G. Zhao, A new moving object detection approach with adaptive double thresholds, in *Fifth International Conference on Computational and Information Sciences (ICCIS)*, June 2013, pp. 102–105

Split-Ring Resonator Multi-band Antenna for WLAN/WIMAX/X Standard

Vinita Sharma, Santosh Meena, and Ritesh Kumar Saraswat

Abstract In this paper, we discussed split-ring resonator inspired monopole antenna design. The proposed design is fabricated on FR-4 substrate (30 × 26.5 × 0.8 mm^3) and also covered the WLAN (5.2 GHz Band), Upper WiMAX (3.5 GHz Band), and X Band (7.5 GHz) application. For these ranges, the respectively achieved bandwidths are 3.56 GHz, 5.19 GHz, 7.5 GHz that covered the WLAN, WIMAX, AND X Band wireless standards, respectively. The proposed antenna shows stable radiation pattern.

Keywords Monopole antenna · Split-ring resonator · WiMAX · WLAN · X band

1 Introduction

The antenna is an appropriate device to cover the various wireless standards which are present in microwave range. Many designs are proposed such as CPW-fed square ring slot antenna [1] to resonate at wireless mode. Another antenna design that has various shaped strips (S and U type) is implemented [2], achieves many wireless communication applications. The Y-shaped monopole antenna design consisting of split-ring slot shows various wireless resonating bands at WLAN (5.2 GHz), WiMAX (3.5 GHz), and X band (7.5 GHz) [3] Many more antennas which are feed in different forms are also designed for wireless applications [4] and microstrip fed slot antenna [5]. Recently, metamaterials loaded design by using SRR [6] and CSRR [7] having application for wireless standards [8]. These metamaterials inspired antenna are having compact size and with simplified geometry.

V. Sharma (✉) · S. Meena
Government Women's Engineering College, Ajmer, Rajasthan, India
e-mail: vinita95sharma@gmail.com

S. Meena
e-mail: smeena@gweca.ac.in

R. K. Saraswat
M.L.V. Textile and Engineering College, Bhilwara, Rajasthan, India
e-mail: ritesh.saraswat9@gmail.com

S. S. Dash et al. (eds.), *Intelligent Computing and Applications*,
Advances in Intelligent Systems and Computing 1172,
https://doi.org/10.1007/978-981-15-5566-4_4

The proposed design having compact size patch indicates triple-band characteristics at wireless standards WLAN (5.2 GHz), WiMAX (3.5 GHz), and X Band (7.5 GHz). The proposed design is showing stable radiation property at respective wireless band.

2 Design Methodology

The proposed design has hexagonal shape which is connected with feed line dimension of $7.2 \times 1.8\ mm^2$ microstrip line as represented in Fig. 1. This design is fabricated on a $30 \times 26.5 \times 0.8\ mm^3$ low-cost FR-4 substrate ($\varepsilon_r = 4.4$, $\tan \delta = 0.002$). This antenna has split-ring radiating element to create multi-band nature for wireless applications. The radiating patch area consists of pair of metallic rings that are in parallel form. The top and side views of the proposed design are represented in Fig. 2.

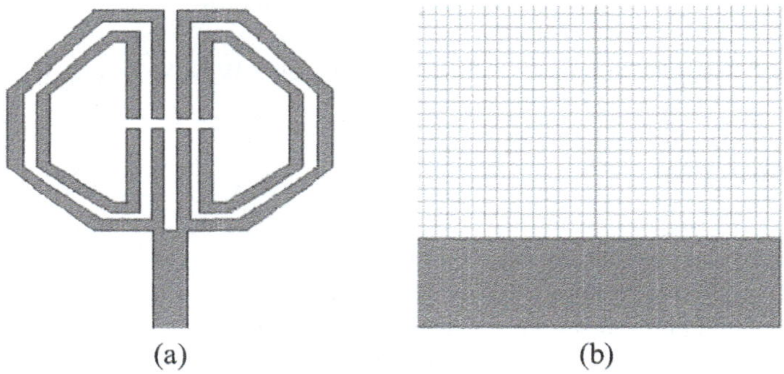

Fig. 1 Antenna design **a** top view **b** bottom view

Fig. 2 Configuration of the designed antenna

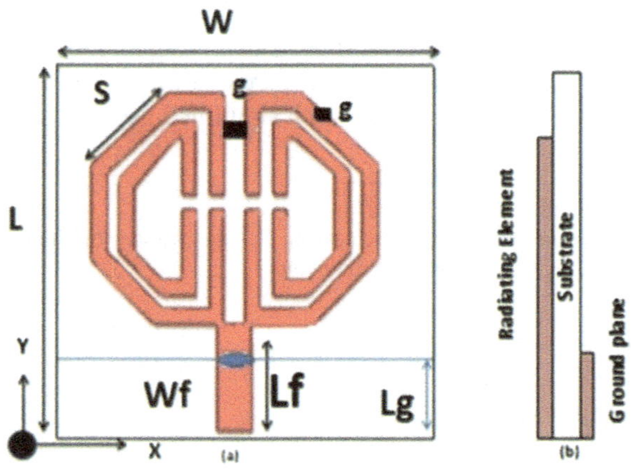

Table 1 Dimension of various proposed antenna parameters

Design parameter	Length (L)	Width (W)	Side length (S)	Gap (G)	Feed line width (W_f)	Feed line length (L_f)	Ground length (L_g)
In mm	30	26.5	0.8	0.75	1.88	7.2	7

As per the antenna design, the dimension width of SRR rings (g) and slot gap of conducting ring are the same. Due to this the proposed design indicates the sufficient bandwidth for wireless standards. Table 1 shows the optimized dimension of the proposed antenna.

3 Result Analysis

With the help of simulating software, we obtained the simulated S parameter as shown in Fig. 3a. The measured S parameter results are obtained in anechoic chamber with the help of vector network analyzer. The proposed design with optimized dimension as mentioned in Table 1 covers three resonant bands at 3.5, 5.2, and 7.5 GHz for wireless standards. By implementation of two concentric spilt rings in radiating patch, it creates the resonant band for WLAN, WiMAX, and X band wireless applications, respectively. The simulated and measured S parameter of the proposed design is shown in Fig. 3.

Figure 4 shows the simulated and measured gain of the proposed antenna at different frequencies. It is noticed that at higher and WLAN/WiMAX frequencies gain is improved with respect to reference level of gain. Figure 5 represents the comparison of radiation efficiency at various resonant frequencies. It is observed that at wireless resonant frequency of WLAN and WiMAX efficiency is increased.

The parametric study regarding width of feed line (W_f) at 1.5 and 2.4 mm is indicated in Fig. 6a, b, respectively. It is identified that wider bandwidth around 3 GHz is obtained for $W_f = 1.88$ mm, i.e., optimized value of width of feed line.

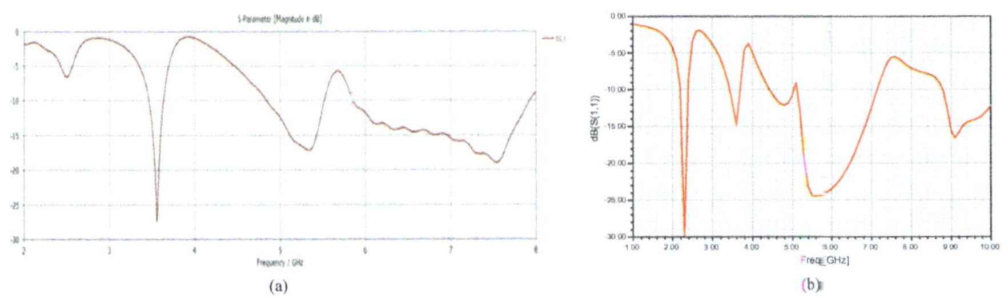

Fig. 3 Simulated and measured S parameter of the proposed antenna

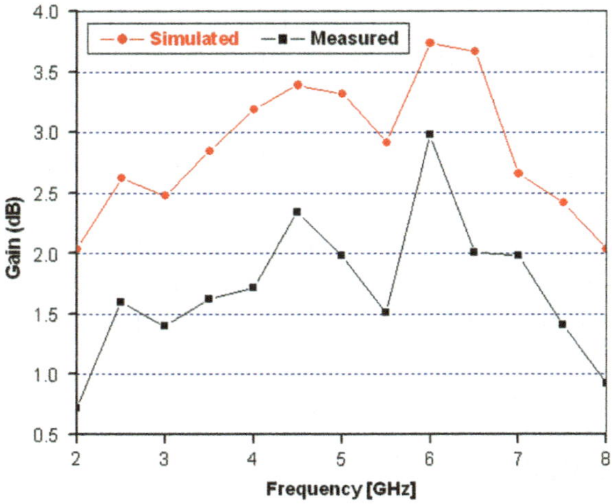

Fig. 4 Simulated and measured gain

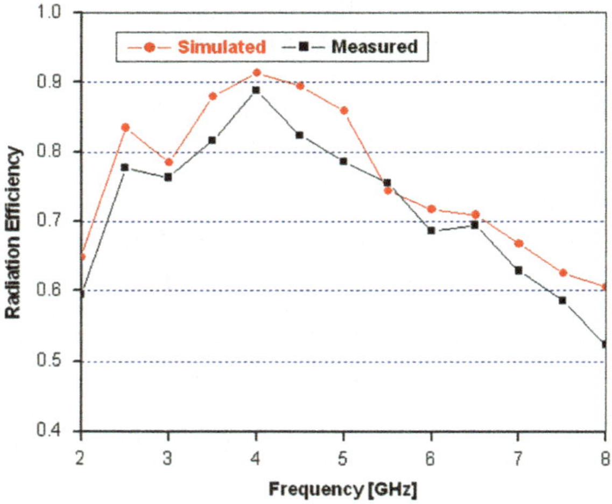

Fig. 5 Simulated and measured efficiency

The simulated surface current distribution at resonant mode 3.5 GHz, 5.2 GHz, and 7.5 GHz are shown in Fig. 7, respectively. It is observed that current density is dense across the periphery of spilt rings at 3.5 GHz as shown in Fig. 7a. Figure 7b, c shows the current distribution characteristics at 5.2 GHz and 7.5 GHz, respectively, indicate maximum distribution around the inner and outer part of split ring that is responsible to create the higher resonant band.

Figure 8 indicates the E and H plane radiation patterns along with 3D polar plot at resonant frequencies 3.5, 5.2, and 7.5 GHz. It is observed that the omnidirectional

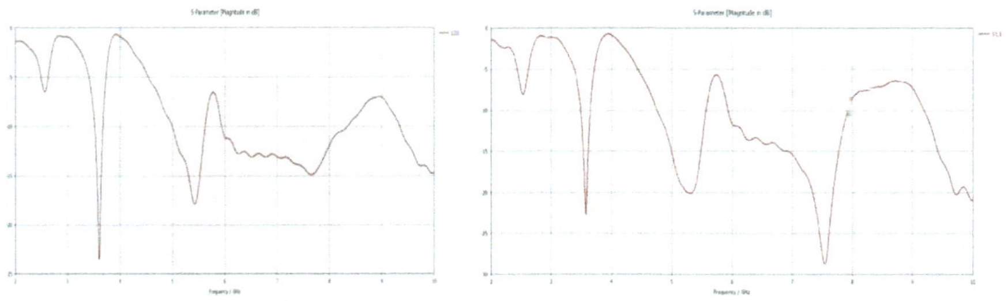

(a) (b)

Fig. 6 Parametric study regarding feed line width (W_f) at 1.5, 2.4 mm

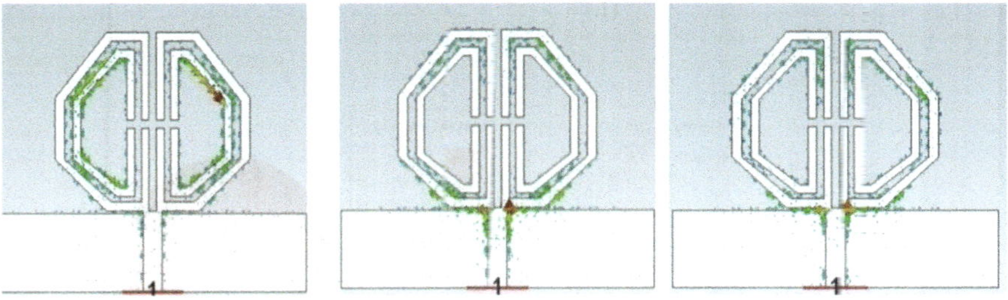

(a) **(b)** **(c)**

Fig. 7 Simulated current distribution **a** 3.5 GHz, **b** 5.2 GHz, **c** 7.5 GHz

radiation patterns are obtained in case of H plane for all the resonant bands, whereas bidirectional patterns are observed in case of E plane. The radiation pattern is stable for all the resonant frequencies. The simulated peak gain of 2.6, 2.36, and 3.03 dBi is obtained across the 3.5, 5.2, and 7.5 GHz resonant frequency,respectively.

4 Conclusion

This antenna shows triple-band characteristics at WILAN (5.2 GHz), upper WMAX (3.5 GHz), and X Band (7.5 GHz) wireless standards. The proposed antenna has compact size with pair of hexagonal shape concentric split ring as a radiating part. The proposed design has appropriate surface current distribution and stable radiation pattern for operating band of wireless applications.

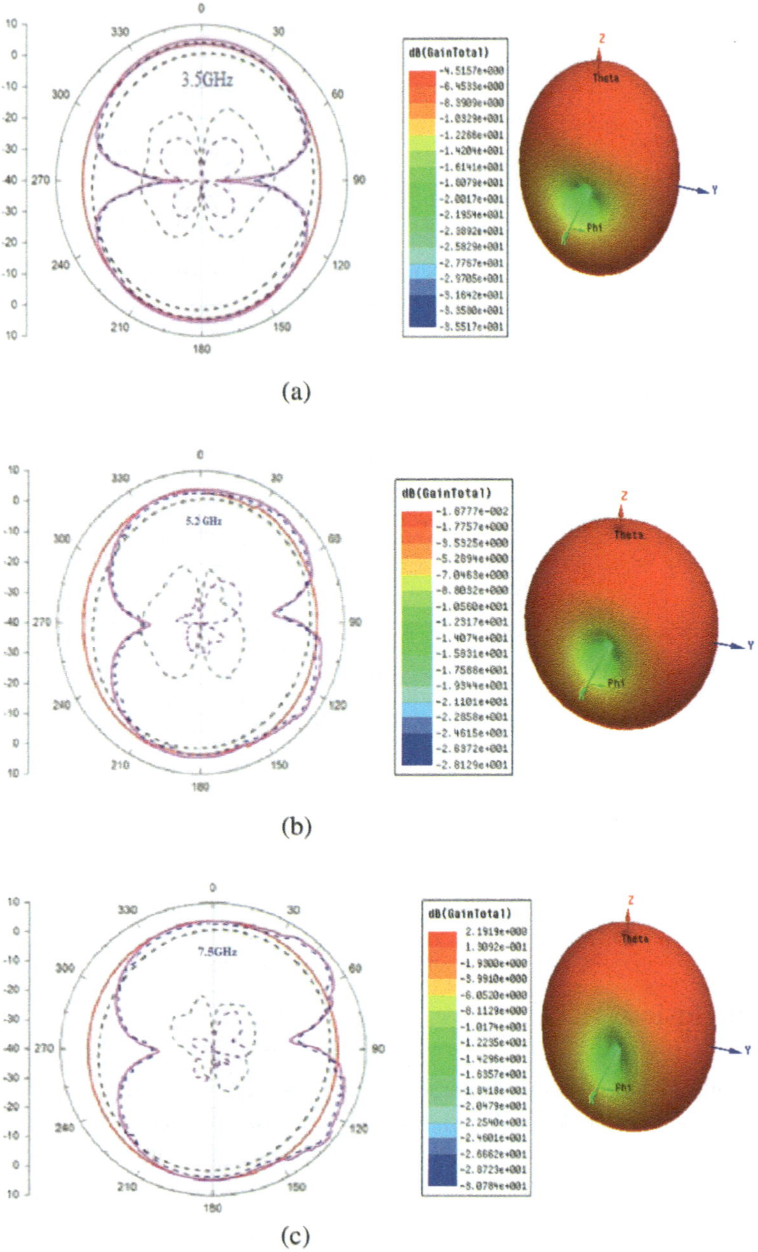

Fig. 8 E-planed H-planed radiation patterns/3-D polar plot at resonant frequency at 3.5, 5.2, 7.5 GHz

References

1. C.Y.D. Sim, H.D. Chen, K.C. Chiu, C.H. Chao, Coplanar waveguide-fed slot antenna for wireless local area network/world-wide interoperability for microwave access applications. IET Microwave Antennas Propag. **6**, 1529–1535 (2012)
2. Y. Xu, Y.C. Jiao, Y.C. Luan, Compact CPW-fed printed monopole antenna with triple-band characteristics for WLAN/WiMAX applications. Electron. Lett. **48**, 1519–1520 (2012)
3. P. Liu, Y. Zou, B. Xie, X. Liu, B. Sun, Compact CPW-fed tri-band printed antenna with meandering split-ring slot for WLAN/ WiMAX applications. IEEE Antennas Wirel. Propag. Lett. **11**, 1242–1244 (2012)
4. X. Li, X.-W. Shi, W. Hu, P. Fei, J.-F. Yu, Compact triband ACS-fed monopole antenna employing open-ended slots for wireless communication. IEEE Antennas Wirel. Propag. Lett. **12**, 388–391 (2013)
5. X.Q. Zhang, Y.C. Jiao, W.H. Wang, Compact wide tri-band slot antenna for WLAN/WiMAX applications. Electron. Lett. **48**, 64–65 (2012)
6. K. Yang, H. Wang, Z. Lei, Y. Xie, H. Lai, CPW-fed slot antenna with triangular SRR terminated feed line for WLAN/WiMAX applications. Electron. Lett. **47**, 685–686 (2011)
7. S.C. Basaran, U. Olgun, K. Sertel, Multiband monopole antenna with complementary split-ring resonators for WLAN and WiMAX applications. Electron. Lett. **49**, 636–638 (2013)
8. Y. Kim, D. Kim, CPW-fed planar ultra wideband antenna having a frequency band notch function. Electron. Lett. **40**, 403–405 (2004)
9. S. Qu, J. Li, Q. Xue, A band-notched ultra wideband printed monopole antenna. IEEE Antennas Wirel. Propag. Lett. **5**, 495–498 (2006)
10. Y. Ding, G. Wang, J. Liang, Compact band-notched ultra-wide-band printed antenna. Microwave Opt. Technol. Lett. **49**, 2686–2689 (2007)
11. Y. Zhang, W. Hong, C. Yu, Z.-Q. Kuai, Y.-D. Don, J.-Y. Zhou, Planar ultra wideband antennas with multiple notched bands based on etched slots on the patch and/or split ring resonators on the feed line. IEEE Trans. Antennas Propag. **56**, 3063–3068 (2008)

Analyses on Architectural and Download Behavior of Xunlei

Bagdat Kamalbayev, Nazerke Seidullayeva, Adilbek Sain, Pritee Parwekar, and Ikechi A. Ukaegbu

Abstract The following paper conducts an extended analysis on different measurement studies performed on popular P2P Chinese application called Xunlei. In this particular research paper, several previous researches have been studied and their arguments were compared, in order to find out finalized reason behind this enormous companies' success. Its packet-level data was evaluated in order to understand applications network architecture and download behavior. Xunlei's core-server layer was analyzed by division of tasks. It was discovered that application uses direct linking and applies multi-source technology in an effective way in order to increase the download speed. Also, it was found that Xunlei uses UDP as the main transport protocol for both IPTV streaming and bulk file transfer. This paper provides a broad study on Xunlei which leads to a better understanding of the Internet.

Keywords Internet measurement · Xunlei · P2P · Bandwidth theft

1 Introduction

The main aim of the following paper is to investigate the P2P application called Xunlei, a file-sharing application, utilizing different existing researches already

B. Kamalbayev · N. Seidullayeva · A. Sain · I. A. Ukaegbu (✉)
School of Engineering and Digital Sciences, Nazarbayev University, Nur-Sultan, Kazakhstan
e-mail: ikechi.ukaegbu@nu.edu.kz

B. Kamalbayev
e-mail: bagdat.kamalbayev@nu.edu.kz

N. Seidullayeva
e-mail: nazerke.seidullayeva@nu.edu.kz

A. Sain
e-mail: adilbek.sain@nu.edu.kz

P. Parwekar
Computer Science & Engineering Department, SRM University, Delhi-NCR, Sonepat, India
e-mail: priteep@srmist.edu.in

S. S. Dash et al. (eds.), *Intelligent Computing and Applications*,
Advances in Intelligent Systems and Computing 1172,
https://doi.org/10.1007/978-981-15-5566-4_5

undertaken toward this company. Such applications have good contribution to the modern internet traffic [1]. Studying this download manager could provide better understanding of the modern Internet [1, 2]. Xunlei is one of the most popular streaming media and file-sharing applications in China. It can be seen from the following statistics which are provided by Wikipedia:

- Over 80 million users installed Xunlei.
- Xunlei's website draws over 50 million hits per day.
- Number of users on the official website exceeded 1880 million.

Xunlei helps people to easily find files and download them at high speed. It uses multi-source technology to deliver all kinds of files and integrates resources on the internet. Also, there is a good contribution of file-sharing and streaming protocols such as eMule, Bit Torrent, FTP, and RTSP [3]. However, Xunlei is not only file-sharing application, but it also affords IPTV functionality which allows people to watch TV programs and movies online. Popularity of Xunlei among users attracts attention of big companies such as Google, IDGVC, and Geyuan Ventures despite that it has not English version and runs only on Windows machines. It is spread outside China due to migrants and soon will be popular worldwide too. Until now, not many researches have been conducted to investigate this popular company.

1.1 Resource Searching and Sharing Mechanisms

In the succeeding paragraphs, Xunlei's resource searching policy will be explained and compared with Client/Server systems and P2P systems [4–8].

(1) Client/Server systems

  - Traditional web downloads: A central server affords content to clients directly.

(2) Peer-to-Peer Systems Peers can exchange content among each other by creating their own overlay networks. Examples:

  - Gnutella: sending many requests to find the needed file
  - eDonkey: uses own servers as querying organs.
  - Bit Torrent: uses torrent files and own servers as querying organs.

(3) Xunlei resource system is like a combination of the previous systems. Xunlei has its own overlay network, but it also uses resources which are located on other servers. There are examples of these resources [6]:

  - Internal server: when a user seeks a file on a website, and the website gets file from a server which belongs to this website.
  - External server: when a user seeks a file on a website, and the website gets file from another server which does not belong to this website.
  - Peer: Xunlei indexes peers who got the file.

1.2 Direct Linking or Bandwidth Leaching

Using external servers is considered as bandwidth theft in the form of direct linking [9]. It is also called as inline linking or leeching. An example of leeching can be usage of linked images, from one website into a webpage which belongs to the second website. If there is no agreement between them, the second website is considered to be leeching to the first one. In Xunlei, a client is called as leecher if it uses Xunlei to download a file from one website and it downloads file from external servers instead of using internal or P2P servers. Direct linking harms external sites earnings from advertisements but allows users to download files at high speed. However, according to the other recent researchers, Xunlei begins to compensate for this so-called bandwidth leaching by providing membership to the Xunlei Union to these external servers. One of the researches reveals that Xunlei also supplies with the bandwidth support to those who are not even Xunlei's servers and sources. For instance, client who uses Xunlei as downloading application, starts to download some files from an external server, and if other Xunlei's clients have already downloaded that file, that client can support the external server as well.

1.3 Contributions of This Paper

First of all, architecture of Xunlei's core-system is studied by revealing its structure and client workflow. Further, its download behavior is evaluated through observation of several download processes for IPTV streaming and bulk file transfer. The main results are [10–12]

- It was revealed that Xunlei's core servers clearly divide tasks between its servers.
- It was found that Xunlei exploits external servers in order to speed up download speeds.
- It was shown that Xunlei's multi-source download manager works effectively.
- It was discovered that Xunlei uses UDP as the main transport protocol for both IPTV streaming and bulk file transfer [13].
- Xunlei uses tracker and cross-protocol system to find the files requested in peers and in other servers easily and quickly. Also, plug-in system details are provided below.

The related work is discussed in the second section. In Sect. 2, experiment and measurements are portrayed and based on these studies Xunlei architecture is explained. The information about download behavior is written and the last section concludes the paper.

2 Results

2.1 Xunlei's Architecture

Xunlei has contributed to the improvement of download session and gave great advantages to the people. The traffic collection of Xunleis sessions has helped to understand the basic working mechanisms of that product, and also to identify the servers which then have been filtered based on IP addresses with the reverse DNS protocol. The defined servers have been analyzed to understand the Xunlei's server infrastructure and to demonstrate the clear advantage over the servers of other popular download software such as eDonkey [6, 7].

Core servers: A Xunlei and eDonkey client/server architectures are different:

(1) The amount and IP addresses of Xunlei's servers are stable and do not alter frequently while those eDonkey servers' features are changed regularly. Thus, Xunlei's client/server architecture aimed to have permanent IP addresses and can be considered as a reliable source of connection. An experiment was conducted to identify the number of Xunlei's server. It was found that 46 IP addresses were used and most of them are located in three provinces of China.
(2) Xunlei servers are organized in a more confounded and systematic way as compared to those of eDonkey [6, 7]. Every server of eDonkey software is primarily responsible for indexing documents for customers. Xunlei servers participate adequately with obvious divisions of work. These categories of work may be listed in such a way:

    a. Client initialization
    b. Session tracking
    c. Nodes server directory
    d. Nodes server
    e. Resource management (providing resource reports with external servers and peers to clients)
    f. Indexing advertisements
    g. Storing advertisements
    h. Picture server
    i. Virus Scan
    j. Statistics
    k. Keep-alive server

Client-Server communication: The Client-server communication in Xunlei has three phases:

(1) **Login** The client uses login session to establish a connection to the Xunlei network. This login session can be initiated with the client request through TCP and session tracking server on UDP/4000, and also the amount of available nodes servers is required at the nodes server directory through UDP. In case of

rejecting request, it continues to link to other nodes in the server directory. This action continues till client gets positive replies and it is then informed to the session tracker, thus login session will be stopped.

(2) **Downloading**: Client starts downloading from one website by requesting an URL, then this URL is translated into a distinctive number and will be provided with resulting resource report where it includes locations of requested file. The translation of URL is conducted by resource management server to identify the needed data. From the provided location, source can download the data in parallel by applying multi-source technology that is provided by Xunlei's technology.

(3) **Idling:** In order to hold idling state, Xunlei client holds communication with server and sends ICMP messages every 10 s and UDP in between 45 and 300 s to the session tracker and node servers. These actions contribute to update and report the client status.

2.2 Download Behavior

The previous analysis only gives general understandings of this huge infrastructure, the more important things are how the resources in this enormous company are allocated between core servers. Also, it is predicted that one of the main reasons for popularity of Xunlei is due to this resource allocation method which increases the rate of downloading. Therefore, in the beginning of the following subsections, the attention was on data redundancy and finally on the protocols that Xunlei mostly uses.

Resource allocation:

As explained in the earlier section, three main resources are used, such as Internal servers, it is when the all the requested files are taken from the specific website which belongs to that specific company; peer-to-peer, is the other type of the resources, where the clients share the requested file between each other; and the last type was external servers, where the requested files are taken from website that does not belong to the specific website. Before analyzing the results, let us understand how the measurements were gained through experimentation. To conduct the experiment, 10 famous files were chosen as well as 4 portal websites. Furthermore, each file was downloaded 5 times from the same website. These totaled as 165 sessions. The number of sessions is higher, due the nonexistence of the requested file in the specific website. In Figs. 1 and 2, the detailed allocation of the resources per session can be seen. The horizontal line on the Figure illustrates the download processes grouped according to websites, meanwhile the vertical line illustrates the types of the resources. According to this figure, it is clearly shown that the most used type of the resources is external servers. These results were collected in 2012 which was the last research that was conducted on this subject. To see if any differences have accrued, this work has conducted another experiment as illustrated in Fig. 2. It is

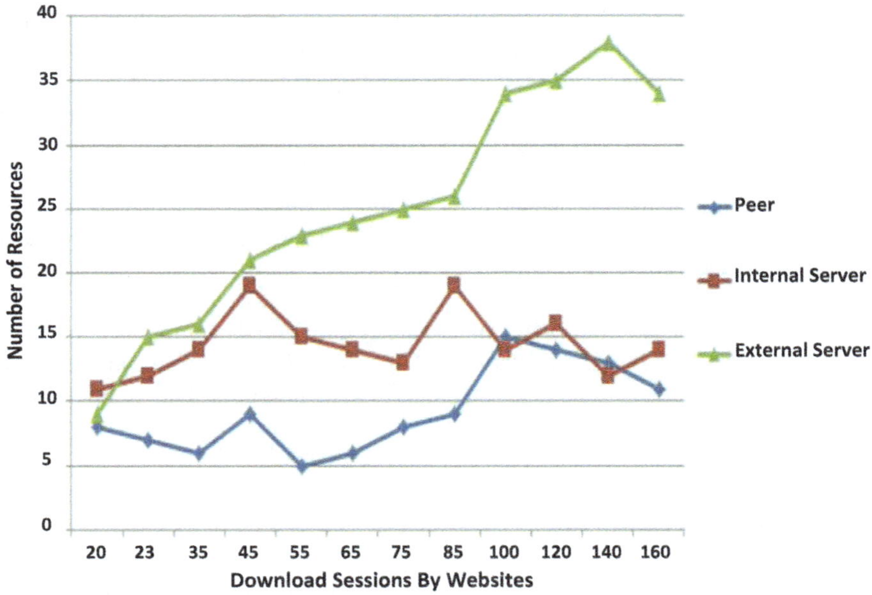

Fig. 1 Experimental result of resources gained from different servers and peers (conducted in 2012)

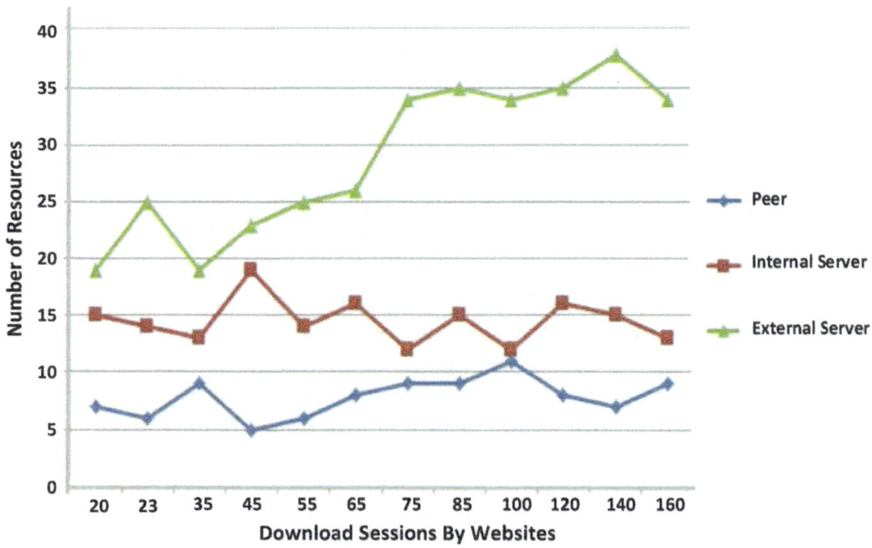

Fig. 2 Experimental result of resources gained from different servers and peers (conducted in 2018)

clearly seen that even after many years behind the Xunlei mainly uses external servers rather than internal servers and peers.

2.3 *Download Redundancy*

Measurements obtained were observed to ascertain how data redundancy behaves. The attention was given especially to this feature, since the large amount of the resources gained does not guarantee that the downloaded files are in good quality. To examine this quality an experiment was conducted in which 210 downloads were examined. In Fig. 3, it is clearly seen that most of the portions of the downloads have less than 1.2 redundancy, and nearly the other rest have only redundancy between 1.2 and 1.5, which is suggested as these downloads have good quality. The results are shown in Fig. 3.

UDP or TCP:

In this last section, bulk-files and IPTV files in Xunlei were examined. The main reason behind it is to find out, what kind of transport protocols Xunlei uses that their downloading rate is very high compared to the other big companies and even has good quality of the downloaded files. It was investigated that Xunlei provides a website where one can not only download the video files but also to watch it online. Therefore, four popular video files were both downloaded and streamed in this experiment. The measurements obtained are shown below on the Table. It is clearly seen that major part of the files or it can be said almost all of them were transferred utilizing UDP rather than TCP. In order to find out the real justification, paper has tried to become Xunlei client and using software Wireshark downloaded several files from this application, meanwhile were capturing files through Wireshark. The results were as was predicted by previous researches, almost all the packets captured were using UDP rather than

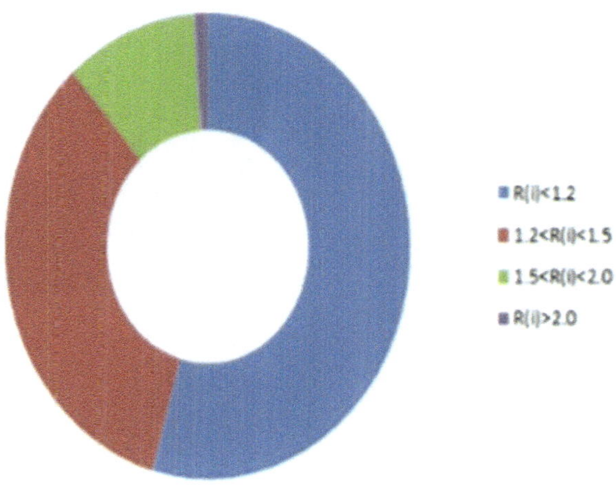

Fig. 3 Redundancy of the different downloaded files

TCP as can be seen from Fig. 4. In Fig. 5, paper has included finalized results gained from packet capturing, it is seen that almost 98 % is UDP. Paper has suggested that the reason behind this good downloading speed might be in the utilization of the right transport protocol. In case of Xunlei, it appears that UDP helps to maintain good velocity.

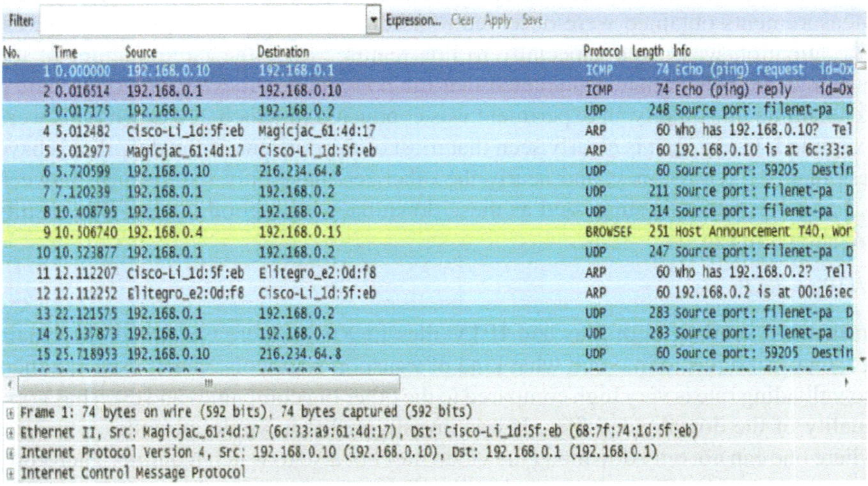

Fig. 4 Packet capturing using Wireshark in order to find out which protocol is used

Fig. 5 UDP is used mainly over TCP protocol in both bulk file-sharing and in IPTV

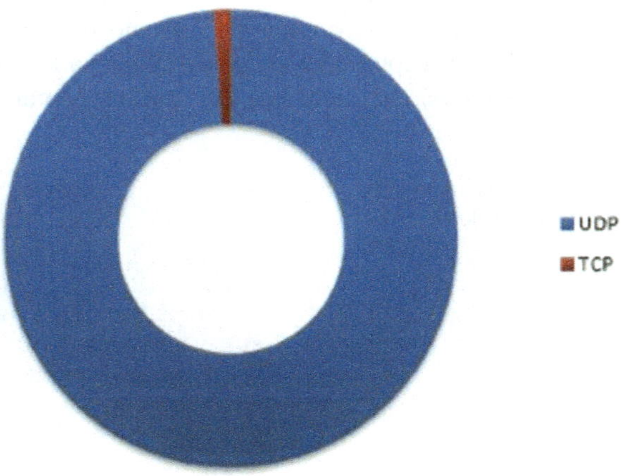

2.4 Plug-in

It was discovered while conducting the experiment that Xunlei utilizes plug-in system. Whenever the client in first case installs the Xunlei client application, immediately plug-in is also installed into the Internet Explorer browser. It allows the Xunlei client to take over the browsers speed from other users who are also downloading the same file. This can be applied to different types of files as well, such as video files, pdf files, also supports different protocols HTTP, FTP, RTSP, and MMS. Also, the suggestions made earlier in this work that Xunlei clients serve as the Bit Torrent and eDonkey client were justified. It might be the other reason, however not the main according to the "Xunlei: Peer-Assisted Download Acceleration on a Massive Scale" article, that Xunlei's downloading acceleration is the highest among Chinese downloading applications [14].

2.5 Tracker

The recent research claims that some interesting analyses have been found during observing the local Xunlei clients traffic. Particularly, the details between the client's message exchange with the Xunlei tracker when it requests the available sources for any file. For instance, when a Xunlei client sends reference to the central tracker, every time central tracker returns 20-byte hash values and 8-byte code for a requested file. Actually, both hash and code values are later used to request resources for the file from other peers and servers. Also, this hash and code values can be received from the tracker in other ways as well, it can be obtained by sending distinctive 20-byte identifier for the file. For Bit Torrent files, the Xunlei client utilizes the info hash of the file by 20-byte identifier which is included to the message which is then send to the tracker. For eDonkey files, this identifier is extracted from its link which is similar to the file size. In case when Xunlei client sends the message to the tracker, particularly the identifier that is not currently tracked by the Xunlei, the tracker does not give any output with two hash and code values. Xunlei's cross-protocol operation, to explain it more accurately let's take a look at an example. For instance, there is a file that exists in HTTP server and in Bit Torrent as well, and also consider that the tracker does not know about the file. Whenever the client downloads this file from Bit Torrent, it begins to map two hash values internal hash within the Bit Torrent info hash. Also, if another Xunlei client begins to download the same file from the server in direct way, it begins to measure the same internal hash values, and then when it informs the tracker, it will know that there is mapping between HTTP URL and internal hash. From this, it can be concluded that the tracker knows that the same file matches both Bit Torrent info hash and HTTP. For that reason, in case Xunlei client begins downloading with Bit Torrent info hash, the tracker will immediately provide the HTTP server link and the Xunlei peers that also have the same file.

3 Conclusion

In this paper, attempt has been made to understand the working principal of China's most famous downloading application, which is known as Xunlei. This software application was chosen since it is used by a large number of people. To find out how the server-side of the application works, several experiments were conducted, however, all of them are according to the packet-level feature, because data transmission in Xunlei is encoded. First, it has suggested to focus on the architectural behavior of the Xunlei, in order to do that paper have studied previous analysis done toward in previously from different researchers, particularly this paper wanted to find out Xunlei's servers and client-server relations. Then, it was decided to investigate Xunlei's downloading behavior, which leads us to figure out that, this company supply with great amount of resources, which definitely makes in its turn the speed rise. However, a major part of the data which is requested is taken from external servers. Furthermore, the great amount of resources does not always guarantee that data downloaded is in good quality, for that reason, downloading redundancy as well was examined. Also, the experiment was conducted to find out what kind of transport layer Xunlei mainly uses. It was revealed that Xunlei despite the file type, bulk file-sharing or IPTV streaming, utilizes 95 % UDP transport protocol rather than TCP. All of these results lead us to conclude our analysis that the key factors about this provider is that it provides a great amount of resources and mainly uses external servers and UDP protocol, and it has great architectural structure where all the task are accordingly divided within the server. However, the external servers are considered as improper, since other external websites gain no advantage from it, but only Xunlei gets the advantage. For that reason, it was suggested that the good way to correct this is to make it an open and business-driven network architecture. However, it was discovered that Xunlei already has solved this problem by introducing the external servers to their Xunlei Union membership.

References

1. H. Schulze, K. Mochalski, Internet study 2008/2009. Ipoque Rep. **37**, 351–362 (2009)
2. W. John, S. Tafvelin, T. Olovsson, Trends and differences in connection-behavior within classes of internet backbone traffic, in *International Conference on Passive and Active Network Measurement* (Springer, Berlin, 2008), pp. 192–201
3. M. Zhang, W. John, C. Chen, A measurement-based study of Xunlei, in *Student Workshop*, PAM (2009)
4. S. Saroiu, P.K. Gummadi, S.D. Gribble, Measurement study of peer-to-peer file sharing systems, in *Proceedings of Multimedia Computing and Networking (MMCN)* (2002)
5. J. Miller, Characterization of data on the gnutella peer-to-peer network. In *First IEEE Consumer Communications and Networking Conference, 2004. CCNC 2004* (IEEE, 2004), pp. 489–494
6. J. Yang, H. Ma, W. Song, J. Cui, C. Zhou, Crawling the edonkey network, in *2006 Fifth International Conference on Grid and Cooperative Computing Workshops* (IEEE, 2006), pp. 133–136

7. O. Heckmann, A. Bock, The edonkey 2000 protocol. Rapport technique, Multimedia Communications Lab, Darmstadt University of Technology (2002), p. 13
8. M. Zhou, Y. Dai, X. Li, A measurement study of the structured overlay network in P2P file-sharing applications, in *Eighth IEEE International Symposium on Multimedia (ISM'06)* (IEEE, 2006), pp. 621–628
9. M. Zhang, C. Chen, N. Brownlee, Behaviors of bandwidth theft in Xunlei, in *2009 2nd IEEE International Conference on Computer Science and Information Technology* (IEEE, 2009), pp. 520–524
10. Y.J. Hall, P. Piemonte, M. Weyant, Joost: A measurement study. School of Computer Science, Carnegie Mellon University, Technical Report (2007); J. Liang, R. Kumar, K.W. Ross, The kazaa overlay: a measurement study. Compu. Netw. J. **49**(6) (2005). (Elsevier)
11. S.A. Baset, H.G. Schulzrinne, An analysis of the skype peer-to-peer internet telephony protocol, in *INFOCOM 2006. 25th IEEE International Conference on Computer Communications Proceedings* (2006)
12. M. Zhang, M. Dusi, W. John, C. Chen, Analysis of UDP traffic usage on internet backbone links, in *2009 Ninth Annual International Symposium on Applications and the Internet* (IEEE, 2009), pp. 280–281
13. P. Dhungel, K.W. Ross, M. Steiner, Y. Tian, X. Hei, Xunlei: peer-assisted download acceleration on a massive scale, in *International Conference on Passive and Active Network Measurement* (Springer, Berlin, 2012), pp. 231–241
14. P. Dhungel, W. Ross, M. Steiner, Y. Tian, X. Hei, Xunlei: Peer-assisted download acceleration on a Massive Scale" in SAINT 09, *9th Annual International Symposium on Applications and the Internet*, 2015

Advanced Driver Assistance System Technologies and Its Challenges Toward the Development of Autonomous Vehicle

Keerthi Jayan and B. Muruganantham

Abstract The automotive industry is more concerned about vehicle and road safety. Nowadays the Govt. has introduced new safety regulations to improve the vehicle and road safety. Most of the accidents are happening because of human errors, violation of traffic rules, etc. To avoid or reduce these accidents and to meet the Govt. safety regulations the automotive industries started implementing new technologies like Advanced Driver Assistance System (ADAS), Passive and Active safety system in vehicles. ADAS is the most important safety feature in modern automobiles. The key functionality of ADAS is to provide more comfort to the driver and to improve road and vehicle safety by minimizing human errors. To achieve these functionalities the system use sensors like Radar (Radio Detection and Ranging), Camera, Light Detection and Ranging (LIDAR), and Ultrasonic sensors. Based on sensor input Electronic Control Unit (ECU) delivers the information to driver in terms of warnings, alarms, etc. or system takes the longitudinal and lateral control to bring the vehicle to safer zone. Focus of this paper is to discuss the hardware and software technologies used for the development of ADAS systems and its challenges.

Keywords ADAS · ECU · LIDAR · RADAR

1 Introduction

Since there is an intensification in popularity of automobiles over the last century, rise in popularity of automobiles has a significant role in road accidents. As per the Global Status Report on Road Safety 2018 statistics [1] which reveals that road accidents on an average cause 20–50 million injuries and 1.3 million fatalities around

K. Jayan (✉) · B. Muruganantham
Department of Computer Science and Engineering, SRM Institute of Science and Technology, Kattankulathur, Chennai, Tamil Nadu 603203, India
e-mail: keerthij@srmist.edu.in

B. Muruganantham
e-mail: muruganb@srmist.edu.in

S. S. Dash et al. (eds.), *Intelligent Computing and Applications*,
Advances in Intelligent Systems and Computing 1172,
https://doi.org/10.1007/978-981-15-5566-4_6

the world. Most of the accidents are due to human errors. The demand for an error-free driving environment is increased. This facilitates the development for Advanced Driver Assistance System (ADAS) technologies. ADAS is the combination of sensors that aid and improve the comfort of driver and safety of passengers on the road. It enhances the machine to human interface by detecting the set of on-road and driver behavior by providing visual and audio information to the driver.

Nowadays most of the modern vehicles are equipped with ADAS functionalities to improve the safety level and provide more comfort to the driver. ADAS systems are capable of sensing environment and analyze, predict, and react to the situation. ADAS functionalities started with providing support to the driver action, for example, Anti-lock Braking System (ABS) [2]. This system helps to avoid wheel locking during braking. Anti-lock breaking systems was introduced in the vehicle about 70 years ago [2], and it was the first driver assistance system [2]. ABS first introduced as an anti-skid system for aircraft was used in the 1950s. In 1970, the first-generation anti-lock braking system was developed by Mercedes-Benz and TELDIX. In the late 1960s, the first Electronic cruise control system was developed. The succeeding developments led to adaptive cruise control systems [3].

Today the ADAS system uses the sensors like Radar, LIDAR, Ultrasonic, and Camera to introduce new functionalities to improve driver comfort and vehicle safety. Research and Development for the ADAS system are focusing to develop a driverless vehicles. The driverless vehicle needs to train with lots of data of different road scenarios and objects. However, the system has its own limitation in terms of technologies to handle the data. The system also needs self-learning capability to improve safety. The Artificial Intelligence (AI) technologies aid the system for self-learning purpose. Society of Automobile Engineers (SAE) levels were established as a classification system for autonomous vehicle in January 2014.

2 Sae Level of Advanced Driver Assistance Systems

In May 2013, National Highway Traffic Safety Administration (NHTSA) established a policy on automated vehicles where it defines the automation levels from zero to four. This level shows the growth of vehicle automation from zero to fourth level, which is the fully automated vehicle. As per the SAE classification, the ADAS systems were classified from level 0 to 5 different stages based on the level of automation. In both set of automation definitions (NHTSA and SAE) the first 3 levels of automations are common, but in SAE automation the fully automated 4th level of NHTSA was subdivided into two, SAE level 4 and level 5. SAE J3016 issued on January 2014 provides the information of all modes of automation in cars. Table 1 shows the summary of levels of driving automation.

Table 1 Summary of levels of driving automation

SAE level	SAE name	SAE narrative definition	Execution of steering and acceleration/ deceleration	Monitoring of driving environment	Fallback performance of dynamic driving task	System capability (driving modes)	BASt level	NHTSA level
Human Driver Monitor the driving Environment								
0	No Automation	The full-time performance by the human driver of all aspects of the dynamic driving task even when enhanced by warning or intervention systems	Human Driver	Human Driver	Human Driver	NA	Driver Only	0
1	Driver Assistance	The driving mode-specific execution by a driver assistance system of either steering or accelerating/deceleration using information about the driving environment arid with the expectation that the human driver performs all remaining aspects of the dynamic task	Human Driver and System	Human Driver	Human Driver	Some Driving Modes	Assisted	1

(continued)

Table 1 (continued)

SAE level	SAE name	SAE narrative definition	Execution of steering and acceleration/ deceleration	Monitoring of driving environment	Fallback performance of dynamic driving task	System capability (driving modes)	BASt level	NHTSA level
2	Partial Automation	The driving mode-specific execution by one or more driver assistance system of both steering or accelerating/deceleration using information about the driving environment and with the expectation that the human driver performs all remaining aspects of the dynamic task	System	Human Driver	Human Driver	Some Driving Modes	Partially Authomated	2
Automated Driving System ("System") Monitor the driving Environment			System	System	Human Driver	Some Driving Modes	Highly Automated	3
3	Conditional Automation	The driving mode-specific performance by an automated driving system of all aspects of the dynamic driving task with the expectation that the human driver will respond appropriately to a request to intervene						

(continued)

Table 1 (continued)

SAE level	SAE name	SAE narrative definition	Execution of steering and acceleration/deceleration	Monitoring of driving environment	Fallback performance of dynamic driving task	System capability (driving modes)	BASt level	NHTSA level
4	High Automation	The driving mode-specific performance by an automated driving system of all aspects of the dynamic driving task, even if a human driver does not respond appropriately to a request to intervene	System	System	System	Some Driving Modes	Highly Automated	3/4
5	Full Automation	The full-time performance by an automatcd driving system of all aspects of dynamic task under all roadway and environmental conditions that can be managed by human driver	System	System	System	All Driving Modes		

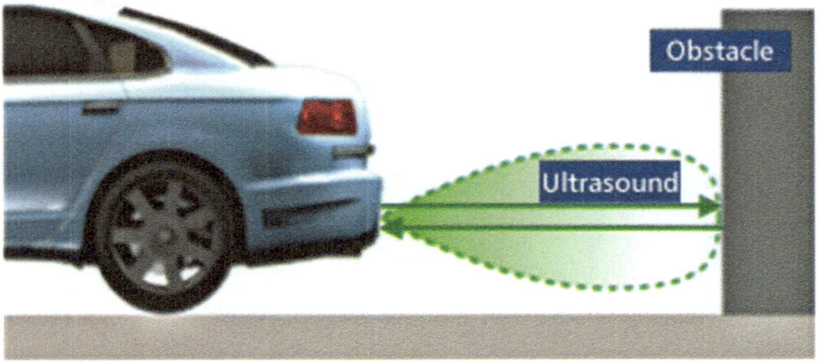

Fig. 1 Overview of parking sensor

2.1 SAE Level Zero: No Automation

Level zero is the basic SAE level of automation where human driver performs all driving tasks like longitudinal control (acceleration or deceleration) and lateral control (steering). However, this system provides information to the driver as a warning that helps driver to monitor the surrounding environment.

2.1.1 Parking Sensor

Parking sensor system provides acoustic warning to the human driver based on the distance between vehicle and detected obstacle that helps the human driver to avoid collision by taking corrective measures. Generally, ultrasonic sensors are used as parking sensors in vehicles. The ultrasonic sensors can detect the presence of neighboring objects without having any physical contact. There are around 4–8 ultrasonic sensors located on front and rear of the vehicle to detect objects.

Ultrasonic sensors determine distance from the sensor to object. The basic principle of distance measurement in ultrasonic sensor is based on Eco [4]. The sensor transmits the sound waves to the environment, after striking on the obstacle those waves return to the origin. The distance between sensor and object can be calculated based on the time of flight of an ultrasonic wave. Figure 1 shows the overview of a parking sensor.

2.1.2 Traffic Sign Recognition (TSR)

In Traffic Sign Recognition (TSR) [5] system, the vehicle is able to recognize traffic signs on the road and provide information to the driver. TSR recognize traffic signs like speed limits, turn, overtaking, slippery road, and other warnings, etc. The system uses a forward-looking camera positioned behind the rearview mirror on the windshield that detects road signs, and TSR system also uses supplementary information from the HD maps. Basic TSR system consists of mainly three parts [5].

1. Knowledge Database: System stores information of traffic sign pattern and representation related to it. This allows fast identification of received pattern
2. Leaning Module: Collect the image data from the received video data and improve the knowledge database of the system.
3. Recognizing Modules: This module process the input image to identify area of traffic sign. This processed traffic sign image is compared with available traffic sign from the knowledge database to get the best matching result and communicate with the driver.

2.1.3 Blind Spot Detection (BSD)

Blind spot detection system processes the sensor information and warns driver in the form of visual audible indicator [2]. Visual indicators often integrated with the outside side-view mirrors. Generally, this system uses two types of sensors like radar and camera [6].

In radar-based blind spot detection system, the sensor usually mounts on rear left and right position on the vehicle bumper fascia. This system uses radio wave to detect the obstacle. The system sends and receives radar signal to the blind zone area and provides visual and audible information to driver. Figure 2 shows the radar-based blind spot detection in vehicle.

Camera-based blind spot detection uses image processing technology to identify object in the blind zone. In this system, camera is located on the rearview mirror or near to the rearview mirror. The system identifies vehicle shape in daytime and light in night-time. Figure 3 shows camera-based blind spot detection.

Fig. 2 Radar-based blind spot detection system

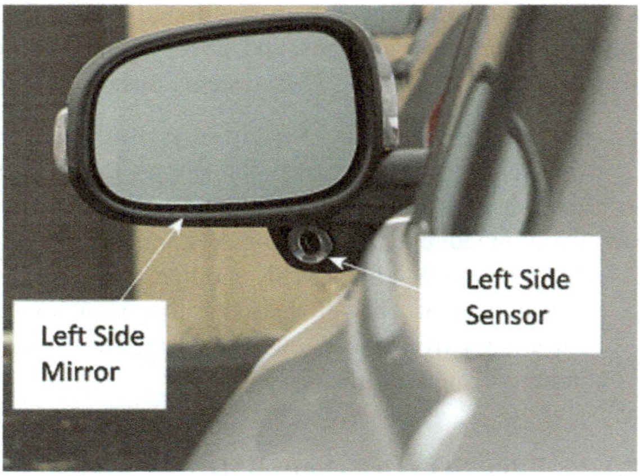

Fig. 3 Camera-based blind spot detection

2.1.4 Surround View

Automotive surround view, also called "around view" or "surround vision monitoring system," [2] is an emerging automotive ADAS technology that provides the driver a 360-degree view of the area surrounded by the vehicle. The vehicle is equipped with four or six fish-eye cameras. These camera information stiches and get complete surround view of the vehicle. This information displayed to the driver in the form of birds-eye view. Surround-view system helps driver to understand the vehicle surroundings in a better way. This system is combined with other driver assistance functions like parking assistance, blind spot detection, etc. and provides more support to the driver. Figure 4 shows illustration of surround-view camera.

2.1.5 Lane Departure Warning (LDW)

The purpose of Lane Departure Warning [7] is to alert the driver with acoustic or haptic steering warnings during the situations where vehicle starts moving out of the lane [7]. LDW systems are equipped with a forward-looking camera to identify the road lane markings and detect when a vehicle starts moving out of its driving lane. Whenever the system detects the vehicle starts moving out of its lane, then the driver is informed about this unintentional lane change in the form of acoustic or haptic steering warnings that helps the driver to bring vehicle back to its lane. Figure 5 shows block diagram of Lane Departure Warning system.

The prediction of a lane departure is based on certain thresholds like time-to-line crossing, actuation of the turn signals, steering wheel angle, and in some cases driver drowsiness and driver inattentiveness monitoring. Figure 6 shows time-to-line crossing threshold for lane departure.

Fig. 4 Illustration of surround view camera

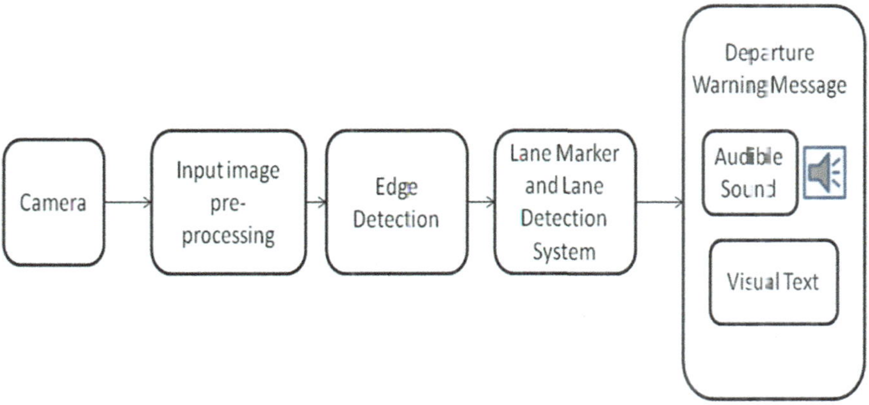

Fig. 5 Block diagram of lane departure warning system

2.1.6 Forward Collision Warning (FCW)

FCW systems are an active safety feature that warns the driver about an upcoming collision with an obstacle ahead that helps the driver to start decelerating the vehicle [8]. FCW systems use a midrange radar sensors and a forward-looking camera to detect objects. When the distance between objects becomes too short that a crash is forthcoming, FCW system alerts the driver so that the driver can vigilant to apply

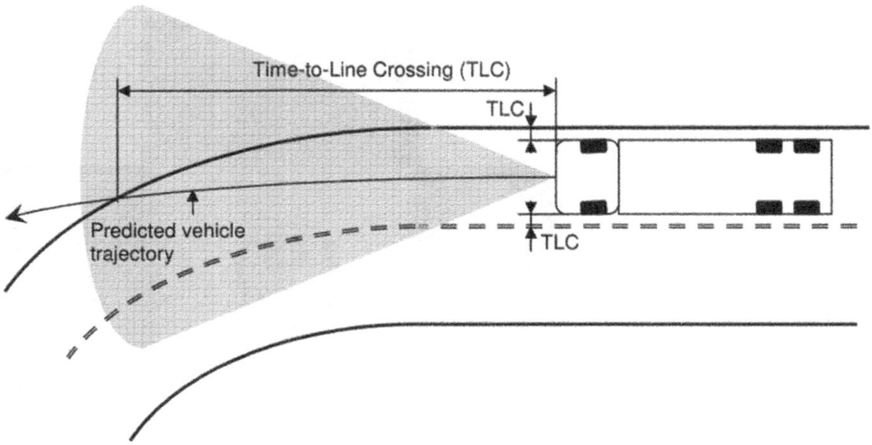

Fig. 6 Time-to-line crossing threshold for lane departure

brakes or do a steering to avoid obstacle and prevent a potential crash. FCW system uses acoustic, visual display, or other warning signals to inform driver about the crash.

2.2 SAE Level One Automation: Driver Assistance

In this level of automation, the driver controls vehicle, but some driving assist features are included in the vehicle design. Small elements of driving such as steering, braking can be done automatically by the vehicle.

2.2.1 Anti-Lock Braking System (ABS)

ABS helps to prevent the wheels from locking up during hard braking situations on low friction surface [9]. ABS brakes increase vehicle safety during braking because it eliminates wheel lockup and avoids vehicle skidding. ABS system allows the vehicle to stop in a straight-line manner. Anti-lock braking system provides flexibility in lateral control of the vehicle during hard braking so that driver can steer the vehicle and avoid the obstacle. The system consists of wheel speed sensors, ABS control module, Hydraulic-modulating unit, Valves, and Pump. Figure 7 shows schematic illustration of Anti-lock Braking System (ABS).

Anti-lock braking system controls tire slip by observing the relative deceleration rates of wheels during braking [10]. One or more wheel speed sensors are used to monitor the wheel speed. If one of the wheels starts to decelerate rapidly than the other wheels, then this shows that the wheel is losing traction and starting to slip.

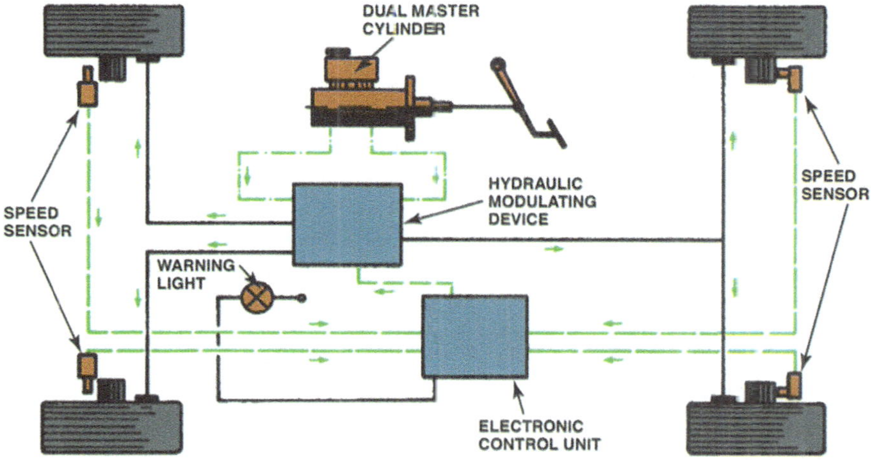

Fig. 7 Schematic illustration of Anti-lock Braking System (ABS)

This leads to wheel locking situation. ABS system quickly responds to this situation by reducing hydraulic brake pressure on the affected wheel. This helps the wheel to accelerate to gain wheel speed and traction. Once traction is regained, the system reapplies brake pressure to decelerate the wheel. This sequence is repeated until the vehicle stops or driver releases the brake pedal.

2.2.2 Traction Control System (TCS)

Traction control system [11] helps to avoid wheel spin during vehicle acceleration. It assists driver in acceleration phase, preventing wheels from spinning by cutting engine torque, thus improving acceleration performance and keeping the vehicle stable. Traction control system continuously monitors wheel speed using wheel speed sensor. If one of the wheels starts to accelerate rapidly than the other wheels, then this shows that the wheel is losing traction and starting to spin. TCS system quickly responds to this situation by increasing hydraulic brake pressure on the affected wheel or by reducing drive torque to the affected wheel. This helps the wheel to decelerate to gain the desired wheel speed and traction.

2.2.3 Electronic Stability Control (ESC)

Electronic Stability Control (ESC) system helps drivers to preserve control of the vehicles in conditions where the vehicle is starting to lose control [12]. It uses computer-controlled technology to apply individual brakes and help bring the car safely back on track. System uses a steering wheel angle sensor to get the desired direction of driver input and use a yaw rate sensor to get the vehicle heading direction.

This information can be used to calculate the desired path. If vehicle is not moving on the desired path, then ESC system controls the individual brake to take vehicle to intended path and improve the vehicle stability [13].

In oversteering situation, the rear end of the vehicle moves outwards, this loses vehicle stability. If this condition happens during a left turn, then the ESC system applies brake on the front right side, which creates a counter torque that brings the vehicle to the desired path.

In understeering situation, the vehicle keeps straight-line path during a turning. If this situation detects during a right turn, the ESC system applies brakes on the rear right side, which creates a counter torque that brings the vehicle to the desired path.

2.2.4 Adaptive Cruise Control (ACC)

Adaptive Cruise Control (ACC) [14] controls vehicle longitudinal velocity. ACC regulates the speed of vehicle to keep a safe distance from the vehicle ahead. This system uses radar and camera sensors for precise control of the vehicle speed. Generally, the data from sensors (radar and camera) will be fused in radar. From this fused data, ACC System identifies the object, its relative velocity, and relative distance. Based on this information, ECU triggers the system to accelerate or decelerate the vehicle.

The main functional enhancement of ACC, compared to standard cruise control, is the ability of sensing forward traffic objects [15]. Dependent on the situation, ACC changes automatically between two basic function modes: set speed control and follow control [15]. Figure 8 shows schematic diagram of Adaptive Cruise Control (ACC) system.

In set speed control, if no preceding vehicle is detected, speed is controlled according to the set speed. In the following control mode, if a slower preceding vehicle is in the same lane, then speed is reduced to that of the preceding vehicle and an appropriate distance is controlled automatically [15]. If necessary, moderate braking by automatic brake actuation is used for deceleration.

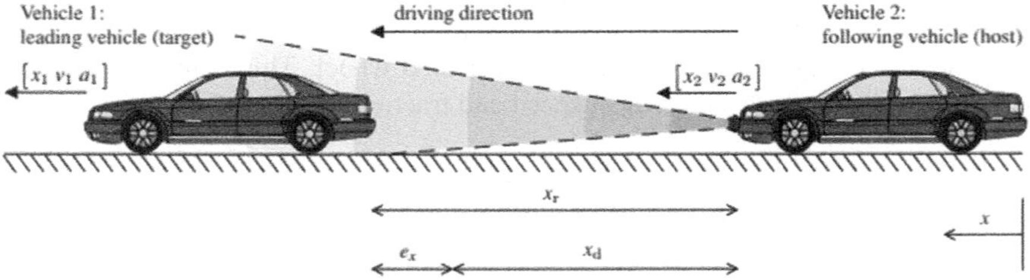

Fig. 8 Adaptive Cruise Control (ACC)

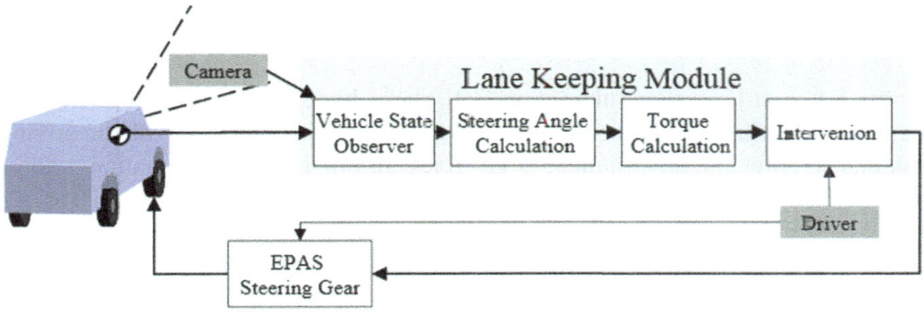

Fig. 9 Lane keeping system

2.2.5 Automatic Emergency Brake Assist (AEB)

AEB is an active safety feature that warns the driver about an imminent collision with an obstacle ahead, which may be either moving or stopped, and suggests decelerating the vehicle. If the driver does not respond to the warning or the driver's response is not sufficient to avoid the crash, then the system takes control and applies brakes to decelerate the vehicle, when an unavoidable collision is identified [16]. The function of the system is to assist the driver to evade or reduce harshness of crash by reducing the vehicle speed. This system uses camera and radar sensor for object identification.

2.2.6 Lane Keeping Assist (LKA)

Lane Keeping Assist system supports the driver to keep e vehicle in the center of the lane with full speed range [17]. System takes lateral control to keep the vehicle in the current lane [18]. This system allows the vehicle to change the lane if driver intentionally changes the lane. The system continuously monitors the driver's action like steering input or turn indicator to identify driver intension. The system uses camera sensor to detect lane markings. It also requires power steering system to provide assist torque. Figure 9 shows block diagram of lane keep assist system [18].

2.2.7 Lane Cantering (LC)

This system is similar to LKA, it not only intervenes on the steering wheel when the vehicle is exceeding lane boundaries but also continuously controls the steering wheel to keep vehicle in center of the lane. Lane Cantering system uses mono camera.

2.3 SAE Level Two Automation: Partial Automation

As per SAE, in level two automation [19], the vehicle is able to take lateral and longitudinal control simultaneously but driver must remain engaged with the driving task and monitor the environment at all times. In this level of automation, two or more automated functions work together to relieve the driver.

2.3.1 Highway Assist (HA)

The Highway Assist provides support to driver by controlling longitudinal and lateral dynamics of the vehicle on highways driving situations. The Highway Assist function can automatically control vehicle acceleration, deceleration, as well as steering to a certain extent. The driver has to monitor the environment and is ready to take over the vehicle control at any time. It is partially automated driving function up to a speed of 180 kmph on highways. The system combines the radar-based Adaptive Cruise Control (ACC) with the video-based lane keeping support.

2.3.2 Autonomous Parking

Autonomous Parking system helps driver to perform parallel, perpendicular, or angular parking. In this function, system takes longitudinal and lateral control of the vehicle [20]. It improves the safety and comfort of driving in a constrained environment. This system controls steering and speed in order to achieve the desired vehicle path with in the available space. The speed and steering angle are controlled with respect to the input from sensor data about the environment. In general, the system uses ultrasonic sensor and camera to get the environment information.

2.3.3 Traffic Jam Assist

Traffic Jam Assist helps the driver in driving in highway traffic jams up to 60 km/h. The Ego vehicle automatically follows the traffic vehicle ahead. The system uses forward-looking radar and camera to sense traffic and road ahead [21]. The fused data from both sensors is used for adapting Ego vehicle speed to the speed of traffic vehicle ahead, with Stop-and-Go functionality. Camera helps to detect the road lane marking that helps to keep vehicle in the canter of occupied lane. The driver needs to monitor both surrounding, traffic, and system status hence driver can override the system at any point of time. Figure 10 shows the Traffic Jam Assist system and it helps dramatically to reduce efforts and stress of highway driving in congested traffic conditions with enhanced vehicle and occupant safety.

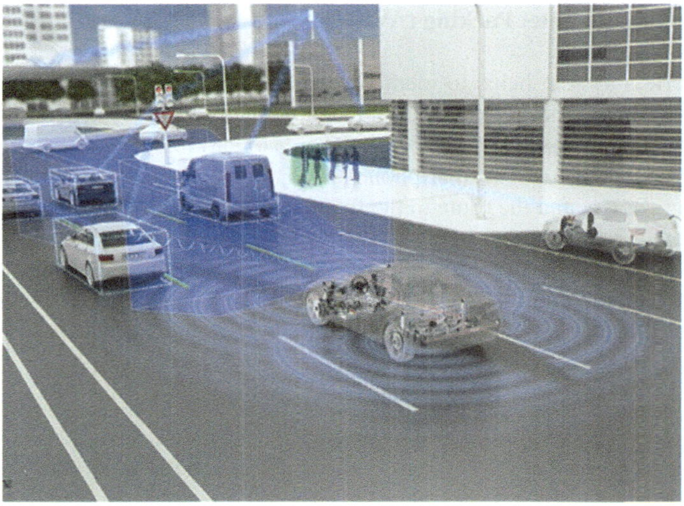

Fig. 10 Traffic jam assist

2.4 SAE Level Three Automation: Conditional Automation

In this level, system does all jobs like monitoring the environment and controlling lateral and longitudinal dynamics like steering acceleration or deceleration of the vehicle but if any dynamic situation comes, then system warns the driver to take the control action [19]. In this level, driver needs to be vigilant to the environment because system transfers driving control to the driver if any dynamic situation arrives. Transfer of vehicle control to driver during dynamic situation leads to safety-related issues. So most of the Original Equipment Manufacturer (OEM) skip this level of automation and directly jump to level four.

2.5 SAE Level Four: High Automation

In level four, vehicle is capable of performing all driving functions under all conditions. The main difference between levels three and four is that the vehicle is capable of handling all dynamic situations without having driver interventions [19]. This level having high automation improves vehicle safety and comfort. Level four also provides manual driving mode so that the automation control can switch between human and machine.

2.5.1 Automatic Valet Parking (AVP)

AVP driverless system helps to identify a free space in parking garage and parks vehicle itself by means of connected car technology [22]. This system allows vehicle to drop driver and passengers at the drop-off area of parking garage and all the driver needs are to be activated via smartphone app. This provides digital connection with the parking garage, and the route to the free recognized parking spot. The vehicle is equipped with a camera that helps system to visualize the driving route and detect obstacles and pedestrian in the route so that vehicle can respond immediately. Camera helps system to identify the free parking slot based on the vehicle size. The system provides a complete automatic parking feature, even if the parking space is so tight.

2.6 SAE Level Five: Full Automation

In level five, vehicle does not have any driver interventions and vehicle can perform all driving functions under all conditions [19]. In this level, vehicle control, environment monitoring, and dynamic responses are handled by vehicle itself. This system helps to improve vehicle and road safety by providing matured driving and avoid human errors. There is no need for pedals, brakes, or a steering wheel, as the autonomous vehicle system controls all critical tasks, monitoring of the environment, and identification of unique driving conditions like traffic jams.

3 Challenges in ADAS System

Vehicles with high levels of ADAS capability considered Autonomous Driving. The objective of active safety or Autonomous Driving is to reduce collisions and ensures safety of vehicles on road. The technology uses sensors and control devices at different levels to assist or fully take over the vehicle controls depending on the levels of automation. In order to build the trust, adequate testing is required and perhaps one fatal incident will actually reduce the feel of safety in autonomous cars. After two fatal incidents involving a Tesla Model X and an autonomous vehicle Volvo XC90, consumers have become less trusting of autonomous vehicles. Since the technology behind autonomous driving is majorly dependent on machine learning, no explicitly specified behavior can be defined.

For tier1 suppliers, the challenges would be offering the best technology with improved performance at lower cost to OEM. Since the base price for cars is remaining relatively stable, semiconductor and other solution providers may face pressure from OEM and customers to keep the ADAS cost low and make the technology as standard.

For OEM, the challenge would be establishing a record of accomplishment of safe cars and increasing customer feeling of safety about driverless cars. It is not easy to reach that level without adequate testing for various environmental and road conditions. Achieving the confidence levels requires more testing time and usually it is limited to overall vehicle development time. Front loading the testing to virtual simulation platforms is one of the practical solution followed across the industry.

4 Conclusion

Improving automotive safety and driver comfort is a major task of Advanced Driver Assistance System. Based on the ADAS technology implemented in the vehicle, the level of automation maturity can be addressed using the SAE level of automation standard. This paper describes SAE levels of automation and ADAS functions implemented at each level of automation. The hardware and software technologies are used for the ADAS function development, and the challenges toward development of autonomous vehicle were addressed. At present, most of the ADAS functions are implemented in high-end vehicles. The suppliers and OEM are putting effort to reduce ADAS system cost and use the best technologies available in the market without compensating the performance and safety. This helps to improve the development of autonomous vehicle in the market. In near future vehicle, the basic SAE Level 1 and Level2 ADAS features like AEB, ACC, Parking Assist, Surround-view systems, etc. will be implemented. As a result, the number of road accidents will be reduced.

References

1. W. H. Organization, Global Status Report On Road Safety 2018 (2018)
2. D. Assistance, M. Galvani, History and future of driver assistance. IEEE Instrum Meas. Mag. **22**(February), 11–16 (2019)
3. A. Shaout, D. Colella, S. Awad, Advanced driver assistance systems—past, present and future, in *ICENCO'2011—2011 7th International Computer Engineering Conference Today Information Society What's Next?* (2012), pp 72–82 (2012)
4. A. Carullo, M. Parvis, S. Member, An Ultrasonic Sensor for Distance Measurement in automotive applications. IEEE Sens. J. **1**(2), 143–147 (2001)
5. K. Dyczkowski, P. Gadecki, A. Kułakowski, Traffic signs recognition system
6. T. National, H. Traffic, S. Administration, Blind Spot Monitoring in Light Vehicles—System Performance, no. July, 2014
7. R.N. Mahajan, A.M. Patil, Mazda lane departure warning system. Int. J. Eng. Tech. Res. 120–123 (2014)
8. E. Dagan, O. Mano, G.P. Stein, Forward collision warning with a single camera modeling acceleration momentary time to contact
9. A.A. Aly, E.-S. Zeidan, A. Hamed, F. Salem, An antilock-braking systems (ABS) control: a technical review. Intell. Control Autom. **02**(03), 186–195 (2011)
10. J.D. Halderman, *Automotive Technology* 4 edn. (2012)
11. Y.A. Ghoneim, W.C. Lin, Y.-K. Chin, D.M. Sidlosky, Enhanced traction stability control system. SAE Tech. Pap. Ser. **1**, 724 (2010)

12. A.T. Van Zanten, Bosch ESP systems: 5 years of experience. SAE Tech. Pap. Ser. **1**, 724 (2010)
13. D. Kim, K. Kim, W. Lee, I. Hwang, Development of Mando ESP (Electronic Stability Program). SAE Tech. Pap. Ser. **1**, 724 (2010)
14. O. Gietelink, J. Ploeg, B. De Schutter, M. Verhaegen, Development of advanced driver assistance systems with vehicle hardware-in-the-loop simulations. Veh. Syst. Dyn. **44**(7), 569–590 (2006)
15. T. Sauer, J. Steinle, O. Tschernoster, The BMW active cruise control ACC. SAE Int. by, no. 724 (2000)
16. T. Henriksson, Driver assistance systems with focus on automatic emergency brake (2011)
17. J. Liu, J. Wu, Y. Su, C. County, Development of an Interactive Lane Keeping Control System for Vehicle, no. 6
18. D. Aspects, O. F. The, U. Interface, S. Vehicle, R. Departures, H. M. Interface, Jochen Pohl Jonas Ekmark Volvo Car Corporation Sweden Paper No. 529 Lane Keeping-Module Lane Tracker EPAS Road Lateral Position, pp. 1–8
19. B. Wendling, First revision of J3016—update on task force activities. SAE Int.
20. A. Paul, R. Chauhan, R. Srivastava, M. Baruah, Advanced Driver Assistance Systems (2018)
21. A. Driving, Automated and Autonomous Driving Regulation under uncertainty Automated and Autonomous Driving Regulation under uncertainty
22. T.J. Titus, S.A.V Balakrishnan, J. Sam, Fully Automated Valet Car Parking System, no. Mar 2017, pp. 69–72

Application of Artificial Intelligence-Based Solution Methodology in Generation Scheduling Problem

Shubham Tiwari, Vikas Bhadoria, and Bharti Dwivedi

Abstract Generation Scheduling commonly known as Unit Commitment Problem is one of the most typical optimization problems in electrical engineering. In this paper, Particle Swarm Optimization Technique with few modifications is implemented as an Artificial Intelligence-based technique for the solution of this convoluted nonlinear optimization problem. The proposed technique is implemented on a standard IEEE ten generator system. The entire work is performed in MATLAB environment. The results have been compared and found better than the other available techniques available in literature.

Keywords Generation scheduling problem (GSP) · Modified particle swarm optimization technique (MPSO) · Priority list method (PLM)

1 Introduction

Generation Scheduling Problem is a highly complex optimization problem which is of great concern for power system engineers and researchers. Proper scheduling of generators can lead to significant profit for generation companies [1]. The nonlinear nature of the problem with involvement of large number of constraints which keep on changing with time makes this problem more severe. The prime objective of UCP is to provide proper scheduling of generators to meet load forecast with minimum cost in a scheduled time frame under the light of various constraints [2–4]. Numerous

S. Tiwari (✉) · V. Bhadoria
Electrical & Electronics Engineering Department, ABES Engineering College, Ghaziabad, India
e-mail: shubham.tiwari@abes.ac.in

V. Bhadoria
e-mail: vikas.bhadoria@abes.ac.in

B. Dwivedi
Electrical Engineering Department, Institute of Engineering & Technology, Lucknow, India
e-mail: bharti.dwivedi@ietlucknow.ac.in

S. S. Dash et al. (eds.), *Intelligent Computing and Applications*,
Advances in Intelligent Systems and Computing 1172,
https://doi.org/10.1007/978-981-15-5566-4_7

efforts have been made to solve UCP with various classical, artificial intelligence-based and hybridized techniques [5–15] but the scope of improvement has always been there [16, 17]. The reason behind choosing PSO was that it was found to be better in comparison to other techniques [18–21].

The main asset of PSO is that it is robust with ease in parameter tuning, moreover, it can be combined with various classical and artificial intelligence-based techniques [21–24]. There are two cases considered in this paper. In the first case the forecasted load is solely satisfied by the thermal generation and the results are compared with other solution methodologies available in literature.

2 Problem Formulation

As the main objective is minimization of operational cost. The objective function [25] is given as Eq. (1).

$$\mathrm{Cost}_N = \sum_{i=1}^{N} \left[\mathrm{FC}_i(P_{ih}) + \mathrm{STC}_i\left(1 - U_{i(h-1)}\right)\right]U_{ih} \tag{1}$$

where, FC_i is the fuel cost of ith thermal generator and P_{ih} is the output power of ith thermal generator at hth hour & U_{ih} is the ON/OFF status of the ith unit at hth hour.

The fuel cost function is given as Eq. (2).

$$\mathrm{FC}_i(P_{ih}) = a_i + b_i P_{ih} + c_i P_{ih}^2 \tag{2}$$

where, a_i, b_i & c_i are the fuel cost constants.

STC_i is the cost incurred during the starting of ith thermal generator. This is given as Eq. (3).

$$\mathrm{STC}_i = \begin{cases} \mathrm{HSc}_i : X_{i(\mathrm{off})} \le (\mathrm{MD}_i + \mathrm{Cs}_i)\mathrm{hrs} \\ \mathrm{CSc}_i : X_{i(\mathrm{off})} > (\mathrm{MD}_i + \mathrm{Cs}_i)\mathrm{hrs} \end{cases} \tag{3}$$

X_i^{off} is the duration in which ith unit is continuously OFF.

A. *Generation Scheduling Constraints*

a. *Equality Constraint (Power Balance Constraint)*

The power balance constraint (generated power is equal to forecasted demand) is given as Eq. (4).

$$\left(\sum_{h=1}^{24}\sum_{i=1}^{N} P_{ih}\right) = \sum_{h=1}^{24} \mathrm{LD}_h \tag{4}$$

where, P_{ih} generated power (MW) of ith unit at hth hour & LD_h is the forecasted load at hth hour.

b. *Spinning Reserve Constraint*

Spinning Reserve is a part of generation (offline) which serves as backup during contingency situations. It is given as Eq. (5).

$$\sum_{i=1}^{N} P_{i(\max)} U_{ih} \geq \mathrm{LD}_h + \mathrm{SR}_h \tag{5}$$

The maximum generation limit and the spinning reserve of individual generator at hth hour are represented by $P_{i(\max)}$ & SR_h, respectively. SR_h is taken as five percent.

c. *Inequality Constraint (Generation Limit Constraint)*

The generation limit constraint is given as Eq. (6).

$$P_{i(\min)} \leq P_{ih} \leq P_{i(\max)} \tag{6}$$

where, $P_{i(\min)}$ & $P_{i(\max)}$ are the minimum and maximum generating limits of individual thermal generator.

d. *Time Constraints (Minimum Up)*

$$X_i^{\mathrm{on}}(t) \geq \mathrm{MU}_i \tag{7}$$

e. *Time Constraint (Minimum Down)*

$$X_i^{\mathrm{off}}(t) \geq \mathrm{MD}_i \tag{8}$$

where, X_i^{on} is the continuous off time ith unit.

f. *Initial Status*

Down time status is taken at the start of the problem. The data regarding thermal generation and load profile is given in Appendix.

3 Solution Methodology

Generation Scheduling Problem (GSP) is a two-stage problem. In the first stage the ON/OFF status of generating units is obtained while in the latter stage economic dispatch is done. In the proposed AI-based hybrid solution methodology the first stage solution is obtained by the initial priority vector given in Eq. (9), this initial vector is updated as [20]. The dispatch is obtained by Modified Particle Swarm Optimization Technique (MPSO) [25]. The second stage solution is obtained from MPSO. The Classical PSO (CPSO) [25] is modified as Eqs. (15, 16).

$$\text{priorityvector} = \frac{P_{(\max),\text{vec}}}{\max.\left[P_{(\max),\text{vec}}\right]} + \frac{\text{MD}_{\text{vec}}}{\max.[\text{MD}_{\text{vec}}]} \tag{9}$$

$$v_{id}^{(k+1)} = \begin{bmatrix} \omega * v_{id}^{k} + \left(\left(c_{1f} - c_{1i}\right) * \frac{\text{iter}}{\text{iter}\max} * c_{1i}\right) * \text{Rand}_1() * \left(P_{\text{best}id} - x_{id}^{k}\right) \\ + \left(\left(c_{2f} - c_{2i}\right) * \frac{\text{iter}}{\text{iter}\max} * c_{2i}\right) * \text{Rand}_2() * \left(G_{\text{best}gd} - x_{id}^{k}\right) \end{bmatrix} \tag{10}$$

where, ω, c_1, c_2 are the inertia weight and acceleration coefficients respectively.

ω is obtained as Eq. (11).

$$\omega_i = [1.1 - (g\text{best}i / p\text{best}i)] \tag{11}$$

where, $p_{\text{best}i}$ and g_{besti} are the local and global best positions of ith particle. The velocity limits are taken as [25].

4 Results and Discussion

The results obtained from the proposed techniques are given in Table 1. The ON/OFF status obtained from stage one results are shown as "*green*" and "*blue*" colors, respectively. The stage two results are given as respective MW values. The convergence characteristics are shown as Fig. 1 and the comparison with other solution methodologies is given in Table 2.

The convergence is obtained in the tenth iteration. The time taken by proposed method is 6 s and overall operational cost obtained is **557,090$**. The comparison of the proposed method to other methods available in literature is given in Table 2.

Table 1 Generation schedule for case one

<table>
<tr><th rowspan="2">Hrs.</th><th colspan="10">Unit No.
Thermal Generators</th><th rowspan="2">Tot. Gen. (MW)</th></tr>
<tr><th>Tg1</th><th>Tg2</th><th>Tg3</th><th>Tg4</th><th>Tg5</th><th>Tg6</th><th>Tg7</th><th>Tg8</th><th>Tg9</th><th>Tg10</th></tr>
<tr><td>H-1</td><td rowspan="24">455</td><td>245</td><td rowspan="6">0</td><td rowspan="5">0</td><td rowspan="3">0</td><td rowspan="8">0</td><td rowspan="9">0</td><td rowspan="10">0</td><td rowspan="11">0</td><td rowspan="24">0</td><td>700</td></tr>
<tr><td>H-2</td><td>295</td><td>750</td></tr>
<tr><td>H-3</td><td>395</td><td>850</td></tr>
<tr><td>H-4</td><td rowspan="3">455</td><td>40</td><td>950</td></tr>
<tr><td>H-5</td><td>90</td><td>1000</td></tr>
<tr><td>H-6</td><td rowspan="16">130</td><td>60</td><td>1100</td></tr>
<tr><td>H-7</td><td>410</td><td rowspan="14">130</td><td>25</td><td>1150</td></tr>
<tr><td>H-8</td><td rowspan="8">455</td><td>30</td><td>1200</td></tr>
<tr><td>H-9</td><td>110</td><td>20</td><td>1300</td></tr>
<tr><td>H-10</td><td rowspan="4">162</td><td>43</td><td>25</td><td>1400</td></tr>
<tr><td>H-11</td><td>80</td><td>25</td><td>13</td><td>1450</td></tr>
<tr><td>H-12</td><td>80</td><td>25</td><td>53</td><td>10</td><td>1500</td></tr>
<tr><td>H-13</td><td>43</td><td>25</td><td rowspan="12">0</td><td rowspan="12">0</td><td>1400</td></tr>
<tr><td>H-14</td><td>110</td><td>20</td><td rowspan="6">0</td><td>1300</td></tr>
<tr><td>H-15</td><td>30</td><td rowspan="5">0</td><td>1200</td></tr>
<tr><td>H-16</td><td>310</td><td>25</td><td>1050</td></tr>
<tr><td>H-17</td><td>260</td><td>25</td><td>1000</td></tr>
<tr><td>H-18</td><td>360</td><td>25</td><td>1100</td></tr>
<tr><td>H-19</td><td rowspan="4">455</td><td>30</td><td>1200</td></tr>
<tr><td>H-20</td><td>162</td><td>43</td><td rowspan="3">25</td><td>1400</td></tr>
<tr><td>H-21</td><td rowspan="4">0</td><td>162</td><td>73</td><td>1300</td></tr>
<tr><td>H-22</td><td rowspan="3">0</td><td>145</td><td>20</td><td>1100</td></tr>
<tr><td>H-23</td><td>425</td><td rowspan="2">0</td><td>20</td><td>0</td><td>900</td></tr>
<tr><td>H-24</td><td>345</td><td>0</td><td>0</td><td>800</td></tr>
</table>

5 Conclusion

It is evident from Table 2 that the proposed AI technique gives minimum operational cost. Although the execution time is given in the table but the comparison would not be justified as the computer efficacies are different. The proposed technique is implemented with the help of a computer having 2 GB RAM and Intel core processor.

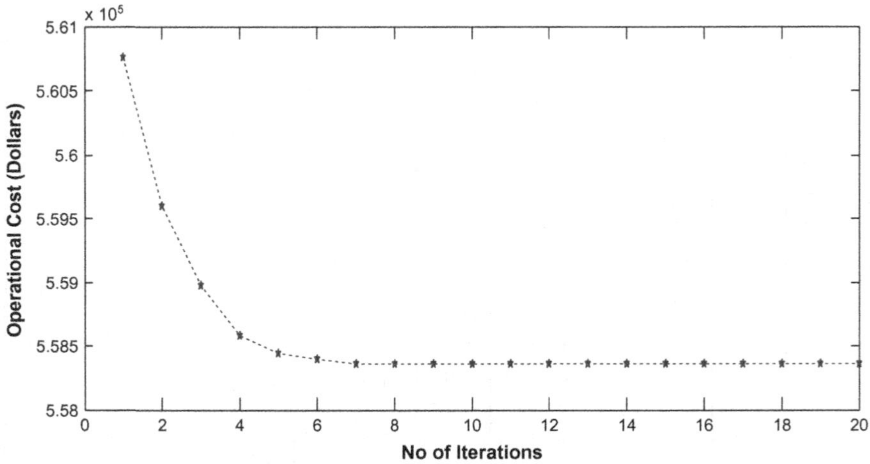

Fig. 1 Convergence characteristic for case one

Table 2 Comparison of operational costs

Method used	Cost ($)	Execution time (s)
BP [15]	565,804	NA
GA[13]	570,781	NA
APSO [24]	561,586	NA
BP [24]	565,450	NA
Advance Three Stage (ATS) PLM + PSO + SMP [13]	557,677	NA
ATS- PSO [21]	557,128	8.82
ATS-WIPSO [21]	557,128	8.36
ATS-CPSO [21]	557,128	7.73
ATS-WICPSO [21]	557,128	6.58
Proposed	557,090	06.00

Appendix

TGs	TG1	TG2	TG3	TG4	TG5	TG6	TG7	TG8	TG9	TG10
P_{max}	455	455	130	130	162	80	85	55	55	55
P_{min}	150	150	20	20	25	20	25	10	10	10
a (\$/h)	1000	970	700	680	450	370	480	660	665	670
b (\$/MWh)	16.19	17.26	16.60	16.50	19.70	22.26	27.74	25.92	27.27	27.79
c (\$/MW^{2}h)	0.00048	0.00031	0.002	0.00211	0.00398	0.0072	0.00079	0.00413	0.0022	0.00173
MD (h)	8	8	5	5	6	3	3	1	1	1
MU (h)	8	8	5	5	6	3	3	1	1	1
HSc (\$/h)	4500	5000	550	560	900	170	260	30	30	30
CSc (\$/h)	9000	10000	1100	1120	1800	340	520	60	60	60
Cs (h)	5	5	4	4	4	2	2	0	0	0
Initial status	8	8	−5	−5	−6	−3	−3	−1	−1	−1

References

1. A.J. Wood, B.F. Wollenberg, *Power Generation, Operation and Control*, 3rd edn. (Wiley, New York, 2013)
2. R.M. Bums, C.A. Gibson, Optimization of priority lists for a unit commitment program, in *Proceedings of the IEEE Power Engineering Society Summer Meeting* (1975)
3. A. Shukla, S.N. Singh, PSO for solving unit commitment problem including renewable energy resources. Electr. India **54**(12), 100–105 (2014)
4. C. Shantanu, T. Senjyu, A.Y. Saber, A. Yona, T. Funabashi, Optimal thermal unit commitment integrated with renewable energy sources using advanced particle swarm optimization. IEEJ Trans. **4**, 617 (2009)
5. W.L. Snyder, H.D. Powell, J.C. Rayburn, Dynamic Programming approach to Unit Commitment. IEEE Trans. n Power Syst. **2**, 339–351 (1987)
6. P.G. Lowery, Generating unit commitment by dynamic programming. IEEE Trans. Power Apparatus Syst. **85**(5), 422–426 (1966)
7. T. Aoki, T. Satoh, M. Itoh, T. Ichimori, K. Masegi, Unit commitment in a large scale power system including fuel constrained thermal and pumped storage hydro. IEEE Trans. Power Syst. **2**(4), 1077–1084 (1987)
8. W.J. Hobbs, G. Hermon, S. Warner, G.B. Sheble, An enhanced dynamic programming approach for unit commitment. IEEE Trans. Power Syst. **3**(3), 1201–1205 (1988)
9. T. Senjyu, K. Shimabukuro, K. Uezato, T. Funabashi, A fast technique for unit commitment problem by extended priority list. IEEE Trans. Power Syst. **18**(2), 882–888 (2003)
10. D. Mukarta, S. Yamashiro, Unit Commitment scheduling by lagrange relaxation method taking into account transmission losses. Electr. Eng. Jpn. **152**, 27–33 (2005)
11. W. Ongsakul, N. Petcharaks, Unit commitment by enhanced adaptive Lagrangian Relaxation. IEEE Trans. Power Syst. **19**(1), 620–628 (2004)
12. B. Xiaomin, S.M. Shahidehpour, Y. Erkeng, Constrained unit commitment by using tabu search algorithm, in *Proceedings of International Conference on Electrical Engineering*, vol. 2 (1996), pp. 1088–1092
13. F. Zhuang, F.D. Galiana, Unit commitment by simulated annealing. IEEE Trans. Power Syst. **5**(1), 311–317 (1990)
14. T. Matsui, T. Takata, M. Kato, M. Aoyagi, M. Kunugi, K. Shimada, J. Nagato, Practical approach to unit commitment problem using genetic algorithm and Lagrangian Relaxation method, in *Intelligent Systems Applications to Power Systems, Proceedings ISAP* (1996)
15. A.Y. Saber, Scalable Unit Commitment by memory bounded ant colony optimization with a local search. Electr. Power Energy Syst. **30**(6–7), 403–414 (2008)
16. S. Tiwari, B. Dwivedi, M.P. Dave, Advancements in unit commitment strategy, in *Proceedings of ICETEESES*, 11–12 Mar 2016, pp. 259–262
17. S. Salem, Unit commitment solution methods, in *World Academy of Science, Engineering and Technology* (2007), pp. 320–325
18. A. Shukla, S.N. Singh, Multi-objective unit commitment using search space based crazy PSO and Normal boundary intersection technique. IET Gener. Transm. Distrib. **10**(5), 1222–1231 (2016)
19. S. Tiwari, A. Kumar, G.S. Chaurasia, G.S. Sirohi, Economic load dispatch using particle swarm optimization. IJAIEM **2**(4), 476–485 (2013)
20. S. Khanmohammadi, M. Amiri, M. Tarafdar Haque, A new three stage method for solving unit commitment method. Energy **35**, 3072–3080 (2010)
21. V.S. Pappala, I. Erlich, A new approach for solving Unit Commitment Problem by adaptive particle swarm optimization, in *Power and Energy Society General Meeting-Conversion and Delivery of Electrical Energy in the 21st Century* (IEEE, USA, 2008), pp. 1–6
22. P. Sriyanyong, Y.H. Song, Unit commitment using particle swarm optimization combined with Lagrangian relaxation, in *Power Engineering Society General Meeting*, vol. 3 (2005), pp. 2752–2759

23. Z.L. Gaing, Discrete particle swarm optimization algorithm for unit commitment, in *IEEE Power Engineering Society General Meeting*, vol. 1 (2003), pp. 13–17
24. Y.W. Jeong, J.B. Park, S.H. Jang, Y.L. Kwang, A new quantum-inspired binary PSO: application to unit commitment problems for power systems. IEEE Trans. Power Syst. **25**(3), 1486–1495 (2010)
25. S. Tiwari, A. Maurya, Particle swarm optimization technique with time varying acceleration coefficients for load dispatch problem. IJRITCC **3**(6), 3878–3885 (2015)

Discomfort Analysis at Lower Back and Classification of Subjects Using Accelerometer

Ramandeep Singh Chowdhary and Mainak Basu

Abstract Lower back pain problems are increasing these days because of a sedentary lifestyle and working habits such as using laptops or computers for long hours and sitting on chairs for continuous duration. Back injury occurrence is also very common in athletes and workers where lifting of heavyweights are required. This research work aims to classify subjects and analyze discomfort at lower back by performing wireless data acquisition using accelerometer sensor. The designed hardware consisting of accelerometer sensor, NRF wireless module, and micro-controller was used to implement node-hub architecture. This research work shows the classification of subjects based on lower back vibration data. Multiple classification algorithms such as decision trees, random forest, naive Bayes, and support vector machine were applied to perform subject classification after performing experiments. The analysis of data shows that the subjects could be classified based on the discomfort level of the lower back using accelerometer data. Such a kind of study could be used for the prediction of the core strength of lower back and treatment of lower back problems by analyzing vibrational unrest.

Keywords Accelerometer · Back analysis · Classification · Discomfort analysis · Micro-controllers · Wearable sensors · Wireless sensor node

1 Introduction

Human lower back discomfort and pain analysis is an important domain for gait researchers as it gives various parameters for classification of subjects and prediction of injuries related to the lower back area. Oliverio et al. [1] stated in their research work that pain in the pelvic region is a very common problem found in almost all

R. S. Chowdhary (✉) · M. Basu
GD Goenka University, Gurgaon, Haryana, India
e-mail: raman85friends@gmail.com

M. Basu
e-mail: mainak.basu87@gmail.com

S. S. Dash et al. (eds.), *Intelligent Computing and Applications*,
Advances in Intelligent Systems and Computing 1172,
https://doi.org/10.1007/978-981-15-5566-4_8

parts of the world specially in developed countries where people had a sedentary lifestyle. This problem raises a huge expense of approximately 2 billion dollars in a developed country like the United States in a year and hence could have a significant impact on the social and economic state of any nation. Taghvaei et al. [2] discussed another important lower back issue which is locomotion for daily activities. Elderly persons face difficulties in performing daily activities due to lack of muscular strength, musculoskeletal injuries, lower back pain, etc. In some of the recent research work, sensor measurement units based on MEMS were used for lower body posture analyses. Urukalo et al. [3] designed a wearable device to find a solution to chronic problems such as back pain and musculoskeletal disorders which had no technological or physiological hurdles. Wenjing et al. [4] stated in their research work that medicines or surgeries are required for persons with acute lower back pain in order to improve the health condition. So to avoid the use of such medicines or undergo surgeries it is always better to analyze the discomfort at the lower back of the patients and opt for corrective measures before it converts to severe back pain.

Some recent systems used to analyze body movements and capture data using sensors were used extensively by researchers for classification and predictions. Molnar et al. [5] used 6D inertial measurement units to access lower back motions. Xu et al. [6] proposed an algorithm which used the angle between two IMU sensors modules and did not use ground reference for angle calculation. Chutatape et al. [7] had implemented a system using one smartphone and measured joint angles at the hip location which led to inappropriate body postures and could result in dislocation of joint. Kam et al. [8] developed a plastic optical fiber-based sensor and used intensity interrogation technique to analyze backbone bending in the sagittal plane. Dobrea et al. [9] designed a wearable warning system which identified an improper sitting posture which could generate cervical pains. This was done by defining several triggering zones and identifying the presence of the system in those zones. Sandag et al. [10] presented machine learning algorithm to classified subjects with back pain using K-Nearest Neighbor (KNN) classification algorithm on non-real-time data and found that Degree Spondylolisthesis had the most important effect on lower back pain syndrome. Further, Galbusera et al. [11] described various techniques of artificial intelligence including machine learning that were being developed with special focus on those used in spine research.

To overcome these problems a single sensor-based instrument was designed in this research work and was used for discomfort analysis of lower back. The collected data was also used to train different classifiers which will eventually perform classification of the test sample. Multiple classifiers were used to compare performance metrics for classification of subjects and hence propose the best-suited classifier for accelerometer related data.

2 Methodology

Complete flow and steps involved in the system architecture are shown and explained in Fig. 1. In this the first step was to collect data from the sensor. The data collection was done when the subject had started performing experiments while wearing wireless sensor node at the lower back. Sensor calibration was carried out by keeping the sensor node stationary before it starts capturing data from the sensor. As soon as the node starts getting data it performs wireless transfer of data to the hub. With the help of a Java software the received data at the hub side was stored in the form of CSV files for creating logs and preprocessing. In the process of preprocessing the concept of data segmentation and data windowing was used. After completion of the preprocessing task, feature extraction was performed to extracts features from the collected data from the subjects. The entire set of data was split into two sets viz. training set and testing set. In the present research work various classifiers were used to classify subjects based on their discomfort level at the lower back. A comparison of performance metrics of different classifiers was carried out to find the best-suited classifier for the given set of data. Both training set and testing set were fed to the classifiers to get the final predicted result as shown in Fig. 1. Shabrina et al. [12] in their research work used visual analogue scales and pain questionnaire methods to analyze lower back pain due to prolonged standing on inclined surface. Some of the

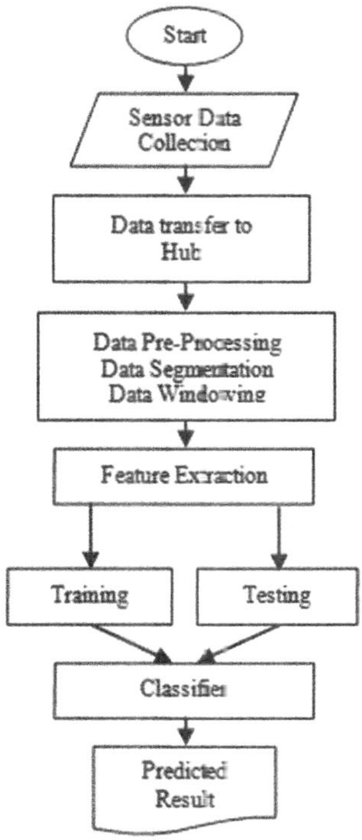

Fig. 1 Flow and steps required for the complete system

important features in such kind of classification process are gender, age, vibrational data in all three axis, nature of job of the subjects, fitness level of the subjects, and hold time of the upper body while performing experiments in the sagittal plane.

3 Hardware Design

The implemented system was designed, fabricated, and was used to perform experiments on 20 different subjects. The complete system consists of 1 NRF based sensor node, which was mounted at the lower back of the subjects while performing experiments. Secondly, it consists of 1 NRF based hub, which was connected with the laptop to creates data logs of the received data. Following sections explain hardware infrastructure and software algorithms used in this research work.

Real-time data acquisition was implemented in this system at the time of performing experiments. Arduino Nano micro-controller board was used in node and hub design. In the case of node architecture the micro-controller board was interfaced with the accelerometer sensor. The data which was transmitted from node to hub was transferred through wireless communication. To perform wireless communication an NRF chip was used as it is a low power device and can operate on battery. The block diagram of wireless sensor node is shown in Fig. 2 where a battery-powered Arduino Nano is interfaced with I2C based inertial sensor and NRF wireless communication chip.

The hub is connected at the remote site with a laptop which received sensor data and created log files in CSV format. The hub architecture consists of Arduino Nano micro-controller board and NRF wireless communication module. It also has a synchronization switch which was used to send synchronization signal to the node to start sending sensor data to the hub for data acquisition.

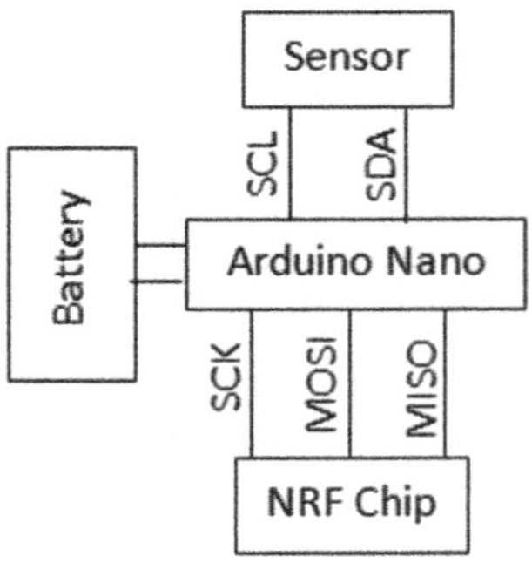

Fig. 2 Block diagram of wireless sensor node

4 Software Algorithms

Python programming language was used to perform discomfort analysis at lower back and classification of subjects based on extracted features. The detailed steps and algorithm are explained in the following sections.

A. Data Preprocessing

Sensor data acquisition frequency is directly proportional to the resolution of the acquired signal value. Typical consumer accelerometer sensor like MPU6050 can support communication using I2C at 400 kHz. It is known that f_{sampling} is approximately equal to 50 Hz is the most suitable value for sampling frequency. The present research work is focused on the detection of accelerometer data at the lower back in sitting upright and forward lean position. The classification algorithms were written in python which detects the subjects with pain in lower back while performing the experiment in forward lean position. It uses various features extracted from the acceleration data after preprocessing. This is possible after segmentation of data and it is stored in with window size. The captured data is processed by various classification algorithms after the data window is filled completely. It is always good to have large window size as it helps in providing stable status. Large window sizes sometimes come with problem of data overlapping for different activities as it was true for sitting upright position and lean forward position while performing experiments. To avoid overlapping problem window size was selected to fit in two experimental positions separately.

B. Classification Method

Classification of subjects based on discomfort analysis of lower back was done in this research work. Various classification algorithms were used in this research to get the predicted output. Four classification algorithms viz. Navie Bayes (NB), Decision Trees (DT), Random Forest (RF), and Support Vector Machine (SVM) were used and their performance metrics were compared. This gives the best-suited classification algorithm for related classification problems. Nezam et al. [13] showed in one of the related research work that support vector machine (SVM) had higher classification accuracy of 83.5% for three pain levels than K-nearest neighbor (KNN) classifier having 80.5% classification accuracy. Abdullah et al. [14] also performed classification of spinal abnormalities and showed that the classification accuracy of KNN algorithm was 85.32% which was better than the percentage of accuracy in case of random forest (RF) classifier which was 79.56%. Estrada et al. [15] also performed posture recognition using cameras and smartphones and had highest classification accuracy with decision trees classifier as 89.83% (for spinal posture) and 95.35% (for head and shoulder posture) as compared to KNN and SVM classifier algorithms.

C. Feature Selection and Extraction

Tian et al. [16] in their research work proposed a system which used three types of feature extraction namely original features, linear discriminant analysis (LDA) features, and kernel discriminant analysis (KDA) features. In this research work several features were extracted from accelerometer data in all three axis (viz. *x*-axis, *y*-axis, and *z*-axis). Features were also extracted which was composed of data from all three dimensions of accelerometer sensor. Detailed explanation of the extracted features is given below where *Axi*, *Ayi*, and *Azi* were the acceleration values in all three axes and the total number of samples in each axis is denoted by N.

- Mean in x-axis for sitting upright position is defined as the summation of acceleration values in x-axis while sitting upright divided by number of samples, i.e., N.

$$\mu(Ax)sit = \frac{1}{N}\sum_{i=1}^{N} Axi \tag{1}$$

- Mean in *y*-axis for sitting upright position is defined as the summation of acceleration values in *y*-axis while sitting upright divided by number of samples, i.e., N.

$$\mu(Ay)\mathrm{sit} = \frac{1}{N}\sum_{i=1}^{N} Ayi \tag{2}$$

- Mean in *z*-axis for sitting upright position is defined as the summation of acceleration values in *z*-axis while sitting upright divided by number of samples, i.e., N.

$$\mu(Az)\mathrm{sit} = \frac{1}{N}\sum_{i=1}^{N} Azi \tag{3}$$

- Variance in *x*-axis, V_{Ax} is given as the spread of the accelerometer data around the mean in *x*-axis,

$$V_{Ax} = \sqrt{\frac{1}{N}\sum_{i=1}^{N}(Axi - \mu(Ax)_{\mathrm{lean}})^2} \tag{4}$$

- Variance in *y*-axis, V_{Ay} is given as the spread of the accelerometer data around the mean in *y*-axis,

$$V_{Ay} = \sqrt{\frac{1}{N}\sum_{i=1}^{N}(Ayi - \mu(Ay)_{\mathrm{lean}})^2} \tag{5}$$

- Variance in z-axis, V_{Az} is given as the spread of the accelerometer data around the mean in z-axis,

$$V_{Az} = \sqrt{\frac{1}{N}\sum_{i=1}^{N}(Azi - \mu(Az)_{\text{lean}})^2} \quad (6)$$

Similarly, mean acceleration values ($\mu(Ax)_{\text{lean}}$, $\mu(Ay)_{\text{lean}}$, and $\mu(Az)_{\text{lean}}$) for lean forward position was also calculated as was done in case of mean acceleration values for sitting upright position in Eqs. 1, 2, and 3.

5 Experimental Setup

The experiments were designed in such a manner that it will be able to capture the accelerometer data from the sensor attached at the lower back of the subjects. A written consent was taken from all subjects before performing experiments. The subjects were addressed to perform predefined experiments calmly and without feeling any pressure. All subjects were counseled about the execution of the experiment in order to collect neutral data from all experiments. Experimental data was collected from 12 subjects (2 females and 10 males) with ages from 24 to 40 years. The device was fastened by a belt at the middle lower back to capture the data. The subjects were asked to sit stationary in an upright position on a chair for 3 s and then lean forward in the sagittal plane and stay there in that position for 10 s. The normal vibrational values of lower back at sitting upright position were collected and stored for analysis. Further the vibrational values from the accelerometer sensor were collected when the lower back is at discomfort position as the spine was under stress in the forward lean position. Figure 3 shows the 2 positions used for getting accelerometer data. The data was collected and transmitted wirelessly to the hub where logs were created.

A. Experiment with Subjects

To perform experiments the subjects were equipped with sensor node which was installed at their lower back. Sensor position was selected in such a manner to avoid thick muscles which could affect the accuracy of classification. Further, the subjects were asked to sit on a chair in an upright position for calibration of the sensor. The calibration process will take 5 s and after that a status LED available on the sensor node blinks to indicate completion of calibration process. Next, the synchronous signal is transmitted from the hub by pressing a synchronization switch The subject will remain in upright position for 5 s so that the lower back acceleration data can be transmitted to the hub. Then the subject was moved in a forward lean position making a posture angle of less than 70° as shown in Fig. 3. Ikegami et al. [17] developed a

Fig. 3 **a** Sitting upright (Position: P1) and **b** lean forward (Position: P2) while performing experiments

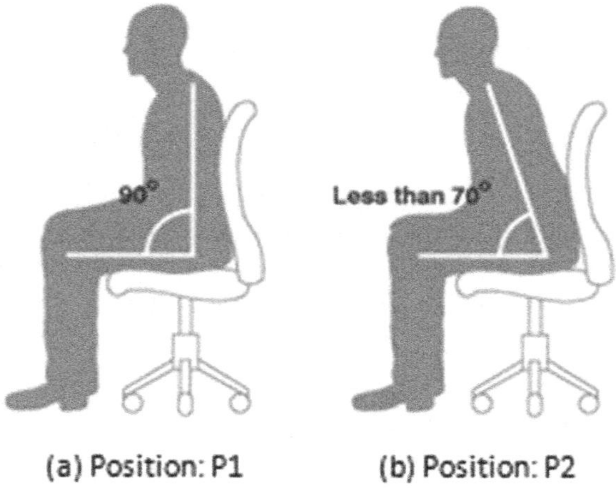

chair which prevents lower back pain while prolonged sitting and doing handwork at the same time. To minimize the effect of sitting for long hours on the lower back, the experiments in the present research work was designed for short duration.

6 Experimental Results

Table 1 shows data values for different features used for discomfort analysis at lower back and classification of subjects. In the experiment conducted, 10 male and 2 female subjects participated who performed an experiment at sitting upright and lean forward position. It was found that mean acceleration value in sitting position around x-axis was 9.893 and in lean forward position was 9.737. This shows that subjects had more vibrations at the lower back while they were in sitting position as compared to lean forward position for a duration of 10 s. This happened because

Table 1 Details of features used for discomfort analysis and classification by different classifiers

Feature	Value	Remark
$\mu(Ax)_{\text{Sit}}$	9.893	Mean of Ax at sitting upright position
$\mu(Ay)_{\text{Sit}}$	0.581	Mean of Ay at sitting upright position
$\mu(Az)_{\text{Sit}}$	1.457	Mean of Az at sitting upright position
$\mu(Ax)_{\text{lean}}$	9.737	Mean of Ax at lean forward position
$\mu(Ay)_{\text{lean}}$	0.445	Mean of Ay at lean forward position
$\mu(Az)_{\text{lean}}$	2.494	Mean of Az at lean forward position
V_{Ax}	0.047	Gives spread of data around $\mu(Ax)_{\text{lean}}$
V_{Ay}	0.036	Gives spread of data around $\mu(Ay)_{\text{lean}}$
V_{Az}	0.089	Gives spread of data around $\mu(Az)_{\text{lean}}$

x-axis is the axis which is passing through the lower back of the subjects in coronal plane. On the other hand, mean acceleration values in sitting position around z-axis was 1.457 and in lean forward position was 2.494. This shows that subjects had more vibrations at the lower back while they were in position P2 (i.e., lean forward position) as compared to position P1 (i.e., sitting upright position). This happened because z-axis is the axis which is passing through the lower back of the subjects in the sagittal plane. Hence more vibrational unrest is there in lean forward position as compared to sitting upright position in the z-axis direction. Mean of acceleration values in sitting and lean forward positions in all three axes were denoted as $\mu(Ax)_{\text{sit}}$, $\mu(Ay)_{\text{sit}}$, $\mu(Az)_{\text{sit}}$, $\mu(Ax)_{\text{lean}}$, $\mu(Ay)_{\text{lean}}$, and $\mu(Az)_{\text{lean}}$. All these values are shown in Table 1 along with variance in all three directions in lean forward position denoted by V_{Ax}, V_{Ay}, and V_{Az}.

Figure 4 shows the accelerometer data for lower back at sitting upright and lean forward position in y-axis direction. It showed the vibrational acceleration data for approximately first 2500 samples in sitting upright position and the vibrational acceleration data for approximately 2000 samples (from sample no. 3000 to sample no. 5000) in lean forward position. The performed experiment showed that the average acceleration had different values and is easily recognizable from Fig. 4. The measured values also indicate that there is more vibrational unrest in the lean forward position as $\mu(Az)_{\text{lean}} > \mu(Az)_{\text{sit}}$. This type of study can be used as meaningful data for predicting posture correction techniques while playing and recovering from back pain problems.

Figure 5 shows the acceleration data (in all three directions) of lower back of the subject while performing experiment. The shift in the acceleration values was observed in the graphs which shows the movement of the subject while sitting and performing experiment in the sagittal plane. The subject was sitting stationary for the initial phase of the experiment and is denoted by the first 2700 sample values in the graph. The subject then moved from position P1 (Sitting upright position) to position P2 (Lean forward position). Position P2 is denoted by sample starting from sample number 4000 and going up to sample number 6500.

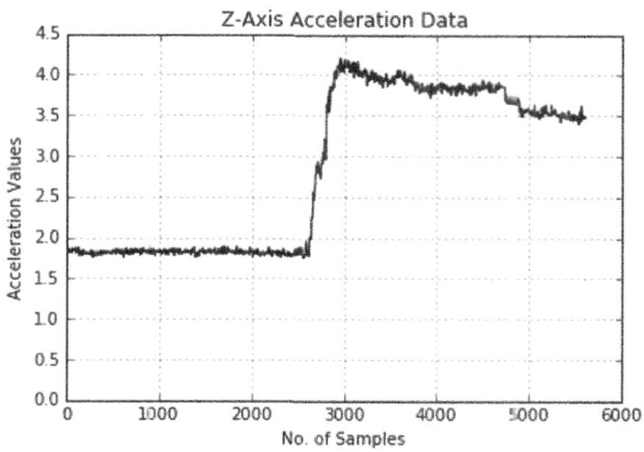

Fig. 4 Acceleration data in Y-axis direction for sitting upright and lean forward position

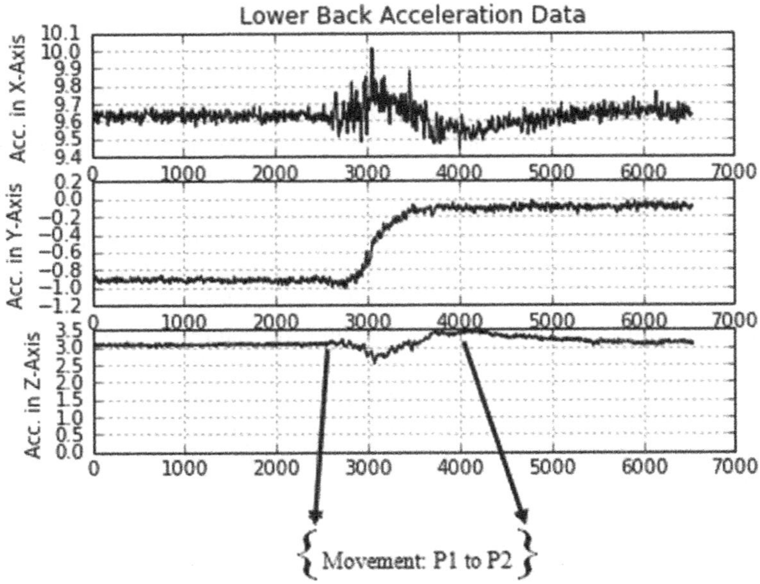

Fig. 5 Acceleration data in Y-axis direction for sitting upright and lean forward position

Table 2 shows the values of different features which were used with various classification algorithms as discussed in Sect. 4 part B. Ax_S, Ay_S, and Az_S denote acceleration values in sitting upright position for x-axis, y-axis, and z-axis, respectively. Similarly, Ax_L, Ay_L, and Az_L denote acceleration values in lean forward position for x-axis, y-axis, and z-axis, respectively. V_{Ax}, V_{Ay}, and V_{Az} denote variance for

Table 2 Experimental values of classification features for sitting upright and lean forward position

Ax_S	Ay_S	Az_S	Ax_L	Ay_L	Az_L	V_{Ax}	V_{Ay}	V_{Az}
9.67	−0.83	2.98	9.57	−0.56	3.34	0.05	0.04	0.08
9.63	−0.92	3.06	9.61	−0.11	3.23	0.05	0.03	0.11
9.45	−0.93	3.57	9.47	−0.76	3.55	0.06	0.04	0.09
9.44	−0.94	3.55	9.46	−0.75	3.53	0.06	0.04	0.09
10.07	−0.80	1.04	9.48	−0.45	3.56	0.04	0.03	0.06
10.04	−0.66	1.38	9.71	−1.18	2.73	0.03	0.04	0.04
10.11	0.01	0.87	10.06	−0.24	1.35	0.03	0.05	0.06
10.10	−1.06	−0.41	9.96	−0.13	1.91	0.05	0.05	0.11
10.09	0.19	1.00	9.90	−0.37	2.22	0.04	0.03	0.06
10.14	0.06	−0.39	10.15	−0.10	−0.04	0.03	0.03	0.05
9.90	−0.26	1.83	9.40	−0.25	3.89	0.08	0.03	0.18
10.07	−0.84	−1.02	10.07	−0.43	0.73	0.04	0.02	0.14

lean forward position in x-axis, y-axis, and z-axis, respectively. It was observed from the table that there were increased values of feature Az_L as compared to feature Az_S hence there is more vibrational unrest in Z-axis direction. A higher vibrational unrest shows more discomfort at lower back due to unstable body posture. This data was analyzed and matched with the subjective discomfort data collected from subjects while performing experiments. The comparison of experimental data with the subjective data helps in assigning target class to the data set. This enables the classifiers to classify the test data into binary class having value as Class_C and Class_D where Class_C signifies the target class which will have subjects with no or less discomfort in lean forward position and Class_D signifies the target class which will have subjects with low or high discomfort in lean forward position.

Figure 6 shows the confusion matrix for predicted classes using random forest (RF) classification algorithm. In this the target class was set to "1" (for Class_D, i.e., subjects with discomfort at lower back in forward lean position) and it was set to "0" (for Class_C, i.e., subjects with no discomfort at lower back in forward lean position). There were true labels (0 and 1) for actual class and predicted labels (0 and 1) for predicted classes. It was observed that random forest classification algorithm was having a classification accuracy of 80% which was highest as compared to decision trees, naïve Bayes, and support vector machine. Figure 6 also shows that there was one false prediction when the target class was falsely predicted as "1". The complete set of data was divided into two sets namely training set and test set. The prediction was carried out the testing set after training the classification algorithms. Two more performance metrics were calculated for all classifiers, i.e., F-score and computation time and was found that F_{Score} is 0.66 and computation time was 0.028 s for random forest classifier.

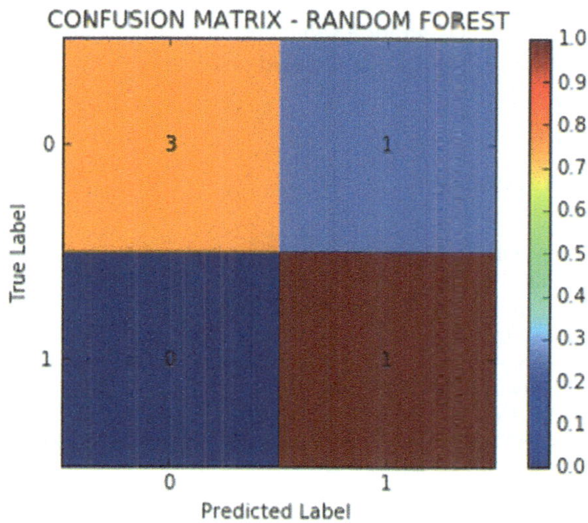

Fig. 6 Classification result in form of confusion matrix for random forest classifier

The classifiers were evaluated by three different performance measures viz. precision, recall, and F_{Score}. The values of various performance metrics can be calculated by using Eqs. 7, 8, and 9.

$$\text{Precision} = \frac{\text{True}_P}{\text{True}_P + \text{False}_P} \tag{7}$$

$$\text{Recall} = \frac{\text{True}_P}{\text{True}_P + \text{False}_N} \tag{8}$$

$$F_{\text{Score}} = \frac{2 * \text{Precision} * \text{Recall}}{\text{Precision} + \text{Recall}} \tag{9}$$

Equation 7, 8, and 9 were used to calculate values of respective performance metrics and the results were calculated as Precision = 0.5, Recall = 1 and F_{Score} = 0.66.

7 Conclusion

The research work resulted in designing of a new device and presented a new way of classification of subjects based on discomfort analysis at lower back using accelerometer data. A wireless sensor node-hub architecture was used for data acquisition and creating logs. This research proved that classification of person having discomfort at lower back could be done from accelerometer data from the sensor mounted at the lower back. It showed that the classification accuracy with random forest classifier was highest with 80% accuracy as compared to naive Bayes, decision trees, and support vector machine classifiers. It was also found that there is higher vibrational unrest around z-axis in lean forward position as compared to sitting upright position. The mean acceleration value in z-axis was 1.457 in sitting upright position and it was 2.494 in z-axis in lean forward position.

References

1. V. Oliverio, O.B. Poli-Neto, Case study: classification algorithms comparison for the multi-label problem of chronic pelvic pain diagnosing, in *IEEE 33rd International Conference on Data Engineering*, pp. 15071509 (2017)
2. S. Taghvaei, Y. Hirata, K. Kosuge, Visual human action classification for control of a passive walker, in *IEEE 7th International Conference on Modeling, Simulation and Applied Optimization (ICMSAO)* (2017), p. 15
3. D. Urukalo, P. Blazevic, S. Charles, J.-P. Carta, The TeachWear healthcare wearable device, in *27th IEEE International Symposium on Robot and Human Interactive Communication* (2018), pp. 638–643

4. W. Du, O.M. Omisore, H. Li, K. Ivanov, S. Han, L. Wang, Recognition of chronic low back pain during lumbar spine movements based on surface electromyography signals. IEEE Access **6**, 65027–65042 (2018)
5. M. Molnar, M. Kok, T. Engel, H. Kaplick, F. Mayer, T. Seel, A method for lower back motion assessment using wearable 6D inertial sensors, in *21st IEEE International Conference on Information Fusion (FUSION)* (2018), p. 799806
6. W. Xu, C. Ortega-Sanchez, I. Murray, Measuring human joint movement with IMUs, in *15th IEEE Student Conference on Research and Development* (2018), pp. 172–177
7. O. Chutatape, K. Naonueng, R. Phoophuangpairoj, Detection of improper postures leading to dislocation of hip prosthesis by a smartphone, in *14th IEEE International Conference on Electrical Engineering/Electronics, Computer, Telecommunications and Information Technology (ECTI-CON)* (2017), pp. 415–418
8. W. Kam, K. O'Sullivan, W.S. Mohammed, E. Lewis, Low cost portable sensor for real-time monitoring of lower back bending, in *25th IEEE International Conference on Optical Fiber Sensors* (2017)
9. D.-M. Dobrea, M.-C. Dobrea, A warning wearable system used to identify poor body postures, in *IEEE International Conference on Advances in Wireless and Optical Communications* (2018), pp. 55–60
10. G.A. Sandag, N.E. Tedry, S. Lolong, Classification of lower back pain using K-Nearest Neighbor algorithm, in *6th IEEE International Conference on Cyber and IT Service Management (CITSM 2018)* (2018), pp. 1–5
11. F. Galbusera, G. Casaroli, T. Bassani, Artificial intelligence and machine learning in spine research. JOR Spine, Jan 2019
12. G. Shabrina, B.M. Iqbal, D.H. Syaifullah, Effect of shoes on lower extremity pain and low back pain during prolonged standing on a sloping medium, in *IEEE International Conference on Intelligent Informatics and Biomedical Sciences (ICIIBMS)*, vol. 3 (2018), pp. 181–187
13. T. Nezam, R. Boostani, V. Abootalebi, K. Rastegar, A novel classification strategy to distinguish five levels of pain using the EEG signal features. IEEE Trans. Affect. Comput. 1–9 (2018)
14. A.A. Abdullah, A. Yaakob, Z. Ibrahim, Prediction of spinal abnormalities using machine learning techniques, in *IEEE International Conference on Computational Approach in Smart Systems Design and Applications (ICASSDA)* (2018), pp. 1–6
15. J. Estrada, L. Vea, Sitting posture recognition for computer users using smartphones and a web camera, in *IEEE Region 10 Conference (TENCON)*, pp. 1520–1525 (2017)
16. Y. Tian, J. Zhang, L. Chen, Y. Geng, X. Wang, Single wearable accelerometer-based human activity recognition via kernel discriminant analysis and QPSO-KELM classifier. IEEE Access **7**, 109216–109227 (2019)
17. K. Ikegami, H. Hirata, H. Ishihara, S. Guo, Development of a chair preventing low back pain with sitting person doing hand working at the same time, in IEEE international conference on mechatronics and automation (2018), pp. 1760–1764

Design of Lyapunov-Based Discrete-Time Adaptive Sliding Mode Control for Slip Control of Hybrid Electric Vehicle

Khushal Chaudhari and Ramesh Ch. Khamari

Abstract This paper has developed discrete-time Fuzzy Adaptive sliding mode control algorithm for controlling the slip ratio of a hybrid electric vehicle. Fuzzy logic algorithm is used to develop controller for controlling slip ratio so as to handle different road conditions. A discrete-time Sliding Mode Observer is designed to observe the vehicle velocity. Furthermore, an adaptive SMC has been designed by employing Lyapunov theory in order to adapt with slip dynamic change for varying or changing road conditions. The performances of designed controller such as ASMC, SMO, FLC, and Fuzzy PID for controlling slip ratio are compared using MATLAB simulation and it is proved that the discrete-time fuzzy ASMC perform most impressively and effectively.

Keywords Discrete-time sliding mode control · SMO · FLC · HEV · Slip ratio · FSMC

1 Introduction

Hybrid electric vehicles have good energy efficiency and reduce emissions. Hybrid electric vehicles (HEVs) are more comfortable and preferable over conventional vehicles (ICVs) [1]. Furthermore, multiple sources that provide power are used in HEVs for driving. Further, with the help of control performance such as TCS and ABS [2], it is easier to get and achieve useful, human favorable, and advanced driving performance. Actuation system (electric motor) of HEV always produces uncertainty error or variation in driving or braking torque as it is not present in

K. Chaudhari
Department of Electronics and Telecommunication, Government Polytechnic, Jintur, Maharashtra 431509, India
e-mail: khushalchaudhari@gmail.com

R. Ch. Khamari (✉)
Department of Electrical Engineering, Government College of Engineering, Keonjhar, Odisha 758002, India
e-mail: rameshkhamari_fee@gcekjr.ac.in

S. S. Dash et al. (eds.), *Intelligent Computing and Applications*,
Advances in Intelligent Systems and Computing 1172,
https://doi.org/10.1007/978-981-15-5566-4_9

that of hydraulic braking system and combustion engine system. Corresponding to the main function of propulsion, because of torque performance and regeneration capability of motors, electric motor will alternatively use the braking option. The main problem during braking on roads according to different road conditions like slippery road, road covered with ice, etc., is that locking condition of the wheels of the vehicle may get occurred which implies that wheel speed nearly tends to approach toward zero and which may result in that vehicle leaving the road. Hence, it creates the possibility of overstepping, collision, or accident of the vehicle. The goal of Antilock Braking System is to maximize wheel traction of the vehicle by preventing the wheels from locking as wheel speed approaches to zero during braking while maintaining vehicle stability on the road. Due to nonlinearity present in system dynamic and time-variant unknown parameters related with wheel characteristic, controlling of a braking system of electric and hybrid electric vehicle are difficult.

A numerous control strategy likes Iterative Learning Control [2], sliding mode control (SMC) [3], fuzzy logic [4], neural networks [5], etc., have been used for controlling the slip ratio of vehicles. Also, advanced controlling strategies such as neural network [6], fuzzy logic control [7], adaptive control [8], hybrid control [9], and PWM technique [10] have been developed for vehicles with IC engine to achieve effective antilock braking performance. Undesirable phenomenon of oscillations having finite frequency and amplitude is known as chattering [11]. Chattering is a problem as it leads to less accuracy and more power losses.

The objective is to develop a controller for achieving the required value of slip ratio such that locking of wheel is avoided and furthermore the controller must handle the nonlinearity and time variation present in HEV dynamics (wheel and vehicle dynamic). Also, observer design is required as vehicle speed is not measurable. Even though in the recent literature, SMC and observer were used for designing for slip ratio controlling of HEV earlier but discrete-time controller is essential for practical implementation of controller on system. Hence unlike [2, 4, 10], this work is focused on development of discrete-time sliding mode observer, discrete-time Adaptive Sliding mode control, Fuzzy logic control, and Fuzzy PID control for slip control of HEV.

The remaining structure of the paper is described as follows. Development of discrete-time Slip Control model and actuator dynamic presented in Sect. 2, Sect. 3 describes the problem development, Sect. 4 describes design of SMO, Sect. 5 describes design of adaptive SMC, design of FLC for SRC is described in Sect. 6, Sect. 7 describes fuzzy PID controller design, Sect. 8 presents simulation results, and Sect. 9 concludes the paper.

2 Slip Control Model and Actuator Dynamic

The dynamics of vehicle in discrete domain is given by [12]

$$\frac{x_1(k+1) - x_1(k)}{T} = f_1(x_1(k)) + d_1\mu(\lambda(k)) \quad (1)$$

$$\frac{x_2(k+1) - x_2(k)}{T} = f_2(x_2(k)) + d_2\mu(\lambda(k)) + d_3\tau_m(k) \quad (2)$$

where T is sampling time, $x_1 = \omega_v$ and $x_2 = \omega_w$,

$$f_1(x_1) = -\frac{c_v r_w}{m}x_1^2,\ f_2(x_2) = -\frac{r_w f_w(x_2)}{I_w} d_1 = \frac{n_w g}{r_w},\ d_2 = -\frac{r_w mg}{I_w}, d_3 = \frac{1}{I_w}$$

Variables information are r_w is radius of wheel, m is vehicle mass, I_v is moment of inertia of wheel, n_w is number of wheel, g is acceleration due to gravity, f_w is viscous wheel friction force, τ_m is braking torque, v is vehicle linear velocity, ω_v is angular velocity of vehicle, ω_w is angular velocity of wheel, c_v is aerodynamic drag coefficient, μ is adhesive coefficient.

The actuator consists of a dc motor for HEV given in [13]. The torque is negative while brake applied on wheel of vehicle. Voltage required to produce braking torque in discrete domain [12] is

$$e(k) = \frac{L}{K_m}\frac{\tau_m(k+1) - \tau_m(k)}{T} + \frac{R}{K_m}\tau_m(k) + K_b\omega_w(k) \quad (3)$$

3 Problem Formulation

Slip ratio is defined as [12]

$$\lambda = \frac{\omega_w - \omega_v}{max(\omega_v, \omega_w)} \quad (4)$$

Equation (4) is rewritten as

$$\lambda(k) = \frac{x_2(k) - x_1(k)}{max(x_1(k), x_2(k))} \quad (5)$$

Deceleration [12]:
For deceleration, $x_1 > x_2$ and hence

$$\frac{\lambda(k+1)-\lambda(k)}{T} = f(\lambda, x) + bu \tag{6}$$

where $f(\lambda, x) = f_2(x_2) - (1+\lambda) f_1(x_1) + [d_2 - (1+\lambda)d_1]\mu(\lambda)$

$$b = d_3, u = \frac{\tau_m}{x_1} x = [x_1, x_2]^T$$

Acceleration [12]:

For acceleration, $x_2 > x_1$ therefore,

$$\frac{\lambda(k+1)-\lambda(k)}{T} = f(\lambda, x) + bu \tag{7}$$

where $f(\lambda, x) = (1-\lambda) f_2(x_2) - f_1(x_1) + [d_2(1-\lambda) - d_1]\mu(\lambda)$

$$b = (1-\lambda)d_3, u = \frac{\tau_m}{x_2}$$

Our main solution is for achieving effective braking and for these, goal of the paper is to get control input u using Eq. (6) as it is known that I_w is directly related with b which is unknown constant gain. Estimated value of b is represented by $\hat{b}$. $\mu - \lambda$ characteristics of surface vehicle [14] and μ is given by

$$\mu(\lambda) = \frac{2\mu_p \lambda_p \lambda}{\lambda_p^2 + \lambda^2} \tag{8}$$

where λ_p is optimal slip and μ_p is optimal adhesive coefficient. Now main goal of the paper is to find the control input u so as to achieve tracking of desired slip ratio to desired value for the Hybrid Electric Vehicle in presence of nonlinearity in $f(\lambda, x)$ due to adhesive coefficient and slip relation.

4 Design of Discrete-Time Sliding Mode Observer

Design of Fuzzy SMC is done in ref. [12]. While driving the vehicle, it is finding to measure vehicle velocity online very difficult. Hence, we suggested the design of a discrete-time SMO observer for estimating or observing vehicle angular velocity. So, observer dynamic chosen in following form

$$\hat{x}_1(k+1) = \hat{x}_1(k) - \frac{T c_v r_w \hat{x}_1^2}{m} - T K_v sgn(\tilde{x}_1) + c_1 \mu T \tag{9}$$

where $\hat{x}_1$ is estimated vehicle velocity, $\tilde{x}_1$ is measurement error and K_v is observer gain. Now $\tilde{x}_1$ is given as

$$\tilde{x}_1(k) = x_1(k) - \hat{x}_1(k) \tag{10}$$

Equation (10) rewritten as

$$\tilde{x}_1(k+1) = x_1(k+1) - \hat{x}_1(k+1) \tag{11}$$

Substituting $x_1(k+1)$ and $\hat{x}_1(k+1)$ from Eqs. (1) and (9) in (11) and solving for $\tilde{x}_1(k+1)$ we get

$$\tilde{x}_1(k+1) = \tilde{x}_1(k) - \frac{T c_v r_w \tilde{x}_1^2}{m} + T K_v sgn(\tilde{x}_1) \tag{12}$$

where $\tilde{x}_1^2 = x_1^2 - \hat{x}_1^2$ SMO observer dynamic represented in Eq. (9) is asymptotically stable if observer gain chosen as

$$K_v \leq \left| \frac{c_v r_w \tilde{x}_1^2}{m} \right| \tag{13}$$

So as prove that, we have chosen the Lyapunov candidate function as

$$V = \frac{1}{2}\tilde{x}_1^2 \tag{14}$$

and

$$\Delta V = V(k+1) - V(k) = \tilde{x}_1(k)\left[\tilde{x}_1(k+1) - \tilde{x}_1(k)\right] \leq 0 \tag{15}$$

Putting Eqs. (9) and (11) in (15), we get

$$T\tilde{x}_1(k)\left[-\frac{c_v r_w \tilde{x}_1^2}{m} + K_v sgn(\tilde{x}_1)\right] \leq 0 \tag{16}$$

Under the assumption, wheel speed and vehicle speed are positive gives the condition as

$$K_v \leq \left| \frac{c_v r_w \tilde{x}_1^2}{m} \right| \tag{17}$$

5 Design of Discrete-Time Adaptive Sliding Mode Control

Equation (6) can also be written as

$$\frac{\lambda(k+1)-\lambda(k)}{T}=f_a(\lambda,x)+\theta h(\lambda,x)+bu \tag{18}$$

where

$$f_a(\lambda,x)=\frac{f_2(x_2)-(1+\lambda)f_1(x_1)}{x_1} \tag{19}$$

$$h(\lambda,x)=\frac{1}{x_1}\frac{2[c_2-(1+\lambda)c_1]\lambda_p\lambda}{\lambda_p^2+\lambda^2} \tag{20}$$

$$\theta=\mu_p \tag{21}$$

Choosing of sliding surface is the same as given in the ref [12].
Choosing Lyapunov candidate function as

$$V=\frac{1}{2}s^2+\frac{1}{2}(\theta-\widehat{\theta})^2 \tag{22}$$

So,

$$\Delta V=V(k+1)-V(k)=s(k)[s(k+1)-s(k)]-\widetilde{\theta}\left[\widehat{\theta}(k+1)-\widehat{\theta}(k)\right]\leq 0 \tag{23}$$

where

$$\widetilde{\theta}=\theta-\widehat{\theta} \tag{24}$$

Substituting Eq. (18) in Eq. (23) leads to

$$\begin{aligned}\Delta V=s(k)T\Big[f_a+\theta h+b\hat{b}^{-1}\Big(-\hat{f}_a-\widehat{\theta}\hat{h}+\lambda_d+(1-qT)s(k)\Big)+b\varepsilon T\,sgn(s(k))\Big]\\ -\widetilde{\theta}\left[\widehat{\theta}(k+1)-\widehat{\theta}(k)\right]\end{aligned} \tag{25}$$

Now, rearranging different terms of Eq. (25), we get

$$\begin{aligned}\Delta V=s(k)T\Big[f_a+\theta h-b\hat{b}^{-1}\hat{f}_a-b\hat{b}^{-1}\widehat{\theta}\hat{h}+b\hat{b}^{-1}\lambda_d+b\hat{b}^{-1}(1-qT)s(k)\\ +b\varepsilon T\,sgn(s(k))]-\widetilde{\theta}\left[\widehat{\theta}(k+1)-\widehat{\theta}(k)\right]\end{aligned} \tag{26}$$

Equation (26) can be rewritten as

$$\begin{aligned}\Delta V=s(k)T(f_a-\hat{f}_a+(1-b\hat{b}^{-1})\hat{f}_a+(1-b\hat{b}^{-1})\widehat{\theta}\hat{h}+b\hat{b}^{-1}\lambda_d\\ +\theta h-\widehat{\theta}\hat{h}+b\hat{b}^{-1}(1-qT)s(k)+b\varepsilon T\,sgn(s(k)))\end{aligned}$$

$$-\tilde{\theta}\left[\hat{\theta}(k+1)-\hat{\theta}(k)\right] \tag{27}$$

Following assumptions are considered for bounds in function f_a and h. $\left|f_a-\hat{f}_a\right| \le F_a$ and $\left|h-\hat{h}\right| \le H$

Using above assumption in Eq. (27) leads to

$$\begin{aligned}\Delta V &= s(k)T(f_a-\hat{f}_a+(1-b\hat{b}^{-1})\hat{f}_a+(1-b\hat{b}^{-1})\hat{\theta}\hat{h}+b\hat{b}^{-1}\lambda_c+\theta\hat{h}\\ &\quad +\theta H-\hat{\theta}\hat{h}+b\hat{b}^{-1}(1-qT)s(k)+b\varepsilon T\,sgn(s(k)))\\ &\quad -\tilde{\theta}\left[\hat{\theta}(k+1)-\hat{\theta}(k)\right]\le 0\end{aligned} \tag{28}$$

Now, rearranging different terms, we get

$$\begin{aligned}\Delta V &= s(k)T(f_a-\hat{f}_a+(1-b\hat{b}^{-1})\hat{f}_a+(1-b\hat{b}^{-1})\hat{\theta}\hat{h}+b\hat{b}^{-}\lambda_d\\ &\quad +\theta H+b\hat{b}^{-1}(1-qT)s(k)+b\varepsilon T\,sgn(s(k)))\\ &\quad -\tilde{\theta}\left[\hat{\theta}(k+1)-\hat{\theta}(k)\right]+s(k)T\tilde{\theta}\hat{h}\le 0\end{aligned} \tag{29}$$

For asymptotically stable,
$\Delta V \le 0$ and hence

$$\begin{aligned}&s(k)T\Big[f_a-\hat{f}_a+(1-b\hat{b}^{-1})\hat{f}_a+(1-b\hat{b}^{-1})\hat{\theta}\hat{h}+b\hat{b}^{-1}\lambda_d+\theta H\\ &\quad +b\hat{b}^{-1}(1-qT)s(k)+b\varepsilon T\,sgn(s(k))\Big]=0\end{aligned} \tag{30}$$

and

$$-\tilde{\theta}\left[\hat{\theta}(k+1)-\hat{\theta}(k)\right]+s(k)T\tilde{\theta}\hat{h}=0 \tag{31}$$

From Eq. (30), we get

$$\begin{aligned}\varepsilon \ge &\Big[f_a-\hat{f}_a+\left(1-b\hat{b}^{-1}\right)\hat{f}_a+\left(1-b\hat{b}^{-1}\right)\hat{\theta}\hat{h}+b\hat{b}^{-1}\lambda_d+\theta H\\ &+b\hat{b}^{-1}(1-qT)s(k)\Big](bT\,sgn(s(k)))^{-1}\end{aligned} \tag{32}$$

Equation (31) provide dynamic for $\hat{\theta}$ given by

$$\hat{\theta}(k+1)=\hat{\theta}(k)+s(k)T\hat{h} \tag{33}$$

Equation (32) provide the value of ε and Eq. (33) provide the estimated value of optimal adhesive coefficient.

Table 1 Rule base for computing control action

λ_e	$\Delta\lambda_e$						
	NB	NM	NS	Z	PS	PM	PB
NB	NB	NB	NM	NM	NS	NS	Z
NM	NB	NM	NM	NS	NS	Z	PS
NS	NM	NM	NS	NS	Z	PS	PS
Z	NM	NS	NS	Z	PS	PS	PM
PS	NS	NS	Z	PS	PS	PM	PM
PM	NS	Z	PS	PS	PM	PM	PB
PB	Z	PS	PS	PM	PM	PB	PB

6 Design of FLC for SRC

The design of an FLC for controlling of the desired value of slip ratio of HEV considers the appropriate membership Functions (MFs). Appropriate MFs will be selected on the basis that chosen MFs will occupy the whole universe of discourse and also the selected membership functions overlap each other. This is to be done so that any kind of discontinuity can be avoided. We have developed Fuzzy Logic controller consisting of two inputs and one output for slip ratio tracking problem of HEV. FLC inputs are λ_e and $\Delta\lambda_e$ and output control action is u. We have chosen 7 numbers of triangular membership function for input and output variable of FLC.

FLC for slip ratio of HEV has the following rules formulated in Table 1.

Rule i: If λ_e is NB and $\Delta\lambda_e$ is NB (negative big), then control action u is NB (negative big).

where $i = 1, 2, 3, \ldots, n$ and n is equal to 49 where n is the number of rules.

7 Fuzzy PID Controller Design

This section provides design of fuzzy PID controller. Two Inputs for Fuzzy PID are λ_e and $\Delta\lambda_e$ and Three outputs of Fuzzy PID are K_p, K_i, K_d. 7 triangular membership functions will be chosen for λ_e, $\Delta\lambda_e$, K_p, K_i, K_d. Rules shown in Table 2.

Rule i: If λ_e is NB and $\Delta\lambda_e$ is NB, then K_p is BS and K_i is BS and K_d is BS.

where $i = 1, 2, 3, \ldots, n$ and n are the number of rules and it is equal to 49.

Table 2 Rule base for computing K_p, K_d, K_i

λ_e	$\Delta\lambda_e$						
	PB	PM	PS	Z	NS	NM	NB
NB	M	MS	MS	S	S	BS	BS
NM	M	M	MS	MS	S	S	BS
NS	MB	M	M	MS	MS	S	S
Z	B	MB	MB	M	MS	MS	S
PS	B	B	MB	MB	M	M	MS
PM	BB	B	B	MB	MB	M	M
PB	BB	BB	B	B	MB	MB	M

8 Simulation Results and Discussion

All the developed controllers namely SMC, ASMC, FLC, and F-PID are simulated in MATLAB considering discrete-time dynamic of HEV [Eqs. (14), (15)]. For simulating purpose, the chosen value is the initial value of wheel speed 88 km/h, vehicle speed 90 km/h. Values of parameter used for simulating the work designed are shown in Table 3.

8.1 *Analysis of Designed Fuzzy Sliding Mode Control and Fuzzy Sliding Mode Control with Observer*

Simulation result of fuzzy SMC is incorporated from ref paper [12]. Figures show performing result for fuzzy SMC and fuzzy SMC with observer. Value of desired slip ratio is selected as λ_d is −0.6, ε is 0.05, q is 300. Necessity of SMO is that observe vehicle speed online because vehicle speed is not measurable practically. Figure 1 shows that the value of slip ratio to controlled has been achieved and settling time of response is nearly 0.18 s. Uncertainty estimation error is shown in Fig. 1 for FSMC and FMSO with observer is zero. Observed vehicle speed (FSMO) closely

Table 3 Vehicle and Actuator parameter

Parameter	Value	Parameter	Value
I_w	0.65 km m^2	λ_p	−0.17
r_w	0.31 m	c_v	0.595 $N/m^2/s^2$
m	1400 kg	T	0.001 s
g	9.8 m/s^2	K_m	0.2073 mkg/h
n_w	4	K_b	2.2 $V/rad/s$
f_w	3500 N	R	0.125 Ω
μ_p	−0.8	L	0.0028 H

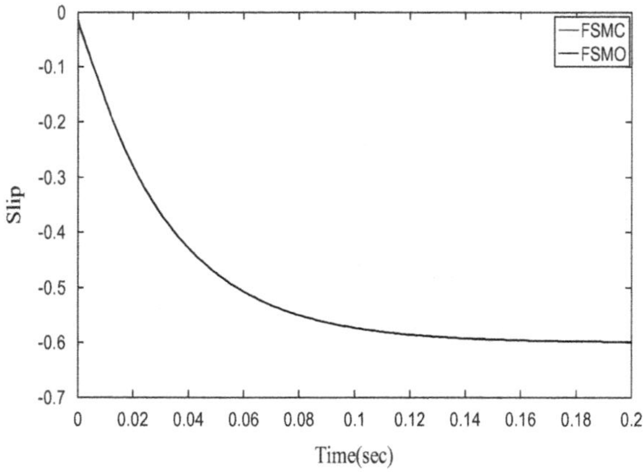

Fig. 1 Slip ratio with FSMC and FSMO

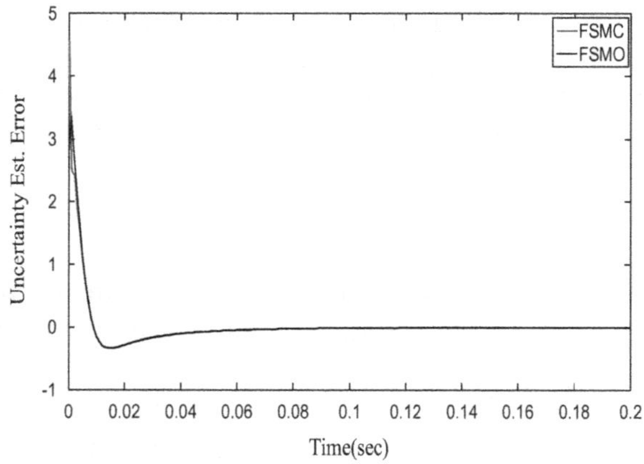

Fig. 2 Estimation error with FSMC and FSMO

resemble vehicle speed as shown in Figs. 3 and 4 provide vehicle speed estimation error approaches to zero. The breaking torque shown in Fig. 5 and voltage required for the actuator to carry out the process shown in Fig. 6.

8.2 *Analysis of Designed Sliding Mode Control, Fuzzy Logic Control, and Fuzzy PID Control*

Simulation result of sliding mode control is incorporated from ref paper [12]. Performing results of SMC, FLC, and fuzzy PID control are shown in figures. Desired slip ratio is selected as λ_d is -0.6, q is 300, ε is 0.05. Controllers such as SMC, FLC, and F-PID are used for tracking the desired value of slip ratio and compared performance are shown in Table 4. Figure 7 shows that the desired slip ratio is being

Fig. 3 Vehicle speed with FSMC and FSMO

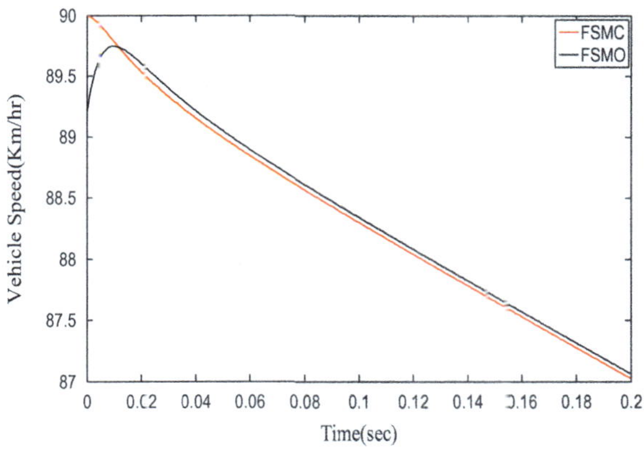

Fig. 4 Vehicle speed estimation error with FSMO

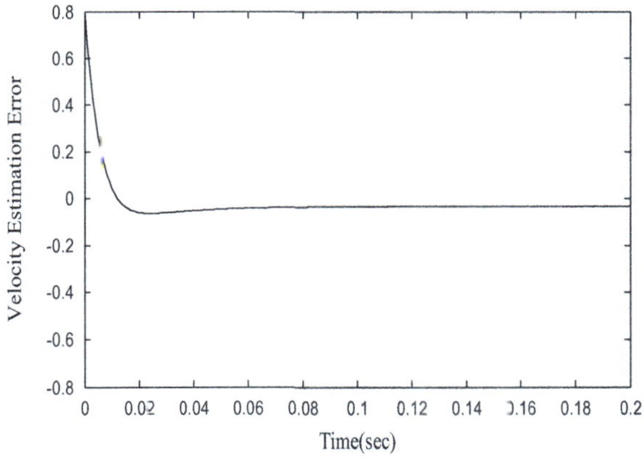

Fig. 5 Braking torque with FSMC and FSMO

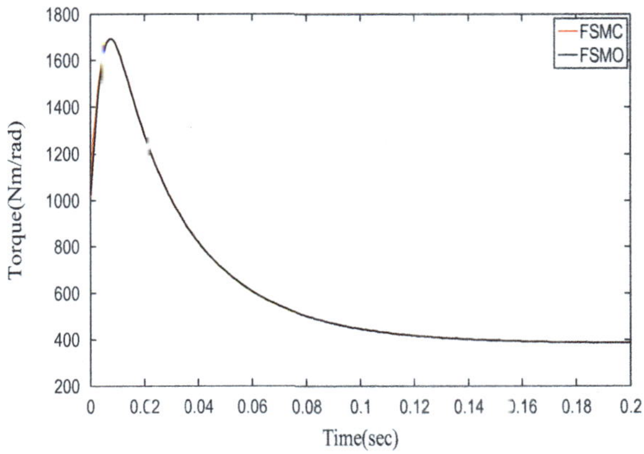

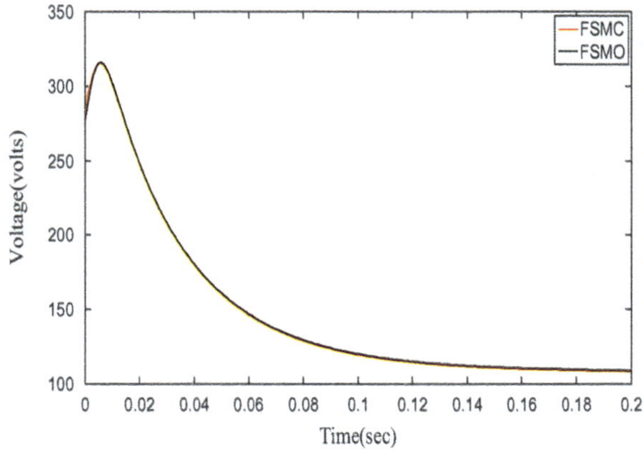

Fig. 6 Required voltage with FSMC and FSMO

Table 4 Tabular comparison

Controller	Chattering	Rise time (s)	Setting time (s)
SMC	Zero	0.0699	0.12
FLC	Zero	0.0066	0.0042
F-PID	Zero	0.0243	0.048

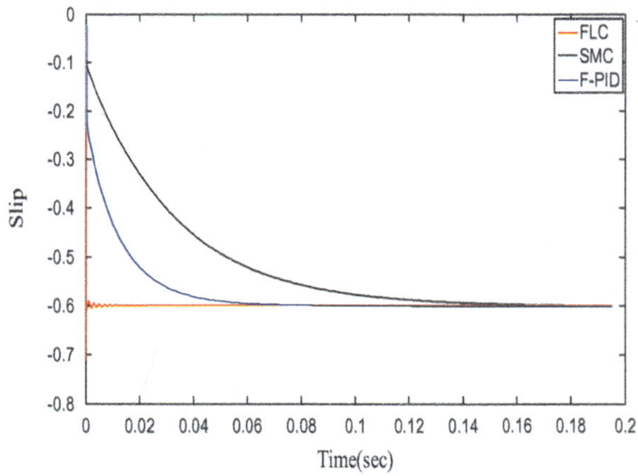

Fig. 7 Slip ratio with FLC, SMC and F-PID

achieved by suggested controllers. Comparison proved that, as compared to SMC and F-PID, response of FLC is impressive and as compared to SMC, the response of F-PID is good. Braking torque required is shown in Fig. 8. By comparing response of braking torque of all suggested controllers, initial value of torque in FLC is higher. Vehicle speed and wheel speed performance shown in Figs. 9 and 10, respectively, and it is given in literature that for deceleration mode both speeds should decrease so that slip ratio is maintained at desired value and it is happening in our figure response also. Voltage excitation required for producing torque is shown in Fig. 11.

Fig. 8 Braking torque with FLC, SMC and F-PID

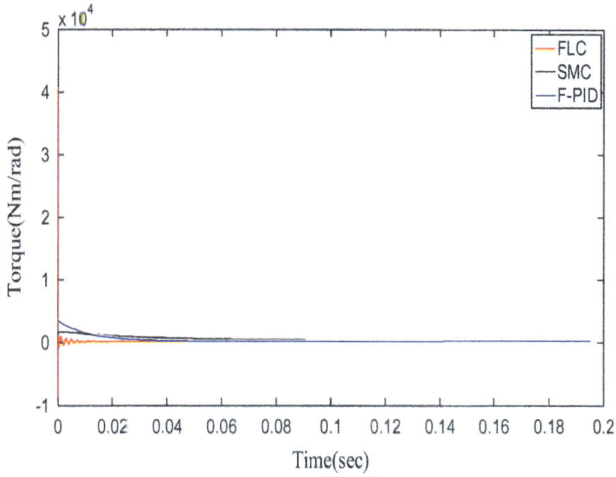

Fig. 9 Vehicle speed with FLC, SMC and F-PID

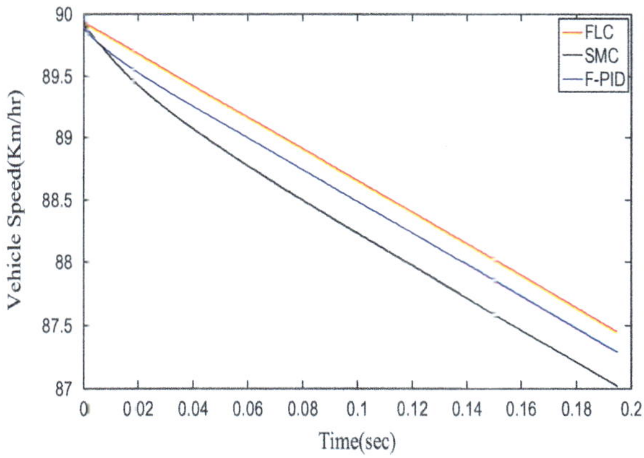

Fig. 10 Wheel speed with FLC, SMC and F-PID

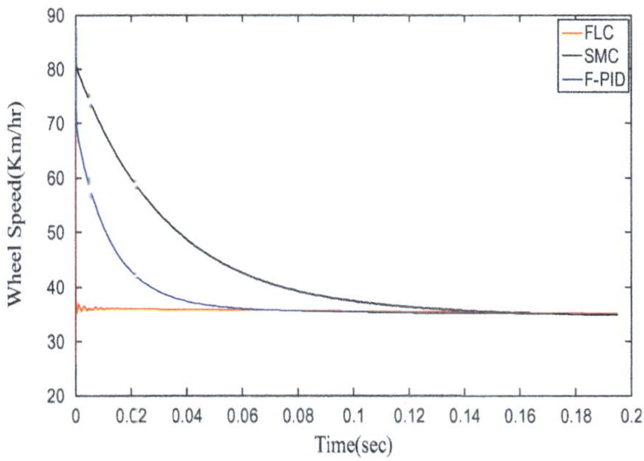

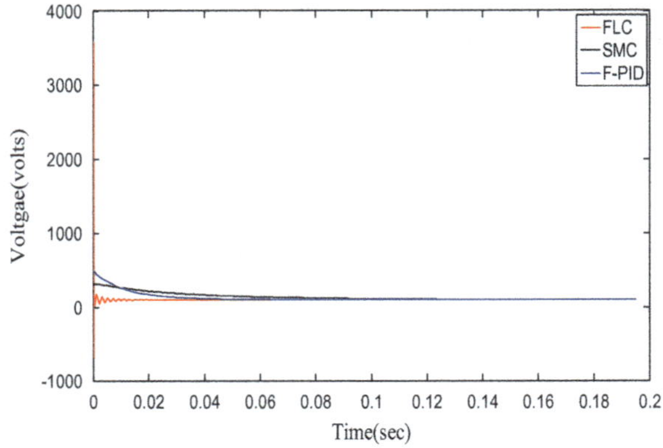

Fig. 11 Required voltage with FLC, SMC and F-PID

In Table 4, if we compare all responses of the controller, then F-PID gives impressive performance in all respects as compared to FLC and SMC.

8.3 Analysis of Designed Adaptive Sliding Mode Control, Fuzzy Logic Control and Fuzzy PID Control

Figures show results of adaptive SMC, FLC, and fuzzy PID control. For controlling purpose, range 0.8 to -0.4 is selected for λ_d. Adaptive SMC, FLC, and fuzzy PID are used for slip ratio (λ_d) tracking and performance of these controllers are observed. Figure 12 shows that λ_d is being achieved by suggested controllers. So, as compared to adaptive SMC and F-PID, response of FLC is impressive and as compared to adaptive SMC, the response of F-PID is good. Comparison provides that initial value of braking torque is higher when FLC is used for controlling slip ratio required is

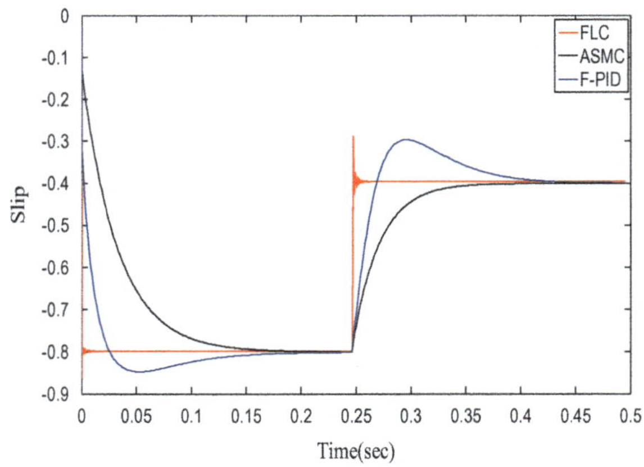

Fig. 12 Slip ratio with FLC, ASMC and F-PID

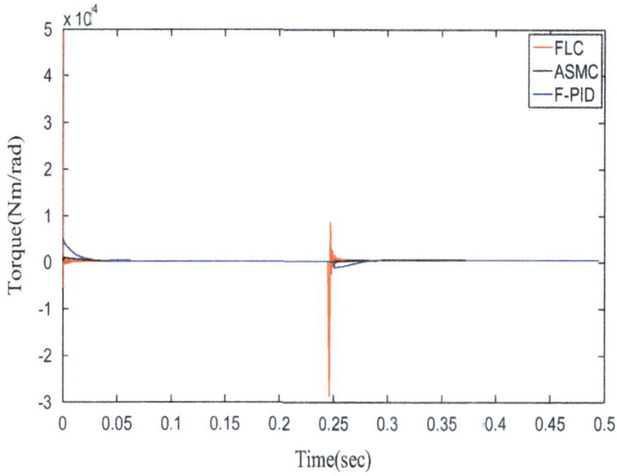

Fig. 13 Braking torque with FLC, ASMC and F-PID

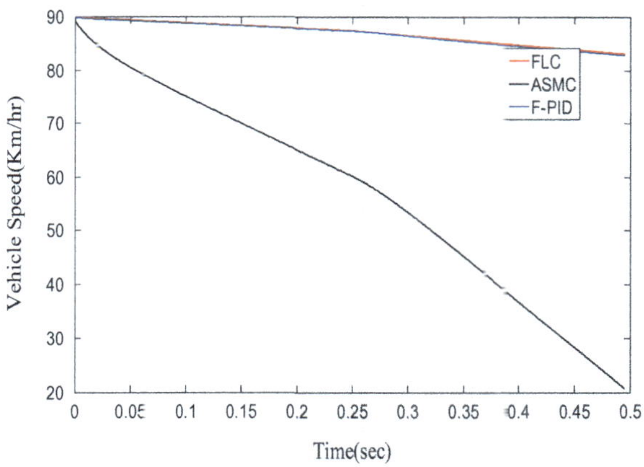

Fig. 14 Vehicle speed with FLC, ASMC and F-PID

shown in Fig. 13. Figures 14 and 15 show vehicle speed and wheel speed profile and it is given in literature that for deceleration mode both speeds should decrease so that slip ratio is maintained at desired value and it is happening in our figure response also. Figure 16 shows the voltage required for actuation purpose which indeed produced torque and response is just following the braking torque response.

9 Conclusion

This paper presented the design of control algorithm schemes like discrete-time Sliding Mode Control, FLC, and Fuzzy PID for achieving control over the desired value of slip ratio. Also, observed control response output of slip ratio of an HEV impressively speeds up using FLC and F-PID. The effectiveness of above-mentioned

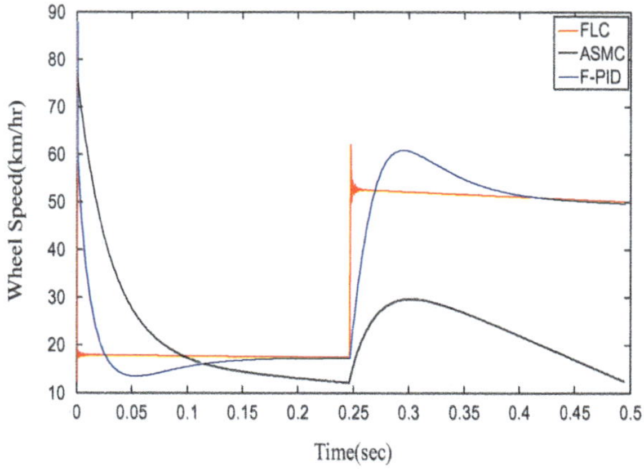

Fig. 15 Wheel speed with FLC, ASMC and F-PID

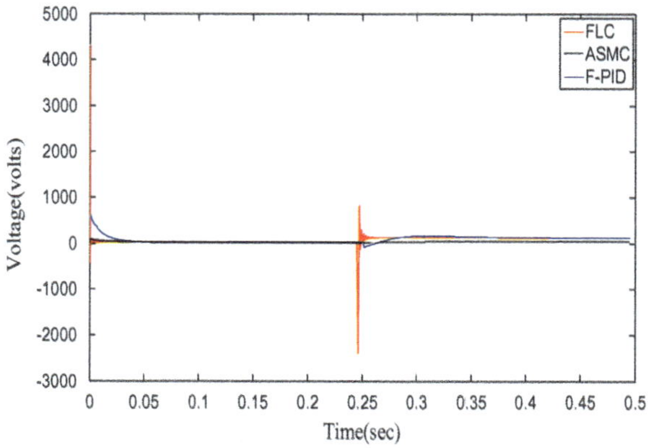

Fig. 16 Required voltage with FLC, ASMC and F-PID

controllers will be observed through MATLAB Simulation results. Problem of uncertainty in the dynamic of slip ratio is addressed using fuzzy logic. Thereafter, designed sliding mode observer using Lyapunov theory successfully estimated the vehicle velocity online. The problems related to the tire, such as road dynamic, slip changes, road changes, etc., are overcomed by designed discrete-time ASMC. Overall, chattering problem is vanished by all designed controllers and hence actuation system is avoided from damage.

References

1. P. Khatun, C.M. Bingam, N. Schofield, P.H. Mellor, Application of fuzzy control algorithms for electric vehicle antilock braking/traction control systems. IEEE Trans. Veh. Technol. **52**(5), 1356–1364 (2003)

2. C. Mi, H. Lin, Y. Zhang, Iterative learning control of antilock braking of electric and hybrid vehicles. IEEE Trans. Veh. Technol. **54**(2), 486–494 (2005)
3. C. Unsal, P. Kachroo, Sliding mode measurement feedback control for antilock braking systems. IEEE Trans. Control Syst. Technol. **7**(2), 271–281 (1999)
4. G.F. Mauer, A fuzzy logic controller for an ABS braking system. IEEE Trans. Fuzzy Syst. Technol. **3**(4), 381–388 (1995)
5. O. Hema Kesavulu, S. Sekhar Dash, N. Chellammal, M. Padma Lalitha, A novel control approach for damping of resonance in a grid interfaced inverter system-fuzzy control approach. Int. J. Ambient Energy, pp. 1–8
6. C.M. Lin, C.F. Hsu, Neural-network hybrid control for antilock braking systems. IEEE Trans. Neural Netw. **14**, 351–359 (2003)
7. C.M. Lin, C.F. Hsu, Self learning fuzzy sliding mode control for antilock braking systems. IEEE Trans. Contr. Syst. Technol. **11**, 273–278 (2003)
8. J.S. Yu, A robust adaptive wheel-slip controller for antilock brake system, in *Proceedings of 36th IEEE Conference of Decision Control*, vol 3 (1997), pp. 2545–2546
9. T.A. Johansen, J. Kalkkuhl, J. Ludemann, I. Petersen, Hybrid control strategies in ABS, in *Proceedings of 2001 American Control Conference*, vol 2 (2001), pp. 1704–1705
10. M. Yoshimura, H. Fujimoto, Slip ratio control of electric vehicle with single-rate PWM considering driving force, in *IEEE International Workshop on Advanced Motion Control* (2012), pp. 738–7432012
11. V. Utkin, H. Lee, Chattering problem in sliding mode control systems, in *International Workshop on Variable Structure Systems* (2006)
12. K. Chaudhari, R. Khamari, Design and comparison of discrete-time sliding mode control and discrete-time fuzzy sliding mode control for slip control of a hybrid electric vehicle. Int. J. Manage. Technol. Eng. **9**, 4199–4203 (2019)
13. B. Subudhi, S.S. Ge, Sliding mode observer based adaptive slip ratio control for electric and hybrid vehicles. IEEE Trans. Intell. Transp. **13**(4), 1617–1627 (2012)
14. F.L. Lewis, A. Yesildirek, K. Liu, Multilayer neural-net robot controller with guaranteed tracking performance. IEEE Trans. Neural Netw. **7**(2), 388–399 (1996)

An Improved Scheme for Organizing E-Commerce-Based Websites Using Semantic Web Mining

S. Vinoth Kumar, H. Shaheen, and T. Sreenivasulu

Abstract In the running of the Internet world, E- commerce industry has its own benchmark in terms of its rapid growth and has made itself an established sector that is indispensable for every industry to trade and do transactions online. As the world is rushing in a rapid manner, India is slogging in the improvisation of the online market, leading to the lack of customized needs of the customers. Bigger companies are trying to put in a different strategic approach taking that into consideration an approach of blended e-mining along with e-commerce has been devised. It would be a design of the semantic- and neural-based page ranking algorithm [2]. This tool upon launching would be a well-defined approach for e-commerce website ranking [1]. It would also facilitate the customers to find the relevant websites on the top of the page during their search for any particular product or business. It would be further customized with all the relevant comparison of the other websites in terms of the product quality and price.

Keywords Neural-based page ranking · Website ranking · E-mining

1 Introduction

One of the fastest growing businesses for the past decade is e-commerce [1]. The customer's needs have taken up the next level for satisfying the demands comparing the competitors and to fetch revenue to the company. Researchers made an analysis

S. Vinoth Kumar (✉)
Department of Computer Science and Engineering, SNS College of Technology, Coimbatore, India
e-mail: profsvinoth@gmail.com

H. Shaheen · T. Sreenivasulu
Department of Computer Science and Engineering, St. Peters Engineering College, Hyderabad, India
e-mail: shaheen66@gmail.com

T. Sreenivasulu
e-mail: sreesmart@hotmail.com

S. S. Dash et al. (eds.), *Intelligent Computing and Applications*, Advances in Intelligent Systems and Computing 1172,
https://doi.org/10.1007/978-981-15-5566-4_10

that in India is also rapidly growing among other countries. The reason behind this sudden growth is due to the increase in the awareness of the Internet-based applications, computer literacy, change of lifestyle, and high rated income [6]. Apart from that are the policies from the companies like trial and exchange feedback of the product, cash upon delivery, and reviews about the product. One of the associate research areas is intelligent neural web based mining would be taking up these features of adaptability and would be able to transit the same to practical application stage where the errors would be minimized to a larger extent and help the customer to extract useful and abstract information from his surfing web pattern [3]. These patterns would help us to do the necessary improvements on the website structure enhancement of the website content making it user-friendly. This paper is going to describe the web mining process deployment to get the maximum benefit of the e-commerce area. It may not only be a user-friendly approach for the customer but also for the data analysts to take a decision on the organizational needs. We have divided the paper into various categories. The phase 2 category is task related. Phase 3 is the identification of the query and analysis. Phase 4 illustrates the objectives of the analysis. Phase 5 details about the analysis method and innovations taken for the page ranking algorithm and website listing tool along with the graphical representation of data [4].

2 Task Related

The semantic- and neural-based web mining technology leads to a better sophisticated ranking of the e-commerce websites. The main objective for taking up such a blend is that it might be able to assist the customer as well as the organizations to take a more clear approach in terms of the transaction as well as on the data-oriented decision-making. Since the old data mining area was not that successful it has ended up in inventing new areas for progressing further [9]. The intention is to perform a discussion of the quality of the e-commerce websites using the data envelopment analysis model (DEA).

Eventually, they also compared the same with the other models like CCR, BCC, and KH. A large amount of data was taken and was automated for obtaining results on e-commerce websites. Various ideas regarding applications are implied in e-commerce [7]. It also emphasized on the handling of customer behavior pattern and feedback reciprocation. This contributed largely on the optimization of the websites. While handling a large volume of data, most of the important information was hidden during retrieving the data. This hidden information would be of great use for the structuring of the web page and to enhance a better ranking [4]. They also added further that the constant watch of the users, their methodology of watch search patterns, etc. would also be helpful for the website optimization. The primary objective of this research is to improvise the search engine algorithm and to minimize the complexities faced by the user. The semantic and neural network methodology is used for an unbiased ranking model of the e-commerce websites. This model has a wider perspective toward the e-commerce industry through an easier way of user navigation and retrieval of specific information [8].

3 Identification of the Query and Analysis

The rapid growth of the e-commerce industry remains untold [5]. Every customer is idealized to use only a search engine instead of the web catalog. Search engine may not be able to fetch the exact requirement, as it is a syntactic-based query. It might match and fetch the results based on the frequency count and proximity fetching the data based on search query and web page. This syntactic match may lack semantics producing wrong results which might fetch either number of unwanted pages or no result at all. Apart from this, the result-producing pages are powered by the search engines which make a very good revenue on the companies listing irrespective of their content, reliability, and relevance to the customer. Few e-commerce companies may not have got the authorization to sell the product. But even then they would have published the same on their websites leading to a confused state of mind for the customer to suffer from not knowing the product in detail along with the other details of warranty, replacement, etc. The reason behind is that the search engines failed to design their structure with reference to the customer's queries and intention of the customer. The other reason is the backpropagation of the errors and retrieval algorithms lead to the biased ranking leading to only the top-based rankings popular.

4 Objectives of the Analysis

The whole aim of the analysis is to enhance the e-commerce website to be ranked in a more better and efficient way using the ranking algorithm through the SNEC process to assist the customer while carrying out online transactions in a more authentic and rational manner. The research implies a semantic- and neural-based approach to deal with the ranking problems. This would optimize the use of web dictionary backpropagation and unbiased ranking process.

4.1 Analysis Method and Innovations

The research comprises backprocessing the retrieved company information using a profiling and dictionary implementation module to improvise the incomplete entries and data cleaning. The dictionary and the candidate web page are then analyzed by another module called as the content priority module. The primary objective of this module is to check for relevance and to remove unwanted data. Then the query is

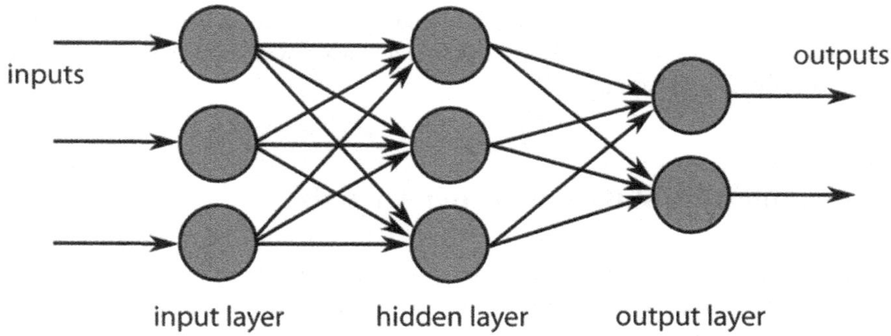

Fig. 1 Analysis method

passed on to the priority module which would check for the priority of the web page based on the customers search and also on the previous searches for the same product by other people. Taking these data the web page is now sent to the next phase of the module called as the semantic module which identifies the user session from other external sources using its algorithm and determines the search to avoid wrong interpretation. One of the most popular NN algorithms is backpropagation algorithm. The BB routine can be fragmented into four key phases followed by which loads of the networks are chosen arbitrarily and the back broadcast routine is employed for estimating the needed modifications. The routine could be disintegrated into the stated four phases as feed onward estimation, back broadcast for the result layers, back broadcast for the concealed layer, and load-based revision. The scheme is halted on experiencing a fault function which is suitably small. It is very irregular and the basic formula for the back broadcast holds some dissimilarity designed by others but it must be precise and much easier for usage. The concluding phase is load revision which occurs for the overall scheme (Fig. 1).

4.2 *Nomenclature*

Semantic- and Neural-based Algorithm

X_i User search product.
Min Minimum length of the keyword X_i.
Max Maximum length of the keyword X_i.
Y Keyword search specific.
ST The web-based e-commerce document to be scanned.
PD Dictionary with reference to the *T*th document.
TXT Words of the document.
A1 Time spent of the browsing by the other visitors.
A2 Time spent on the web page creation.

F Frequency of the number of keywords found.
NF Not found keywords.
tanθ Linear activation function for the training of the neural network.
MT Mass of the input.

4.3 Module

Module 1:

Step 1: Input from the user.
Step 2: Filtering of unwanted terms from the user.
Step 3: Track the movement of the pattern sequence of the user data.
Step 4: Track the web pages through the search engine.
Step 5: Divide the strings into various words like: $Y_1, Y_2, \ldots, Y_n$.
Step 6: Determination of minimum and the maximum length of the words

Min := StrLen ($Y1$), Max : = StrLen ($Y1$)
For $k = 1$ to n do
Initialize $F := 0$ and NF:= 0
If ST found in PD then
$F: = F + 1$
Else $F: =$ NF + 1.

Step 11: Remove those web pages where NF > F.

Module 2

Step 12: To evaluate the timestamp $A2$ for the creation of web page.
Step 13: On the beginning of the user session, determine $a1$ which is session duration of current page and determine new value of $A1$ as follows:

If $A1 = 0$ then $A1 = a1$
Else A1 = $(A1 + a1)/2$.

Step 14: Assign a higher priority to web page if $A2$ is low and $A1$ is high.
Step 15: Update the time database of tool with keywords, page address, and $A1$.

Module 3

Step 16: Identify navigation session by comparing user search query with each of the search query present in user profile database as

LCS $[c, d] = 0$ if $c = 0$, or $d = 0$ OR
LCS $[c, d] =$ LCS $[c - 1, d - 1] + 1$, if c, d <>0 and $X1c = X2d$ OR
LCS $[c, d] = \max($LCS$[c - 1, d]$, LCS$[c, d - 1]$, if $c, d > 0$ and $X1c$ <>$X2d$.

Step 17: Class generation using Web Ontology Language.

Module 4

Step 18: Normalize all the priority inputs from module 2, 3, and 4.
Step 19: Train the network using various sets of inputs and outputs with linear activation function as

$$\{O\} = \tan\theta\ \{I\}.$$

Step 20: Use sigmoidal function for output evaluation in hidden and output layers as $\{O\} = [(1/1 + e - 1)]$ and summation function as $\sum(C1\mathrm{MT}1 + C2\mathrm{MT}2 + C3\mathrm{MT}3 + C4\mathrm{MT}4 + C5\mathrm{MT}5 + B)$.

Step 21: The error rate is determined by adjusting the weight age of the synapses.
Step 22: At last the web pages are displayed in a decreasing manner in order of the ranking priority.

4.4 Priority on Time Spent

SBPP algorithm mentioned in this exploration research implies the usage of the importance to website priority under ASP.NET framework. Through this research, we would be able to explore more than five e-commerce websites using the design done. The tool will allow the number of entries based on the design. After entering the data the tool would search in accordance with the content and the statistical data (like the number of times the page has been visited, the product specification, etc.) (Fig. 2).

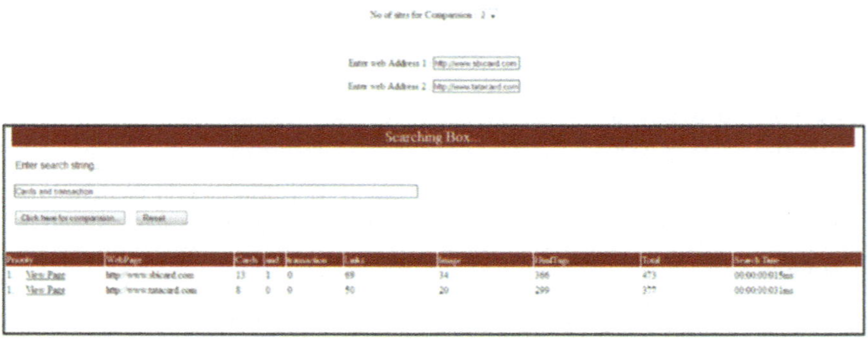

Fig. 2 Priority on time spent

4.5 A Statistical Approach Toward the Research

This segment shows the comparative outcomes of parameters precision and efficiency, respectively. Outcomes were obtained while acting of weighted page rank algorithm on the dataset of Internet pages. Diverse iterations have been completed to test the consistency of the version. Precision: it approaches how properly the website precedence tool (WPT) is working. Internet site precedence device allows evaluation of websites, the use of drop-down container, and seek box to specify a string of unique product. The drop-down field provides as many URL's (uniform resource locator) of the website and after contrast, WPT tool assigns precedence to every candidate website based on the calculation of content priority module, time spent priority module, advice module, and neural priority module. Subsequently, precision is used to measure the consistency of the consequences for each and every time the system runs. Greater the relevancy of the fetched web pages better might be the consistency of the machine. The better consistency of the results implies that the website priority tool is operating correctly. As an end result, better accuracy of net site priority tool leads to higher precision. Relevancy is calculated with the aid of measuring the space of the statistics. Statistics has been stored in array/matrix shape. Distance will be calculated for every row by comparing it with all different rows. For each row, lesser the gap among rows more relevant will be records and vice versa. Precision values of the proposed machine had been received by way of making use of more than one testing rounds (iterations) about 25 on the statistics set (Fig. 3).

Precision primarily based evaluation of the designed version. The foreseen graph represents the correct values for Internet site precedence tool, Google, and the proposed WPR. The line graph here truly shows that the proposed weighted page rank set of rules has high precision values for all the iterations. The graphical layout of both the model's page rank and stepped forward weighted page rank showcases the

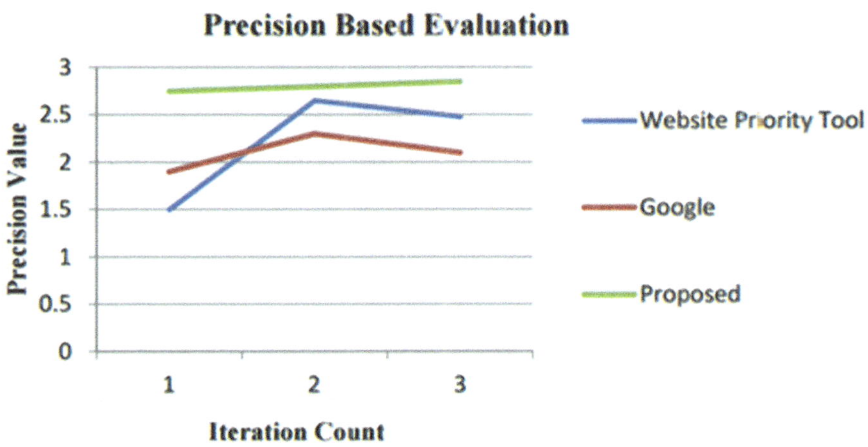

Fig. 3 Precision-based evaluation

comparative evaluation which has been evaluated on the premise of the precision of the simulated outcomes. The upgrades are stated which validates the better conduct of the designed WPR scheme than the prevailing schemes.

5 Conclusion and Future Work

We see a developing interest in the use of the semantic and neural community for solving net programs inside the approaching years. The inherent capability of neuro-semantic techniques in managing indistinct, big-scale, and unstructured information is a great fit for Internet-related problems. Earlier studies have a tendency to consciousness on a way to extract wanted facts from unstructured net facts. Recently, we have seen the use of neural methodologies in building a based web [10]. The semantic Internet is one instance of such. The perception of a structured web can be made extra practical while the concept is employed due to the fact that net records have a tendency to be unpredictable in nature. We count on to peer an integration of gentle computing techniques in semantic Internet methodologies within the near destiny. A genetic set of rules for net software ought to also turn out to be extra famous as Internet applications get large in scale.

References

1. N. Verma, Improved web mining for e-commerce website restructuring, in *IEEE International Conference Computational Intelligence & Communication Technology*, UP, India (2015), pp. 155–160
2. N. Verma, D. Malhotra, M. Malhotra, J. Singh, E-commerce website ranking using semantic web mining and neural computing, in *Proceedings of International Conference on Advanced Computing Technologies and Applications*, Mumbai, India (2015). Procedia Comput. Sci., Science Direct. Elsevier, **45**, 42–51. https://doi.org/10.1016/j.procs.2015.03.080
3. D. Malhotra, N. Verma, An ingenious pattern matching approach to ameliorate web page rank. Int. J. Comput. Appl. **65**(24), 33–39 (2013). https://doi.org/10.5120/11235-6543
4. D. Malhotra, Intelligent web mining to ameliorate web page rank using back propagation neural network, in *Proceedings of 5th International Conference, Confluence: The Next Generation Information Technology Summit*, Noida, India (IEEE, 2014), pp 77–81. https://doi.org/10.1109/confluence.2014.6949254
5. S. Sharma, P. Sharma, Use of data mining in various fields. IOSR J. Comput. Eng. IOSR-JCE **16**(3) (2014)
6. N. Padhy, P. Mishra, R. Panigrahi, The survey of data mining applications and feature scope. Int. J. Comput. Sci. Eng. Inf. Technol. IJCSET **2**(3) (2012)
7. H. Kaur, A review of applications of data mining in the field of education. Int. J. Adv. Res. Comput. Commun. Eng. **4**(4), 2319–5940 (2015)
8. S. Bagga, G.N. Singh, Applications of data mining. Int. J. Sci. Emerg. Technol. Latest Trends (2012)

9. N. Jain, V. Srivastava, Data mining techniques. Int. J. Res. Eng. Technol. eISSN: 2319-1163 | pISSN: 2321-7308
10. D. Kumar, D. Bhardwaj, Rise of data mining: current and future application areas. Int. J. Comput. Sci. Issues **8**(5) (2011). ISSN (Online): 1694-0814

Performance Estimation and Analysis Over the Supervised Learning Approaches for Motor Imagery EEG Signals Classification

Gopal Chandra Jana, Shivam Shukla, Divyansh Srivastava, and Anupam Agrawal

Abstract In this paper, a comparative analysis has been done to estimate a robust classifier to classify motor imagery EEG data. First, segment detection and feature extraction have been done over the raw EEG data. Then the frequency domain features have been extracted using FFT. Six classifiers DNN, SVM, KNN, Naïve Bayes,' Random Forest, and Decision Tree have been considered for this study. The DNN model configured with four layers and used the binary cross-entropy loss function and sigmoid activation function for all layers. The optimizer used is "Adam" having the default-learning rate of 0.001. In this experiment, for the purpose of the estimation of the performance of various classifiers, the experiment used dataset IVa from BCI Competition III, which consisted of EEG signal data for five subjects, namely 'aa,' 'al,' 'av,' 'aw,' and 'ay.' The highest average accuracy of 70.32% achieved by the DNN model, whereas the model achieved an accuracy of 80.39% over the subject 'aw.' The objective of this experiment encompasses the different models for the classification of various motor tasks from EEG signals.

Keywords Electroencephalogram (EEG) · Brain–computer interface (BCI) · Motor imagery · Deep neural network (DNN) · SVM · KNN · Naive Bayes · Random forest · Decision tree

---

G. C. Jana (✉) · A. Agrawal
Interactive Technologies and Multimedia Research Lab, Department of Information Technology, Indian Institute of Information Technology-Allahabad (IIITA), Allahabad, India
e-mail: go.gopal.ch.jana@gmail.com

A. Agrawal
e-mail: anupam@iiita.ac.in

S. Shukla · D. Srivastava
Research Intern, Interactive Technologies & Multimedia Research Lab, Department of Information Technology, Indian Institute of Information Technology-Allahabad (IIITA), Allahabad, India
e-mail: create.shivamshukla@gmail.com

D. Srivastava
e-mail: srivastavadivyansh98@gmail.com

S. S. Dash et al. (eds.), *Intelligent Computing and Applications*,
Advances in Intelligent Systems and Computing 1172,
https://doi.org/10.1007/978-981-15-5566-4_12

1 Introduction

Motor Imagery based on Brain–Computer Interface (BCI) technologies is a constantly evolving field due to the immense researches being conducted and novel methods being constantly brought out and improved. Motor Imagery (MI) is a type of mental process which involves motor actions. In this particular process, the person simulates a given action mentally rather than performing them. It may involve different limb movements or body part movements, but all these tasks are accomplished only mentally, using imagination to do that particular task. Examples include the right hand or foot movement. The development in this field may bring about significant changes in the lives of persons with neurological disorders. The BCI employs pathways between the brain and any other external actuators or devices for performing the action without the actual limb movement. These systems can be extremely beneficial for neural rehabilitation for the persons with neurological and neuromuscular disorders. MI-based BCI is now evolving as a domain due to the improvement in the performance of different classification techniques and the introduction of non-invasive BCI techniques like EEG, MEG, etc., which are easy and risk-free in terms of internal organ damage and cost. The electroencephalograph (EEG)-based signals depend upon the neuron's activity inside the brain of the subjects. The main challenges in the field are due to signal temporal and spatial resolutions and considering an efficient algorithm to work on EEG signal features with its extraction strategies that are available. Other than this, the data collection using invasive methods are less cost-effective and accurate in comparison to non-invasive ones, while risky than non-invasive. This work has been done in two phases, one is data preprocessing and another one is classification. Data preprocessing phase have been done in parts; one is segment detection and another one is feature extraction. To estimate a robust classifier for this proposed approach, we consider different classifiers for getting the best performance for the EEG signal-based motor imagery classification for the different subjects. The extensive analysis for different classifiers—Deep Neural Networks (DNN), Support Vector Machines (SVM), K-Nearest Neighbor (KNN), Naïve Bayes algorithm, and Random Forests have been applied to the EEG signal-based motor imagery dataset, which consists the signal data of five different subjects. These models estimate the prediction accuracy of different motor imagery tasks. This paper focusses on the performance analysis of the different classifiers with a supervised approach for the classification of EEG signals for either right hand or foot which are taken from BCI Competition III dataset IVa. The further improvement in the techniques to classify the EEG-based signals will bring more accuracy in the analysis of various disorders and the development of many different devices and applications for the same.

Organization of paper is done in various sections: Next Section (Sect. 2) discusses the related works and literature review, Furthermore, Sect. 3 describes the dataset used for classification models. Sect. 4 discusses methodologies and classifiers model. Sec. 5 discusses the results and analysis of classification results. In Sect. 6, the conclusions are provided and different references for the paper are shown.

2 Related Works

This approach is about MI classification among two classes of EEG signals generated due to the right hand and right foot activity of the subjects. There has been continuous work going in this field with the intent of making the classifiers more and more accurate with different combinations of the feature extraction strategies and different classifiers too. In [1], the motor imagery classification has been done using traditional classifier models like SVM and MLP are employed on handcrafted features extracted from EEG signal data and both of classifier model achieved a considerable accuracy with of 85% and 85.71%, respectively. With the purpose of designing, a reliable and robust model of BCI, an enhanced method using A DNN-based approach had been proposed in [2] for motor imagery classification. Similarly, an improved approach has been proposed in [3], where the authors use SVM for classification and Particle Swarm Optimizations (PSO) for the selection of kernels and penalty parameters for motor imagery classification based on EEG signals. Different other classifiers like KNN have also been used in the work of [4] for motor imagery EEG signal classification, and the proposed model achieved a maximum classification accuracy of 91% for left and right hand movements and extraction of features have been done using the wavelet-based techniques. A comparative strategy for finding the effectiveness of the different feature extraction techniques has been proposed in [5], where the raw data samples are compared with DWT-based feature extraction. It demonstrated that the improvement with their strategy came out significantly in terms of different performance measures. With the same consideration, a hybrid method based on Discrete Wavelet Transform (DWT) and Artificial Neural Network (ANN) [6, 7] has been proposed in [8, 9] for the classification of physical action, while some other motor imagery classification technique has been proposed in [10] using Gaussian Naive Bayes algorithm on the feature of EEG signal data extracted using the Common Spatial Pattern (CSP) technique. Other than this some more powerful models like CNN (Convolutional Neural Networks) are also being considered to make the generalization procedure even better. In [11], the paper uses the five-layer CNN for the classification of different motor imagery tasks, while there is also a comparative analysis that is performed with SVMs and other composite feature extraction techniques. The CNNs have made the accuracy level even higher for motor imagery data. In [12], the CNNs were used along with the time–frequency methods using STFT (Short-Term Fourier Transform) and CWT (Continuous Wavelet Transform) were employed as preprocessing stage. Also, the most popular Neural Network (NN) techniques were used by the authors of [13] over the BCI Competition Data. NN techniques provide an efficient and less cost-effective classification approach, and they are continuously being applied on the different classification problems by tuned the hyperparameters. This literature review show that still there is a scope of analyzing the performance of an approach using DNN and FFT with respect to other supervised classification approaches for motor imagery EEG signal classification for a significant breakthrough in the area of BCI.

3 Experimental Data Description

The dataset used for this experiment has been taken from BCI III Competition. Dataset IVa [14] has been used for the estimation of the performance of different models in the proposed methodology. As mentioned in the BCI III Competition weblink, the data were acquired using a non-invasive technique from five healthy participants or subjects whose data were labeled as: 'aa,' 'al,' 'av,' 'aw,' and 'ay.' An EEG cap of 128 Ag/AgCl electrodes is used, out of which only 118 electrodes were considered for data acquisition purposes. In the duration of 3.5 s, which is the duration of the visual cue for each subject, the subjects have to perform one of imagery, namely Left Hand (L), Right Hand (R), and Right Foot (F). In this experiment, we consider two classes: one is right hand and another one is right foot. The data were available in two different formats, it will be either in zipped ASC II format or mat format. In this experiment, 100-Hz ASCII format of data has been used which contains two different txt files, one for EEG values and another one is for the corresponding class label.

4 Methodologies

Experiment of this proposed approach has been done in several stages. An illustration of the proposed approach is shown in Fig. 1.

Experimental Environment Setup: Different environmental parameters are considered and setup for this experiment. The experimental procedure assumes the usage of the following libraries and tools

1. Anaconda (Python 3.6 or above)
2. Jupyter notebook (for implementation)
3. Virtual Environment (for Python) is enabled and workable with following libraries: sklearn, sklearn.metrics (for accuracy_score, confusion_matrix), sklearn.svm (for svm implementation), sklearn.datasets, sklearn.naive_bayes(for Naïve Bayes implementation), sklearn.neighbors (for K-nearest Neighbour implementation), sklearn.ensemble (for Random Forest implementation),

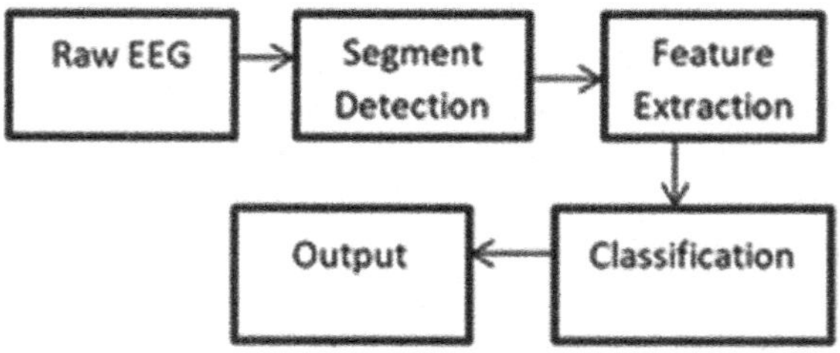

Fig. 1 Illustration of the proposed approach

sklearn.tree (for Decision Tree implementation), Numpy (basic numeric operations on matrix, vectors, load txt file), scipy.fftpack, keras.utils (up_utils), matplotlib.pyplot (for visualization purposes), keras.layers (dense), keras.models (sequential).

The methodology used in this paper encompasses some important key components which are as follows:

A. **Segment Detection**

The input raw EEG signals data are provided by the BCI Competition III—Dataset IVa. This dataset has few segments and these are belonging to class label NaN which are not important for our classification process. So, in this phase, the segment detection has been performed to extract the correct segment of signal from the original signal. The extracted segment is now passed for the feature extraction stage.

B. **Feature Extraction**

In the feature extraction phase, transform EEG values from the segmented signal are taken as representative for that original signal. Fast Fourier Transform has been used for this phase which is mentioned below

Fast Fourier Transform: The Fast Fourier Transform has been used in this phase for the transformation time-domain EEG signal into the frequency domain. This conversion is crucial for this experiment to extract signal features in terms of frequency from the time-domain segmented samples. The FFT transform [15] the N points time-domain signal samples into the separated N different frequency-domain signals points. Finally, from these N point spectra, the N frequency spectrum is extracted. This N point frequency spectrum is meant to be extracted into a single spectrum for the synthesis of the overall signals into the spectrum. FFT is one of the prominent ways to apply Discrete Fourier Transform (DFT). The basic equation governing the DFT (which is the main process behind DFT for efficient working) is:

$$F(n) = \sum_{0}^{N-1} x(k) e^{-j2\pi k \frac{n}{N}}.$$

The value of $F(n)$ is the amplitude of the signal in the frequency domain at value n, while N is the total discrete samples of signals taken.

In this phase, we have applied FFT on all segmented signal samples for extracting frequency-domain features. Extracted frequency-domain features pass into the classifier for the classification of right hand or foot.

C. **Classification**

Deep Neural Networks (DNN): DNN [16, 17] is one of the classifiers that have been used for this experiment. The DNN configured with four layers having input layer, hidden layer-1, hidden layer-2, and output layer. The hidden layer-1 has seven

Table 1 Layer configuration of DNN

Layers	Input shape	Output shape
Dense	(118, 1)	(None, 12)
Dropout	(None, 12)	(None, 12)
Dense	(None, 12)	(None, 8)
Dropout	(None, 8)	(None, 8)
Dense	(None, 8)	(None, 6)
Dense	(None, 6)	(None, 3)

Table 2 Subject-wise training epochs and batch size

Subject	Epochs	Batch size
aa	150	32
al	100	32
av	115	16
aw	150	32
ay	25	4

neurons, while the hidden layer-2 has six neurons. The final output layer is provided with the two output neurons. The classification into two classes is accomplished by this network.

The DNN model uses the loss function as binary cross-entropy, activation function as sigmoid for all layers. The optimizer used is "Adam" having the default learning rate of 0.001. The input layer takes an input of 118 frequency-domain features. Layer-wise details of DNN have been mentioned in Table 1.

The following Table 2 describes the number of epochs and batch size of each subject data have been used in the training session for the DNN.

Support Vector Machines (SVM): SVM [1] are one of the popular classification models which have a pretty simple algorithm. In this experiment, SVM have been used for classification. Different kernel functions have been used to estimate the best kernel function for which SVM achieves high classification accuracy over the motor imagery EEG data. Two kernels have been considered which are mentioned below.

1. RBF Kernel (Radial Basis Function)
2. Sigmoid Kernel

The input data for the SVM is kept the same as that of other classifiers and no change has been done. The results of this particular technique have been shown in the result and analysis section.

Naive Bayes Classifier: Naive Bayes classifiers [10] are another category of classifiers that depends upon the probabilistic analysis for their classification algorithm. The foundation of this algorithm is based on Bayes Theorem, these are very powerful and faster models that perform better on some class of data. It has been used in this experiment to compare the classification capability w.r.t other classifiers over the same input data. The results of this particular technique have been shown in the result and analysis section.

K-Nearest Neighbor (KNN): K nearest neighbor [4] is also one of the renowned supervised algorithms in which the vicinity of a particular data point w.r.t to other available points. It does not make up any structure of the overall data during training. Rather all the training examples are kept at the same time for the test example for the nearest neighbor.

In this experiment, KNN with three nearest neighbors has been taken to classify the test data correctly. The results of this particular technique have been shown in the result and analysis section.

Random Forest Method: Random Forest models [15] work by preparing multiple decision trees at the time of their training. The output of particular test data is determined by either mode or mean of the different decision trees that are prepared. This is a type of ensemble-learning algorithms. For this experiment, the number of estimators for this method is ten for the overall implementation of this paper. The results of this particular technique have been shown in the result and analysis section.

Decision Tree: Decision Tree (DT) [18] which is a powerful supervised learning algorithm. DT classification techniques have the capability of representing the information in terms of trees. These are able to provide decisions with the help of this underlying structure. It works like a flowchart, a flow goes from one point to another in flowchart w.r.t different conditions and parameters.

In this experiment, we have also considered decision trees as classifiers for comparative analysis with different models that have been used over the same input data. The maximum depth of decision trees for this experiment has been considered as 3. The results of this particular technique have been shown in the result and analysis section.

5 Results and Analysis

With the intention of finding out the best classifiers from the six different types of supervised classification approaches, we hereby estimated the performance of six different kinds of classifiers to classify the motor imagery dataset. The performance of the classifiers has been recorded subject-wise. Classification performance has been shown in section in terms of confusion matrix and accuracy. In this section, we try to show the results of all the considered classifiers separately, and this section ends with the results of the DNN classifier which has been achieved the highest accuracy for all subjects.

Support Vector Machines (SVM): SVM with Radial Basis kernel function (RBF) and sigmoid kernels have been tested over the input data. These model performances have been recorded subject-wise. Performance variation has been found w.r.t the kernel and datasets, whereas both models achieved the highest accuracy of 57.14% for the subject 'ay' while in case of other subjects' performance is less than 57.14%.

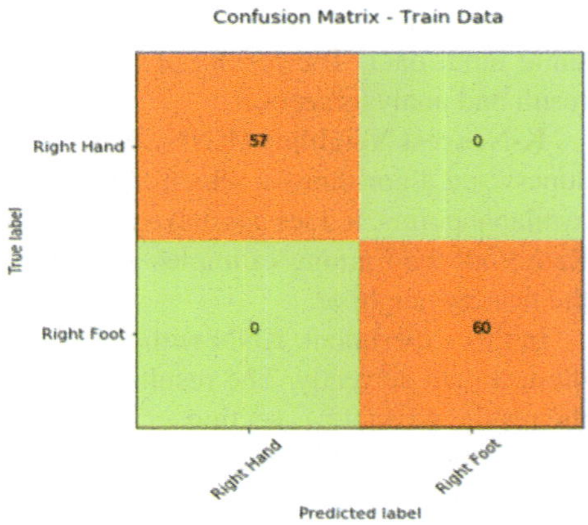

Fig. 2 Confusion matrix over the training data of subject 'aa' for SVM (RBF)

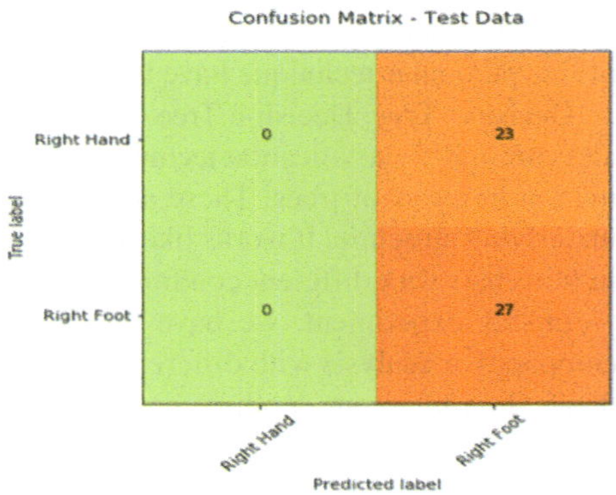

Fig. 3 Confusion matrix over the testing data of subject 'aa' for SVM (RBF)

Figures 2 and 3 depict the confusion matrix of the SVM classifier with RBF kernel over the training data and testing data of subject 'aa,' whereas Fig. 4 shows subject-wise accuracy (%) achieved by the SVM with RBF and sigmoid kernels. It is pretty evident that SVM with both kernels achieved similar accuracy for the individual subjects.

Naive Bayes Classifier:

The NB classifier had shown the best accuracy over the subject 'av' of 52.17%, which is pretty lower than the other classifiers' performance. Figures 5 and 6 show the depiction of the confusion matrix plotted the training data and the testing data of subject 'aa' for NB, whereas Fig. 7 shows the subject-wise performance (accuracy) achieved by the classifier.

Evidently from Fig. 7, the NB classifier achieved an accuracy of 52.17% over subject 'av.' Other than this, the NB classifier achieved even less accuracy than 50% over the other subject's data.

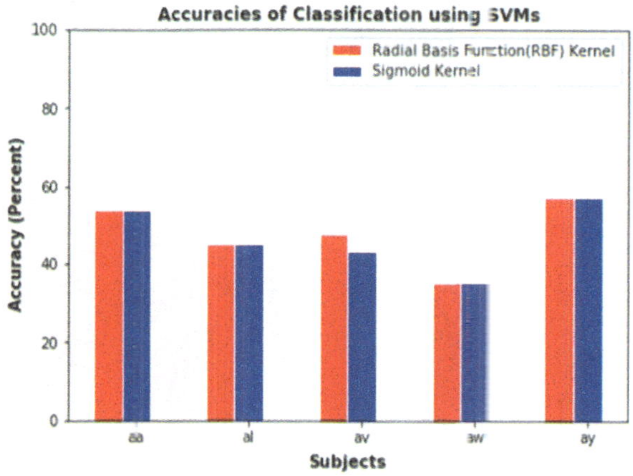

Fig. 4 Subject-wise accuracy (%) achieved by the SVM with RBF and sigmoid kernels

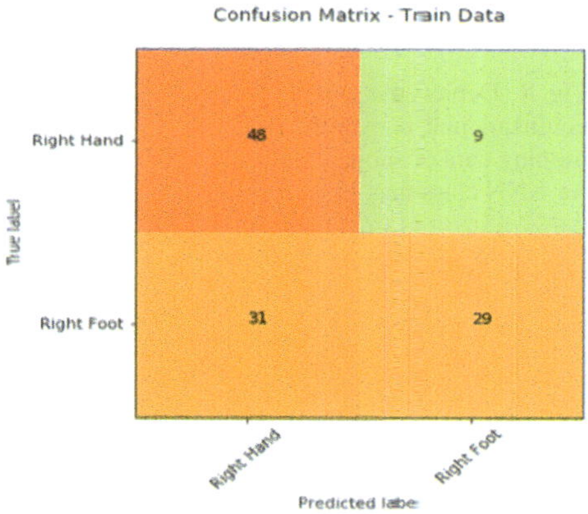

Fig. 5 Confusion matrix over the training data of subject 'aa' for Naïve Bayes classifier

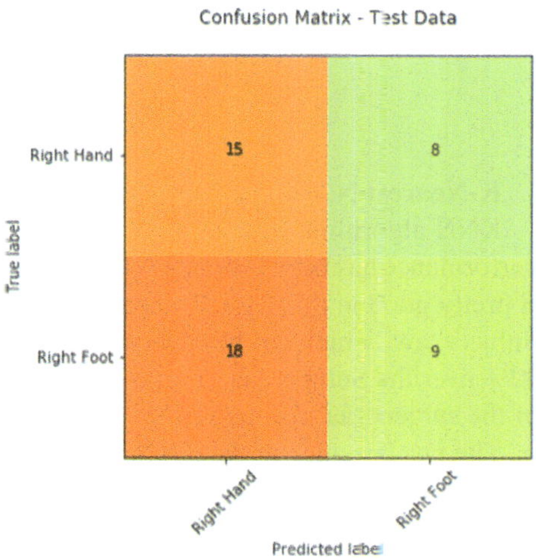

Fig. 6 Confusion matrix over the testing data of subject 'aa' for Naïve Bayes classifier

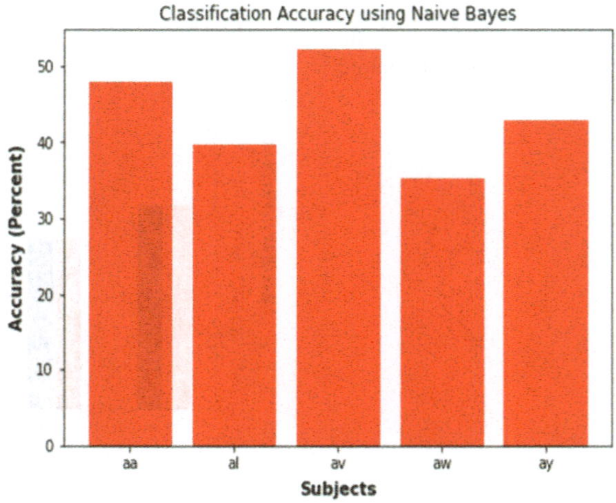

Fig. 7 Subject-wise accuracy (%) achieved by the Naïve Bayes classifier

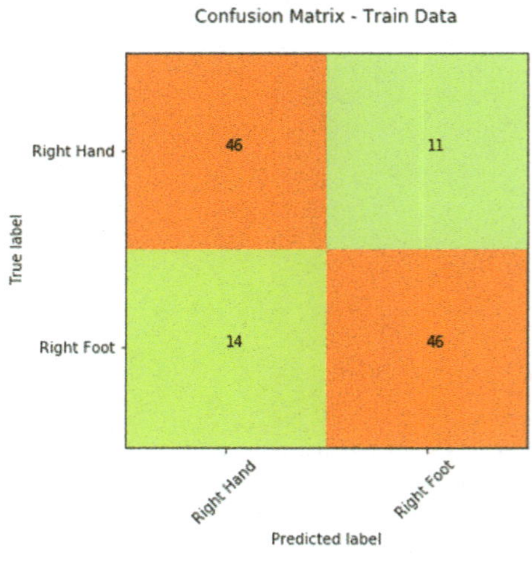

Fig. 8 Depicts the confusion matrix over the training data of subject 'aa' for KNN classifier

K-Nearest Neighbour (KNN):

KNN algorithm has been successfully applied over the motor imagery data and performance has been estimated for all subjects. The overall performance of the KNN is pretty good in this case. The best accuracy of 64.70% has been achieved over the Subject 'aw' which has been shown in Fig. 10, where KNN achieved an accuracy of 42% over the Subject 'aa.' Figures 8 and 9 are the depiction of the confusion matrix of the subject 'aa' for the KNN-based classification.

The performance of the KNN over all subjects has been shown in Fig. 10, which indicated that the KNN cannot be granted as the robust model for the motor imagery based on EEG data with achieved accuracies of 42%, 47.16%, 52.17%, 64.70%, and 57.14% over the subjects, namely 'aa,' 'al,' 'av,' 'aw,' and 'ay,' respectively.

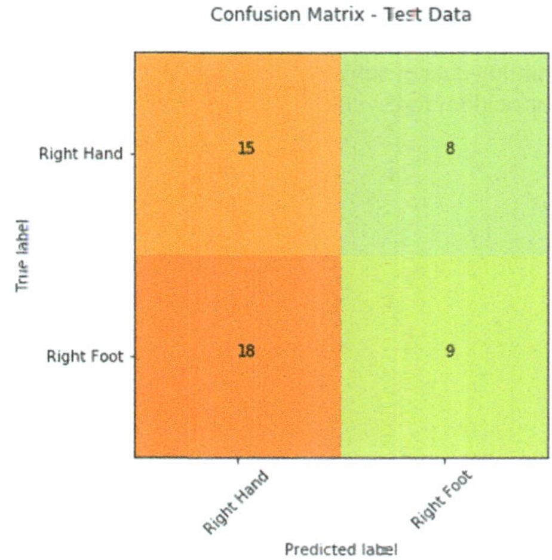

Fig. 9 Depicts the confusion matrix over the testing data of subject 'aa' for KNN classifier

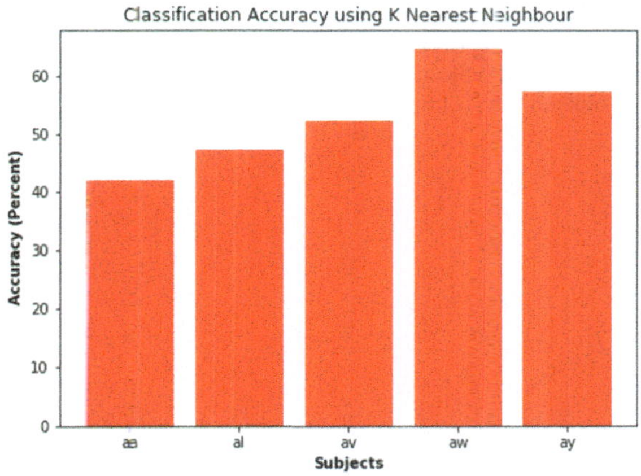

Fig. 10 Depicts subject-wise accuracy (%) achieved by the KNN classifier

Random Forest Method:

In this experiment, Random Forest classification approach has been applied over all subjects. Following Figs. 11 and 12 are the confusion matrixes over training data and testing data of subject 'aa' of the subject 'aa' for classification using random forest approach.

The accuracy of 50%, 24.52%, 56.52%, 23.52%, and 57.14% achieved by the Random Forest approach over the subjects, namely 'aa,' 'al,' 'av,' 'aw,' and 'ay,' respectively. The random forest approach achieved at most 57.14% accuracy over the subject 'ay.' While the accuracy was pretty lower with respect to other subjects. Following Fig. 13 is the depiction of the same.

Figure 13 shows the inconsistent performance with respect to the different subjects. Moreover, the performance was significantly less for the subject 'al,' and 'aw,' in comparison to the subject 'aa' and 'ay.' The overall performance of Random

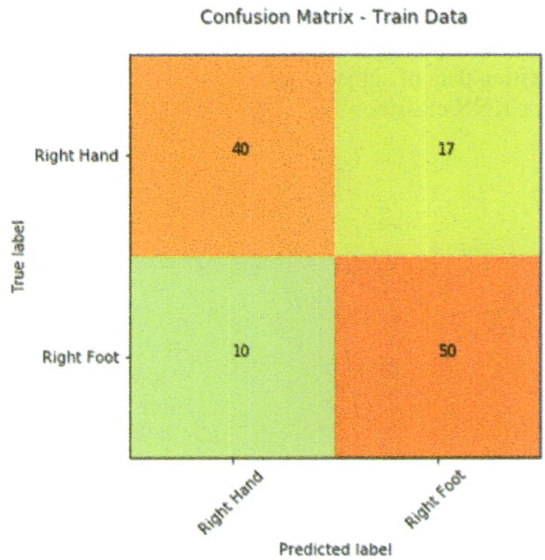

Fig. 11 Depicts the confusion matrix over the training data of subject 'aa' for random forest classifier

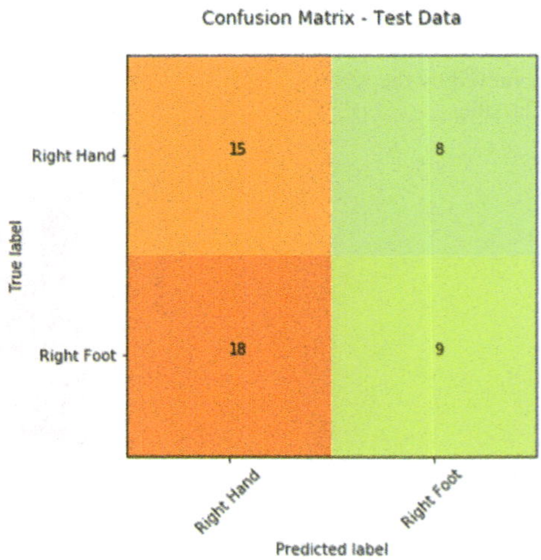

Fig. 12 Depicts the confusion matrix over the testing data of subject 'aa' for random forest classifier

Forest classifier is significantly less than that the KNN which indicated that the Random Forest classifier cannot be granted as the robust model for the motor imagery EEG data classification.

Decision Tree:

In this experiment, DT approach has been used over all subjects. The performance has been estimated for all subjects. Following Figs. 14 and 15 are the confusion matrix over the training data and testing data of subject 'aa.'

For DT, the performance of classifier attained a maximum accuracy of 71.42% for the subject 'ay,' while for others, it achieved an accuracy of 44.0%, 11.32%, 69.56%, and 11.76% for the subjects aa, al, av, aw, respectively, which has been shown in Figure 16. The classifier performed very poorly with an accuracy of 11.32%.

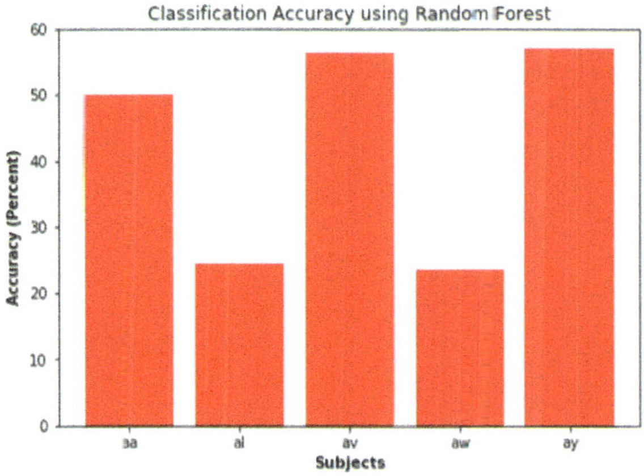

Fig. 13 Shows subject-wise accuracy (%) achieved by the random forest classifier

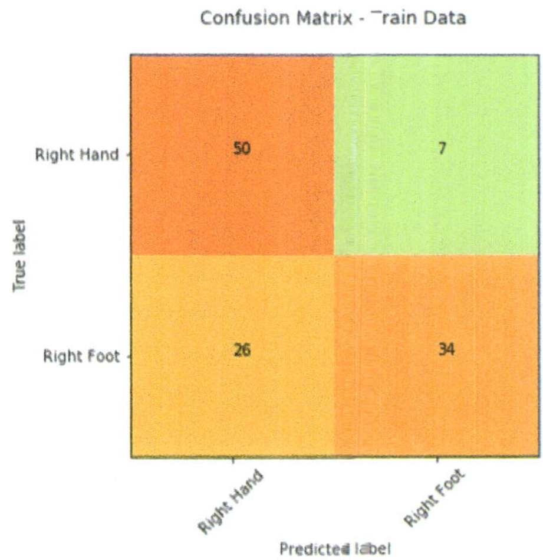

Fig. 14 Depicting the confusion matrix over the training data of subject 'aa' for the decision tree classifier

It is evident that DTs although performed well over the subjects 'aa,' 'av,' and 'ay,' while for other two subjects, DT classifier achieved very less accuracy. This is pretty evident from the overall data that decision trees cannot perform very well with the present combination of feature extraction techniques.

Deep Neural Networks:

For this experiment, DNN and their different architectures have been applied over the pre-processed motor imagery data. In this approach, the Fast Fourier Transform technique over the segmented data and then the DNN has been applied to classify. The performance of DNN has been estimated over all subjects which have been shown in Figure 19. It shows subject-wise performance in terms of accuracy of 63.99%, 67.30%, 63.77%, 80.39%, and 76.19% achieved by the DNN classifier over the subjects, namely 'aa,' 'al,' 'av,' 'aw,' and 'ay,' respectively. The DNN model

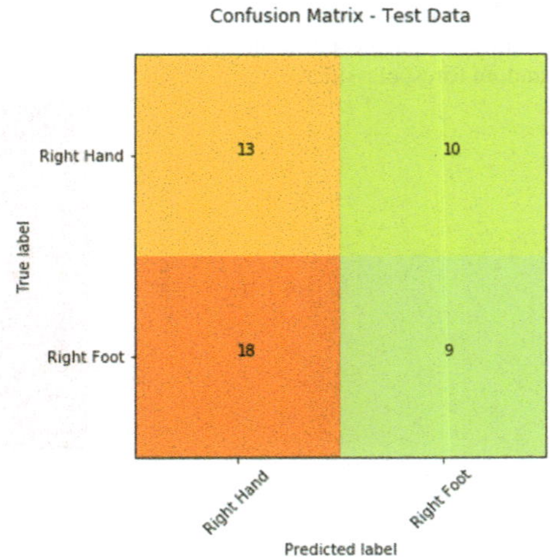

Fig. 15 Depicting the confusion matrix over the testing data of subject 'aa' for decision tree

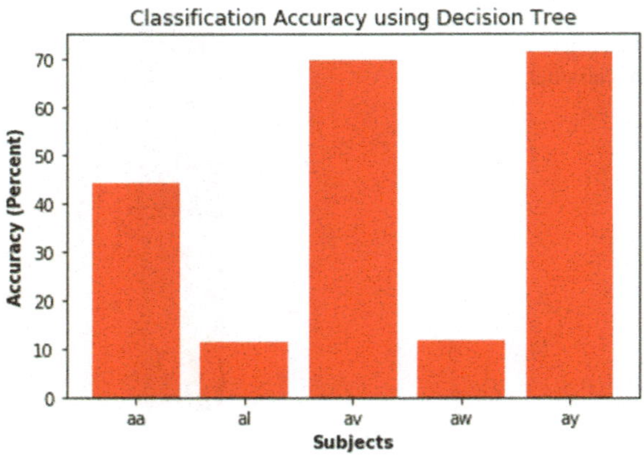

Fig. 16 Shows subject-wise Accuracy (%) achieved by decision tree classifier

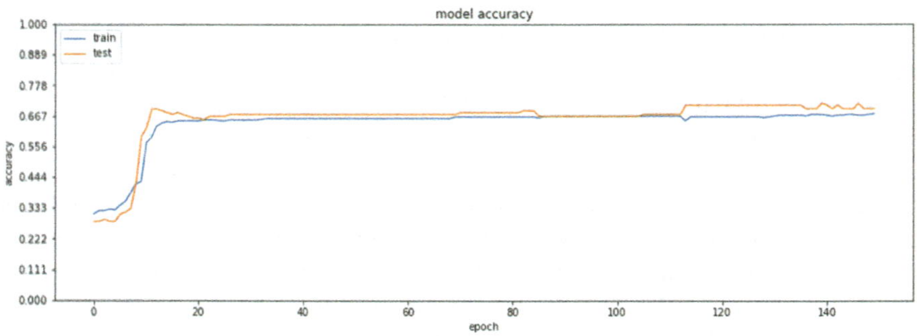

Fig. 17 Trend of training and testing accuracy w.r.t epochs over the subject 'aa'

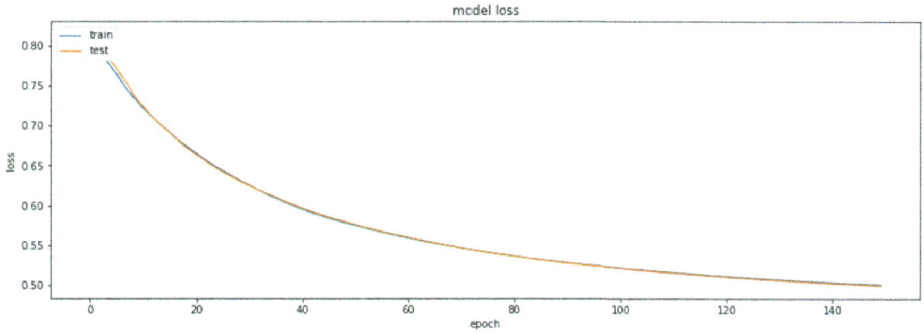

Fig. 18 Loss trend of the DNN model over the training and testing of subject 'aa'

Fig. 19 Shows subject-wise accuracy (%) achieved by the DNN Classifier

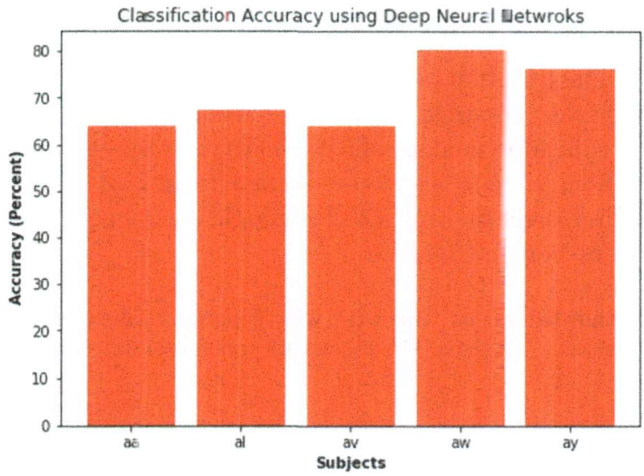

has achieved better accuracy than the SVM, Naïve Bayes, KNN, Random Forest, and Decision Tree classifiers over all subjects. The variation of DNN accuracy with respect to the number of epochs over the subjects 'aa' has been shown in Fig. 17. Also, the loss trend of the DNN model has been estimated with respect to the number of epochs over all subjects. Figure 18 shows the loss trend of the DNN model over the subjects 'aa.'

Finally, an average accuracy has been calculated for the all considered classifier over all subjects, and it has been found that average accuracy of 70.32%, 52.63%, 47.90%, 47.03%, 43.58%, 42.34%, and 41.61% achieved by DNN, KNN, SVM with RBF, SVM with sigmoid, Naive Bayes, Random Forest, Decision Tree classifier, respectively. It indicates the robustness of the DNN to classify the motor imagery EEG over all subjects. This comparative analysis shows the considered classifiers performed well at some of the subjects with the motor imagery data. Thus, the considered DNN architecture with feature extraction techniques is one of the robust approaches for motor imagery EEG data classification. The further analysis can be done with a different set of hyperparameters to enhance the overall classification accuracy.

6 Conclusion with Future Scope

In this paper, a comparative analysis has been done upon the performance of some selective classifiers using a supervised learning technique over the motor imagery EEG data. The classification approaches have been applied to the frequency-domain features that have been extracted using FFT. The performance of the seven classifiers has been estimated over the five subjects of motor imagery EEG data. The estimated average accuracy indicates the considered DNN architecture achieved the highest classification performance over the extracted features. Thus, the estimated result shows DNN and further architecture is preferably one of the suitable approaches for classification of the motor imagery based EEG data. Several feature extraction and selection strategies can be considered which will enhance the robustness of the classification models which will serve as the future work.

Source code and experimental results of this paper can be found from the author's website (https://sites.google.com/site/gcjanahomepage/publications/Publications-Source-Codes).

Acknowledgments This work was supported by the Indian Institute of Information Technology Allahabad, UP, India. The authors are grateful for this support.

References

1. R. Chatterjee, T. Bandyopadhyay, EEG based motor imagery classification using SVM and MLP, in *2016 2nd International Conference on Computational Intelligence and Networks (CINE)*, Bhubaneswar (2006), pp. 84–89
2. G.C. Jana, A. Swetapadma, P.K. Pattnaik, Enhancing the performance of motor imagery classification to design a robust brain computer interface using feed forward back-propagation neural network. Ain Shams Eng. J. **9**(4), 2871–2878 (2018)
3. Y. Ma, X. Ding, Q. She, Z. Luo, T. Potter, Y. Zhang, Classification of motor imagery EEG signals with support vector machines and particle swarm optimization. Comput. Math. Methods Med. **2016**, Article ID 4941235 (2016)
4. R. Aldea, M. Fira, A. Lazăr, Classifications of motor imagery tasks using k-nearest neighbors, in *12th Symposium on Neural Network Applications in Electrical Engineering (NEUREL)*, Belgrade (2014), pp. 115–120
5. L. Vega-Escobar, A.E. Castro-Ospina, L. Duque-Munoz, DWT-based feature extraction for motor imagery classification, in *6th Latin-American Conference on Networked and Electronic Media (LACNEM 2015)*, Medellin (2015), pp. 1–6
6. M. Wairagkar, Motor imagery based brain computer interface (bci) using artificial neural network classifiers, in *Proceedings of the British Conference of Undergraduate Research*, vol 1 (2014)

7. H. Yang, S. Sakhavi, K.K. Ang, C. Guan, On the use of convolutional neural networks and augmented CSP features for multi-class motor imagery of EEG signals classification, in *Conference Proceedings of IEEE Engineering in Medicine and Biology Society* (2015)
8. G.C. Jana, A. Swetapadma, P.K. Pattnaik, An intelligent method for classification of normal and aggressive actions from electromyography signals, in *1st International Conference on Electronics, Materials Engineering and Nano-Technology (IEMENTech)* (2017), pp. 1–5
9. G.C. Jana, A. Swetapadma, P.K. Pattnaik, A hybrid method for classification of physical action using discrete wavelet transform and artificial neural network. Int. J. Bioinf. Res. Appl. (IJBRA) **17**(1), xx–xx (2021). (In press)
10. S.S.R.J. Rabha, K.Y. Nagarjuna, D. Samanta, P. Mitra, M. Sarma, Motor imagery EEG signal processing and classification using machine learning approach, in *2017 International Conference on New Trends in Computing Sciences (ICTCS)* (2017), pp. 61–66
11. Zhichuan Tang, Chao Li, Shouqian Sun, Single-trial EEG classification of motor imagery using deep convolutional neural networks. Optik **130**, 11–18 (2017)
12. S. Chaudhary, S. Taran, V. Bajaj, A. Sengur, Convolutional neural network based approach towards motor imagery tasks EEG signals classification. IEEE Sens. J. **19**(12), 4494–4500 (2019)
13. L. Ming-Ai, W. Rui, H. Dong-Mei, Y. Jin-Fu, Feature extraction and classification of mental EEG for motor imagery, in *2009 Fifth International Conference on Natural Computation* (2019), pp. 139–143
14. [DataSet] http://www.bbci.de/competition/iii/desc_IVa.html
15. D. Steyrl, R Scherer, G.R. Müller-Putz, Using random forests for classifying motor imagery EEG, in *Proceedings of TOBI Workshop IV* (2013), pp. 89–90
16. N. Lu, T. Li, X. Ren, H. Miao, A deep learning scheme for motor imagery classification based on restricted Boltzmann Machines. IEEE Trans. Neural Syst. Rehabil. Eng. **25**(6), 566–576 (2017)
17. S. Kumar, A. Sharma, K. Mamun, T. Tsunoda, A deep learning approach for motor imagery EEG signal classification, in *2016 3rd Asia-Pacific World Congress on Computer Science and Engineering (APWC on CSE)* (2016), pp. 34–39
18. S. Guan, K. Zhao, S. Yang, Motor imagery EEG classification based on decision tree framework and Riemannian geometry. Comput. Intell. Neurosci. **2019**, Article ID 5627156 (2019)

Fully Automated Digital Mammogram Segmentation

Karuna Sharma and Saurabh Mukherjee

Abstract The "Computer-Aided Detection and Diagnosis" (CADx) system plays a vital role as second look to identify and analyze the breast carcinoma. The functioning of CADx can be degraded due to some factors like the appearance of impulse and speckle noise, artifacts, and low contrast both in CC and MLO views and pectoral muscles appears in mammogram's MLO view. For this reason, noise elimination, artifacts and pectoral muscles, mammogram image enhancement, and breast profile extraction are significant prior process stages in the CADx system for breast carcinoma analysis. The research is aimed to propose a precise and effective method for completely automated mammogram image segmentation which includes CC and MLO views. In this paper, median filter and Wiener filter are used for noise removal, threshold based on otsu's method for artifact removal, "Contrast limited histogram equalization" method for mammogram enrichment in both views and multilevel threshold with canny edge detection for pectoral muscle segmentation and breast parenchyma extraction. The proposed work was examined on mammographic images containing both views from CBIS-DDSM and MIAS databases. Jaccard similarity index, Dice similarity index, and Score Matching methods have been employed to estimate the performance of segmentation results that represents the proposed work's effectivity and usability.

Keywords Computer-aided detection and diagnosis · Noise · Image enhancement · Thresholding · Otsu's method · Edge detection · Segmentation

K. Sharma (✉) · S. Mukherjee
Department of Computer Science, Faculty of Mathematics and Computing, Banasthali Vidyapith, Banasthali, Tonk 304022, Rajasthan, India
e-mail: karuna4frenz@gmail.com

S. Mukherjee
e-mail: mukherjee.saurabh@rediffmail.com

S. S. Dash et al. (eds.), *Intelligent Computing and Applications*, Advances in Intelligent Systems and Computing 1172,
https://doi.org/10.1007/978-981-15-5566-4_13

1 Introduction

The breast carcinoma utmost continual ill health spotted among women, in both emerging and established countries' death. As stated by "World Health Organization" (WHO) and "National Carcinoma Institute" [1] in 2004, breast carcinoma reported as the cause of 13% of total demises in the globe [2] and one among eight women faces breast carcinoma in some phase throughout her lifetime in the "United States" (US). The radiologists use "Computer-Aided Detection and Diagnosis" (CADx) system extensively as an investigative and assessment tool for breast carcinoma detection at the primary phase. It is an extremely trustworthy method for the primary detection of breast carcinoma, decreasing life-threatening rates up to 25%. But the performance of CADx can be degraded due to some factors like the appearance of impulse and speckle noise, artifacts, and low contrast both in CC and MLO views and pectoral muscles appear in mammogram's MLO view. For this reason, noise elimination, artifacts and pectoral muscles, mammogram image enhancement, and breast profile extraction are significant preprocessing stages in the CADx system to analyze breast carcinoma. [3] proposed an algorithm to remove artifacts and pectoral muscle by using a region description, split and merge method to extract pectoral muscle. In [4], authors studied the several filters as mean, median, and Wiener filters by applying various window sizes using DDSM ("Digital Database for Screening Mammography") database, evaluated with PSNR ("Peak Signal to Noise Ratio'). [5] proposed an approach to remove pectoral muscles by using ROR ("Robust Outlying Ratio") method and to remove Gaussian and impulse noise by using DCT ("Discrete Cosine Transform") filter ROR-NLM. [6] performed histogram equalization [7], and used histogram equalization with Grey relational analysis for the enhancement of mammogram. Agarwal et al. [8] used homomorphic filtering (MH-FIL) method to modify the histogram-based contrast enhancement method. CLAHE enhances smaller regions in mammograms better. [9, 10] proposed CLAHE ("Contrast Limited Adaptive Histogram Equalization")-based mammogram image enhancement, mathematical morphology, and multiscale laplacian Gaussian pyramid transform. [11] proposed Otsu threshold and multiple regression analysis-based pectoral muscle segmentation. Shape-based mask with morphological operators is employed to the mammographic image by [12] and fitting function with cubic polynomial for the segmentation of pectoral muscle region [13]. In this paper, the clustering technique based on K-Means to eliminate pectoral muscle morphology-based operations, and "seeded region growing" (SRG) techniques to remove noise and artifacts are proposed. Kwok et al. [14] used iterative "cliff detection" to detect the pectoral muscle region. The pectoral muscle detection with morphology-based operations and the "random sample consensus" method named "RANSAC" are proposed in [15]. The region of pectoral muscle is segmented by using geometrical shapes and CLAHE for mammogram image enhancement is proposed in [16]. In [17], histogram-based 8-neighborhood connected component labeling to remove pectoral muscle is proposed. [18] proposed binary threshold-based pectoral muscle segmentation. In [19], a hybrid approach to delineate the pectoral region border by applying Hough transform and to segment pectoral muscle active contour is proposed. In [20], threshold with an active contour model for breast boundary extraction and canny edge detection for pectoral muscle removal are used. In [21], the authors proposed a two-phase novel "Margin Setting Algorithm" (MSA)-based breast and pectoral region segmentation. [22] proposed a low-contrast pectoral region segmentation by using local histogram equalization and estimation of polynomial curvature for the selected areas. In [23], an adaptive gamma correction-based pectoral muscle detection with 98.45% accuracy is proposed. [24] proposed linear enhancement mask and global threshold-based technique to detect the pectoral region boundary. In [25], the author presented a review of the segmentation method for pectoral muscle, breast boundary, micro-calcification, and mass lesions. [26] proposed analysis on particle swarm optimization and Ant bee colony optimization methods to search optimal multilevel threshold value for segmentation.

2 Research Gaps

After going through a literature review, we found the following research gaps:

1. There is little research work available for segmenting mammographic images including both CC and MLO views using a single system.
2. There is only a little research work available for integration of artifact removal, noise suppression, pectoral muscle segmentation, breast boundary extraction, and mammogram image enhancement.

3 Objectives

1. To develop a comprehensive algorithm for completely automated segmentation of mammograms on CC and MLO views.
2. To create a GUI for the implementation of a fully automated segmentation of mammograms on CC and MLO views.

4 Research Methodology

a. **Proposed Research Methodology**
   See Fig. 1.
b. **Tools and Techniques**
   Image Database: The secondary data are used for this study from the "Mammographic Image Analysis Society" (MIAS) and "Curated Breast Imaging Subset" of DDSM ("CBIS-DDSM"). The MIAS database includes 320 films of only mediolateral oblique (MLO) views with 1024 × 1024 pixels and investigated and

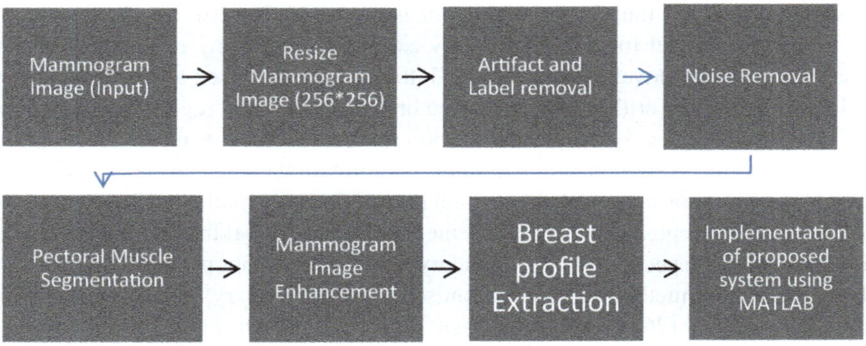

Fig. 1 Sequence of methods to be used in this research

labeled by specialized radiologists. The DDSM consists of 2620 mammograms of normal, benign, and malignant cases with verified pathology information.

Techniques used in proposed Algorithm

Normalization of Grayscale Image: Normalization is a technique to transform a gray image with intensities in the (Min, Max) range, into a new image with intensities in the (IMin, IMax) range. The normalization of a digital grayscale image is performed according to the formula equation:

$$\text{INormalized} = ((\text{I} - \text{Min}) * ((\text{IMax} - \text{IMin}) \div (\text{Max} - \text{Min}))) + \text{IMin} \quad (1)$$

Global Image Thresholding: Otsu developed a discriminate analysis-based method to determine the optimum thresholds to separate the level of the classes in grayscale image at its maximum [27]. Otsu suggested the between-class variance which is the sum of sigma functions of each class is given by Eq. (2):

$$f(t) = \sigma_0 + \sigma_1 \quad (2)$$

$$\sigma_0 = \omega_0(\mu_0 - \mu_t)^2, \sigma_1 = \omega_1(\mu_1 - \mu_t)^2 \quad (3)$$

Here, (μT) is the gray image's mean intensity. The mean level $(\mu 1)$ of two classes for the case of bi-level threshold is given by Eq. (4) [28]:

$$\sigma_0 = \sum_{i=0}^{t-1} \frac{iP_i}{\omega_0} \quad \sigma_1 = \sum_{i=0}^{t-1} \frac{iP_i}{\omega_1} \quad (4)$$

The optimal value of threshold to maximize the function for a between-class variance can be specified using Eq. (5) [28]:

$$\sigma_\omega^2(t) = \omega_0(t)\sigma_0^2(t) + \omega_1(t)\sigma_1^2(t) \quad (5)$$

Optimal Multilevel Threshold: The multilevel threshold method subdivides the image pixels into several distinct groups having similar gray levels within a specific range. The Otsu's method can be applied for multiple threshold [29] segmentation. The optimum threshold values t_1^* and t_2^* can be computed as Eq. (5).

$$f(x, y) = \begin{Bmatrix} a\ if\ g(x, y) > t_2 \\ b\ if\ t_1 < g(x, y) \le t_2 \\ c\ if\ g(x, y) \le t_1 \end{Bmatrix} \tag{6}$$

"Canny Edge Detection": Canny method for detection of an edge is the summation of four complex exponentials calculated by Gaussian's first derivative. As stated by Canny, detection of edges can be done by convolution of the noisy image with a function $f(x)$ and denoting the edges in the output at the maxima of convolution. The procedure to compute edges using a Canny edge detector is given as follows:

- Compute the image gradient $f(a, b)$ by convolution of image with Gaussian's first derivative in x and y directions which is expressed by Eqs. (7) and (8):

$$fa(a, b) = f(a, b) * \left(-\frac{a}{\sigma^2}\right) e^{-\frac{a^2+b^2}{2\sigma^2}} \tag{7}$$

$$fb(a, b) = f(a, b) * \left(-\frac{b}{\sigma^2}\right) e^{-(a^2+b^2)/2\sigma^2} \tag{8}$$

- Apply Non-Maxima Suppression on the result of above step.
- Perform hysteresis threshold on the resultant gradient of above step.

Canny edge detection method works on two thresholds by scanning image form leftward to rightward and upward to downward. If the non-maxima-suppressed gradient magnitude at the pixel is larger than the highest threshold, then it is declared as an edge point. On the other hand, if the non-maxima-suppressed gradient magnitude at the pixel is larger than the lowest threshold, then it is also declared as an edge point.

"Contrast-Limited Adaptive Histogram Equalization": The "Contrast-Limited Adaptive Histogram Equalization" computes the contrast-limited transformation function for each small region and, as a result of this operation, each small area's contrast is enhanced as the 'Distribution' value histogram matches approximately with the histogram of the outcome region. The neighborhood regions are then combined to remove falsely inferred boundaries using bilinear interpolation. The contrast is limited to avoid amplify any noise in homogeneous areas that may exhibit in the input image.

Performance evaluation of segmentation result: Jaccard similarity index, Dice similarity index, and Score Matching methods are employed to estimate the effectiveness of segmented region results.

(a) The Jaccard Similarity index J computes the likeness of real ("ground truth") image and segmented image as follows:

$$J(I_g, I_s) = \frac{|I_g \cap I_s|}{|I_g \cup I_s|} \tag{9}$$

where I_g is the real ("ground truth") image segmented manually and I_s is the segmented image by proposed work.

(b) The Dice Similarity index D computes the likeness of real ("ground truth") image and segmented image as follows:

$$D(I_g, I_s) = \frac{2 * |I_g \cap I_s|}{|I_g| + |I_s|} \tag{10}$$

where I_g is the real ("ground truth") image segmented manually and I_s represents the segmented image by proposed work.

(c) The Score matching method measures the closeness of the segmented image boundary and the ground truth image boundary as follows:

$$sm = 2 * p * r/(r + p) \tag{11}$$

where p is the precision and r is the recall value and the score match is the harmonic mean of p and r with a distance error tolerance.

c. **Implementation of Proposed Algorithm**
See Fig. 2.

Algorithm 1. Pseudocode for the Proposed Algorithm START

1. Read the mammogram image

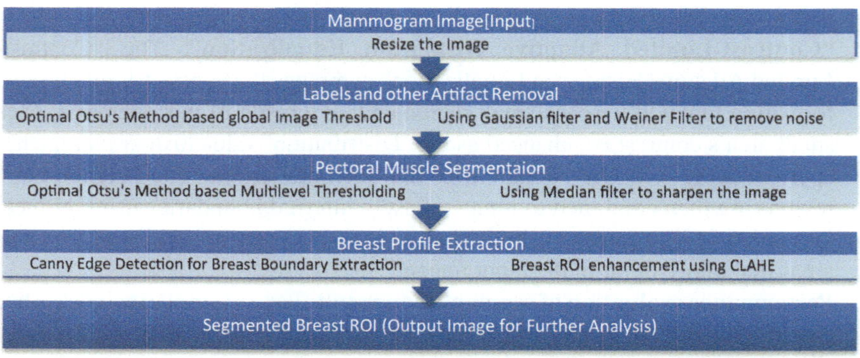

Fig. 2 Steps for implementation of proposed work

2. Resize the grayscale image to 256*256 pixels
3. IF orientation is right
   a. Flip the image left to right

   End IF
4. Apply "Contrast-Limited Adaptive Histogram Equalization" to enhance the breast ROI.
5. Apply optimal Otsu's method-based global image threshold
6. Filter the mammogram image using a Gaussian filter and Wiener filter to remove impulse and speckle noise.
7. IF MLO view image then
   a. Apply optimal Otsu's method-based multilevel Threshold
   b. Filter the image using the median filter to sharpen the image.

   End IF
8. Detect the edges of breast boundary using edge detection method developed by Canny.
9. Filtering the mammogram using Gaussian filter to smoothen the edges of image.

END

5 Result and Discussion

We have implemented the proposed approach using GUI created in Matlab2018b. A result of the complete algorithm is presented with a number of input and output mammogram images of both MLO and CC views with both breasts. In our work, we used a global threshold technique based on Otsu's method to eliminate artifacts and labels. Level five Multi-threshold, morphological structuring, and region fill are used for pectoral muscle segmentation and extraction. Breast structure excluding pectoral region is extracted using an edge detector developed by Canny and Gaussian filtering. The proposed work's result exhibited the capability to eliminate artifacts and label, remove and extract pectoral muscle, and extract breast profile and to enhance the contrast of image without losing any information. The outcome exhibit reducing the size of image will reduce the computational time of proposed work (Figs. 3, 4, 5, 6, 7, 8, 9, 10, 11, 12, 13, 14, 15, 16, 17, 18, 19, 20, 21, 22, 23, 24, and 25).

6 Conclusion

This research paper dealing with the identification and segmenting of the pectoral muscle region and breast profile from both MLO and CC view of mammogram

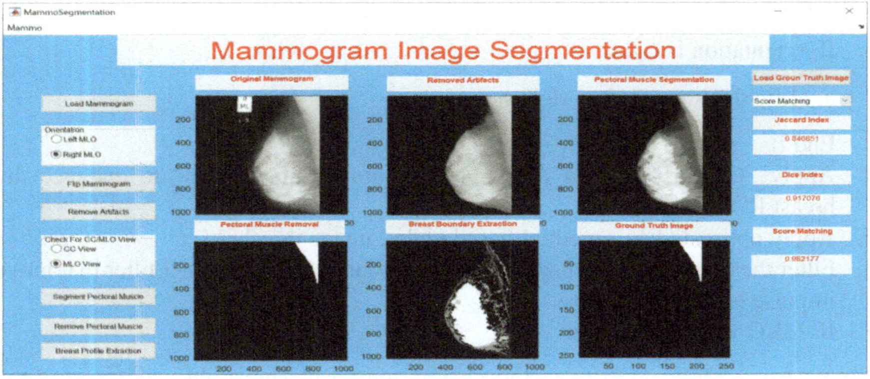

Fig. 3 Mdb004.pgm

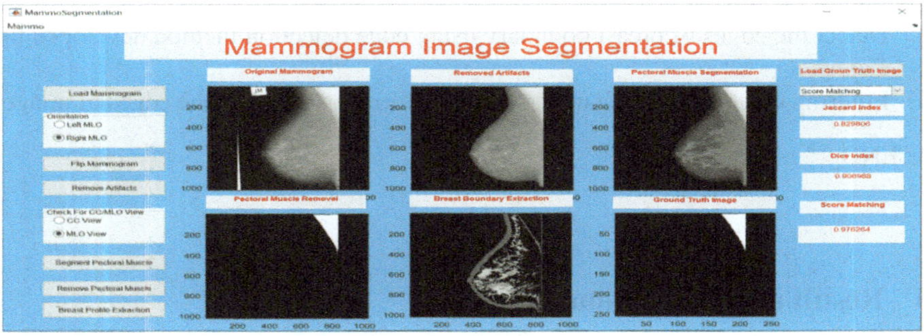

Fig. 4 Mdb009.pgm

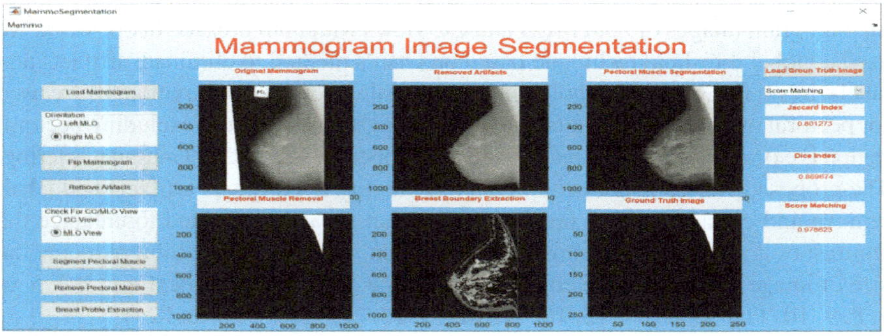

Fig. 5 mdb012.pgm

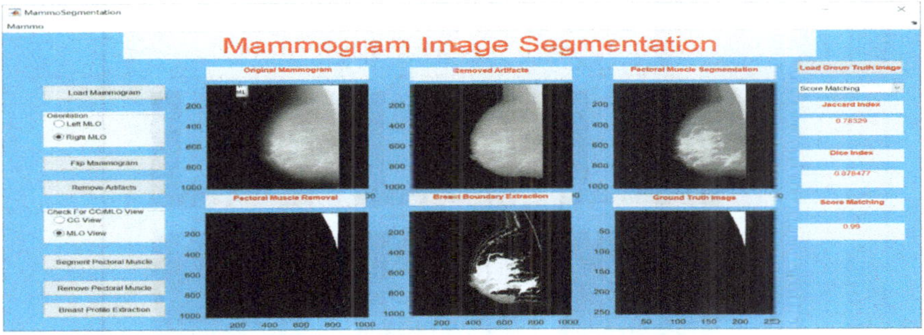

Fig. 6 Mdb014.pgm

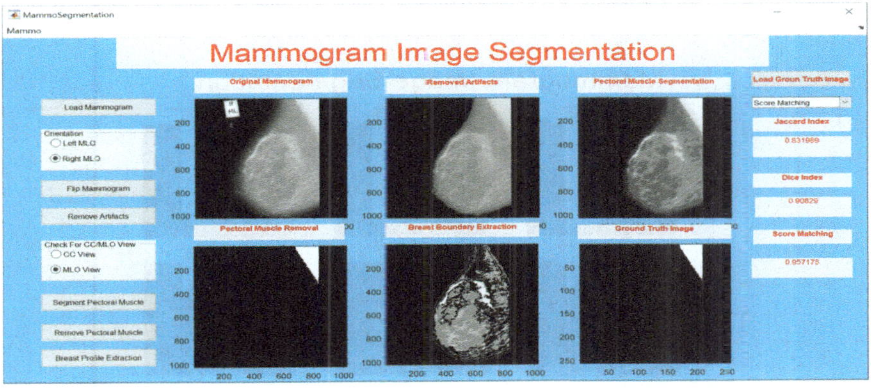

Fig. 7 Mdb020.pgm

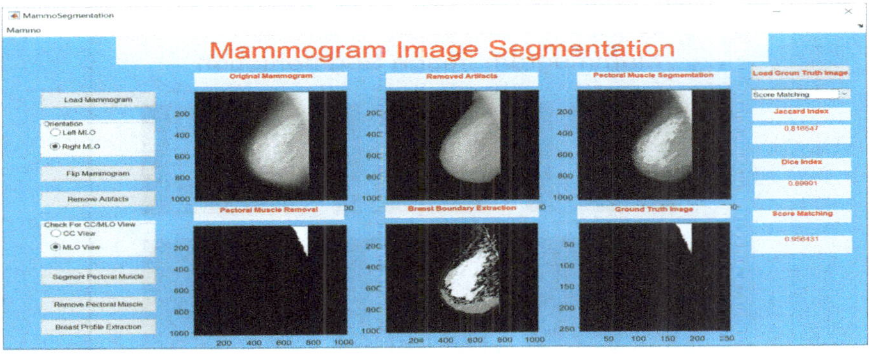

Fig. 8 Mdb022.pgm

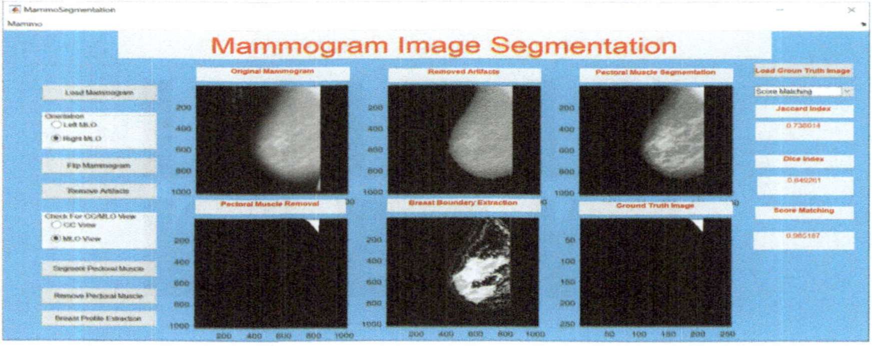

Fig. 9 Mdb024.pgm

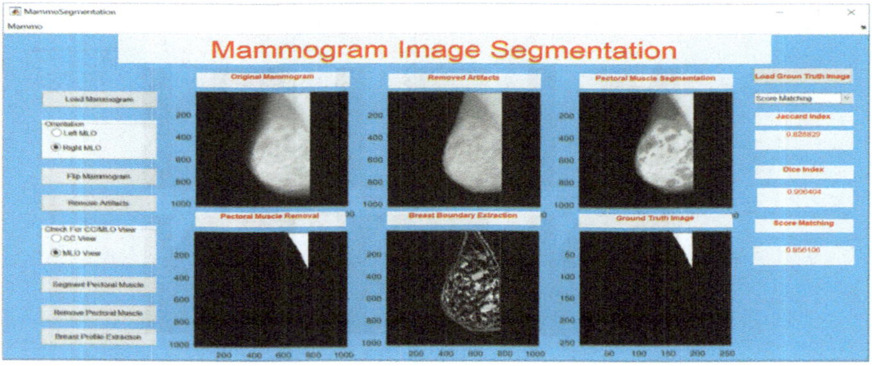

Fig. 10 Mdb030.pgm

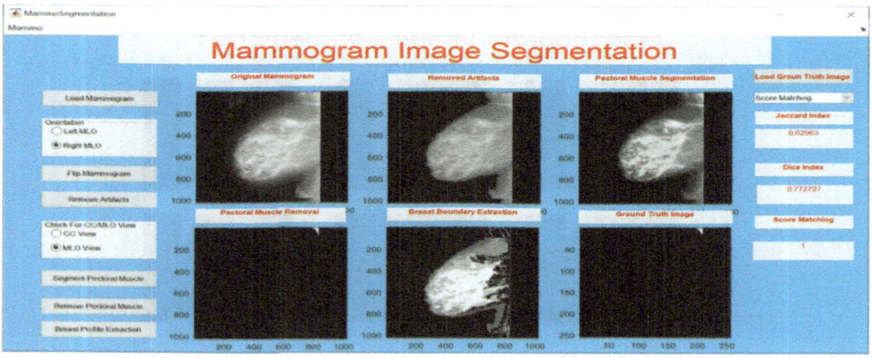

Fig. 11 Mdb032.pgm

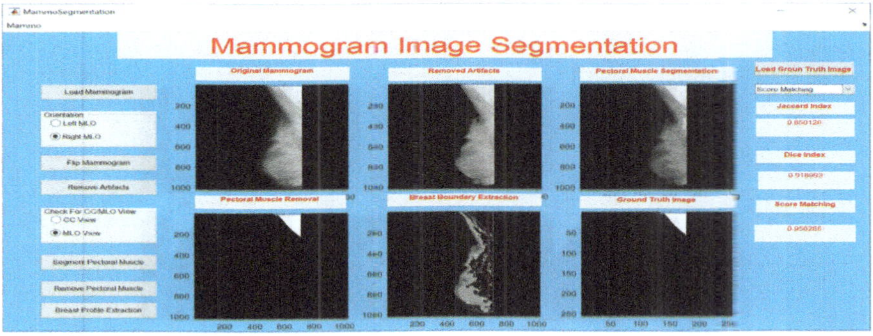

Fig. 12 Mdb034.pgm

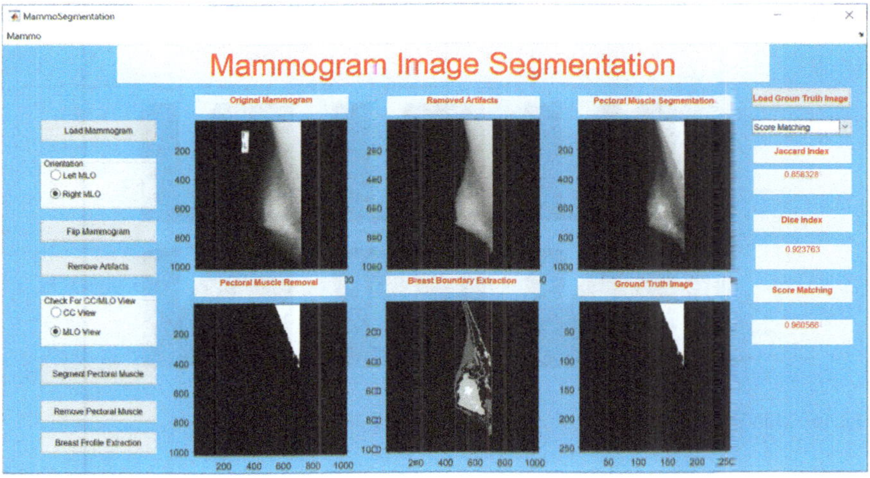

Fig. 13 Mdb036.pgm

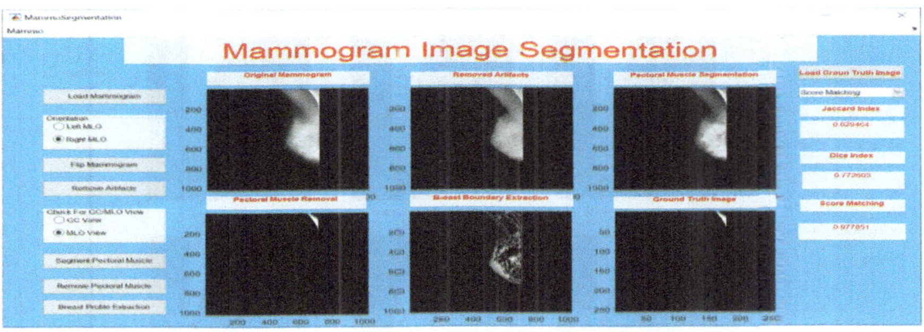

Fig. 14 Mdb040.pgm

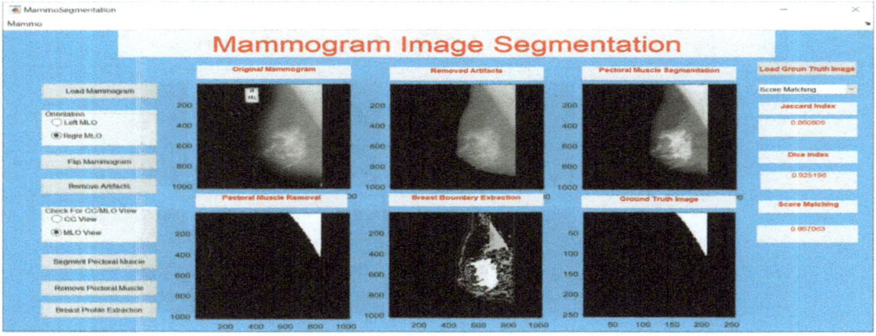

Fig. 15 Mdb042.pgm

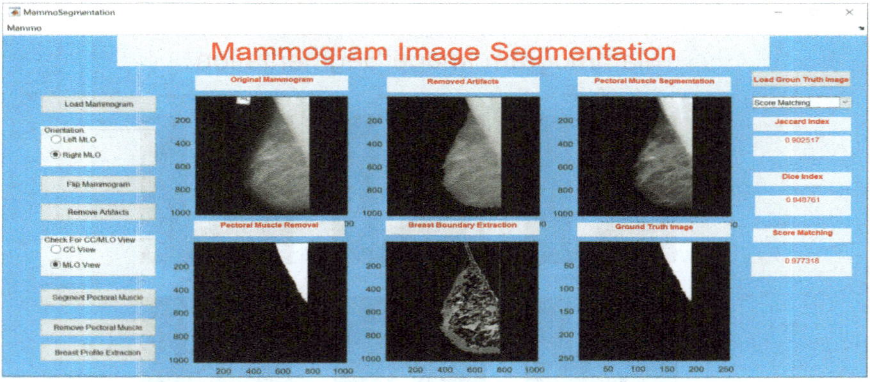

Fig. 16 Mdb046.pgm

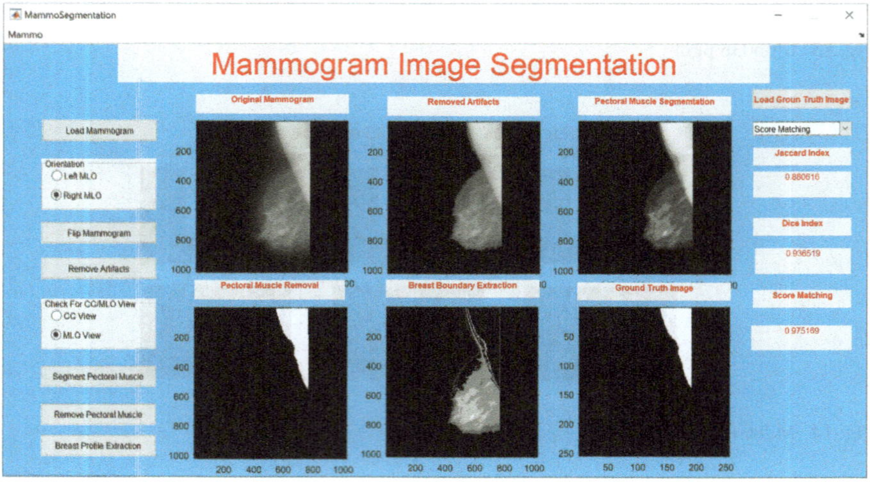

Fig. 17 Mdb047.pgm

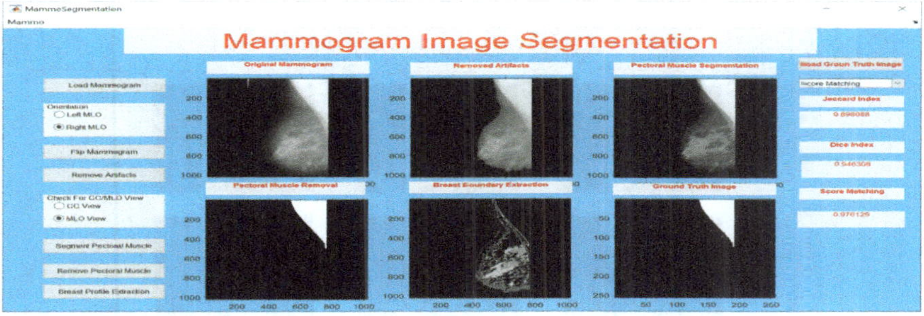

Fig. 18 Mdb048.pgm

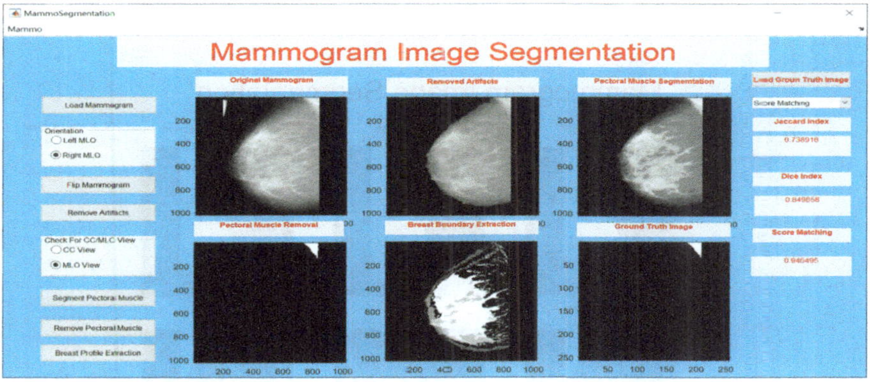

Fig. 19 Mdb050.pgm

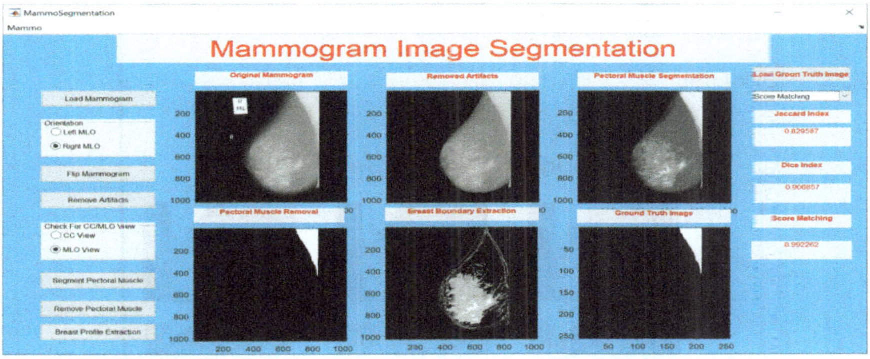

Fig. 20 Mdb058.pgm

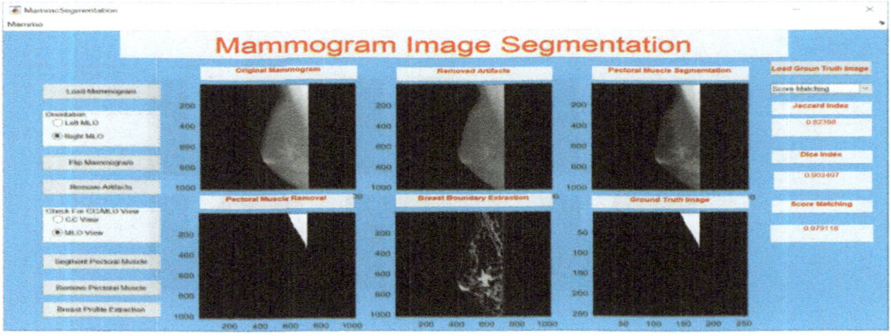

Fig. 21 Mdb060.pgm

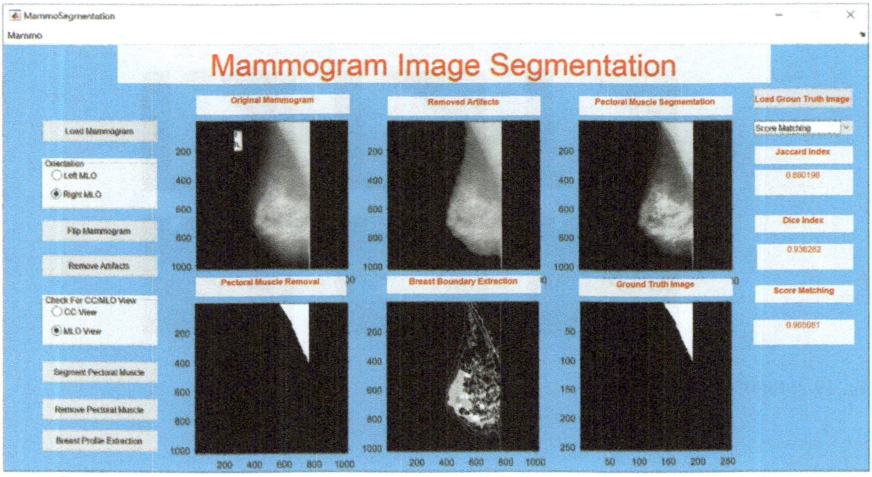

Fig. 22 Mdb064.pgm

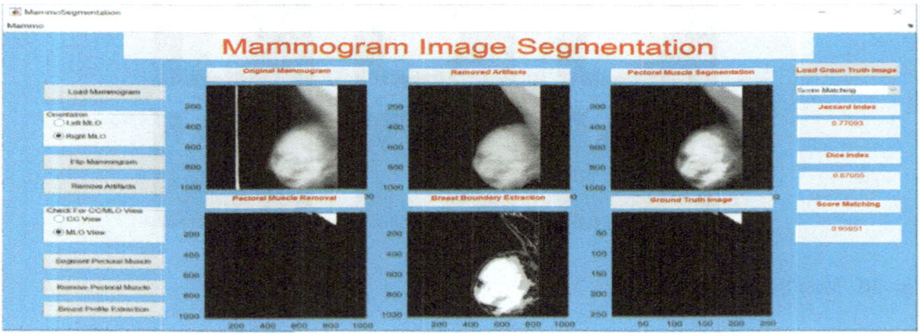

Fig. 23 Mdb068.pgm

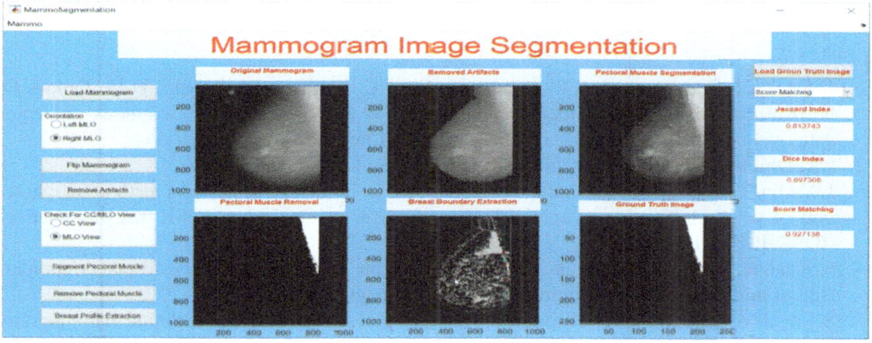

Fig. 24 Mdb069.pgm

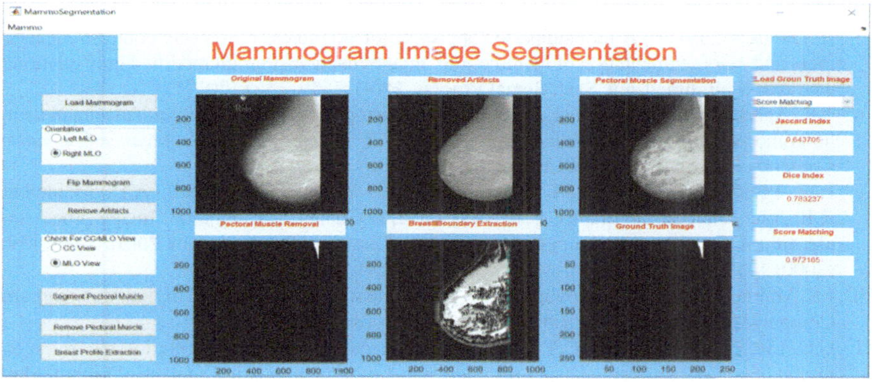

Fig. 25 Mdb070.pgm

contains the left as well right breast. The proposed approach performs Level Five Multilevel threshold to segment pectoral muscle which does not deal with the detection of a straightline. The local histogram equalization performs enhancement of pectoral muscles. Then the pectoral region border is found with morphological structuring and breast region, receptively, using Canny edge detection and threshold technique. The investigated outcomes obtained for the proposed work are tested on 100 mammograms, 50 from the MIAS database mammograms and 50 from the CBIS-DDSM database consisting of both views of breasts. The future work of this research will involve a comparative study of the proposed approach with other approaches. The results obtained on 100 images of CBIS-DDSM and MIAS database have shown an excellent output. The outcome mammogram images can be used further for the next CADx stage to assist for discrimination, identification, and diagnosis of breast carcinoma.

References

1. J. Ferlay, I. Soerjomataram, R. Dikshit, S. Eser, C. Mathers, M. Rebelo, D.M. Parkin, D. Forman, F. Bray, Carcinoma incidence and mortality worldwide: sources, methods and major patterns in GLOBOCAN 2012. Int. J. Carcinoma **136**(5), E359–E386 (2015)
2. P.K. Dhillon, Breast carcinoma fact sheet. Mortality **11**(5), 10–18 (2009)
3. J.A. Ojo, T.M. Adepoju, E. Omdiora, O. Olabiyisi, O. Bello, Pre-processing method for extraction of pectoral muscle and removal of artefacts in mammogram. IOSR J. Comput. Eng. (IOSR-JCE) e 16(3), 6–9
4. S. Kannan, N.P. Subiramaniyam, A.T. Rajamanickam, A. Balamurugan, Performance comparison of noise reduction in mammogram images. Image, 1(3 x 3), 5 x 5
5. S. Sreedevi, E. Sherly, A novel approach for removal of pectoral muscles in digital mammogram. Procedia Comput. Sci **46**, 1724–1731 (2015)
6. E.D. Pisano, S. Zong, B.M. Hemminger, M. DeLuca, R.E. Johnston, K. Muller, M.P. Braeuning, S.M. Pizer, Contrast limited adaptive histogram equalization image processing to improve the detection of simulated spiculations in dense mammograms. J. Digit. Imaging **11**(4), 193–200 (1998)
7. B. Gupta, M. Tiwari, A tool supported approach for brightness preserving contrast enhancement and mass segmentation of mammogram images using histogram modified grey relational analysis. Multidimension. Syst. Signal Process. **28**(4), 1549–1567 (2017)
8. T.K. Agarwal, M. Tiwari, S.S. Lamba, Modified histogram-based contrast enhancement using homomorphic filtering for medical images, in *2014 IEEE International on Advance Computing Conference (IACC)* (IEEE, 2014), pp. 964–968
9. E.D. Pisano, E.B. Cole, B.M. Hemminger, M.J. Yaffe, S.R. Aylward, A.D. Maidment, R.E. Johnston, M.B. Williams, L.T. Niklason, E.F. Conant, L.L. Fajardo, Image processing algorithms for digital mammography: a pictorial essay 1. Radiographics **20**(5), 1479–1491 (2000)
10. I.K. Maitra, S. Nag, S.K. Bandyopadhyay, Technique for preprocessing of digital mammogram. Comput. Methods Programs Biomed. **107**(2), 175–188 (2012)
11. C.C. Liu, C.Y. Tsai, J. Liu, C.Y. Yu, S.S. Yu, A pectoral muscle segmentation algorithm for digital mammograms using Otsu thresholding and multiple regression analysis. Comput. Math Appl. **65**(5), 1100–1107 (2012)
12. C. Chen, G. Liu, J. Wang, G. Sudlow, Shape-based automatic detection of pectoral muscle boundary in mammograms. J. Med. Biol. Eng. **35**(3), 315–322 (2015)
13. N. Alam, M.J. Islam, Pectoral muscle elimination on mammogram using K-means clustering approach. Int. J. Comput. Vis. Sign. Process. **4**(1), 11–21 (2014)
14. S.M. Kwok, R. Chandrasekhar, Y. Attikiouzel, M.T. Rickard, Automatic pectoral muscle segmentation on mediolateral oblique view mammograms. IEEE Trans. Med. Imaging **23**(9), 1129–1140 (2004)
15. W.B. Yoon, J.E. Oh, E.Y. Chae, H.H. Kim, S.Y. Lee, K.G. Kim, Automatic detection of pectoral muscle region for computer-aided diagnosis using MIAS mammograms. BioMed Res. Int. (2016)
16. S.A. Taghanaki, Y. Liu, B. Miles, G. Hamarneh, Geometry-based pectoral muscle segmentation from mlo mammogram views. IEEE Trans. Biomed. Eng. **64**(11), 2662–2671 (2017)
17. R. Boss, K. Thangavel, D. Daniel, *Automatic Mammogram Image Breast Region Extraction and Removal of Pectoral Muscle* (2013). arXiv preprint arXiv:1307.7474
18. A.K. Mohideen, K. Thangavel, Removal of pectoral muscle region in digital mammograms using binary thresholding. Int. J. Comput. Vis. Image Process. (IJCVIP) **2**(3), 21–29 (2012)
19. A.L. Pavan, A. Vacavant, A.F. Alves, A.P. Trindade, D.R. de Pina, Automatic identification and extraction of pectoral muscle in digital mammography, in *World Congress on Medical Physics and Biomedical Engineering 2018* (Springer, Singapore, 2019), pp. 151–154
20. A. Rampun, P.J. Morrow, B.W. Scotney, J. Winder, Fully automated breast boundary and pectoral muscle segmentation in mammograms. Artif. Intell. Med. **79**, 28–41 (2017)

21. H. Al-Ghaib, Y. Wang, R. Adhami, A new machine learning algorithm for breast and pectoral muscle segmentation. Eur. J. Adv. Eng. Technol. **2**, 21–29 (2015)
22. M. Mustra, M. Grgic, Robust automatic breast and pectoral muscle segmentation from scanned mammograms. Sig. Process. **93**(10), 2817–2827 (2013)
23. S.J.S. Gardezi, F. Adjed, I. Faye, N. Kamel, M.M. Eltoukhy, Segmentation of pectoral muscle using the adaptive gamma corrections. Multimedia Tools Appl. **77**(3), 3919–3940 (2018)
24. P.S. Vikhe, V.R. Thool, Detection and segmentation of pectoral muscle on MLO-view mammogram using enhancement filter. J. Med. Syst. **41**(12), 190 (2017)
25. J. Dabass, S. Arora, R. Vig, M. Hanmandlu, Segmentation techniques for breast carcinoma imaging modalities-a review, in *9th International Conference on Cloud Computing, Data Science & Engineering* (2019)
26. B. Akay, A study on particle swarm optimization and artificial bee colony algorithms for multilevel thresholding. Appl. Soft Comput. **13**(6), 3066–3091 (2013)
27. T. Kurban, P. Civicioglu, R. Kurban, E. Bescok, Comparison of evolutionary and swarm based computational techniques for multilevel color image thresholding. Appl. Soft Comput. **23**, 128–143 (2014)
28. A.K. Bhandari, A. Kumar, G.K. Singh, Modified artificial bee colony based computationally efficient multilevel thresholding for satellite image segmentation using Kapur's, Otsu and Tsallis functions. Expert Syst. Appl. **42**(3), 1573–1601 (2015)
29. ChH Bindu, K.S. Prasad, An efficient medical image segmentation using conventional OTSU method. Int. J. Adv. Sci. Technol. **38**, 67–74 (2012)

Empirical Study of Computational Intelligence Approaches for the Early Detection of Autism Spectrum Disorder

Mst. Arifa Khatun, Md. Asraf Ali, Md. Razu Ahmed, Sheak Rashed Haider Noori, and Arun Sahayadhas

Abstract The objective of the research is to develop a predictive model that can significantly enhance the detection and monitoring performance of Autism Spectrum Disorder (ASD) using four supervised learning techniques. In this study, we applied four supervised-based classification techniques to the clinical ASD data obtained from 704 patients. Then, we compared the four machine learning (ML) algorithms performance across tenfold cross-validation, ROC curve, classification accuracy, F1 measure, precision, recall, and specificity. The analysis findings indicate that Support Vector Machine (SVM) achieved the uppermost performance than the other classifiers in terms of accuracy (85%), f1 measure (87%), precision (87%), and recall (88%). Our work presents a significant predictive model for ASD that can effectively help the ASD patients and medical practitioners.

Keywords Autism spectrum disorder (ASD) · Machine learning · Classification · Screening tools · SVM

1 Introduction

Autism spectrum disorder (ASD) is a complex neurobehavioral syndrome that refers to impairments in different social skills and developmental, repetitive activities combined with nonverbal communication. In 2018 alone, according to the Autism and Developmental Disabilities Monitoring (ADDM) Network, a division of Centers

Mst. A. Khatun · S. R. H. Noori
Department of Computer Science and Engineering, Daffodil International University, Dhaka, Bangladesh

Md. A. Ali · Md. R. Ahmed (✉)
Department of Software Engineering, Daffodil International University, Dhaka, Bangladesh
e-mail: razu35-1072@diu.edu.bd

A. Sahayadhas
Artificial Intelligence Research Lab, Vels Institute of Science, Technology and Advanced Studies, Chennai, India

S. S. Dash et al. (eds.), *Intelligent Computing and Applications*,
Advances in Intelligent Systems and Computing 1172,
https://doi.org/10.1007/978-981-15-5566-4_14

for Disease Control and Prevention (CDC), 1 in 59 is recognized with ASD [1]. Moreover, males are four times more likely to be diagnosed with ASD than females [1]. Studies have shown that 31% of adolescents affected with ASD have a severe psychological impairment (i.e., intelligence quotient less than 70), another 25% adolescents are in the marginal range (i.e., IQ: 71–85), and only 44% adolescents have IQ in the normal range (i.e., IQ greater than 85) [2]. The existing scientific evidence indicates that the possible factors causing ASD in children may be due to ecological and genomic factors [3]. These possible factors indicate that the primary signs of autism usually appear at early age (i.e., 2 or 3 years). Consequently, a study shows that early diagnosis within the beginning stage of life can lead to positive outcomes of autism-affected people [4].

Researchers have presented different autism diagnosis tools and techniques which are based on some scripted questionnaires with scoresheets that yield results. Based on the inference, the clinical practitioner can take their decision. Example of different autism screening tools and techniques are the "Modified Checklist for Autism in Toddlers (M-CHAT) [5], Autism Diagnostic Interview (ADI) [6], Autism Diagnostic Observation Schedule Revised (ADOS-R) [7], and Child Behavior Checklist (CBCL)" [8], etc. The existing ADI-R and ADOS have shown that competitive performance as inputs to train autism screening algorithms was clinically confirmed in several experimental research findings [9–12]. Cognoa [13] is another data-driven machine learning (ML)-based tool and it has been validated by multiple research studies [12, 14, 15]. The Cognoa ASD screening tool has been developed with the aim of helping the practitioners to diagnose autism in children between the age of 18 and 72 months [16].

ML algorithms have been used to improve the diagnosis process of ASD as well as to obtain information to deliver faster access to clinical services for the medical practitioner so that an effective decision based on the ASD diagnosis can be made [17, 18]. Moreover, ML classifiers are the most operational and effective methods for low-cost screening ASD tools including different clinical contexts. Most of the time, clinical data are generated from different clinical environments but the data consists of an imbalance, unstructured format making it difficult to be applied to the screening tools. The consistency and reliability of ML models may vary conditionally based on the real-time trained data. In this work, we highlight these challenges and current practical tools and strategies for early diagnosis using several ML algorithms. The main aspect of this work is to examine different classifiers' performance through different performance measurement techniques and attain more effective decisions from clinical data. Many of the studies focused on only accuracy and classification based on ASD data. Therefore, the outperform classification techniques have been considered for the predictive model.

The rest of the paper is presented as follows, the materials and methods of this study are presented in Sect. 2. In Sect. 3, the experimental results are described. Finally, conclusions and future recommendations of this work are illustrated in Sect. 4.

2 Materials and Methods

A. Dataset Collection

In this work, we used the ASD dataset created by Fadi Fayez Thabtah and which is provided by the "UCI Machine Learning Repository" [19, 20]. We considered 704 instances including 21 attributes for our predictive model. Table 1 shows the original attributes from the dataset which is collected from the UCI Machine Learning Repository.

B. Classification Techniques for clinical tools

In the last decades, ML algorithms have been performing a significant role and it is recognized to solve the medical problem and clinical diagnosis [21, 22]. Studies show that ML approaches can perform a significant role in improving the consistency and effectiveness of clinical ASD screeners [23–25]. In this work, four ML classifiers have been considered to perform the ASD screening model: "k-nearest neighbors (KNN), Logistics Regression (LR), Random Forest (RF) and Support Vector Machine (SVM)".

Table 1 Dataset attribute from UCI machine learning repository

#ID	Feature	Data type
1	Age of patients	Numerical
2	Gender of patients	string_value
3	Ethnicity of patients	string_value
4	Born alongside jaundice	Yes/no
5	Family_member has a PDD	Yes/no
6	Completed test (i.e. doctors/parents)	String_value
7	Residence	String_value
8	Used_the_screening app previous	yes/no
9	Screening techniques	Numerical
10	The answer to Q1	binary_value (0/1)
11	The answer to Q2	binary_value (0/1)
12	The answer to Q3	binary_value (0/1)
13	The answer to Q4	binary_value (0/1)
14	The answer to Q5	binary_value (0/1)
15	The answer to Q6	binary_value (0/1)
16	The answer to Q7	binary_value (0/1)
17	The answer to Q8	binary_value (0/1)
18	The answer to Q9	binary_value (0/1)
19	The answer to Q10	binary_value (0/1)
20	The screening test results	Numerical
21	Target class	Yes/no

Table 2 The confusion matrix of actual and projected class

Projected class		
Actual class	True_positive (TP_{sample})	False_positive (FP_{sample})
	False_negative (FN_{sample})	True_negative (TN_{sample})

TP_s: the amount of true_positive samples; FP_s: the amount of false_positive samples
FN_s: the amount of false_negative samples; TN_s: the amount of true_negative samples

Table 3 Performance measurement benchmark

Performance metrics	Mathematical equation
Accuracy	$\frac{(TP+TN)}{(TP+FP+TN+FN)}$
Precision	$\frac{TP}{(TP+FP)}$
Recall	$\frac{TP}{(TP+FN)}$
F1 measure	$\frac{2*(Recall*Precision)}{(Recall+Precision)}$
Specificity	$\frac{TN}{(TN+FP)}$

C. *Performance Measurement*

In this paper, performances have been validated using the technique of tenfold cross-validation [26] and different performance measurement approaches. The confusion matrix is an operational tool for examining how good a ML classification technique can recognize of dissimilar classes. Table 2 shows a confusion matrix for actual and projected classes.

In order to evaluate and compare these supervised ML techniques for ASD screening, we used different performance measurement techniques as shown in Table 3.

D. *Experimental Setup*

In this experiment, the ASD dataset has been considered to develop the ASD predictive model. Then we performed different data pre-processing methods on the ASD dataset to make a tidy dataset, such as correlation analysis to find the redundant values, missing values analysis, feature selection analysis, etc. The detailed workflow of the ASD predictive model is presented in Fig. 1.

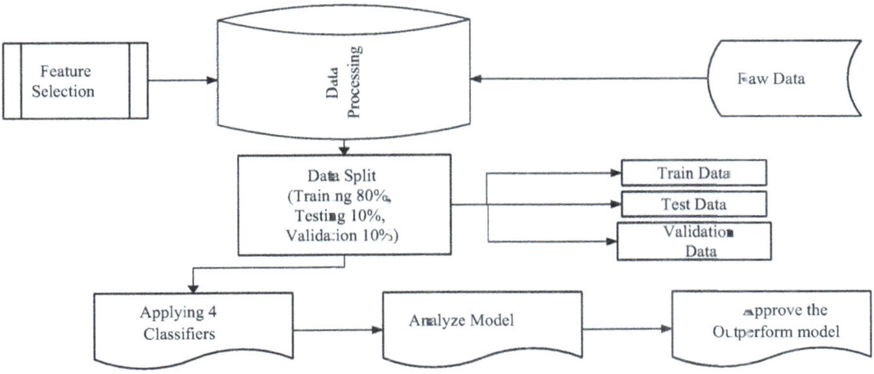

Fig. 1 Workflow of ASD screening model

3 Result and Discussions

A. Data Preprocessing

In this section, we considered various evaluations to investigate the ASD dataset. Figure 2 shows that the white and European country-wise distribution of ASD and it is looking likely to similar for topmost five countries such as the United States, the United Kingdom, Australia, New Zealand, and Canada as top providers of positive ASD. Moreover, Fig. 3 the observation shows that the adults and toddlers have an extreme risk of being an ASD patient in which people are based on white and Europeans Ethnicities. The Black and Asian people are following the next risk factor with a smaller ratio. We conclude that our study has presented a possible genetic link for ASD positive.

Here, Fig. 4 shows that the ASD positive ratio is more familiar among boys than girls. But our study presents a different scenario of adults. Whereas in adults, it is

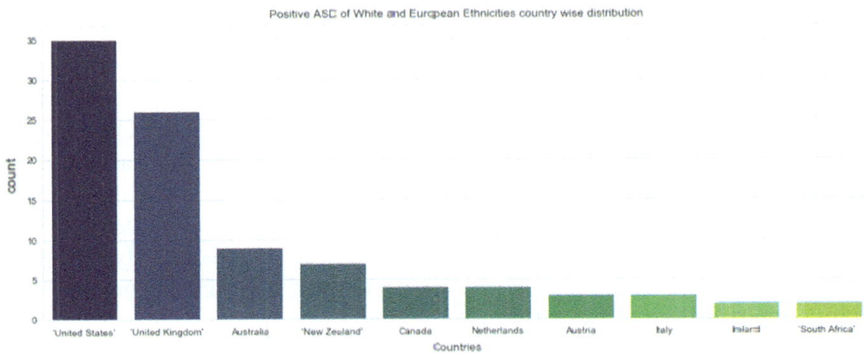

Fig. 2 The ratio of ASD based on white and European country

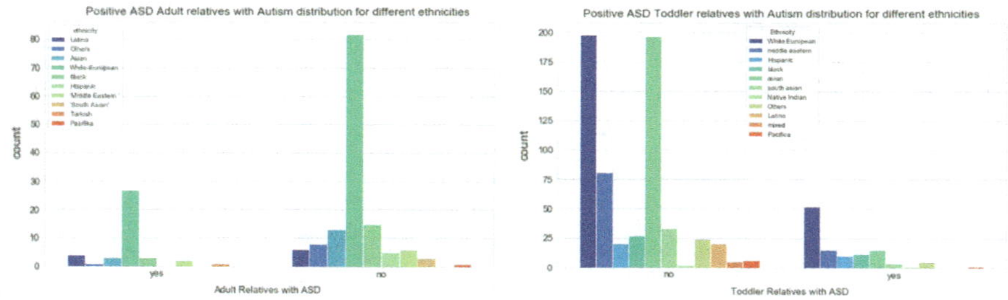

Fig. 3 ASD positive relatives with autism distribution for different ethnic peoples

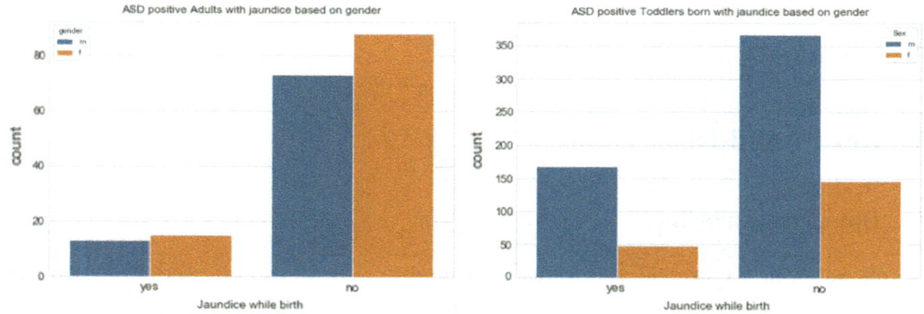

Fig. 4 ASD positive with jaundice based on gender

lower than the female gender but, considering toddlers, it is likely four times higher ratio than girls. From the ASD dataset, 704 samples and 21 attributes were taken into the analysis of the predictive ASD model. We splitted the ASD dataset into three chunks, whereas the training set comprises 80%, the test set comprises 10%, and another 10% of the splitted data for validation test. Hence, the ASD dataset was as well examined to validate the redundant values. we have used heatmap to find the redundant and correlated columns in the ASD dataset. Our analysis shows that there are no columns correlated with one to one in Fig. 5.

B. **Analysis of the performance**

The prediction performance of four ML classifiers was examined for the classification of ASD. Figure 6 shows the performance of four supervision-based classification techniques. With respect to precision, recall, and f1 measure, SVM achieved the highest performance than other classifiers. Moreover, when considering the accuracy (Fig. 7), SVM obtained the highest (i.e., 85%). In addition, LR achieved the lowest performance in terms of accuracy, precision, recall, and f1 measure. By looking at KNN and RF classification techniques, we can observe that their performance was almost similar. However, the performance results suggesting that the SVM is more effective and reliable in the prediction of ASD models.

Another measurement for classifiers is ROC (receiver operating curve) [27], which is based on "false positive rate (x-axis) and true positive rate (y-axis)". However, the

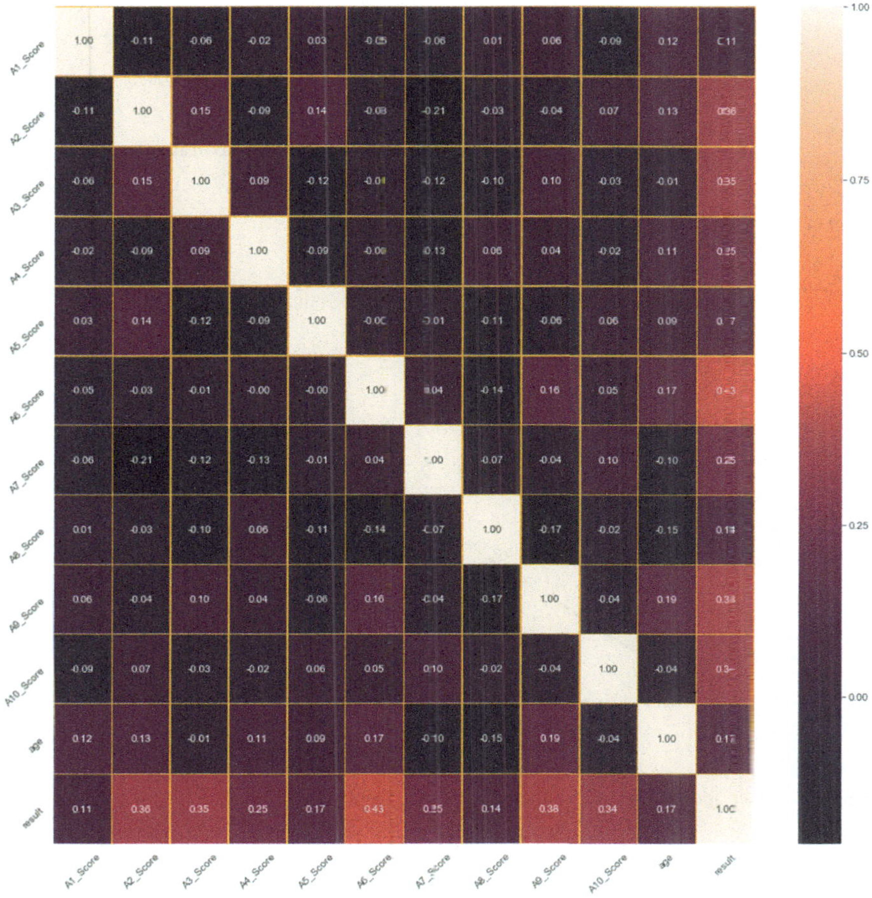

Fig. 5 Heat map for checking correlated columns

Fig. 6 The figure shows the performance of four classifiers

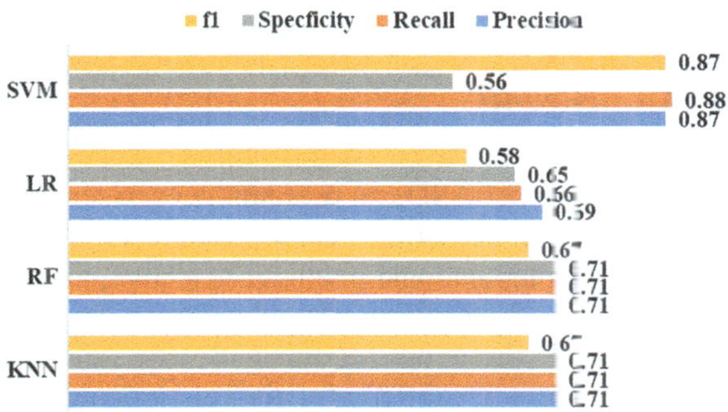

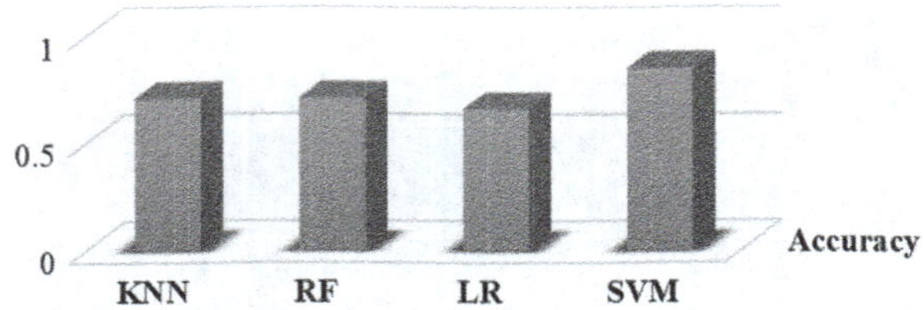

Fig. 7 Classification accuracy for ASD models

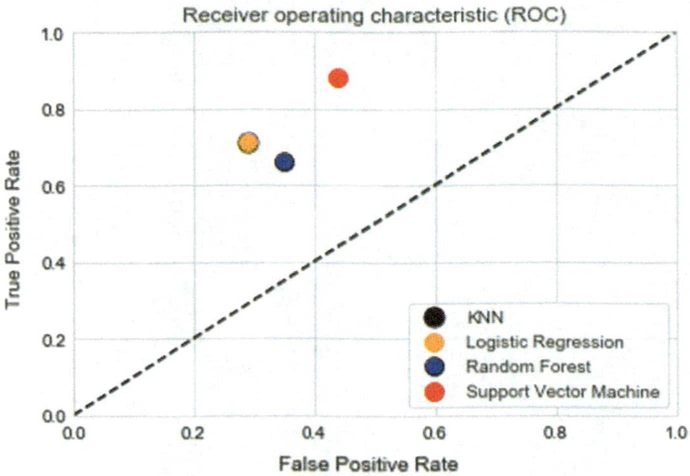

Fig. 8 Receiver operating curve for the classification techniques

ROC curve is unbiased of both classes for measuring the capability of the predictive classifier. Figure 8 shows that SVM performs better than other classifiers to predict ASD.

4 Conclusion

Early diagnosis of ASD can prevent the occurrences of ASD patients and can have a significant influence on its clinical treatment. This work presents a workflow that is based on computational intelligence techniques for the forecast and diagnosis of ASD. This work used four classification techniques in the early identification of ASD patients. These four ML methods were validated with tenfold cross-validation techniques including different statistical measurement techniques. The performance result shows that the SVM achieved the highest performance (i.e., 85%). Therefore, this ML-based predictive application can be used for early diagnosis of ASD patients and which will be helpful for clinical practitioners and the health-care research community.

Conflict of Interest None.

References

1. J. Baio, Prevalence of Autism spectrum disorder among children aged 8 years—autism and developmental disabilities monitoring network, 11 Sites, United States, 2014. MMWR. Surveill. Summ. **67**(6), 1–23 (2018)
2. Autism and Health: A Special Report by Autism Speaks | Autism Speaks. [Online]. Available: https://www.autismspeaks.org/science-news/autism-and-health-special-report-autism-speaks. Accessed 7 Sept 2019
3. What Is Autism? | Autism Speaks. [Online]. Available: https://www.autismspeaks.org/what-autism. Accessed 9 Sept 2019
4. M.S. Durkin et al., Socioeconomic inequality in the prevalence of autism spectrum disorder: evidence from a US cross-sectional study. PloS one **5**(7), e11551 (2010)
5. C. Chlebowski, D.L. Robins, M.L. Barton, D. Fein, Large-scale use of the modified checklist for autism in low-risk toddlers. Pediatrics **131**(4), e1121–e1127 (2013)
6. C. Gillberg, M. Cederlund, K. Lamberg, L. Zeijlon, Brief report: 'The autism epidemic'. The registered prevalence of autism in a Swedish Urban Area. J. Autism Dev. Disord. **36**(3), 429–435 (2006)
7. C. Lord, S. Risi, L. Lambrecht, E.H. Cook, B.L. Leventhal, P.C. DiLavore A. Pickles, M. Rutter, The Autism diagnostic observation schedule—generic: a standard measure of social and communication deficits associated with the spectrum of Autism. J. Autism Dev. Disord. **30**(3), 205–223 (2000)
8. T.M. Achenbach, L. Rescorla, Manual for the ASEBA school-age forms & profiles: an integrated system of multi-informant assessment. ASEBA (2001)
9. D.P. Wall, R. Dally, R. Luyster, J.-Y. Jung, T.F. DeLuca, Use of artificial intelligence to shorten the behavioral diagnosis of autism. PLoS ONE **7**(8), e43855 (2012)
10. E. Ruzich et al., Measuring autistic traits in the general population: a systematic review of the Autism-Spectrum Quotient (AQ) in a nonclinical population sample of 6,900 typical adult males and females. Mol. Autism **6**(1), 2 (2015)
11. D.P. Wall, J. Kosmicki, T.F. DeLuca, E. Harstad, V.A. Fusaro, Use of machine learning to shorten observation-based screening and diagnosis of autism. Transl. Psychiatry **2**(4), e100–e100 (2012)
12. M. Duda, J. Daniels, D.P. Wall, Clinical evaluation of a novel and mobile autism risk assessment. J. Autism Dev. Disord. **46**(6), 1953–1961 (2016)
13. Cognoa | Home. [Online]. Available: https://www.cognoa.com/. Accessed 9 Sept 2019
14. S.M. Kanne, L.A. Carpenter, Z. Warren, Screening in toddlers and preschoolers at risk for autism spectrum disorder: evaluating a novel mobile-health screening tool. Autism Res. **11**(7), 1038–1049 (2018)
15. A. Sarkar, J. Wade, A. Swanson, A. Weitlauf, Z. Warren, N. Sarkar, *A Data-Driven Mobile Application for Efficient, Engaging, and Accurate Screening of ASD in Toddlers* (Springer, Cham, 2018), pp. 560–570
16. Cognoa autism devices obtain FDA breakthrough status. [Online]. Available: https://www.medicaldevice-network.com/news/cognoa-autism-devices/. Accessed 9 Sept 2019
17. F. Thabtah, Machine learning in autistic spectrum disorder behavioral research: A review and ways forward. Inf. Heal. Soc. Care **44**(3), 278–297 (2019)

18. J.L. Lopez Marcano, *Classification of ADHD and non-ADHD Using AR Models and Machine Learning Algorithms* (2016)
19. F. Thabtah, Autism spectrum disorder screening, in *Proceedings of the 1st International Conference on Medical and Health Informatics 2017 - ICMHI '17* (2017), pp. 1–6
20. "UCI Machine Learning Repository: Autism Screening Adult Data Set." [Online]. Available: https://archive.ics.uci.edu/ml/datasets/Autism+Screening+Adult. Accessed 9 Sept 2019
21. M.R. Ahmed, S.M. Hasan Mahmud, M.A. Hossin, H. Jahan, S.R. Haider Noori, A cloud based four-tier architecture for early detection of heart disease with machine learning algorithms, in *2018 IEEE 4th International Conference on Computer and Communications (ICCC)* (2018), pp. 1951–1955
22. S.M.H. Mahmud, M.A. Hossin, M.R. Ahmed, S.R.H. Noori, M.N.I. Sarkar, Machine learning based unified framework for diabetes prediction, in *Proceedings of the 2018 International Conference on Big Data Engineering and Technology - BDET 2018* (2018), pp. 46–50
23. H. Abbas, F. Garberson, E. Glover, D.P. Wall, Machine learning approach for early detection of autism by combining questionnaire and home video screening. J. Am. Med. Inf. Assoc. **25**(8), 1000–1007 (2018)
24. K.K. Hyde et al., Applications of supervised machine learning in autism spectrum disorder research: a review. Rev. J. Autism Dev. Disord. **6**(2), 128–146 (2019)
25. E. Stevens, D.R. Dixon, M.N. Novack, D. Granpeesheh, T. Smith, E. Linstead, Identification and analysis of behavioral phenotypes in autism spectrum disorder via unsupervised machine learning. Int. J. Med. Inform. **129**, 29–36 (2019)
26. M. Duda, R. Ma, N. Haber, D.P. Wall, Use of machine learning for behavioral distinction of autism and ADHD. Transl. Psychiatry **6**(2), e732–e732 (2016)
27. C.E. Metz, Basic principles of ROC analysis. Semin. Nucl. Med. **8**(4), 283–298 (1978)

Intelligent Monitoring of Bearings Using Node MCU Module

Saroj Kumar, Shankar Sehgal, Harmesh Kumar, and Sarbjeet Singh

Abstract This paper discusses the application of NodeMCU to intelligent monitoring of bearings via an online method using an accelerometer to detect the vibration level. An accelerometer was used to detect the vibration level and NodeMCU module for sending a message to the end-user regarding excessive vibration levels. NodeMCU module serves as a low-cost industrial-internet-of-things setup for online monitoring of bearings. In the experiment, the set-up had a motor (to provide torque to the shaft), two ball bearings set, a shaft coupling (to connect main shaft to motor shaft), a NodeMCU (for sending a warning message), an accelerometer (to detect the vibration level), and Blynk app (to control the NodeMCU). The experimental setup was designed to detect the vibration level in time domain as well as in frequency domain and the setup was able to send the warning message in both the cases. By using this type of experimental setup, the unwanted breakdown and uncertain failure of machines due to bearing failure can be avoided. The setup helped in alerting the user about any failure in real time whenever the magnitude of vibrations exceeded its predetermined threshold limit. This experimental setup is found to be very relevant for applications in small- and medium-scale industries due to its low-cost, ease of operation, and good accuracy.

Keywords Accelerometer · Bearings · Blynk app · Industrial-internet-of-things · NodeMCU

1 Introduction

In the current scenario, the unwanted shutdown and breakdown problems in industries as well as other organizations is most common. Due to the ever-changing nature of

S. Kumar · S. Sehgal (✉) · H. Kumar
Mechanical Engineering Department, UIET, Punjab University, Chandigarh 160014, India
e-mail: sehgals@pu.ac.in

S. Singh
Mechanical Engineering Department, CCET, Jammu 181122, India

S. S. Dash et al. (eds.), *Intelligent Computing and Applications*,
Advances in Intelligent Systems and Computing 1172,
https://doi.org/10.1007/978-981-15-5566-4_15

the working environment, it is desired to put into practice an effective e-maintenance strategy for evolving proper usage of existing assets and reliability and safety [1, 2]. The control of these types of failures in real-time can be possible by the application of industrial-internet-of-things (IIoT). IIoT can be used in real-time monitoring by integrating different components like Arduino UNO, GSM SIM900A module (for sending a message to the user), and also Fast Fourier Transformation (FFT) setup. In the current paper, the NodeMCU module has been used to send the warning message in real time and this module is also cheaper than Arduino UNO and SIM GSM900A developed earlier.

Hoshi, in 2006, developed a simple test setup for monitoring of damages produced on rolling surfaces in a ball bearing-based setup of a machine tool spindle [3]. This setup was able to monitor the initiation as well as the progress of the damage occurring on rolling surfaces of the bearing and could also be used to predict the life of a bearing assembly. In 2008, Cao and Jiang [4] developed a service-oriented architecture-based system for supporting the decision-making process in the maintenance of machines. Yang et al. [5], in 2009 used the vibration signal produced by a motor along with phase currents for detecting the presence of a fault in the system and for further diagnosis purposes. Later in 2010, Zhao et al. [6] developed service-oriented architecture-based remote machine health-monitoring system which can be used to diagnose the faults present in industrial machinery located at remote places by using the concepts of web services, smart client, extensible mark-up language technologies, and visual studio. Lazar et al. [7] in 2014 used vision-based robot predictive control in flexible automatic manufacturing systems. In 2019, Goyal et al. [8, 9] developed a laser-based non-contact vibration measurement system for monitoring the condition of the machine in rea -time. It verified the effectiveness and practicality of the system. Shelkeet al. [10] used the time domain as well as frequency-domain methods for monitoring the health of ball bearing by measuring their high-amplitude dynamic responses arising out of wear, corrosion, fatigue or crack, etc., and also by extracting their features to reduce the downtime of machines. In 2017, Hassin [11] used modulation signal bispectrum method to analyze the dynamic responses. Recently, in 2019, Pesch and Scavelli [12] performed the health monitoring of active magnetic bearings using an online monitoring method.

Although several methods have been proposed and implemented for condition monitoring of various types of industrial machines. But it is observed that NodeMCU cum Blynk mobile app-based online condition monitoring of bearings assembly is not used earlier. The proposed NodeMCU-based method is low cost and easy to use due to its linking with the android-based mobile app Blynk. This paper deals with the development of such low-cost user-friendly online condition monitoring system that can be adopted easily by small- and medium-scale industries.

2 Materials and Methods

An experimental system as shown in Fig. 1 was developed which consisted of a steel table containing a motor, two bearing-set housings, and a shaft connected to a coupling. On the left-hand side bearing housing, an ADXL335 accelerometer was mounted for detection of vibration level of that bearing. The experimental setup also contained NodeMCU microcontroller, a low-cost module for sending real-time notifications. NodeMCU system can transfer real-time data to user mobile through Blynk app-based IIoT system. Initially, the experimental data were collected for correct as well as faulty bearings. The collected data were then analyzed and the program coding was done for NodeMCU and Blynk app. The entire work could be divided into two prototypes. First prototype comprised of collection and analysis of data in the time domain, while the second prototype was based on collection and interpretation of data in the frequency domain.

The whole procedure of the experiment using NodeMCU can be summarized as follows. First of all, vibration data were collected in the time domain by the accelerometer. An Arduino sketch (program) was made to convert real time data from time-domain format into the frequency domain format. Analyzed data points (time-domain format in the first prototype; frequency-domain format in the second prototype) were processed statistically to find the mean and standard deviation. A threshold value was set to keep a check for faulty bearing readings. Blynk app was developed to send a warning message to the end user in case the vibration signal exceeds the threshold value. Analyzed values were compared with a threshold value

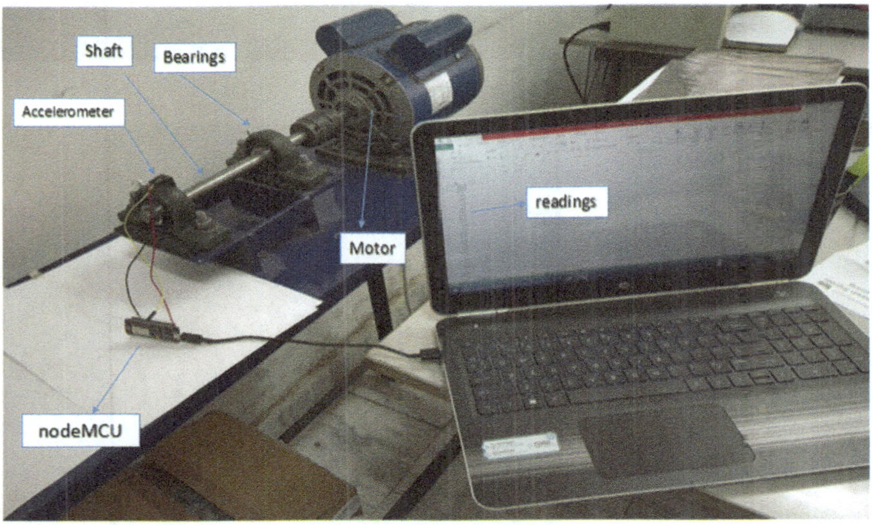

Fig. 1 NodeMCU-based system for online monitoring of vibration responses

and appropriate warning notifications were sent to the end-user through the Blynk mobile app.

3 Results and Discussion

The time-domain signals for correct and faulty bearings signals were captured consecutively by using the ADXL335 accelerometer. Blynk app was linked with the NodeMCU microcontroller and was used to send the warning messages to the end-user as shown in Figs. 2 and 3 for time domain and frequency domain, respectively. Blynk app was also used to send the warning message on the email id of the end-user as shown in Fig. 4.

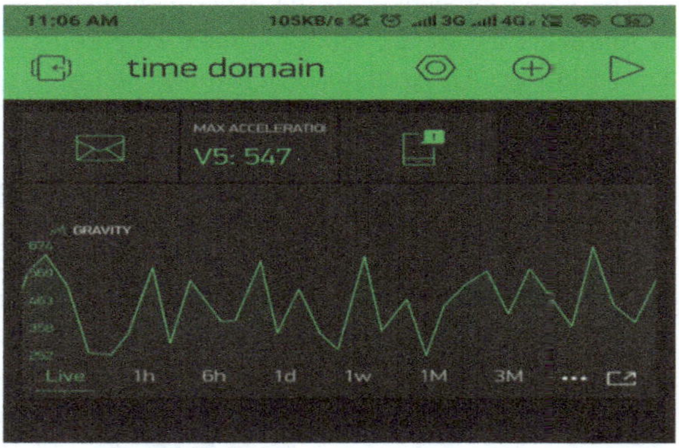

Fig. 2 Blynk App graphical user interface in time domain

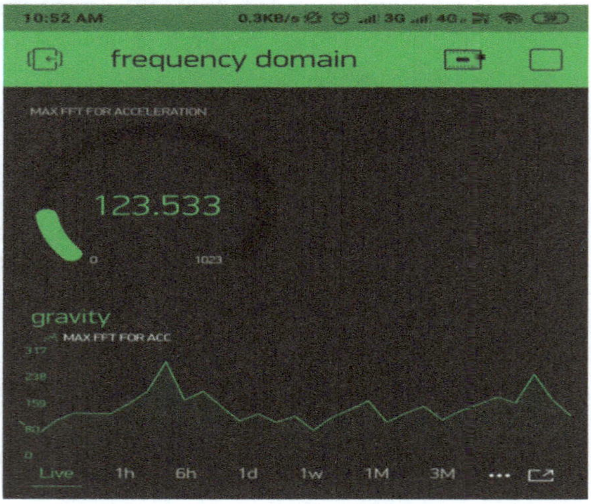

Fig. 3 Blynk App graphical user interface in frequency domain

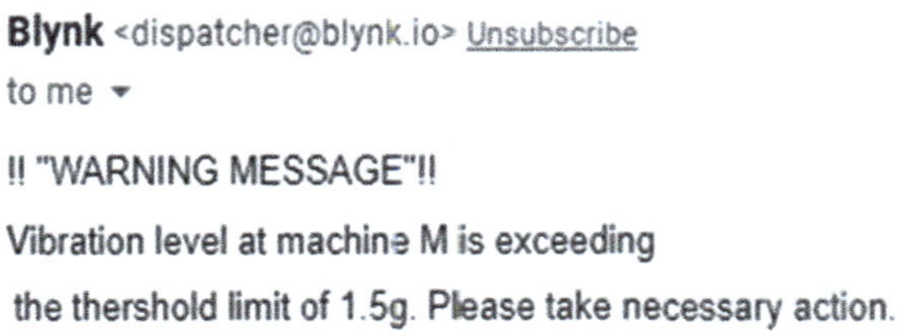

Fig. 4 Warning message of Blynk app in time domain

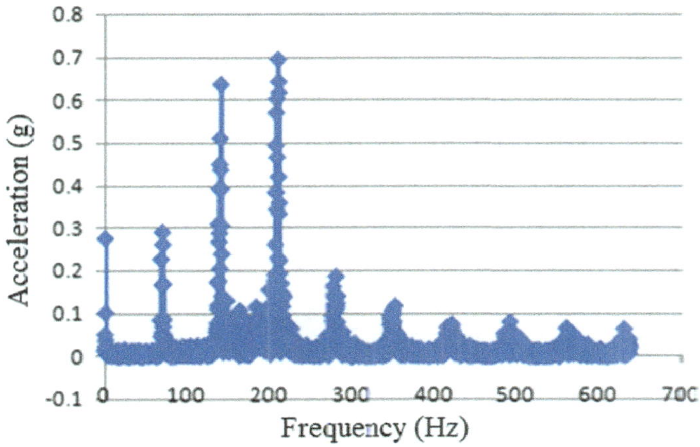

Fig. 5 Acceleration signal for correct bearing

After the study of time-domain signal, FFT function was applied to the time domain signals to convert it into frequency-domain format. The experimental plot for the FFT of a healthy bearing is shown in Fig. 5 and for the faulty bearing is shown in Fig. 6. The peak results obtained in the frequency domain for healthy bearing signals were observed to be 0.7 g and, in case of faulty bearing, it was 0.85 g. Based on these results, the threshold limit for Blynk app was set accordingly to the maximum peak signal of the correct bearing. Figure 7 shows the corresponding experimental warning message received in the inbox email of end-user.

4 Conclusion

The IIoT has significantly attracted the attention of researchers throughout the most recent couple of years. With the progress in sensor hardware technology and cost-effective materials, sensors are expected to be attached to all the items around us, so that these can interact with each other with minimal human intervention. The results of NodeMCU and Blynk app-based setup show that the proposed low-cost,

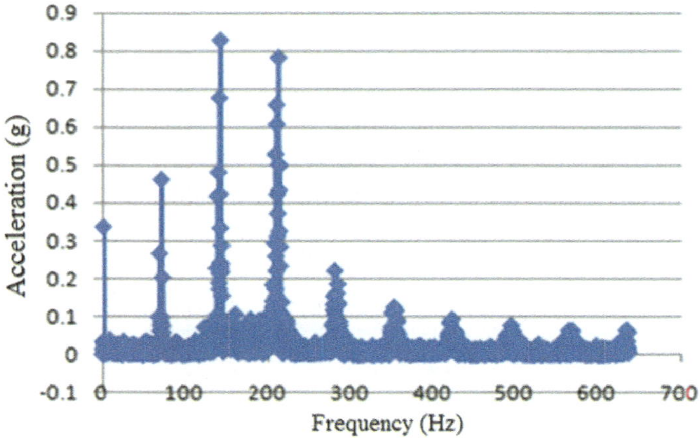

Fig. 6 Acceleration signal for faulty bearing

Fig. 7 Warning message of Blynk app in frequency domain

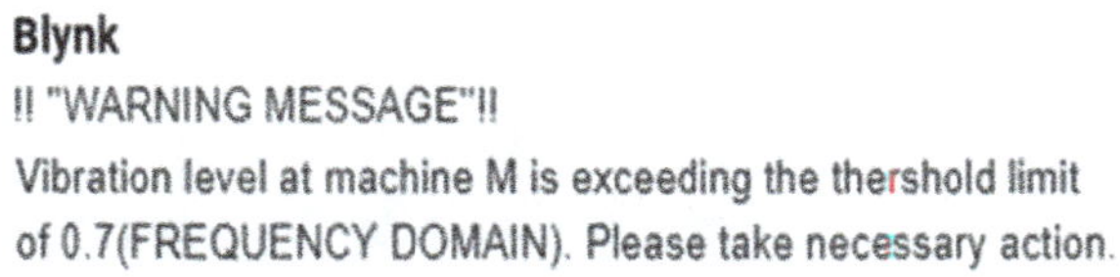

user-friendly system can be successfully utilized in online condition monitoring of the bearings for sending a warning message to the end-user regarding any unwanted high-vibration signals arising due to fault in the bearing setup. The proposed setup has the potential for small- and medium-scale enterprises to avoid unplanned shutdowns due to unexpected failure of bearings.

Acknowledgments Authors are thankful to the team members Pankaj Tiwari, Prachi Kaura, Harshita Sharma, Muskan Goyal, Raunak Sharma, and Anamika Saini, all from UIET, Panjab University, Chandigarh, India, for their support during the execution of this project.

References

1. X. Cao, P. Jiang, Development of SOA based equipments maintenance decision support system, ed. by C. Xiong, et al., in *International Conference on Intelligent Robotics and Applications. Lecture Notes in Computer Science* (Springer, Berlin, Heidelberg, 2008), pp. 576–582. https://doi.org/10.1007/978-3-540-88518-4_62
2. D. Goyal et al., Non-contact sensor placement strategy for condition monitoring of rotating machine-elements. Eng. Sci. Technol. an Int. J. **22**(2), 489–501 (2019). https://doi.org/10.1016/J.JESTCH.2018.12.006
3. D. Goyal et al., Optimization of condition-based maintenance using soft computing. Neural Comput. Appl. **28**(Suppl 1), S829–S844 (2017). https://doi.org/10.1007/s00521-016-2377-6

4. D. Goyal, B.S. Pabla, Development of non-contact structural health monitoring system for machine tools. J. Appl. Res. Technol. **14**(4), 245–258 (2016). https://doi.org/10.1016/J.JART.2016.06.003
5. D. Goyal, B.S. Pabla, The vibration monitoring methods and signal processing techniques for structural health monitoring: a review. Arch. Comput. Methods Eng. **23**(4), 585–594 (2016). https://doi.org/10.1007/s11831-015-9145-0
6. O. Hassin, et al., Monitoring mis-operating conditions of journal bearings based on modulation signal bispectrum analysis of vibration signals, in *First Conference on Engineering Sciences and Technology* (Libya, 2018), pp. 509–517. https://doi.org/10.21467/proceedings.4.18
7. T. Hoshi, Damage monitoring of ball bearing. CIRP Ann. **55**(1), 427–430 (2006). https://doi.org/10.1016/S0007-8506(07)60451-X
8. C. Lazar, et al., Vision-guided robot manipulation predictive control for automating manufacturing, ed. by T. Borangiu, et al., in *Service Orientation in Holonic and Multi-Agent Manufacturing and Robotics. Studies in Computational Intelligence* (Springer, Cham, 2014), pp. 313–328. https://doi.org/10.1007/978-3-319-04735-5_21
9. A.H. Pesch, P.N. Scavelli, Condition monitoring of active magnetic bearings on the internet of things. Actuators **8**(17), 1–13 (2019). https://doi.org/10.3390/act8010017
10. S.V. Shelke et al., Condition monitoring of ball bearing using vibration analysis and feature extraction. Int. Res. J. Eng. Technol. **3**(2), 361–365 (2016)
11. Z. Yang, et al., A study of rolling-element bearing fault diagnosis using motor's vibration and current signatures, in *7th IFAC Symposium on Fault Detection, Supervision and Safety of Technical Process* (IFAC, Barcelona, Spain, 2009), pp. 354–359. https://doi.org/10.3182/20090630-4-ES-2003.0307
12. F. Zhao et al., SOA-based remote condition monitoring and fault diagnosis system. Int. J. Adv. Manuf. Technol. **46**(9–12), 1191–1200 (2010). https://doi.org/10.1007/s00170-009-2178-5

Image Denoising Using Various Image Enhancement Techniques

S. P. Premnath and J. Arokia Renjith

Abstract The main aim of the image enhancement technique is to process any image given as input and to obtain the resultant outcome more accurately than the existing image. The level of accuracy of an image can be restored in different forms using image enhancement techniques. The choices of choosing different image enhancement techniques may vary depending upon the quality of the picture, task, and atmospheric conditions. The different algorithms used for enhancement and the concepts are discussed in this paper. The frequency and spatial domain of the image can be enhanced by using these processes.

Keywords Image enhancement · Spatial domain enhancement · Frequency domain-based enhancement · Histogram equalization

1 Introduction

Image enhancement is the most predominant and not easy technique in image research. The image enhancement technique intends to provide a high-quality clear image such as satellite images, real-life photographs, medical images experiencing low disparity, and unpleasant noise. To increase the quality of an image, it is important to enhance the contrast and remove the noise (Fig. 1).

Different techniques are available to refine the standard of the digital image without diminishing it. The two broad categories, which are utilized to intensify the clarity of the image are as follows:

(a) Spatial Domain

S. P. Premnath (✉)
Department of Electronics and Communication Engineering, Sri Krishna College of Engineering and Technology, Coimbatore, India
e-mail: premnathsp@yahoo.com

J. Arokia Renjith
Department of Computer Science and Engineering, Jeppiaar Engineering College, Chennai, India
e-mail: arokiarenjith@gmail.com

S. S. Dash et al. (eds.), *Intelligent Computing and Applications*,
Advances in Intelligent Systems and Computing 1172,
https://doi.org/10.1007/978-981-15-5566-4_16

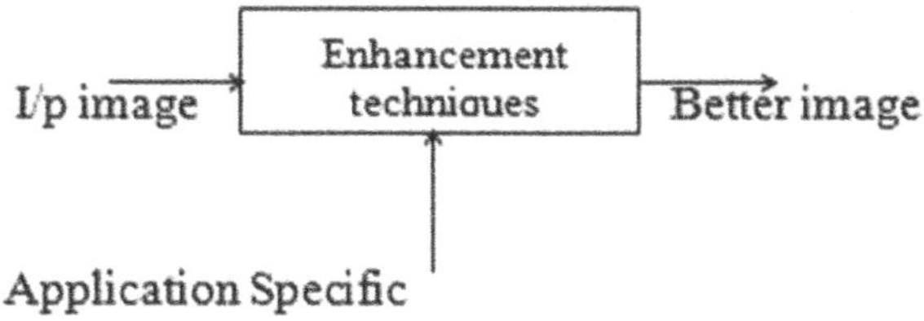

Fig. 1 Enhancing technique

(b) Frequency Domain.

The spatial domain technique defines that the picture or an image is composed of the grouping of pixels. In this method, the quality of an image can be enriched by directly operating on the pixels of an image [1]. The expression for the spatial domain process can be given as follows:

$$g(x, y) = T[f(x, y)]$$

where $g(x, y)$ denotes the given input image, in this g indicates the gray-level value and the coordinates are denoted by (x, y). T be the transformation applied on the input image, to produce a newly improved image $f(x, y)$. The two different ways in which the image can be spatially enhanced are as follows:

(a) Point Processing
(b) Neighborhood Processing.

The next approach to improve the quality of the image is based on the frequency-domain method. In this approach when an image is given as input, the first stage is to convert the input image into a frequency domain, during the conversion of the image into the frequency domain, Fourier transform of the image is quantified first [2]. In order to retrieve the resultant image, the inverse Fourier transform is performed over Fourier transform. In order to enrich the clarity of an image, the gain frequency-domain method is categorized into three types. They are as follows: Image smoothing, Image Sharpening, and Periodic noise reduction (Fig. 2).

To enrich the quality of an image (i.e., based on image brightness, contrast, etc.), the above-said image enhancing techniques are performed. The resultant image will

Fig. 2 Effect of image enhancement

be more improved in quality due to the application of transformation function in the given input values.

2 Point Processing Operations

Point processing operation is one of the simplest spatial domain operations that deals with the individual pixel intensity values. The intensity values are altered using transformation techniques as per the requirement [3]. Gray level at any point of the image plays an important role in magnification of the image at any point (Fig. 3).

Point processing can be represented as follows:

$$S = T(r)$$

where S represents the pixel value after processing and r denotes the original pixel value of an image.

(A) **Negative Transformation**

The primary and simple operations in image processing are to compute the negative image. The Gray (or) White features deeply in a surrounding mass of an image in the darkish point can be magnified through negative images. The negative transformation can be defined by

$$S = (L - 1) - r$$

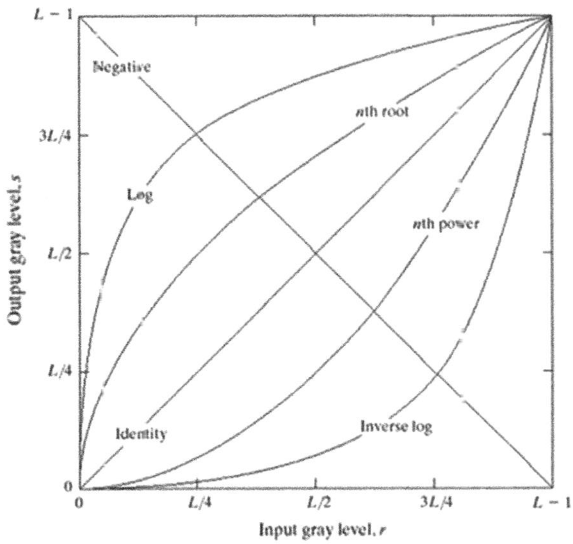

Fig. 3 Some basic gray-level transformation function

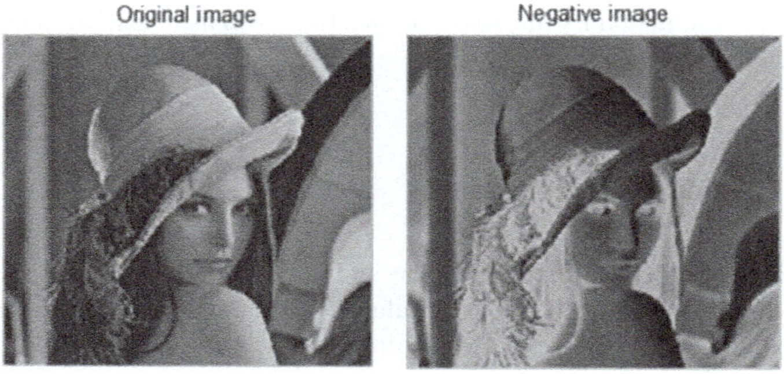

Fig. 4 Effect of negative transformation

Fig. 5 Identity transformation representation

Image intensity level in negative transform lies between in the range of [0, $L-1$]. $\mathrm{L}-1$ denotes the maximum pixel values; r is the pixel value of an image [4] (Fig. 4).

(B) **Identity Transformation**

In identity transformation, the point of the pre-image and the image pixel is same, the point is united (Fig. 5).

If all the points of pre-image and image are same, the entire figure is united.

(C) **Log Transformation**

In certain cases, the effective variation between upper and lower limits of an image may be greater in size than the desired device capacity. Due to the variation in the image, low-pixel values get masked. To overcome this, effective variation needs to be compressed, i.e., log transformation method enlarges the values of darkish pixels and compresses the bright pixel [5]. The high quality of an image compression ratio can be obtained by the log operator. So, this log operator is used to compress the effective range of an image. Log transformation can be given as follows:

$$S = C.\log(1 + |r|)$$

Normalization constant can be denoted by C and r denotes the input intensity (Fig. 6).

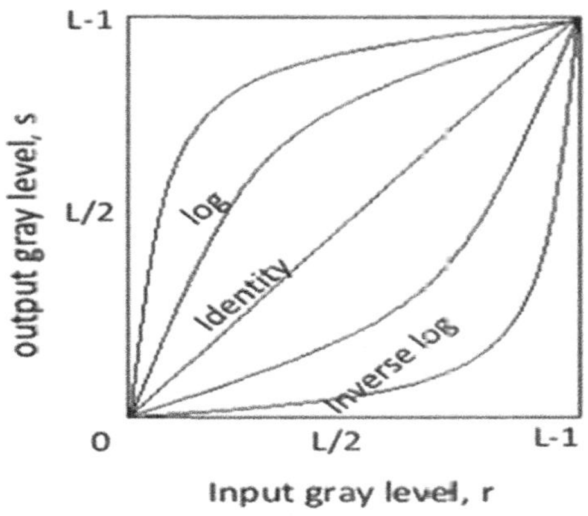

Fig. 6 Log transformation representation

The shape of log curve indicates the variation of low gray-level pixel values in the input image to a bright value in the outcome and vice versa.

(D) **Contrast Stretching**

Due to a lack in lighting effect, effective range (or) improper line aperture consequence in the formation of low-contrast images. In order to magnify the image, the effective range of the gray level needs to be enhanced. So, the contrast stretching technique is used to enrich the quality of an image [6]. Contrast stretching can be represented as follows:

$$k = \frac{1}{1 + (p + r)^z}$$

where r represents the values of a given input image, k denotes the value of the resultant image, and p be the thresholding value and be the slope (Fig. 7).

The above diagram shows the outcome of the variable z. If $z = 1$, then stretching becomes edge transformation. If $z > 1$, then the range of intensity value of an image is defined by the curve which is smoother, and when $z < 1$, the transformation makes the negative and also stretching.

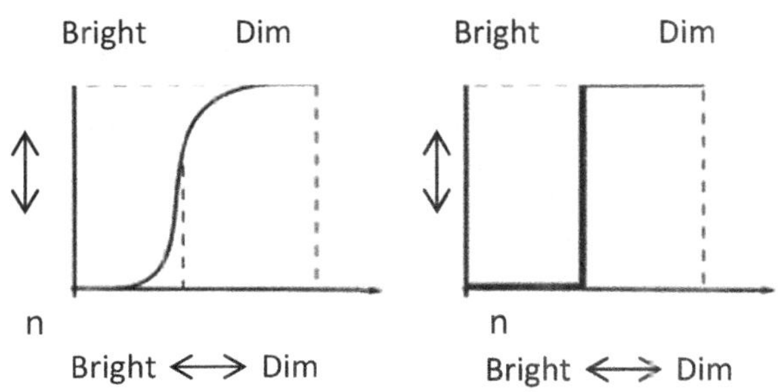

Fig. 7 Contrast stretching

(E) Intensity-Level Slicing

Intensity-level slicing method is used to represent the specific range of the image. For example, if the figure does not contain the values below 40 (or) above 255, then the contrast of the image decreases [7]. If it is remapped by increasing the brightness of the image in the range of [0, 255], then the contrast stretching of the image can be increased.

Figure 8a shows the variation features a range of gray levels and makes other range levels as a less desirable state. Figure 8b shows the variation features range and it maintains a further level of an image.

$$v = \begin{cases} L, & a \le u \le b \\ 0, & \text{otherwise} \end{cases} \quad v = \begin{cases} L, & a \le u \le b \\ u, & \text{otherwise} \end{cases}$$

(F) Bit Plane Slicing

This transformation is useful in determining the number of visually significant bits in an image. Consider a 256*256 image with 256 gray levels. Suppose each pixel is represented by 8 bit, it is desired to extract the nth most significant bit (Fig. 9).

In this transformation, higher order bits contain visually sufficient data and low-order bits contain suitable details of an image.

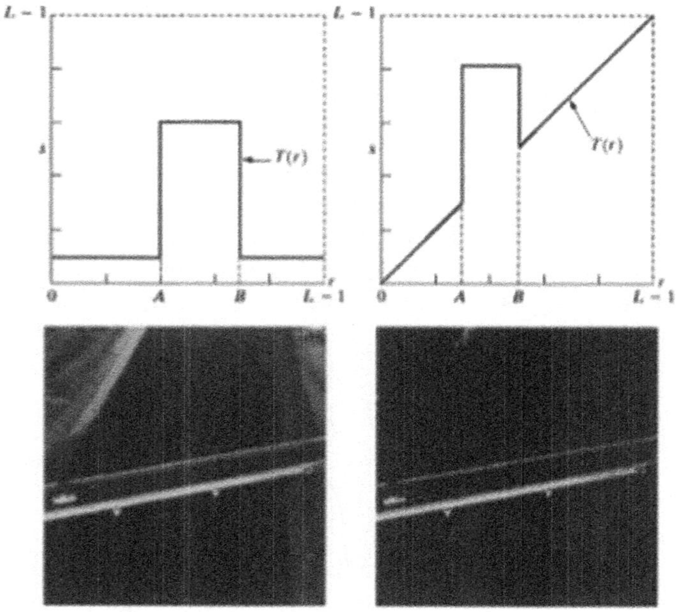

(a) (b)

Fig. 8 Specific range of gray level

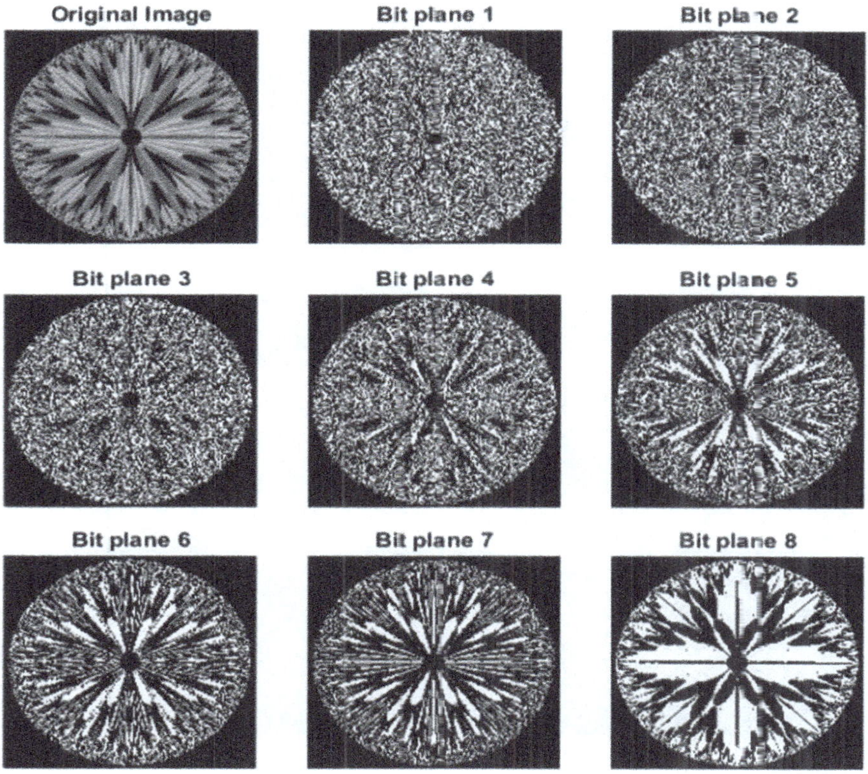

Fig. 9 Bit plane slicing

(G) Power Law Transformation

The quality of any image can be retained by applying power law transformation. Power law transformation can also be called as Gamma Correction. It can be used to rectify the errors that occurred while taking the photo and processing the image further.

The power law transformation can be given as follows:

$$f(x, y) = c.g(\mathrm{x}, \mathrm{y})^{\mu}$$
$$s = c.r^{\mu}$$

where c and μ are the non-negative constants. Figure 10a is the real image, for a positive constant $c = 1$, then the positive γ is 3, 4, 5, respectively, for the image B, C, and D.

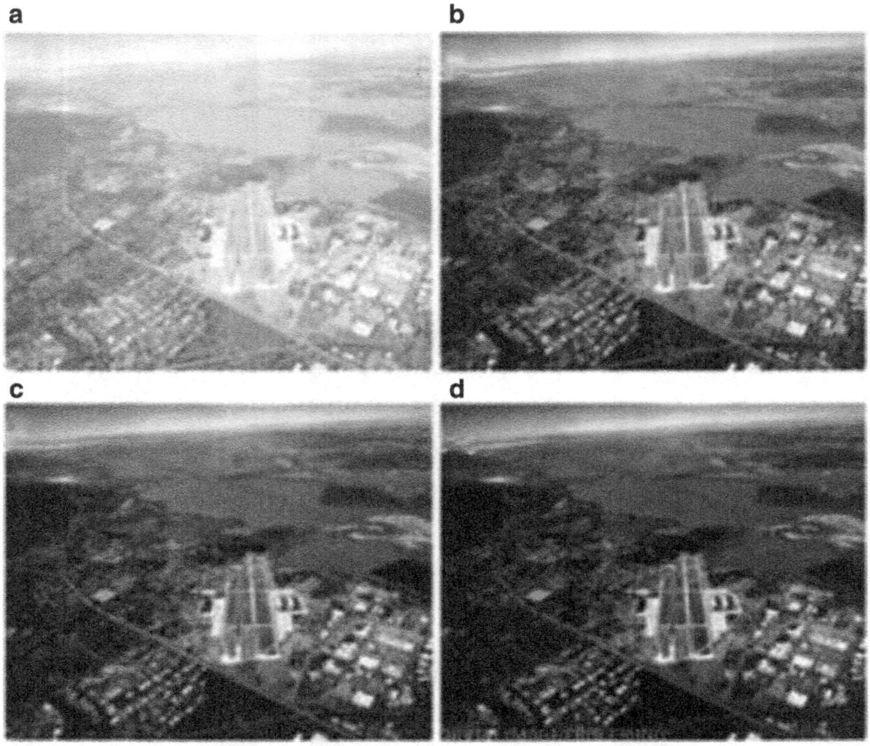

Fig. 10 Gamma correction

3 Neighborhood Pixel Processing

Neighborhood pixel processing technique is one of the spatial domain approaches to enrich the quality of an image. In this technique, one pixel is considered at a time and it is modified accordingly. The adjacent pixel values are also taken into consideration. Pixel value can be changed based on 8 neighbors [1]. The quality of the image can be increased by using neighborhood pixel processing when compared with point processing.

Figure 11a shows the original image and Fig. 11b after filtering the image.

(A) Spatial Averaging And Spatial Low-Pass Filtering

Spatial averaging method is used here to reduce the noise produced by the isolated and random pixels. The pixel value by range from small to high range value. In order to retain the image, the small range values are weighted to the average of adjacent pixel values.

Fig. 11 **a, b**

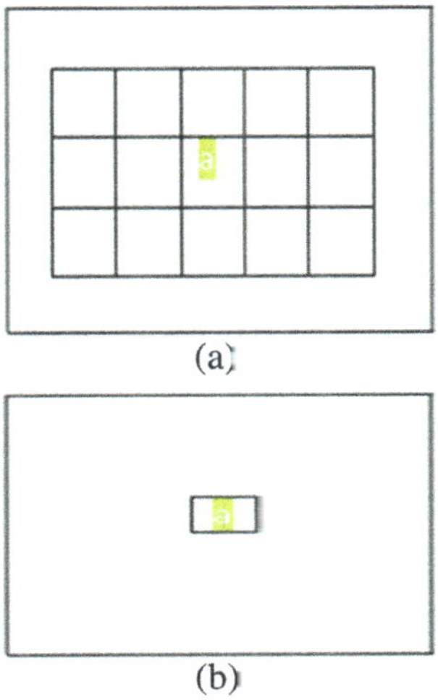

$$Ts(s, z) = \sum \sum_{(i,j)\in I} a(i, j)b(s - i, z - j)$$

where *b*(*s*, *j*) represents the output image and *T*(*s*, *z*) represents the input image. *W* is the windowing technique, *a* (*i*, *j*) is the filter weights. Low-pass filtering technique is used to reduce the noise present in the pixels.

(B) Directional Smoothing

To protect edges from blurring while smoothing, spatial averages are calculated in several directions, and the direction giving the smallest changes before and after filtering is selected.

$$s(k, r : \theta) = \frac{1}{N_\theta} \sum_{(U,V)\in Z_\theta} \sum n(k - u, r - v)$$

The direction θ is found in a manner such that $|n(k, r) - s(k, r : \theta|$ is minimum.

(C) Directional Smoothing

Generally, there is a desire to enlarge the specified location of the image [8]. This requires taking an image and displaying it as a larger image.

- **Zero-Order Hold**: Zero order is performed by considering the pixel values of an image, i.e., the previous pixel values are also used repeatedly to perform this function. In this, (*n***n*) image size can be varied to (2*n*)*(2*n*). Then convolve with

$$H = \begin{bmatrix} 1 & 1 \\ 1 & 1 \end{bmatrix}$$

$K\,(i, j) = c\,(x, y)$ x and y are given as $int\left(\frac{m}{2}\right)$, respectively.

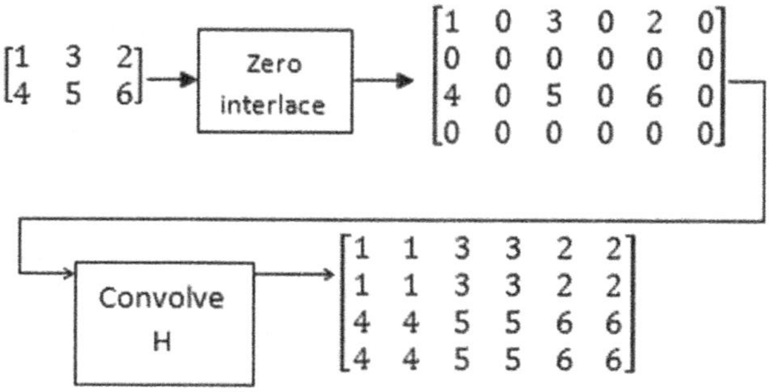

The straightline is fitted between pixels along rows and columns. It can be obtained by interlacing the image by rows and columns of zero and convolve with

$$\mathrm{M} = \begin{bmatrix} \frac{1}{4} & \frac{1}{2} & \frac{1}{4} \\ \frac{1}{2} & 1 & \frac{1}{2} \\ \frac{1}{4} & \frac{1}{2} & \frac{1}{4} \end{bmatrix}$$

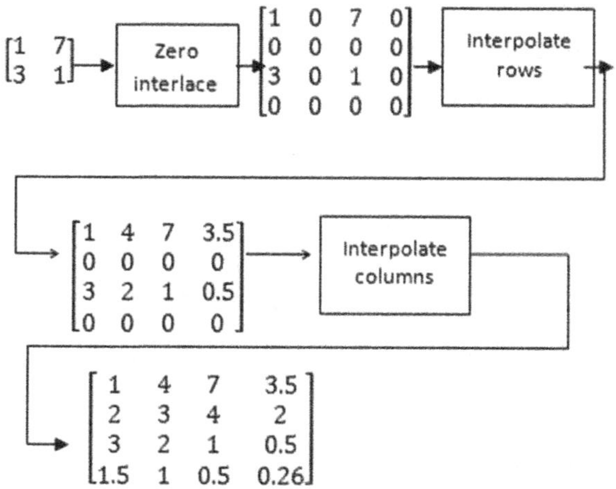

4 Histogram Processing

Histogram processing is the act of reconstruction of an image by modifying its histogram [9]. The gray of a histogram digital image lies in the range of $[0, L - 1]$. The discrete function can be given as follows:

$$h(\gamma_k) = n_k$$

where r_k denotes the gray level of an image and n_k denotes the pixels of an gray-level image r_k. Histograms regularly produce the new pixel values of an image by shifting the values of the existing image. The normalized histogram is given by, $p(\gamma_k) = \frac{n_k}{n}$, where $K = 0, 1, \ldots, L - 1$.

(A) **Histogram Equalization**

Histogram equalization is a simple method used in enhancing the image. In the enhancing processes, histogram equalization is used to reduce the gray level pixel values, so that the brightness level of an image can be enhanced. The resultant image will be more accurate.

(B) **Frequency-Domain Techniques**

In frequency-domain approach, Fourier transform of an image can be determined first by converting an image to a frequency domain. In order to retrieve the desired quality of an image in the resultant, the intensity pixel values of the image are modified by taking inverse Fourier transform (Fig. 12).

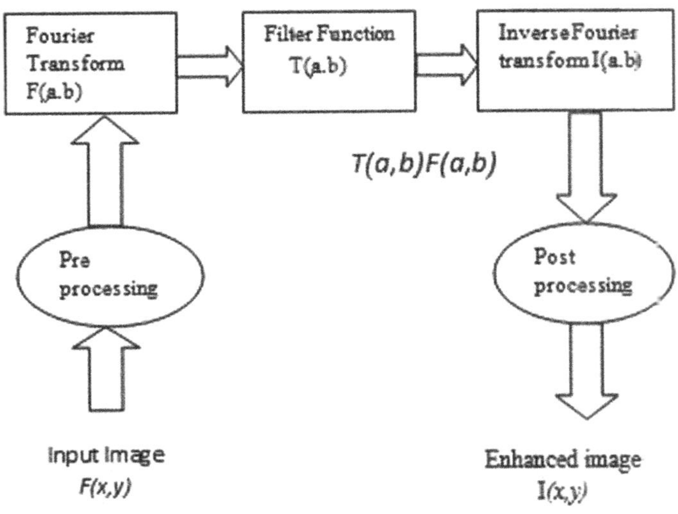

Fig. 12 Frequency domain filtering operation

The transfer function can be given as follows:

$$I(a, b) = T(a, b)F(a, b)$$

where *I(a, b)* is the enhance image, *T(a, b)* represents the filter function, and *F(a, b)* is the original image before processing. Since it depends on the frequency-domain approach, parameters such as high-frequency components can be enhanced easily.

(C) **Low-Pass Filtering**

The simple filtering techniques used in image enhancement are low-pass filtering. The low-pass filter reduces noise content present in the image. Noise present in each pixel may vary and it affects the quality of the image. Enabling low-pass filter, it allows selectively smooth image. Due to this, resultant image obtained is a quality image [10]. The cutoff frequency of the low-pass filter can be represented as follows:

$$H(i, j) = \begin{cases} 1, \ if \sqrt{(i^2 + j^2)} \leq r_0 \\ 0, \ if \sqrt{(i^2 + j^2)} \geq \gamma_0 \end{cases}$$

The cutoff frequency γ_0 of the ideal low-pass filter denotes the amount of noise gets suppressed. Contrast and blurring of the image can be reduced when the value of the γ_0 tends to reduce.

(D) **High-Pass Filtering**

Image enhancement can be performed by various image processing tasks. Among the various filtering techniques, high-pass filtering makes a more smoothened image on resultant by allowing the high frequency to pass and attenuates the low frequencies.

$$H(i, j) = \begin{cases} 1, \ if \sqrt{(i^2 + j^2)} \leq r_0 \\ 0, \ if \sqrt{(i^2 + j^2)} \geq \gamma_0 \end{cases}$$

5 Conclusion

Image enhancement algorithms offer various enhancing approaches for retrieving the images to obtain quality as well as clarity of an image. These image enhancement techniques in spatial domain and frequency domain have been successfully discussed. Depending on the quality of the image and noise with which it is corrupted by varying the methods, visual quality can be improved. The computational cost of enhancement algorithms is not discussed here. It plays a major responsibility in picking accurate algorithms for different conditions. The combination of the above algorithms can be used to produce a more enhanced image.

References

1. R. Maini, H. Aggarwal, *A Comprehensive Review of Image Enhancement Techniques*
2. A.K. Jain, *Fundamentals of Digital Image Processing* (Prentice Hall, Englewood Cliffs, NJ, 1989)
3. S.M. Pizer et al., Adaptive histogram equalization and its variations. Comput. Vis. Graph. Image Process. **39**, 355–368 (1987)
4. R.C. Gonzalez, R.E. Woods, *Digital Image Processing* (Prentice Hall, Upper Saddle River, New Jersey, 2002)
5. R. Jain, R. Kasturi, B.G. Schunck, *Machine Vision* (McGraw-Hill International Edition, 1995)
6. H. Lidong, Z. Wei, W. Jun, S. Zebin, *Combination of Contrast Limited Adaptive Histogram Equalisation and Discrete Wavelet Transform for Image Enhancement* (2014)
7. H.C. Andrews, Digital image restoration: a survey. Computer **7**(5), 36–45 (1974)
8. Y. Yao, B. Abidi, M. Abidi, *Digital Imaging with extreme Zoom: System Design and Image Restoration*
9. R. Hummel, Histogram modification techniques. Comput. Graph. Image Process. **4**, 209–224 (1975)
10. G.L. Anderson, A.N. Netravah, Image restoration based on a subjective criterion. IEEE Trans. Syst. Man Cybern. **SMC-6**, 845 (1976)

Energy Consumption Analysis and Proposed Power-Aware Scheduling Algorithm in Cloud Computing

Juhi Singh

Abstract Cloud computing is a large-scale paradigm using remote servers hosted on the internet. Servers over the data centers provide services in the form of SaaS, PaaS, and IaaS. The concept of cloud computing is the virtualization of resources that provides services. Cloud is an infinite resource pool along with the availability of these resources to multiple users. The energy consumption in a cloud environment is due to the number of parameters that have fixed values or it is variable. In the paper, we had rectified a model for energy consumption and an algorithm is proposed for the power awareness scheduling algorithm. Also, the paper shows a relationship between workload, utility, power, and energy consumption is presented as a model.

Keywords Cloud computing · Cost · Energy consumption · Green computing · Power · Resource management · Scheduling · Utility · Workload

1 Introduction

Cloud Computing is an environment that provides services on users' requests as per demand. From the consumer's point of view, the cloud is a scalable service and consumers can get as many as resources as per the usage. So, from the provider's point of view, it is important to manage resources with both points that it can serve the consumer by providing resources in a better way and gets profit without compromising the quality of services or violating the SLA. There are a number of parameters taken into consideration to provide better services. Complications of these parameters may lead to a penalty, time limitation, unavailability, less utilization, etc. The provider needs to first keep the track with the limitations of time and energy resources availability. Also, services should be optimized with minimum cost.

Various resources like compute, infrastructure, applications, storage, network, platform, and database are used in the cloud environment. The resources can be classified into two categories.

J. Singh (✉)
Faculty of Computer Science and Engineering, SRMU, Lucknow, UP, India
e-mail: juhisingh.srmcem@gmail.com

S. S. Dash et al. (eds.), *Intelligent Computing and Applications*,
Advances in Intelligent Systems and Computing 1172,
https://doi.org/10.1007/978-981-15-5566-4_17

Physical Resources → Computer, Disk, Database, Network, Scientific Instruments.
Logical Resources → Execution, Monitoring, Communicate applications.

These resources need to manage efficiently and in an optimized way over the cloud environment. Cloud computing has concepts of virtualized resources. Resource management is done with the objectives to reduce overhead, to reduce latency, and to improve throughput. Resource management also includes scalability, optimal utility, specialized environment, and cost-effectiveness. Several challenges of resource management can be categorized into broad categories of hardware and software.

1. Resource Management—Challenges (Hardware resources)
   - CPU (central processing unit)
   - Memory
   - Storage
   - Workstations
   - Network elements
   - Sensors/actuators
2. Resource Management—Challenges (Logical resources)
   - Operating system
   - Network throughput/bandwidth
   - Energy Consumption
   - Load balancing mechanisms
   - Protocols
   - Delays
   - Security of Information
   - Application Programming Interfaces

Taking into consideration these points, the term resource management indicates the methods or steps that control the capabilities provided by cloud resources. Resource management also includes how services can be made available in the best way to other entities; these entities could be users, applications, services in an optimized and efficient manner [1]. It can be efficient in energy optimization, maximizing in profit, efficient in providing the best quality or efficient with respect to SLA. Cloud service providers have huge data centers such as Google, Microsoft, Yahoo, Amazon, and IBM. These resources are outsourced from these providers. It is calculated that the server consumes 0.5% of total electricity usage over the world [2]. There are different approaches and methods done by researchers for resource management.

Resource provisioning → Resources are assigned to the workload given.
Resource allocation → Resources are distributed among the workload
Resource Adaptation → Resources adjust themselves automatically to fulfill the user's workload requirements.

Resource Mapping → Resources that are required by the workload and resources that are provided by the cloud infrastructure are co-related correspondingly.

Resource Modeling → A structure is designed to find the resource requirements of a workload with attributes.

Resource Estimation → Approximate prediction of resources required is done for the execution of workload.

Resource Discovery → Resources identified for execution that are available.

Resource Brokering → A third party agent negotiate the resources for the assurance of resource availability.

Resource Scheduling → A proper schedule of resources and events is determined to depend on the number of factors, i.e., its duration, resources allocated, predecessor activities, and predecessor relationships [3].

Green Computing can be taken as advanced scheduling schemas to reduce energy consumption, power-aware, thermally aware, data center designs to reduce Power Usage Effectiveness, cooling systems rack design. There are various reasons to introduce Green Data Centers for economic research with the aim to reduce costs, since many facilities are at their peak operating stage and we need to reduce cost at feasible availability of resources without expanding new power sources. Also, in terms of energy, the majority of energy sources are fossil fuels and a huge volume of CO_2 emitted each year from power plants. Thus, green computing is the advanced technology that works on power awareness methodology (Fig. 1).

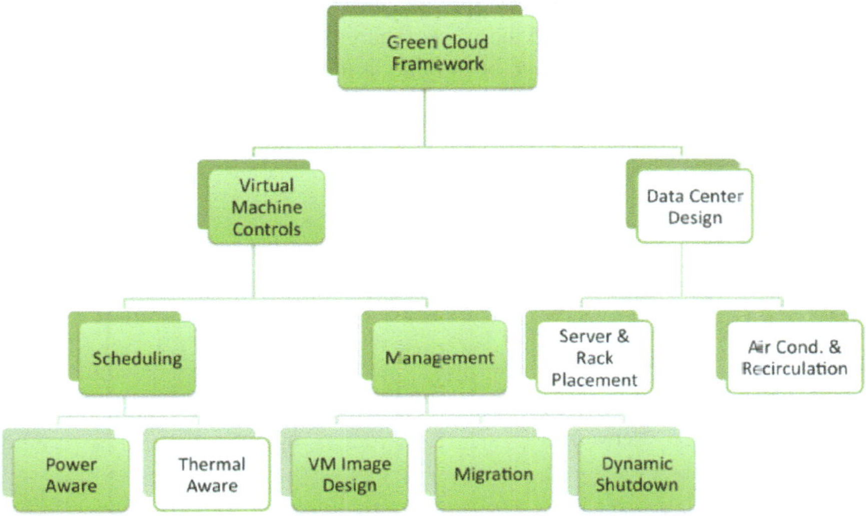

Fig. 1 Green cloud framework

2 Related Work

Management of resources over the cloud environment affects a lot on power consumption so we need to use the resources more efficiently. Various scheduling algorithm in the cloud environment is done for a heterogeneous system to optimize the cost by using the resources more efficiently [4–6]. Data center plays a vital role in power consumption. Dynamic methods are used for managing the utilization of energy and resources over data centers [7–9]. Initially, the work was done for a single user to optimize energy consumption in the research paper [7]. Further, the research continued for power control by adjusting the frequency at the processing level [8]. Later, the research work was done to find relation power and frequency for optimal power allocation [9].

Further to make data centers eco-friendly and to save energy, a new technique came into role, i.e., Green Computing. Moving toward optimal power consumption as introduced in green computing advance research, work is proposed as Powernap, i.e., an approach of energy conservation, where the migration of servers is done between active state and a near-zero power idle state [10]. A unique state of server called "Somniloquy" is introduced to save power more effectively [11]. Li et al. proposed a cost model based on total cost and utilization cost [12]. There was a limitation in the work that the calculation was based on a single hardware component. So, on further research work, Jung et al. focused on physical host power consumption [13]. The work–energy consumption had no effect on workload and hardware specifically. Li et al. introduced a new infrastructure for CPU utilization based on dynamic console VMs and Verma et al. enhanced the work, where the characteristics of VMs were used [14–16]. Various works have been done in terms of energy optimization and power consumption [17–19] (Fig. 2).

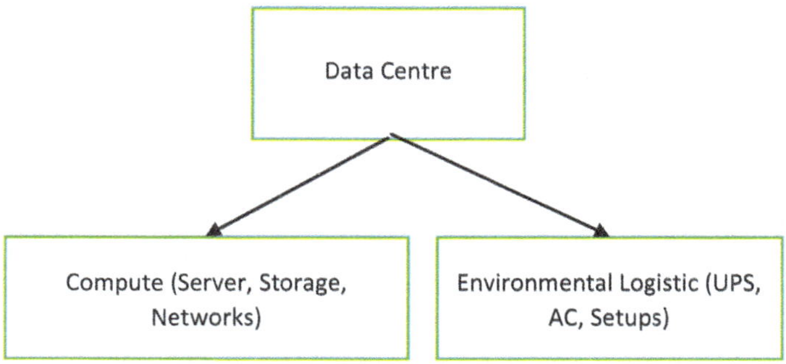

Fig. 2 Datacenter resources

3 Energy Consumption Analysis Model

In other research papers [20], it is discussed that total energy consumption is taken as a summation of all fixed energy consumed and variable energy consumed referred to as E_{fix} and E_{var}. The total consumption defined as E_{total} is shown by the formula given below:

$$E_{\text{total}} = E_{\text{fix}} + E_{\text{var}} \tag{1}$$

where, in the cloud environment, variable energy consumption can be classified in storage, computation, communication resources, and other resources.

1. E_{storage} referred to as the energy consumption of storage resources or memory.
2. E_{comp} referred to as the energy consumption of computation resources or servers.
3. E_{comm} referred to as the energy consumption of communication resources or networks.
4. E_{other} referred to as other resources as environmental logistic or UPS. AC, Setups.

$$E_{\text{total}} = E_{\text{fix}} + \left(E_{\text{storage}} + E_{\text{comp}} + E_{\text{comm}} + E_{\text{other}}\right) \tag{2}$$

For task scheduling algorithm, we can count total energy consumption as energy consumed by all tasks. And, energy consumed by all tasks is not just the addition of all tasks. Scheduling overhead will generate energy consumption referred to as E_{sche}. So further for scheduled task total energy consumed can be taken as follows:

$$E_{\text{total}} = E_{\text{fix}} + \left(\sum E_{\text{storage}} + \sum E_{\text{comp}} + \sum E_{\text{comm}} + \sum E_{\text{other}}\right) - E_{\text{sche}} \tag{3}$$

where $\sum$ represents the summation of each task 1 to a number of task n. For each task, energy consumption is tightly bound with workload, so we say energy consumption is directly proportional to the workload of each task, and the energy consumption increases with the number of processes in execution with the task [21].

4 Power Model

In concern of power consumption, data center are classified into two categories:

1. Compute
2. Environmental Logistic

To calculate energy, the information requires power efficiency and task resource demands of each VM, as in general, power is defined as the energy consumed per unit time. And, resource demand is taken static parameters to compile a task, i.e., size of data through a disk, number of instructions, size of data through network, and job id where it is generated from. Contrast to this dynamic parameter is taken as energy consumption and execution time of the task [22]. Power and energy of CPU in the cloud are modeled as follows:

$$P(u) = k \times P\text{max} + (1 - k) \times P\text{max} \times u \tag{4}$$

where P_{max} is the maximum power consumed when the server is fully utilized, u is the CPU utilization, and k is the fraction of power consumed by the idle server. The utilization of CPU changes with time since workload changes. Thus, the CPU utilization is taken as a function of time which is represented by $u(t)$. Therefore, the total energy consumed by a physical host is taken as an integral of the power consumption function over a period of time:

$$E = \int P(u(t)) \tag{5}$$

Energy consumption by hosts in data centers includes that of disk storage, CPU, and network interfaces. The dynamic power consumption of cloud data centers is mainly generated by the workload on each running server, while the main sources for power consumption are resource demands of tasks that drive server workloads. From the above-derived formulas, we find that there is a tight bonding between CPU utilization and power consumption of system [22].

5 Proposed Algorithm

Power-Aware Scheduling of VM

1. Initialize Buffer
2. Start the loop till resource pool and assign processing elements.

```
for i = 1 to i <=|pool| do
pe_i = num cores in pool_i
end for
```

3. While queue is not empty calculate power and energy

```
while(true)
for i = 1to i <=|queue| do
  vm = queue_i
  for j = 1 to j <=|pool| do
    if pe_j >=1 then
    if check capacity vm on pe_j then
```

schedule vm on pe_j
$P(u) = k \times P_{max} + (1 - k) \times P_{max} \times u$
$E = \int P(u(t))$
pe_j-1
end if
end if
end for
end for
wait for internal t
end while

6 Cost and Penalty Cost

In the virtualization technique, physical resource is assigned a logical name, and when there is a request made physical resource is provided a pointer Resource management helps to maintain SLA i.e., a contract between service provider and service consumer. Resources are provided on availability. KPI of SLA are:

- Uptime
- Downtime
- Inbytes.

Cloud computing finds the availability on two parameters, namely service uptime and service downtime with a formulation

$$\text{Availability} = 1-(\text{service downtime}/\text{service uptime}) \tag{6}$$

For example
Service uptime = 128 min
Service downtime = 32 min
Availability = 1 – (32/128) = 0.75

Cloud is cheaper when utility premium is less than the ratio of peak demand and average demand. Since demand is an essential parameter related to utility, we can calculate utility pricing details as follows:

$$\begin{aligned} CT &= U \times B \times Dtdt \\ &= A \times U \times T0\ B \times T\ BT \\ &= P \times B \times T \end{aligned} \tag{7}$$

While owning resources, there is a penalty whenever your resources do not match the instantaneous demand, which leads to pay for unused resources, or suffer the penalty of missing service delivery that is formulated as follows:

$D(t)$—Instantaneous Demand at time t
$R(t)$—Resources at time t
Penalty Cost $\alpha|D(t) - R(t)|dt$
If demand is flat, penalty $= 0$
If demand is linear periodic provisioning is acceptable.

7 Conclusions and Future Work

Energy consumption in the cloud computing model is analyzed for static and dynamic consumption in the cloud environment. Dynamic consumption of energy at data centers is optimized. The optimized energy consumption improves energy efficiency according to workload at datacenters. The energy consumption model is analyzed and derived the relationship between workload, utility, power, and energy consumed. In this paper, the proposed power scheduling algorithm for a virtual machine in the cloud environment is elaborated. Also, the papers represent the derivation of the resource management parameters, i.e., availability and utility. Resource management maintains the service-level agreement to reduce the penalty cost. Although a set of parameters is considered to model energy consumption, there are a number of environmental parameters that affect and can be taken to rectify energy consumption. Further, the research directions include how to conserve energy within a cloud environment more accurately.

References

1. http://www.cse.hcmut.edu.vn/~ptvu/gc/2012/GC-pp.pdf
2. A.J. Younge, G. von Laszewski, L. Wang, S. Lopez-Alarcon, W. Carithers, Efficient resource management for cloud computing environments
3. N.M. Gonzalez, T.C.M. de Brito Carvalho, C.C. Gonzalez et al, Review open access cloud resource management: towards efficient execution of large-scale scientific applications and workflows on complex infrastructures. J Cloud Comput. Adv. Syst. Appl. (2017)
4. S. Abrishami, M. Naghibzadeh, Epema, Deadline-constrained workflow scheduling algorithms for infrastructure as a service clouds. Futur. Gener. Comput. Syst. **29**(1), 158–169 (2013)
5. A. Bala, I. Chana, A survey of various workflow scheduling algorithms in cloud environment, in *IJCA Proceedings on 2nd National Conference on Information and Communication Technology*, Nagpur vol. NCICT (2011), pp 26–30
6. H. Arabnejad, J.G. Barbosa, List scheduling algorithm for heterogeneous systems by an optimistic cost table. IEEE Trans. Parallel Distrib. Syst. **25**(3), 682–694 (2014)
7. V. Sharma, A. Thomas, T. Abdelzaher, K. Skadron, Power-aware QoS management in web servers, in *Proceedings of the Real-Time Systems Symposium*, 2003
8. X. Wang, M. Chen, Cluster-level feedback power control for performance optimization, in *Proceedings of the IEEE 14th International Symposium on High Performance Computer Architecture*, 2008
9. A. Gandhi, M. Harchol-Balter, R. Das, and C. Lefurgy, "Optimal Power Allocation in Server Farms", Proc. Of SIGMETRICS Conference on Measurement and Modeling of Computer Systems, 2009

10. D. Meisner, B.T. Gold, T.F. Wenisch, PowerNap: eliminating server idle power. in *Proceedings of ASPLOS*, 2009
11. Y. Agarwal et al., Somniloquy: augmenting network interfaces to reduce PC energy usage, in *Proceedings of NSDI*, 2009
12. X. Li, Y. Li, T. Liu, J. Qiu, F. Wang, The method and tool of cost analysis for cloud computing, in *The IEEE International Conference on Cloud Computing (CLOUD 2009)*, Bangalore, India, 2009, pp. 93–100
13. G. Jung, M.A. Hiltunen, K.R. Joshi, Mistral: dynamically managing power performance, and adaptation cost in cloud infrastructures, in *The International Conference on Distributed Computing Systems*, Genova, Italy, 2010, pp. 62–73
14. L. Liu, H. Wang, X. Liu, X. Jin, W. He, Q. Wang, Y. Chen, Greencloud: a new architecture for green data center, in *IEEE Conference on Autonomic Computing (ICAC)*, Barcelona, Spain, 2009, pp. 29–38
15. A. Verma, P. Ahuja, A. Neogi, Power-aware dynamic placement of HPC applications, in *The 22nd Annual International Conference on Supercomputing (ICS 2008)*, Island of Kos, Aegean Sea, Greece, 2008, pp. 175–184
16. R. Nathuji, K. Schwan, Virtualpower: coordinated power management in virtualized enterprise systems, in *The 21st ACM Symposium on Operating Systems Principles (SOSP 2007)*, Stevenson, Washington, USA, 2007, pp. 265–278
17. J. Stoess, C. Lang, F. Bellosa, Energy management for hypervisor-based virtual machines, in *The USENIX Annual Technical Conference (USENIX)* 2007, pp. 1–14
18. Q. Chen, P. Grosso, K.v.d. Veldt, C.d. Laat, R. Hofman, H. Bal, Profiling energy consumption of VMs for green cloud computing, in *The 9th IEEE International Conference on Dependable, Autonomic and Secure Computing (DASC)*, Sydney, Australia, 2011, pp. 768–775
19. A. Kansal, F. Zhao, N. Kothari, A.A. Bhattacharya, Virtual machine power metering and provisioning, in *The 1st ACM Symposium on Cloud Computing (SoCC 2010)*, Indianapolis, Indiana, USA, 2011, pp. 39–50
20. F.F. Chen, J.-G. Schneider, Y. Yang, J. Grundy, Q. He, An energy consumption model and analysis tool for cloud computing environments, in *1st ICSE Workshop on Green and Sustainable Software Engineering*, Zurich, Switzerland, 3rd June 2012
21. Z. Zhang, S. Fu, Characterizing power and energy usage in cloud computing systems, in *The 3rd IEEE International Conference on Cloud Computing Technology and Science (CloudCom 2011)*, Athens, Greece, 2011, pp. 146–153
22. P. Mathiyalagan, M. Rajshree, S. Shobika, M. Sowmiya, An efficient dynamic scheduling based on greedy algorithm for heterogeneity clouds, Int. J.Adv. Res. Comput. Commun. Eng. **6**(3), 2017. ISO 3297:2007 Certified

Relationship Between Community Structure and Clustering Coefficient

Himansu Sekhar Pattanayak, Harsh K. Verma, and A. L. Sangal

Abstract Overlapping community is a phenomenon often observed in numerous real-world networks. Fire-spread (Pattanayak et al. in Swarm Evol Comput. 44: 1–48 (2019) [11]) community detection algorithm is an efficient algorithm to detect overlapping community structures. In this work, the Fire-spread algorithm is modified to establish a relationship between community structure and clustering coefficient. By using different networks and executing the modified Fire-spread algorithm, it is found that the clustering coefficient is highly correlated with community structure. Finally, a simpler community detection algorithm, derived from the fire-spread algorithm, is proposed, where the clustering coefficient is used as a threshold value. To validate the proposed algorithm, it is compared with some state of art community detection algorithms based on the NMI score.

Keywords Community · Detection · Complex network · Probabilistic computing · Clustering · NMI score · Fire-spread algorithm

1 Introduction

Real-world networks vary in sizes and structures. These networks include both naturally occurring networks and human-made networks. In a network, there are few actors (participants) and the interaction between those actors [1]. The actors are generally represented as nodes of the network and interactions are represented as edges. The edges can be directional or non-directional. A social network such as a friendship network, the edges are non-directional whereas an email network [2]

H. S. Pattanayak (✉) · H. K. Verma · A. L. Sangal
Department of CSE, NIT Jalandhr, Jalandhr, India
e-mail: himansusekharpattanayak@gmail.com

H. K. Verma
e-mail: vermah@nitj.ac.in

A. L. Sangal
e-mail: sangalal@nitj.ac.in

S. S. Dash et al. (eds.), *Intelligent Computing and Applications*,
Advances in Intelligent Systems and Computing 1172,
https://doi.org/10.1007/978-981-15-5566-4_18

representing email communication between the actors is necessarily directional. Similarly, edges of a network can either be weighted or unweighted. Edges of a friendship network can be unweighted, whereas edges of a co-purchase network [2] are weighted. Networks include social groups, families, villages [3, 4], World Wide Web (web pages containing similar topics are connected more densely among themselves) [4], biological networks [5].

The real-world networks have been studied extensively by the researchers. These networks exhibit scale-free [6] and community structure [7] properties. Community structure is an organization of nodes where some of the nodes are tightly connected with each other compared to the outside nodes [7, 8]. Community detection techniques are used for the detection of suspicious events in the telecom network, detection of terrorist networks, refactoring of software, lung cancer detection [9].

Community detection is the process of identifying the functional groupings without any other information except for the structural information. In community detection, usually, the Modularity [9] and Conductance [10] scores are used. As described in [10], the clustering coefficient is not used as a parameter in community detection. However, the clustering coefficient has a very strong co-relation with community structure, which will be discussed in the following sections.

The organization of the paper is given in the following order. A literature survey related to the subject is presented in Sect. 2. The proposed algorithm, which is a simplified version of Fire-spread community detection algorithm [11], is discussed in Sect. 3. Implementation of the proposed algorithm and datasets are described in Sect. 4, a comparison of the proposed algorithm with other community detection algorithms is presented in Sect. 5 and the paper is concluded in Sect. 6.

2 Related Work

A brief review of the related research and literature is presented in this section. Some earliest efficient community detection algorithms are based on optimization of modularity score [12, 13]. The algorithm proposed in [14] is based on a similar technique. However, it is capable of detecting hierarchical communities. Similar hierarchical community detection algorithms, which use an incremental method by combining sub-graphs, are presented in [15–19].

The authors of [20] have the set of weights for each node of the network and have applied the K-means clustering algorithm. However, the value of k, i.e., the number of communities needed to be predefined. The proposed algorithm in [21], is an enhancement over the previous work, in which the authors have used a nonlinear Lagrangian programming model that reduces the complexity of the previous algorithm. The authors have used the K-means technique to get an approximate solution and then refined the solution using simulated annealing in [22]. The community detection algorithm in [23] is also relied on K-means clustering.

The work described in [24] has applied a supervised learning technique. The algorithm needs prior knowledge about the network. The authors in [25] have used unsupervised division algorithm and verified their technique using 408 networks.

The authors of [26] have used multimodal optimization based on the firefly algorithm. The work described in [27], is an enhancement of the previous work by adding a multi-objective Pareto-optimal solution to the original. The authors of [28] have used PSO (particle swarm optimization technique) for community discovery which is efficient even for isolated nodes. There are some other community detection algorithms based on swarm intelligence described in [29, 30]. The authors of [31, 32] have used genetic algorithms with multi-objective solutions for community detection. The research presented in [33], uses ant colony optimization technique by defining each node as an ant. The algorithm proposed in [11], is a novel nature-inspired community detection algorithm that relies on local community detection using the Fire-spread algorithm.

The authors of [34] have proposed a community detection algorithm based on Integer programming and computation of the capacity of different communities. The author of [35] has designed a community detection algorithm that can able to work in a noisy environment and responsive to the dynamic behaviour of the community structures.

3 Proposed Work

The research work presented in this paper is divided into two parts; (1) study on the relationship between community structure and clustering-coefficient, (2) designing a simple and efficient community detection algorithm derived from the Fire-spread algorithm [11]. The original Fire-spread algorithm is a community detection algorithm, based on the propagation of fire in real scenarios. It takes its inspiration from naturally occurring fire. Neighbouring points of the source of fire are always at a higher risk of catching fire. We have modified the fire-spread community detection algorithm to simplify its application and investigate the relationship between community structure and clustering coefficient [10].

(A) Fire-spread Community Detection Algorithm

The probability to catch fire (P_i) at a node i is proportional to fire transmission coefficient p_i. The fire propagation probability between node i and j is denoted as $P_{i|j}$ and is defined as follows:

$$p_{i|j} = \frac{1}{G \cdot \text{degree}(j)} \tag{1}$$

where j is a neighbour node of i in graph G.

The total probability of catching fire at node i can be defined in terms of sum of the fire transmission coefficient of all the neighbours.

$$P_i = P_1 * p_{i|1} + P_2 * p_{i|2} + P_3 * p_{i|3} \ldots + P_j * p_{i|j} + \cdots \tag{2}$$

where $P_1, P_2, P_3\ldots$ and P_j are the probabilities of catching fire at 1st, 2nd, 3rd and jth neighbours, respectively.

The algorithm starts with a seed node. An *R-radius* subgraph (g) containing all the neighbours of *seed* which are less than or equal to *R-edge* distance from *seed* are selected to localize the search of community nodes. Initially, the fire catching probability of seed is set to 1.0, and for other nodes, it is set to 0. In each step, the fire catching probability (P_j) of any node j, with an edge (i, j) form node i, is updated as $P_j = P_j + P_i * p_{i|j}$. The number of steps required for the algorithm is R, as the maximum distance from the seed node to any other node is R.

After the completion of the updating of the fire catching probability for all the nodes, the nodes are compared with threshold values. Nodes having greater fire catching probability than the threshold are selected as member nodes along with the seed node. The value of threshold varies for each node depending on the distance from seed node, average node density, degree of the node and number of nodes of the subgraph g. This process is continued repeatedly until no node is left without being a part of any community. The seed nodes are the nodes that are not part of any communities discovered so far. This algorithm is efficient for overlapping community detection as it allows a node to be part of more than community does. The disadvantage of the algorithm is the complexity in calculating the threshold values for each node. Therefore, a simpler version of the algorithm is proposed, in which the threshold values are calculated, based on the clustering coefficient.

(B) The R-radius Neighbourhood

Let, two nodes A and B are part of community C which has n nodes. Let, A has node degree d_1 and B has node degree d_2. Since both A and B belong to the same community C, the probability that both nodes connected to each other by an edge can be calculated.

Probability of immediate neighbour =

$$\left({}^{n-2}C_{d1-1} - {}^{n-2}C_{d2-1}\right)/\left({}^{n-1}C_{d1} + {}^{n-1}C_{d2}\right) \tag{3}$$

Probability that both nodes are connected through a single intermediate node=

$$\frac{(d_1 + d_2)}{(n-2)} \tag{4}$$

Probability that both nodes are connected with 2-radius neighbourhood (P) = Probability of immediate neighbour + Probability that both nodes are connected by single intermediate node =

$$\left({}^{n-2}C_{d1-1} +{}^{n-2} C_{d2-1}\right)/\left({}^{n-1}C_{d1} +{}^{n-1} C_{d2}\right) + \frac{(d_1 + d_2)}{(n-2)} \quad (5)$$

Without losing generality, it is assumed that $d_1 = d_2 = d =$ average node degree inside community C. So, replacing d in Eq. (5)

$$P = \left({}^{n-2}C_{d-1} +{}^{n-1} C_{d-1}\right)/\left({}^{n-1}C_{d} +{}^{n-1} C_{d}\right) + \frac{2d}{(n-2)} \quad (6)$$

$$= \frac{(n-2)!}{(d-1)! * (n-d-1)!} * \frac{(n-d-1)! * d!}{(n-1)!} + \frac{2d}{n-2} = \frac{d}{(n-1)} + \frac{2d}{n-2} \quad (7)$$

$$P = \frac{d}{(n-1)} + \frac{2d}{n-2} \geq \frac{d}{(n-1)} + \frac{2d}{n-1} = \frac{3d}{(n-1)} \geq \frac{3d}{n} \quad (8)$$

If $d \geq \frac{n}{3}$ then $P \geq 1$

Therefore, it can be concluded that it is highly probable that any two nodes of the community C will be in 2-*radius* neighbourhood, if the average degrees of nodes inside C are greater than or equal to $n/3$. Therefore, for dense graphs the R-value can be set 2. However, depending on the density of the graph the value of R may be higher.

C. **The Proposed Algorithm**

In the original FS algorithm, the *fire propagation probability* and threshold values are variable depending upon nodes and their edges. However, in the proposed algorithm the *fire propagation probability* is kept constant for all the edges. The threshold value is also kept constant for each node. By changing the threshold values from 0.0 to 1.0, with a fixed *fire propagation probability* value, the *recall* and *precision* are calculated. The best values of *recall* and *precision* corresponding to the chosen threshold value are recorded and compared with the proposed algorithm with the *clustering coefficient* as a threshold. For this experiment, the network with a known community structure is available.

4 Implementation

The Fire-spread working of algorithm is consisting of two parts. The first algorithm is a local fire-spread algorithm that discovers a community around a seed node. The second part of the algorithm is the community detection algorithm which selects random seed nodes, calls the local Fire-spread algorithm and divides the network into overlapping communities. For all the networks, the *fire propagation probability* is set to 0.2. In this section, the original Fire-spread algorithm analysed by modifying it to work with fixed *propagation probability* and a varying *threshold* value. Community membership of all the datasets are already available, hence we are able to calculate the *recall* and *precision* values by executing local community detection algorithm with specific seed nodes. The best values of recall and *precision* will be used to investigate the effect of community structure on the clustering coefficient, which will be subsequently used to set a fixed *threshold* value for the proposed modified algorithm.

(A) **Dataset**:

The datasets that are used for the experiments are Zackary's Karate Club network [36] (Fig. 1). Bottlenose Dolphins network [37] (Fig. 2), American Political Book network [38] (Fig. 3), Adjective–Noun Network [39] (Fig. 4), Relaxed Caveman Network [40] (8 clusters containing 10 nodes each with 0. 1 being the probability of inter-cluster edge, Fig. 5) and Shared Community Network (A synthetic network with 5 communities with 20 nodes in each community, 25% nodes of each community are shared with other communities, Fig. 6).

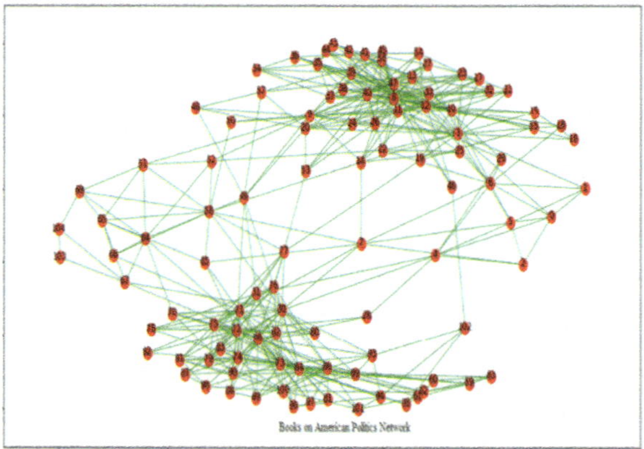

Fig. 1 Books on American politics network graph

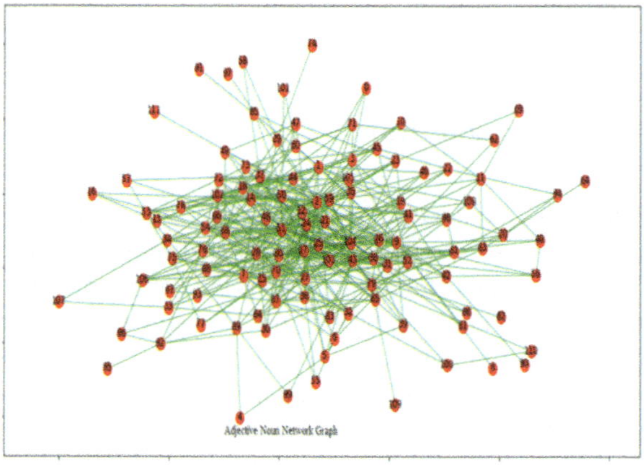

Fig. 2 Adjective–noun network graph

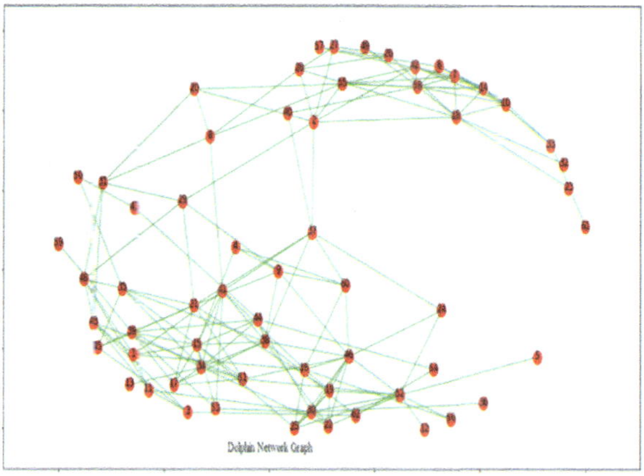

Fig. 3 Dolphin network graph

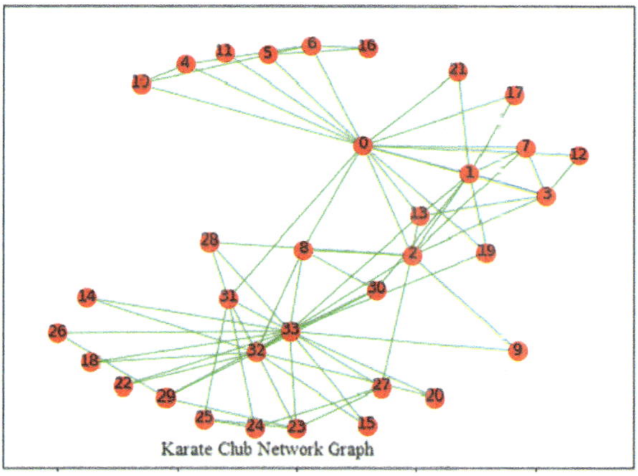

Fig. 4 Karate club network graph

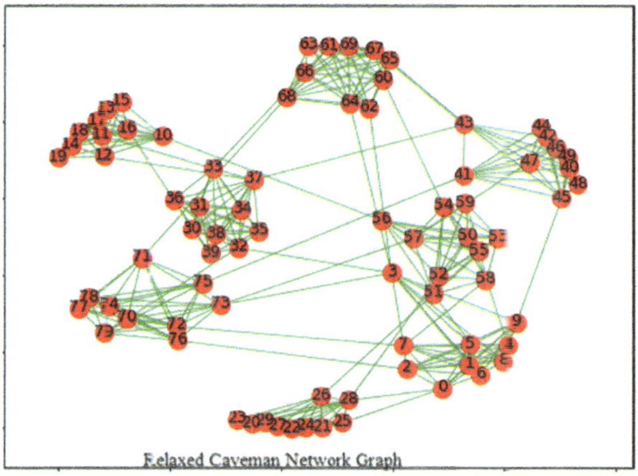

Fig. 5 Relaxed Caveman N/W graph

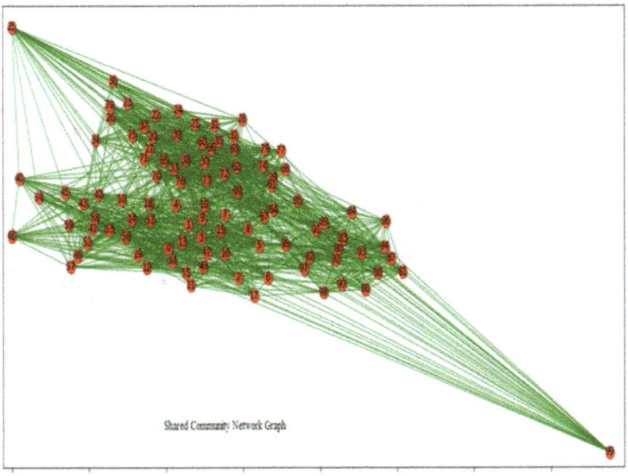

Fig. 6 Shared community graph

(B) **Recall and Precision**

The following formulas are used for calculations of precision and recall.

$$\text{Recall(C)} = \frac{\text{No.of Nodes of Community } C \text{ found by Algorithm}}{\text{Actual no of nodes of commmunity}} \tag{9}$$

$$\text{Precision}(C) = \frac{\text{No.of Nodes of Community } C \text{ found by Algorithm}}{\text{No of nodes found as community by the Algorithm}} \tag{10}$$

The probability value is fixed at 0.2 for all the networks and we calculate the value of precision and recall for all threshold ranging from 0.0 to 1.0 with 20 different values. The *precision* and *recall* values against the different *threshold* values are plotted in the following graphs. The six plots are *recall* and *precision* values for six datasets. The aim is to find the *optimal threshold* value which will maximize both recall and precision. The American political book graph is shown in Fig. 7. The best value of *threshold* for this network is 0.15 with $recall = 0.93$ and $precision = 0.95$.

Figure 8 represents the adjective–noun network. The best value of $recall = 0.97$ and $precision = 1.0$ is found *at a threshold* value of 0.30. For the Dolphin network plot given in Fig. 9, the performance is best with $recall = 0.94$ and $precision = 0.94$ at $threshold = 0.05$. Figure 10 is a plot for Karate club network. The FS algorithm performs best when $threshold = 0.2$. *Recall* and *precision* values are found to be 0.89 and 0.85, respectively. As it can be seen in Fig. 11, FS algorithm performs best with recall = 1.0 and $precision = 1.0$ when the *threshold* value is set to 0.35 for the Relaxed Caveman network. The shared network plot is shown in Fig. 12. The FS algorithm has best $recall = 1.0$ and $precision = 1.0$ when the *threshold* is set at 0:2.

For most of the networks, best values of recall and precision are obtained with $threshold = 0.2$ expect for the Dolphin network where the best *threshold* is found to be 0.05. The best values of thresholds found are used for further experimental works. By observing all the six networks, it can be concluded that with fixed probability

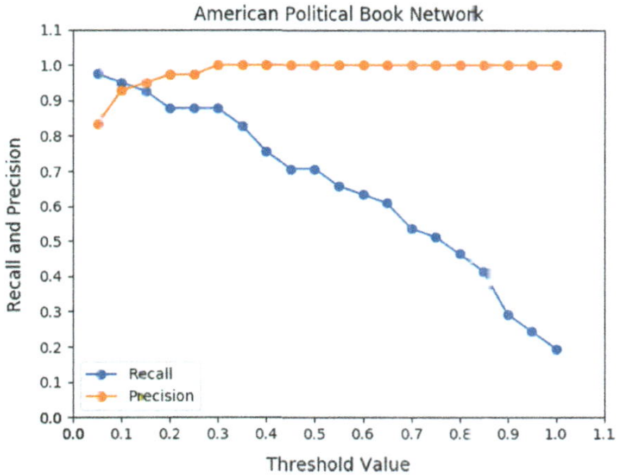

Fig. 7 Books politics network

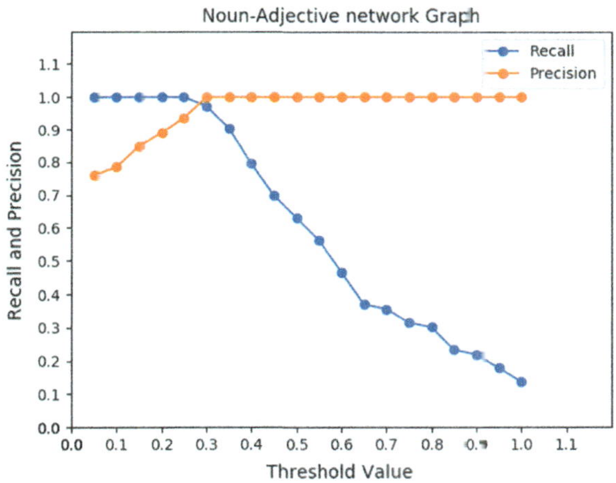

Fig. 8 Adjective–noun network

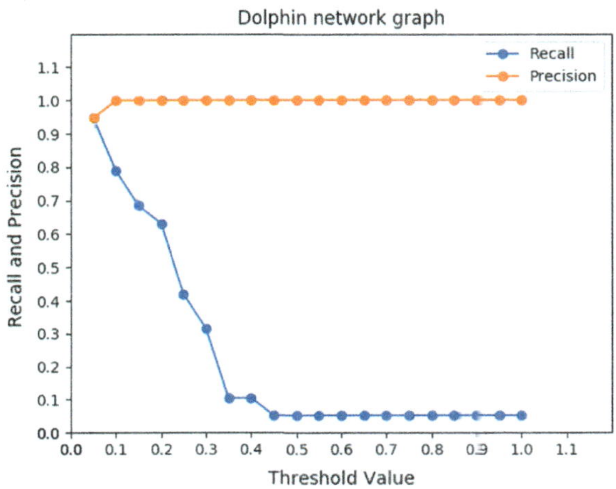

Fig. 9 Dolphin network graph

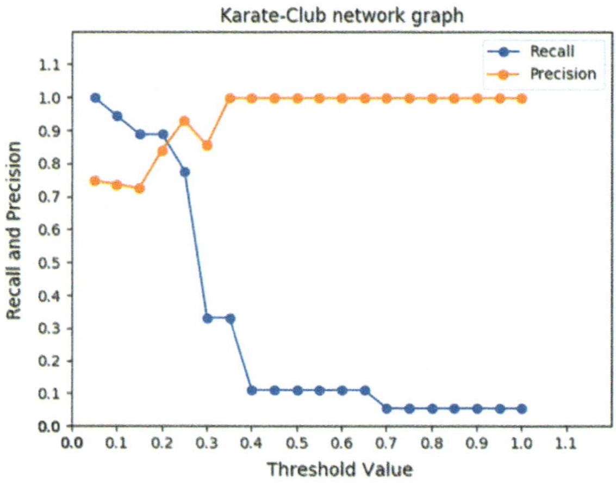

Fig. 10 Zachary's Karate club graph

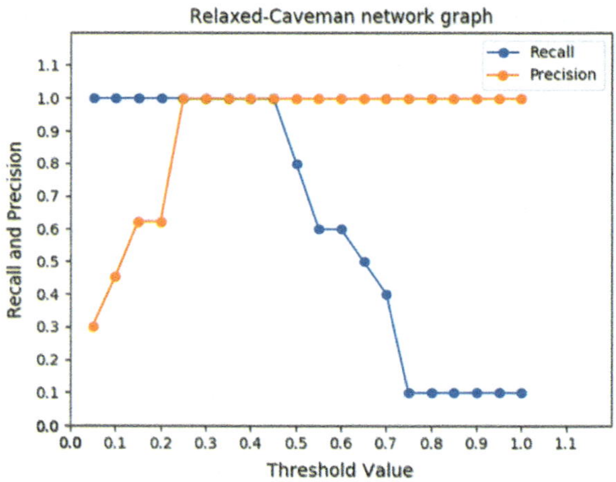

Fig. 11 Relaxed Caveman graph

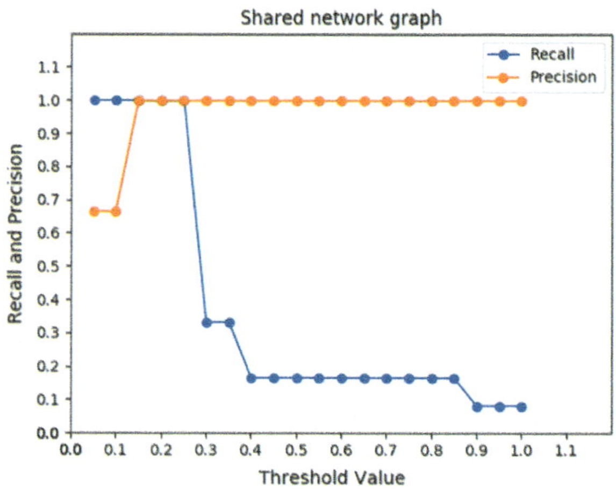

Fig. 12 Shared community network

Table 1 Threshold value of networks using clustering coefficient

Network graph	Avg node degree	Clustering coefficient	$2.0\frac{\text{Clustering Coefficient}}{\text{Avg Node degree}}$
Political book	8.204	0.515	0.1254
Noun–adjective	8.0625	0.195	0.0482
Dolphin	4.18	0.325	0.1554
Karate	4.75	0.529	0.2228
Relaxed Caveman	9	0.71	0.1578
Shared	4.222	0.783	0.3704

value; an increase in *threshold* results in a decrease in *recall* but an *increase* in *precision*. Setting higher *threshold* value results in the detection of high-resolution hierarchical structures. Lower threshold values result in the detection of communities with a higher degree of sharing of nodes.

For calculating threshold, we have used the local clustering coefficient which is a measure to quantify how close a node along with its neighbour to be a clique. To calculate the local clustering coefficient for a node i the following formula is used

$$\text{Clustering Coefficient} = \frac{2 * |e_{jk}|}{|N_i| * (|N_i| - 1)} \tag{11}$$

where $|N_i|$ = number of neighbour nodes of node i and e_{jk} is set of all possible pairs of nodes such that j, k are neighbours of i. The subgraph, which has a higher clustering coefficient with a lesser average node degree, has a higher threshold value than a subgraph with a lesser clustering coefficient and higher average node degree. Table-1 shows the threshold value for the networks based on clustering coefficients.

$$\text{Threshold} = \text{constant} * \frac{\text{Clustering Coefficient}}{\text{Avg Node degree}} \tag{12}$$

The *constant* = 2.0 is set for all the networks. We have experimentally set the value after repeated experiments.

The recall and precision of the proposed algorithm are compared with the best recall and precision of the original algorithm.

C. **Local Fire-Spread Algorithm:**

Input: Graph ***G***, *fire propagation probability* ***p***, *radius* ***R***
For each ***node*** in ***G***

If ***node*** does not belong to any community

Use ***node*** as *seed node*
Find ***Ego-Graph*** for the ***node*** with *radius* ***R***
Find ***threshold*** = **(2****Clustering Coefficient*)/*Average node degree***

Call *local Fire spread* Algorithm with input $\boldsymbol{g}, \boldsymbol{p}$ & $\boldsymbol{t}$
Store the list of nodes returned by the *Local Fire Spread* algorithm.

return list of overlapping communities

Fire-Spread Algorithm:

Input*: R-radius subgraph* **g**, *fire propagation probability* **p**, *Threshold* ***t****, seed* ***s***
Initialize each ***node*** in ***g***

Initialize ***seed*** Probability to catch fire $\boldsymbol{P_s}$**= 1.0**
Initialize nodes other than ***seed*** with $\boldsymbol{P_i}$**= 0.0**

For each ***node*** in ***g***

For $\boldsymbol{i \in Neighbor\ (node)}$

Update $\boldsymbol{P_i = P_i + P_{node} * p}$
Update $\boldsymbol{P_{node} = P_{node} + P_i * p}$

For each ***node*** in ***g***

Check if a node belongs to the community using *threshold* ***t***

return list of community nodes

D. **Comparison of Proposed Algorithm with Fire-Spread Algorithm**

The proposed algorithm is compared with the Fire-spread algorithm in terms of *precision* and *recall* and presented in Table 2. The best precision and recall values are taken for the Fire-spread algorithm.

As it can be seen from Table 3, the performance of the proposed algorithm and the original algorithm are same for the synthetic networks with recall and precision values being 1.0. For the real-world networks, both the algorithms perform very similarly baring a few cases. The above experimental analysis is used to set a simpler threshold value for the network based on local clustering coefficient.

Table 2 Comparison between proposed algorithm and fire-spread algorithm

Network graph	Fire-spread algorithm		Proposed algorithm	
	Recall	Precision	Recall	Precision
Political book	0.9268	0.95	0.8536	1.0
Noun–adjective	0.9726	1.0	1.0	0.9634
Dolphin	0.9473	0.9473	0.9473	1.0
Karate	0.8889	0.8421	1.0	0.8571
Relaxed Caveman	1.0	1.0	1.0	1.0
Shared	1.0	1.0	1.0	1.0

Table 3 NMI score: books on American politics

Algorithms	Max	Mean	Min	Standard deviation
Proposed algorithm	0.6005	0 5996	0.5982	0.0005
Enhanced firefly	0.5953	0 5927	0.5915	0.0004
Firefly	0.5829	0 5819	0.5803	0.0007
Blondel	0.5954	0 5937	0.5924	0.0012
Ronhovde-Nussinov	0.5948	0 5941	0.5878	0.0029
Clauset-Newman	0.5427	0 5386	0.5304	0.00327

5 Results and Discussion

In this section, NMI (Normalized Mutual Information) score is used for all the networks and the proposed algorithm is compared with some well-known algorithms, such as Enhanced Firefly algorithm [26], Firefly algorithm [27], Blondel algorithm [12], Ronhovde–Nussinov [14] algorithm and Clauset–Newman [13]. All the experiments are repeated ten times for accuracy.

For any two clusters C_1 and C_2, $NMI\ (C_1, C_2)$ is a measure to find the amount of mutual information, the two clusters share with each other for the networks, where communities or clusters may overlap. The version of NMI we used for our comparisons is based on [41]: Given two clusters C_1 and C_2

$$\mathrm{NMI(C1, C2)} = \frac{H(C_1) + H(C_1) - H(C_1, C_2)}{((H(C_1) + H(C_1))/2)} \tag{13}$$

where $H(C_1)$= entropy of cluster C_1, $H(C_2)$= entropy of cluster C_2 and $H(C_1, C_2)$ joint entropy of cluster C_1 and C_2. $H(C)$ for any cluster C, $H(C) = -\sum_{c\in C} p(c) * \lg(p(c))$ (14)

$$H(C_1, C_2) = \sum_{c_1\in C_1} \sum_{c_2\in C_2} P(c_1, c_2) * \lg\left(\frac{P(c_1, c_2)}{p(c_1) * p(c_2)}\right) \tag{15}$$

where $P\ (c)$ for any node c is the probability of and node $c. \in$ community C.

To calculate probability $P\ (c)$ for any node in community C, we have used the formula

$$P(c) = \frac{\text{No.of edges of c connected to nodes inside community C}}{\text{Degree}(c)} \tag{16}$$

Table 3 represents the NMI score for Books on American politics network. We have calculated the maximum, minimum, average NMI score for each algorithm along with standard deviation with 10 runs of the algorithms. The proposed algorithm gives the best maximum = 0.6005, minimum = 0.5982 and average = 0.5996 NMI scores among all the algorithms. Enhanced firefly algorithm gives second best result with maximum = 0.5953, minimum = 0.5915 and average = 0.5927 NMI score. Clauset-Newman algorithm performs worst with maximum = 0.5427, minimum = 0.5304 and average = 0.5386 NMI score. Enhanced firefly algorithm gives the least S.D. (standard deviation) with S.D. value = 0.0004. However, the proposed algorithm give second best result with S.D. value = 0.0005.

Table 4 contains the NMI score for Adjective Noun network. The proposed algorithm performs best with maximum = 0.8635, average = 0.8619, minimum = 0.8584 NMI score. Enhanced firefly algorithm performs second best in terms of NMI score. In this network, proposed algorithms perform best in terms of S.D. with S.D. value = 0.0012. Blondel algorithm performs second best in terms of S.D. value with S.D. = 0.0013.

Table 5 represents the Dolphin network. Among all the algorithms our proposed performs best with 0.9907, 0.9898, and 0.9884 as maximum, average and minimum NMI scores, respectively. In this network Firefly and Ronhovde–Nussinov algorithms perform best in terms of S.D. with S.D. value = 0.0001. The proposed algorithm along with the Enhanced Firefly algorithm performs second best in terms of S.D. value with S.D = 0.0002.

Table 6 represents the famous Karate club network data. In this network, the proposed algorithm performs best in terms of maximum, minimum and average NMI score and Enhanced firefly algorithm performs second best. In terms of SD, the proposed algorithm performs best with S.D = 0.0001.

Relaxed Caveman network data is stored in Table 7. The proposed algorithm has the best maximum = 0.3803, average = 0.3791 and minimum = 0.3786 NMI

Table 4 NMI score: adjective noun network

Algorithms	Max	Mean	Min	Standard deviation
Proposed algorithm	0.8635	0.8619	0.8584	0.0012
Enhanced firefly	0.8592	0.8566	0.8512	0.0021
Firefly	0.8498	0.8466	0.8396	0.0025
Blondel	0.8535	0.8517	0.8487	0.0013
Ronhovde–Nussinov	0.8526	0.8488	0.8410	0.0028
Clauset–Newman	0.8321	0.8276	0.8189	0.0034

Table 5 NMI score: Dolphin network

Algorithms	Max	Mean	Min	Standard deviation
Fire-spread	0.9907	0.9898	0.9884	0.0002
Enhanced firefly	0.9857	0.9838	0.9825	0.0002
Firefly	0.9832	0.9826	0.9817	0.0001
Blondel	0.9837	0.9822	0.9809	0.0003
Ronhovde–Nussinov	0.9868	0.9863	0.9855	0.0001
Clauset–Newman	0.5910	0.5881	0.5807	0.0025

Table 6 NMI score: Karate club network

Algorithms	Max	Mean	Min	Standard deviation
Fire-spread	0.9975	0.9959	0.9964	0.0001
Enhanced firefly	0.9969	0.9956	0.9961	0.0001
Firefly	0.9792	0.9753	0.9683	0.0021
Blondel	0.9897	0.9847	0.9812	0.0015
Ronhovde–Nussinov	0.9868	0.9860	0.9856	0.0002
Clauset–Newman	0.7019	0.6941	0.6883	0.0261

Table 7 NMI score: Caveman network

Algorithms	Max	Mean	Min	Standard deviation
Fire-spread	0.3803	0.3791	0.3786	0.0004
Enhanced firefly	0.3765	0.3741	0.3723	0.0009
Firefly	0.3739	0.3691	0.3658	0.0018
Blondel	0.3743	0.3712	0.3687	0.0009
Ronhovde–Nussinov	0.3716	0.3675	0.3624	0.0023
Clauset–Newman	0.3484	0.3409	0.3357	0.0092

score and Enhanced Firefly has the second best NMI scores. In terms of SD, the Fire-spread algorithm performs best with value 0.0004. Both Enhanced Firefly and Blondel algorithm have second best S.D. with value 0.0009.

Table 8 represents data for Shared Community Network. This network was specially designed; with 25% nodes are shared between communities, to evaluate the proposed algorithm for overlapping communities. As it is given in the table, proposed algorithm performs best in terms of maximum, minimum, average NMI score along with S.D. with values 0.6165, 0.6127, 0.6145 and 0.0007, respectively.

Table 8 Shared community network

Algorithms	Max	Mean	Min	Standard deviation
Fire-spread	0.6165	0.6145	0.6127	0.0007
Enhanced Firefly	0.6153	0.6138	0.6104	0.0008
Firefly	0.6127	0.6086	0.6049	0.0016
Blondel	0.6119	0.6095	0.6072	0.0009
Ronhovde–Nussinov	0.6111	0.6063	0.6016	0.0019
Clauset–Newman	0.5732	0.5574	0.5421	0.0057

6 Conclusion

From the data presented in the result section, it can be concluded that the proposed algorithm has the best performance with both real networks and synthetic networks. The algorithm is capable of detection overlapping community structure as it permits any node to be part of more than community does.

Since the proposed algorithm is derived from the Fire-spread community detection algorithm, it inherits some of the best features of its parent algorithm. The proposed algorithm can be executed at local levels for finding the community around a seed node and can able to assign a single node to multiple communities. However, the algorithm has a simpler implementation in comparison with its parent algorithm, as the threshold value can be calculated from the clustering coefficient.

From the experiments presented in Sect. 4, it is also clear that the clustering index is co-related to community structure, hence can be used as a threshold value. In this work, we also analysed the behaviour of Fire-spread algorithm in great detail with different propagation probabilities and threshold values.

The proposed algorithm is used only on an undirected and unweighted network. The algorithm can be applied for a directed and weighted network as a future work.

References

1. S. Wasserman, K. Faust, *Social Network Analysis: Methods and Applications* (Cambridge University Press, 1994)
2. J. Leskovec, A. Krevl, https://snap.stanford.edu/data/
3. M.S. Granovetter, The strength of weak ties. Am. J. Sociol. **78**, 1360–1380 (1973)
4. G. Flake, S. Lawrence, C. Giles, Efficient identification of web communities, in *KDD'00: Proceedings of the 6th ACM SIGKDD International Conference on Knowledge Discovery and Data Mining* (2000), pp 150–160
5. G. Palla, I. Derényi, I. Farkas, T. Vicsek, Uncovering the overlapping community structure of complex networks in nature and society. Nature **435**(7043), 814–818 (2005)
6. R. Pastor-Satorras, A. Vespignani, Epidemic spreading in scale-free networks. Phys. Rev. Lett. **86**(14), 3200 (2001)
7. M. Girvan, M.E. Newman, Community structure in social and biological networks. Proc. Natl. Acad. Sci. **99**(12), 7821–7826 (2002)
8. A. Gog, D. Dumitrescu, B. Hirsbrunner, Community detection in complex networks using collaborative evolutionary algorithms. Adv. Artifi. Life 886–894 (2007)
9. M.S. Ahuja, J. Singh, Neha practical applications of community detection. Int. J. Adv. Res. Comput. Sci. Softw. Eng. **6**(4) (2016)
10. H.S. Pattanayak, H.K. Verma, A.L. Sangal, Community detection metrics and algorithms in social networks, in *2018 First International Conference on Secure Cyber Computing and Communication (ICSCCC)*. IEEE (2018, December), (pp. 483–489)
11. H.S. Pattanayak, A.L. Sangal, H.K. Verma, Community detection in social networks based on fire propagation. Swarm Evol. Comput. **44**, 31–48 (2019)
12. V.D. Blondel, J.L. Guillaume, R. Lambiotte, E. Lefebvre, Fast unfolding of communities in large networks. J. Stat. Mech: Theory Exp. **2008**, P10008 (2008)
13. A. Clauset, M.E.J. Newman, C. Moore, Finding community structure in very large networks. Phys. Rev. E **70**, 066111 (2004)

14. P. Ronhovde, Z. Nussinov, Multiresolution community detection for mega scale networks by information-based replica correlations. Phys. Rev. E **80**, 16109 (2009)
15. S. Boettcher, A.G. Percus, Extremal optimization for graph partitioning. Phys. Rev E **64** (2001). (Article ID: 026114)
16. D. Fisher, J. Artif, Iterative optimization and simplification of hierarchical clusterings. J. Artif. Intell. Res. **4**, 147–180 (1996)
17. D. He, J. Liu, D. Liu, D. Jin, Z. Jia, Ant colony optimization for community detection in large-scale complex networks, in *Proceedings of 7th International Conference on Natural Computation (ICNC 2011)*, vol. 2 (IEEE Press, 2011), pp. 1151–1155
18. T. Herlau, M. Mørup, M.N. Schmidt, L.K. Hansen, Detecting hierarchical structure in networks, in *Proceedings of 3rd International Workshop on Cognitive Information Processing (CIP 2012)* (IEEE Press, 2012), pp. 1–6
19. Z. Zhao, C. Li, X. Zhang, F. Chiclana, E.H. Viedma, An incremental method to detect communities in dynamic evolving social networks. Knowl. Based Syst. **163**, 404–415 (2019)
20. Z. Gao, N. Jin, Detecting community structure in complex networks based on k-means clustering and data field theory, in *Proceedings of 20th Chines Control and Decision Conference (CCDC 2008)* (IEEE Press, 2008), pp. 4411–4416
21. W. Li, 229 (2013) Revealing network communities with a nonlinear programming method, Inform. Sci. 18–28
22. J. Liu, T. Liu, Detecting community structure in complex networks using simulated annealing with k-means algorithms. Phys. A: Stat. Mech. Appl. **389**(11), 2300–2309 (2010)
23. J.B. MacQueen, Some methods for classification and analysis of multivariate observations, in *Proceedings of the Fifth Berkeley Symposium on Mathematical Statistics and Probability*, vol. 1 (1967), pp. 281–297
24. Z.Y. Zhang, Community structure detection in complex networks with partial background information, EPL **101** (4), 6 (2013). Article No. 48005
25. A. Ghasemian, H. Hosseinmardi, A. Clauset, Evaluating overfit and underfit in models of network community structure. IEEE Trans. Knowl. Data Eng. (2019)
26. B. Amiri et al., Community detection in complex networks: multi–objective enhanced firefly algorithm. Knowl. Based Syst. **46**, 1–11 (2013)
27. X.S. Yang, *Firefly Algorithms for Multimodal Optimization* (Foundations and Applications, Stochastic Algorithms, 2009), pp. 169–178
28. X. Duan, C. Wang, X. Liu, Y. Lin, Web community detection model using particle swarm optimization, in *Proceedings of IEEE Congress on Evolutionary Computation (CEC 2008)* (IEEE Press, 2008), pp. 1074–1079
29. R.J. Kuo, Y.D. Huang, C.C. Lin, Y.H. Wu, F.E. Zulvia, Automatic kernel clustering with bee colony optimization algorithm. Inform. Sci. **283**, 107–122 (2014)
30. Q. Cai, M. Gong, L. Ma, S. Ruan, F. Yuan, L. Jiao, Greedy discrete particle swarm optimization for large-scale social network clustering. Inform. Sci. **2014** (2014). Available online 7 October
31. C. Pizzuti, GA-Net: a genetic algorithm for community detection in social networks, in *Proceedings of 10th International Conference on Parallel Problem Solving from Nature (PPSN 2008)*, Lecture Notes in Computer Science, vol. 5199 (2008) pp. 1081–1090
32. A.I. Hafez, N.I. Ghali, A.E. Hassanien, A.A. Fahmy, Genetic algorithms for community detection in social networks, in *Proceedings of 12th International Conference on Intelligent Systems Design and Applications* (ISDA 2012) (IEEE Press, 2012), pp. 460–465
33. V. Estivill-Castro, Why so many clustering algorithms: a position paper. ACM SIGKDD Explor. Newslett. **4**(1), 65–75 (2002)
34. D.J. Watts, P.S. Dodds, M.E.J. Newman, Identity and search in social networks. Science **296**(5571), 1302–1305 (2002)
35. L. Wang, J. Wang, Y. Bi, W. Wu, W. Xu, B. Lian, Noise-tolerance community detection and evolution in dynamic social networks. J. Comb. Optimiz. **28**(3), 600–612 (2014)
36. W.W. Zachary, An information flow model for conflict and fission in small groups. J. Anthropol. Res. 452–473 (1977)

37. D. Lusseau, K. Schneider, O.J. Boisseau, P. Haase, E. Slooten, S.M. Dawson, The bottlenose dolphin community of Doubtful Sound features a large proportion of long-lasting associations. Behav. Ecol. Sociobiol. **54**, 396–405 (2003)
38. V. Krebs. http://www.orgnet.com/
39. M.E.J. Newman, Phys. Rev. E **74**, 036104 (2006)
40. D.J. Watts, *Small Worlds* (Princeton University Press, 1999)
41. L. Danon, A. Diaz-Guilera, J. Duch, A. Arenas, Comparing community structure identification. J. Stat. Mech.: Theory Exp. **2005**, P09008 (2005)

Digital Image Forensics-Image Verification Techniques

Anuj Rani and Ajit Jain

Abstract In the present scenario of the digital era, we are on the brink of marginal transformation in digital imaging technology. Developments of high computational and artificial intelligence techniques have produced wonderful image editing techniques that create imposing results in stint frames. And, doing this will affect the original artifacts present in digital images. This refers to the manipulation or forgery of digital data with the help of image editing applications and this also erodes our trust in digital images. To rely on the semantics of digital images, there is significant research contribution in terms of various image forgery detection techniques. These techniques help to re-evaluate the authenticity of digital images. In this article, an examination of various research contributions is conducted. The primary goal of this work is to give a glimpse of various current existing techniques related to digital image forensics. These techniques are assessed as per their proficiencies and boundaries. Whenever possible, a comparison between similar domain techniques is presented. The main focus is on pixel and physics-based techniques. Our analysis discusses the challenges that are to be addressed in this area of forensic science and provides insight into various datasets available for researchers to develop and test new approaches. Finally, the work recommends possible future research directions.

Keywords Content authentication · Forgery detection · Image forensics · Passive techniques

A. Rani (✉)
Department of Computer Science, G L Bajaj Institute of Technology and Management, Greater Noida, Utter Pradesh, India
e-mail: anuanuj1989@gmail.com

A. Jain
Department of Computer Science, Banasthali Vidyapith, Tonk, Rajasthan 304022, India
e-mail: ajitjain_2k@yahoo.co.in

S. S. Dash et al. (eds.), *Intelligent Computing and Applications*,
Advances in Intelligent Systems and Computing 1172,
https://doi.org/10.1007/978-981-15-5566-4_19

1 Introduction

Today, there barely exists in any field where digital images are not used. Some of the popular areas, where digital images are used, are digital media, e-media, military, law and jurisdiction, science and technology, medical imaging science, newspapers, social media, and many more. As per the usage of different types of digital images in various fields, the importance of their originality is a major concern. We believe in what we see rather than verbal content. Therefore, give our consent to this digital data. The widespread use of this data is commonly tampered with digital imaging techniques and hence misrepresents the meaning of information to us [1]. Image tampering changes the meaning of image data for a kind of malevolent purpose. On the other side, these manipulations are detected using digital image forensics. Tampering with images is not a new concept even first forgery is reported in the 1860s. Many types of forgery making techniques are available in the literature. An example of image forgery is depicted in Fig. 1. In this figure, the photomontage forgery is shown in which different images are joined to create a new image using pasting, overlapping, and rendering operations. These operations are carried out using freely available and accessible image editing applications such as Photoshop, GIMP, Paintshop, and many more. Under such conditions of image manipulations, it becomes essential to find whether the produced image is real or fake because of their usage in many secret domains. Therefore, this brings the importance or motivation to design new image counterfeit recognition methods that can identify and prove the authenticity of digital images. Many researchers are currently working in this cutting-edge domain of digital forensics to develop new techniques to combat image forgeries. And, here it becomes important to state that many research papers are published by IEEE and Elsevier from the year 2000 to 2017 which shows the increased pattern of publications in this domain. An analysis is shown in Fig. 2 which confirms that digital image forensic techniques are growing at a high pace. To find the image vulnerabilities, forgery detection methods are majorly divided into two types, namely active and passive/blind approaches. In an active approach, a kind

Fig. 1 A forged image created using kiwi fruit and lemons using GIMP application software [2]

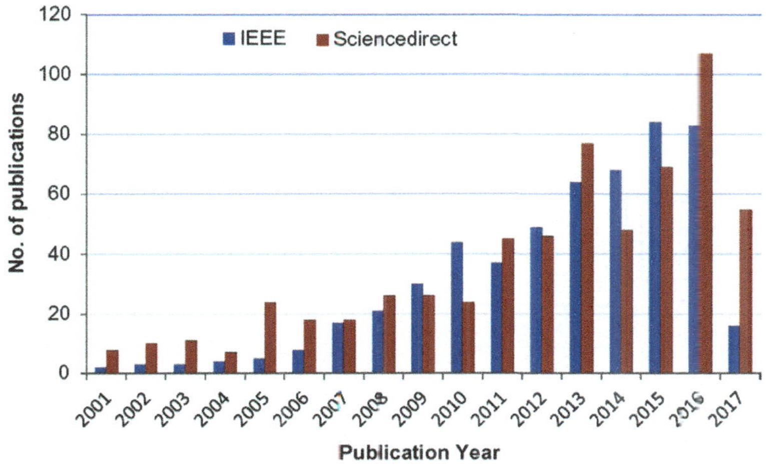

Fig. 2 A pattern of published papers by IEEE and Elsevier from 2000 to 2017. The analysis is done using query "Image forgery Detection"

of preprocessing over data is required and examples of this approach are signature generation and watermarking (hiding of data in host signal).

On the other side, passive techniques do not have such steps and, therefore, more vigorous compares to active methods. In these techniques, a significant challenge is to detect the image tempering because there exists no concealed data.

Many approaches are used to recognize the traces of manipulations. The passive methods are divided into five types as explained below [3].

- Pixel-Based Techniques:

These techniques are used to detect pixel-level geometric anomalies during the tempering process. Inconsistencies are identified at pixel levels. Techniques included in this are mainly copy-move and splicing-based forgery detection approaches.

- Format (Compression)-Based Techniques:

Format-based methods use the anomalies generated at the time of compression of the image. The generally used methods in this category are double JPEG.

- Camera or source-based Technique:

This technique changes the facts by sensor noise, camera lens properties, and on-chip processing in the cameras.

- Physics-based Techniques:

Such approaches detect manipulations in 2D or 3D lighting parameters with reflections and shadows of intrinsic lighting features.

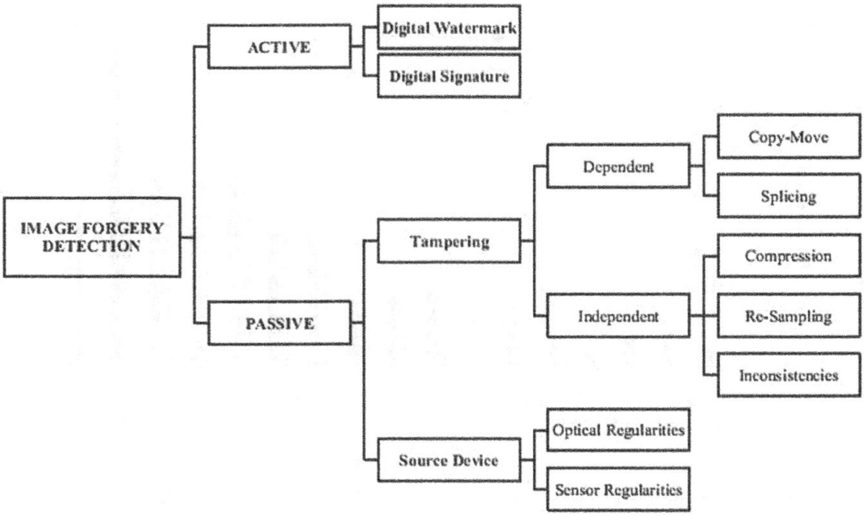

Fig. 3 Classification of various image forgery detection techniques

- Geometric Techniques:

These approaches implement the concept of geometrical analysis of object location related to camera location for forgery detection. A detailed description of these approaches is shown in Fig. 3. We attempted the following research questions in this study.

RQ1: How to analyze data to find image forgeries.
RQ2: Provide availability of popular image forensics techniques for researchers.
RQ3: A comparative study for various image forensics techniques.

This paper focuses on passive forensic approaches. The paper is segregated into three sections. Sec. 2 is about the related comprehensive study of various forgery detection techniques followed by Sect. 3 about the conclusion and future aspects of this domain.

2 Critical Review of Various Forgery Detection Techniques

A critical review of various good quality publications in the domain of image forensics is presented in this section. Our major target is to bring various types of current image manipulation-detecting techniques to the researchers, which can help them to advance in this field.

(A) **Pixel-Based Forgery Detection Techniques**

In pixel-based techniques, the copy-move-based forgery detection methods are mainly discussed. The copy-move-based approaches are alienated into keypoint-based, block-based, hybrid-based, and splicing-based approaches. Copy-move forgery is the most common type of manipulation where a portion from the image is copied and inserted on other regions of a similar image. This will hide the existing object from the image and create a duplicate forged version of the image [4]. There are many ways to make forgery using copy-move. Some of these techniques are known as keypoint- and block-based copy-move forgeries. In the keypoint-based forgeries, keypoints or features are extracted using feature extraction algorithms. Distinct features such as corners, blobs, and contours/edges/boundaries are estimated form images, and later descriptors and features are matched to identify the forged copied regions. These algorithms are mainly implemented using three categories like SIFT (scale-invariant feature transform), SURF (speed up robust features), and Harris Corner points.

Silva et al. [5] proposed a copy-move-based image tampering detection approach using a multi-scale analysis and voting process of the image. They extracted interest points and clustered them into regions based on geometrical constraints. These features are robust against operations like rotation and scaling. Using such points finally, a multi-scale image representation is generated and the final decision of the process is taken by a voting process for various image regions. They compared their approach to 15 other techniques and produced batter results. Cozzolino et al. [6] implemented a copy-move-based forgery algorithm for 44 precisely detection and localization of forgery using rotation-invariant features. They applied the fast approximate nearest-neighbor search algorithm and PatchMatch. Their experimental results are robust and quicker than other state-of-art techniques. Yu et al. [7] suggested a method to solve the problem of redundant feature points and feature fusion to detect image forgery. The two-stage process enhances the matching performance by merging MROGH and HH descriptors. Their technique is effective in detecting highly precise copy-move forgeries. Zandi et al. [8] suggested a high computation cost-efficient technique to identify copy-move forgery. A new nearest point detector is recommended using the block-based and keypoint-based approaches. The estimated keypoint from the image is considered even from the low-contrast areas of the image so that a unique matrix is computed. A filtering algorithm is also used to filter the wrongly matched areas. Their process is iterated with keypoint and, finally, the copied areas are estimated. Their results are satisfactory. Wang et al. [9] suggested a keypoint-based copy-move forgery technique for small even image portions. In the first step of their approach, the tampered part of the image is segmented into non-overlapping and superpixels. Using the local entropy information, these superpixels are classified into smooth, textured, and strong textures. Then, further, in the second step, the stable image keypoint is obtained from each superpixel by using their content-based adaptive feature point detector. Further, the local visual features such as exponent moments magnitudes are generated for every key point and the best is generalized using 2NN (nearest-neighbor) algorithms for matching keypoints. In the last step of the algorithm, the false matched points are removed and duplicated areas are estimated using zero-mean normalized cross-correlation measure. These techniques produce better results against geometric transformations, compression, additive Gaussian noise, etc.

On the contrary, the block-based copy-move forgeries extract the features implemented on segmented images. These segmented images can have many overlapping and non-overlapping image blocks for which feature vectors are estimated and matched using the block-based procedures. There exist many block-based matching algorithms in previous works; some of them are discussed here. Lee et al. [10] implemented a blind forensics technique to detect block-based copy-move forgeries. In their work, the input image is isolated into various overlapping blocks, and HoG gradients are used for every block. Similarities of these blocks are estimated using the statistical features. In the last step of their algorithm, the feature vectors are lexicographically arranged and cloned image areas are identified. This technique has the capability to detect copy-move forgeries and also deals with various image distortion features. Liang et al. [11] presented an approach for object removal by exemplar-based inpainting integrates principal pixel mapping, greatest zero-connectivity component labeling, and fragment splicing detection. Central pixel mapping speeds up the copied areas by matching such blocks with similar hash values. The zero-connectivity component is basically required to identify the manipulated pixels in the blocks estimated using labeling. Finally, fragment splicing detection is used to discretize the forged regions from the best match image areas. This approach is capable of producing 85% forgery detection accuracy. Huang et al. [12] proposed a method to find copy-move forgery using JPEG compression and Gaussian noise *i*th blurring attacks. Their technique involves steps such as feature extraction, matching, and copied block identification. For feature extraction, they used Fast Fourier Transformation (FFT), SVD, and Principal Component Analysis (PCA). The combination of FFT, SVD, and PCA is used to extract similar features. In the last step of their approach, pixels on the upper left corners of duplicated regions are used for visual examination. Their results are satisfactory. Zhou et al. [13] suggested a method based on color moments and the other five descriptors. Forged image is divided into fixed-size overlapped blocks, and clustering shares whole search space into regions of alike color distribution. Tampered blocks reside in the same cluster in which copied and similar blocks have the same color distribution. A group of deep compositional (PNN) pattern neural network is trained with estimated features. Finally, the feature vector from various clusters indicated possible forged areas. Their accuracy for image forgery detection is good even against various blurring attacks.

In splicing-based image forgery detection, two or more number of images are compounded to make a new copy of the image. Splicing operation is somehow close to copy-move forgery. Various authors already published good research papers based on splicing-based approaches. Zhou et al. [14] proposed a 2D noncasual Markov model for detecting splicing-based image forgeries. They estimated discriminative model parameters to differentiate the spliced regions. The model is applied in block DCT domain and cross-domain features are picked as final features for the classification purpose. They also used SVM to classify the spliced images on the public image splicing dataset. The results are superior compared with other techniques in their domain. Bahrami et al. [15] proposed a splicing detection based approach using image blur estimation. In their work, after block-based image partitioning, the local blur approximation is applied to find the features from the local blur kernel. Image blocks are categorized in the motion blur-based features to produce the invariant blur forgeries. Finally, a fine splicing localization is performed to enhance the accuracy of region edges. And, the differences in these image areas are detected as a sign of forgeries. Their results are able to estimate blur-type image splicing manipulations. Zhan et al. [16] suggested a splicing detection technique that is based on irrelevant areas between spliced and original image patches. These areas are computed by minimum eigenvalues estimated using PCA without knowing any prior knowledge. The similarity matrix for PCA is estimated using a similar pixel strategy so that local minimum eigenvalue is obtained for the separation of spliced portions of images. Cozzolino and Verdoliva [17] implemented a method in which expressive local features are obtained from noise residue of image and provide to the autoencoder as input to generate the implicit model of the data. This autoencoder finds the spliced region as an anomaly. Their demonstrated results are able to identify spliced-based forgery detection. Abrahim et al. [18] proposed a technique to find spliced regions using image texture properties. They used various texture features to capture the edge of objects and color features. Using these features, a feature vector is generated to find the spliced object. Three features are obtained and trained using ANN. In the second model, the majority voting for three features is performed. By combining different features and information the NN classifies the spliced image. Song et al. [19] also proposed motion blur-based forgery detection. Their approach is divided into three parts. In the first step, a CNN-based motion blur kernel estimation method is implemented to find image patches available in the image. In the second step, a shared motion blur kernel-based image manipulation detection technique is suggested to detect whether a cluster of motion blur kernels is estimated from the same 3D camera trajectory. In the last part, a consistency propagation approach is implemented to specify the forged image portions. Their results are compelling and detect motion blur based-image splicing.

There exist some hybrid approaches to detect pixel-based forgeries. Li et al. [20] proposed a method to enhance the functioning of forgery localization using tampering possibility maps. They estimated statistical feature-based identifier and copy-move forgery detector and, further, adjusted these results to find tampered possibility maps. Afterward, exploring these maps and comparing them with many fusion schemes, they finally localized their results. Their findings are improved over other techniques. Korus and Huang [21] investigated a multi-scale analysis technique that uses many candidates tapering maps to analyze the windows which are more reliable and produce better localization resolution of images. They combined three techniques for multi-scale fusion and verified their results *i*th other state-of-art techniques. This hybrid technique can detect image forgeries. For more work related to pixel-based techniques can be referenced from [22–26]. A comparative study between various pixel-based approaches is shown in Table 1.

Table 1 Comparative study between various pixels based approaches

Author citation	Features estimated	Classifier used	Performance measure
[5]	Speed up robust features	Nearest neighbour distance ratio	85.35%
[6]	Zernike moments, principal components transformations, Fourier Mellin transform	Dense linear fitting	95.92%
[7]	MROGH and hybrid using MROGH and HH	G2NN	94%
[8]	Polar cosine transform	RANSAC	90%
[9]	Speed up robust features	Rg2NN	93.3%
[11]	Hash values	GZCL	96%
[12]	Cellular automata and local binary pattern	FL ANN	Not defined
[13]	Color moments, color layout features, color and edge directive descriptive	Neural network	Not defined
[15]	Local blur kernals	EM algorithm	92.8%
[16]	Local minimum eigenvalues	Threshold-based clustering	94%
[17]	Image residuals	Auto encoder network	Not defined
[26]	Speed up robust features	Threshold-based clustering	98%

(B) Source And JPEG Compression-Based Forgery Detection

Identification of camera source using which photos are captured has a significant importance in digital image forensics. Using source-based techniques, the identification of the source camera can be identified. These techniques identify and extract the features of image acquisition devices using the source processing elements such as lens aberration, sensor deficiencies, and color filter array (CFA), etc. By analyzing the irregularities in these parameters, source-based forgeries can be detected. On the other side, the compression-based techniques find the irregularities in the JPEG compression format. The primary purpose of these techniques is to find the area of images where the differences of JPEG compression are detected. These approaches for JPEG-based methods use block-based artifacts caused due to JPEG compression. These types of inconsistencies are used to locate the compression- and resampling-based forgeries in images. The techniques related to source and compressions are discussed in this section.

Choi et al. [27] proposed a source identification approach using intrinsic lens aberration. They focused on the features if lens radial distortions and obtain such features for all the pixels of images. They used a classifier to classify the source camera. Their results are producing 87–91% of source identification accuracy. Lukas et al. [28] implemented a novel method to detect the source using the sensor's pattern noise. For each camera, they estimated the reference pattern noise as a sign of uniqueness of the camera. They referred to the reference noise as a spectrum of identity for multiple images. Their experiment shows that their technique is able to detect the source of the image effectively. Chan et al. [29] proposed a confidence map and pixel-weighted correlation method to recognize the camera model. A denoizing approach is implemented to find the photo response non-uniformity (PRNU) between real and fake images. Non-linear regression model is used to test the noise in images and, subsequently, a confidence map for pixels is estimated. This identifies the PRNU noise and therefore the camera model. In a similar kind of work, Tapinar et al. [30] show that for a given numerous seam carved images of a camera, the source identification is possible using PRNU. Source identification is possible for uncarved blocks of 50 by 50. They estimated PRNU from the images using these blocks and identification of the camera model is to be carried out. A recent work by Yang et al. [31] suggested a solution to identify the camera using content-adaptive fusion residual networks. This image is divided into three sub-sets, namely saturation, smoothness, and others to find image content. Then, they trained the fusion network for saturated images through transfer learning. The fusion network is formed with three parallel residual networks, and the differences in these networks are recorded. These are obtained as per the presented feature data. The convolution is performed in the initial stage to improve the features. Finally, these features find the differences in image data to justify the source-based forgery. There are other approaches also available in the literature which identified CFA-based inconsistencies and, therefore, camera models.

Nguyen and Katzenbeisser [32] proposed a resampling detection-based technique. They combined various resampling techniques and hybrid median filtering for covering the hints of resampling. But, their technique is robust to identify resampling artifacts to find the traces of forgery. Their results show that their results are better even against various attacks on images. Kirchner and Gloe [33] find resampling detection using re-compression in images. The blocking artifacts rise during the compression and the previous compression helps to locate the forgery They applied affine transformation on compressed images to identify the forgeries Su et al. [34] derived a method based on image resampling detection algorithm for blind deconvolution by proposing an inverse filtering process. Their method is effectively avoiding the interfaces that arise due to JPEG block artifacts. The results of the technique are convincing. Zach et al. [35] suggested an approach for powering the identification of JPEG ghosts. These ghosts are expended to differentiate the single and twofold JPEG compression and this could be a common clue for detection image forgeries. They estimated jpeg ghosts automatically and, finally, manipulation in images. The outcomes of this technique are comparatively better to other such methods. Wang and Zhang [36] propose a double JPEG compression detection approach using CNN. CNN is used to design and classifies the histograms of DCT coefficients. These coefficients differ as per single and double compression. The localization results are obtained according to classification and tampered areas are identified. Their results are able to find tampering in double compression and forgery localization. Pasquini et al. [37] proposed a technique for detecting JPEG compression artifacts stored in various image formats. They computed DCT coefficients from 8 by 8 blocks and applied a hypothesis of no compression and compression with specific quality features. Further to classification, tests are conducted using statistics relying on the discriminative threshold. This statistical analysis is based on the Gaussian distribution of the coefficients which provides better detectors for any uncompressed image. These identify the differences in image data find the manipulations. Other relevant studies in these areas can be referred to [1, 2, 38–40].

(C) **Physics- and Geometric-Based Forgery Detection**

Physics-based approaches included the analysis of inconsistencies in lighting environments. These methods further include 2D and 3D lighting and lighting environments. These methods believe in the supposition that any change in the image will change the geometrical and physical characteristics of images and the changed aspects can be used to find the traces of forgeries. In lighting-based techniques, the direction of light sources are estimated to locate the forged regions or objects in the images for various kind of lighting environments including from simple to complex. Very few researchers explored this area of forgery detection and very limited literature is available based on this type of forgery detection techniques. Some of the excellent works in this area are discussed here in this paper.

Johnson and Farid [41] proposed a technique to detect forgeries using 2D and 3D lighting. They estimated intensity parameters and respective lighting differences in image areas to localize the forgeries. Their results can find the directions for infinite light sources. The same researchers in [42] described a technique for finding forgeries by detecting lighting inconsistencies in the complex lighting environment. They show a technique to estimate the complex lighting with low-dimensional model and estimated the parameters of this model using spherical coefficients. The irregularities in these lighting coefficients are used as an indication of image manipulation. Kee and Farid [43] proposed a technique to estimate 3D lighting using Lambertian reflectance model. Their technique is based on the estimation of spherical harmonics and 3D is estimated for the face profile of the objects. The limitation of the technique is that it works only for the face profiles and respective 3D models but not for other generalized objects. 3D surface normals are estimated from this model to identify the image forgery. Peng et al. [44] implemented an improved 3D lighting environment based approach using a general surface reflection model. The considered local geometry and texture information for their reflection model. Their method is more applicable for objects like human faces which are non-convex and bumpy. Their results of the technique are effective for finding forgery detection. Kumar and Srivastava [45] designed a lighting based technique to detect light source direction. They estimated lighting features from various objects of the image using the surface normal and intensity estimation. They selected patches areas from the same intensity portion and estimated light source directions. For a change in the intensity, their technique performs poorly. Their result detects forgeries for similar kinds of image intensity patches using lighting estimation. Riess [46] proposed a method to optimize the constraints of surface albedo and shape from shading. Riess detected image forgery using illumination estimation from an image. The author considered objects boundary to estimate the direction of the light environment. Their results are able to detect lighting using illumination profiles. Kumar and Srivastava [47] proposed a technique to identify the inconsistencies in lighting direction. They estimated angle of incidence for each object present in the image and further calculations for surface normal are done. They considered the RED band to access the surface properties for illumination estimation. Their results are satisfactory and detect forgery for various types of objects available in images. Same authors also proposed lighting-based forgery detection in [48, 49].

Complex lighting-based work is very limited in this domain and very few authors explored this area. The complex lighting is defined when the lighting environment is lighted by more than one light source. That is if more than one light source is available in the scene then that scene environment is considered a complex lighting environment. Some of the works which can be refereed in this direction are [50–54]. A comparison between various physics-based approaches is shown in Table 2.

Table 2 Comparison between various physics-based approaches

Citation	Features used	Classifier used	Performance/error measure
[41]	Localization of occluding boundaries	Kind of least square estimation	Error percentage 18%
[44]	Surface reflectance model for illumination estimation and spherical coefficients	Position-dependent transfer function	89.2%
[45]	Surface geometry	–	Angle error threshold 8°–10°
[46]	Illumination and color features	–	–
[47]	Surface texture and color features	Least square estimation	95%
[50]	Intensity, curve, and recursive refinement fittings	Approximation technique	Mean error of 8.55° and 8.84°
[52]	Calibration object parameters and critical point estimation	–	Error threshold 5°
[53]	Modified critical point estimation and intensity	–	Threshold for an error of 2.5°–4°
[54]	PCA	–	AUC of 98.1% and a DR of 97.4%

3 Conclusion and Future Direction

A thoughtful speculative study is presented in this paper, which highlights image forgery detection techniques. The paper aims to provide a classification on various state-of-art methods so that researchers can explore more and gain a better understanding of these techniques in this area. Some of the valuable and recent researches in this direction are discussed in this paper. Various works state that knowledge of detection features and approaches, modeling techniques, and tools are highly required to work in this domain. Even there is no technique that can identify all types of image forgeries, which is a challenging task. In the future direction of this work, we need to design more powerful approaches that can estimate forgeries in complex lighting environments and a technique that can provide a higher percentage of successful detection of various types of image manipulations. Also, the requirement of a dataset is much needed as most of the researcher establishes their work using either a limited number of forged images or small datasets for a kind of forgeries.

References

1. M. Kumar, S. Srivastava, Image forgery detection based on physics and pixels: a study. Aust. J. Forensic Sci. **51**(2), 119–134 (2019)
2. S. Walia, K. Kumar, Digital image forgery detection: a systematic scrutiny. Aust. J. Forensic Sci. (2018). https://doi.org/10.1080/00450618.2018.1424241
3. H. Farid, Image forgery detection. IEEE Signal Process Mag. **26**(2), 16–25 (2009)
4. M. Dilshad, S. Ghrera, V. Tyagi, Pixel based image forgery detection: a review. IETE J. Educ. **55**(1), 40–44 (2014)
5. E. Silva, T. Carvalho, A. Ferreira, A. Rocha, Going deeper into copy-move forgery detection: exploring image telltales via multi-scale analysis and voting processes. J. Vis. Commun. Image Represent. **26**, 16–32 (2015)
6. D. Cozzolino, G. Poggi, L. Verdoliva, Efficient dense-field copy-move forgery detection. IEEE Trans. Inf. Forensics Secur. **10**(11), 2284–2297 (2015)
7. L. Yu, Q. Han, X. Niu, Feature point-based copy-move forgery detection: covering the non-textured areas. Multimedia Tools Appl. **75**(2), 1159–1176 (2016)
8. M. Zandi, A. Mahmoudi-Aznaveh, A. Talebpour, Iterative copy-move forgery detection based on a new interest point detector. IEEE Trans. Inf. Forensics Secur. **11**(11), 2499–2512 (2016)
9. X.-Y. Wang, S.L.Y.-N. Liu, Y. Niu, H.-Y. Yang, Z.-L. Zhou, A new keypoint-based copy-move forgery detection for small smooth regions. Multimedia Tools and Applications **76**(22), 23353–23382 (2017)
10. J.-C. Lee, C.-P. Chang, W.-K. Chen, Detection of copy–move image forgery using histogram of orientated gradients. Inf. Sci. **321**, 250–262 (2015)
11. Z. Liang, G. Yang, X. Ding, L. Li, An efficient forgery detection algorithm for object removal by exemplar-based image inpainting. J. Vis. Commun. Image Represent. **30**(2015), 75–85 (2015)
12. D.-Y. Huang, C.-N. Huang, W.-C. Hu, C.-H. Chou, Robustness of copy-move forgery detection under high JPEG compression artifacts. Multimedia Tools Appl. **76**(1), 1509–1530 (2017)
13. H. Zhou, Y. Shen, X. Zhu, B. Liu, Z.F.N. Fan, Digital image modification detection using color information and its histograms. Forensic Sci. Int. **266**, 379–388 (2016)
14. X. Zhao, S. Wang, S. Li, J. Li, Passive image-splicing detection by a 2-D noncausal markov model. IEEE Trans. Circuits Syst. Video Technol. **25**(2), 185–199 (2015)

15. K. Bahrami, A.C. Kot, L. Li, H. Li, Blurred image splicing localization by exposing blur type inconsistency. IEEE Trans. Inf. Forensics Secur. **10**(5), 999–1009 (2015)
16. L. Zhan, Y. Zhu, Z. Mo, An image splicing detection method based on PCA minimum eigenvalues. J. Inf. Hiding Multimedia Signal Process. **7**(3), 610–619 (2016)
17. D. Cozzolino, L. Verdoliva, Single-image splicing localization through autoencoder-based anomaly detection, in IEEE International Workshop on Information Forensics and Security (WIFS), Abu Dhabi, United Arab Emirates 2016
18. A.R. Abrahim, M.S.M. Rahim, G.B. Sulong, Splicing image forgery identification based on artificial neural network approach and texture features. Cluster Comput. **22**(S1), 647–660 (2019)
19. C. Song, P. Zeng, Z. Wang, T. Li, L. Qiao, L. Shen, Image forgery detection based on motion blur estimated using convolutional neural network. IEEE Sens. J. (2019). https://doi.org/10.1109/jsen.2019.2928480
20. H. Li, W. Luo, X. Qiu, J. Huang, Image forgery localization via integrating tampering possibility maps. IEEE Trans. Inf. Forensics Secur. **12**(5), 1240–1252 (2017)
21. P. Korus, J. Huang, Multi-scale fusion for improved localization of malicious tampering in digital images. IEEE Trans. Image Process **25**(3), 1312–1326 (2016)
22. F. Akhtar, H. Qayyum, Two fold image forgery detection system using combined keypoint based method and block based method. J. Inf. Commun. Technol. Robotic Appl. **9**(2), 62–70 (2018)
23. X.-Y. Wang, L.-X. Jiao, X.-B. Wang, H.-Y. Yang, P.-P. Niu, A new keypoint-based copy-move forgery detection for color image. Appl. Intell. **48**(10), 3630–3652 (2018)
24. X.-Y. Wang, Y.-N. Liu, H. Xu, P. Wang, H.-Y. Yang, Robust copy–move forgery detection using quaternion exponent moments. Pattern Anal. Appl. **21**(2), 451–467 (2018)
25. H.-Y. Yang, S.-R. Qi, Y. Niu, P.-P. Niu, X.-Y. Wang, Copy-move forgery detection based on adaptive keypoints extraction and matching. Multimedia Tools Appl. 1–28 (2019). https://doi.org/10.1007/s11042-019-08169-w
26. P. Srivastava, M. Kumar, V. Deep, P. Sharma, A technique to detect copy-move forgery using enhanced SURF. Int. J. Eng. Adv. Technol **8**(6S), 676–680 (2019)
27. K.S. Choi, E.Y. Lam, K.K. Wong, Source camera identification using footprints from lens aberration, in *Proceedings of SPIE-IS&T Electronic Imaging*, SPIE, vol. 6069 2006
28. J. Lukas, J. Fridrich, M. Goljan, Digital camera identification from sensor pattern noise. IEEE Trans. Inf. Forensics Secur. **1**(2), 205–214 (2006)
29. L.-H. Chan, N.-F. Law, W.-C. Siu, A confidence map and pixel-based weighted correlation for PRNU-based camera identification. Digit. Invest. **10**(2013), 215–225 (2013)
30. S. Taspinar, M. Mohanty, N. Memon, PRNU-based camera attribution from multiple seam-carved images. IEEE Trans. Inf. Forensics Secur. **12**(12), 3065–3080 (2017)
31. P. Yang, R. Ni, Y. Zhao, W. Zhao, Source camera identification based on content-adaptive fusion residual networks. Pattern Recogn. Lett. **119**, 195–204 (2019)
32. H. C. Nguyen, S. Katzenbeisser, in Robust resampling detection in digital images, ed. by B. De Decker, D.W. Chadwick. *Communications and multimedia security. CMS 2012*. Lecture Notes in Computer Science, vol. 7394 (Springer, Berlin, Heidelberg, 2012)
33. M. Kirchner, T. Gloe, On resampling detection in re-compressed images in *First IEEE International Workshop on Information Forensics and Security (WIFS)*, London, UK, 2009
34. Y. Su, X. Jin, C. Zhang, Y. Chen, Hierarchical image resampling detection based on blind deconvolution. J. Vis. Commun. Image Represent. **48**, 480–490 (2017)
35. F. Zach, C. Riess, E. Angelopoulou, in Automated image forgery detection through classification of JPEG ghosts, ed. by A. Pinz, T. Pock, H. Bischof, F. Leberl. *Pattern Recognition. DAGM/OAGM 2012. Lecture Notes in Computer Science*, vol. 7476 (Springer, Berlin, Heidelberg, 2012)
36. Q. Wang, R. Zhang, Double JPEG compression forensics based on a convolutional neural network. EURASIP J. Inf. Secur. **23**(2016), 1–12 (2016)
37. C. Pasquini, G. Boato, F. Pérez-González, Statistical detection of JPEG traces in digital images in uncompressed formats. IEEE Trans. Inf. Forensics Secur. **12**(12), 2890–2905 (2017)

38. P. Korus, Digital image integrity—a survey of protection and verification techniques. Digit. Signal Proc. **71**(2017), 1–26 (2017)
39. S. Mandelli, P. Bestagini, S. Tubaro, D. Cozzolino, L. Verdoliva, Blind detection and localization of video temporal splicing exploiting sensor-based footprints, in *26th European Signal Processing Conference (EUSIPCO)*, Rome, Italy, 2018
40. B. Gupta, M. Tiwari, Improving source camera identification performance using DCT based image frequency components dependent sensor pattern noise extraction method. Digit. Invest. **24**(2018), 121–127 (2018)
41. M. K. Johnson, H. Farid, Exposing digital forgeries by detecting inconsistencies in lighting, in ACM Multimedia and Security Workshop, New York USA, 2005
42. M. K. Johnson, H. Farid, Exposing digital forgeries in complex lighting environments, IEEE Trans. Inf. Forensics Secur. **2**(3), 450–461 (2007)
43. E. Kee, H. Farid, Exposing digital forgeries from 3-D lighting environments, in IEEE international workshop on information forensics and security, Seattle, WA, USA, 2010
44. B. Peng, W. Wang, J. Dong, T. Tan, Improved 3D lighting environment estimation for image forgery detection, in IEEE International Workshop on Information Forensics and Security (WIFS), 2015
45. M. Kumar, S. Srivastava, Identifying photo forgery using lighting elements. Indian J. Sci. Technol. **9**(48), 1–5 (2016)
46. C. Riess, in Illumination analysis in physics-based image forensics: a joint discussion of illumination direction and color, ed. by A. Piva, I. Tinnirello, S. Morosi. *Digital Communication. Towards a Smart and Secure Future Internet. TIWDC 2017. Communications in Computer and Information Science*, vol. 766 (Springer, Cham, 2017)
47. M. Kumar, S. Srivastava, Image authentication by assessing manipulations using illumination. Multimedia Tools Appl. **78**(9), 12451–12463 (2019)
48. M. Kumar, S. Srivastava, in Image tampering detection based on inherent lighting fingerprints, ed. by D. Hemanth, S. Smys. *Computational Vision and Bio Inspired Computing*. Lecture Notes in Computational Vision and Biomechanics, vol. 28 (Springer, Cham, 2018)
49. M. Kumar, A. Rani, S. Srivastava, Image forensics based on lighting estimation. Int. J. Image Graph. **19**(3):195004:1–195004:14 (2019)
50. J. Lopez-Moreno, E. Garces, S. Hadap, E. Reinhard, D. Gutierre, Multiple light source estimation in a single image. Comput. Graph Forum **32**(8), 170–182 (2013)
51. S. Xu, A. Wallace, Recovering Surface reflectance and multiple light locations and intensities from image data. Pattern Recogn. Lett. **29**(11), 1639–1647 (2008)
52. Y. Wang, D. Samaras, Estimation of multiple directional light sources for synthesis of mixed reality images, in 10th Pacific Conference on Computer Graphics and Applications, Beijing, China, 2002
53. M. Kumar, S. Srivastava, N. Uddin, Forgery detection using multiple light sources for synthetic images. Aust. J.Forensic Sci. **51**(3), 243–250 (2019)
54. A. Mazumdar, P.K. Bora, Estimation of lighting environment for exposing image splicing forgeries. Multimedia Tools Appl. **78**(14), 19839–19860 (2019)

Using Automated Predictive Analytics in an Online Shopping Ecosystem

Ruchi Mittal

Abstract This paper is based on a UCI dataset analysis, which includes data generated through Google analytics. The data attributes describe the shopping intent of consumers and their likelihood of exiting the e-commerce website during browsing. This study has selected five attributes out of a total of 18 that were generated and seeks to capture the shopping intent based on the website browsing behavior of shoppers. All selected attributes are in numeric form, and the dependent variable is pagevalue which represents the average value for the user-visited webpage before the completion of a transaction. The proximity of a forthcoming special day is checked for its mediation effect on other independent variables based on web browsing behavior. The findings give some very interesting and relevant insights into the use of analytics on human online behavior.

Keywords Predictive analytics · Structural model · Shopping intent · Browsing behavior

1 Introduction

The online shopping industry is growing rapidly and customers are increasingly preferring to buy online because of varied factors such as ease, variety, schemes, and saving their time and cost. It is thus a promising domain for the researchers to make use of the technology to understand the changing requirements of the online shoppers to help online retailers to make appropriate decisions to meet the competition and to have an edge over their competitors. For this purpose, many different types of demographical, behavioral, and other information can be collected from the customers, who are making online purchases that are quite significant to be analyzed critically in order to provide solutions to many business problems. Many studies have been conducted to analyze the impact of demographic factors of customers on online purchases and one of such study [1] indicated that in context to online shopping, only

R. Mittal (✉)
Chitkara University Institute of Engineering and Technology, Chitkara University, Punjab, India
e-mail: ruchi.mittal@chitkara.edu.in

S. S. Dash et al. (eds.), *Intelligent Computing and Applications*,
Advances in Intelligent Systems and Computing 1172,
https://doi.org/10.1007/978-981-15-5566-4_20

a few of the demographic factors like age, gender, income, and education are significant for the researchers to analyze the predictive power of the online purchase corresponding to particular products. Previous research extended the study to understand the important factors contributing to the success of the online purchase experience other than the basic demographics and concluded that social connections and prior online shopping experience also attribute to various online shopping categories [2]. Reference [3] recognized the problem faced by the online customers while making a decision for the product just on the basis of a given image on the website, and thus some of the other associated information such as price, brand, and previous customers' reviews on price, reliability, durability, and quality and shopping experience can be vital in building the faith in the new customer and in increasing their online shopping convenience. They proposed a ranking allocation mechanism to the previous customers' reviews as it is difficult for the new shoppers to get through all the reviews [3].

Data mining is the inclusion of many disciplines such as database technologies, statistics, artificial intelligence, machine learning, etc., that advance the analytical capabilities of the managers in uncovering the hidden facts, important for the decision making of an organization. Data mining is quite popular in all the business sectors (online and onsite) to optimize their cost and benefits. Business analysts and data scientists can choose among various functionalities of data mining such as classification and prediction, clustering, association analysis, class description, outlier or time-series analysis, etc., based on the type of data and the and the requirement of the application and can suggest intelligent insights for the growth and sustainability of the organization.

Also with the availability of high-speed and low-cost internet facility, selling and buying online is widely becoming popular and convenient. Incorporation of the appropriate data mining tools and technologies can help online retailers to collect, store, process and analyze the large amount customers' data and can exploit the benefits in important decision-making such as identifying the potential customers; market basket analysis; future sales prediction; product planning; segmenting and profiling customers for target marketing; and improved customer relationship management for online shopping organizations [4].

This study takes a different and unique approach to online shopping. An existing open-source dataset was selected from the UCI repository and the following research question was framed based on the web browsing behavior of customers: How can we predict the "pagevalue" of an e-commerce site based on select attributes and how do certain attributes mediate the relation among these different attributes? Here "pagevalue" refers to the average value for a webpage that a shopper visited before completing an e-commerce transaction.

To achieve the objectives the following attributes were selected as predictors of "pagevalue": "product-related information", "bounce rate", "exit rate". An upcoming "special day" was used as a mediator in the hypothesized relationships between the predictors and the criterion attribute (or independent and dependent attributes).

The paper is organized as follows; in Sect. 2 the review of the literature is discussed. This is followed by research methodology in Sect. 3. Section 4 deals with the results and analysis Sect. 5 deals with discussions and interpretation. The final section is the conclusion and future research.

2 Review of the Literature

A prior study emphasized the need for various data mining techniques such as segmentation, association, and classification on the vast amount of data collected from the emerging style of online shopping instead of the conventional shopping approach and also described that how such mining and analysis can help to have better customer relationship management (CRM) and to further optimize the profits of an organization [5]. In order to analyze the customers' requirements and preferences, a novel method based on online customers' reviews was examined by first clustering the customers and then by identifying the relevant and salient attributes of a product in these clusters based on the customer choice criteria [6]. Looking at customer feedback as an integral part of CRM, a previous study, after identifying the importance of feedback system, not only to analyze the customer satisfaction post-purchase/usage, but also to help the app developers in future developments and modifications, the authors presented an ordered review of different studies in the domain of online product reviews and customers' feedback on mobile app store and also acknowledged the effectiveness of applying data mining techniques for the same [7]. In a separate study, the authors tried to aid the business decision-making by using data mining functionalities such as decision trees to discover the type of buyers in the times when discounts were offered by the stores and also, to devise the strategies to increase the customer base for the organization [8]. In another interesting study, a prototype of an application was discussed which provides instant product suggestions to potential shoppers based on even a few preferred product attributes. Based on footwear retailing, the researchers collected the data related to the parameters of foot and shows description of the ease on feet and then applied data mining functionalities such as decision tree, neural network and KNN to come up with a predictive model giving the relevant parameters such as the length of the shoe and also stated that decision tree algorithm has given the best predictive accuracy [9]

Another study on e-commerce recognized the potential of an automated and intelligent system to extract knowledge from the tremendously growing data in e-commerce and applied and compared different data mining classification techniques using Weka machine learning software on a credit-card dataset for identifying possible frauds during online payments to sustain the possibility of paying online which is very important in e-commerce systems. The authors also described that J48 outperforms other classifiers on the mentioned dataset [10]. Another study categorized online shoppers based on their demographics and consumption patterns to further understand the behavioral aspects of shoppers belonging to these identified segments to help the retailers to devise appropriate strategies to cater to the needs of different types of customers. The study used a soft-clustering method for segmentation as it gives better cluster quality over other approaches [11]. In another study, researchers identified the importance of customers' reviews for the products purchased online to not only help the retailers to make more accurate business decisions but also the new buyers to decide on the purchase of a particular product. The authors used Weka machine learning tool to mine such reviews generated from online purchases and conducted a sentiment analysis (SA) followed by classification of these reviews using different classification algorithms [12]. In a different study, the authors used TANGARA machine learning tool in order to investigate the reviews of buyers on different products and brands and for this purpose, the authors applied various mining techniques of classification, clustering, association and correlation to get online business insights like the relationship between customers and sellers and also to classify customers based on their remarks on different products [13]. In a separate study, the authors provided a comparative analysis of 11 data mining techniques of clustering and classification on the online customers' shopping dataset to identify the most appropriate classifier and proposed that decision tree classification scheme gives better results over clustering for customer segmentation on e-commerce websites [14].

Many studies have been carried out emphasizing the use of tools and technologies to analyze the vast amount of data being generated in various domains of business organizations existing online and offline, so based on the review of previous studies, this study aims to get the insights for online shopping behavior of customers based on select attributes drawn from the web browsing behavior of customers.

3 Research Methodology

3.1 Dataset

The dataset has been adopted from the UCI machine learning repository. The dataset comprises a total of 12,330 records of transactions and deals with the intention of shoppers towards purchases while shopping online [15]. The original dataset includes 18 attributes (10 numerical and 8 categorical attributes) extracted threw aggregated page views, which were collated by Google Analytics. Select attributes have been shortlisted for the purpose of this study, namely: (1) product-related duration, (2) bounce rate, (3) exit rate, (4) pagevalue, and (5) special day.

3.2 Description of Attributes

Product-related duration refers to the duration of time spent by the potential consumer on the product-related page of the e-commerce portal. *Bounce rate* refers to the feature for a web page that calculates the percentage of visitors, who enter the site from that page and then leave without triggering any other requests to the Google

Analytics server during that session. The value of *Exit Rate* feature for a specific web page is calculated as for all page views to the page, the percentage that was the last in the session. The *Pagevalue* feature represents the average value for a web page that a user visited before completing an e-commerce transaction. The *Special Day* feature indicates the closeness of the site visiting time to a specific special day (e.g., Mother's Day, Valentine's Day) in which the sessions are more likely to be finalized with transaction.

3.3 Predictive Model

The data analysis was done using AMOS version 18—structural equation modeling (SEM) and tested a total of seven hypotheses drawn from the conceptual model as per Fig. 1.

"Structural equation modeling (SEM) is a collection of data analyses techniques that allow a set of relationships between one or more independent variables (IVs), either continuous or discrete, and one or more dependent variables (DVs), either continuous or discrete, to be examined. Both IVs and DVs can be either factors or measured variables. Structural equation modeling is also referred to as causal modeling, causal analysis, simultaneous equation modeling, analysis of covariance structures, path analysis, or confirmatory factor analysis" [16]. SEM has been very effectively used for prediction and model testing in earlier studies [17–20]. Based on the model as per Fig. 1, eight hypotheses have been formulated. The SEM analysis performs the hypothesis testing and the results are presented in the next section.

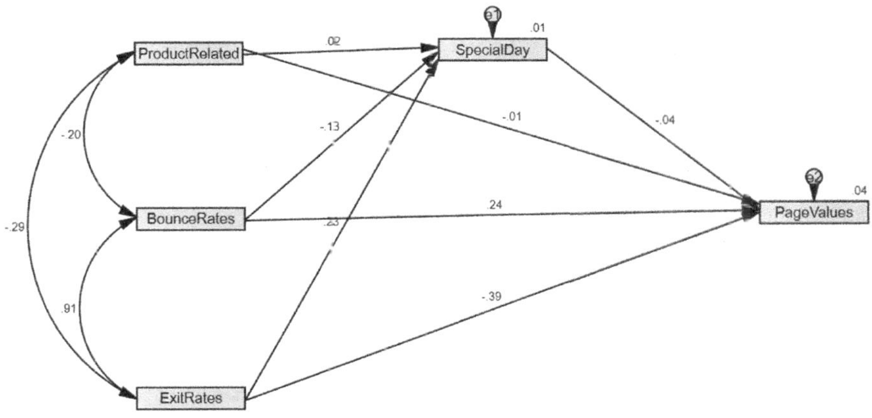

Fig. 1 Conceptual model

4 Results and Analysis

See Tables 1, 2, 3, 4, 5 and 6.

Table 1 Regression weights: (group number 1—default model)

			Estimate	S.E.	C.R.	*P*	Label
Special day	<—	Product related	0.000	0.000	1.623	0.105	
Special day	<—	Bounce rate	−0.533	0.091	−5.844	***	
Special day	<—	Exit rate	0.922	0.093	9.906	***	
Pagevalue	<—	Special day	−3.840	0.828	−4.636	***	
Pagevalue	<—	Product related	−0.004	0.004	−1.105	0.269	
Pagevalue	<—	Bounce rate	91.415	8.395	10.889	***	
Pagevalue	<—	Exit rate	−149.497	8.597	−17.390	***	

Table 2 Standardized regression weights: (group number 1—default model)

			Estimate
Special day	<—	Product related	0.015
Special day	<—	Bounce rate	−0.130
Special day	<—	Exit rate	0.225
Pagevalue	<—	Special day	−0.041
Pagevalue	<—	Product related	−0.010
Pagevalue	<—	bounce rate	0.239
Pagevalue	<—	Exit rates	−0.391

Table 3 Standardized indirect effects (group number 1—default model)

	Product related	Exit rate	Bounce rate	Special day
Special day	0.000	0.000	0.000	0.000
Pagevalue	−0.001	−0.009	0.005	0.000

Table 4 Standardized indirect effects—two-tailed significance (BC) (group number 1—default model)

	Product related	Exit rates	Bounce rates	Special day
Special day	...	...	...	...
Pagevalue	0.035	0.001	0.001	...

Table 5 Standardized direct effects (group number 1—default model)

	Product related	Exit rate	Bounce rate	Special day
Special day	0.015	0.225	−0.130	0.000
Pagevalue	−0.010	−0.391	0.239	−0.041

Table 6 Standardized direct effects—two-tailed significance (BC) (group number 1—default model)

	Product related	Exit rate	Bounce rate	Special day
Special day	0.039	0.001	0.001	...
Pagevalue	0.211	0.001	0.001	0.001

5 Discussions and Interpretation

H1 There is a significant relationship between *special day* and *product-related duration.*

Table 1 shows the regression weight in the relationship between special day and product-related duration. The regression coefficient value of 0.000, standard error value of 0.000, critical ratio 1.623 which is below 1.96, and p value greater than <0.01 show that there is no significant relationship between special day and product-related duration.

H2 There is a significant relationship between *special day* and *bounce rate.*

Table 1 shows the regression weight in the relationship between special day and bounce rate. The regression coefficient value of −0.533, standard error value of 0.091, critical ratio −5.844 which is below 1.96, and p value <0.01 show that special day has a significant relationship with bounce rate. Table 2 also indicates a negative −0.130 standardized beta value of between the relationship of special day and bounce rate. This indicates that the bounce rate is a negative significant predictor of special day. Hence, it is concluded that bounce rate significantly affects the special day in a negative manner, which means that more the bounce rate lower the probability of a forthcoming special day.

H3 There is a significant relationship between *special day* and *exit rate.*

Table 1 shows the regression weight in the relationship between special day and exit rate. The regression coefficient value of 0.922, standard error value of 0.093, critical ratio 9.906 which is above 1.96, and p value<0.01 show that special day has a significant relationship with exit rate. Table 2 also indicates a positive standardized beta value of 0.225 between the relationship of special day and exit rate. This indicates that exit rate is a positive significant predictor of special day. Hence, it is concluded that exit rate significantly affects the special day in a positive manner which means that more the exit rate higher the probability of a forthcoming special day.

H4 There is a significant relationship between *pagevalue* and *special day.*

Table 1 shows the regression weight in the relationship between pagevalue and special day. The regression coefficient value of −3.840, standard error value of 0.828, critical ratio −4.636, and *p* value <0.01 show that special day has a significant relationship with pagevalue. Table 2 also indicates a negative standardized beta value of −0.041 between the relationship of special day and pagevalue. This indicates that special day is a negative significant predictor of pagevalue. Hence, it is concluded that special day significantly affects the special day in a negative manner, which means that higher the probability of a forthcoming special day lower the pagevalue.

H5 There is a significant relationship between *pagevalue* and *product related duration.*

Table 1 shows the regression weight in the relationship between pagevalue and product-related duration. The regression coefficient value of −0.004, standard error value of 0.004, critical ratio −1.105, and *p* value >0.01 show that product-related duration has a nonsignificant relationship with pagevalue.

H6 There is a significant relationship between *pagevalue* and *bounce rate.*

Table 1 shows the regression weight in the relationship between pagevalue and bounce rate. The regression coefficient value of 91.415, standard error value of 8.395, critical ratio 10.889 which is above the threshold of 1.96, and *p* value <0.01 show that bounce rate has a significant relationship with pagevalue. Table 2 also indicates a positive standardized beta value of 0.239 between the relationship of bounce rate and pagevalue. This indicates that bounce rate is a positive significant predictor of pagevalue. Hence, it is concluded that bounce rate significantly affects the pagevalue in a positive manner, which means that more the bounce rate higher the pagevalue.

H7 There is a significant relationship between *pagevalue* and *exit rate.*

Table 1 shows the regression weight in the relationship between pagevalue and exit rate. The regression coefficient value of −149.497, standard error value of 8.597, critical ratio of −17.390 which is above the threshold of 1.96, and *p* value <0.01 show that exit rate has a significant relationship with pagevalue. Table 2 also indicates a negative standardized beta value of −0.391 between the relationship of exit rate and pagevalue. This indicates that exit rate is a negative significant predictor of pagevalue. Hence, it is concluded that exit rate significantly affects the pagevalue in a negative manner which means that more the exit rate higher the pagevalue.

H8 An upcoming special day mediates the relationship between (a) product-related duration and pagevalue, (b) bounce rate and pagevalue, (c) exit rate and pagevalue.

Results of the mediation analysis show that special day mediates the relationship between product-related duration and pagevalue, bounce rate and pagevalue, and exit rate and pagevalue (please refer to Table 3, 4, 5, and 6).

6 Conclusion and Future Research

The following select attributes were shortlisted for the purpose of this study, namely, product-related duration, bounce rate, exit rate, pagevalue and special day. The data were subjected to structural equation modeling (SEM). As per the data analyses, there is no significant relationship between special day and product-related duration. Bounce rate significantly affects the special day in a negative manner which means that more the bounce rate lower the probability of a forthcoming special day. Exit rate significantly affects the special day in a positive manner which means that more the exit rate higher the probability of a forthcoming special day. Special day significantly affects the special day in a negative manner which means that higher the probability of a forthcoming special day lower the pagevalue. Product-related duration has a nonsignificant relationship with pagevalue. Bounce rate significantly affects the pagevalue in a positive manner, which means that more the bounce rate higher the pagevalue. Exit rate significantly affects the pagevalue in a negative manner which means that more the exit rate higher the pagevalue. Special day shows an effective mediation effect as hypothesized in the conceptual model.

In a future study, the balance 13 attributes may be included and different structural models may be created and tested. Different attributes may be checked for mediation effect and some attributes may also be evaluated for moderation effect.

References

1. M.K. Chang, W. Cheung, V.S. Lai, Literature derived reference models for the adoption of online shopping. Inf. Manag. **42**(4), 543–559 (2005)
2. M.B. Naseri, G. Elliott, Role of demographics, social connectedness and prior internet experience in adoption of online shopping: applications for direct marketing. J. Target. Meas. Anal. Mark. **19**(2), 69–84 (2011)
3. K. Zhang, R. Narayanan, A.N. Choudhary, Voice of the customers: mining online customer reviews for product feature-based ranking. *WOSN*, *10*, 11
4. M. Ismail, M.M. Ibrahim, Z.M. Sanusi, M. Nat, Data mining in electronic commerce: benefits and challenges. Int. J. Commun. Netw. Syst. Sci. **8**(12), 501 (2015)
5. M. Riaz, A. Arooj, M.T. Hassan, J.B. Kim, Clustering based association rule mining on online stores for optimized cross product recommendation, in *The 2014 International Conference on Control, Automation and Information Sciences (ICCAIS 2014)*, 2014 (pp. 176–181). IEEE
6. R. Rai, Identifying key product attributes and their importance levels from online customer reviews, in *ASME 2012 international design engineering technical conferences and computers and information in engineering conference*, 2013 (pp. 533–540). American Society of Mechanical Engineers Digital Collection
7. N. Genc-Nayebi, A. Abran, A systematic literature review: opinion mining studies from mobile app store user reviews. J. Syst. Softw. **125**, 207–219 (2017)
8. S. Kumar, R.K. Arora, Analyzing customer behaviour through data mining. IJCATR (Online) **4**, 12 (2015)
9. P. Sudta, K. Kanchan, C. Chantrapornchai, Children shoes suggestion system using data mining. Int. J. Database Theory Appl. **5**(1), 21–36 (2012)
10. F. Alam, S. Pachauri, Detection using weka. Adv. Comput. Sci. Technol. **10**(5), 1731–1743 (2017)

11. R.S. Wu, P.H. Chou, Customer segmentation of multiple category data in e-commerce using a soft-clustering approach. Electron. Commer. Res. Appl. **10**(3), 331–341 (2011)
12. A.A. Kshirsagar, P.A. Deshkar,. Review analyzer analysis of product reviews on WEKA classifiers, in *2015 International Conference on Innovations in Information, Embedded and Communication Systems (ICIIECS),* 2013 (pp. 1–5). IEEE
13. V. Jain, Narula, G.S., M. Singh, Implementation of data mining in online shopping system using tanagra tool. Int. J. Comput. Sci. Eng. (2013)
14. R.A.E.D. Ahmeda, M.E. Shehaba, S. Morsya, N. Mekawiea, Performance study of classification algorithms for consumer online shopping attitudes and behavior using data mining, in *2015 Fifth International Conference on Communication Systems and Network Technologies,* 2015 (pp. 1344–1349). IEEE
15. C.O. Sakar, S.O. Polat, M. Katircioglu, Y. Kastro, Real-time prediction of online shoppers' purchasing intention using multilayer perceptron and LSTM recurrent neural networks. Neural Comput. Appl. 1–16
16. J.B. Ullman, P.M. Bentler, Structural equation modeling. Handb. Psychol. 607–634 (2003)
17. R. Mittal, V. Rattan, Evaluating rule based machine learning classifiers for customer spending in a shopping mall. J. Adv. Res. Dyn. Control Syst. **11**(09), 716–719 (2019)
18. A.S. Kumar, P.K. Chand, A. Mittal, Emotional intelligence and its relationship to employability skills and employer satisfaction with fresh engineering graduates. Int. J. Qual. Res. **13**(3), 735–752 (2019)
19. Vikas Uppal, Amit Mittal, Antecedents of post purchase consumer dissatisfaction in household white goods. J. Adv. Res. Dyn. Control Syst. **11**(5), 1958–1966 (2019)
20. Ruchi Mittal, Predictive modeling for bank direct marketing: a sem based approach on select attributes. Int. J. Adv. Sci. Technol. **27**(1), 283–288 (2019)

Design and Development of Home Automation System

Anjali Pandey, Yagyiyaraj Singh, and Medhavi Malik

Abstract Internet of things is based on the integration of various real-world objects by interconnecting them using the Internet. This concept can be applied to build an automated smart home. A phone application will be used to control various electrical and electronic appliances such as fans, lights, fire alarms, automatic doors, smoke detectors, etc., through various sensors and actuators. The microcontroller used in the proposed prototype is node MCU. It is Wi-Fi equipped and can be programmed using Arduino IDE. The working principle of the proposed system is based on different types of wireless communication techniques such as ZigBee, Wi-Fi, Bluetooth, etc. This enables us to remotely control manage and monitor home appliances. An automated home is smarter, safer, more convenient, time saving as well as energy efficient.

Keywords Home automation · Internet of Things · Server · Mobile devices

1 Introduction

Internet of things is a technology that connects devices, machines, and tools to the Internet by means of wireless technologies. Over 9 billion things are connected to the Internet as of now. It also involves the unification of technologies such as low-power embedded systems, cloud computing, big data, machine learning, and networking. Internet of things contains a network of physical objects which are interconnected to one another through the Internet. Some areas identified as IoT enablers are RFID, nanotechnology, sensors, and smart networks [1].

A. Pandey · Y. Singh · M. Malik (✉)
SRM IST Delhi NCR Campus Ghaziabad, Ghaziabad, India
e-mail: medhavimalik28@gmail.com

A. Pandey
e-mail: Anjalipandey868@gmail.com

Y. Singh
e-mail: yagyiyaraj@gmail.com

S. S. Dash et al. (eds.), *Intelligent Computing and Applications*,
Advances in Intelligent Systems and Computing 1172,
https://doi.org/10.1007/978-981-15-5566-4_21

Benefits of Using Internet of Things

- Optimization: It is mainly required in industries, for example, strict quality control, supply chain management, inventory management.
- Increase in productivity: With the help of IoT, employees can have improved decision-making and efficient communication.
- Better business models: IoT enables companies to provide service based on real-time sensor data and information [2].

Modern Day IoT Applications

- Traffic congestion,
- Smart home,
- Medical applications,
- Smart city,
- Weather prediction,
- Smart parking, and
- Soil moisture monitoring.

IoT Market Share

- Business manufacturing,
- Health care,
- Retail, and
- Security [3].

Home Automation

A smart home includes the interconnection of various devices using a variety of wired and wireless technologies. "It is a home that is equipped with lighting, heating, electronic devices that can be controlled remotely by smart phone or computer". For example, you can switch on the air condition of your room while sitting in any other part of the home or installing an alarm system that sends a notification to your phone the moment alarm rings. An automated home can provide a safe home environment and personalization according to our needs. A smart home optimizes structures, systems, service, and management [4].

Home area network (HAN): It can be referred to as a network within a home that allows controlling the devices remotely. HAN elements are as follows:

- **Internet protocol**—It is a set of protocols and rules for exchanging messages and information among devices in a network connected through the Internet [5].
- **Wired HAN**—In this, we can use various types of wires like optical fiber, power lines, coaxial cables, and telephone lines. It is a low-cost method and provides efficient use of already existing house infrastructure.
- **Wireless HAN**—It is less complex as compared to wired HAN. It uses technologies such as Bluetooth, ZigBee, Wi-Fi, etc.

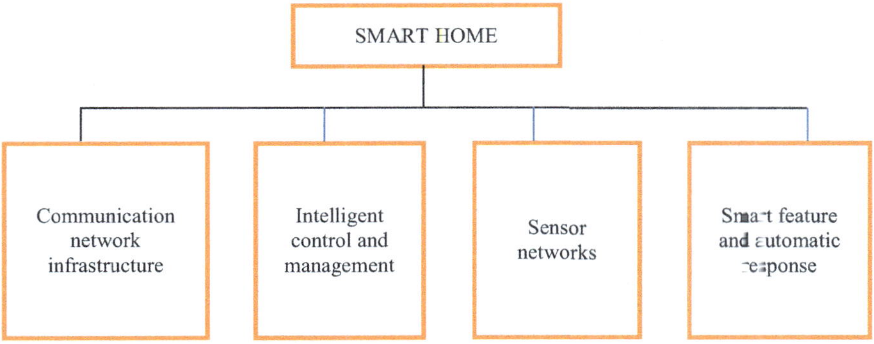

Advantages of Home Automation System

The ability to manage your home's electronic system from one main control system can make your household run smoother, feel better, and save energy. The trick is to find a system that will meet all the demands of your household now and in the future.

1. Energy Saving: It is one of the most important advantages of home automation and can be done in various ways such as automatically switching off the air conditioner when the temperature is optimum or generating energy using photovoltaic cells (solar cells) installed on the rooftop of the house [6].
2. Remote access: This feature will allow you to monitor your home's safety and alter the appliance's settings remotely using your phone.
3. Interoperability: The beauty of a home automation system lies in its ability to tie diverse electronics together so that they can work as one unified system.
4. Variety of user interface: There are many different ways you can control your automated home for example by touching the colorful icons on a portable touch panel of sliding fingers across the screen of your smartphone [7].

2 Proposed Work

The fundamental purpose of this paper is to develop an environment in which one can control home appliances through a mobile application on a smartphone device. All the electronic devices connected to the Internet in the system will run without any human interference.

For building a smart home, the following are the basic requirements:

1. A microcontroller such as Node MCU, Arduino UNO, etc.
2. Various Sensors such as light sensor, temperature sensor, gas sensor, water flow, motion sensor, and many other sensors.
3. Electronics appliances such as lights, fans, AC, television, speakers, surveillance cameras, etc.

Microcontroller (Node MCU):

The information about various sensors is gained by using a microcontroller.

Node MCU, technically isn't a device but it is firmware. There are two types of Node MCU boards, version 0.9 and 1.0.

1. Version 0.9: This board is blue and loaded with ESP-12 chip and it comes with 16 pins
2. Version 1.0: This board is black and comes with ESP-12E (enhanced). And this version comes with 22 pins [8].

ESP8266 is more recent release than the Arduino. There is a 32-bit RISC processor clocked at 80 MHz, along with generous RAM complement and storage for up to 16 Mb of external flash storage.

It has a USB connection for easy interface with the computer.

"Node MCU is based on ESP8266-12E Wi-Fi system on chip loaded with an open source firmware. It is perfect for IoT applications and various other situations where wireless connection is required" [9].

Sensors:

A sensor senses some kind of change in a system and forwards or processes this gathered information in a certain manner.

> "A device which detects or measures a physical property and records, indicates, otherwise responds to it."
>
> —**Oxford dictionary** [10]

When a sensor senses some physical change in a system then it performs input functions and responds to stimuli.

Operation of the sensor is very similar to the sensors in the human body. For example, skin is one of the sensors of the human body. When it senses or receives a pain trigger, then it causes the organism to "be aware"'. Similarly, sensors sense the changes and respond to these changes accordingly.

There are two categories of sensors: Analog and Digital.

1. Analog Sensors: Analog sensors produce a continuous output signal or voltage, which is generally proportional to the quantity being measured.
   Physical quantity such as temperature, speed, pressure, displacement, strain, etc. are all analog quantities and they tend to be continuous in nature.
2. Digital Sensors: Digital sensors produce discrete digital output signals and voltage that are in digital representation of the quantity being measured [3].
   Digital sensors produce a binary output signal in the form of logic "1" or logic "0", ("ON" or "OFF")

Sensor Types:

1. Light sensor: It converts light energy into electrical signals. Example: photodiode, light-dependent sensor.

2. Temperature sensor: It is used to measure temperature. It uses a converter to convert the temperature value to an electrical value. Example: thermocouple, thermistor.
3. Sound sensor: It uses diaphragm which vibrates when hit by sound and this vibration is converted into an electrical signal by the sensor. Example: carbon microphone and piezoelectric crystal [3].

There are a number of such examples of sensors.

Actuators:

An actuator is a component of a machine or a system that moves or controls the mechanism of the system.

An actuator requires a control signal and a source of energy.

Upon receiving a control signal, the actuator responds by converting the energy into mechanical motion [3].

Planning:

There are various technologies that can be incorporated into a Smart Home:

1. Smart lights:
   Light is a very critical component of any home. It affects almost every part of the household. So replacing lights with smart lights in homes can save both our time and money.
   Smart lights are efficient as they can be automatically switched ON and OFF according to the usage.
2. Smart fans:
   Smart fans can be automatically turned ON and OFF after a certain temperature has reached.
3. Smart security system:
   A security system can detect when someone has entered your property or home and sound an alarm. The automation system can send images captured by surveillance cameras to your smartphone.
4. Heating and cooling system:
   It detects the temperature of a room. If the temperature is high then it turns ON the cooling system of the room.

Implementation:

Software Architecture:

To implement home automation, two software are used that are Arduino IDE and mobile app (Blynk app).

Arduino IDE:

The Arduino IDE is an open-source software that is mainly used for writing and compiling the code into the Arduino module (board).

A range of modules available are Arduino UNO, Arduino Mega, Node MCU, and many more.

Each of these modules consists of a microcontroller that is programmed and accepts the information in the form of code.

The IDE environment mainly contains two basic parts: Editor and Compiler.

Editor: It is used for writing the required code.

Compiler: It is used for compiling and uploading the code into the given board.

It supports both C and C++ languages [11].

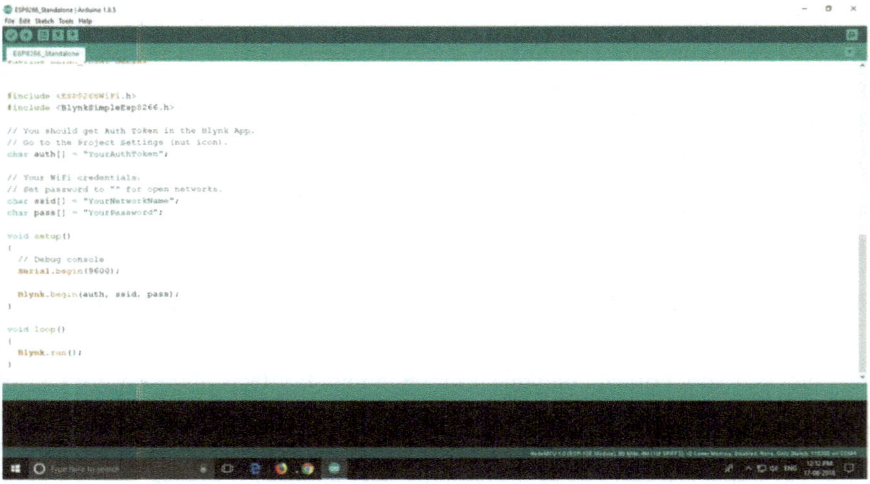

Mobile Application:

Home automation demands a proper home automation mobile app to coordinate all the system components and create an efficient infrastructure. All the home appliances can be controlled using a mobile application.

The operation becomes very easy by using a mobile application.

You can control each and every appliance in the system by just one tap on your mobile screen.

We can create our own mobile application for our system or we can use different pre-made mobile applications.

There are different mobile applications for home automation available on the play store. We just need to download it and set it up according to our requirements. Blynk app is one of the pre-made mobile applications for home automation. In this paper, we will be focusing on building a smart home using Blynk app [12].

Blynk App:

> Blynk app is a new platform that allows you to quickly build interface for controlling and monitoring your hardware projects from your iOS and Android devices. After downloading the Blynk app, you can create a project dashboard and arrange buttons, sliders, graphs and other widgets onto the screen.

Blynk Architecture:

The Blynk platform includes the following components:

1. Blynk app builder: Allows you to build apps for your project using various widgets. It is available for android and iOS platforms.
2. Blynk Server: Responsible for all the communications between your mobile (device) that is running the Blynk app and the hardware.
3. Blynk libraries: Enables communication with the server and processes all the incoming and outgoing commands from your Blynk app and the hardware.
4. Connection types: Blynk app supports the following connection types to connect your microcontroller board (hardware) with Blynk cloud and Blynk'sserver: Ethernet, Wi-Fi, Bluetooth, Cellular, Serial [13].

Steps to Setup Blynk App:

Blynk app is used to visualize data, control equipment remotely, set notifications and rules, manage multiple devices, etc. Since app development is an expensive pre-made module and the feature reduces cost.

Step 1: Select the project name and choose the device.

Step 2: After selecting the project name and board, the token is sent to your registered id by the Blynk app.

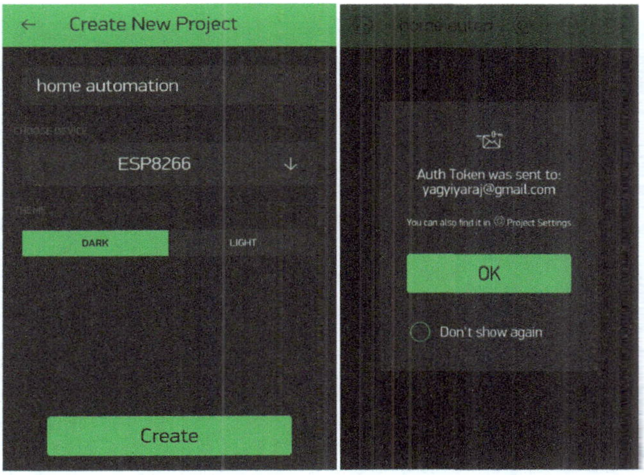

Step 3: Now select the widgets from the widget box. You can select buttons, sliders, timer, vertical slider, and various other widgets.

Step 4: Select the required type and number of pins.

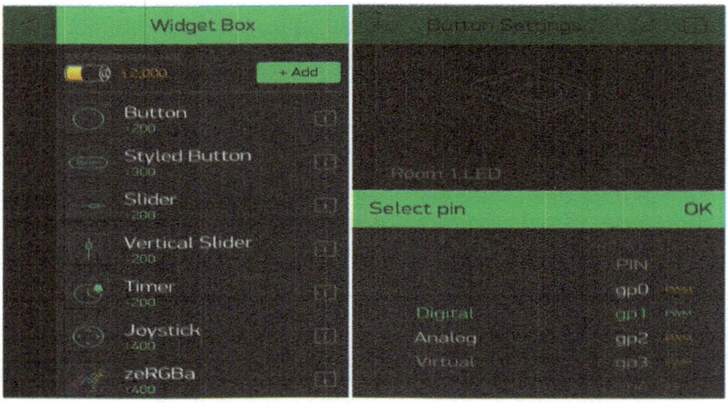

Step 5: Six buttons have been selected for controlling lights and fans and gauge that will indicate temperature. You can add widgets according to your requirement.

Step 6: The status is displayed on the top, when you are online you can control devices using the buttons.

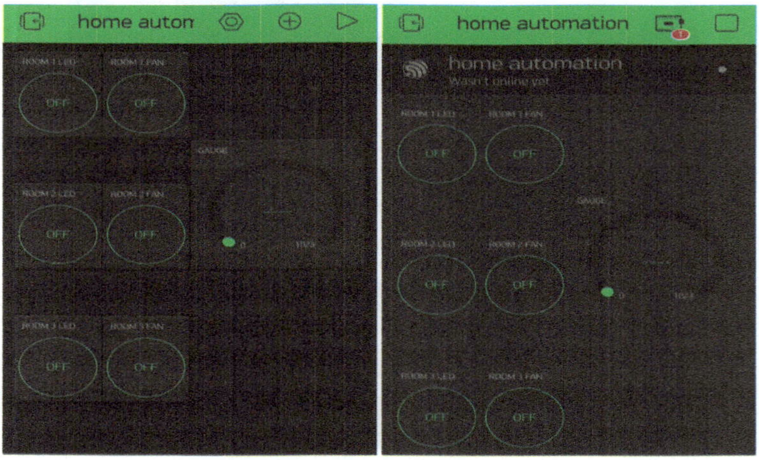

Design:

The system proposed above is designed using an Android-based smartphone that has touch screen operation and an application for controlling the elements of an automated home. This application will use the MQTT protocol for the transmission of the signal to node MCU or any other microcontroller that has been used in the system. MQTT stands for message queuing telemetry transport which uses publish-subscribe messaging pattern. Thus by using our smartphone, we can send the ON/OFF signal to the microcontroller which is in turn connected to various sensors and home appliances that we need to monitor, control, and manage. Hence a home automation system can be designed with the help of an Android application that can be controlled

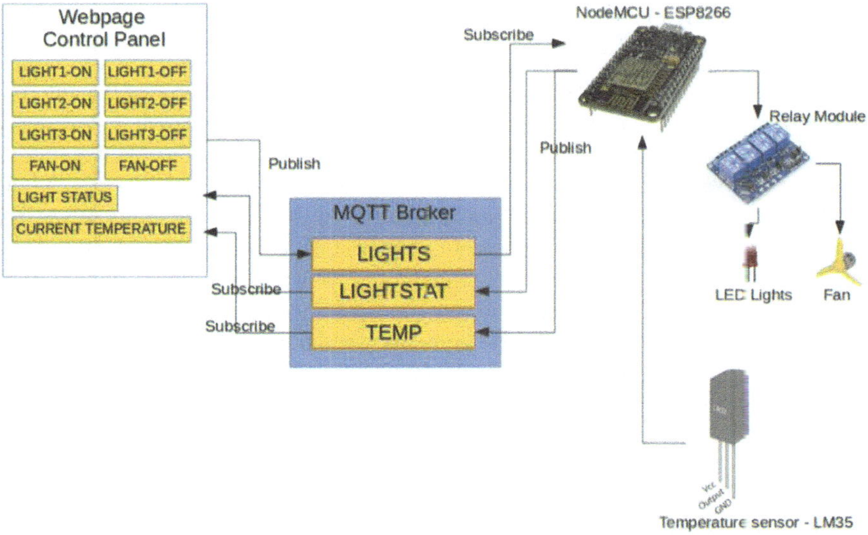

Fig. 1 Design of a basic home automation system

remotely by touching a particular button on the smartphone which will turn on/off an appliance or sensor through a wireless network [14] (Fig. 1).

3 Scope of Home Automation System

1. Security: A smart home can have various devices such as security alarm, smoke detectors, and security cameras. These devices can be controlled easily with the help of a home automation system and hence improving the security of the house.
2. Saving time: An efficient home automation system can simplify our daily routines by providing the feature to easily control various elements of our home [3].
3. Comfort and remote access: By providing the ability to remotely manage and monitor various home appliances, a home automation system provides a great deal of comfort and peace of mind to people living in that house as it can easily adapt and fit according to the lifestyle of people and enable easy customization options.
4. Boosting efficiency: The efficiency of a household can be increased in various ways by using a home automation system as it provides interoperability, energy saving, cost saving, and adding to the convenience of user [15].

4 Conclusion

Using node MCU is cost-effective as it will decrease the overall cost of the automation system since it as cheaper than other microcontrollers. Node MCU has many things in common with Arduino like both of them are microcontroller equipped prototyping board which can be programmed using Arduino IDE. Hence node MCU is more compact and Wi-Fi-equipped alternative of Arduino. By using Wi-Fi for connectivity rather than using other technologies such as Bluetooth and ZigBee we can overcome the problem of short-range connectivity. Since Wi-Fi will provide wireless connectivity it will be more efficient than wired connections because if devices are connected through wires then data will be transferred at comparatively low speed and finding out any fault in wiring is not easy.

References

1. https://www.sap.com/india/trends/internet-of-things.html
2. https://readwrite.com/2019/03/07/how-iot-is-optimizing-costs-for-industrial-sectors/
3. NPTEL online course- introduction to internet of things by Prof. Sudip misra
4. https://www.xfinity.com/hub/smart-home/home-automation
5. https://www.techopedia.com/definition/5366/internet-protocol-ip
6. https://www.wespeakiot.com/how-smart-homes-help-saving-energy/
7. https://www.smarthouseintegration.com/smarthouse-blog/10-key-features-of-home-automation-systems/
8. https://www.instructables.com
9. https://www.electronicwings.com/nodemcu/basics
10. Oxford dictionary
11. https://www.theengineeringprojects.com/2018/10/introduction-to-arduino-ide.html
12. https://www.sam-solutions.com/blog/why-is-a-mobile-app-important-for-an-efficient-smart-home-solution/
13. https://subscription.packtpub.com/book/application_development/9781788995061/1/ch01lv11sec10/what-is-blynk
14. Kodali RK, Jain V, Bose S, Boppana (2016) IoT based smart security and home automation system. In: International conference on computing, communication and automation (ICCCA2016)
15. https://www.slideshare.net/DanKinsella/smart-home-systems-52056045

Performance Analysis Study of Stochastic Computing Based Neuron

A. Dinesh Babu and C. Gomathy

Abstract Deep learning systems are computationally more expensive and their performance depends strongly on the underlying hardware platform. The basic requirements of a deep neural network (DNN) are simple and high-precision computations that can be performed faster. A deep neural network involves enormous data manipulations, of which basic arithmetic operations such as addition and multiplication form the building blocks. But implementing these operations in DNN is a challenging task with less hardware requirements. In this paper, a study has been made on how DNN effectively processes the diverse multi-sensor data. All these can be taken into consideration for the applications which work on constrained resource environment, e.g., IoT devices.

Keywords Stochastic computing · DNN · Stochastic neuron · Deep learning

1 Introduction

With the advent of smart mobile phones in the past decade, our day-to-day life has become easier. Security has become mandatory for these devices. Speech recognition, visual recognition are playing a key role. For all these, a large amount of sensory inputs are taken and processed to give the final expected output. Moreover, a major part of these processes takes place in the warehouse computer which is communicated through the cloud. Moving certain processes from cloud into the device would save a lot of power and make the system an intelligent one. All these require enhancing the efficiency of the algorithms used and also the efficiency of the mobile devices which perform these functions.

A. Dinesh Babu (✉) · C. Gomathy
Department of ECE, SRM Institute of Science and Technology, Vadapalani, Chennai, India
e-mail: dineshba@srmist.edu.in

C. Gomathy
e-mail: hod.ece.vdp@srmist.edu.in

S. S. Dash et al. (eds.), *Intelligent Computing and Applications*,
Advances in Intelligent Systems and Computing 1172,
https://doi.org/10.1007/978-981-15-5566-4_22

In recent years, many improvements have been made in the algorithmic predictions which are more accurate. This is mainly because of the introduction of deep neural networks for these applications. These networks are nowadays used in object classification, detections. Past few years, various researches have been made in the deep neural network which is inspired from biological systems [1]. Restricted Boltzmann machine in one such deep learning system, which has created a great impact because of its higher performance in solving complex function [2].

It is classified into two major domains; learning phase and inference phase. The learning phase comprises mapping the input data to their respective desired outputs, while the inference phase makes use of this configuration to calculate the outputs for the data. Deep belief networks have produced desirable results for vision recognition processes. DBN are made by peeling the RBM one above the other to form a deep neural network [3]. However, power consumption is very high for all these processes. With the advent of the IoT for mobile devices have pushed the researchers to build lower power machine learning pervasive devices and also to make the maximum use of the onboard power supply [4].

Stacked layers of RBMs are used to construct the Deep belief network, with the output layer at the final level as seen in Fig. 1. These networks comprise of various matrix multiplications in every levels. These multiplications are very costly in terms of hardware implementation since they require a large silicon portion and high-power consumption. Moreover, the nonlinearity functions are performed with the help of lookup tables, which require a large amount of memory [5]. Thus DBN implementation using VLSI efficiently is a big task [6].

Stochastic computing, nowadays has been seen as an alternative solution for the above-stated problem. Stochastic computing (SC) methods may result in fast computation, low power consumption during hardware implementation. With the help of

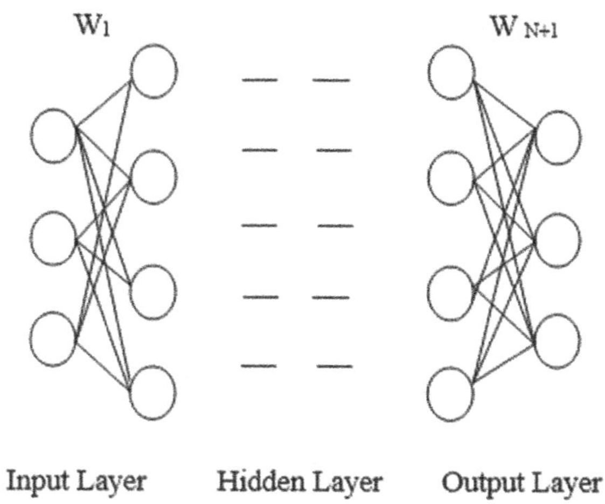

Fig. 1 DBN with N-layers where weights are denoted using *W* and number of layers are denoted by *N*

SC, various complex computational units can be easily implemented. Various multiplication circuits and adder circuits can be implemented with the help of unipolar SC, i.e. using AND gate and multiplexer.

Since the DNN consist of a numerous number of multipliers, various approaches have been taken into consideration for minimizing the computational speed, power, and area [7]. All these are required for end IoT devices since they have a limited onboard power supply, as well as the size of these devices is very small. Literature survey has also been done, and a study has made on the possible solutions to overcome the previously existing problems.

2 Literature Survey and Related Works

2.1 *Structure of a Neuron*

The scope of DNN is to find a solution to complex problems. DNN is a group of neurons with processing capability and connection-related algorithms. The neuron acts as a processing unit for information [8]. The neurons, are similar to the biological neurons, since both of them have inputs, output, and hidden layers, which produce an output with respect to the given inputs. The artificial neurons (Fig. 2) are formed by [9]:

- Inputs: Direct inputs or the outputs from the other neuron's activation unit.
- Weights and Bias: Can be either positive or negative. These play an important role in determining the function of the neural network.
- Summation Unit: It is used for adding up the input signals, which were previously multiplied with their corresponding weights. Finally, bias is added to it.
- Activation Unit: It controls the previous stage output and normalizes it between 0 and 1 or −1, 1.
- Output: The output of the neuron which drives the next neuron.

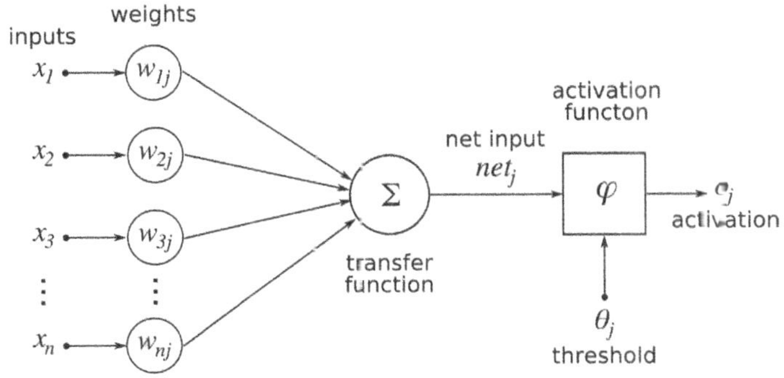

Fig. 2 Basic neuron architecture

2.2 Neural Network and its Types

The neural network comprises a number of neurons that are arranged in a series of sequential layers [10, 11]. The neural network (Fig. 3) comprises different layers:

- Input layer—This layer receives real-time inputs, and starts processing the input. The input layer is a visible layer.
- Hidden layer—This layer is sandwiched between the input and output layers. The role of this layer is to convert the inputs by computing it with their respective weights and bias and then process them to the output layer
- Output layer—This layer computes the output of the neural network.

Some of the neural network types includes the following:

- Single-layer perceptron: Network with two inputs and one output without any hidden layers [12]. It is also known as the Basic Perceptron model (Fig. 4a).
- Multilayer perceptron: Network contains many hidden layers between the input and the output layers. It is also known as Feedforward Deep Neural Network (Fig. 4b) [13].
- Restricted Boltzmann machine neural network: The nodes in the network are with the next layers but not within the same layer. The neurons in this network make stochastic decisions, i.e., to turn on or turn off. (Fig. 5)

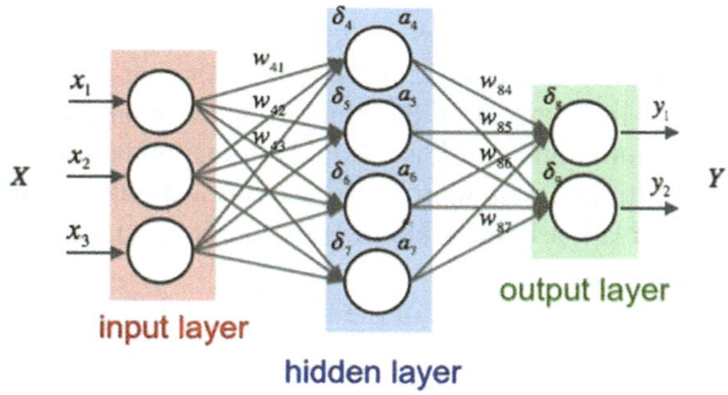

Fig. 3 Basic neural network architecture

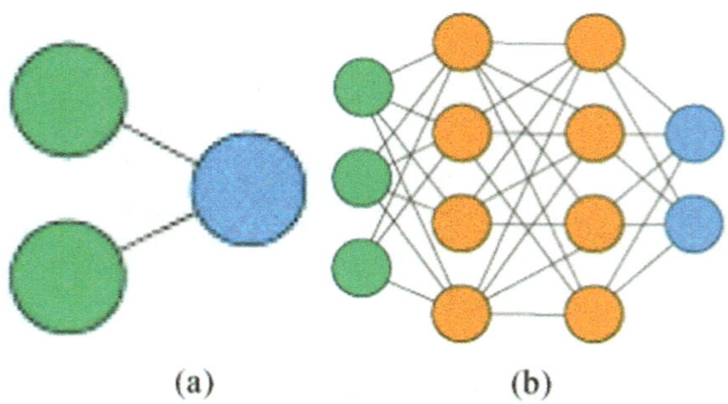

Fig. 4 **a** Structure of single-layer perceptron **b** structure of multilayer perceptron

Fig. 5 Structure of restricted boltzmann machine

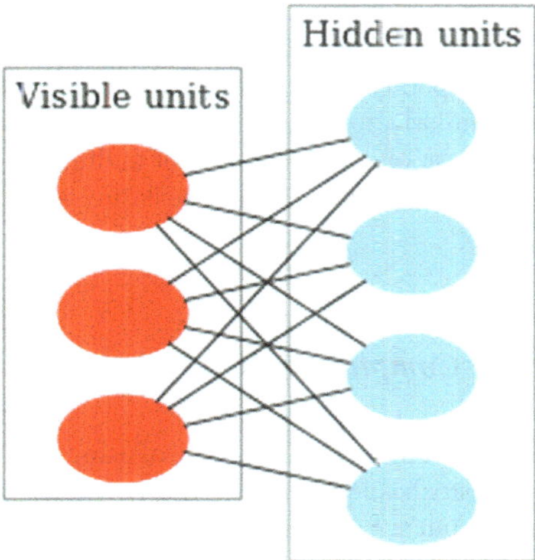

3 Stochastic Computing (Sc) for DNN

The computational functionality of the neuron mainly depends on the weights and the bias, which are multiplied and added, respectively, with the given inputs. Moreover, the implementation of the deep belief networks in VLSI is more expensive in terms of computation because it needs a large number of matrix multiplications and there is no direct representation for activation unit (sigmoid function, tanh function, etc.) [14]. It is generally generated with the help of Lookup Tables which in turn occupy some more additional memory. In order to overcome this, a better functionality is needed. The following are the challenges and ways to overcome them.

3.1 Computions in Each Neuron

Each neuron performs some computational functions in which the input value of input nodes or the visible nodes are multiplied with their corresponding respective weighs. The weights are assigned to each neuron in the first layer. These weights are multiplied with their respective activations from the input layer and their weighted sum is computed [15]. After this, bias is added to the weighted sum.

$$X_i = \sum_{k=1}^{N} W_{ik} V_k + B_i$$

$$G_i = 1 = \sigma(X_i)1 + \exp(-X_j)$$

where N represents the number of input visible nodes, V represents the activation value of the input nodes, B represents the bias, W represents the weights, X represents the temporary midway value, and G represents the output value of the neuron [16].

Similarly, for a neural network, many such computations take place which results in the use of a large number of multipliers and adder circuits. In order to minimize such computational load, stochastic computing can be seen as a better alternative [17].

3.2 Computation Using Stochastic Computing

- One of the major advantage of using stochastic computing is that the multiplication of stochastic numbers needs only a single AND gate (Fig. 6). These numbers are a sequence of random bits. For stochastic multiplication, bit-wise AND operation is performed between the two stochastic streams.
- $X = \text{AND(C, D)} = \text{C} \cdot \text{D}$
- Similarly, addition operations can be performed with the help of OR gate or scaled adder (Figs. 7 and 8).
- $\text{Z} = \text{C} \cdot \text{S} + \text{D} \cdot (1 - S)$ for MUX and
- $\text{Y} = \text{A} + \text{B} - \text{A} \cdot \text{B}$ for OR gate

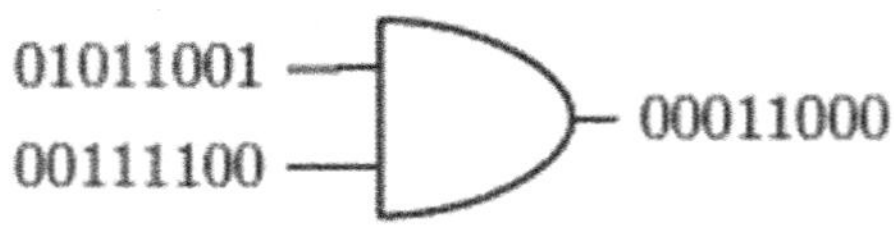

Fig. 6 Structure of stochastic AND gate used for multiplication

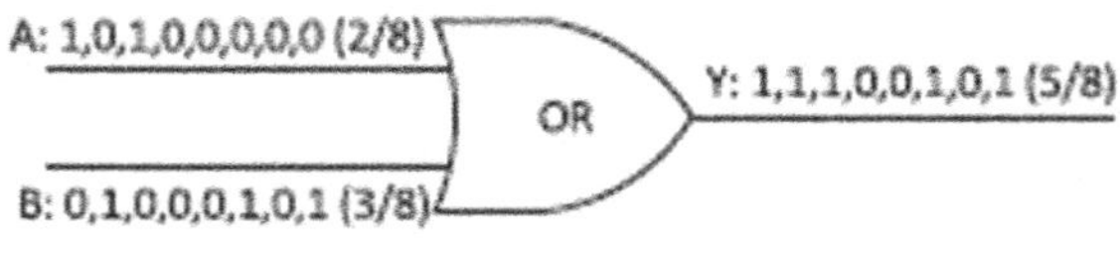

Fig. 7 Structure of stochastic OR gate used for addition

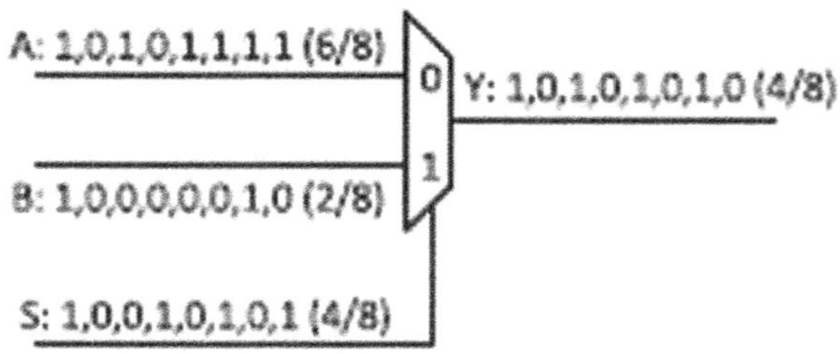

Fig. 8 Structure of stochastic MUX used for addition

- Using stochastic computing, the number of multiplication and addition processes in the neural network gets reduced to a large extent. Since in stochastic computing these multiplication and addition operations can be performed by using bitwise AND gates in unipolar format and addition operations can be performed with the help of scaled adders or an OR gate.

4 Neuron Design Using Stochastic Computing

Neuron can be built using stochastic computing [18]. For this integer, stochastic stream is used. It is built by using the summation of the two binary to stochastic (B2S) stream inputs (Fig. 9). The binary to stochastic unit consists of LFSR, constant number generator (Fig. 10), comparator (Fig. 11), and a counter unit (Fig. 12).

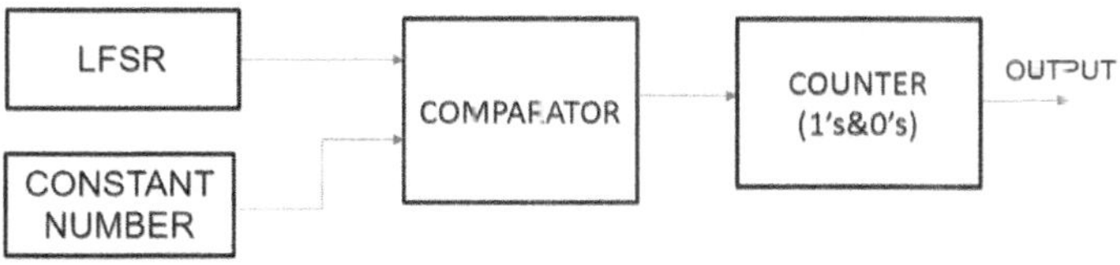

Fig. 9 Structure of binary to stochastic unit (B2S)

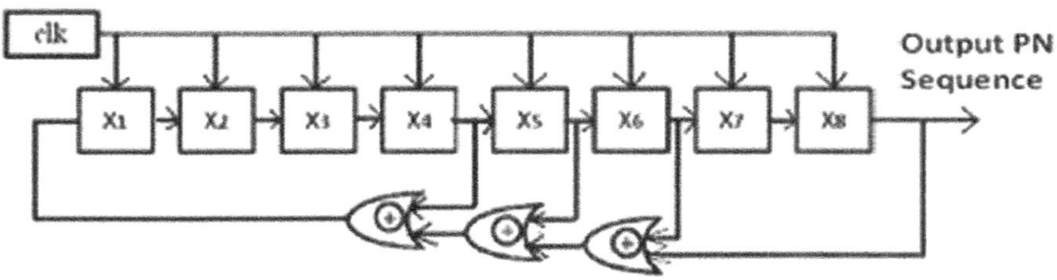

Fig. 10 Structure of LFSR to generate random numbers

Fig. 11 Structure of comparator to produces 1's and 0's binary output

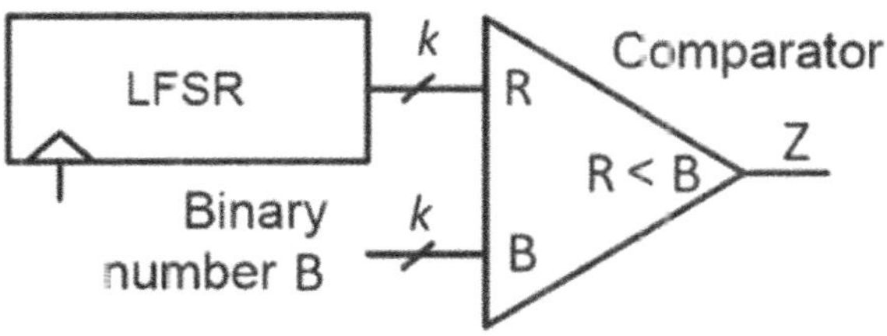

COUNTER

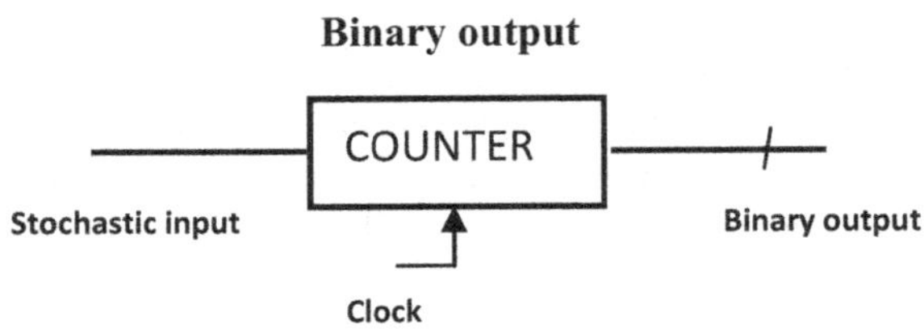

Fig. 12 Structure of counter to count the number of 1's and 0's

The LFSR unit is used to generate the random numbers, these generated random numbers are then compared with the constant number. The constant number is generally denoted by the count of LFSR numbers which are taken into consideration. These two values are then compared using a comparator. Such that if the value of the LFSR number is less than the constant number then a high output (1's) is produced by the comparator and if the value of the constant number is lesser than the value of the LFSR number then a low output (0's) is produced.

The counter is used to count the number of 1's and 0's which are produced by the comparator unit. The final output from the counter is the output of the binary to the stochastic unit. The next stage is to construct the binary to integer stochastic unit (B2IS) (Fig. 13). This is constructed with the help of two binary to stochastic units [19].

Integer addition is performed between the two B2S, and the output of this unit produces the output of the B2IS unit. This output acts as one of the inputs for the bitwise AND operator, which performs the multiplication function in the stochastic

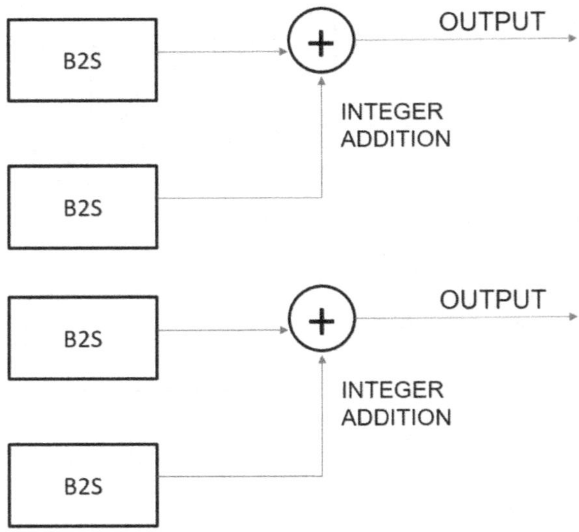

Fig. 13 Structure of binary to integer stochastic unit (B2IS)

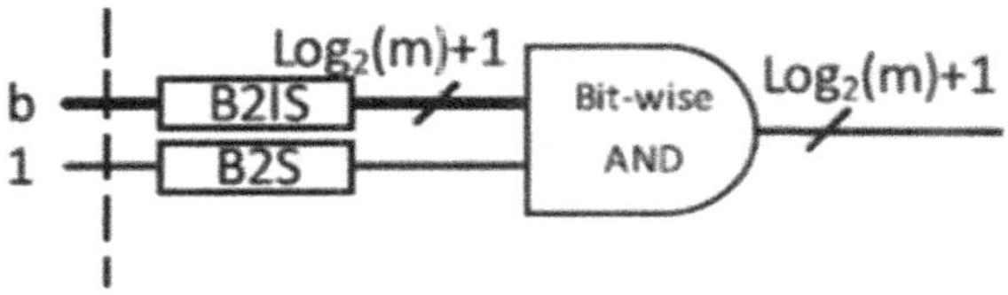

Fig. 14 Structure of stochastic multiplier using bitwise AND operator

computing process. The other input for the bitwise AND operator is the corresponding weighs (Fig. 14).

Similarly, many such inputs node which performs bitwise operators are fed into a binary adder, where the summation of the values from various activation units takes place. Along with that, the bias value is added to it. The final output from the adder unit is fed into an activation unit. The activation unit can be a sigmoid function or tanh function [20]. The activation unit determines the on or off state of each node (Fig. 15).

Thus, when a neuron is built using stochastic computing logic, the number of multiplier and adder used is replaced with the help of bitwise AND operator and scaled adder or an OR gate, respectively.

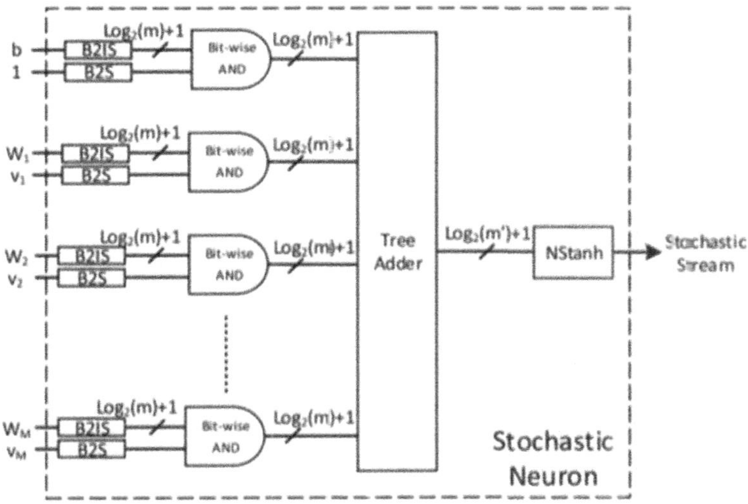

Fig. 15 Structure of neuron using stochastic computing

Table 1 Comparison between the normal neural network and stochastic computing based neural network

	No of multiplication process
No. of multiplier used in normal neural network	540 + 120 + 100 + 10 = 780 780 multipliers contains 1520 multiplication process
No. of used bitwise AND operators in SC	780 stochastic multiplication processes

5 Results and Discussions

With the advent of the stochastic computing process, the number of multiplication processes and addition processes reduces to a large extent during hardware implantation. Consider a neural network with 540 visible nodes, 120 hidden nodes in the first hidden layer, 100 hidden nodes in the second layer, and 10 visible nodes at the output. Therefore to compute the output of the first layer, 540 such multiplication operations are required, along that their respective weighs occupies a large memory space. Moreover simultaneous implementation of this network requires a large area in the silicon wafer.

All these can be reduced with the help stochastic computing logic. Instead of performing these many numbers of multiplications, in stochastic computing, these multiplication operations can be computed using stochastic multiplication where bitwise AND operations are used. And the summation of the multiplication process is done with the help of an OR gate (Table 1).

6 Conclusion

Thus, implementation of a deep neural network using stochastic computing will reduce more than half of the area which was previously required for the conventional neural network [21]. Moreover when it is implemented for a complex function, precision is more since stochastic computing uses the random probabilistic value to determine the output. Since the area consumption is reduced their corresponding power consumption is also reduced to a large extent. Such a type of neural network using stochastic computing can be used for end IoT devices, which result in better performance under a constrained resource environment. The processing capability of the DNN for diverse sets of data can be increased efficiently using the stochastic computing process. Thus, stochastic computing based DNN performs better than conventional DNN models.

References

1. J.A. Bullinaria, *Introduction to neural networks, Lectures*, University of Birmingham, UK, 2004
2. R.F. Molanes, K. Amarasinghe, J.J. Rodriguez-adina, M. Manic, Deep learning and reconfigurable platforms in the internet of things, 2018
3. Y. Chen, T. Luo, S. Liu, S. Zhang, L, He, J Wang, L. Li, T. Chen, Z. Xu, N. Sun, O. Temam, DaDianNao: a machine-learning supercomputer, 2014
4. H. Li, K. Ota, M. Dong, Learning IOT in edge: deep learning for the internet of things with edge computing, 2018
5. S. Han, H. Mao, W.J. Dally, Deep Compression: compressing deep neural network with pruning, trained quantization and huffman coding (2015)
6. P. Maechler, C. Studer, D.E. Bellasi, A. Maleki, A. Burg, N. Felber, H. Kaeslin, R.G. Baraniuk, VLSI design of approximate message passing for signal restoration and compressive sensing, 2012
7. T. Rzayev, S. Moradi, D.H. Albonesi, R, Manohar, Deeprecon: dynamically reconfigurable architecture for accelerating deep neural networks, 2015
8. L. Ranganath, J. Kumar, P.S.N. Reddy, Design of MAC unit in artificial neural network architecture using verilog HDL, 2013
9. C. Sheng, Brodersen, Weste & Harris 3rd ed. Hardware Multiply/Accumulate (MAC) Unit, Motorola, 1992
10. L. Cavigelli, L. Benini, A 803 GOp/s/W convolutional network accelerator. IEEE Trans. Circ. Syst. Video Technol **99**, 1–1 (2016)
11. N.K. Bose, P. Liang, Neural network fundamentals with graphs, algorithms, and applications, 1996
12. T. Theocharides, G. Link, N. Vijaykrishnan M.J. Irwin, V. Srikantam, A generic reconfigurable neural network architecture implemented as a network on Chip, 2004
13. S. Haykin, *Neural networks a comprehensive foundation*, 2nd edn. (Prentice Hall Publishing, New Jersey, VoU, 1999), pp 6–7
14. Y.H. Chen, T. Krishna, J. Emer, V. Sze, Eyeriss: an energy-efficient reconfigurable accelerator for deep convolutional neural networks, 2016
15. Y.S. Anshika, N. Goee, S. Indu, Neuronal logic gates realization using vedic mathematics, 2015
16. L.A. Basheer, M. Hajmeer, Artificial neural networks: fundamentals, computing, design and application. J. Microbiol. Methods **43**, 3–31 (2000)
17. O.C., Yolcu, E. Bas, E. Egrioglu, U. Yolcu, Single multiplicative neuron model artificial neural network with autoregressive coefficient for time series modelling, 2018
18. E. Nurvitadhi, J. Sim, D. Sheffield, A. Mishra, S. Krishnan, D. Marr, Accelerating recurrent neural networks in analytics servers, 2016
19. I. Kouretas C. Basetas, V. Paliouras. IEEE, Low-power logarithmic number system addition/subtraction and their impact on digital filters, 2012
20. A. Amaricai, O. Boncalo, C.-E. Gavriliu, Low-precision DSP-based floating-point multiply-add fused for field programmable gate arrays (2013)
21. G. James, D. Witten, T. Hastie, R. Tibshirani, An introduction to statistical learning: with applications, in *R. Springer Texts in Statistics*, 2014

Scheduling of Parallel Tasks in Cloud Environment Using DAG MODEL

Sakshi Kapoor and Surya Narayan Panda

Abstract The scheduling of tasks in a heterogeneous multiprocessor system in the cloud is still a demanding problem that is being explored by many researchers. Parallel computing which itself is a research area is now integrated with cloud computing. In this paper, we present a systematic study of a directed acyclic graph (DAG) model in a parallel multiprocessor system. Basic concepts of parallel computing along with issues like task scheduling have been discussed in detail. Computational solutions can be demonstrated as directed acyclic graphs (DAG) model having edges and nodes with weights. Each task or job in the DAG has its own execution time, which incorporates into various processors. This paper gives the basic idea of parallel processing using DAG. The main objective behind scheduling using the DAG model is to reduce the finish time or completion time of parallel applications through proper assignment of tasks among multiprocessors. Most of these schemes consider the precedence constraints among tasks. Up to now, Direct Acyclic Graph (DAG) is considered a prominent approach used for modeling the precedence constraints between the nodes or tasks. There are various scheduling algorithms that use DAG for scheduling jobs. The most popular heterogeneous earliest finish time (HEFT) is one of them. A literature review on various task scheduling schemes in combination with artificial intelligence concepts is also presented.

Keywords Task scheduling · Cloud computing · Parallel computing · Directed acyclic graph · Multiprocessor · Heterogeneous environment

1 Introduction

Cloud Computing is the delivery of services over the Internet. It is a kind of structure that provides software, hardware, and platform in the form of SaaS (Software-as-a-service), IaaS (Infrastructure-as-a-service), PaaS (Platform-as-a-service), known as service models of cloud. The service provider is the one, who provides all these

S. Kapoor (✉) · S. N. Panda
Chitkara Institute of Engineering and Technology, Chitkara University, Punjab, India
e-mail: sakshi.kapoor@chitkara.edu.in

S. S. Dash et al. (eds.), *Intelligent Computing and Applications*,
Advances in Intelligent Systems and Computing 1172,
https://doi.org/10.1007/978-981-15-5566-4_23

services to the clients via the Internet [1–3]. The parallel execution of jobs is needed to get the work done more speedily. Parallel computing is basically an architecture in which various processors execute applications or tasks simultaneously. Parallel computing can be called parallel processing [2, 3]. It divides the workload of large computations between multiple processors, at the same time. Parallel computing principles are followed by most of the supercomputers in order to operate properly and efficiently. Parallel computing makes the execution of various portions of tasks or applications possible simultaneously in a heterogeneous multiprocessor environment. A high degree of parallelism enhances the high conduct of existed systems [4]. Not only parallel, but serial or sequential execution of the jobs or tasks and automated interprocess synchronization become efficient through parallel processing. Scientific problems such as modeling of human bones, environmental modeling, statistical mechanics as well as genetic engineering are very complicated if solved through simulation as it needs powerful computers. Parallel computing architectures solve these problems with high efficiency. The division of parallel tasks is analytical for the indication of distributed computing systems [5–7].

The main purpose behind task scheduling is proper assignment of jobs or tasks to multiprocessors in order to reduce the finish time of parallel applications [6]. There are two forms of scheduling problems: one is static and the other is dynamic. Static scheduling is the one in which characteristics or properties of a parallel program like task communication, synchronization requirements as well as data dependency are known before the actual execution of the program. Therefore, a parallel program can be symbolized by a weighted *directed acyclic graph* (DAG) consisting of weighted edges and nodes [7].

Nowadays, scheduling in the multiprocessor system or environment is an active research area, in which various assumptions are freely recommended. Shockingly, some assumptions are neither obviously expressed nor reliably utilized by the vast majority of the analysts. Therefore, it is hard to welcome the benefits of different scheduling algorithms and quantitatively to assess their efficiency and performance. To keep away from this issue, we present directed acyclic graph (DAG), understood portrayal of parallel applications is a DAG (Directed Acyclic Graph) in which jobs or tasks are symbolized by nodes in the DAG. Inter-task dependencies are represented by arcs. The actual finish time of the last job or node in the DAG gives the schedule length or makespan [3, 6]. DAG is a task precedence model, where directed edges give the communication time between the nodes or the tasks and is popular for static scheduling of tightly coupled tasks or jobs on multiprocessors in a heterogeneous environment. A task precedence graph can model the program more efficiently in most of the parallel applications as it conquers all the temporal dependencies between the jobs or tasks.

When a system is demonstrated as a group of homogeneous processors, where one task is executed or computed at a time, the goal is to limit the schedule length or finish time or makespan. Sometimes, even the communication cost is explicitly modeled. Subsequently, on assigning two neighboring jobs directly to different processors, the subsequent one is postponed by a measure of time relative to the weight of the directed edge between them, yet if allocated to a similar processor, no such delay in task happens. Many limited instances of scheduling using DAG have been demonstrated to be NP-complete, implying that it is unrealistic to search for an ideal schedule. This is genuinely notwithstanding for very limited exceptional cases, for example, at the point when communication is free and all the jobs have unit execution times [8].

The paper is systematized in this sequence: Firstly, the introduction of Parallel Computing in the cloud with a brief idea about DAG is given. In Sect. 2, we focus on the need and issues of parallel computing like task scheduling along with their solutions like DAG. In Sect. 3, the DAG Model along with an explanation about DAG scheduling and performance parameters are discussed. Lastly, a literature review on DAG scheduling using different artificial intelligence techniques is presented, followed by the heterogeneous earliest finish time (HEFT) algorithm for efficient task scheduling is discussed. The conclusion is discussed in Sect. 4.

2 Parallel Computing in Cloud

It is the set of multi-processors or multiple processing elements that solves problems concurrently. Problems are decomposed into subproblems which are to be distributed among multiprocessors and to be solved in parallel or simultaneously [9].

2.1 Need for Parallel Computing

- Larger and complex problems are harder to solve through serial computing.
- The advantage of all nonlocal resources can be taken in case local resources are finite [10].
- In serial computing, there is a huge wastage of potential computing power which makes a parallel computing environment utilize better work of the hardware.
- The entire real world is dynamic in nature like numerous things occur at a specific time, however at different places simultaneously. This information is broadly immense to handle [7].
- There is a need for dynamic simulation and modeling in real-world data and for accomplishing the equivalent, parallel processing is the key.
- Complex and enormous datasets and their management can be sorted out by utilizing parallel processing methodology [11].
- With parallel computing, there will be effective and efficient utilization of the assets. Hardware is destined to be utilized adequately though in serial computing, less portion of the hardware was utilized and the rest rendered inactive.

2.2 Issues in Parallel Computing

- **Task scheduling of Parallel Tasks**: Task scheduling in a heterogeneous multiprocessor environment is found to be a difficult issue [6, 8, 10, 12]. Achievement of high performance in a multiprocessor system is a key factor in scheduling parallel tasks. The main objective of task scheduling is to map tasks in parallel on multiprocessors and order their execution in such a way that a reduced schedule length is given under the point of confinement of undertaking priority necessities. Scheduling is considered to be a highly important issue since improper planning of jobs can neglect to abuse the genuine capability of the framework and can balance the increase from parallelization [12]. The goal of scheduling is to limit the completion or finish time of a parallel application by appropriately assigning the jobs to the processors. There are so many scheduling schemes that are found to be efficient in parallel task scheduling in a multiprocessor environment. The vast majority of these plans consider the priority limitations among assignments. Up to now, Direct Acyclic Graph (DAG) is the popular and significant model utilized as displaying the precedence constraints among the jobs [13].
- **Load Balancing**: Partitioning of work correspondingly between several processors, so that no processor becomes under-loaded or overloaded.
- **Reducing Execution time**: Noted executed time of last executed job should be reduced in order to overcome high energy consumption.
  Computation time—It is associated with the resources used by the nodes in the DAG.
  Idle time—It is the waiting time for the data from other processors.
  Communication time—It is time taken by the processors to transmit communications among them (Fig. 1).

3 Task Scheduling

Task scheduling can be characterized as relegating the job or task to a processor for executing at a specific time. The process of task scheduling relies on the accompanying components:

- Total number of processors.
- Performance or efficiency of processors.
- Mapping of tasks to processors.
- Arrangement of tasks in order to execute on a specific processor.

The above components profoundly rely upon one another and figure out the advanced outcomes. They all cooperate together and nobody considered the individual. Static scheduling is generally done at compile time, in which the qualities of a parallel program like communication, synchronization prerequisites are known before the execution of the program [7]. In contrast to static scheduling, dynamic scheduling takes a few assumptions about the parallel program before the execution

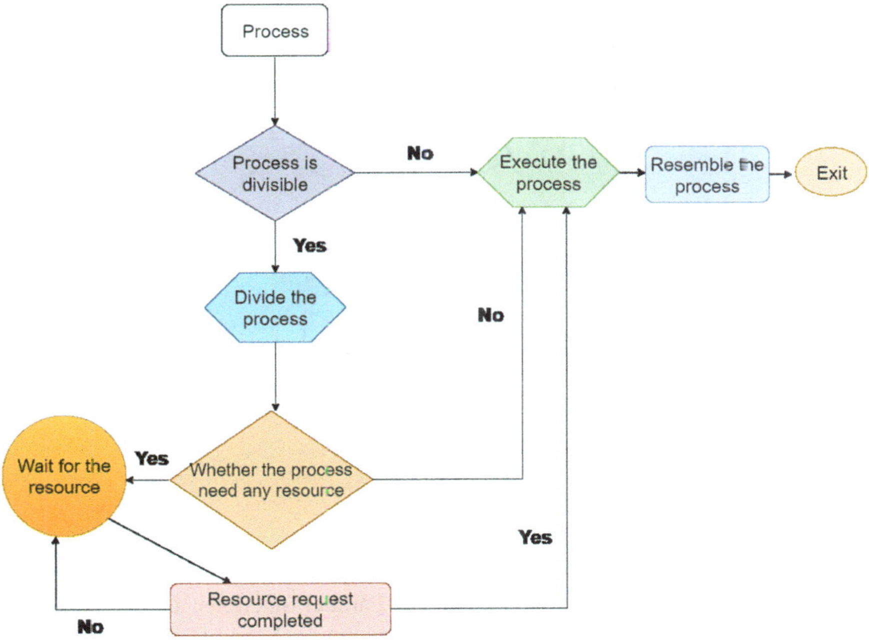

Fig. 1 Working of parallel processing [3]

takes place and thus, scheduling choices must be made on-the-fly. The objective of dynamic scheduling not only includes a reduction in the program finish time yet, also, the minimization of the planning overhead which comprises a noteworthy segment of the cost paid for running the scheduler. Parallel computing is an enthusiastic type of data processing, in which assignments or tasks are separated into components, which are executed at the same time each on its individual processor [9, 11]. The processors might be orchestrated in homogeneous or heterogeneous conditions. Representation of task scheduling is efficiently done by Directed Acyclic Graphs (DAG).

3.1 Model for Task Scheduling Problem

Task scheduling helps to limit the overall or total finish time by properly assigning tasks to the processors and provisioning of execution sequencing of all the given tasks. Due to efficient scheduling, the precedence constraints among the tasks are saved. The general completion time of a parallel program is called the makespan. The objective of efficient scheduling is to reduce or lessen the makespan.

Fig. 2 DAG model [5]

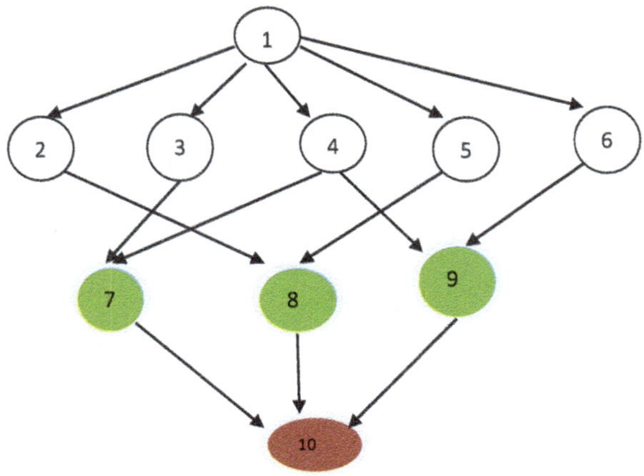

3.2 DAG Model

The Direct Acyclic Graph (DAG) Model Direct Acyclic Graph is the authentic model of a parallel framework or parallel system [7]. Parallel functions, as well as parallel programs, can be described in the form of DAG, where V and E are set of vertices and edges. Tasks are represented by the nodes which are basically set of instructions that must be executed sequentially without preemption in the same processor [14]. Every node contains some weight known as computation cost whereas the edge in the graph gives communication cost between tasks and nodes. The starting node is the source node or the parent node while the sink node is known as the child node. A node having no parent is the entry node and a node having no child is known as an exit node of the DAG [15].

Rules to be followed in an efficient DAG Scheduling:

- Execution will only be as per the level.
- Next-level jobs cannot be executed if there is any job left at the previous level.
- No child can be executed before its parent.
- Level jobs will be ranked as per some rule architecture (Fig. 2).

3.3 Literature Review on Task Scheduling Using DAG

The recent trends in Cloud Computing involve efficient energy utilization for an optimal number of jobs in the least span of time. Therefore, various researchers have implemented in different algorithms and artificial intelligence concepts in order to reduce makespan and energy consumption. A literature survey on various task scheduling schemes is given in Table 1.

Through Literature Survey, it is cleared that with the use of artificial intelligence algorithms and techniques with the DAG Model, the energy consumption, as well

Table 1 Literature review on task scheduling using DAG

S. No.	Year	Paper title	Scheduling parameters	Tool used	Improvement
1.	2016	Task scheduling algorithms with multiple factors in cloud computing environment [16]	Cost metrics, load balancing	CloudSim	Improving performance in comparison to conventional algorithms
2.	2016	Enhanced bee colony algorithm for efficient load balancing and scheduling in Cloud [17]	Schedule length, cost	MATLAB	The decrease in schedule length which enhances the overall performance
3.	2014	A task scheduling algorithm based on genetic algorithm and ant colony optimization in cloud computing [18]	Execution time	MATLAB	Increase in the efficiency of the proposed algorithm
4.	2014	Improved ant colony algorithm based on PSO and its application on cloud computing resource scheduling [19]	Execution time	MATLAB	Decrease in schedule length
5.	2014	Task scheduling algorithm based on improved Min-Min algorithm in a cloud computing environment [20]	Schedule length, load balancing	CloudSim	High performance
6.	2015	Credit-based scheduling algorithm in cloud computing environment [21]	Schedule length	CloudSim	Decrease in the schedule length
7.	2012	Dynamic task scheduling algorithm with load balancing for heterogeneous computing system [22]	Resource utilization, schedule length	Distributed algorithm simulator	The concept of clustering is used for resolving load balancing as well as decreasing schedule length

as schedule length in the DAG, will be minimized to a great extent. In parallel computing, the task scheduling process becomes energy efficient in applying genetic algorithms and artificial intelligence concepts.

3.4 *Task Scheduling Algorithm Using DAG*

Heterogeneous Earliest Finish Time (HEFT)

HEFT is an algorithm which considers dependent tasks onto a network of processors in a heterogeneous environment and taking communication time or cost between the tasks into consideration. HEFT takes a DAG consisting of jobs or tasks as input, processors, execution time of tasks on each processor as well as the time to communicate results from each job to each of its children between each pair of processors [23–25].

1. Creation of DAG consisting of tasks T_j in the Cloud.
2. Set computation cost (T_j) and communication cost (T_j) between the nodes or tasks.
3. **For** j **= 1 to** T_i
   (Calculation of the order of execution of tasks)
4. **if (**T_j**= last task), then**
5. O (T_j) = Avg (T_j) on all processors, where O is the order of tasks and Avg is the average of tasks.
6. **else**
7. O (T_j) = Avg (T_j) + max (order of task value of predecessor task of current task) + communication cost (T_j).
8. **end if**
9. **end for**
10. Arrangement of tasks in decreasing order in a list based on O (T_j).
11. **for (*T* in list)**
12. Assign T to the processor which has the min execution time.
13. **end for**
14. **End.**

4 Conclusion

Larger and complex problems are harder to solve through Serial computing. To overcome the problems of serial computing, parallel computing comes into existence. In parallel computing, task scheduling is a very difficult and challenging issue. DAG Model can be used for easy and efficient scheduling of parallel jobs or tasks. The goal of scheduling is to limit the completion or finish time of a parallel application by appropriately assigning the jobs to the processors. A literature review on DAG

Scheduling using an artificial intelligence approach is also given. A popular DAG-based algorithm HEFT is also discussed in the paper.

References

1. J. Gibson, R. Rondeau, D. Eveleigh, Q. Tan, Benefits and challenges of three cloud computing service models, in *2012 Fourth International Conference on Computational Aspects of Social Networks (CASoN)*. IEEE (2012, November), pp. 198–205
2. J. Ekanayake, G. Fox, High performance parallel computing with clouds and cloud technologies, in *International Conference on Cloud Computing* (Springer, Berlin, Heidelberg, 2009), pp. 20–38
3. T. Hagras, J. Janecek, A high performance, low complexity algorithm for compile-time task scheduling in heterogeneous systems, in *Proceedings of 18th International Parallel and Distributed Processing Symposium*. IEEE (2004), p. 107)
4. C. Cai, L. Wang, S.U. Khan, J. Tao, Energy-aware high performance computing: a taxonomy study, in 2011 IEEE 17th International Conference on Parallel and Distributed Systems. IEEE (2011), pp. 953–958
5. G.P. Gibilisco, M. Li, L. Zhang, D. Ardagna, Stage aware performance modeling of DAG based in memory analytic platforms, in 2016 IEEE 9th International Conference on Cloud Computing (CLOUD). IEEE (2016), pp. 188–195
6. A.R. Arunarani, D. Manjula, V. Sugumaran, Task scheduling techniques in cloud computing: a literature survey. Future Gener. Comput. Syst. **91**, 407–415 (2019)
7. S. Baskiyar, K.K. Palli, Low power scheduling of dags to minimize finish times, in *International Conference on High-Performance Computing* (Springer, Berlin, Heidelberg, (2006, December), pp. 353–362
8. R. Ghosh, Y. Simmhan, Distributed scheduling of event analytics across edge and cloud. ACM Trans. Cyber Phys. Syst. **2**(4), 24 (2018)
9. F. Suter, F. Desprez, H. Casanova, From heterogeneous task scheduling to heterogeneous mixed parallel scheduling, in *European Conference on Parallel Processing* (Springer, Berlin, Heidelberg, 2004), pp. 230–237
10. F. Guo, L. Yu, S. Tian, J. Yu, A workflow task scheduling algorithm based on the resources' fuzzy clustering in cloud computing environment. Int. J. Commun Syst **28**(6), 1053–1067 (2015)
11. X. Zhou, G. Zhang, J. Sun, J. Zhou, T. Wei, S. Hu, Minimizing cost and makespan for workflow scheduling in cloud using fuzzy dominance sort based HEFT. Future Gener. Comput. Syst. **93**, 278–289 (2019)
12. Z. Tang, L. Qi, Z. Cheng, K. Li, S.U. Khan, K. Li, An energy-efficient task scheduling algorithm in DVFS-enabled cloud environment. J. Grid Comput. **14**(1), 55–74 (2016)
13. S. Yassa, R. Chelouah, H. Kadima, B. Granado, Multi-objective approach for energy-aware workflow scheduling in cloud computing environments. Sci. World J. (2013)
14. M. Mezmaz, N. Melab, Y. Kessaci, Y.C. Lee, E.G. Talbi, A.Y. Zomaya, D. Tuyttens, A parallel bi-objective hybrid metaheuristic for energy-aware scheduling for cloud computing systems. J. Parallel Distrib. Comput. **71**(11), 1497–1508 (2011)
15. S. Selvarani, G.S. Sadhasivam, Improved cost-based algorithm for task scheduling in cloud computing, in *2010 IEEE International Conference on Computational Intelligence and Computing Research*. IEEE (2010, December), pp. 1–5
16. N. Bansal, A. Awasthi, S. Bansal, Task scheduling algorithms with multiple factor in cloud computing environment, in *Information Systems Design and Intelligent Applications* (Springer, New Delhi, 016), pp. 619–627
17. K.R. Babu, P. Samuel, Enhanced bee colony algorithm for efficient load.balancing and scheduling in cloud, in *Innovations in Bio-inspired Computing and Applications* (Springer, Cham, 2016), pp. 67–78

18. C.Y. Liu, C.M. Zou, P. Wu, A task scheduling algorithm based on genetic algorithm and ant colony optimization in cloud computing, in 2014 13th International Symposium on Distributed Computing and Applications to Business, Engineering and Science. IEEE (2014, November), pp. 68–72
19. H. Yang, Improved ant colony algorithm based on PSO and its application on cloud computing resource scheduling, in *Advanced Materials Research*, vol. 989. (Trans Tech Publications, 2014), pp. 2192–2195
20. G. Wang, H.C. Yu, Task scheduling algorithm based on improved Min-Min algorithm in cloud computing environment, in *Applied Mechanics and Materials*, vol. 303 (Trans Tech Publications, 2013), pp. 2429–2432
21. A. Thomas, G. Krishnalal, V.J. Raj, Credit based scheduling algorithm in cloud computing environment. Procedia Comput. Sci. **46**, 913–920 (2015)
22. D.M. Abdelkader, F. Omara, Dynamic task scheduling algorithm with load balancing for heterogeneous computing system. Egypt. Inform. J. **13**(2), 135–145 (2012)
23. H. Topcuoglu, S. Hariri, M.Y. Wu, Task scheduling algorithms for heterogeneous processors, in *Proceedings. Eighth Heterogeneous Computing Workshop (HCW'99)*. IEEE, (1999, April), pp. 3–14
24. H. Topcuoglu, S. Hariri, M.Y. Wu, Performance-effective and low-complexity task scheduling for heterogeneous computing. IEEE Trans. Parallel Distrib. Syst. **13**(3), 260–274 (2002)
25. Z. Cao, L. Zhou, B. Hu, C. Lin, An adaptive scheduling algorithm for dynamic jobs for dealing with the flexible job shop scheduling problem. Bus. Inf. Syst. Eng. **61**(3), 299–309 (2019)

GREENIE—Smart Home with Smart Power Transmission

N. Noor Alleema, S. Babeetha, V. L. Hariharan, S. Sangeetha, and H. Shraddha

Abstract The energy consumption is increasing in day-to-day life. Nowadays, the growth of smart devices has increased, which consumes more energy so people are shifting to renewable resources for producing energy. Modern homes are shifting from conventional switches to a centralized control system. Home automation refers to handling and controlling home. This project mainly focusses on transmitting energy wirelessly using GREENIE in home automation. The energy from various sources and it is distributed to the devices through WPT, where these devices are controlled using IoT, by using Arduino the various sensor are connected to the devices and controlled using a web server and smartphones.

Keywords WPT—Wireless Power Transmission · IoT—Internet of things

1 Introduction

Nowadays, wires are playing an important role in the transmission of electricity in many sectors such as homes, industry, medical sector, etc. Transmitting electric current leads to various problems, for example, when the overload current is passed through wires it leads to the short circuit and wire is easily set to fire, this shows the highly inflammable nature of wires, secondly the skilled persons like electrician

N. Noor Alleema (✉) · S. Babeetha · V. L. Hariharan · S. Sangeetha · H. Shraddha
SRM Institute of Science and Technology, SRM University, Ramapuram, Chennai 603203, India
e-mail: noor25nrs@gmail.com

S. Babeetha
e-mail: babeeths14@gmail.com

V. L. Hariharan
e-mail: hariharanvl04@gmail.com

S. Sangeetha
e-mail: sankarmanickam1965@gmail.com

H. Shraddha
e-mail: shraddhamutha9940@gmail.com

S. S. Dash et al. (eds.), *Intelligent Computing and Applications*,
Advances in Intelligent Systems and Computing 1172,
https://doi.org/10.1007/978-981-15-5566-4_24

is required to work or else leads to shocks, leakage of current compared to all the problems occurred due to the usage of wires the complexity of wires is a major one in homes, industries. The idea is mainly focusing on the problem that is to reduce the complexity of wire in home, to solve this problem a powerful technology is used that is called WPT to transmit the current wirelessly.

2 Methodology

2.1 Wireless Power Transmission (WST)

It is the technology used to transfer electric current without the usage of wire by the principle of magnetic induction. Here the primary coil is connected to a power source, which produces the magnetic field that is received by the secondary coil in the form of electric current by this way the transmission is done more efficient and low cost, there are three ways to transmit which depend on the range of the transmission such as Short range, Medium range, and Long range.

(a) **Short-Range Wireless Power Transmission**

The short-range wireless electric transmission works on the principle based on inductive coupling, for example, transformer passed electricity in the form of electromagnetic fields. When the current is passed to the primary coil which is connected to the power source directly, it produces a magnetic field inside and outside the coil. The field produced inside the coil is a strong magnetic field and outside is the weak magnetic field. The secondary coil (as shown in Fig. 1) receives the fields produced by the primary coil. The electromagnetic flux produced between two coils induces the secondary coil.

(b) **Medium-Range Wireless Powerless Transmission**

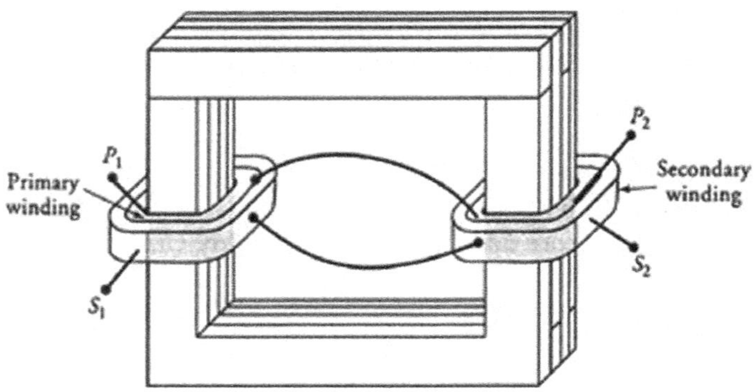

Fig. 1 Construction of transformer

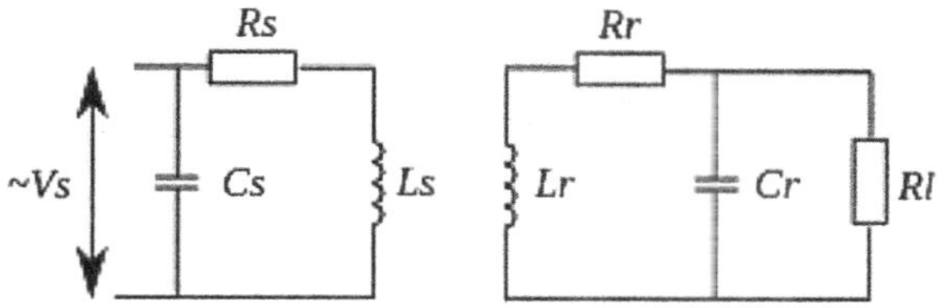

Fig. 2 Resonance inductive circuit

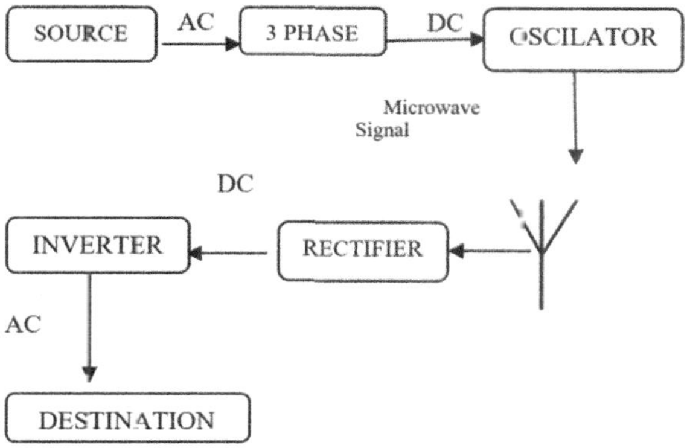

Fig. 3 Microwave power transmission

This approach works on the combination of resonance and inductive coupling (shown in Fig. 2), where resonance is used to send the magnetic field from primary coil to secondary coil more efficiently and inductive coupling is used to increase the magnetic flux between two coils. The transmission of power in this method is possible only if both the coils are having the same frequency. This type of power transmission is mostly used in homes.

(c) **Long-Range Wireless Power Transmission**

In this transmission, a beam of microwaves or laser waves are used to accomplish the wireless transmission of electric current. Antennas are used to transmit the electric current from source to destination. First, the electric current is converted into a microwave signal which is transmitted through an antenna, signals are received at destination and it is converted back to the electric current. For the conversion of an electric current into microwave signals, AC signal is converted into DC signal at the source, at the receiver end the DC signal is converted to AC signal (Fig. 3).

3 Architechture of Smart Home with GREENIE

The renewable energy is collected from various resources then it is transmitted to GREENIE Board wirelessly, which is also attached with the main power supply separately shown in Fig. 3.1. From the GREENIE Board, the power is transmitted to

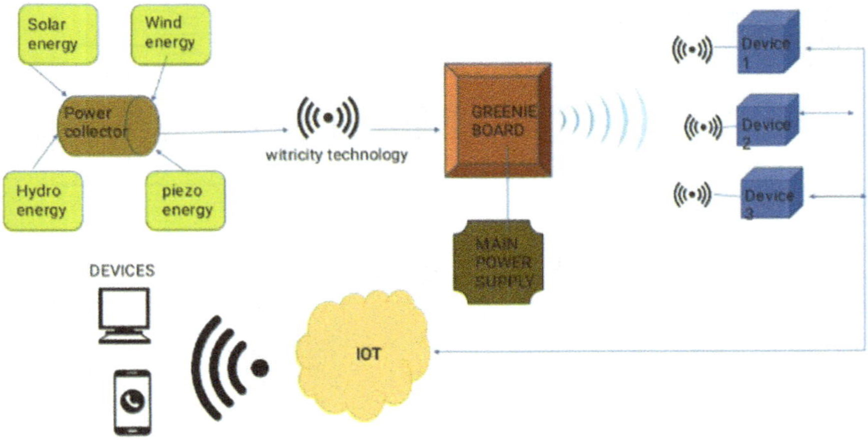

Fig. 4 Architecture

devices in the Smart Home. The devices are controlled using smartphones with the help of IoT (Fig. 4).

4 Implementation

4.1 GREENIE Board Architechture

It is divided into two segments: sender and receiver. The sender consists of the source coil attached to the main power supply. LC circuit is used for resonance frequency. The receiver side consists of a destination coil along with the LC circuit. The sender board is connected with the main power supply and receiver board is connected with the respective devices (as shown in Fig. 3.2) each board is tuned with the resonance frequency. Once the power starts flowing through board it produces magnetic field lines which induce the receiver board in this way the power supply for the house is transmitted. The whole house is connected with various sensors and they are controlled using a single application each device will be assigned with the corresponding port number in Arduino board that can be operated using the application Fig. 5.

4.2 Smart Home Construction

The construction of a smart home is divided into three segments:

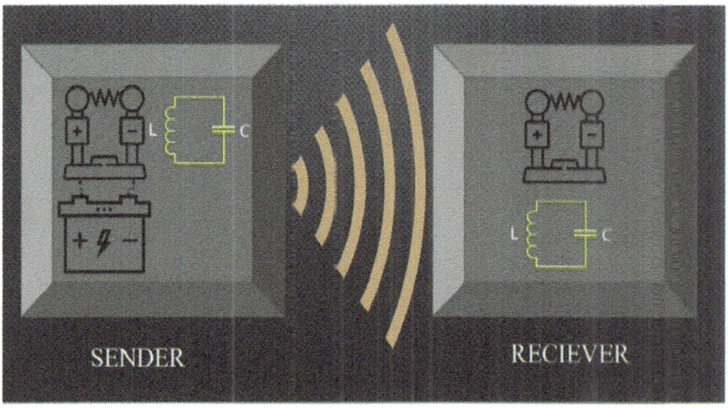

Fig. 5 GREENIE Board

- Construction,
- Connectivity, and
- Application.

(a) **Construction**

The smart home consists of three rooms, where Arduino board controls each room separately which acts as a slave. Raspberry Pi acts as a master. The I2C signal is passed to the Arduino boards with specified slave addresses (Figs. 6 and 7).

The devices are connected to the Arduino board with a respective sensor. For example, the living room consists of devices like lights, fans, etc. which is connected

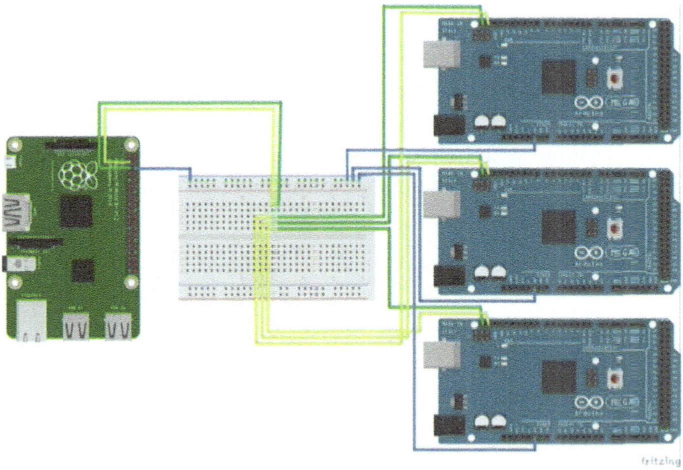

Fig. 6 I2C connection between Raspberry Pi and multiple Arduino boards

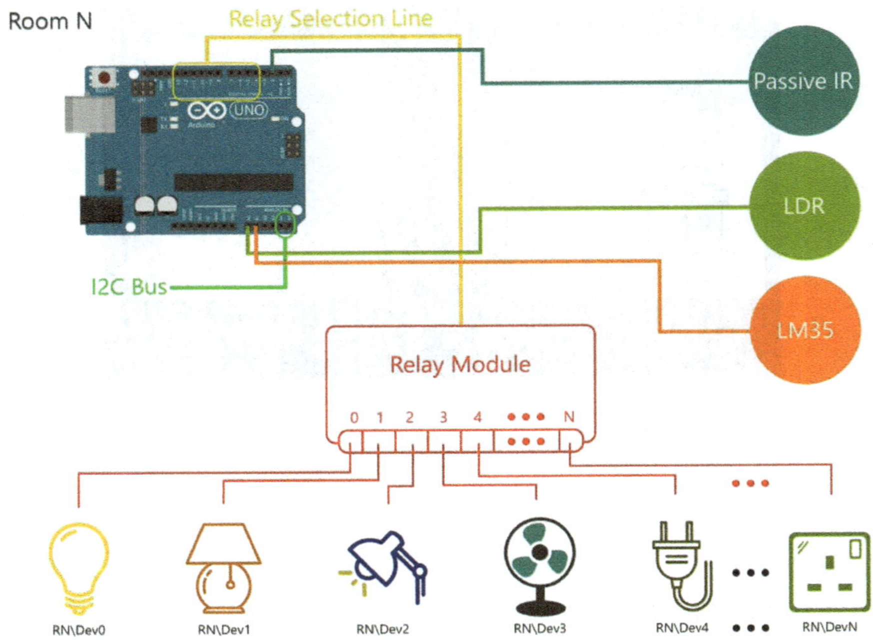

Fig. 7 Room configuration

to the Arduino board using relay module. Passive IR sensor is used to detect the motion. LDR sensor is used to maintain the intensity of light. LM35 sensor is used to monitor the temperature and humidity of the room.

(b) **Connectivity**

The SDA and SLC pins in Raspberry Pi are connected to the A4 and A5 pins of the Arduino board for the I2c connection. Sensor is connected to the digital I/O pins. Here 2. 2-channel relay connections are used to connect the devices. When Raspberry Pi requests the Arduino with help of slave address on the other side Arduino board runs the slave, program, which fetches the inputs from the sensors and sends the signal to the devices (Fig. 8).

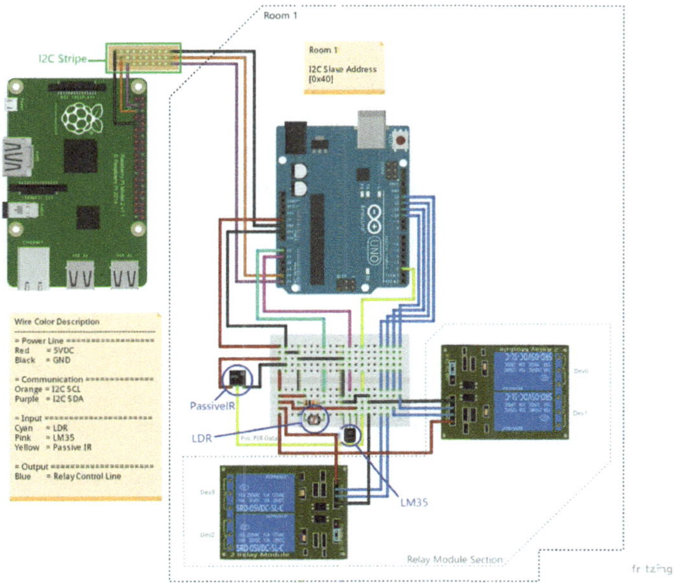

Fig. 8 Overall connectivity of a room

(c) Application

The application is developed and deployed in the Raspberry Pi, which is used to control the devices. Here we should enter the Arduino board slave address so that it can be operated (Fig. 9).

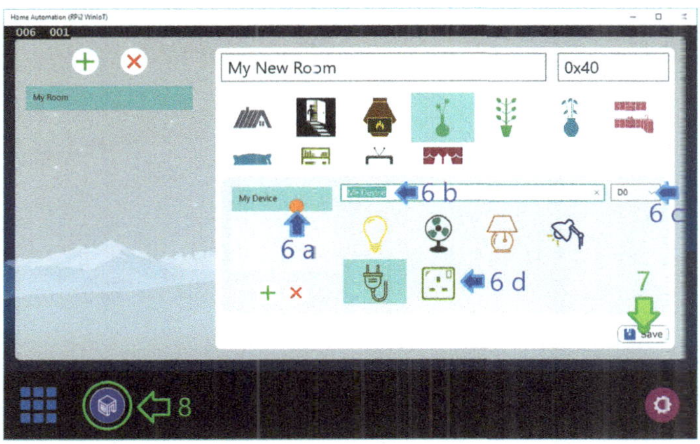

Fig. 9 Application interface

5 Conclusion

The Smart Home is built using wireless transmission of electric current, which enhances the feature of a smart home. The usage of wires can be reduced and the efficiency of power transmission can be increased. The maintenance cost of the smart home can be reduced and the wires are replaced by coils, which helps in preventing the wire's hazards.

References

1. J. Han, C.-S. Choi, W.-K. Park, I. Lee, S.-H. Kim, Smart home energy management system including renewable energy based on ZigBee and PLC. IEEE Trans. Consum. Electron. **60**(2) (2014)
2. A. Singh, H. Mehta, A. Nawal, O.V. Gnana Swathika, Arduino based home automation control powered by photovoltaic cells, in *Proceedings of the Second International Conference on Computing Methodologies and Communication (ICCMC 2018) IEEE Conference Record # 42656*, IEEE Xplore ISBN 978-1-5386-3452-3 (2018)
3. A. Sharma, Application of wireless power transfer for home appliances using inductive resonance coupling. Int. J. Eng. Trends Technol. (IJETT) **16**(4) (2014)
4. Y. Wang, J. Qiao, J. Du, F. Wang, W. Zhang, A view of research on wireless power transmission. J. Phys. Conf. Ser. **1074**, 012140 (2018)
5. B. Che, *Research on Omni-Direction Wireless Power Transfer Based on Rotation Magnetic Field* (Harbin Institute of Technology, 2014)
6. A. Kurs, A. Karalis, R. Moffatt et al., Wireless power transfer via strongly coupled magnetic resonances. Science **317**(5834), 83–86 (2007)
7. T. Nikola, Apparatus for transmitting electrical energy US1119732, 1914
8. Z. Zhao, Y. Zhang, K. Chen, New progress of mangetically-coupled resonant wireless power transfer technology. Proc. CSEE, **33**(3), 1–13 + 21 (2013)
9. Z. Zhao, Y. Zhang , K. Chen, New Progress of magnetically-coupled resonant wireless power transfer technology. Proc. CSEE **33**(3), 1–13 + 21 (2013)
10. L. Chen, R. Zeng, Study of wireless power transmission system based on magnetic resonance communications power supply technology **34**(4), 10–12 (2017)

ANAVI: Advanced Navigation Assistance for Visually Impaired

Arjun Sharma, Vivek Ram Vasan, and S. Prasanna Bharathi

Abstract The advancement in technology over the decades has provided for exceptional solutions and aid for the visually impaired. However, they are standalone solutions that target specific problems encountered by visually impaired individuals. This paper addresses some of the most common problems faced by the visually impaired and to provide an innovative and inexpensive solution that collectively addresses the problems. The solution proposed in this paper makes use of RF communication, AWS (Amazon Web Services) IoT (Internet of Things) core, and Computer Vision to aid visually impaired people to detect an object and use public transportation independently. Computer vision labels identify the obstacles or objects in front to aid the visually impaired individual to navigate through rough terrain. The problem faced by the visually impaired is access to public transport, which a majority of them use to commute daily as it is the only viable mobility option in our country. The visually impaired are forced to always rely on other individuals to help them travel making them feel dependent. The other problem is when navigating the streets or in their home, it is hard to recognize the type of object and the distance at which it is in front of them. Hence, there is a system that is required to allow them to travel independently and help them feel empowered. The handheld unit will be connected to the bus stop modules using RF module and Py-camera. The bus stop modules will have an ESP 8266 which provides the Internet connection. The modules and the buses will be connected using the services of AWS IoT core. Even though there are many ways for the visually impaired individuals to navigate the streets, as the environment and the objects around us change rapidly, it is close to impossible for them to spot using conventional methods and that is where this system comes in handy. For one

A. Sharma (✉)
ESP-AS13, Robert Bosch Engineering and Business Solutions, Adugodi, Bangalore, India
e-mail: arj.sharma1997@gmail.com

V. R. Vasan
ESP-AS2, Robert Bosch Engineering and Business Solutions, Adugodi, Bangalore, India
e-mail: vivek435@gmail.com

S. Prasanna Bharathi
Department of ECE, SRM Institute of Science and Technology, Vadapalani, Chennai, India
e-mail: prasannss@srmist.edu.in

S. S. Dash et al. (eds.), *Intelligent Computing and Applications*,
Advances in Intelligent Systems and Computing 1172,
https://doi.org/10.1007/978-981-15-5566-4_25

such instance, an open drainage, which is 1 by 1 foot wide, cannot be detected using the tap and go method or sound sensing methods while the system has a pretrained dataset which can send a voice alert saying there is an open drain at 20 m in front and a vibration-based alert stimulation is given at the stick handle to help circumnavigate the open drain.

Keywords RF · AWS · Computer vision · Py-camera · Visually impaired · IoT · ESP8266

1 Introduction

Eye Blindness or the number of visually impaired people around the world stands at about 1.3 billion. The studies have found that visually impaired people suffer from secondary road traffic accidents and there are times when they get lost in a crowded place, in some instances they fall due to objects placed in the middle of walk paths. The main factor of such untoward incidents is the lack of guidance or help provided by the general public. There have been a lot of emerging technologies but never has there been a combination of such principles used for the lesser fortunate in our nation. With emerging technologies like machine learning and artificial intelligence, there is a huge scope to put such technology into work in a much crucial situation and give the impaired with an eye of technology to assist and guide them [1]. With the advent of IoT and the widespread availability of internet connectivity, we propose a system that is easy to implement over the existing infrastructure with minimal changes. The need of an established connected system is to connect the unconnected and offer people with visual impairment a mean to see things [1], instead of walking towards their destination they can afford a public transport with a click of a button [2]. The system is ideated keeping in mind the development of technology, which caters to the needy and gives a solution to a problem statement or a purpose.

The existing solutions for this particular problem include the usage of RFID tags to identify the required bus or its user, a bus arrival based detection by the use of RF waves by the visually impaired person [3]. Though this method has been proved to work, it has a lot of disadvantages such as cost and change in existing infrastructure to serve a select group of people. RFID is a complex but trivial method to cater to the visually impaired person to move from point *A* to point *B* [4]. The use of sensor and artificial based system to provide assistance to visually impaired people can be seen from the plethora of works in the reference section below [5–11]. However, technology has not served large communities to solve the common problem such as obstacle avoidance, safe and convenient transportation to the visually impaired people. The common problems found by studies indicate that the price of this life-changing technology is higher or the device gets bulky and difficult to manage.

Public transportation is the only viable mode of transport for the majority of the visually impaired. Hence, it is imperative that there exists a system that is inclusive of them as well. The current scenario forces the visually impaired to rely on another individual for guidance. We offer a device and an infrastructure that will empower them and make them feel independent. We propose to design a system that integrates all the buses deployed and the bus stops through IoT. The visually impaired individual will be provided with a Handheld Unit that allows them to enter the bus number that they wish to board. All other solutions revolve around the similar concept of some sort of usage in the RFID field of technology. Clearly, none of these measures have been scaled or implemented on a national level to serve the users.

Our solution involves the use of a handheld device which will be given to the visually impaired user to input the number of the bus through voice recognition or manual input and can use the onboard Computer Vision based module to detect whether there is an object in the front. The distance and the type of object will be captured and analyzed to give a voice-based alert. When a bus-based request is requested, a request is sent to the bus stop from which they wish to board the bus through radio frequency. The modules placed at the bus stops (with internet connectivity) will be connected to a gateway device (server). This module will publish the message to the gateway, where the buses will be at the receiving end of the gateway, enabled through an Android application at the disposal of every bus driver. The bus driver will now know the location of the visually impaired individual and will be given the choice to accept or clear the request and an auditory signal will be heard to indicate the location of the door. The particular devices and technologies used to provide the aforementioned system will be explained further.

2 Methodology

The method which is proposed deals with the identification of real-time objects when a visually impaired person uses the walking stick. The walking stick is multifunctional as it can be used to communicate with local transport to board the vehicle and identify objects while walking to their destination.

The working of the product is split into three parts:

1. Handheld Unit—User to Bus stop.
2. Bus Stop Module—Bus stop to Amazon Server.
3. ANAVI Driver app—Amazon Server to Bus driver.

2.1 Handheld Unit

This subsystem consists of a Matrix Keypad, Ultrasonic Sensor, NRF module, and a Py-camera to obtain the inputs and transfer the data to the Bus stop receiver as well as to detect objects and measure the distance.

2.1.1 Matrix Keypad and NRF Module

The Matrix Keypad will be present on the top of the module. The connector lines enable the keypad to be used directly. We check the logic levels of all the rows and columns infinitely. Any changes in the state of the rows or columns will be detected in the code. Each key has a unique combination of a row and a column. Hence by obtaining the row and column whose states have changed, we can obtain the key pressed by the user. The visually impaired individual can identify all the characters by use. This is required as it act as an easy method of obtaining the required bus number which the user wishes to board. This will be processed by the microcontroller and the data will be transmitted to the bus stop utilizing RF communication.

We utilize a cost-effective NRF module to transmit the data (bus number) to the bus stop receiver. The module is an 8-pin device, with each pin having a unique function to implement the communication protocol. The module requires a constant supply of 3.3 V. The module runs on a 2.4 GHz bandwidth and provides a range of 100 m. It can be programmed as a transmitter or a receiver based on the requirements. The module can open up to 6 different channels for communication. We are employing this module at two locations. Firstly, it is programmed as a transmitter on the HHU module. Secondly, it is programmed as a receiver at the bus stop module.

2.2 Py-camera and SENIC Ultrasonic Sensor

The Py-camera and the ultrasonic sensor module is connected to a Raspberry Pi for Image and Distance Processing. Object detection is done using OpenCv and a pretrained dataset from the open-source Tensorflow Detection Model Zoo. This has some restrictions due to the speed and accuracy of the data detection. To counter such discrepancies, SSDLite-MobileNet is used to train the model and to get fast image recognition. To support high accuracy, any model with small and faster quantized kernels that allows fixed-point matching is implemented and datasets are trained in the PC and an export is used in the Pi to make the model optimized for faster detection.

The Ultrasonic Sensor can measure a distance that has a range of 70 feet and an accuracy of 0.833 inches. In this project, we use it to tag the image detected so that it enables the visually impaired person to know the distance of objects, buildings, and any other obstacle in front of them. An earphone can be connected to the onboard Raspberry Pi module to get Audio feedback which the visually impaired person can use to navigate. A vibrator is placed on the stick to send an alert if an object is detected too close to the individual (Fig. 1).

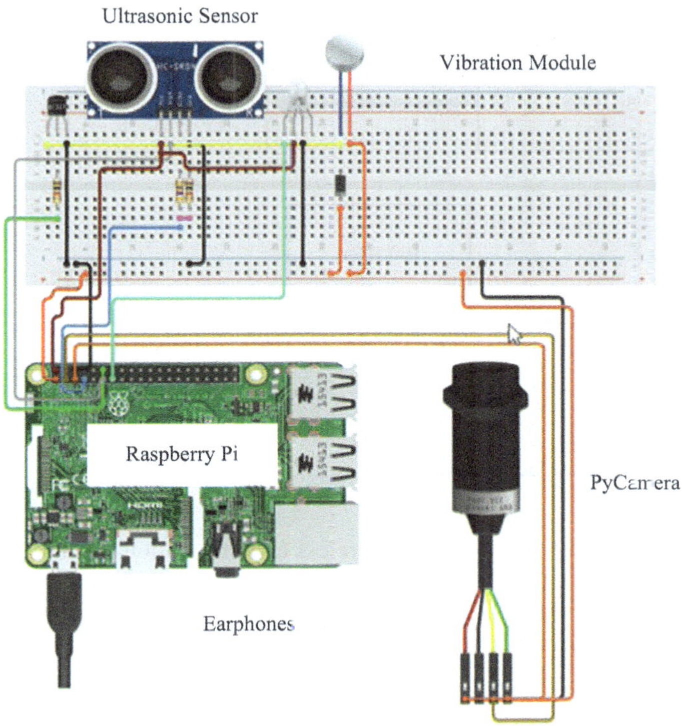

Fig. 1 Raspberry pi interfaced with Py-camera for computer vision and ultrasonic sensor for distance calculation

2.3 Bus Stop Module

The bus stop module acts both as a receiver and a transmitter. It receives data from the HHU and it is sent through serial communication to the Wi-Fi module. This is provided with an internet connection to transmit data through a wireless connection to the server which will be accessed by the bus.

2.3.1 Wi-Fi Connectivity Board-ESP8266

The ESP8266 is a cost-efficient Wi-Fi-enabled microchip with a full TCP/IP stack and computational capability, embedded onto the NodeMCU board. We require this device to access the Amazon server to publish the data received from the previous subsystem. As it has internet capability, it will wirelessly transmit data to AWS IoT gateway. It has storage capability and 9 digital GPIOs. It is powered with a supply voltage of 3.3 V. The code to identify, connect and transmit data to the gateway is run on this device.

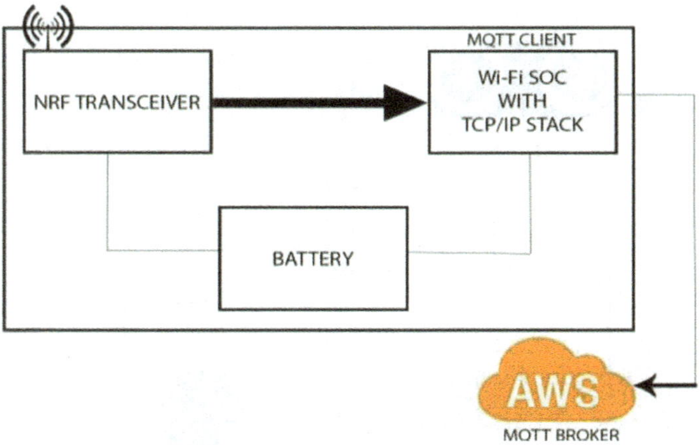

Fig. 2 Bus stop module block diagram

This module will be enabled with wireless communication capability. Hence, it can transfer the data from the client to the gateway server endpoint provided by Amazon. The bus stop receives and transmits data. Accommodations are made to store data in a buffer in case of multiple entries simultaneously (Fig. 2).

2.4 *ANAVI Driver App*

The final subsystem will be on the mobile phone of every bus driver. An Android application will be deployed on their devices. This app will run a Graphical User Interface. It will also serve as a subscriber to the server gateway and will be run in the background. The incoming data will be received from the server vis-à-vis AWS IoT services. This data will present to the driver about the presence of a visually impaired person along with the name of the bus stop. The driver needs to indicate his arrival at the bus stop. This is enabled by providing the ACCEPT or CLEAR command to the driver.

The ACCEPT push button on the GUI will result in giving an auditory output to the user indicating the exact location of the entrance to the bus (Fig. 3).

2.5 *IoT Enabler*

Our IoT enabler provides these services through AWS. The details of AWS implementation will be mentioned ahead. AWS IoT cloud platform provides a secure, two-directional communication between hardware objects (such as smart sensors,

Fig. 3 ANAVI driver app

digital actuators, embedded devices with edge connectivity, and smart appliances), the AWS Cloud has encryption of MQTT. We have implemented the data transfer using the MQTT protocol.

Devices acknowledge the sent commands by sending ACK messages through a JSON format when an MQTT protocol is used. There can be multiple MQTT-based Connection Topics relating to a single group name that identifies the unique device which is updated based on its state and it sends back an acknowledgment for the state change. When a message is published, the message is dispatched to the message broker (MQTT), which ensures sending the messages published to that particular MQTT group to reach all the clients subscribed to that group. The IoT broker uses topics that routes messages to receiving clients from the sender. The topic to publish to or subscribe from can be entered in the dialog box as shown and the MQTT broker can transfer the messages from end to end. This process of subscribing and publishing is included in our code at every bus stop module exclusively (Fig. 4)

3 Implementation and Result Analysis

In the proposed work, the use of a handheld device was implemented which will be given to the visually impaired user to input the number of the bus through voice

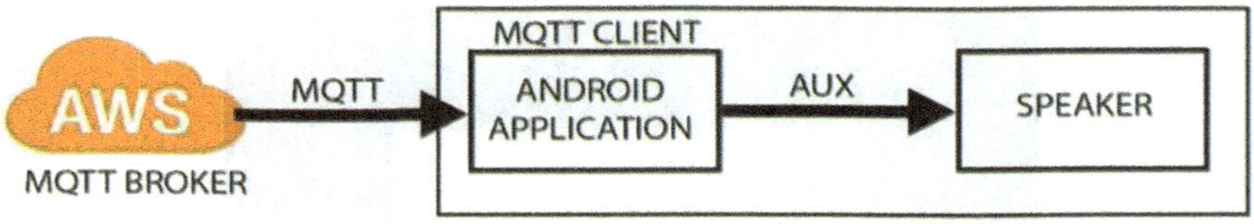

Fig. 4 Module block diagram

recognition or manual input. The use of an onboard Computer Vision based module to detect whether there is an object in front is implemented. The distance and the type of object will be captured and analyzed to give a voice-based alert. When a bus-based request is processed, the request is sent to the bus stop from which they wish to board the bus through radio-frequency communication. The modules placed at the bus stops (with internet connectivity) will be connected to a gateway device (server). This module will publish the message to the gateway, where the buses will be at the receiving end of the gateway, enabled through an Android application at the disposal of every bus driver. The bus driver will now know the location of the visually impaired individual and will be given the choice to accept or clear the request and an auditory signal will be heard to indicate the location of the door of the bus once the bus has arrived at the corresponding bus stop. The particular devices and technologies used to provide the aforementioned system will be explained further.

3.1 Design Verification

To verify the design of the blind stick and the camera position along with the ultrasonic sensor placement for optimal performance and to avoid low lying objects and wrong calibration, simulations were run to study various scenarios that could be encountered by the visually impaired individuals. The developed prototype was also tested on the field with visually impaired individuals handling the prototype equipment. Figure 5 shows the placement of the aspects being calibrated. The collective data from the simulations as well as the user experience from the visually impaired individuals was processed and the following results were tabulated for optimal performance (refer Table 1).

The app is tested for the range of requests processed by the AWS IoT core system. The driver's app must come under a range of 10 km of the visually impaired person to get a request to accept or clear. Figure 6 shows the request from a visually impaired individual being transmitted from the server to the bus driver's app. The notification is displayed on the app with the bus stop name details and the options to Accept or Clear the request. The reliability of internet connection and the latency for request processing were also tested which resulted in improving our Android app code to handle unexpected exceptions to increase the performance. The design and architecture were verified under various scenarios and the experiments concluded with the

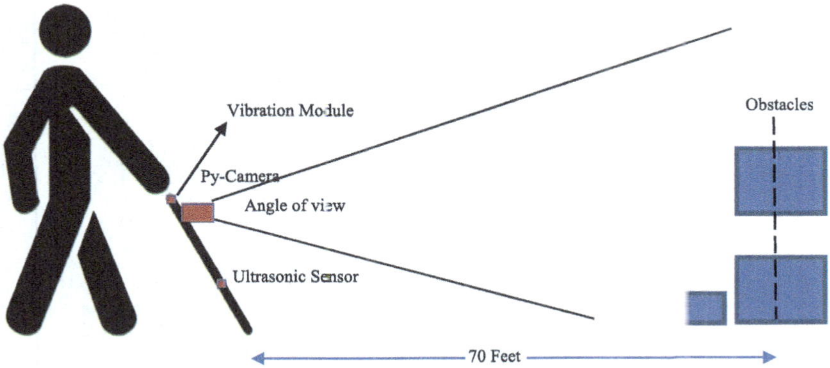

Fig. 5 Walking stick attached with the modules

Table 1 Blind stick calibration with mean stick length 130 cm

Average precision	Intersection over union	Area	Max detections
Average precision	0.49:0.95	All	0.453
Average precision	0.50	All	0.0668
Average precision	0.75	All	0.4721
Average precision	0.49:0.95	Small	0.267
Average precision	0.49:0.95	Small	0.470
Average precision	0.49:0.95	Medium	0.589
Average recall	0.49:0.95	Large	0.337
Average recall	0.49:0.95	All	0.531
Average recall	0.49:0.95	All	0.577
Average recall	0.49:0.95	Small	0.368
Average recall	0.49:0.95	Large	0.702
Average recall	0.49:0.95	Medium	0.604

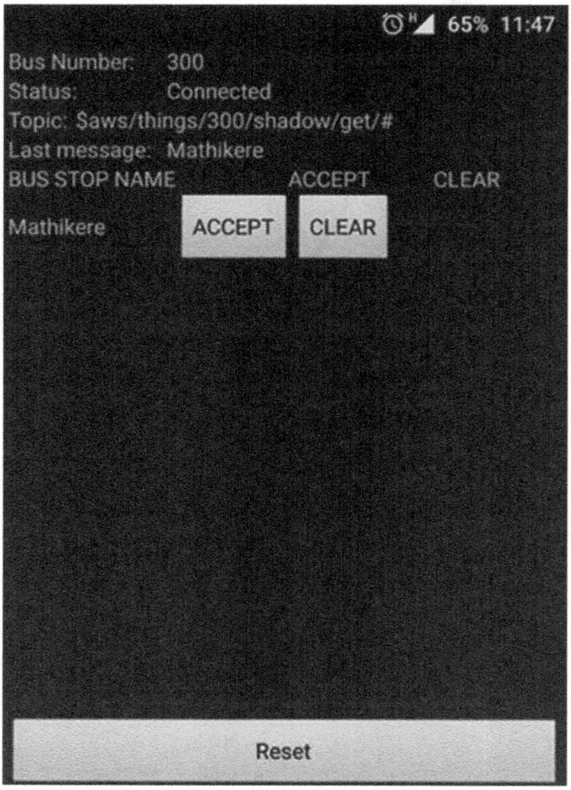

Fig. 6 ANAVI interface

successful working of the driver app to process requests as well as provide auditory signals at the acceptance of corresponding requests.

3.2 Result Analysis

SSD_Mobilenet_V2_fpn_Coco is fast compared to other models by giving about 525 ms. The drawback of Raspberry Pi running on such powerful datasets is it heats the Raspberry Pi, though a heat shrink can solve this problem. Images trained and tested were of the resolution 420 * 240 in RBG though the size is big, it would improve the model accuracy and reduce the recall. The camera was running on a 4–7 FPS and this can simultaneously capture about 25 objects and label them with an accuracy of 85.69% according to the testing datasets (Table 2).

Table 2 6 data object experiment results

Position calibration		Remarks
Ultrasonic sensor	24.6 cm	From below
Vibration sensor	4.5 cm	From the handle
Camera AOV	62.2 × 48.8°	Front angle
Camera position	14.8 cm	From the top
Microcontroller module	30 cm	From the top

4 Summary

ANAVI is a user-friendly device that works primarily on computer vision and RF transmission to send crucial information to the user and cloud. This solution has solved the most important aspect which is providing essential visual and navigation assistance to the visually impaired individuals in a simple and cost-effective way.

Field trials show that the camera has a wide focus area, which leads to multiple object detection which gives the user more data to process. For better user decision-making, in future developments, we can narrow down the focus area of the camera. This solution can also be scaled up by improving the infrastructure of the connectivity of the device to the cloud and expanding its reach to other transport sectors. This serves as a precursor to the Intelligent Transport System toward which the world is progressing.

From the results of our experiment, we can observe that the accuracy of image processing to detect obstacles is improved to 85.69% resulting in better performance compared to other developments in the field. Further improvements can be made by learning more image data sets to recognize real-life obstacles with improved accuracy. With this solution, the objectives achieved include enabling the visually impaired to access public transportation independently with minimal changes to the existing infrastructure.

References

1. M. Hersh, M. Johnson, *Assistive Technology for Visually Impaired and Blind People* (Springer Publishing Company, 2008)
2. D. Dakopoulos, N.G. Bourbakis, Wearable obstacle avoidance electronic travel aids for blind: a survey. IEEE Trans. Syst. Man Cybern. Part C (Appl. Rev.) **40**(1), 25–35 (2010)
3. M.Z.H. Noor, I. Ismail, M.F. Saaid, *Bus Detection Device For The Blind Using RFID Application* (IEEE Xplore, 2009)
4. K.N. Sushma, B.C. Smitha, B.N. Kiran, Implementation of RFID for blind bus boarding system, in *A Treatise on Electricity and Magnetism*, ed. by J.C. Maxwell, ICCSP, vol. 2, 3rd edn. (Clarendon, Oxford, 1892), pp. 68–73
5. S. Sivan, G. Darsan, Computer vision based assistive technology for blind and visually impaired people. https://doi.org/10.1145/2967878.2967923

6. S. Abbouda, S. Hanassya, S. Levy-Tzedek, S. Maidenbaum, A. Amedi, Eyemusic: introducing a visual colorful experience for the blind using auditory sensory substitution. Restorative Neurol. Neurosc. **32**(2) (2014)
7. A. Geiger, M. Roser, R. Urtasun, Efficient large-scale stereo matching, in *Asian Conference on Computer Vision (ACCV)* (2010)
8. R. Tapu, B. Mocanu, A. Bursuc, T. Zaharia, A smartphone-based obstacle detection and classification system for assisting visually impaired people, in *2013 IEEE International Conference on Computer Vision Workshops (ICCVW)* (2013), pp. 444–451
9. L. Chen, B.-L. Guo, W. Sun, Obstacle detection system for visually impaired people based on stereo vision, in *2010 Fourth International Conference on Genetic and Evolutionary Computing (ICGEC)* (2010), pp. 723–726
10. F.L.M. Milottta, D. Allegra, F. Stanco, G.M. Farinella, An electronic travel aid to assist blind and visually impaired people to avoid obstacles, in *International Conference on Computer Analysis of Images and Patterns* (2015), pp. 604–615
11. A. Morar, F. Moldoveanu, L. Petrescu, A. Moldoveanu, Real time indoor 3D pipeline for an advanced sensory substitution device, in *International Conference on Image Analysis and Processing* (2017)

THIRD EYE—Shopfloor Data Processing and Visualization Using Image Recognition

S. Prasanna Bharathi, Vivek Ram Vasan, and Arjun Sharma

Abstract This paper describes how superimposed data using an Augmented Reality based system is appropriate in manufacturing and warehousing locations to improve operations. Access to Data is penultimate in the manufacturing industry, and giving the required data in real-time based on the component or a machine can help in increasing productivity with an added insight of using, a handheld mobile phone, thus avoiding any interaction with computers. Production and machine details give the shop floor managers the flexibility to analyze machines at the shop floor, which aids on time defect resolution and machine status. This paper focuses on object detection based data analytical application, which can differentiate between two or more similar machines using feature extraction methods and texture mapping to enhance object detection in the used game engine and also to develop the application and map those to the database of that machine, hence provide an analytical information in a hands-free manner. The effectiveness of the algorithm was tested on machines that look similar or near to identical was differentiate using our methodical image processing algorithm and the data pertaining to the machine is processed on the cloud. Improvements to process the data on the device to improve overlay efficiency and speed was experimented using various studies. Compared to various image processing software, THIRD EYE was rated 93.6% efficient in image recognition and data overlay to the user was one new feature we provide as an application to the user.

Keywords Augmented reality · Object detection · Feature extraction · Texture mapping

---

S. Prasanna Bharathi (✉)
Department of ECE, SRM Institute of Science and Technology, Vadapalani, Chennai India
e-mail: prasanns@srmist.edu.in

V. R. Vasan
ESP-AS2, Robert Bosch Engineering and Business Solutions, Adugodi, Bangalore, India
e-mail: vivek435@gmail.com

A. Sharma
ESP-AS13, Robert Bosch Engineering and Business Solutions, Adugodi, Bangalore, India
e-mail: arj.sharma1997@gmail.com

S. S. Dash et al. (eds.), *Intelligent Computing and Applications*,
Advances in Intelligent Systems and Computing 1172,
https://doi.org/10.1007/978-981-15-5566-4_26

1 Introduction

Augmented Reality (AR) is a game-changing technology in the manufacturing industry which are used as part of the Industry 4.0 solutions. This technology gels with automation and digitizing the industries by making the process simpler and easily understood. Mobile augmented reality can be accessed on smartphones as it extends and enhances the user experience, thus making it more customer friendly. The current use cases in the manufacturing industries are for process training and shop floor employees are replacing maintenance manuals for service technicians. This methodology is adopted by industries to ensure process training is appreciated and effortless to explain, giving the operator a hands-on experience, which proves to be more efficient than the current training practices. Service technicians extensively use AR devices instead of manuals, as the steps are elementary, exclusively because of the crunch of operating space and visual aid which helps in identification of the correct part to be serviced, apart from the step by step explanation. This is a game-changing technology since there's no necessity for the operator to have prior experience in doing such tasks and provides a visual guide. Augmented Reality also holds several applications in other industries and functions. Product advertisements can also adapt to this technology to provide the user an interactive platform. The user can perceive the product in an uncomplicated virtual environment.

AR technology superimposing data onto the real world the application is yet to be explored for manufacturing usage and needs extensive research [1]. Virtual Reality uses CAD-based 3D objects that allow the operator to experience and interact with virtual environments [2–4]. The operators will be able to interact with the virtual world which could be a simulation of a complex process or the viewing of volumetric data [5]. Augmented Reality (AR) aids the user to visualize the reality with virtual objects superimposed on the live feed. Virtual data can provide significant information that the user cannot perceive with their own processing capability. Augmented Reality based applications are used in various areas such as training and entertainment (Thomas et al. [6]), social topics (Dähne et al. [7]) and product assembly, machine downtime and repair of manufacturing equipment (Curtis et al. [8]; Navab et al. [9]). AR technologies have been implemented in a few scientific visualization tasks such as HUD, ultrasound scans, and in the healthcare sector for the visualization of CT scans or MRI. In the above-mentioned applications, Augmented Reality is a unique platform for data visualization for enhancing the understandability of data among the users, but it has not been employed to embed insightful information pertaining to the real object itself [10]. From the above-cited paper, it can be observed that AR visualization increases the understandability and interaction of the displayed data, however, there is no implementation of the AR technology in the industrial and manufacturing field [11], in which effective ways are there to view and analyze data.

However, the problem of image/object identification in real time has been extensively studied. Although there are many platforms to do such processing, there is limited or no special package for AR-based applications, which can do recognition on a very precise scale. Nevertheless, applications have come out with superimposed images on real-time video feed which is popular despite several restrictions in terms of 3D plane and the quality of image or object it tracks. Currently, there are no systems which can give real-time information of a process in the manufacturing industry, and simultaneously display data when the detected object or machine changes, even though they look alike. The main motive of object detection and recognition is to discover the scene and find the target. Firstly, for the classification of information, we have to rely on the enhanced image information for better detection and ordering of objects. Research shows there was a lack in parallelism strategy to develop a mobile-based image processing algorithm to deal with limited CPU-GPU memory utilization without compromise of memory [12], in this work the ability to parallelly process data and perform image recognition on the target. Image processing using OpenGL ES 1.1 is studied [13] to save computing space by the efficiency of such platform for processing decreases the accuracy of recognition, and hence advanced yet compact versions of OpenCV paired with Vuforia engine is studied for compute efficiency.

In the present paper, a method is proposed that identifies similar machines and gives the current production data of the machine which is crucial to line managers at a particular interval of time. The recognition and detection of the image targets are done on OpenCV, which runs in parallel to the Vuforia game engine to classify the images. The image targets are matched with respect to the feature value. This method is put into Microsoft HoloLens and an AR-enabled mobile phone which the operator uses to get insights into the production data as he scans around the production floor using his AR glass. Studies have shown that users are well adapted when provided with enriching data which is spatially aligned thus not affecting the view of the user [14]. The system exploits a cyber-physical system, which enables the user to seamlessly be connected with the database in a basic way, to view data of different machines instantly. AR devices are popular, yet expensive to be scaled to such business verticals since it gives possibilities for exploring Augmented Reality using a mobile device to guide and visualize data.

2 Methodology

The method which is proposed deals with image identification of similar objects, predominantly used for industrial manufacture (focuses mainly on production machines on the production floor). The application is designed in such a way that, it can differentiate two similar objects depending upon various parameters which tend to differ within a similar equipment/machine when placed in a dynamic environment such as a production floor. An example of such similar machines is depicted in Fig. 1.

The first step for image processing is the application should be able to process two resembling objects thereby and able to differentiate based on a few features. The design should also be able to cater to fast-moving object/camera feed in order to deliver accurate data based on the image detected. Various methods are incorporated into the design such as contrast-based feature extraction, contour detection and

Fig. 1 Similar datasets for image training

texture mapping. The algorithm is divided into 2 stages to simplify the detection of similar objects. Contour detection and texture mapping are used to identify objects based on the contour and the texture, which helps to narrows down the detection process. Mobile platform architecture is shown in Fig. 2.

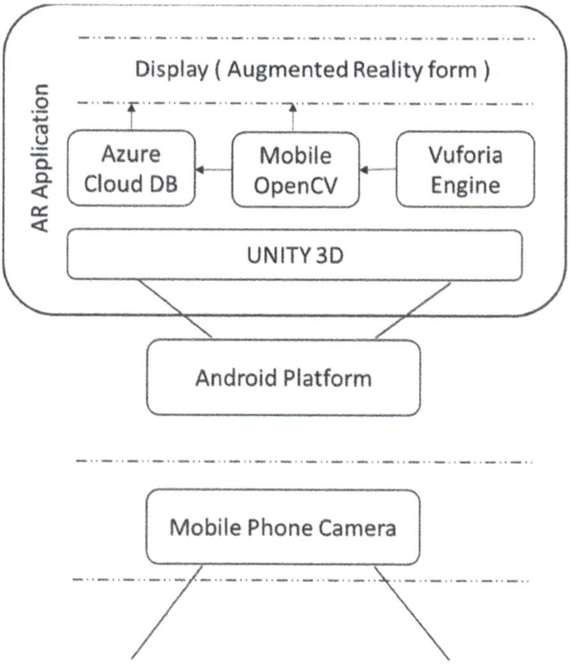

Fig. 2 Mobile app architecture on iOS/Android

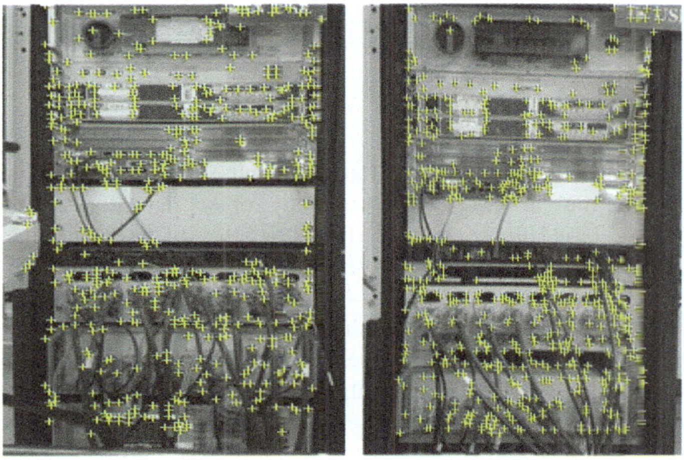

Fig. 3 Classification of feature points of similar images

Once the machine or object is detected the images with a similar algorithm is run to check if there is any matching object and further it is narrowed down to one unique object. Feature extraction uses a search engine that checks for unique elements in an image, which can be based out of the high-contrast spots due to the lighting condition in that area or certain stickers or markers for that particular machine. There can also be some wiring density, which makes it different from a similar machine. These feature points are stored once the engine is trained with the image sets and forms a defined constellation based on the feature points it detects. Once the application is narrowed down on the object, the maker is recognized by the constellation. The above process is depicted in Fig. 3.

The detection algorithm and the application are tested against the conditions as follows:

- The detection algorithm should be able to operate in real time.
- The algorithm must be able to classify two similar objects based on certain feature points.
- The efficiency is tested against varying environmental conditions such as background change, light variations and wiring density changes.
- The detection speed should operate within 800 ms.
- The algorithm should work efficiently with human or object disturbance at permissible levels.

3 Proposed System Design

3.1 Unity 3D and OpenCV

Augmented reality requires a powerful and efficient game engine like unity 3D, to put into use, the mobile's webcam for real-time object detection using its inbuilt computer vision-based libraries. Unity host flexibility of the environment where the application is developed, every layer can have a script to dedicate a certain functionality. Game objects are tagged for the text values which are going to be displayed post-image detection. The application will have communication protocols for simultaneous communication with the cloud database, which is Azure cloud. For unity instance and the image detection algorithm which works on unity platform to communicate with each other, UDP (User Datagram Protocol) is used over TCP as it doesn't expect any acknowledgment token before sending the next packet. This method is opted to optimize speed for the application to identify the image within the stipulated time frame. This offers lower bandwidth and latency (Figs. 4 and 5).

OpenCV in the unity platform uses Python API to detect the image based on the algorithm implemented. An asset identification will be sent in terms of image targets that are compatible with Vuforia inbuilt along with a unity platform to identify the object to a predefined port using sockets. This is used to trigger a game object, which syncs with the cloud and displays information pertaining to that object (Fig. 6).

The detection and processing phases performed by OpenCV is as follows:

1. Frames are captured from the camera on board the device as images.
2. The image is stored and a Gaussian blur is added to reduce the noise of the image and for further processing.
3. A color filter is added by replacing the image with black and white so that an outline of the image is obtained. Contours are detected and are marked on the image.

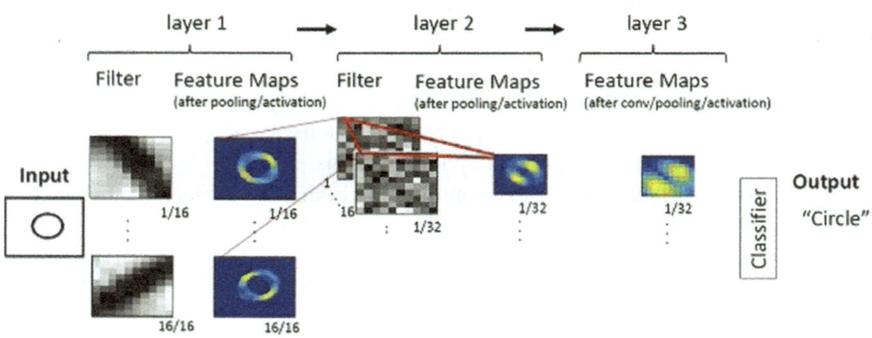

Fig. 4 Feature extraction and classification

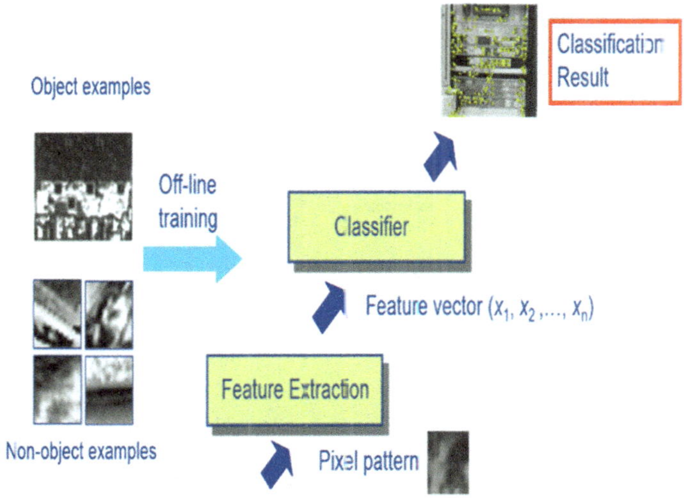

Fig. 5 Software architecture

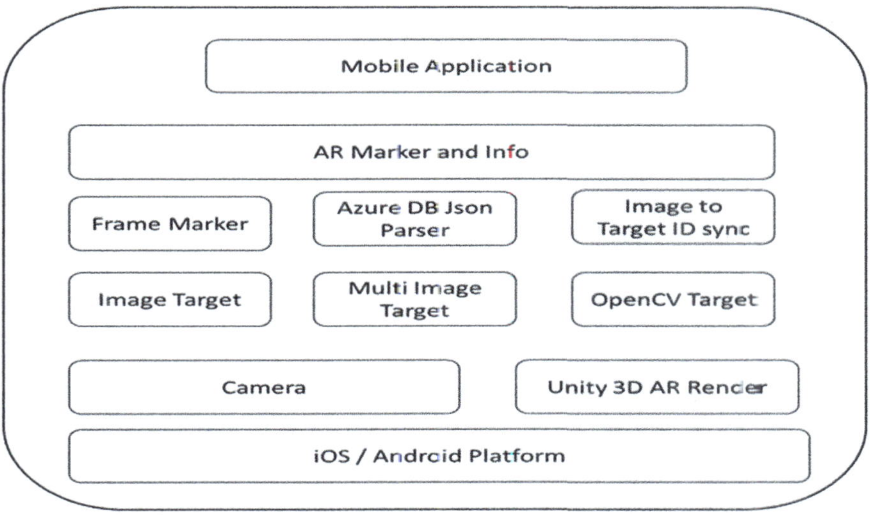

Fig. 6 Object detection—basic scheme

4. Image is then scanned for feature points and these points are later represented and stored as a constellation for identification of the object.
5. Image target script is invoked to match the feature and contour detection points to a specific image target stored in the application (Fig. 7).
6. Image target ID is communicated through the UDP protocol to unity platform.

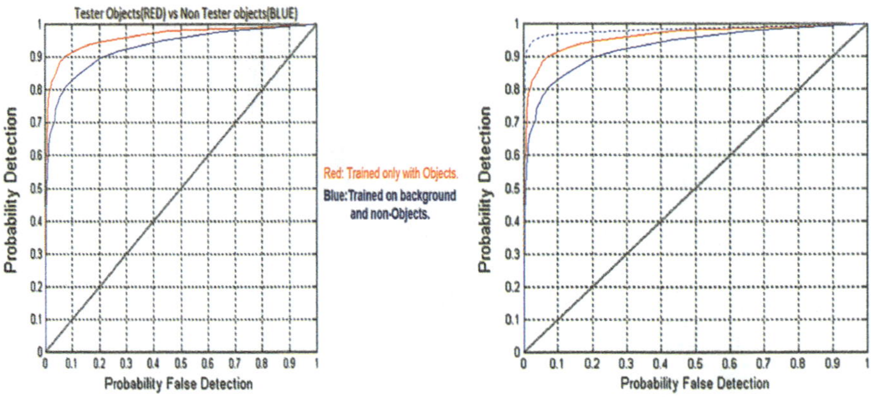

Fig. 7 Probability of false detection feature and texture-based biasing—blue dotted line

3.2 Port Forwarding and Game Object Setup

Unity platform handles the inputs via UDP and creates a trigger to generate the instances of the game object. A game object is a component that can consist of text, 3D models, and other functions that can also create legacy components using the Scripting API in unity. An outline marker is marked on the identified object to give a visual representation, so that the object is detected. The function which is optional can be implemented via script. To receive data from the UDP layer go to the project hierarchy under managers tabs –> player controller, the script named Player Controller. CS is modified to receive the commands (Script is shown below). A Boolean variable is added to check if the variable is initialized on each frame. Based on that trigger action, the flag will be set. The trigger enables a handshake technique to receive data from the database, which uses the object ID to parse data back to the text game object.

Commands to be Added in the Script File

```
Thread recieveThread;
UdpClient client;
Int port;
Public Gameobject Player;
InputFielddataViz;
bool data;
```

Back in the Text Game Object, a script file is added to invoke the script to receive the trigger from the UDP script. Boolean variable is initialized to false, and it sets to true when a trigger is received from the OpenCV Python code. A text game object is attached to the player controller, which stores the data and is parsed back from the database when the trigger is set.

Script File
The port is initialized in this script.

```
Void start()
{
port = 5066;
data = false;
dataViz = gameObject.GetComponent < InputField > ();
InitUDP();
}
```

3.3 *Scene Creation and Database Connection*

The scene can be created using a real-time feed of the camera as well as for the representation of the object, to sense if it has been detected or not. There are panels that change color from red to green when the object is detected. Enabling the back camera in the project properties' settings so that the camera opens when the application starts. Text game objects are mapped to the 3D plane accordingly. The text game object also has a script file that parses data from the cloud database based on the objectID. For example, JSON parsing is used in terms of object name/objectID, which is parsed from DB to the application. Data is superimposed on the realtime feed from the camera, giving it an augmented reality feel.

For the information to be displayed on the screen 2D or 3D, text objects can be used based on the software complexity. The detection panels can be downloaded from the unity asset store or can also be used from PNG images. The text object and image target must be destroyed using command Destroy GameObject() so that it saves space and avoids rendering of many image targets which is active in the background and may slow down the application (Fig. 8).

The data parsed uses the code as follows:

```
The data parsed uses the code as follows:
void Start()
{
stringurl="http://10.10.0.25:8080/TestRESTService/MachineData.svc/Yield/{NET0500206}";
WWWFormformDate = new WWWForm();
formDate.AddField ("name    ", "product");
    WWW www = new WWW(stringurl, formDate);
StartCoroutine(request(www));
}
IEnumerator request (WWW www)
  {
    yield return www;
displayText.text = www.text;
  }
```

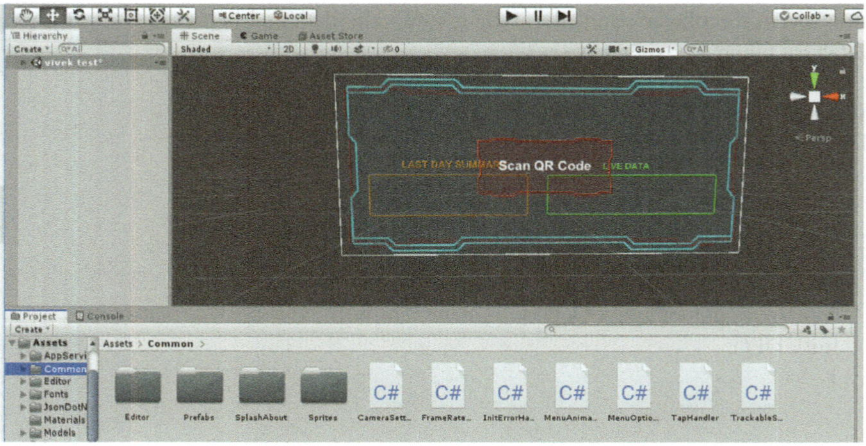

Fig. 8 Unity scene setup with script files in the background

4 Implementation and Results

The final application consists of three layers, the primary layer of image detection followed by, the network layer which transports the data between both interfaces and initiates a trigger to the third layer of cloud computing. The speed and efficiency are tested layer-wise and, the object detection efficiency is marked to 94% while, at the network layer there is 98% of packet delivery. In cloud computation, the later speed of fetch and display is about 300 ms. Various environmental parameters are accounted and tested during the application testing. From these tests, light and materials of objects which cause reflection and color of the object whether it's light or dark doesn't take the features of the object, which leaves it undetectable by the algorithm. To counter these abnormalities, we have adopted the match criteria threshold based on the number of feature points for feature matching to accommodate these changes. The final application has to be compressed due to the heaviness of the package due to OpenCV embedded into the application. Graphics and other display texture were limited to few 3D objects and mostly 2D objects were applied.

The object detection algorithm consists of feature matching, followed by contour and texture mapping to increase the match probability. However, rules are evaluated to make the algorithm flexible in case of 90% match is not possible due to environmental impacts such as light conditions that have the ability to change the image feature extraction properties. Unity platform has also been tweaked not only to serve as a game and AR platform but also to be used as a gateway to view vast operation critical data in industries (Fig. 9).

Fig. 9 Scene after object detection and data is parsed

5 Conclusion

Finally, the basic goal of the design is achieved very closely. The application is able to identify similar machines and is also able to display operation crucial information on a superimposed way to the user. AR devices can be integrated with this solution or it can also fit into the phone segment as an Android/iOS app. We have presented a flexible and extendable AR-based environment for the implementation of complex image classification environments in this application.

The system logs which were taken during the image processing when running on the smartphone shows parallelism strategy was used to balance CPU–GPU computational needs and does not overload the device. The test was run on Samsung mid-range smartphones. The application was found to be more efficient than other similar applications done by the referenced researchers and thus giving an operational efficiency of the application stood at 92.3% for image recognition and data computing stood at 86%.

Due to the complexity of the image detection package in the application, it tends to crash frequently when the background process runs simultaneously which the authors are focusing on the further scope of improvement. The scope for implementation is integrating single-shot detectors and Mobile Nets for fast and efficient object detection. To achieve high detection rates, feature point extraction can be used for better contrast and texture pattern differentiation to get significant feature points, hence improving the image classification. To reduce software complexity, we can also use single-shot detectors and send the captured image over the cloud to a cloud-based GPU for image processing and parse the machine information back to the scene.

In comparison to the object recognition, the game engine operates by using a scanned 3D model to detect the object. The conventional combined platform to recognize and classify images using a combination of features, contour and texture can achieve higher than 94% efficiency under normal conditions, while we are extending the capabilities to recognize objects under bright light using other techniques such as Gaussian blur after grayscale image conversion. Post the erosion and dilation of the image, the extract information has been applied to label the image sets.

References

1. R. Azuma, *A Survey of Augmented Reality* (ACM SIG-Graph, 1997), pp. 1–38
2. A. Van Dam, A. Forsberg, D. Laidlaw, J. LaViola, R. Simpson, Immersive VR for scientific visualization: a progress report. IEEE Comput. Graph. Appl. **20**(6), 26–52 (2000)
3. W. Sherman, A. Craig, *Understanding Virtual Reality: Interface, Applications and Design* (MorganKaufmann Publishers, 2003)
4. G. Giraldi, R. Silva, J.C. Oliveira, *Introduction to Virtual Reality* (LNCC, Brazil, 2003). Technical Report: 06/2003
5. A. Fuhrmann et al., Occlusion in collaborative augmented environments. Comput. Graph. **23**(6), 809–819 (1999)
6. B. Thomas, B. Close, J. Donoghue, J. Squires, P.D. Bondi, M. Morris, W. Pieraski, ARQuake: an outdoor/indoor augmented reality first person application, in *Proceedings of International Symposium on Wearable Computers* (2000)
7. P. Dähne, J.N. Karigiannis, Archeoguide: system architecture of a mobile outdoor augmented reality system, in *Proceedings of IEEE and ACM International Symposium on Augmented Reality* (Darmstadt, Germany, 2002)
8. D. Curtis, D. Mizell, P. Gruenbaum, A. Janin, Several devils in the details: making an ar application work in the airplane factory, in *Proceedings of IEEE Workshop on Augmented Reality* (San Francisco, 1998–2011)
9. N. Navab, M. Mitschke, A. Bani-Hashemi, Merging visible and invisible: two camera-augmented mobile C-arm (CAMC) applications, in *Proc. IEEE and ACM Int. Workshop on Augmented Reality*. San Francisco, CA, pp. 134–141 (1999)
10. A. Fuhrmann, H. Löffelmann, D. Schmalstieg, M. Gervautz, Collaborative visualization in augmented reality. IEEE Comput. Graphics Appl. **18–4**, 54–59 (1998)
11. S. Boivin, A. Gagalowicz, *Image-Based Rendering for Industrial Applications* (ERCIM News, 2001)
12. R. Thabet, R. Mahmoudi, M.H. Bedoui, Image processing on mobile devices: an overview, in *IEEE IPAS'14: International Image Processing Applications And Systems Conference* (2014). 978-1-4799-7069-8/14
13. C. Xiao, Z. Lifeng, Implementation of mobile augmented reality based on Vuforia and Rawajali, in *2014 IEEE 5th International Conference on Software Engineering and Service Science*, 978-1-4799-3279-5/14 (2014)
14. Z. Szalavari, D. Schmalstieg, A. Fuhrmann, M. Gervautz, Studierstube—an environment for collaboration in augmented reality. Virtual Reality Syst. Dev. Appl. **3**(1), 37–49 (1998)

A Hybrid Search Group Algorithm and Pattern Search Optimized PIDA Controller for Automatic Generation Control of Interconnected Power System

Smrutiranjan Nayak, Sanjeeb Kar, and Subhransu Sekhar Dash

Abstract A hybrid Search Group Algorithm (SGA) and Pattern Search (hSGA-PS) technique with the PIDA controller, to deal with Automatic Generation Control (AGC) of power system is presented. In the first stage, three nonlinear power systems with PID controller is considered and the controller parameters are tuned by SGA. The supremacy of the SGA -tuned PID-controlled AGC system is demonstrated by comparing the published Firefly Algorithm (FA) optimization procedure for the same interconnected power system. Then in the second stage, the PID controller is replaced with Proportional–Integral–Derivative and Acceleration (PIDA) controller and the optimum gains of the PIDA controller are optimized employing the SGA technique. It has been demonstrated that SGA-tuned PIDA controller improves the performance significantly compared with the SGA -tuned PID controller. Pattern Search (PS), a local optimization method is used in the third stage to fine-tune the PIDA controller parameters delivered by the SGA. The advantage of the hSGA-PS-tuned PIDA controller over the SGA-tuned PIDA controller, SGA-tuned PID controller, FA-tuned PID controller is demonstrated. Furthermore, in the sensitivity analysis, the system parameters, operation load conditions, and the location of disturbance are changed and the results are analyzed. The performance and results from the sensitivity analysis reveal the effectiveness of the hSGA-PS-tuned PIDA controller aimed at AGC of the power system.

S. Nayak · S. Kar
Department of Electrical Engineering, ITER, Sikha 'O' Anusandhan University, Bhubaneswar, Odisha, India
e-mail: smrutikiit40@gmail.com

S. Kar
e-mail: sanjeebkar@soa.ac.in

S. S. Dash (✉)
Department of Electrical Engineering, Government College of Engineering Keonjhar, Keonjhar, Odisha, India
e-mail: subhransudash_fee@gcekjr.ac.in

S. S. Dash et al. (eds.), *Intelligent Computing and Applications*,
Advances in Intelligent Systems and Computing 1172,
https://doi.org/10.1007/978-981-15-5566-4_27

Keywords Automatic generation control (AGC) · Governor dead band (GDB) · Generation rate constraint (GRC) · PIDA controller · Search group algorithm (SGA) · Pattern search (PS)

1 Introduction

In the multi-area power system, units of individual areas are interconnected with another area via transmission lines. If real power load changes abruptly, the frequency of the power system and the power flow in tie line changes from its nominal value. The deviations are due to the difference between the electrical load demand and generation and cause unwanted effects. Automatic Generation Control (AGC) in each area regulates the generator set point automatically for the corresponding load change by calculating and driving Area Control Error (ACE) to zero [1–4].

Therefore, an effort is made for the design of the PIDA controller for AGC which is tuned by the hSGA-PS technique. The aim of this work is of threefold: (i) to exhibit the superiority of new powerful computational intelligence technique like SGA over FA-tuned PID controller for AGC, (ii) to exhibit the advantage of PIDA structure over PID structure, (iii) to show the effectiveness of hybridization of SGA and PS, where PS is used to improve the results of SGA-tuned PIDA values [5–9]. Here, three-area reheat thermal system with different capacities with nonlinearities is considered for investigation.

2 System and Controller

2.1 System Investigated

A three-area test system is considered which contains three thermal units of different capacities of each area with GDB and GRC as shown in Fig. 1. The most commonly specified value of GRC in the reheat unit is around 3% per min. Governor dead band (GDB) is a physical constraint resulted due to mechanical friction and backlash and due to overlap of valves in the hydraulic relay. The speed of GDB affects the dynamics of the power system. It causes an increase in undershoot/overshoot and steady-state apparent speed regulation. The typical width of GDB is 0.06% (0.036 Hz). In the current research, a GDB of 0.036 Hz and GRC of 3%/min are considered.

2.2 Controller Structure

In Fig. 2, PID controller structure used for AGC is shown with usual notation. In Fig. 3, the structure of proposed PIDA controller is shown, where K_P, K_I, K_D are

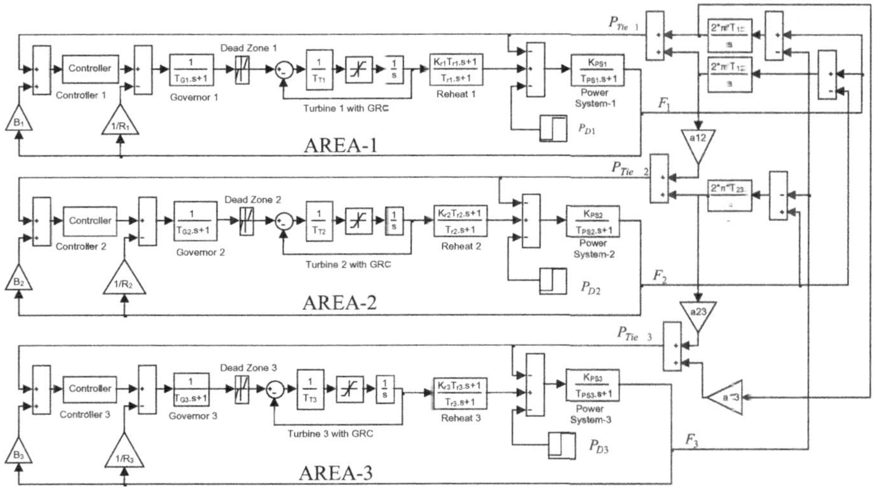

Fig. 1 Three-area test system with reheat, GDB, and GRC

Fig. 2 Structure of PID controller

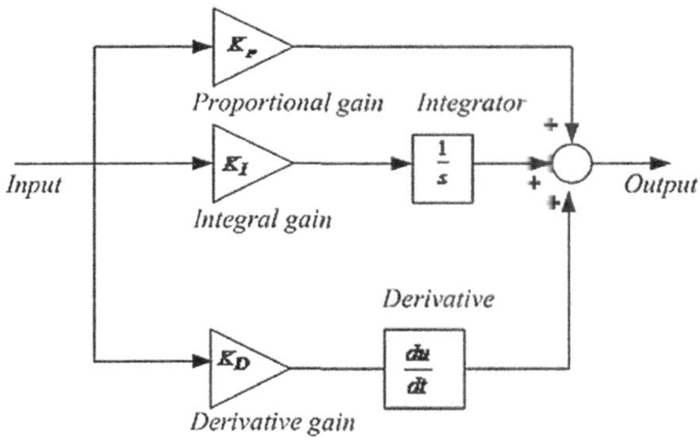

Fig. 3 Structure of PIDA controller

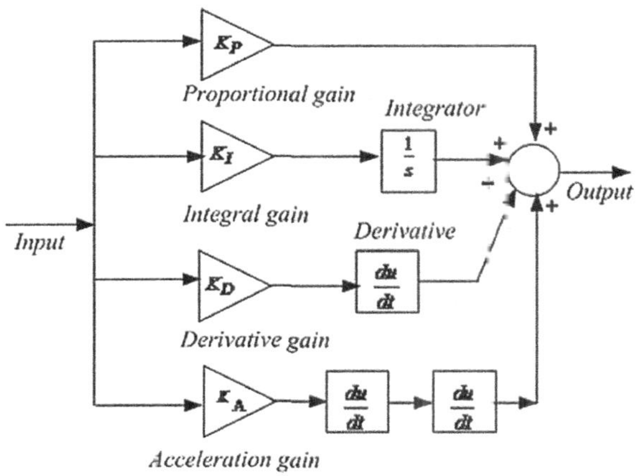

the conventional PID controller gains and K_A is acceleration gain. The addition of new acceleration term to PID controller makes the system response faster with less overshoot.

2.3 Objective Function

The performance criteria suitable for AGC studies is ITAE as reported in the literature [6] and hence the same is selected to tune PID/PIDA controller parameters. ITAE is expressed in Eq. (1).

$$J = \text{ITAE} = \int_{0}^{t_{\text{sim}}} (|\Delta F_i| + |\Delta P_{\text{Tie}-i}|) \cdot t \cdot \mathrm{d}t \tag{1}$$

where ΔF_i frequency variation of area i; $\Delta P_{\text{Tie}-i}$ net transmission line power variation in area i; t_{sim} is the simulation time range.

2.4 Search Group Algorithm

The key feature of SGA is that it establishes a balance among the exploitation and exploration stage of the algorithm run. SGA is comprised of five steps; each step is explained as follows [10, 11].

2.4.1 Initial Population

Initial population P is generated randomly as per Eq. (2)

$$P_{ij} = X_j^{\min} + \left(X_j^{\max} - X_j^{\min}\right)U[0,1], \quad \text{where } j = 1, \ldots, n, i = 1, \ldots, n_{\text{pop}} \tag{2}$$

P_{ij} is jth design parameter of the ith individual of the population P, n represents the total number of search parameters, n_{pop} represents the total no of population, U[0,1] is a random number from 0 to 1 uniformly distributed, $X_j^{\min}$ and $X_j^{\max}$ are the minimum and maximum limits of the jth search parameter.

2.4.2 Selection of Initial Search Group

Once population is created, each individual population's objective function is calculated and a search group R is constructed by choosing n_g points from the population P. The selection is accomplished by a standard tournament selection procedure and each row of R represents an individual. The members of rare ranked after each iteration. If R_i represents the ith row of R, then R_1 represents the best design and R_{n_g} is the worst design in R, n_g is the number of members in the search group [12].

2.4.3 Mutation of the Search Group

To enhance exploration capability, n_{mut} individuals are replaced from R by new individuals generated according to Eq. (3)

$$X_j^{\text{mut}} = E\left[R_j\right] + t\varepsilon\sigma\left[R_j\right], \quad \text{for } j = 1, \ldots, n \tag{3}$$

where X_j^{mut} is the jth design parameter of a mutated individual, σ and E are standard deviation and mean operators, ε represents a suitable random number, t represents a control parameter which decides how far a mutated individual is produced from the average value of the population. The individual which is to be substituted is decided by an "inverse tournament" selection procedure in which the design having worst objective function cost is chosen to be replaced.

2.4.4 Creation of Families of Individual Search Group Member

Perturbation defined by Eq. (4) is used by each member of the search group to generate a set of individuals known as family

$$X_j^{\text{new}} = R_{ij} + \alpha\varepsilon \quad \text{for } j = 1, \ldots, n \tag{4}$$

where perturbation size is controlled by α The value of α reduces after each iteration k of the search process. α is updated by Eq. (5):

$$\alpha^{k+1} = b\alpha^k \tag{5}$$

where b is a parameter that defines the way that the value of α^k reduces as the iterations pass by.

The distance of a newly generated individual from its search group member is controlled by α^k and is accountable for the exploitation and exploration ability of the SGA. In the first stage of iterations, α^k value is chosen such a high value such that, it allows exploring the maximum of the design regions. As the iteration lapses by, α^k reduces and allows exploiting the design regions. The range of α^k is given by

$$\alpha^0 \leq \alpha^k \leq \alpha_{\min}$$

where $\alpha_{\min}$ is the minimum value of α^k, guaranteeing minimum mobility of the new points up to the last iterations of the SGA, α^0 is the value of α the first iteration.

Let each family be denoted by F_i, where $i = 1$ to n_g. The objective function value of each population of search group decides the number of individuals to be generated by that member. Better the quality of objective function, the higher is the number of individuals it generates. The total number of individuals generated by search group members at each iteration are kept constant, i.e. equals to $\left(n_{\text{pop}} - n_g\right)$.

2.4.5 Formulation of New Search Group

In the first $\text{it}^{\max}$ iterations, the best member from each family is taken into account to formulate a new search group and this is called the global phase of algorithm. When the iteration number is greater than $\text{it}^{\max}_{\text{global}}$, a new search group is created by taking fittest n_g points from all the families and this stage is known as local phase of SGA. Where $\text{it}^{\max}$ is maximum iterations, $\text{it}^{\max}_{\text{global}}$ is the maximum iteration of the global phase.

2.5 Pattern Search

Pattern Search (PS) commences with the initial point X_0 that is the best solution delivered by the SGA. At the first iteration, the direction vectors or patterns are created as [0, 1], [1, 0], [−1, 0] and [0, −1] using a scalar = 1 known as mesh size [13–16].

3 Results and Discussions

3.1 Implementation of SGA/HSGA-PS Algorithm

MATLAB/Simulink platform is employed to create the model of the test system. A simultaneous 10% step load is applied in area −1 and 2 at $t = 0$ s for objective function calculation. The parameters of SGA play a significant contribution to the performance of SGA and the range of these parameters is chosen as per literature [11]. The SGA control parameter values used in the algorithm are presented in Table 1. The PID and PIDA parameters are tuned separately by using SGA. The tuned parameters of PID/PIDA are noted down after the completion of a maximum iteration $\text{it}^{\max}$. The SGA was repeated for 50 times and the best solutions among 50 runs of the optimization process are presented in Table 2. The best solutions provided

Table 1 SGA parameter values

Parameters	Value/expression
n_{pop}	100
α^0	2
α_{min}	0.01
it^{max}	50
it^{max}_{global}	$0.5 \times it^{max}$
n_g	$0.2 \times n_{pop}$
n_{mut}	$0.03 \times n_{pop}$
t	1, 2, 3 respectively for each muted individual

by SGA-tuned PIDA controller parameters are then fine-tuned by PS algorithm. The best solution for PIDA gains provided by SGA are used as PS algorithm's initial points. In PS algorithm, the values of mesh size, mesh contraction factor, and mesh expansion factor are taken as 1, 0.5, and 2, respectively. The number of generations and objective function calculations are set as 10 and 100, respectively. The flowchart of the proposed hSGA-PS approach is illustrated in Fig. 4. The final best value of hSGA-PS-tuned PIDA controller parameters are recorded in Table 2.

3.2 Analysis of Results

For a simultaneous application of a 10% step rise in area 1 and area 2, load demand at $t = 0.0$ s, the performance of the system using SGA-tuned PID/PIDA controller and hSGA-PS-tuned PIDA controller is presented in Table 3. The performances of the SGA-tuned PID/PIDA controller and hSGA-PS-tuned PIDA controller are compared with recently published FA-tuned PID controller parameters [9] for the same test system. It is observed from Table 3 that a reduced ITAE value is achieve during SGA-tuned PID controller (ITAE = 45.0126) as compared to FA-tuned PID controller (ITAE = 90.8448). This shows SGA performs better than FA. In [9], the superiority of FA was demonstrated compared to some other recently proposed optimization technique. Again the performance is improved by decreasing ITAE value by using SGA-tuned PIDA controller (ITAE = 36.2459) compared to SGA-tuned PID controller. Finally, by the application of proposed hSGA-PS algorithm to PIDA controller outperform the SGA-tuned PIDA controller. Minimum ITAE value of 35.4556 is obtained by hSGA-PS-tuned PIDA controller as stated in Table 3. The system dynamic responses for variation in frequency and variation in tie-line power are shown in Figs. 5, 6. It is observed from the Figs. 5, 6 that best dynamic performance is achieved by hSGA-PS-tuned PIDA controller, significant enhancement is witnessed with proposed SGA-tuned PIDA controller than SGA optimized PID controller, also SGA-optimized PID controller gives better response than FA optimized PID controller (Table 4).

Table 2 Tuned PID/PIDA controller parameters

Technique/controller	Proportional gain			Integral gain			Derivative gain			Acceleration gain		
	K_{P1}	K_{P2}	K_{P3}	K_{I1}	K_{I2}	K_{I3}	K_{D1}	K_{D2}	K_{D3}	K_{A1}	K_{A2}	K_{A3}
FA PID [9]	−1.5293	−1.6612	−0.1738	−1.6021	−0.2000	−0.1257	−1.1051	−1.3578	−0.2705	–	–	–
SGA PID	−0.6413	−0.9158	−0.6726	−0.0998	−0.0955	−0.1211	−0.7065	−0.9549	−1.5670	–	–	–
SGA PIDA	−0.3602	−0.4726	−0.7399	−0.1881	−0.0856	0.0043	−0.5519	−0.3621	−1.0458	−0.0264	−0.0682	−0.1331
hSGA-PS PIDA	−0.3602	−0.4726	−0.7399	−0.1881	−0.0856	0.0043	−0.5519	−0.3621	−1.0458	−0.0264	−0.0682	−0.2581

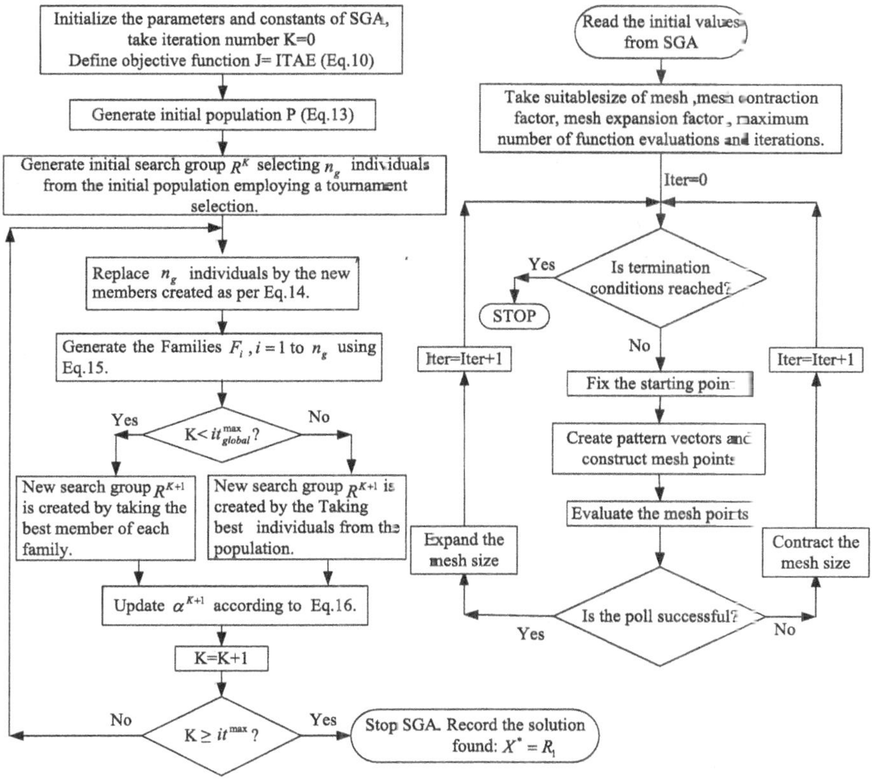

Fig. 4 Flowchart of the hSGA-PS algorithm

Table 3 ITAE value and % improvement for each controller for 10% step load demand in area-1 and area-2

Optimization technique/Controller	ITAE	
	Value	% Imp.
FA: PID controller [9]	90.8448	–
SGA: PID controller	45.0126	50.45
SGA: PIDA controller	36.2459	60.10
hSGA-PS: PIDA controller	35.4556	60.97

Further, only a 10% step rise in area 1 load is considered at $t = 0$ s to analyze the influence of disturbance location on the performance of system. The system dynamic response for single disturbance are presented in Figs. 7, 8. It is obvious from Figs. 7, 8 that the proposed hSGA-PS-optimized PIDA controllers are robust and delivers satisfactory performance when disturbance location changes. Also improved results are achieved with the proposed SGA-tuned PID controller in comparison with FA-optimized PID [9], SGA-tuned PIDA controller has better responses compared to SGA-optimized PID controller and finally hSGA-PS-optimized PIDA controllers

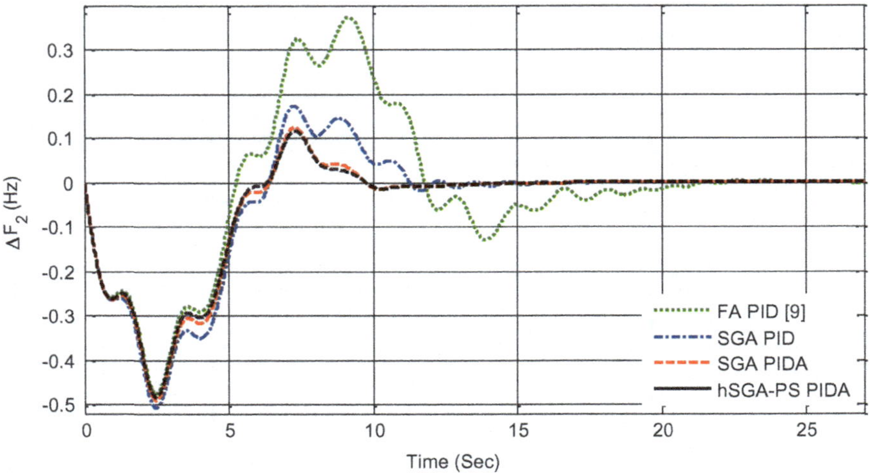

Fig. 5 Area-2 frequency deviance for simultaneous 10% step load demand in area-1 and area-2

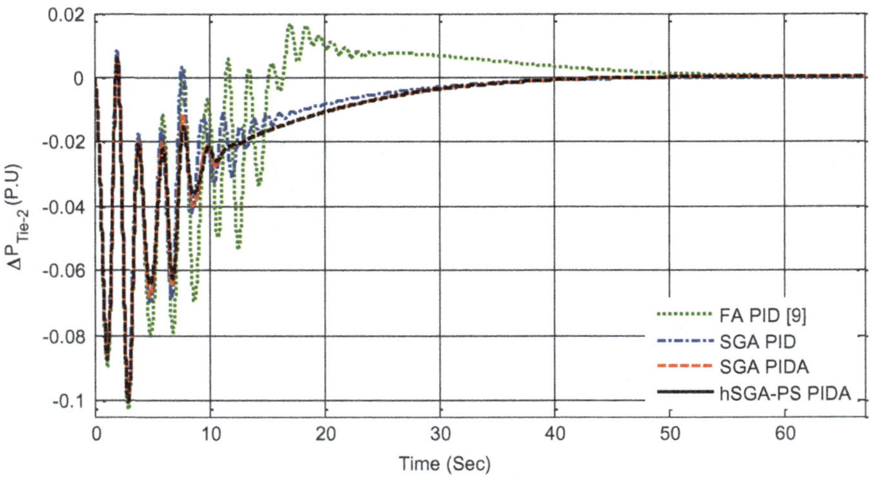

Fig. 6 Area-2 tie line power deviance for simultaneous 10% step load demand in area-1 and area-2

have best results compared to SGA-tuned PIDA controller. Hence, hSGA-PS-tuned PIDA controller parameters obtained from double disturbance are also effective for a single disturbance. So the proposed design of PIDA controllers is robust and performs satisfactorily regardless of load disturbance location. Figure 9 shows the system variation with change in load.

Table 4 Sensitivity analysis under varied conditions

Parameter variation	% Change	Settling time T_s (s)						ITAE
		ΔF_1	ΔF_2	ΔF_3	ΔP_{tie-1}	ΔP_{tie-2}	ΔP_{tie-3}	
Nominal	0	14.96	14.49	15.59	35.98	34.01	20.74	35.4556
Loading condition	−50	24.09	27.68	24.47	35.55	32.73	20.49	47.5721
	−25	14.54	22.67	15.07	35.8	33.49	20.64	40.5453
	+25	15.48	14.87	16.05	36.13	34.38	20.79	32.9753
	+50	16.17	15.48	16.75	36.26	34.86	21.13	31.9027
T_G	−50	14.81	14.32	15.41	36.1	34.05	20.57	34.4011
	−25	14.91	14.42	15.52	36.03	34.02	20.66	34.899
	+25	14.86	14.5	15.61	35.97	34.01	20.79	36.1316
	+50	14.71	14.47	15.61	35.96	34.01	20.83	36.8923
T_T	−50	15.13	14.59	15.71	36.23	34.07	20.7	35.0917
	−25	15.07	14.55	15.65	36.1	34.03	20.72	35.2686
	+25	14.76	14.35	15.45	35.88	33.97	20.72	35.7341
	+50	14.73	14.25	15.28	35.79	33.94	20.68	36.0302
H	−50	30.75	31.73	31.48	35.45	25.2	12.07	62.9446
	−25	23.06	26.11	23.63	35.94	31.96	18.44	42.5669
	+25	17.47	16.6	18.03	36.77	36.9	22.76	35.9119
	+50	20.05	18.79	20.55	37.39	38.44	25.1	37.7925

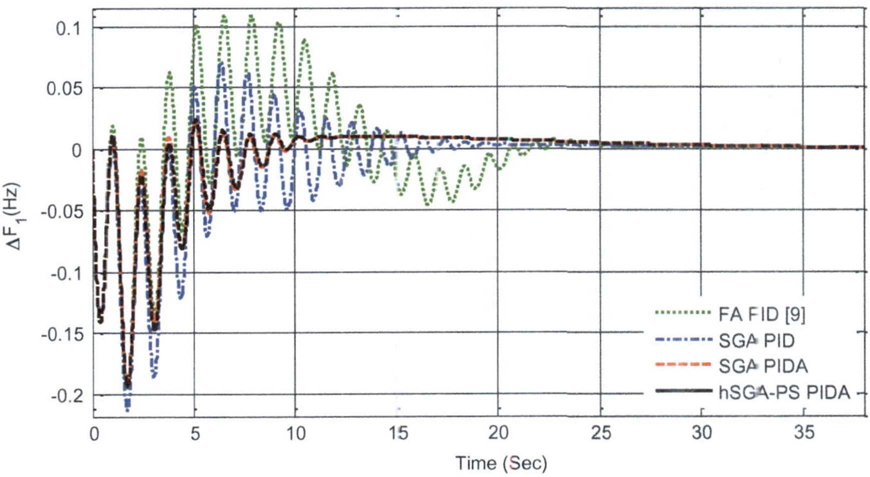

Fig. 7 Area-1 frequency deviance for 10% step load demand in area-1

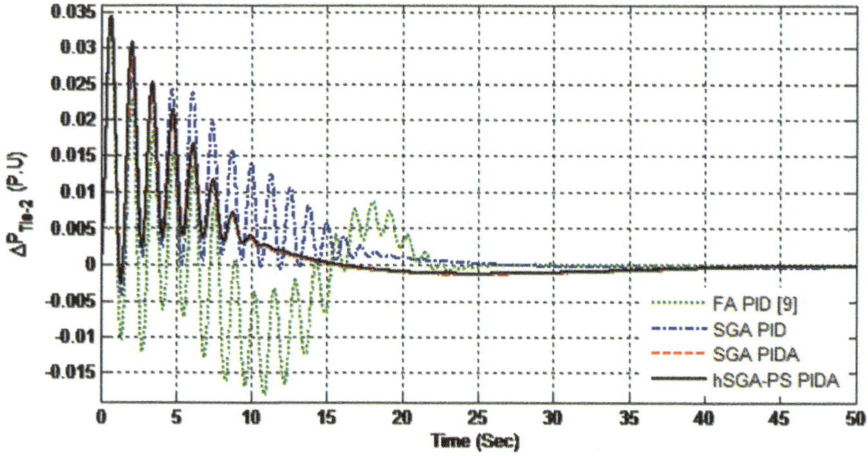

Fig. 8 Area-2 tie line power deviance for 10% step load demand in area-1

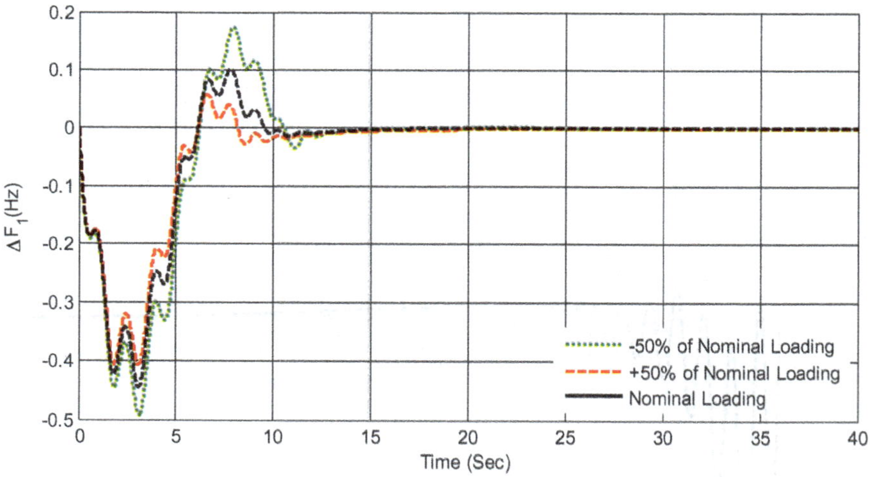

Fig. 9 Area-1 frequency deviance with the variation of loading

4 Conclusion

In this study, a hybrid SGA and PS algorithm has been applied to tune PIDA gains for AGC of the power system. For the first time, SGA is used in this area to optimize the PID/PIDA controller gains using ITAE objective functions. Firstly, SGA is used to tune PID parameters in a three unequal area interconnected thermal power system with GDB and GRC nonlinearities and the superiority of SGA technique is demonstrated by comparing the results with a published optimization method which was proved superior performance compared to GA, DE, BFOA, PSO, hBFOA-PSO, and Ziegler–Nichols-based controllers. Then in the second stage, PIDA controller gains are tuned by SGA and the result reveals that SGA-optimized PIDA controller gives a significant improvement in the response than the SGA-tuned PID controller. In the third stage, to take advantage of local search capabilities of PS, PS is used to fine-tune the PIDA controller parameter, in which SGA-tuned values are taken as the initial starting point. The proposed hSGA-PS-based PIDA controller provides improved performance compared to SGA tuned PIDA controller. So, it can be concluded that the proposed hSGA-PS-based PIDA controller is very effective, robust and gives better system performance compared to FA-tuned PID controller.

Appendix

Nominal parameters of the investigated system are

Three unequal areas thermal system [9, 20]:

$f = 60$ (Hz); $D_1 = D_3 = 0.015$, $D_2 = 0.016$ (p u Hz); $B_1 = 0.3483$, $B_2 = 0.3827$, $B_3 = 0.3692$ (p u Hz); $2H_1 = 0.1667$, $2H_2 = 0.2017$, $2H_3 = 0.1247$, (p u s); $T_{G_1} = 0.08$, $T_{G_2} = 0.06$, $T_{G_3} = 0.07$ (s); $R_1 = 3.0$, $R_2 = 2.73$, $R_3 = 2.82$ (Hz/p u); $T_{T_1} = 0.4$, $T_{T_2} = 0.44$, $T_{T_3} = 0.3$ (s); $T_{r1} = T_{r2} = T_{r3} = 10$ (s); $K_{r1} = K_{r2} = K_{r3} = 0.5$; $T_{12} = 0.2$, $T_{23} = 0.12$, $T_{31} = 0.25$ (p u/Hz), $P_{r1} = 2000\,\text{MW}$, $P_{r2} = 4000\,\text{MW}$, $P_{r3} = 8000\,\text{MW}$.

References

1. P. Kundur, *Power System Stability and Control*, 8th edn. (Tata Mc-Graw Hill, New Delhi, 2009)
2. O.I. Elgerd, *Electric Energy Systems Theory: An Introduction* (Tata Mc-Graw Hill, New Delhi, 2007)
3. H. Bevrani, *Robust Power System Frequency Control* (Springer, 2009)
4. L.C. Saikia, J. Nanda, S. Mishra, Performance comparison of several classical controllers in AGC for multi-area interconnected thermal system. Int. J. Electr. Power Energ. Syst. **33**, 394–401 (2011)
5. S. Mishra, G. Mallesham, P.C. Sekhar, Biogeography based optimal state feedback controller for frequency regulation of a smart micro grid. IEEE Trans. Smart Grid **4**, 628–637 (2013)
6. H. Shabani, B. Vahidi, M.A. Ebrahimpour, Robust PID controller based on imperialist competitive algorithm for load-frequency control of power systems. ISA Trans. **52**, 88–95 (2012)
7. R.K. Sahu, S. Panda, U.K. Rout, D.K. Sahoo, Teaching learning based optimization algorithm for automatic generation control of power system using 2-DOF PID controller. Int. J. Electr. Power Energ. Syst. **77**, 287–301 (2016)
8. U.K. Rout, R.K. Sahu, S. Panda, Design and analysis of differential evolution algorithm based automatic generation control for interconnected power system. Ain Shams Eng J. **4**, 409–421 (2013)
9. S. Padhan, R.K. Sahu, S. Panda, Application of firefly algorithm for load frequency control of multi-area interconnected power system. Elect. Power Comp. Syst. **42**, 1419–1430 (2014)

10. R.K. Sahu, S. Panda, S. Padhan, A hybrid firefly algorithm and pattern search technique for automatic generation control of multi area power systems. Int. J. Electr. Power Energ. Syst. **64**, 9–23 (2015)
11. S.G. Matheus, R.H. Lopez, L.F.F. Miguel, Search group algorithm: a new metaheuristic method for the optimization of truss structures design. Comput. Struct. **153**, 165–184 (2015)
12. Y. Bao, Z. Hu, T. Xiong, A PSO and pattern search based memetic algorithm for SVMs parameters optimization. Neurocomputing **117**, 98–106 (2013)
13. A.K. Al-Othman, N.A. Ahmed, M.E. Al Sharidah, A. Hanan, A hybrid real coded genetic algorithm-pattern search approach for selective harmonic elimination of PWM AC/AC voltage controller. Int. J. Electr. Power Energ. Syst. **44**, 123–133 (2013)
14. U. Thaiwasin, S. Sujitjorn, Y. Prempraneerach, J. Ngamwiwit, Torsional resonance suppression via PIDA controller, in *2000 TENCON Proceedings of Intelligent Systems and Technologies for the New Millennium* (2000), pp. 498–503
15. D.Y. Ha, I.Y. Lee, Y.S. Cho, Y.D. Lim, B.K. Choi, The design of PIDA controller with pre-compensator, in *ISIE 2001. 2001 IEEE International Symposium on Industrial Electronics Proceedings (Cat. No.01TH8570)* (2001), pp. 798–804
16. S. Sornmuang, S. Sujitjorn, GA-based PIDA control design optimization with an application to ac motor speed control. Int. J. Math. Comput. Simul. **4**, 67–80 (2010)

Fractional Order PID Controlled PV Fed Quadratic Boost Converter TZ Source Inverter Fed Permanent Magnet Brushless Motor Drive

N. K. Rayaguru and S. Sekar

Abstract Permanent magnet brushless motors are industry friendly motors with a significant issue with respect to power quality. There is always a necessity to solve dynamic response related issues in a converter inverter system with less power devices and minimal complexities. This work proposes combination of QBC and TZSI for the control of PV fed PMBLDCM. DC from-PV is stepped up utilizing QBC and the yield of QBC is applied to TZSI. This effort covenants with comparison of closed-loop proportional resonant and fractional order PID controlled quadratic boost converter with TZ source inverter fed permanent magnet brushless motor drive systems. The effort incorporates the control strategies in both open and closed-loop control of an industry friendly permanent magnet brushless motor drives system incorporating a conventional FOPID controller and an advanced proportional resonant controller. The FOPID and PR controllers regulate the speed of revolution of the motor and thereby reduces the ripples associated in the torque of the motor. Finally the improvement time response is obtained which is depicted through the results on simulation performed indicates reduced steady-state error and enhanced dynamic response.

Keywords Permanent magnet brushless motor · MPPT · PWM inverter · Quadratic Boost Converter (QBC) · T-Z source inverter (TZSI) · Motor load · PR-proportional resonant · FOPID-Fractional order PID · Steady-state error · Rise time · Peak time and settling time

1 Introduction

The solar-based energy source had taken the best opportunity among other ways of generation of electricity from renewable sources because the energy produced is

N. K. Rayaguru (✉) · S. Sekar
Department of EEE, Hindustan Institute of Technology and Science, 1, Rajiv Gandhi Salai OMR, Padur, Chennai, Tamilnadu 603103, India
e-mail: nkrayaguru29@gmail.com

S. Sekar
e-mail: ssekar.pt@hindustanuniv.ac.in

S. S. Dash et al. (eds.), *Intelligent Computing and Applications*, Advances in Intelligent Systems and Computing 1172,
https://doi.org/10.1007/978-981-15-5566-4_29

cordial. Moreover had a longer existence with acceptable payback period, exceptionally versatile and compact. A basic method to structure the output voltage of sun-based energy sources was to associate several solar panels in use and an ordinary VSI. However, the design had some disadvantages like constant low quality and low efficacy. A simulation model was created that used the essential circuit conditions of the photovoltaic (PV) cells, along with the impacts of temperature change and irradiation.

The existing various DC/DC converter topologies are buck, boost, and buck-boost; Sepic, Luo and C'uk [1–3].

Among these power converters, a boost converter that raises the input voltage to an appropriate level is most utilized because of a few favorable circumstances, for example, low ripple current waves, high proficiency, quick transient reaction, and an improved unwavering quality [4–6].

This work likewise investigates the closed-loop execution of the QBC framework by utilizing different PI, FOPID controller [7–9]. Closed structure which incorporates two cascaded control circles; inside current control and outside speed control are utilized.

The above papers do not deal with QBC-TZSI fed PMBLDC Drive system. This work suggests the QBC for the control of PMBLDC Drive system. The exceeding writing does not deal with evaluation of FOPID and PR-based TZSI fed PMBLDC drive systems. This effort mainly presents closed-loop simulation and execution of QBC-TZSI fed PMBLDC Drive. PRC is proposed for the closed loop control of QBC-TZSI fed PMBLDC Drive system. The objectives of this work are to minimize settling time and steady-state error of QBC-TZSI fed PMBLDC Drive system using FOPID/PR controller.

2 Quadratic Boost Converter

2.1 *Circuit-Description*

In single QBC has a single switch M_1, two inductors L_1 and L_2, two Capacitors C_1, C_2, three diodes named D_1, D_2, D_3, and a motor load. A single QBC same as cascaded boost converter but it uses Single switch. The QBC system consists of two single QBC in parallel connection (Fig. 1) [10, 11].

2.2 *Analysis of Single QBC*

Mode 1-Operation

This mode begins when MOSFET switch M_1 is turned on. During this mode D_1 and D_3 Reverse biased connection and D_2 forward biased connection. Now the inductors

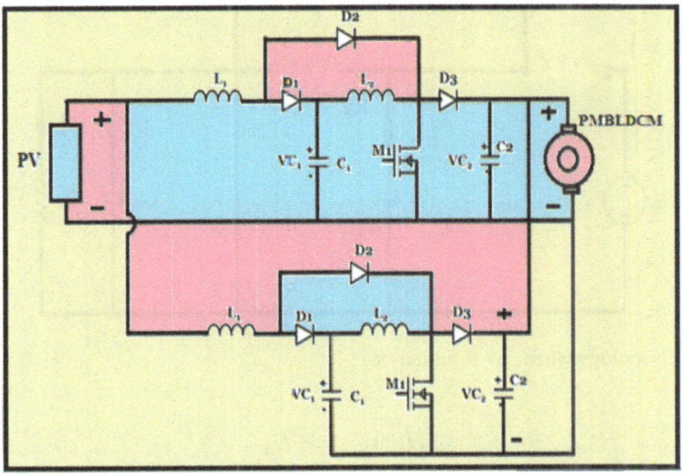

Fig. 1 The proposed circuit diagram of QBC

L_1 and L_2 get energized by V in and C_1 separately. Subsequently Inductor current, I_{L_1} and I_{L_2} increment in size. Capacitor C_2 supplies energy to the load (Figs. 2, 3) [12].

$$(\Delta I_L)_{\text{On}} = \frac{V_s K T}{L_1} \tag{1}$$

$$(\Delta I_{L_2})_{\text{On}} = \frac{V_S K T}{L_1}$$

Meanwhile the capacitor currents are expressed by

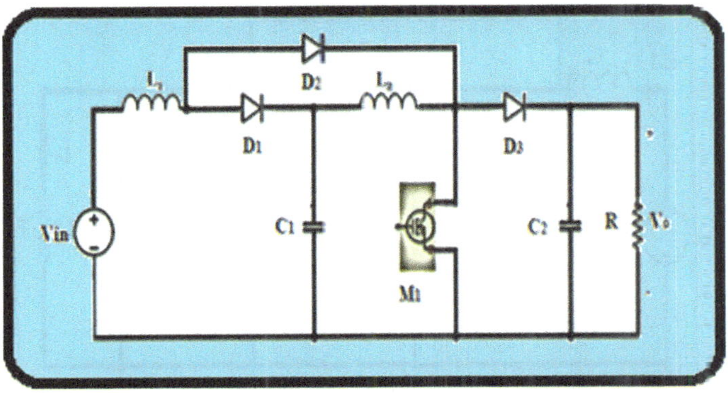

Fig. 2 Single quadratic boost converter: topology

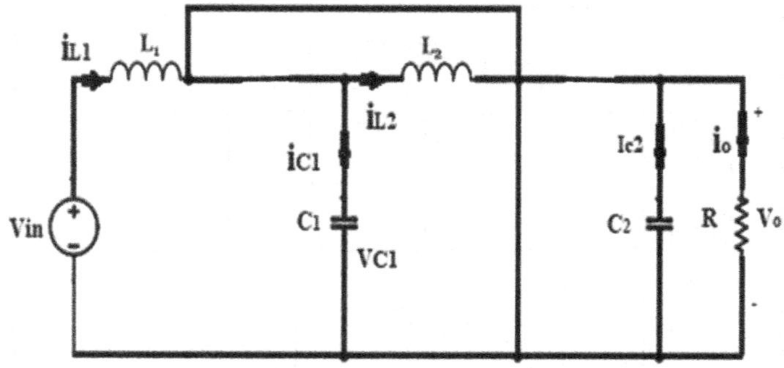

Fig. 3 QBC yopology when SW is turned-on

$$\left(I_{C_1}\right)_{\text{On}} = -I_{L_2} \tag{3}$$

$$\left(I_{C_2}\right)_{\text{On}} = -I_0 \tag{4}$$

where ΔI_{L_1} and ΔI_{L_2} are the change in inductor current 1 and 2, respectively. K, the duty cycle of switch, V_{in} is the input voltage, and T the switching period [13, 14].

Mode 2 Operation

During this mode switch M_1 is in off state. Diodes D_1 and D_3 are forward biased and diode D_2 is in reverse biased condition. The current in the inductor L_1 and L_2 now begins diminishing as the energy put away in the inductor L_1 is presently moved to capacitor C_1 and energy put away in the inductor L_2 is currently moved to the load side. Capacitor C_1, C_2 gets energized during this mode. During this time voltage over the inductor turns around as the pace of progress of current is insignificant, i.e., decreasing current. The capacitor current for this state as (Figs. 4, 5) [15].

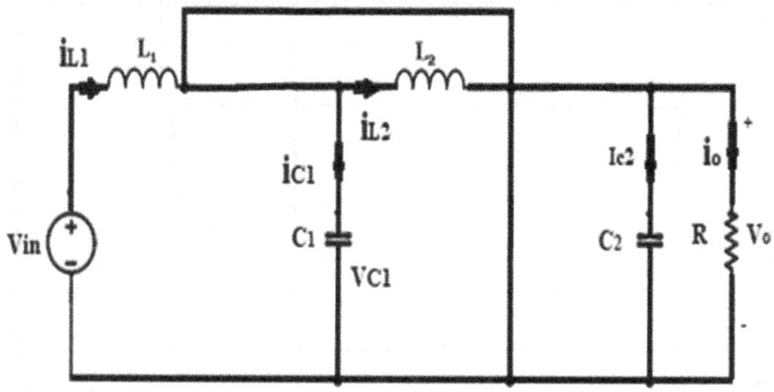

Fig. 4 QBC topology when SW is turned-off

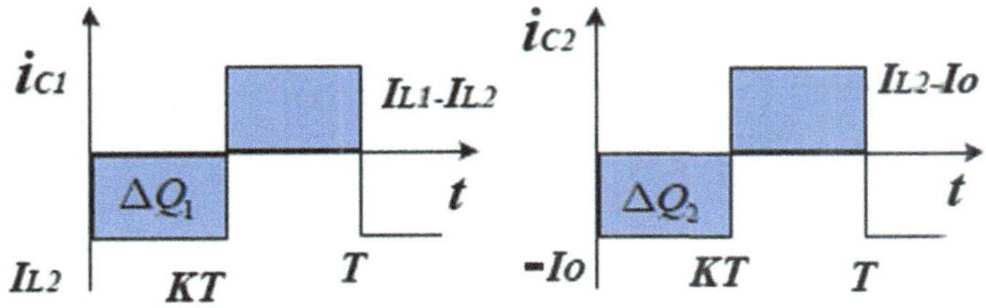

Fig. 5 Capacitor current waveform

$$(\Delta I_{L_1})_{\text{Off}} = \frac{V_s - VC_1(1-K)T}{L_1} \tag{5}$$

$$(\Delta I_{L_2})_{\text{Off}} = \frac{VC_1 - V_0(1-K)T}{L_2} \tag{6}$$

$$(I_{C_l})_{\text{Off}} = I_{L_1} - I_{L_2} \tag{7}$$

$$(I_{C_2})_{\text{Off}} = I_{L_2} - I_0 \tag{8}$$

3 System Configuration

Figure 6 delineates the block representation of open-loop QBC TZSI fed PMBLDCM. The PV Panel is characterized as the connection of PV cells in series/parallel to create the necessary voltage and current. Produced by PV is DC. DC is stepped up using QBC. The output of QBC is converted to AC using TZSI. The TZSI feeds PMBLDCM.

FOPID/PR is outlined in Fig. 7. Motor speed is measured and it is related with the reference speed to get error in speed. The speed error is implied to the speed FOPID/PR controller. The output of speed FOPID/PR is utilized to create current reference. The reference current is compared with the actual current and the current

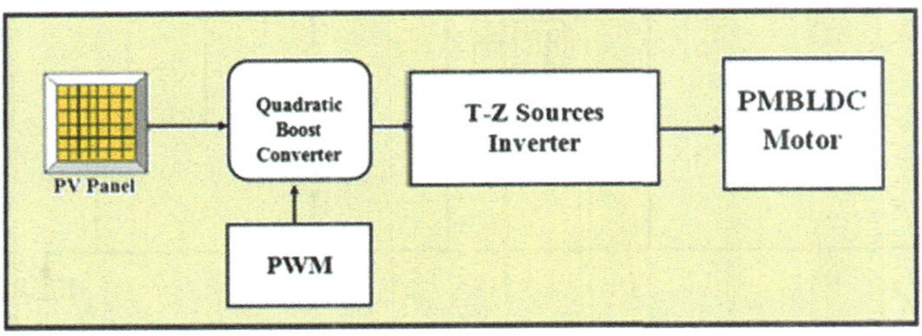

Fig. 6 Block diagram of open-loop QBC TZSI fed PMBLDCM

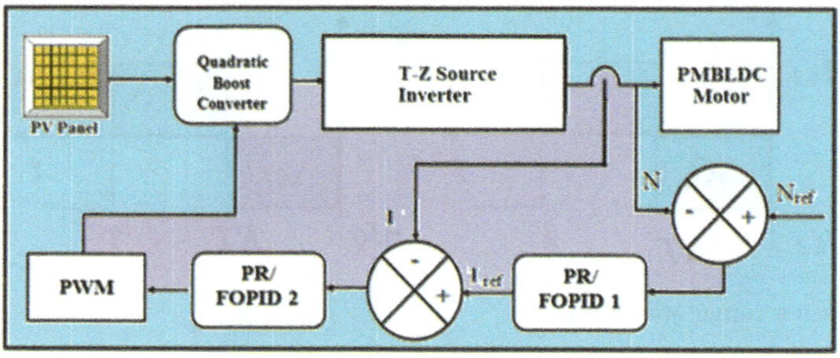

Fig. 7 Block diagram of closed-loop QBC-T-ZSI fed PMBLDCM with PR and FOPID controllers

error is applied to current FOPID/PR and its output is exploited to update the PWM of QBC [16–18].

4 Simulation Results

4.1 QBC–TZSI-PMBLDC with Source Disturbance

Simulink results of QBC–TZSI-PMBLDC with source disturbance are outlined here. Circuit diagram of QBC with T-Z source is appeared in Fig. 8. The pulses for inverter are generated based on the information from the hall-sensor. Scopes are connected to measure the QBC voltage, inverter voltage, motor speed, and motor torque.

Voltage across PV of QBC-TZSI-PMBLDC is outlined in Fig. 9 and its value enhances to 95 V at $t = 1$ s. The augment in PV voltage is due to enhance in insolation. Voltage across QBC is outlined in Fig. 10 and its value enhances to 75 V. The QBC voltage is enhanced due to enhancement in PV voltage.

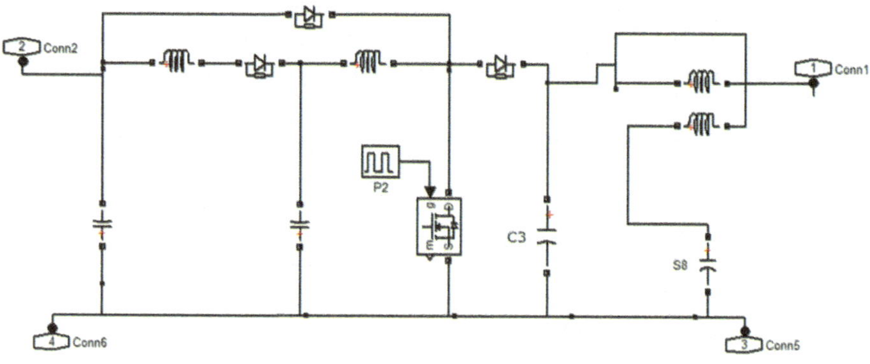

Fig. 8 Circuit diagram of QBC with T-Z source

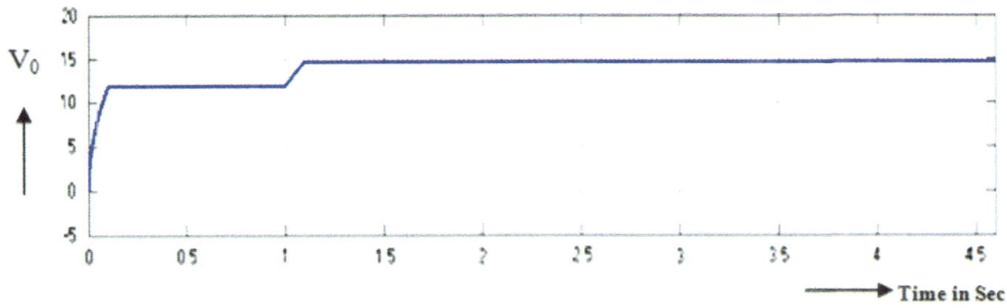

Fig. 9 Voltage across PV of QBC–TZSI-PMBLDC

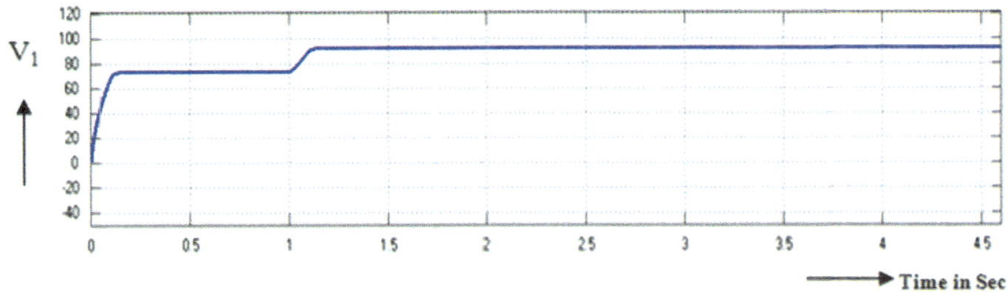

Fig. 10 Voltage across QBC

Voltage across inverter is outlined in Fig. 11 and its value enhances to 75 V. The augment in inverter voltage is due to enhancement in PV voltage. Motor speed of QBC-TZSI-PMBLDC is appeared in Fig. 12 and it's value enhances to 390 RPM. The augment in speed is due to enhance in PV voltage.

Motor Torque of QBC-TZSI-PMBLDC is outlined in Fig. 13 and its value enhances to 1.5 N m. The enhancement in torque is due to boost in voltage.

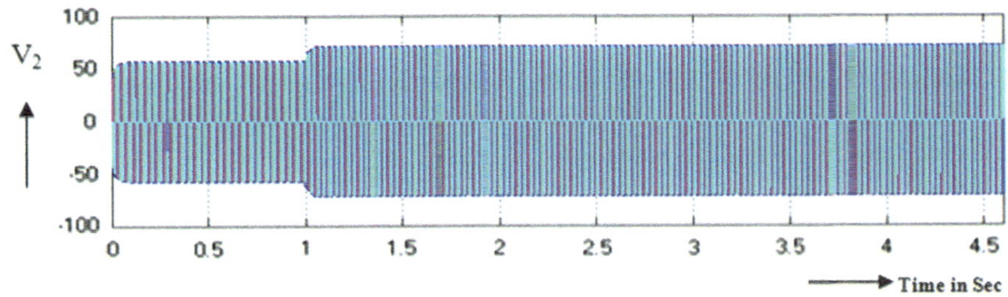

Fig. 11 Voltage across inverter of QBC-TZSI-PMBLDC

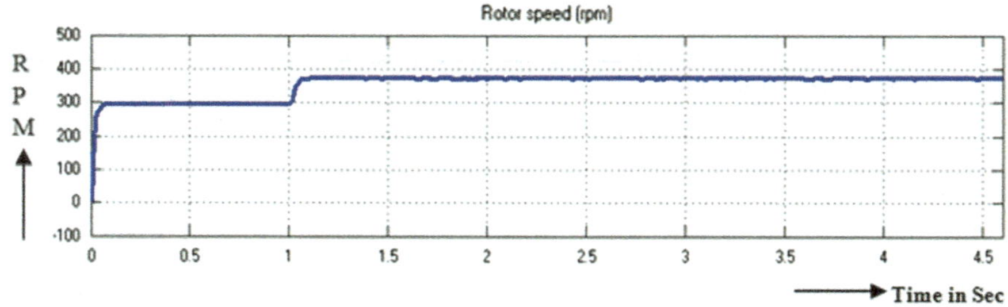

Fig. 12 Motor speed of QBC-TZSI-PMBLDC

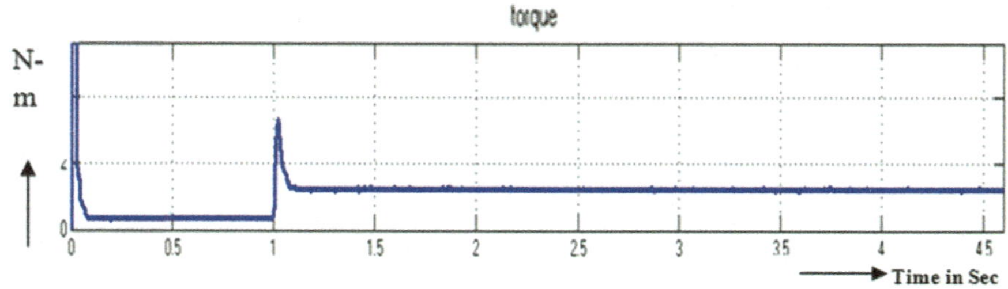

Fig. 13 Motor Torque of QBC-TZSI-PMBLDC

4.2 Closed-Loop PR Controlled QBC-TZSI-PMBLDC

Simulink diagram of PR controlled closed-loop QBC-TZSI-PMBLDC is outlined in Fig. 14. First PR controller is used for both voltage control mode as well as current control mode. Then PR controllers are replaced with FOPID controllers and the speed, torque, and power are measured. Here the dynamic performance of the controllers is compared. The error signal is coursed by FOPID as well as PR to sustain the speed of QBC-TZSI-PMBLDC constant and diminish the steady-state errors [19, 20].

Voltage across QBC-TZSI-PMBLDC is presented in Fig. 15 and it's value is 70 V. Voltage across motor load of PR controlled closed-loop QBC-TZSI-PMBLDC is presented in Fig. 16 and it's value is 60 V.

Motor speed of PR controlled closed-loop QBC-TZSI-PMBLDC is presented in Fig. 17 and it's value is 300 RPM. Motor torque of PR controlled closed-loop QBC-TZSI-PMBLDC is outlined in Fig. 18 and it's value is 1.7 N m.

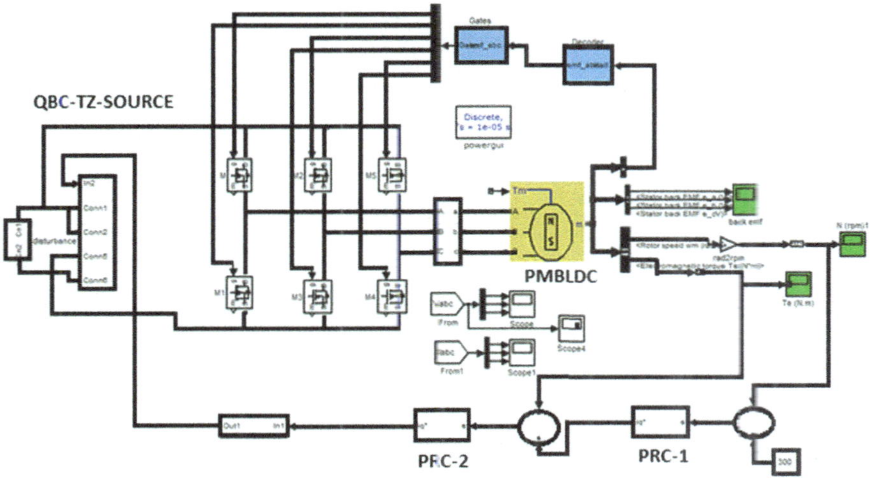

Fig. 14 Circuit diagram of PRC closed-loop QBC-TZSI-PMBLDC

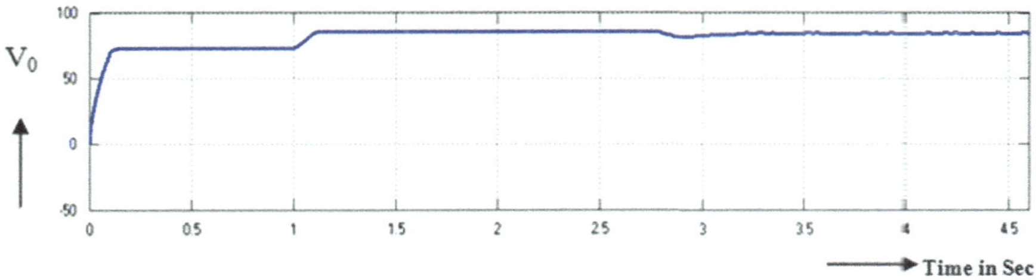

Fig. 15 Voltage across QBC-TZSI-PMBLDC

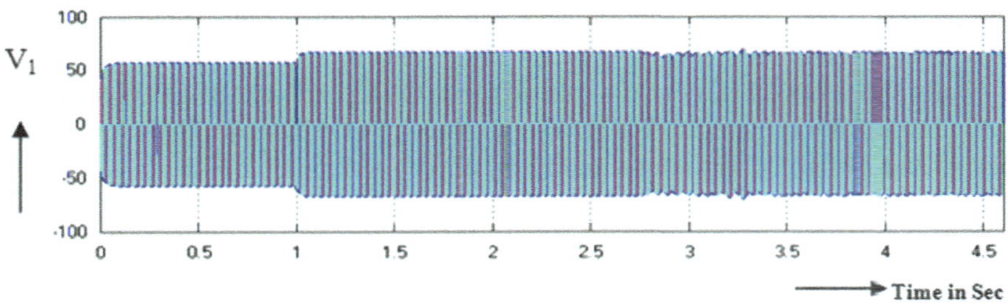

Fig. 16 Voltage across motor load of PRC closed-loop QBC-TZSI-PMBLDC

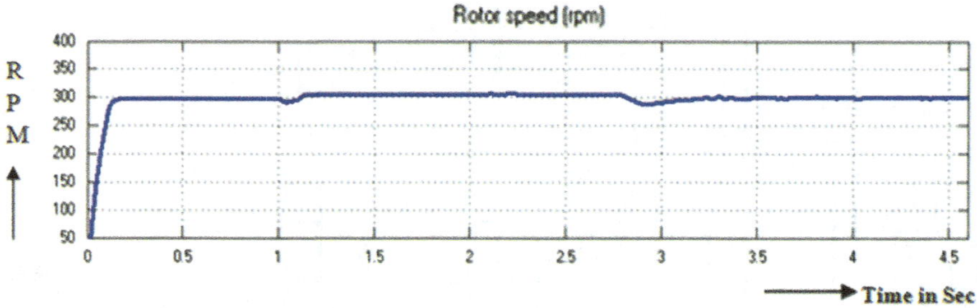

Fig. 17 Motor speed of PRC closed-loop QBC-TZSI-PMBLDC

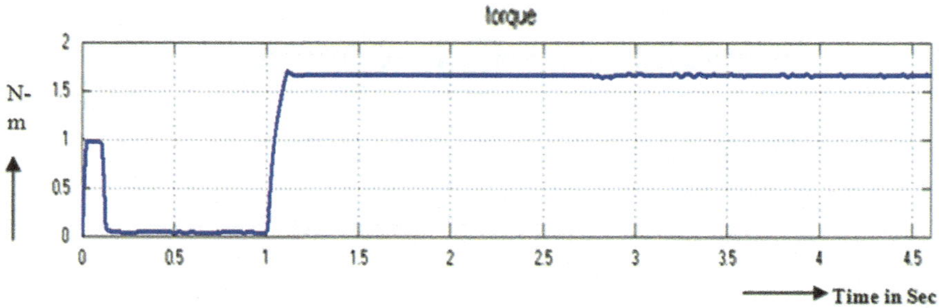

Fig. 18 Motor Torque of PRC closed-loop QBC-TZSI-PMBLDC

4.3 Closed-Loop FOPID controlled QBC-TZSI-PMBLDC

Voltage across QBC-TZSI-PMBLDC is depicted in Fig. 19 and it's value is 100 V. Voltage across motor load of FOPID controlled closed-loop QBC-TZSI-PMBLDC with TZSI is presented in Fig. 20 and it's value is 60 V.

Motor speed of FOPID controlled closed-loop QBC-TZSI-PMBLDC is presented in Fig. 21 and it's value is 300 RPM. Motor torque of FOPID controlled closed-loop QBC-TZSI-PMBLDC is outlined in Fig. 22 and it's value is 1.2 N m.

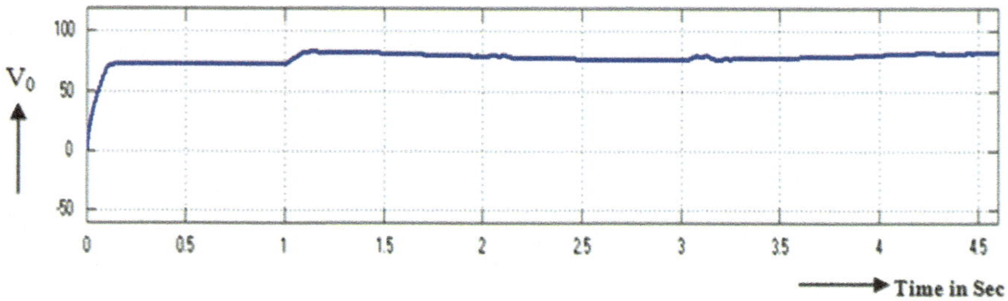

Fig. 19 Voltage across QBC

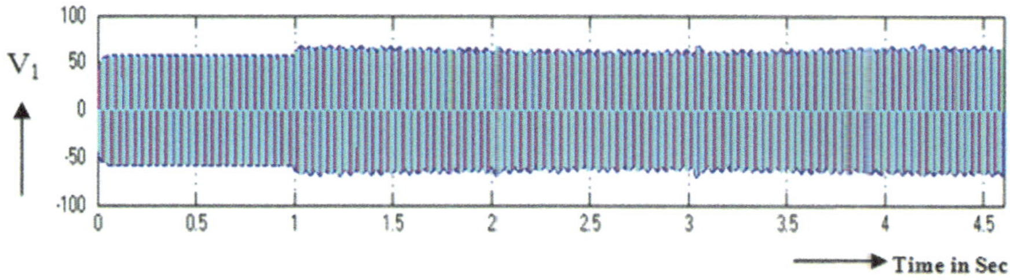

Fig. 20 Voltage across motor load of FOPID controlled closed-loop QBC-TZSI-PMBLDC

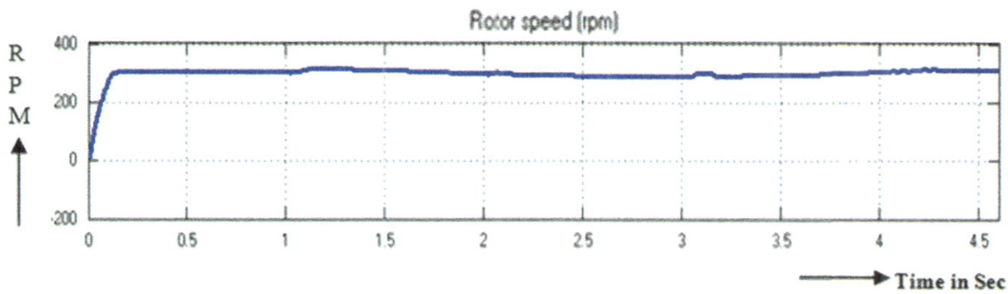

Fig. 21 Motor speed of FOPID controlled closed-loop QBC-TZSI-PMBLDC

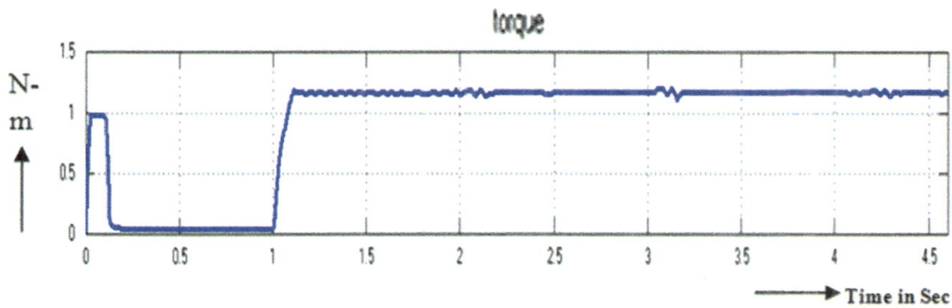

Fig. 22 Motor Torque of FOPID controlled closed-loop QBC-TZSI-PMBLDC

Comparsion of time domain parameters for motor speed using PR-PR and FOPID-FOPID is presented in Table 1. By using FOPID-FOPID controller, 'RT (rise time)' is diminished from 1.15 to 1.14 s; 'PT (peak time)' is diminished from 2.45 to 1.34 s;

Table 1 Comparsion of time domain parameters for motor speed using PR and FOPID

Controller	Rise time (s)	Peak time (s)	Setting time (s)	Steady-state error (RPM)
Dual-PR	1.15	2.45	3.02	3.6
Dual-FOPID	1.14	1.34	2.23	2.8

Table 2 Comparison of time domain parameters for motor torque using PR-PR and FOPID-FOPID

Controller	Rise time (s)	Peak time (s)	Setting time (s)	Steady-state error (N m)
Dual-PR	1.13	2.82	2.93	0.8
Dual-FOPID	1.11	1.29	1.82	0.7

'ST (settling time)' is diminished from 3.02 to 2.23 s; 'SSE (steady-state-error)' is diminished from 3.6 to 2.8 RPM for motor speed (Table 2).

Comparsion of Time domain parameters for motor torque using PR-PR and FOPID-FOPID is presented in Table 1. By using FOPID controller, 'RT (rise-time)' is diminished from 1.13 to 1.11 s; 'PT (peak time)' is diminished from 2.82 to 1.29 s; 'ST (settling time)' is diminished from 2.93 to 1.82 s; 'SSE (steady-state error)' is diminished from 0.8 to 0.7 N m for motor torque.

5 Conclusion

Closed-loop proportional resonant and fractional order PID controlled QBC-TZSI-PMBLDC fed permanent magnet brushless motor drive systems are designed, analyzed, and simulated utilizing MATLAB-simulink. By using FOPID controller, 'ST (settling time)' is diminished from 3.6 to 2.8 s and 'SSE (steady-state error)' is diminished from 0.8 to 0.7 N m. Hence, the outcome represents that the dual-loop FOPID controlled closed-loop QBC-TZSI-PMBLDC is superior to dual-loop PR controlled closed-loop QBC-TZSI-PMBLDC. The benefits of proposed QBC-TZSI-PMBLDC system are diminished number of switches and enhanced-response. The drawback of QBC-TZSI-PMBLDC is that the number of passive components is increased.

The present effort covenants with the comparison of closed-loop proportional resonant and fractional order PID controlled QBC-TZSI-PMBLDC. The comparison of closed-loop Fractional PID and Fuzzy logic controlled QBC-TZSI-PMBLDC can be done in future.

References

1. A. Mudhol, A.J.P. Pius, Design and implementation of boost converter for photovoltaic systems. IJIREEICE **4**(2) (2016)
2. P. Wang, L. Zhou, Y. Zhang, J. Li, M. Sumner, Input-parallel output series DC-DC boost converter with a wide input voltage range, for fuel cell vehicle. IEEE Trans. Veh. Technol. **66**(9), 7771–7781 (2017)
3. C.C. Chan, A. Bouscayrol, K. Chen, Electric, hybrid, and fuel-cell vehicles: architectures and modeling. IEEE Trans. Veh. Technol. **59**(2), 589–598 (2010)
4. T.R. Choudhury, B. Nayak, Comparison and analysis of cascaded and quadratic boost converter. IEEE. 978-1-4799-7455-9/15/$31.00 (2015)

5. M.A. Al-Saffar, E.H. Ismail, A.J. Sabzali, High efficiency quadratic boost converter. IEEE (2012). 978-1-4577-1216-6/12/$26.00 (2012)
6. Y. Wang, Y. Qiu, Q. Bian, Y. Guan, D. Xu, A single switch quadratic boost high step up DC-DC converter. IEEE Trans. Ind. Electron. https://doi.org/10.1109/tie.2018.2860550 (2018)
7. L.S. Vilefort, W.G. Silva, D.A. de Andrade, F.V.R. da Silva, L.C.G. de Freitas, E.A.A. Coelho, J.B. Vieira Jr, Quadratic boost converter using a soft single switch. IEEE. 978-1-4799-0272-9/13/$31.00©2013
8. P. Priya, G.S. Basha, S.V.S. Niranjan, R. Seyezhai, Analysis and experimentation of quadratic boost converter for photovoltaic applications. Indian J. Sci. Technol. **10**(37) (2017). https://doi.org/10.17485/ijst/2017/v10i37/117553
9. M. Forouzesh, Y.P. Siwakoti, S.A. Gorji, B. Lehman, Step-up DC–DC converters: a comprehensive review of voltage-boosting techniques topologies, and applications. IEEE Trans. Power Electron. **32**(12) (2017)
10. A.M.S.S. Andrade, L. Schuch, M.L. de S. Martins, High-stepup PV-module-integrated converter for PV-energy harvest in-FREEDM-systems. IEEE Trans. Ind. Appl. **53**(2), 1138–1148 (2017)
11. N.R. lingam, S. Priyanga, Implementation of PV-based boost converter using PI controller with PSO algorithm (2017)
12. K.B. Tawfiq, A.S. Mansour, H.S. Ramadan, M. Becherif, Wind energy conversion system topologies and converters: comparative review. **162**, 38–47 (2019)
13. C.-C. Hua, C.-H. Cheng, Design and implementation of power converters for wind energy conversion system. IEEE (2010)
14. J. Bauman, M. Kazerani, Acomparative study of fuel-cell battery, fuel cell-ultra capacitor, and fuel-cell-battery-ultra capacitor vehicles. IEEE Trans. Veh. Technol. **57**(2), 760–769 (2008)
15. C.-M. Lai, Y.-H. Cheng, M.-H. Hsieh, Y.-C. Lin, Development of a bidirectional DC/DC converter with dual-battery energy storage for hybrid electric vehicle system. IEEE Trans. Veh. Technol. **67**(2), 1036–1052 (2018)
16. Y. Zhang, L. Zhou, M. Sumner, P. Wang, Single-switch, wide voltage gain range, boost DC–DC converter for fuel cell vehicles. IEEE Trans. Veh. Technol. **67**(1), 134–145 (2018)
17. F. Zhang, X. Zhang, M. Zhang, A.S.E. Edmonds, Literature review of electric vehicle technology and its applications, in *2016 5th International Conference on Computer Science and Network Technology (ICCSNT)*, Changchun, China, Dec. 2016, pp. 10–11
18. Y.B. Koca, A. Yonetken, Y. Oğuz, Boost converter design and analysis for photovoltaic systems (2019)
19. A. Khaligh, Z. Li, Battery, ultra capacitor, fuel cell and hybrid energy storage systems for electric, hybrid electric, fuel cell, and plug-in hybrid electric vehicles: state-of-the-art. IEEE Trans. Veh. Technol. **59**(6), 2806–2814 (2010)
20. Z. Chen, P. Yang, G. Zhou, J. Xu, Z. Chen, Variable duty cycle control for quadratic boost PFC converter. IEEE Trans. Ind. Electron. **63**(7) (2016)

Peformance Analysis of Joysticks Used in Infotainment Control System in Automobiles

Vivek Ram Vasan, S. Prasanna Bharathi, Arjun Sharma, and G. Chamundeeswari

Abstract Distraction while driving can be attributed as an important cause for accidents and the use of smartphones while driving is identified as the major reasons for distracting the driver. In order to reduce distractions due to the usage of mobile phones in automobiles, latest technologies are available to operate the mobile phones through the dashboard HUD's and steering wheel housed controls. Android Auto is one such technology developed at Google that allows the android phones with android version 5 and above to be interfaced and controlled by the Car dashboard controls. Mirror link is another standard available to interface smart phones to a car's infotainment system. The driver and passengers can interact and use them for assistance while driving by using the steering wheel controls and the dashboard buttons which is part of the automotive infotainment system. However, these technologies are available only in some of the latest high-end premium cars released after 2015. There are many Pluggable Hands-free devices available, that enable the users to control the mobile phones in the car. These devices are mountable on cars' components such as steering wheel and are connected to the phone using Bluetooth protocol. They provide key-based input interfaces using which the user can control specific features of the mobile phone.

Keywords Android Client-Server application · Atmega microcontroller · HC-05 module · HUD

V. R. Vasan (✉)
ESP-AS2, Robert Bosch Engineering and Business Solutions, Adugodi, Bangalore India
e-mail: vivek435@gmail.com

S. Prasanna Bharathi
Department of ECE, SRM Institute of Science and Technology, Vadapalani, Chennai, India
e-mail: prasanns@srmist.edu.in

A. Sharma
ESP-AS13, Robert Bosch Engineering and Business Solutions, Adugodi, Bangalore, India
e-mail: arj.sharma1997@gmail.com

G. Chamundeeswari
Saveetha School of Engineering, SIMATS, Chennai, India
e-mail: easwarig@gmail.com

S. S. Dash et al. (eds.), *Intelligent Computing and Applications*,
Advances in Intelligent Systems and Computing 1172,
https://doi.org/10.1007/978-981-15-5566-4_30

1 Introduction

Distraction of Driver is considered as one of the primary causes of accidents and the use of smartphone is identified as a common practice that can potentially distract the driver during driving. Distraction caused by mobile phones includes talking on the mobile phone, chatting, voice recognition, use of maps for navigation and tuning the sound system. It would be very tedious and not realistic to estimate how many distracted drivers drive on a particular section of the road because surveys and cameras do not aid to come to a conclusion of such distraction. However, it is clear that distraction of driver is a major cause of road traffic accidents because a distracted driver tends to fail spotting a stop sign, or see a red light at a traffic junction, and also could violate the permitted speed limit that might put their own safety or that of others at risk.

Research literature has highlighted that during driving speaking on the phone is a very risky in terms of road safety and it contributes to increase in road accidents [1], thus, presenting as a significant threat to the public. Smartphone usage while driving is predominant in adolescent and amateur drivers. Recent studies shows [2] that approximately 50% of the young drivers of the age in between 18 and 25 years, indulge in the practice of the use smartphone while driving, in which about 25% of them read emails and browse the World Wide Web, and around 65% of them send text messages.

It was observed from a study report that among 2500 distracted driver-related incidents is the considered the highest of smartphone caused harmful accidents corresponded to young drivers [3], Horberry et al. [4]. Another report shows that more than 65% of drivers indulging in smartphone usage while driving constituted of drivers less than 40 years of age. A conversation over the mobile phone distracts the driver by immersing the individual into the contents of the phone rather than concentrating on the task which is driving. Thus, particular research was undertaken to study the reaction times of drivers as a measure of the crash probability due to distraction from mobile phone usage. The research considered various study conditions which include a test drive in a simulator, laboratory and field trials.

An experiment conducted [5] to observe the braking maneuvers of drivers who use a mobile phone at the blink of a red light in a Test bench setup deduced that hand-held as well as hands-free smart phone tasks results in slower response times to perform the task of braking. A desktop simulator experiment conducted where the distracted driver has to commence driving after a green light while conversing over a smartphone deduced that the distracted drivers took almost 1/3 of a second more to begin driving [6].

Many studies have targeted to study the negative effects of mobile phone usage on Car-tailing maneuvers. Car-following is the behavior of a driver to follow a vehicle in the front. It is a predominant driving situation as well as a prerequisite for safe driving [7]. In car simulation, [8] the adverse effect of following a car while the use of smartphones and in-car infotainment system was studied. A study of 40 active drivers performed a driving maneuver for 80 km where a total of 20 car-tailing scenarios occurred arbitrarily. The drivers were tested in simulated driving conditions where they are exposed phone conversation task in 8 of these car-following situations. It showed that there was an increase response time for mobile phone conversations while driving. Furthermore, it was found that the drivers did not compensate for the increase in response time by increasing the spacing between cars while conversing over the mobile. However, recent reports show a decrease in driving speed when a phone call is received by the driver, an act known as driving risk compensation and avoidance [9]. For example, significant decrease in speed when driving on call with the phone on the driver's hand was observed [10], compared to other types of mobile conversations.

A slow braking response time was observed for the driver conversing over a mobile phone when compared to driving without a phone [11]. The distracted drivers additionally took more time to recuperate their lost speed after braking. It was also observed that the participants indulging in conversation over the mobile phones appeared to have a more cautious driving profile and maintained lower speed and higher spacing with respect to following distance compared to non-distracted drivers [12]. However, the crash rates were still higher for distracted driving compared to non-distracted driving [13]. Particularly within the distracted driving conditions, there was no significant difference observed between driving with hands-free and hand-held mobile phone conversations.

Driver response time to an unforeseen event, the speed and the spacing are the key factors to ensure stability and flow of road traffic. Talking over the phone while driving can greatly affect such factors, thus, causes the driver to drive poorly while following the leading vehicle [14]. Studies have tried to summarize the danger of mobile phone usage in a car in the above-mentioned scenarios, however, the data is limited and the knowledge we have on this critical issue remains elusive.

2 Feasible Mechanical Design for the Novel Car Accessory

2.1 Button/Key Pad-Based Input Interface

The Bluetooth kits available in market have buttons as the input interface through which the driver can interact with the device to control the phone's apps. These devices would require the driver to use his/her index finger or the thumb to press the buttons and use the device in order to make or receive a call.

The button size is expected to be big enough to enable the driver to distinguish between the various button options available. Also, key-based input interfaces would require the driver's attention for at least a fraction of second which is sufficient to cause another distraction which might prove fatal. The devices are mounted on the steering wheel. Hence the size of the device is the function of available number of input buttons. This would limit the number of controls the device can provide. For example, a device with four buttons can provide four functionalities like

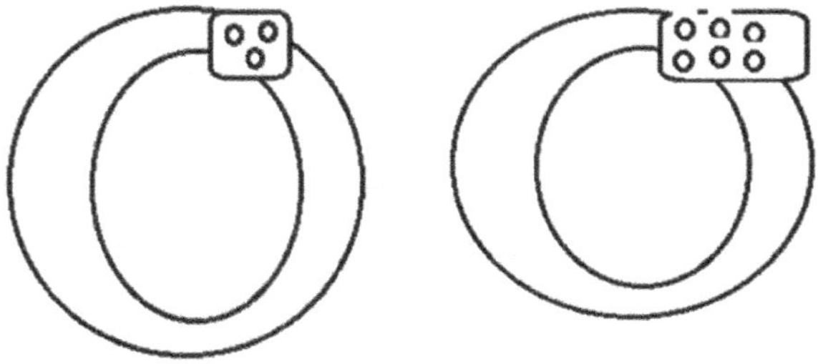

Fig. 1 2012 steering wheel button system [R] to the 2018 steering wheel [L]

- Receive Incoming Call
- End a Call
- Increase Volume
- Decrease Volume.

In order to provide extra features, extra buttons have to be added which would in turn increase the size of the device. As the size of the device increase, it becomes more of a obstruction to the driver handling the steering wheel. Therefore, it could be clearly concluded that currently available devices with button input interface have a series of limitations and disadvantages. Figure 1 illustrates the increase in the size of the device as the number of buttons increases.

2.2 *Joy Stick-Based Input Interface*

A joystick is a lever that has the ability to move in all directions. It is flexible in nature and provides many degrees of motion. The idea to use a joystick is to translate the movement of a flexible vertical stick which is made of plastic into electronic signal that a controller can process. To put into perspective, translating physical movement into electric signal. The current work proposes a joystick-based design for the car accessory using which mobile phone can be controlled. The design is illustrated in Fig. 2.

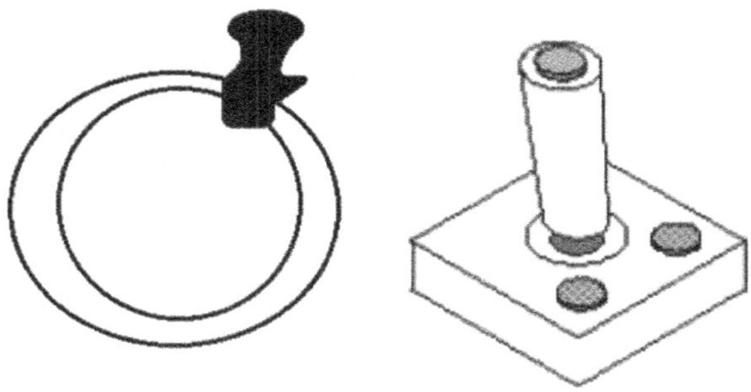

Fig. 2 Joystick device mounted on steering wheel and a conventional joystick

3 Advantages of Selecting Joystick-Based Design

Joystick has a set of advantages over the keypad/button-based input interface. The advantages are listed below:

1. A joystick occupies relatively less space compared to the latter. The base of the joystick is fixed and hence the design can be made as optimal as possible to make the entire device occupy less space on the steering wheel.
2. It is easy to train the users on the usage. The joystick movement is either towards the x-axis or y-axis. The human cognitive ability enables the user to pick up such actions easily. The drivers need not have to spend even a fraction of second in identifying the key to be operated in order to provide a specific Input.
3. The ease of handling a joystick, the design can be provided as slender and flexible as possible. This would enable the driver to control the joystick with the help of the region between the thumb and index finger without having to move the finger from the steering wheel. This would ensure perfect grip while handling the device.
4. The joysticks can be designed in such a way that they are able to move in z dimension as well. Hence the number of Input controls that can be provided by a joystick is higher. A minimum of 4 to a maximum of 9 input signals can be generated using a single joystick. This is a space saving factor and definitely overcomes the disadvantage of keypad-based input as the size of the device increases with increase in keys.
5. The joystick can be used as a base for keypad as well. The column and the head of the joystick can be embedded with keypads if required in order to increase the number of input signals that can be generated using a single device.
6. The joystick inputs can be designed for a particular pattern, for example, moving the joystick towards x-axis twice within a small-time interval can be decoded as another input signal thus increasing the number of input signals that can be generated.
7. The joystick occupies less space and hence the manufacturing material cost would be relatively lesser when compared to a device with keypad providing the same set of features.

3.1 CAD Model Using Solid Works

A CAD model has been designed for the prototype device using solid works. The mechanical design proposes a device with a flexible head and rectangular base. The rectangular base can be used to house the electronic components. The dimension of the rectangular base is 5 cm × 3 cm × 3 cm. It is 5 cm wide, 3 cm long and has a thickness of 3 cm. This would enable the device to occupy very less space on the steering wheel. The rectangular base is provided with facility to attach a flexible rubber strap by means of which the device can be wrapped around the steering wheel or one of the rims of the wheel or the connector between the hub and the rim as per the user's preference. The length of the Joystick head is about 2.5 cm (Figs. 3, 4, 5, 6).

Fig. 3 A CAD model for the prototype

Fig. 4 The mounting of the device on the steering wheel is illustrated

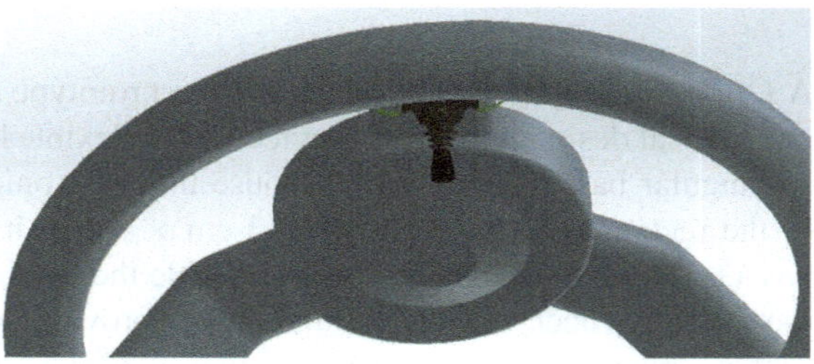

Fig. 5 Front view of the mounted device

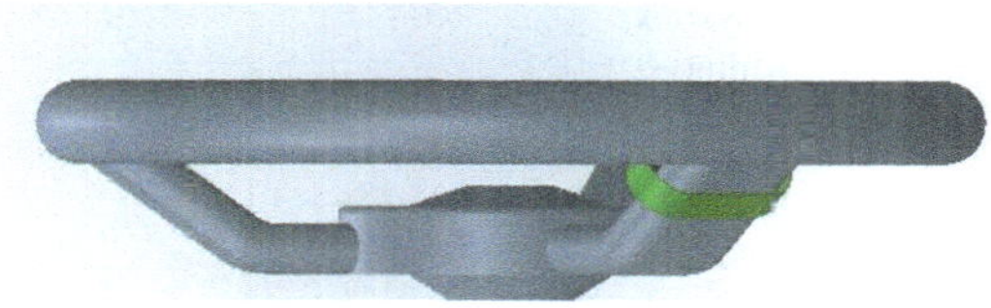

Fig. 6 The side view of the mounted device

4 Electronic System Design of the Novel Car Accessory

This chapter is dedicated to designing the electronic system of the car accessory proposed as part of the work. The device can be built as a microcontroller-based device. The joystick is to be interfaced to the input pins of the controller The inputs should be decoded based on the signal strength obtained at the input lines. A Bluetooth module is used to transmit Bluetooth messages that are intercepted by the android phone. These messages are used to control the features in the phone using an android service (Fig. 7).

4.1 *Interfacing BT and Joystick to Arduino*

The joystick was interfaced to Analog Pin A_0 (x-axis) and A_1 (y-axis) on Arduino board. The power of 5 V required by the joystick was drawn from the power rails available on the Arduino board. The GND was connected to GND point on the board.

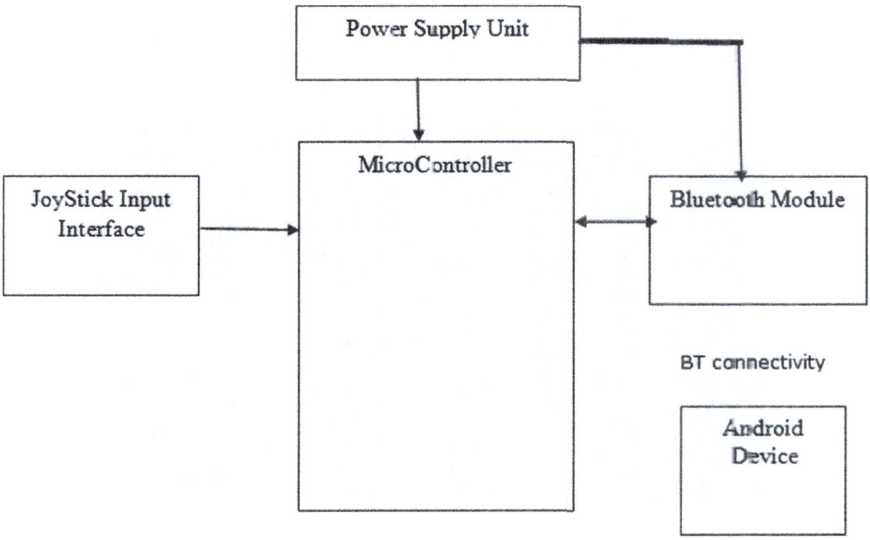

Fig. 7 The system is represented using the block diagram

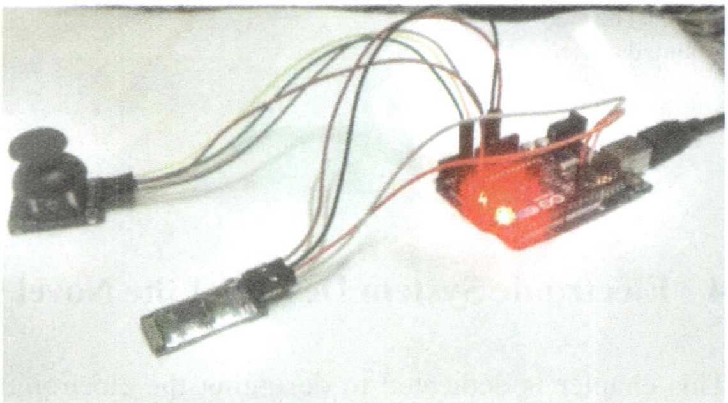

Fig. 8 BT and Joystick interfaced to Arduino

The HC-05 Bluetooth module was powered using the onboard power supply from the Arduino board, while Arduino board drew its power from USB port of the Laptop or can be given through a battery pack. The TXD pin of HC05 was interfaced to PD_0(RXD), and the RXD pin to PD_1(TXD) of the Arduino board. The GND point of HC-05 was grounded.

The Arduino board was programmed to send BT data. Tera Term Software was used to verify if the data was sent by the BT module accurately. The data transmitted by the BT module based on the inputs from joystick is observed using Tera Term emulator. 1 is sent for Positive *X*-axis, Data 2 is sent for Negative *X*-axis, Data 3 is sent for Positive *Y*-axis, and Data 4 is sent for Negative *Y*-axis. The Tera term emulator is shown in Fig. 9 (Fig. 8).

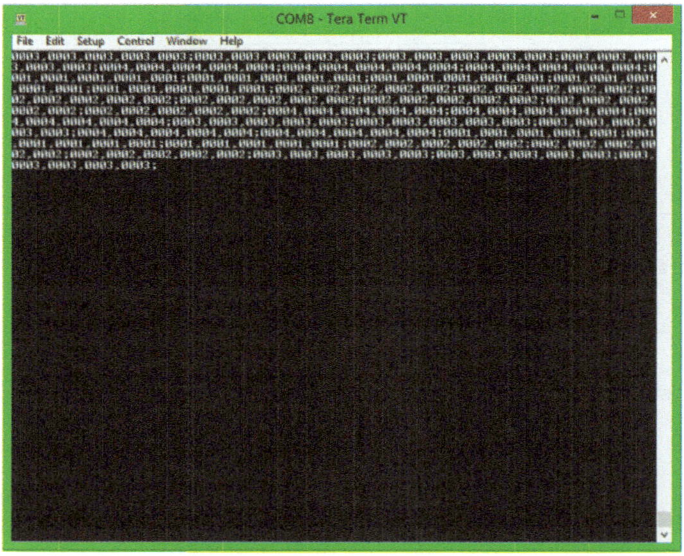

Fig. 9 Bluetooth message received on Tera term emulator

4.2 The Bluetooth Android Service

The android platform supports Bluetooth network stack, using which a device can exchange data wirelessly with another Bluetooth device. Android Bluetooth APIs which are part of the application framework are used to build Bluetooth-based services and applications in android. The use of Bluetooth function call's (API's), an application can perform the following:

- Search for accessible Bluetooth-enabled devices
- RFCOMM channels Establishment
- Connect to other Bluetooth devices
- Transfer data and control apps.

Bluetooth-enabled devices should first form a communication channel to transmit data between each other. This is called Pairing Process. One device being discoverable makes it available for incoming requests for connection. The application then finds the detected device using a SDP process. Once the discoverable device accepts the request for pairing, they exchange the security keys.

5 Experiments and Result Analysis

In the proposed work, a joystick has been used as an input device instead of a traditional key-based input. The space advantage is obvious in the mechanical design of the device. However, in order to test the effectiveness of the usage of the device, an experiment was designed. The time taken by a user to respond to a request, for example, the user has a hands-free device plugged in his car and a call arrives, the user gets distracted and makes an effect to receive the call by pressing the key on the hands-free device that is mounted on the steering wheel. If the number of keys available on the device is more, the user needs to spend some effort to identify the right key to be pressed. This adds on to the response time. The response time is directly proportional to the amount of distraction a user experience

5.1 Experiment Design

To verify the difference in response time while using a key-based device and Joystick-based device, two MATLAB GUI applications were created. The first GUI simulated the behavior of a key-based device (Fig. 10).

The key GUI consisted of 6 keys labeled randomly as A key, B key, M key, T key, V key, and Y key. A Start key and End key were provided to start the experiment. On

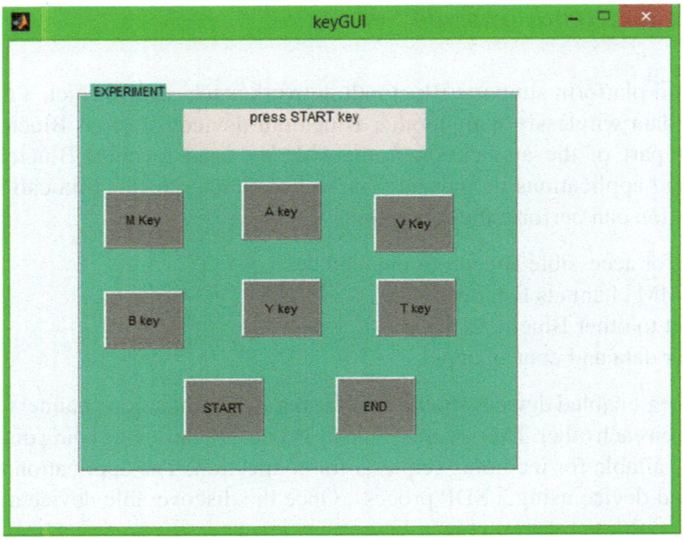

Fig. 10 Key GUI experiment

pressing the Start key, a message was displayed to press Key A and a sequence of keys randomly. The message was displayed at random time intervals and sequences (Fig. 11).

Ten people of two different age groups 18–35, 35–60 response time was measured while using the GUI. Each User was given three trials and their response time, the time from when the message was displayed to the time when the key was pressed for each of the keys was measured and average response time was recorded (Fig. 12).

Another GUI to simulate joystick behavior was designed. It consisted of only four input points one at north, south, east, and west directions, the input points did not have any labels. The same group of people who used Key Experiment GUI was

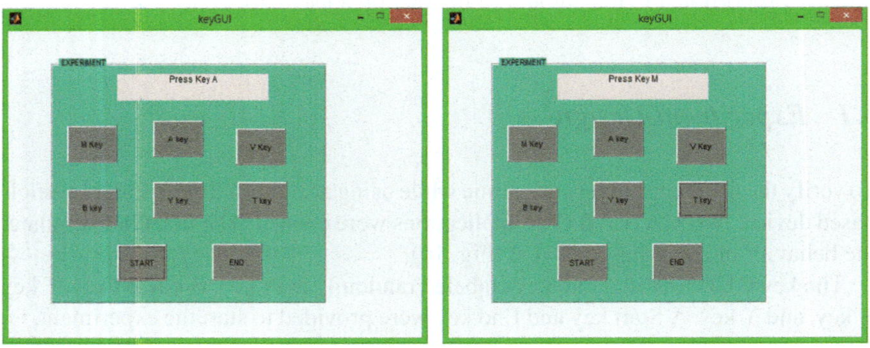

Fig. 11 Key experiment sequence

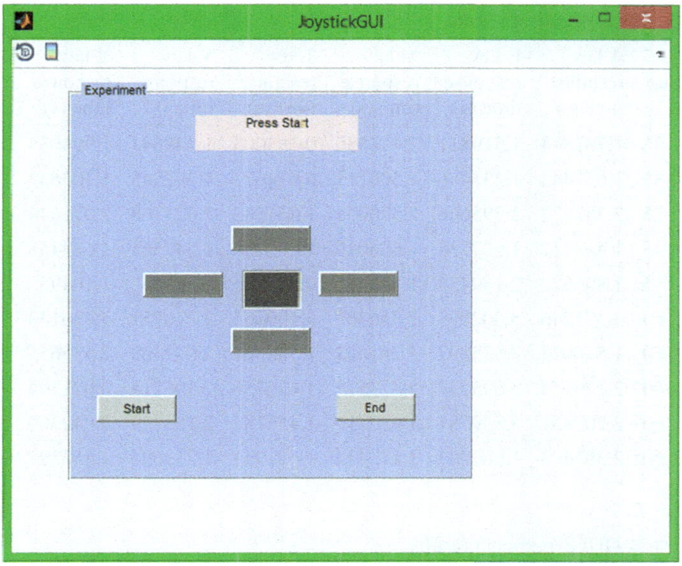

Fig. 12 Joystick GUI experiment

made to use the joystick experiment GUI. Commands, like up, down, left, right, were displayed in a random sequence as shown in Fig. 13 and the time taken by the user to respond to those messages were recorded. Each user was again given three trials and the average response time was recorded (Tables 1, 2, 3, and Fig. 14).

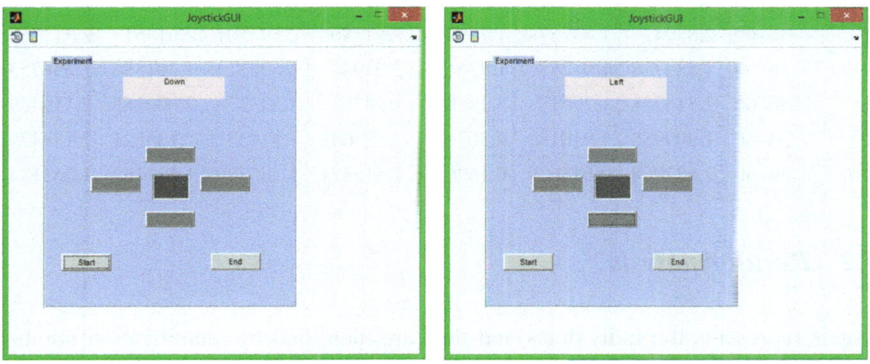

Fig. 13 Joystick experiment sequence

Table 1 Key GUI experiment results

Person	Age group	A key response time (s)	B key response time (s)	M key response time (s)	T key response time (s)	V key response time (s)	Y key response time (s)	Average response time (s)
1	18–35	1.701834	1.519811	1.474536	0.483003	1.248641	1.994034	1.4036431
2	18–35	1.657443	1.255486	1.500717	0.614673	1.392545	1.403633	1.3040828
3	18–35	2.357872	2.295666	2.570059	1.231183	2.023168	2.071644	2.0915986
4	18–35	1.463217	1.607738	1.686016	1.063738	1.549565	1.643458	1.5022887
5	18–35	1.896672	2.159435	1.584615	0.930515	0.31873	1.10495	1.3324862
6	36–60	3.371246	5.602098	2.356469	3.190361	2.276751	3.436133	3.3721764
7	36–60	1.576033	3.575897	3.281622	1.558561	3.072469	2.535655	2.600039
8	36–60	2.129173	3.695737	2.423518	1.626736	2.506214	3.021394	2.5671287
9	36–60	2.202831	3.839781	1.932216	1.458182	2.153239	7.042369	3.1047698
10	36–60	2.301473	1.663203	1.112434	1.621941	1.259696	1.453597	1.5687241

Table 2 Joystick GUI experiment results

Person	Age group	Up response time (s)	Down response time (s)	Left response time (s)	Right response time (s)	Up twice response time (s)	Up and right response time (s)	Average response time (s)
1	18–35	1.694997	1.245673	1.131581	1.214475	1.196757	1.016823	1.250051
2	18–35	1.82407	1.172241	0.703676	0.759536	1.114881	1.031267	1.100945
3	18–35	1.719633	1.054821	0.930883	1.503861	1.552299	1.832409	1.432318
4	18–35	1.839778	1.03758	1.876124	1.039421	1.698226	1.100322	1.431909
5	18–35	1.001523	1.113212	0.891723	0.711242	1.27988	1.823234	1.136802
6	36–60	2.83941	1.837578	1.769216	2.823801	2.592501	2.01351	2.312669
7	36–60	1.839778	1.03758	1.876124	2.039421	1.698226	1.701386	1.698753
8	36–60	2.316238	1.29518	3.198072	1.58362	2.323278	2.000185	2.119429
9	36–60	1.999988	1.910189	2.563862	2.583041	1.91427	2.135521	2.184479
10	36–60	2.687908	0.648679	0.169645	1.615324	1.280389	1.117162	1.253184

5.2 *Result Analysis*

X-axis represents the individuals and they are identified by identification number 1–10. The *Y*-axis represents the response time of the device in seconds.

Red-colored line indicates each individual's response time in Key GUI Experiment and Blue colored line indicates each individual's response time in Joystick GUI Experiment.

From the data statistics it is clearly evident that the response time in case of key-based inputs is relatively higher when compared to joystick-based Input interface.

Table 3 Data statistics for each of the experiment

Joystick GUI response Time (s)	
Min	1.1009
Max	2.3127
Mean	1.5921
Median	1.4321
Mode	1.1009
Std	0.4584
Range	1.2117
Key GUI response Time (s)	
Min	1.3041
Max	3.3722
Mean	2.0847
Median	1.8302
Mode	1.3041
Std	0.7775
Range	2.0681

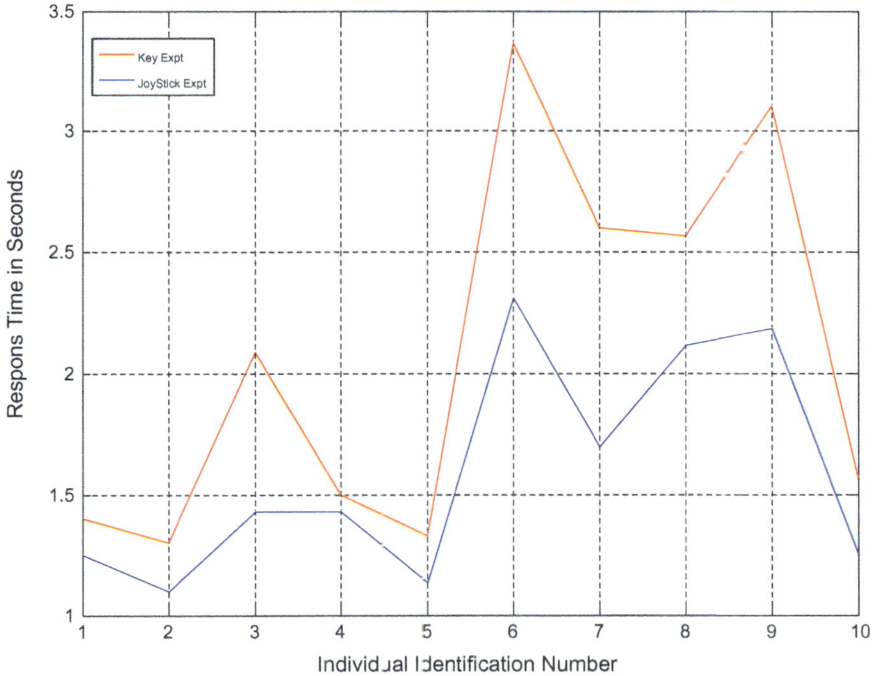

Fig. 14 A comparison of the response time

The maximum response time for joystick was 2.3127 s and that for key input was 3.3722 s almost 50% more than the response time of joystick experiment.

6 Summary

Many Hands-free devices are available in the market that could be mounted in the car and they enable us to control some of the features of the Android phone like make/receive call, play music, etc. Keys are provided on the joystick by means of which a user can perform a particular action. Some of these devices have built-in speakers/displays that play/display the phone data. However, the smartphones these days are big and powerful enough that eliminates the need for separate speaker/display units in hands-free device. The current work proposes a hands-free device that would use a unique design—A joystick-based Input Interface.

A CAD model has been designed for the prototype device using solid works. The mechanical design proposes a device with a flexible head and rectangular base. The rectangular base can be used to house the electronic components. The dimension of the rectangular base is 5 cm $\times$ 3 cm $\times$ 3 cm. It is 5 cm wide, 3 cm long and has a thickness of 3 cm. This would enable the device to occupy very less space on the steering wheel. Due to the Axial movement provided by the joystick, it is possible to generate many input actions using a single joystick. This would offer the user to control many functionalities on the phone using different input actions. Also, the distraction level is directly proportional to time taken by the driver in performing a specific action using the hands-free, say connect a call by pressing a key. An experiment was conducted to measure the response time of the users with respect to handling key-based device and joystick-based devices. It was observed that the response time reduces by almost 50% in case of the joystick device as the human brain finds it much easier to respond to repetitive common actions rather than the search action.

The design and production cost of the proposed device is relatively cheaper when compared to other devices available in the market as it uses minimal manufacturing material due to its size. The electronic components chosen: ATMEGA 328 Microcontroller, HC05 Bluetooth are relatively cheaper and hence the device can be introduced to the market at a price almost equal to Rupees 500. The prototype for the device was developed on an Arduino board. A two-axis joystick was used as the input device. A HC-05 Bluetooth module was used to communicate to the android device. An Android Bluetooth service was created which received the commands from the Bluetooth module and controlled certain features of the phone which in turn reduces the driver's distraction level and helps him/her drive safely on the road.

References

1. D.A. Redelmeier M.D., R.J. Tibshirani, *Association between Cellular-Telephone Calls and Motor Vehicle Collisions*. https://doi.org/10.1056/NEJM199702133360701
2. AAMI, *AAMI young driver's annual road safety index, AAMI*, Melbourne (2010). Retrieved 27 September 2012, from http://aami.com.au/sites/default/files/fm/news/2010-AAMI-Young-Drivers-Index.pdf
3. L.T. Lam, Distractions and the risk of car crash injury: The effect of drivers' age. J. Saf. Res. https://doi.org/10.1016/S0022-4375(02)00034-8
4. T. Horberry, Driver distraction: The effects of concurrent in-vehicle tasks, road environment complexity and age on driving performance. Accid. Anal. Prev. **38**, 185–191 (2006)
5. W. Consiglio, *Effect of cellular telephone conversations and other potential interface on reaction time in a braking response*. https://doi.org/10.1016/S0001-4575(02)00027-1
6. K.E. Beede, Engrossed in conversation: The impact of cell phones on simulated driving performance. https://doi.org/10.1016/j.aap.2005.10.015
7. M. Saifuzzaman, Z. Zheng, Incorporating human-factors in car-following models: a review of recent developments and research needs. Transp. Res. Part C Emerg. Technol. (2014). https://doi.org/10.1016/j.trc.2014.09.008
8. H. Alm, L. Nilsson, The effects of a mobile telephone task on driver behaviour in a car following situation. Accid. Anal. Prev. **27**, 707–715. http://dx.doi.org/10.1016/0001-4575(95)00026-V
9. J. Törnros, Mobile phone use—effects of conversation on mental workload and driving speed in rural and urban environments. Transp. Res. Part F Traffic Psychol. Behav. (2006). https://doi.org/10.1016/j.trf.2006.01.008
10. T. Ranney, G. Watson, E.N. Mazzae, Y.E. Papelis, O. Ahmad, J.R. Wightman, *Examination of the Distraction Effects of Wireless Phone Interfaces Using the National Advanced Driving Simulator-Preliminary Report on Freeway Pilot Study* (2004). Paper No. DOT 809 737
11. D. Strayer, J. Watson, F. Drews, Cognitive distraction while multitasking in the automobile. Psychol. Learn. Motiv. **54**, 29–58 (2011). https://doi.org/10.1016/B978-0-12-385527-5.00002-4
12. R. Matthews, S. Legg, S. Charlton, The effect of cell phone type on drivers subjective workload during concurrent driving and conversing. Accid. Anal. Prev. **35**(4), 451–457 (2003)
13. L. Nilsson, Behavioural research in an advanced driving simulator—experiences of the VTI system, in *Paper Presented at the 37th Annual Meeting of the Human Factors and Ergonomic Society*, Seattle, USA, 11–15 October 1993
14. C. Patten, A. Kircher, J. Östlund, L. Nilsson, Using mobile telephones: cognitive workload and attention resource allocation. Accid. Anal. Prev. **36**, 341–350 (2004)

Probabilistic Principal Component Analysis (PPCA) Based Dimensionality Reduction and Deep Learning for Cancer Classification

D. Menaga and S. Revathi

Abstract Cancer is a threatening disease that causes the death of major humans. The DNA microarray-based gene expression profiling is an effective technique employed for cancer classification, diagnosis, and treatment. The early detection creates a need for a better and accurate method that offers the information of cancer in the patient that enables better decision-making by the clinicians and treating them. Hence, in this work, a cancer classification technique is devised by utilizing the gene expression data collected from the patients. Since the input gene expression data is huge, the dimensionality of the data is reduced, undergoing transformation using Probabilistic Principal Component Analysis (PPCA), for the efficient computation of the maximum likelihood estimates. Once the dimensionality of the data is reduced, it is classified using Deep Belief Network (DBN) to perform the cancer classification. DBN classifies the dimensionality reduced gene data, providing two classes, namely, normal and abnormal. The comparative performance analysis of the proposed technique is evaluated based on the three metrics, such as sensitivity, specificity, and accuracy. The effectiveness of the devised methodology was deliberated through the achieved accuracy value of 0.8085, specificity value of 0.8333 and sensitivity of 0.7981.

Keywords Cancer classification · Gene expression data · Dimensionality reduction · PPCA · DBN

1 Introduction

The cancerous tumors can be treated appropriately only when the exact type of cancer is identified earlier. The conventional cancer diagnosis was done on the basis of the

D. Menaga (✉) · S. Revathi
Department of CSE, B.S. Abdur Rahman Crescent Institute of Science and Technology, Chennai, India
e-mail: dev.menaga@gmail.com

S. Revathi
e-mail: srevathi@crescent.education

S. S. Dash et al. (eds.), *Intelligent Computing and Applications*,
Advances in Intelligent Systems and Computing 1172,
https://doi.org/10.1007/978-981-15-5566-4_31

clinical and morphological appearance [1]. Since cancer is caused by various factors, these traditional methods were not capable of identifying the cancer effectively [2, 3]. The advancement in cancer cell classification has gained a lot of research interest due to the emerging need for cancer treatment. The cancer classification is also based on the tumor's morphological appearance, but it is affected by the limitation of incorrect prediction of the cancer cell. The incorrect prediction is mainly due to the nature of cancer cell, which portrays distinct symptoms at distinct conditions. With the aim of resolving this limitation, the DNA microarray technology was adopted for performing the gene expression data-based classification as it provided highly accurate classification [4, 5]. It is an essential platform widely utilized for analyzing the gene expression data of millions of genes in various experimental studies [6]. The microarray technology was evolved estimating the variation in all the genes available in the provided biological sample by the utilization of the hybridization mechanism. The gene expression patterns in the microarray data provide the useful information related to the significant advancement in the fields of biology and clinical medicine for the medical prognosis and diagnosis [7]. The exploration of the molecular variation existing in the cancer cells for the categorization of the tissue into cancerous or normal tissues is one of the prevalent applications of the microarray technology [8].

The DNA microarray methodology performs the identification, classification, and prediction of the distinct cancer types [9, 10]. The proliferation of gene expression is done for estimating the enormous genes available within a single RNA by practicing hybridization process. The DNA microarray is created by integrating the healthy DNA reference and the testing DNA [11, 12]. It is utilized for expressing and comparing the additional insight about the genes at specified time and status for carrying out the identification of cancer or any other disease [13, 15]. The color spot matrix comprising of the gene expression level's quantitative values is generated by the utilization of the laser and fluorophores [14]. The obtained gene expression level is defined as a signature beneficial for the diagnosis of the distinct diseases, which occur mainly due to the alteration in the gene expression level [16]. The DNA microarrays when concatenated with the other computational approaches become capable of providing an efficient diagnosis. The DNA microarray classification comprises three steps, they are prediction of the gene, discovery of class, and prediction of class [17, 19]. The basic task of microarray data classification is the determination of the class for the concerned unknown samples. The millions of genes together constitute the DNA microarray samples and the selection of best gene set is a tough task, so the modeling of a trustworthy classification methodology is an immediate need. The presence of unwanted genes increases both the noise and the dimensionality of the obtained gene expression matrix; this further leads to high computational complexity [14, 18].

Various classification methodologies based on the gene expression data were developed for performing the cancer identification. Anyhow, the gene expression data-based cancer classification suffers from the issues, like high dimensionality and multi-class imbalance [8]. Hence, in this paper an effective cancer cell classification methodology and feature selection scheme are to be devised for the classification of the given microarray data with the intention of identifying cancer. The distinct research works related to the microarray dataset-based cancer cell classification were not able to correctly identify the reduced number of informative genes with maximum classification accuracy. Initially, the dimensionality of the gene expression data is reduced by employing the PPCA [21] through the effective estimation of maximum likelihood. This reduces the dimensionality related complexity issues during classification. Subsequently, the gene expression data with the reduced dimension is classified utilizing the DBN classifier. From the obtained classification results, the cells under examination can be either declared as a normal cell or abnormal cell.

The rest of this paper is organized as follows: The brief introduction on the gene expression-based cancer cell classification is presented in Sect. 1 and the survey on the relevant works is discussed in Sect. 2. Section 3 elaborates the proposed method of cancer cell classification by adopting the PPCA and DBN classifier and the simulation results of the proposed classification methodology are demonstrated in Sect. 4, and finally, in Sect. 5, the conclusion of the research is presented.

2 Motivation

2.1 Literature Survey

Here, four literature works related to the gene expression-based cancer cell classification are presented.

Liu et al. [2] developed a simple Principal Component Accumulation (PCAcc) scheme for resolving the issues in the gene expression data-based cancer classification. The main intention was to reduce the dimensionality through the utilization of multiple Principal Component (PC) subspaces rather than the first few PCs. In spite of this advantage; this method could not provide the biological interpretations. Patricia Melin Oscar Castillo [3] presented type-2 fuzzy logic for resolving the issues faced by clustering, pattern recognition, and classification. The type-2 fuzzy rule's structure is the same as the structure of the type-1 fuzzy rule case; because the only difference between the type-1 and type-2 is the membership functions nature. The devised scheme was applicable in a wide range of application with high complexity and the maximum degree of uncertainty and is computationally inexpensive.

Alshamlan et al. [15] presented a Bio-Inspired evolutionary gene selection methodology for performing the microarray-based cancer classification. The developed algorithm was an effective wrapper gene selection scheme; it has the ability to search the near-optimal or optimal solutions in the large dimensional space. This scheme achieved maximum classification accuracy with a limited number of selected genes. Van De Vijver et al. [14] developed a complementary DNA (cDNA) microarray analysis for the treatment of breast cancer. The prognosis profile is a strong development predictor for performing the patient's distant metastases with the lymph-node positive disease. This method suffered from the risks of under treatment and overtreatment because of the improper patient selection.

2.2 *Challenges*

The distinct challenges faced by the gene expression-based cancer cell classification are elaborated as follows,

- The microarray dataset is affected by number of irrelevant genes and the level of noisy genes; these made the classification task highly challenging [15].
- Cancer classification is a tedious process since it is historically dependent on the biological knowledge rather than the unbiased and systematic schemes for the effective recognition of the tumor subtypes [4].
- The availability of unwanted genes increases both the noise as well as dimensionality to the gene expression data; these leads to hike in the complexity of computation during clustering and classification [19].
- The numbers of features present are comparatively larger than the available number of cancer samples, this leads to the risk of overfitting during the classifier's training phase [20].

3 Gene Expression-Based Cancer Cell Classification by Adopting the Proposed PPCA-Based DBN Classifier

The proposed PPCA-based DBN classifier for classifying the cancer cell based on the gene expression data is elaborated in this section. Figure 1 portrays the block diagram of the cancer cell classification employing the proposed PPCA-based DBN classifier. As a first step; the features are extracted from the provided input gene data. In general, the extracted features will be fed to the classifier for training; but here the dimensionality reduction phase is included for reducing the dimensionality of the

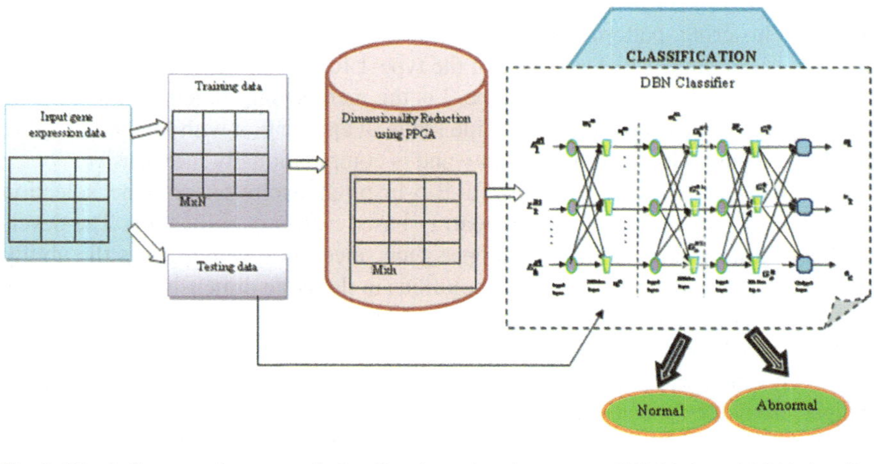

Fig. 1 Block diagram of cancer cell classification using the proposed PPCA-based DBN classifier

feature space. The methodology adopted for the dimensionality minimization is the PPCA. Now, the feature vector with the reduced dimensionality will be trained and classified using the DBN classifier. Eventually, the provided input gene expression data will be classified as either normal cell or abnormal cell conveying the absence or presence of cancer accordingly.

3.1 Input Gene Expression Data

Let the input gene expression data be indicated as and the gene data's dimension carrying the gene information be. The considered gene expression data comprises gene information from number of individuals. The millions of genes in a person together constitute the gene expression data; the mechanism of learning the gene characteristics supports in better understanding of human life. The classification of gene expression data must be done for the earlier diagnosis and prevention of diseases.

3.2 Dimensionality Reduction Using PPCA

The provided input data is subjected to feature extraction for the extraction of relevant features. As the input database is of large size, probably there will be a traffic hike in the feature space dimensionality. This further leads to the upsurge in the complexity during the classifier's training phase. Hence, PPCA [21] is employed in this work, for reducing the dimensionality of the feature space. The PPCA utilizes the minor algebraic manipulation for the dimensionality reduction; it is denoted as *B*. The dimensionality reduction is done by practicing the conditional probability on the time-space; it is expressed as shown in the below equation,

$$\langle r_u | s_u \rangle = P^{-1} B^T (s_u - \mu) \tag{1}$$

Here, the mean value is denoted as μ, the conditional probability over the time s_u is denoted as r_u, and the projection measure having the value of $\left(B^T B\right)^{-1}$ is indicated as P^{-1}. Eventually, the feature space dimensionality is reduced and is provided to the classifier. Let *S* be the selected genes obtained as a result of dimensionality reduction and is of size $M \times h$, such that $h < N$, and the database is represented as

$$S = \left\{S_{xy}\right\}\ 1 < x \le M; 1 < y \le h \tag{2}$$

3.3 DBN-Based Classification of Cancer

The next step after the dimensionality reduction is the classification of the provided feature vector, for the identification of cancer. The classification scheme adopted for the cancer cell classification is the DBN.

(i) DBN Architecture

The DBN [22] is a generative neural network comprising of many layers. Every layer is composed of the input layer represented by the visible neurons and the output layer represented by the hidden neurons. One of the extraordinary characteristics of the DBN is that there are no links among the visible neurons and no links among the hidden neurons. The connections are symmetric and the visible neurons are completely interlinked with the hidden ones. The employed DBN comprises Multi-Layer Perceptron (MLP) layer and multiple layers of Restricted Boltzmann Machines (RBM). The MLP layer comprises the input layers, hidden layers, and output layers; and the RBM's hidden layer acts as the output layer of the successive layers. Here, the DBN architecture is designed by considering two RBMs and an MLP layer, and the architecture is sketched in Fig. 2. The DBN is provided with the feature vector with reduced dimension from the PPCA for the effective cancer cell identification. Because of the dimensionality reduction through PPCA, the developed classification scheme achieves maximum classification accuracy.

The input layer and hidden layer of the RBM1 is given as

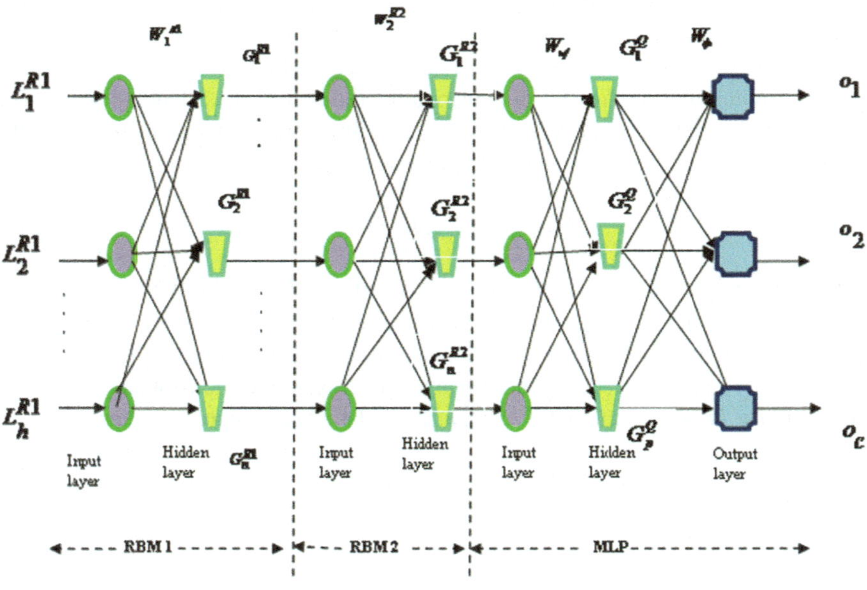

Fig. 2 DBN architecture

$$L_1^{R1} = \left\{L_1^{R1}, L_2^{R1}, \ldots, L_j^{R1}, \ldots, L_h^{R1}\right\} = \left\{S_{xy}\right\}; 1 < j \leq h \quad (3)$$

$$G_1^{R1} = \left\{G_1^{R1}, G_2^{R1}, \ldots, G_\phi^{R1}, \ldots, G_n^{R1}\right\}; 1 < \phi \leq n \quad (4)$$

where the number of input neurons in RBM1 of DBN is denoted as h and the number of hidden neurons in RBM1 of DBM is denoted as n. The RBM1's weight is denoted as $W_1^{R1} = \left\{W_{j\phi}^{R1}\right\}$ and the RBM1's weight dimension is indicated as $h \times n$. The output of RBM1 layer is given by the below equation,

$$G_\phi^{R1} = \varepsilon\left[\psi_\phi^{R1} + \sum_j L_j^{R1} \times W_{j\phi}^{R1}\right] \quad (5)$$

Here, ε is the activation function, the bias of the ϕth hidden layer of RBM1 is denoted as ψ_ϕ^{R1}, and the jth neuron in the RBM1 layer is indicated as L_j^{R1}.

Likewise, RBM's input layer and hidden layer are represented as

$$L_1^{R2} = \left\{L_1^{R2}, L_2^{R2}, \ldots, L_a^{R2}\right\} = \left\{G_\phi^{R1}\right\}; 1 < \phi \leq n \quad (6)$$

$$G_1^{R2} = \left\{G_1^{R2}, G_2^{R2}, \ldots, G_\phi^{R2}, \ldots, G_n^{R2}\right\} \quad (7)$$

The total number of hidden neurons in the RBM1 layer is same as the total number of input neurons in the RBM2 layer. The RBM2 layer's weight is indicated as $W_2^{R2} = \left\{W_{\phi\phi}^{R2}\right\}$. The existing neuron weight between the RBM1's ϕth visible neuron and the RBM2's ϕth hidden neuron is denoted as $W_{\phi\phi}^{R2}$. The RBM2 layer's output is expressed by the below equation

$$G_\phi^{R2} = \varepsilon\left[\psi_\phi^{R2} + \sum_j L_j^{R2} \times W_{\phi\phi}^{R2}\right] \forall L_j^{R2} \approx G_\phi^{R1} \quad (8)$$

Now, the output from the RBM2's hidden layer is fed as the input to the MLP layer of the DBN classifier. The MLP's input layer and hidden layer is represented as shown in Eqs. (9) and (10), respectively.

$$Q = \left\{Q_1, Q_2, \ldots, Q_\phi, \ldots, Q_n\right\} = \left\{G_\phi^{R2}\right\}; 1 \leq \phi \leq n \quad (9)$$

$$G^Q = \left\{G_1^Q, G_2^Q, \ldots, G_u^Q, \ldots, G_p^Q\right\}; 1 \leq u \leq p \quad (10)$$

The MLP layer's output is represented as

$$o = \left\{o_1, o_2, \ldots, o_f, \ldots, o_c\right\} \quad (11)$$

Here, the total count of output is denoted as c and the fth output neuron's output is denoted as o_f. The MLP layer's output is expressed as shown below

$$o_f = \sum_{u=1}^{p} W_{uf} * G_u^Q; 1 \leq u \leq p; 1 < f \leq c \tag{12}$$

Here, the weight between the fth output neuron and uth hidden neuron is denoted as W_{uf} and the dimension of which is given by $(p \times c)$. The output at the hidden layer is expressed as shown in Eq. (13)

$$G_u^Q = \left[\sum_{\phi=1}^{b} W_{\phi u} * Q_\phi \right] K_u \forall Q_\phi = G_\phi^{R2} \tag{13}$$

Here, the hidden neuron's bias is denoted as K_u and the weight between the ϕth input neuron and the uth hidden neuron is denoted as $W_{\phi u}$ and its dimension is given by $(b \times p)$.

(ii) Training Phase

The training phase in the MLP layer of DBN is carried out by practicing the supervised steepest descent algorithm. The RBMs are trained by utilizing the unsupervised learning of steepest descent algorithm. The series of steps followed during the training phase of DBN are explained below:

Step 1 Training of RBM1 layer and RBM2 layer
The first step is the training of the RBM layer. The DBN classifier employed in this research comprises two RBMlayers. The training sample with the selected genes of reduced dimension is provided as the input to the RBM1 layer for estimating the data's probability distribution. Then, the computed probability distribution is encoded with the weights and the determined output is provided as the input to the RBM2 layer. The similar procedure is carried out by the RBM2 layer and the obtained output is fed to the MLP layer.

Step 2: Training Phase of MLP layer
The steps followed in the DBN's MLP layer training are given below

(b) **Read the input sample**: The output from the RBM2 layer indicated as G_ϕ^{R2} is provided as the input to MLP layer.
(c) **Estimation of MLP layer's output**: The MLP layer's outputs are given by G_u and o_f and the outputs are dependent on the steepest descent algorithm's weight update equation.

(a) **Initialization**: The first step is the random initialization of neuron weights of the MLP layer. Let the visible layer weight be indicated as W_{uf} and the hidden layer weight be indicated as $W_{\phi u}$.

Step 3: Error computation
The network's error calculated on the basis of the average Mean Square Error (MSE) is expressed using the below equation

$$E_{\text{Average}} = \frac{1}{h} \times \sum_{j=i}^{h} \left(o_f^j - o_{\text{grd}}^j \right); 1 \le f \le c \tag{14}$$

Here, the total number of data samples is denoted as h, the output of the DBN classifier is denoted as o_f^j, and the desired output of the DBN classifier is indicated as o_{grd}^j.

Step 4: Updating the weights in the MLP layer By concerning the partial derivative of the average error E_{Average}, the weights of the visible neurons and the hidden neurons are updated. The incremental weights are determined by utilizing the derivatives of Eq. (14) are given in the below equations

$$\Delta W_{\phi u}^{SD} = -\xi \times \frac{\alpha E_{Average}}{\alpha W_{\phi u}} \tag{15}$$

$$\Delta W_{\phi j}^{\text{SD}} = -\xi \times \frac{\alpha E_{\text{Average}}}{\alpha W_{uj}} \tag{16}$$

Here, the learning rate is denoted as ξ. The steepest descent algorithm's weight update equation is expressed as

$$W_{\phi u}^{\text{SD}}(T + 1) = W_{\phi u}^{\text{SD}}(T) + \Delta W_{\phi u}^{\text{SD}} \tag{17}$$

Step 5: Re-computation of the average error of MLP outputs The average error of the MLP output is determined by utilizing Eq. (14). If the obtained error value is high, then the above steps are repeated. If the obtained error value is low, then the neuron weight is updated using Eq. (17), which is the weight update equation of the steepest descent algorithm.

Step 6: Termination The above steps are iteratively repeated for a maximum number of iterations, denoted as T. Eventually, the optimal neuron weights are found and on the basis of the obtained best solution, either the absence or presence of cancer is identified. The decisions are made by the genes, which vary from the normal behavior of genes.

4 Results and Discussion

The experimentation results of the proposed PPCA-based DBN classifier for the cancer cell classification are demonstrated in this section.

4.1 Experimental Setup

The experimentation is done in a PC configured with Intel core-i3 processor, Windows 10 OS, and 4 GB RAM. The implementation tool adopted in this research for carrying out the experiment is JAVA.

4.2 Dataset Description

The Standard databases, like Colon dataset [23] and Leukemia dataset [24], are utilized in this research for performing the experimentation. The Leukemia dataset contains gene expression measurements on 72 leukemia patients. The Leukemia dataset comprises 7128 genes; it is stored as a matrix of dimension 7128×72 (10 MB) with the column names representing the class labels. There is also a smaller subset of these data, consisting of 3571 genes, which is stored as the 3571×72 matrix (5 MB), with the column names representing the class labels. The colon dataset designed for studying colon tumors has 6500 genes comprising of 22 samples of normal colon tissue and 40 samples of colon tumor.

4.3 Evaluation Metrics

The evaluation metrics concerned for the analysis of the devised PPCA-based DBN are sensitivity, specificity, and accuracy; these metrics are defined in this part as follows:

Sensitivity: The sensitivity metric estimates the true positives attained by the developed classifier. It is formulated as

$$\text{Sensitivity} = \frac{\text{TP}}{\text{TP} + \text{FN}} \tag{18}$$

where the true positive is denoted as TP, the true negative is indicated as TN, the false positive is denoted as FP, and the false negative is indicated as FN.

Specificity: The specificity metric computes the true negatives attained by the devised PPCA-based DBN classifier, and is expressed as

$$\text{Specificity} = \frac{\text{TN}}{\text{TN} + \text{FP}} \tag{19}$$

Accuracy: The accuracy is defined as the percentage of accurate classification for the identification of the presence/absence of cancer. It is computed as

$$\text{Accuracy} = \frac{\text{TN} + \text{TP}}{\text{TN} + \text{TP} + \text{FN} + \text{FP}} \tag{20}$$

4.4 Methods Taken for the Comparison

The different classifiers chosen here for the comparative performance analysis are NN [9], DBN [22], PPCA + NN [21, 9]. The performance of these comparative techniques will be compared with that of the proposed method to evaluate its effectiveness.

4.5 Comparative Analysis

The results of comparative analysis of the devised PPCA-based DBN classifier with the comparative models are demonstrated in this section by varying the percentage of training data. When the 50% of data is used for training, remaining 50% of data is used for testing. Similarly, for 60, 70, 80, and 90% of the training data, the testing data is 40, 30, 20, and 10%.

(i) Using Colon dataset

The comparative analysis results of comparative classifiers on the colon dataset are described here.

The comparative analysis based on varying percentage of training data (50–90%) on the colon dataset is depicted in Fig. 3. Figure 3.a displays the comparative analysis of the distinct classifiers based on sensitivity. At 90% training data, sensitivity measure attained by the NN classifier is 0.7384, the DBN classifier is 0.74, the PPCA + NN classifier is 0.7977, and the PPCA + DBN classifier is 0.7984. Figure 3.b sketches the comparative analysis of the distinct classifiers based on specificity. For 90% training data, the specificity values of the NN classifier, DBN classifier, PPCA + NN classifier, and PPCA + DBN classifier are 0.5416, 0.6521, 0.6994, and 0.7672, respectively. The comparative analysis of the distinct classifiers based on accuracy is pictorized in Fig. 3.c. The accuracy measure attained at 90% training data by the NN classifier is 0.5416, the DBN classifier is 0.6521, the PPCA + NN classifier is 0.6994, and the PPCA + DBN classifier is 0.7672. The obtained maximum accuracy, high specificity, and sensitivity deliberate the effectiveness of the proposed PPCA + DBN classifier.

Figure 4 depicts the comparative analysis based on varying number of hidden neurons on the colon dataset. The comparative analysis of the different classifiers based on sensitivity is displayed in Fig. 4.a. The sensitivity values for the number of hidden neurons = 500 are 0.5803 for NN classifier, 0.6782 for DBN classifier, 0.7973 for PPCA + NN classifier, and 0.7981 for PPCA + DBN classifier. Figure 4.b

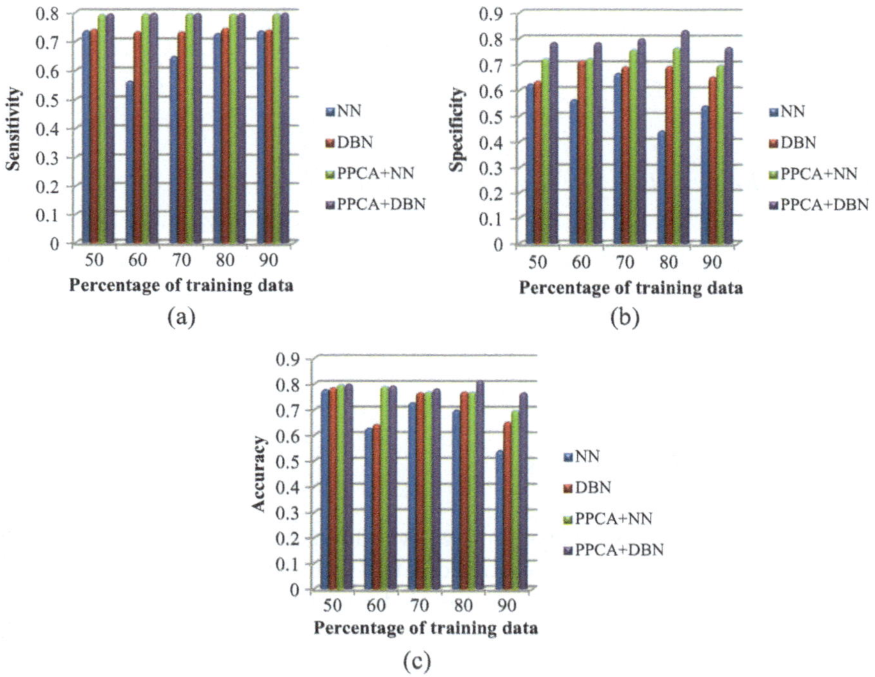

Fig. 3 Comparative analysis using colon dataset based on training data percentage **a** sensitivity, **b** specificity, and **c** accuracy

portrays the comparative analysis of the different classifiers based on specificity. As the number of hidden neurons = 500, the achieved specificity values are 0.7058 for NN classifier, 0.7328 for DBN classifier, 0.7487 for PPCA + NN classifier, and 0.8333 for PPCA + DBN classifier. The comparative analysis of the distinct classifiers based on accuracy is sketched in Fig. 4.c. The accuracy measures attained when the number of hidden neurons = 500 are 0.6714 for NN classifier, 0.7285 for DBN classifier, 0.7686 for PPCA + NN classifier, and 0.8085 for PPCA + DBN classifier. The superiority of the proposed PPCA + DBN classifier over the other classifiers is conveyed through the obtained values.

4.5.1 Using Leukemia Dataset

The comparative analysis results of the distinct classifiers on the Leukemia dataset are elaborated in this part.

The comparative analysis based on varying percentages of training data on the Leukemia dataset is depicted in Fig. 5. Figure 5a displays the comparative analysis of the distinct classifiers based on sensitivity. For 90% training data, sensitivity measure attained by the NN classifier is 0.3826, the DBN classifier is 0.6, the PPCA + NN

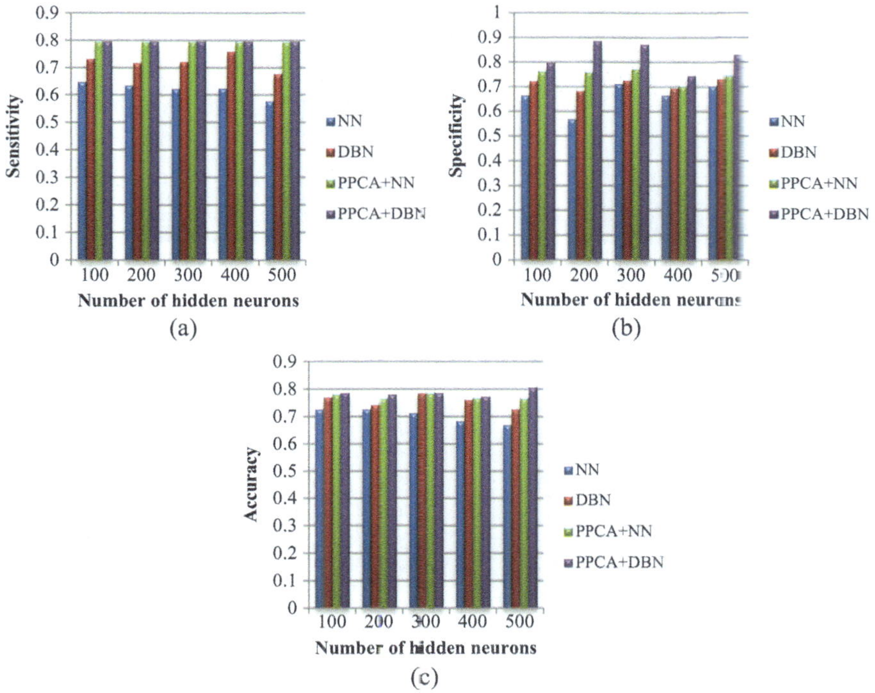

Fig. 4 Comparative analysis using colon dataset based on the number of hidden neurons **a** sensitivity, **b** specificity, and **c** accuracy

classifier is 0.7946, and the PPCA + DBN classifier is 0.7951. Figure 5b sketches the comparative analysis of the distinct classifier based on specificity. For 90% training data, the specificity value of the NN classifier, DBN classifier, PPCA − NN classifier, and PPCA + DBN classifier is 0.6395, 0.7246, 0.7374, and 0.75, respectively. The comparative analysis of the distinct classifier based on accuracy is pictorized in Fig. 5c. The accuracy measures attained at 90% training data by the NN classifier is 0.4438, the DBN classifier is 0.6286, the PPCA + NN classifier is 0.6534, and the PPCA + DBN classifier is 0.7057. The determined accuracy, specificity, and sensitivity elucidate the effectiveness of the proposed PPCA + DBN classifier over the other classifiers.

Figure 6 depicts the comparative analysis based on varying number of hidden neurons on the Leukemia dataset. The comparative analysis of the different classifiers based on sensitivity is displayed in Fig. 6a. The sensitivity values at number of hidden neurons = 500 are 0.3826 for NN classifier, 0.6 for DBN classifier, 0.7926 for PPCA + NN classifier, and 0.7938 for PPCA + DBN classifier. Figure 6b portrays the comparative analysis of the different classifiers based on specificity. At the number of hidden neurons = 500, the achieved specificity values are 0.6309 for NN classifier, 0.7355 for DBN classifier, 0.75 for PPCA + NN classifier, and 0.8550 for PPCA +

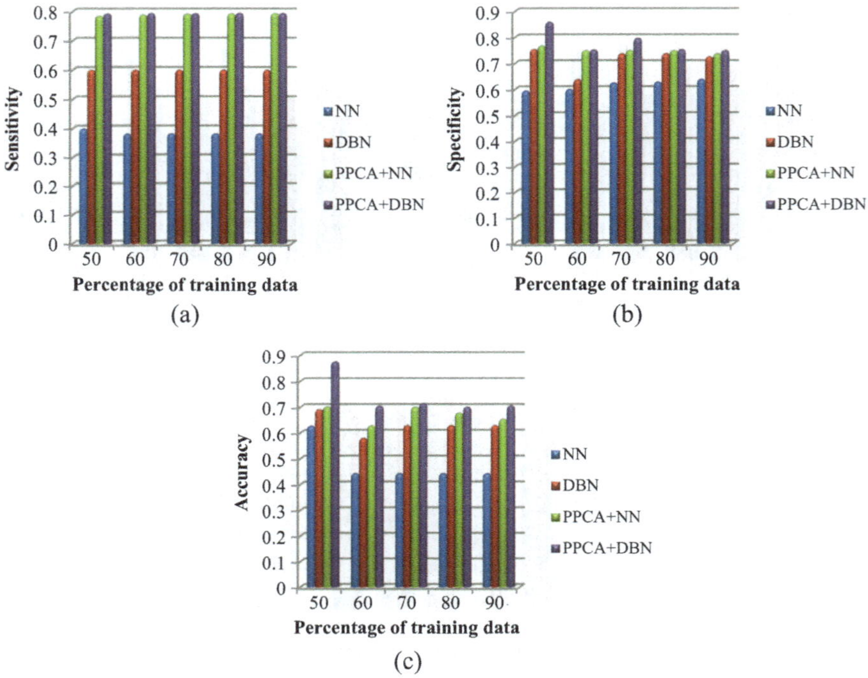

Fig. 5 Comparative analysis using leukemia dataset using based on training data percentage **a** sensitivity, **b** specificity, and **c** accuracy

DBN classifier. The comparative analysis of the distinct classifiers based on accuracy is sketched in Fig. 6c. The accuracy measures attained when the number of hidden neurons = 500 for NN classifier, DBN classifier, PPCA + NN classifier, and PPCA + DBN classifier are 0.4438, 0.6283, 0.6978, and 0.7643, respectively. The excellent performance of the proposed PPCA + DBN classifier over the other classifiers is conveyed through the determined values.

5 Conclusion

In this paper, an effective cancer cell classification methodology, entitled PPCA-based DBN, is developed for the classification of cells. Initially, for the gene expression data provided as the input, the dimensionality of the feature space is reduced by adopting the PPCA. Then, the dimensionality reduced features are fed to the DBN classifier for the classification of the data into normal and abnormal. Eventually, the presence or absence of the cancer is discovered by the proposed PPCA-based DBN classifier. The experimentation of the proposed PPCA-based DBN is evaluated by utilizing the metrics like sensitivity, specificity, and accuracy. The simulation results

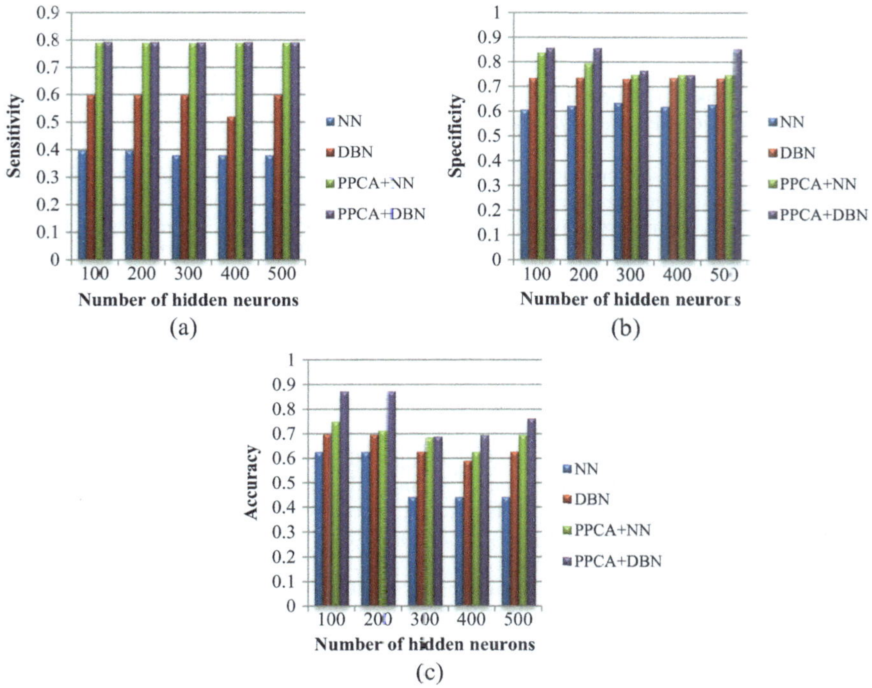

Fig. 6 Comparative analysis using leukemia dataset based on number of hidden neurons **a** sensitivity, **b** specificity, and **c** accuracy

based on the number of hidden neurons deliberated that the proposed PPCA-based DBN outperformed the other classifiers by achieving the sensitivity value of 0.7981, specificity value of 0.8333, and accuracy measure of 0.8085.

References

1. Z. Liu, D. Tang, Y. Cai, R. Wang, F. Chen, A hybrid method based on ensemble WELM for handling multi-class imbalance in cancer microarray data. Neurocomputing **266**, 641–650 (2017)
2. J.J. Liu, W.S. Cai, X.G. Shao, Cancer classification based on microarray gene expression data using a principal component accumulation method. Sci. China Chem. **54**(5), 802–811 (2011)
3. Patricia Melin, Oscar Castillo, A review on type-2 fuzzy logic applications in clustering, classification and pattern recognition. Appl. Soft Comput. **21**, 568–577 (2014)
4. T.R. Golub, D.K. Slonim, P. Tamayo, C. Huard, M. Gaasenbeek, J.P. Mesirov, H. Coller et al., Molecular classification of cancer: class discovery and class prediction by gene expression monitoring. Science **286**(5439), 531–537 (1999)
5. Z. Zainuddin, P. Ong, Reliable multiclass cancer classification of microarray gene expression profiles using an improved wavelet neural network. Expert Syst. Appl. **38**(11), 13711–13722 (2011)

6. V. García, J.S. Sánchez, Mapping microarray gene expression data into dissimilarity spaces for tumor classification. Inf. Sci. **294**, 362–375 (2015)
7. R.M. Luque-Baena, D. Urda, J.L. Subirats, L. Franco, J.M. Jerez, Analysis of cancer microarray data using constructive neural networks and genetic algorithms, in *Proceedings of the International Work-Conference on Bioinformatics and Biomedical Engineering (IWBBIO)* (2013), pp. 55–63
8. A. Antoniadis, S. Lambert-Lacroix, F. Leblanc, Effective dimension reduction methods for tumor classification using gene expression data. Bioinformatics **19**(5), 563–570 (2003)
9. B.A. Garro, K. Rodríguez, R.A. Vázquez, Classification of DNA microarrays using artificial neural networks and ABC algorithm. Appl. Soft Comput. **38**, 548–560 (2016)
10. C. Sotiriou, M.J. Piccart, Taking gene-expression profiling to the clinic: when will molecular signatures become relevant to patient care? Nat. Rev. Cancer **7**(7), 545 (2007)
11. A. Abderrahim, E.-G. Talbi, M. Khaled, Hybridization of genetic and quantum algorithm for gene selection and classification of microarray data. Int. J. Found. Comput. Sci. **23**(02), 431–444 (2012)
12. E. Alba, J. Garcia-Nieto, L. Jourdan, E.-G. Talbi, Gene selection in cancer classification using PSO/SVM and GA/SVM hybrid algorithms, in *The Proceedings of the IEEE Congress on Evolutionary Computation (CEC 2007)* (2007), pp. 284–290
13. C.J. Alonso-González, Q.I. Moro-Sancho, A. Simon-Hurtado, R. Varela-Arrabal, Microarray gene expression classification with few genes: criteria to combine attribute selection and classification methods. Expert. Syst. Appl. **39**(8), 7270–7280 (2012)
14. M.J. Van De Vijver, Y.D. He, L.J. Van't Veer, H. Dai, A.A.M. Hart, D.W. Voskuil, G.J. Schreiber et al., A gene-expression signature as a predictor of survival in breast cancer. N. Engl. J. Medicine. Eng. J. Med. **347** (25), 1999–2009 (2002)
15. H.M. Alshamlan, G.H. Badr, Y.A. Alohali, The performance of bio-inspired evolutionary gene selection methods for cancer classification using microarray dataset. Int. J. Biosci., Biochem. Bioinform. **4**(3), 166 (2014)
16. M. Dashtban, M. Balafar, Gene selection for microarray cancer classification using a new evolutionary method employing artificial intelligence concepts. Genomics. **109**(2), 91–107 (2017)
17. T.K. Karakach, R.M. Flight, S.E. Douglas, P.D. Wentzell, An introduction to DNA microarrays for gene expression analysis. Chemometr. Intell. Lab. Syst. **104**(2), 28–52 (2010)
18. S. Ghorai, A. Mukherjee, S. Sengupta, P.K. Dutta, Multicategory cancer classification from gene expression data by multiclass NPPC ensemble, in *The Proceedings of the IEEE International Conference on Systems in Medicine and Biology (ICSMB)* (2010), pp. 41–48
19. H.M. Alshamlan, G. H. Badr, Y. Alohali, A study of cancer microarray gene expression profile: objectives and approaches, in *Proceedings of the World Congress on Engineering*, vol 2 (2013), pp. 1–6
20. H. Hijazi, C. Chan, A classification framework applied to cancer gene expression profiles. J. Healthc. Eng. **4**(2), 255–283 (2013)
21. M.E. Tipping, C.M. Bishop, Probabilistic principal component analysis. J. Roy. Stat. Soc. Series B (Statistical Methodology) **61**(3), 611–622 (1999)
22. M.A. Salama, A.E. Hassanien, A.A. Fahmy, Deep belief network for clustering and classification of a continuous data, in *Proceedings of the The 10th IEEE International Symposium on Signal Processing and Information Technology* (2010), pp. 473–477
23. U. Alon, N. Barkai, D.A. Notterman, K. Gish, S. Ybarra, D. Mack, A.J. Levine, Broad patterns of gene expression revealed by clustering of tumor and normal colon tissues probed by oligonucleotide arrays. PNAS. **96**(12), 6745–6750 (June 8, 1999)
24. Leukemia data. https://web.stanford.edu/~hastie/CASI_files/DATA/leukemia.html. Accessed on 07 Nov 2017

Blockchain Technology for Data Sharing in Decentralized Storage System

D. Praveena Anjelin and S. Ganesh Kumar

Abstract In Cloud architecture, the storage systems can be centralized and decentralized environments. Centralized storage systems stored data in private cloud, maintained in a single location, can be accessed by one or more user, whereas in decentralized storage system, the data is stored on more than one or multiple servers. Users or companies are being a part of the decentralized cloud storage cloud to host the servers. The data may be stored on any server, i.e., no dedicated server for data storage and which can be accessed easily. In decentralized storage system, the files are stored and protected with the help of blockchain technology. Blockchain is nothing but a chain of blocks (computers or servers), which are connected by using cryptography. Each block contains a set of transactions which has been encrypted and shared in a secured way among multiple servers. Blockchain technology ensures confidentiality and integrity, which can be implemented by using peer-to-peer network. In this paper, we discussed how the technology improves security used by different applications in distributed environment. Further, we proposed adaptive encryption algorithm to improve security and access control.

Keywords Adaptive encryption · Cryptography · Decentralized storage · Blockchain

1 Introduction

In the past few years, most of the organizations have shown concern on outsourcing, it outsources the data and as well as gives functional services to clouds. Actually, the classical cloud storage is a centralized storage system acted as a trusted third-party storage provider to store and transfer the data. This system consumes high operational

D. Praveena Anjelin (✉) · S. Ganesh Kumar
Department of Computer Science and Engineering, SRM Institute of Science and Technology, SRM University, Kattankulathur, Chennai 603203, India
e-mail: pd6231@srmist.edu.in

S. Ganesh Kumar
e-mail: ganeshk1@srmist.edu.in

S. S. Dash et al. (eds.), *Intelligent Computing and Applications*,
Advances in Intelligent Systems and Computing 1172,
https://doi.org/10.1007/978-981-15-5566-4_32

cost, less security, poor performance, and lack of availability to transfer. To overcome the drawbacks of centralized cloud storage, a decentralized storage system (DSS) has been introduced with many features. Blockchain-based cloud storage solutions allow users to store data safely and give access to all participants in digital activities with most effect. Filecoin, storj, ppio, ochain, opacity, bittorrent, neo, and Dfinity are the various platforms of decentralized storage solutions.

Instead of central server, [1] InterPlanetary File System (IPFS) is a distributed file system which stores data on large number of computers that are connected by peer-to-peer networks. IPFS can communicate through Transmission Control Protocol (TCP) which has certain rules for data storage and transmission among connected nodes in the network. Node addressing system identifies nodes and content addressing system constructs the index of the content. Both node address and content address are stored in InterPlanetary Naming System. Files stored on IPFS can be accessed and shared by the users connected to an ipfs node. Files are split into small blocks, those blocks are hashed and distributed to storage nodes in the network. In order to consider privacy and data availability, DSS is better than centralized cloud storage systems, in which operational cost is much lesser than centralized storage. Existing decentralized storage systems supports easy implementation of encryption and storage of data, but the difficulty of data sharing is done secretly. Hence, we proposed a distributed framework to control access and search data using Ethereum Block Chain blockchain technology. Further to ensure security, we used ELGAMAL for encryption which protects the data in cloud storage.

The rest of the paper is sectioned as follows: Sect. 2 consists of literature review of existing technologies that have been used for data sharing in DSS. Section 3 presents reviews storage solutions of blockchain technology. Section 4, elaborate the user access control policy in Blockchain technology. Section 5 presents comparative analysis of various security algorithms in blockchain with DSS. Section 6 presents review of application areas of blockchain. Section 7 concludes the chapter.

2 Literature Review

Decentralized Cloud Storage In Decentralized Cloud storage system, the data is stored on multiple computers or servers connected by P2P network. These servers are hosted by a general user or an organization which can contribute to this decentralized cloud. The files are encrypted by cryptography and protected with the help of blockchain technology. The participants in this system can also earn money in cryptocurrency by sharing their unused space. Each computer in this network stores encrypted form of user data, the user is the only authenticated person to access and manage the files through their own public or private key. There are so many advantages in this technology,

- No dedicated servers for data storage.
- Peer-to-peer system network.

- Always availability of data.
- User can earn extra money by allowing free space to share.
- Since files are encrypted, it achieves higher rate of data security.
- Operational cost is lesser.
- Load balancing is very flexible.
- It requires less computing power and bandwidth.

Existing Technologies in DSS Distributed storage systems have huge volume of structured data spread over many nodes in the network, which provide highly obtainable service with no distinct end of failure. Some organizations provide distributed storage systems such as Sia, Storj, Maidsafe, and Ethereum, which are based on blockchain technology and a peer-to-peer architecture.

SIA: Sia is a decentralized peer-to-peer cloud storage scheme with blockchain technology. When the user chooses to store data on Sia blockchain. Sia first splits whole data into blocks, encrypts and distributes the data into various nodes connected in the network. The user can retrieve data by making request with a private key. User can get paid in Siacoin by renting out the extra space on their PC to the Sia decentralized network. Similarly, users who wish to use storage space need to pay Siacoins to a host. Sia does this by creating a marketplace for hosts and users via the token economics of Siacoin. Siacoin currency is used to execute a file storage contracts on the Sia blockchain and aim to reduce the cost of cloud storage. A host in this network puts a file contract agreement for keeping files, space required, and to invest Siacoins as collateral. A host can be punished if he does not serve to the user who made the data request. Once the agreement is over, the sum will be credited to the host in Siacoin account by the file contractor.

Storj: This storage platform is built on the Ethereum network best suite for cryptocurrency, store data in decentralized in asecure manner. [2] The user can upload the file with their own private key in order to ensure authenticate person and file is encrypted before sharing in the network. It separated the files from the user and stored in decentralized storage model. When user wants to access the file, he has to make a request, then storj uses distributed hash tables to locate all the shards and piece them together. Storj rents its own network to many users and puts charges for those who had used the network. This technology creates revolution in file sharing due to availability of data among connected nodes. Storj has partnerships with Microsoft Azure and Heroku to deploy its own development tools which creates great initiative for the open-source developer ecosystem. User authentication with direct payment is a major success in storj.

Swarm: Swarm is another decentralized storage system built on Ethereum network. It is one of the best content distributor and provides redundant store for web application. It provides various base layer services for web3 such as media streaming, none-to-node messaging, database services, etc. Actually the content is hosted on peer-to-peer network instead of individual computers or servers. This network is responsible for user authentication and allows them to access resources with payment. It directly integrated with Ethereum blockchain for service payments and data availability.

Cassandra: Cassandra is a distributed storage system used for storing and running a huge volume of structured data across various commodity servers. Cassandra supports dynamic control over data layout for clients with a simple data model [3]. The major distributed system techniques used in Cassandra are partitioning, scaling, replication, creating membership, and failure handling. To handle, either read or write requests, all these modules work under synchrony mode. Partitions the data dynamically, encodes it by hashing method, and assigns it to all nodes in the network. This distributed system uses replication algorithm to improve data availability and durability. Among multiple nodes, one can act as coordinator node which is responsible for keeping redundant data files whenever failure takes place in the network. It provides various replication polices such as Rack Aware, Rack Unaware, and Datacenter Aware to their clients stating that how and when data needs to be replicated. In Bootstrapping module, whenever a node wants to join in the group, first that node has to read its configuration file and membership identity card will be issued. The configuration file contains a list of linking points in it, with the help of this a node can join in the group and share files. It achieves high throughput, high performance, and providing better scalability.

Data Storage systems
Meta Product manufacturers always use their own private cloud to store data due to lack of security and privacy issues. Manufacturers can't trust another network and avoid data sharing [4]. Shrestha, A. K. and Vassileva, J proposed decentralized storage platform for Meta products, such as smartphones, wearable sensor devices, and smart cars. Meta Products must store group user data, later these data can be reused. It is a trusted system for sharing user information across various domains deployed by different organizations. It maintains distributed file account which holds the following information, who can contribute with sharing systems, what file may access, when that file retrieved, etc. This storage system has inbuilt trust mechanism, so it easily identifies the threat whenever a user tries to access an unwanted file. It supports personalized and content-oriented services to their clients.

Replication is a key technology of DSS [5] Yijie wang and Sijun Li proposed indirect replication algorithm. In the proposed model, the data is split into different data blocks, these blocks are encoded and distributed among several storage systems. Since data is redundant among data blocks, bipartite graph is used to encode the data blocks. The major advantage of using this algorithm is it provides security, durability, and availability of data. This model can be implemented with less operational cost and storage cost.

3 Blockchain Technology

Cloud storage is a centralized database system, in the extension of a centralized one, a decentralized storage framework is designed with blockchain technology. As we know that blockchain is a distributed database, a user can store any kind of information across different blocks (personal computer or server) connected in the network. Blocks are connected in chronological chain model, hence it is named as blockchain. It can be divided into three categories.

- Cryptocurrency bitcoin
- Smart contracts
- Application.

Cryptocurrency bitcoin: It is nothing but digital cash and as well as currency paid through online modes. It was established in the year 2008 with a group of people under the pseudonym Satoshi nakamoto, effectively solved the Byzantine (cryptographic research) problem. It is the most useful tool for value transfer underlying blockchain technology. The blockchain technology replicated all transactions done in the Bitcoin network. Bitcoin uses proof-of-work mechanism to prevent double-spending in the network; i.e., one who may spend same funds twice. The proof-of-work solved by miners in the bitcoin. Miners are the bitcoin nodes which checks its blockchain history and verifies all transaction made in the network. If anyone in the network tries to change the history, it takes more computational power in the network in order to verify. Actually proof-of-work mechanism is very expensive but, it is the only prevention method against Sybil attack. Sybil attack means, a node in network claims multiple fake identities and gain resources without spending currency. This attack highly influenced on following applications, voting systems, location-aware routing, data aggregation, and reputation evaluation.

Smart contracts: Smart contract is nothing but crypto contract, maintained by peer-to-peer network. In 1994, smart contracts idea was first proposed by Nick Szabo in 1994. He was interested in starting digital currency. Then later, in 2008, the cryptocurrency bitcoin was developed via a blockchain platform, this technology enabled smart contract. This tool can provide some coordination and assigned agreements among nodes who participate in the network. This tool can directly control digital payment, currency transfer, or sharing assets between nodes in the network. It enforces the contractor to follow all rules and obligations related to the agreement as the same way what traditional contract did by generating tokens. Blockchain managed the self-enforcing agreement which is embedded in some other computer code. The code contains smart contract agreement policy which parties need to interact; what file needs to be shared and cost details. The agreement is enforced automatically when the rules are met, i.e., it unlocks the access and manages tokenized assets. In technical aspects, the merit of the smart contract is self-verifying, self-executing, and tamper resistant. In economic aspects, decentralized ledger initiated transactions. So, low expenses and maintains higher transparency. In legal aspects, security is the major issue that needs to be resolved.

Blockchains: In general, blockchain requires a huge computational resource to run, since it has been available in public and accessed by many nodes in the network. So, the system is almost utilized its maximum potential but the user should have limitations to access. Sometimes an unauthorized user may access the network, he can download the file, i.e., confidentiality of data, secure sharing, and redundancy are major issues in blockchain. In distributed systems, private data must be shared by all nodes connected on the network. So, data may be completely exposed. In centralized storage systems, data can be stored securely with less transparency. This is one of the primary contradiction between centralized storage system and decentralized public blockchain. Based on access mechanism, blockchain can be classified as,

Public blockchain: A blockchain grants permissions to issue transactions and to read access for all data of the chain to all users.

Private blockchain: A blockchain is available for a predefined list of entities to access stored data and create a transaction on the chain. There are two types of blockchains.

Permissionless blockchain: This type of blockchain does not restrict the identities of transactions.

Permissioned blockchain: This type of blockchain accepts only the predefined entities performed transactions with known identifications [6, 7].

Both private and permissioned blockchains are not the same based on storage mechanism. For example, a permissioned blockchain can be available to the public to read the data from blockchain, but the general users might not have permission to create transactions. It has distributed ledgers to maintain storage details, access details and protocols can be used to provide security and increase the availability of data. But, in private blockchains may have full permission to access the blockchain by preconfigured entities. Moreover blockchain may be any type, must have multi-level permissions such as connection establishment, create storage block, send and receive transactions, read data from decentralized storage systems. All these technologies support security, data consistency, and easy access. The following table shows comparison among various blockchains (Table 1).

Blockchain-based storage solutions: Blockchain technology provides decentralized storage structure with a secure manner when compared to traditional cloud storage. There has been significant growth in technology with respect to storage, security, authentication, and access control. Whenever a node in network wants to upload a file, first that must be encrypted before sharing and also ensure availability

Table 1 Comparisons among public blockchain, consortium blockchain and private blockchain [6]

S. No	Property	Public blockchain	Consortium blockchain	Private blockchain
1	Consensus determination	All miners	Selected set of nodes	One organization
2	Read permission	Public	Could be public or restricted	Could be public or restricted
3	Immutability	Nearly impossible to tamper	Could be tampered	Could be tampered
4	Efficiency	Low	High	High
5	Centralized	No	Partial	Yes
6	Consensus process	Permissionless	Permissioned	Permissioned

of data [8]. Metadisk is an open-source software to provide decentralized storage system which is more secure and efficient. Its main objective to establish a constant platform for P2P cloud storage network (Storj). This application provides an interface for nontechnical users also. Using this API, users can upload and download files in a secure way. Client-side encryption has taken place on it. Users can generate private key and assign to files while uploading files on the network. Multiple nodes in a network may have the same file according to data availability.

Once the file has been encrypted, the SHA-256 hash serves a unique identity and way to detect file tampering. [9] In blockchain distributed storage, first data storage provides decentralized platform where, ana organization can register in order to access distributed data and further it can be shared. Anonymous access control component ensures the client access is provided by data owner itself. The data owner may not aware of who will be the data user and how many times the data consumer has to show the credential proof. Private keyword search mechanism helps a data user to recognize the encrypted information. The consumer can directly interact with blockchain node by giving a single keyword to get the fingerprint of specific data content. At last, data consumer retrieves encrypted documents from decentralized storage system. Keyword search cryptosystem consists of KeyGen, Trapdoor, Encrypt, and test algorithms. KeyGen generates private search key, encrypt algorithm produces ciphertext for keyword which are performed by the data owner. KeyDerieve algorithm is used by the owner to produce secret search key for data consumer. Trapdoor is run by data user for keyword. Test algorithm is executed by smart contracts on blockchain.

4 Access Control Mechanism in BlockChain Technology

Samarati and de Vimercati [10] defined Access Control as "The process of mediating every request to resources and data maintained by a system and determining whether the request should be granted or denied." The primary goal of access control mechanism is to verify and limit the actions that authorized users can perform within a computer system. Access control has set certain limitations to users, as what a user could perform directly with the system, what programs can be executed on behalf of the users who are permitted to execute those programs with approval An access control community should have an object, subject, operations, permissions, control list, and matrix.

Access control Models: The access control system has been developed based on security policy and security model. Security mechanism is a model which comprises of low-level implementation of both hardware and software. Access control models can be categorized as,

1. Mandatory Access Control
2. Discretionary Access Control
3. Role-based Access Control

4. Organization-based Access Control
5. Attribute-based Access Control
6. Centralized and Decentralized Access Control.

Laurent et al. [11] They proposed a new access control method. In that, a smart auditable contracts deployed in blockchain infrastructures which is a transparent one and provides controlled access to outsourced data, i.e., Malicious entities should not access data without the knowledge of the data owner. They would implement this proposed new solution on Ethereum blockchain network. The proposed access control mechanism has four entities which are defined as Data Storage Provide (DSP): It is not an active client having read access alone and governs distributed remote servers and hosts application services. Data Owner (DO): Ensures authorized entities and makes uses of resources provided by data storage provider in order to store and share data in network. Data retriever (DR): DR is a client of blockchain, he can access the content stored in remote servers. Blockchain Infrastructure (BC): It is a mediator, permits to authenticate DRs. It ensures authenticated access control with respect to efficient whitelist definitions and preserves privacy among entities. Unlinking properties can support one-time access control which ensures unlinkability between different access sessions by users.

Ronghua and Chen[12] A blockchain-enabled decentralized access control (BlendCAC), is suitable for decentralized storage in a secured manner. Its major aim is to provide effective access control mechanism to devices, services, and information especially in large scale IoTs (Internet of Things). It efficiently does process related to IoT and also provides granularity, scalability, and dynamicity of AC strategies for IoTs. The proposed system, IoT devices acts as a master, controls all the resources used in the network instead of centralized authority. The operations are carried out in Ac as listed below. Registration: Entities that participate in blockchain network must create an account with a pair of keys for authentication purpose. Smart Contract Deployment: Domain owner developed and deployed capability tokens on blockchain network, those tokens are managed by smart contract. Capability Propagation: Initially, an entity sends access rights request to the domain owner to get the capability token. Then the system checks for authorization, if it is found to be right entity, then capability token is issued by data owner. Capability token encodes the access rights, then initiates transactions in order to update token data in the smart contract. Authorization Validation: Service providers performed authorization validation on receiving service request from an entity.

Steichen et al. [13] IPFS uses Ethereum smart contract to provide access-controlled file sharing. The proposed mechanism allows users to do all financial transactions in its own cryptocurrency area and custom currencies called tokens. Access control system has some privileges regarding addition, modification, and removal of permissions that are recorded on blockchain. The access control contract contains four functions such as, AMIOwner, MIOwnerMultiple, CheckAccess, and CheckAccessMultiple. AdBlock function used to register an IPFS file chunk in control list. Cryptographic hashes identify IPFS file chunks. This hash value is used as a key stored along with file chunks. GrantAccess function accepts request given by data owner, check for hash value. If value exists, then it sends the key to owner. Remove access function used to remove the user whenever hash value mismatched or different being posted by data owner. AMIOwner: It accepts the request and checks the hash value of file chunks. If it has the same value what a data owner already registered, then it gives access rights to data owner to get access of file chunks of storage blockchain. Its values are mismatched, then it returns false message. DeleteBlock function reverts if the hash value is empty, soon it removes files from data storage since the file has not been requested so far from the owner's side. *f* hash value resides, it knows that file may be accessed and retrieved soon.

Thwin and Vasupongayya [14] Personal health record system (PHR system) stores individual health-related information. This system allows data owners to share and manage data among selected individuals. Information stored in database may be incorrect because of immutability, irreversibility properties, and originality of content may be inequitable. Fortunately, blockchain technology gives potential solutions to solve this issue. The proposed system supports tampering resistance feature which includes fine-grained access control, auditability, tamper resistance, and revocability of consent. To preserve privacy, proxy reencryption and cryptographic techniques are used. Proxy reencryption scheme is an asymmetric cryptosystem; enables its users to share their decryprion capabilities with others. The ciphertext is encrypted with user public key, that file can be decrypted by another user by using their own private key but the content cannot be retrieved fully during transactions. Since the encrypted file can be constructed in such a way, ensure secure data sharing scheme. The data owner must send reencryption key to proxy in order to share data. The reencryption key is generated with the combination of owners secret key and users public key. Moreover the proxy cannot read any information from original data with the help of reencryption key.

5 Privacy/Security in Blockchain

(i) **Existing Algorithm**:

Decentralized storage solutions create massive attraction among storage providers. Data can be stored in different nodes in the network and being shared with entire network. However, uploading, downloading, and sharing of files may not be secure in a distributed platform. Since, authentication and security can be a big challenge in distributed architecture. Existing decentralized architectures are constructed to support huge volume of data storage and shard, but, still there are no effective solutions for security issues. P2P network has been widely implemented over the distributed systems. We compared different blockchain technologies so far that had been implemented, security algorithms used, and their performance with evaluation metrics. The comparison of different blockchain platforms along with network performance has been shown in Table 2.

Table 2 Comparisons of blockchain technologies, security algorithm used

S. No	Platforms/applications	Business logic	Configuration mode	Encryption	Workload	Transaction data	evaluation metrics
1	Ethereum	Smart contract	Private network	Single public key	Random	Deployment time, completion time	Execution time, throughput
2	Hyperledger Fabric	Chaincode	Private network	Different private key	Random	Completion time	Latency and throughput
3	Smart home systems	Bitcoin	Private blockchain	Signcryption	Asynchronous	Completion time	Execution time
4	Peer-to-peer	Medibchain protocol (smart contract)	Private blockchain	Elliptic curve cryptography			
5	P2P network	Bitcoin	Private	Digital signature algorithm	Random	Transaction time	Latency

(ii) **Adaptive Encryption Algorithm**:

Adaptive encryption techniques can be applied on decentralized storage systems because of data confidentiality. This algorithm permits the cloud server to do a large set of SQL operations over encrypted data. The below listed encryption schemes are used in it.

- Deterministic
- Order Preserving encryption
- Random
- Search
- Plain
- Sum.

Each client can participate in DSS to get services by direct execution of SQL operations. This algorithm guarantees the same level of scalability and data availability. Initially, data is encrypted using key, then it is stored in cloud server. Whenever the client wants to get data, he has to decrypt with key value after downloading a file from the server. This algorithm is well-suited for encryption since it does not allow even a trusted party to manage encryption details. It also simplifies database server configuration because of operation automation.

6 Application Areas

Blockchain technology is very proficient and gives copious reimbursement such as decentralization, persistency, auditability, and anonymity. Cryptocurrency, financial services, Internet of Things, risk management are various sectors where blockchain technology can be applied.

Finance:

(i) **Financial Services**: Bitcoin is one of mostly widely used cryptocurrency method. Blockchain technology can be used for financial transactions, clearing and settlement of financial assets, and reducing cost and risk. Microsoft Azure, IBM are tremendous software companies offering blockchain services to the general public.

(ii) **P2P financial market**: Blockchain builds a P2P financial network to establish financial services among network in a secure and reliable way. Blockchain creates shared computation protocols to create a P2P financial shared computational market. This MPC market allows a P2P network to do computational task in offload mode.

(iii) **Enterprise transformation**: The traditional organization has completed all its transactions smoothly with help of blockchain technology. For example, the traditional postal operators (Pos) acts as an intermediator between customers and merchants. Pos make use of cryptocurrency techniques to extend both financial and nonfinancial services.

(iv) **Risk management**: Blockchain technology provides risk-management framework used to decide the investments plan, collaterals, and analyze investment risk also. Smart contract enables decentralized autonomous organizations to manage with all its business activity collaborations.

Internet of things (IoT): IoT is nothing but integration of objects (electronic devices and smart objects) with internet to provide different services to users. Some of the applications of IoTs as logistic management with Radio-Frequency Identification (RFID), smart homes, smart grids, e-health,maritime industry, etc. Blockchain with IoT potentially improves e-business model. This model has DAC, the people who trade with DAC to get transactions quickly, get coins, and the people can also exchange data with intermediary. Blockchain can improve privacy in IoT applications. Hardjono and Smith (2016) proposed a privacy-preserving method for IoT device into cloud ecosystem. This system ensures security by not allowing unauthorized access of devices. IBM ADEPT system developed to build a distributed platform, where devices are connected.

Public and Social Services: Blockchain can be widely used for public services such as patent management, income tax generation and monitoring system, digital signatures, marriage registrations, etc. Decentralized storage access provided to all users connected in network which reduces hard copy system and easy access to user.

(i) **Education**: Blockchain learning is proposed by Devine (2015) which achieves online educational marketing strategy. In this learning model, the teachers can segregate the learning materials into blocks; those blocks are packed and placed into the blockchain. Any user can access blocks on blockchain by paying coins through cryptocurrecy.

(ii) **Energy saving**: Gogerty and Zitoli proposed solarcoin. Solarcoin encourages the usage of renewable energies by rewarding digital currency to energy producers.

(iii) **Land Registration**: Blockchain technology highly contributes in public services and as a proof land registration is major application of it. The land information such as physical status, registered details, authorized person, landmark, cost details are publicized on blockchains. Whenever there is any changes in details, that are updated lively on blockchains to improve effectiveness of communal services consequently.

Reputation System: Reputation is nothing but trustworthy. Simply saying, an Individual's reputation can be evaluated based on the previous transaction done by him. Individuals overall interactions can be taken place within community. In e-commerce applications, to achieve high reputation, many service providers enrolled large number of fake customers. This creates problem for vendors also. This issue can be solved with the support of blockchain. Reputation is an important thing for an academician [15]. Blockchain-based distributed system is proposed to manage educational record and reputation. In this model, initially each institution and their professionally qualified workers will be awarded (educational reputation currency). An institution also awards the intellectual worker by transferring record status to them. All transactions are stored dynamically. So, changes in reputations can be detected easily. In web community, members can be reputed based on their qualities. Carboni proposed a reputation model based on blockchain, the service providers in web community may be reputed by getting feedback from customers. Reputation values are stored on blocks across a distributed network.

7 Conclusion

Decentralized storage system is created a revolution in storage area where the public can interact with internet. It provides distributed data storage, though there may issues on security, no safe on stored data, trustless environment. In this cyber world, no security, difficult to predict third party access, no privacy of sharing data. Blockchain is emerging technology which gives promising potential solutions for these issues. Data stored on this blockchain is more secure and ensures confidentiality by using cryptographic techniques. It provides high-quality services to all connected in network and set check constraints for third party access. The aim of this paper is to show a comprehensive view of blockchain technology and its merits. This paper includes study of existing technologies in DSS, storage solutions of DSS with blockchain technology, various access control mechanisms implemented for blockchain. In this paper, we have also discussed about security algorithms used in blockchain, and comparisons among various public blockchain used, consortium blockchain used and private blockchain used in network. Further we discussed about various application areas, where blockchain technology could be implemented.

References

1. J. Benet, IPFS—Content addressed, versioned, p2p file system (draft 3), (2014). https://ipfs.io/ipfs/QmR7GSQM93Cx5eAg6a6yRzNde1FQv7uL6X1o4k7zrJa3LX/ipfs.draft3.pdf
2. Storj—A Peer-to-Peer Cloud Storage Network. https://storj.io/storj.pdf
3. A. Lakshman, P. Malik, Cassandra—A Decentralized Structured Storage System, from Facebook Org
4. A.K. Shrestha, J. Vassileva, Towards decentralized data storage in general cloud platform for meta-products, in *Proceedings of the International Conference on Big Data and Advanced Wireless Technologies (BDAW '16)* (ACM, New York, NY, 2016)
5. Y. Wang, S. Li, Research and performance evaluation of data replication technology in distributed storage systems. Comput. Math. Appl.
6. Z. Beaven, D. Neilson, R. Osborne, P. Pacifico, Blockchain for creative industries music on the blockchain (2016)
7. Public versus Private Blockchains Part 1: Permissioned Blockchains (2015)
8. S. Wilkinson, J. Lowry, Metadisk: Blockchain-Based Decentralized File Storage Application. http://storj.io. August 20, 2014

9. H. Giang, W. Keong, Blockchain-based system for secure data storage with private keyword search, in *on 13th World Congress on Services* (IEEE, 2017)
10. P. Samarati, S. De Capitani, D. Vimercati, Access control: policies, models, and mechanisms (2000)
11. M. Laurent, N. Kaaniche, C. Le, M. Vander Plaetse, An access control scheme based on blockchain technology
12. E.B. Ronghua, G. Chen, BlendCAC: a Blockchain–Enabled Decentralized Capability-based Access Control for IoTs (April 2018)
13. M. Steichen et al., Blockchain-based, Decntralized Access Control for IPFS. IEEE conf. Internet of Things (2018)
14. T.T. Thwin, S.Vasupongayya, Blockchain-based access control model to preserve privacy for personal health record systems. Secur. Commun. Netw. Article ID 8315614, (2019)
15. Z. Zheng, H.-N Dai, Bloakchain challenges and opportunities: a survey. Article ID 101504, (Oct2018)

Understanding Concepts of Blockchain Technology for Building the DApps

P. Shamili, B. Muruganantham, and B. Sriman

Abstract Blockchains are the digital ledgers with tamper resistance and are implemented in the distributed networks with the consensus protocols which were introduced by Satoshi Nakamoto. A high-level secured technology for users both in the public and private networks. The peers in the network who share their transactions with the other peers are recorded and once recorded transactions cannot be modified by any of the peers in the blockchain technology. Here the digital signatures are used to provide the trusted third party to prevent the double-spending attacks. Due to the longest chain generated by the user by using the high computational power, there may be possibility of an attacker to access the whole network and this is prevented by the proof of work consensus protocol as forking concept. This paper provides a review of the blockchain network and helps the readers to understand the working of blockchain technology.

Keywords Blockchain · Consensus protocols · Smart contracts · Distributed network · Ethereum · Hyperledger composer

1 Introduction

The signed transactions that are arranged into the blocks and distributed and stored in the digital ledgers are named as the blockchains. After the consensus decision and the validation of the blocks linked to one another, they are linked through the

P. Shamili (✉) · B. Muruganantham · B. Sriman
Department of Computer Science and Engineering, SRM Institute of Science and Technology, Kattankulathur, Chennai, Tamil Nadu 603203, India
e-mail: shamilip@srmist.edu.in

B. Muruganantham
e-mail: muruganb@srmist.edu.in

B. Sriman
e-mail: srimanb@srmist.edu.in

S. S. Dash et al. (eds.), *Intelligent Computing and Applications*,
Advances in Intelligent Systems and Computing 1172,
https://doi.org/10.1007/978-981-15-5566-4_33

cryptographic value of the previous block [1]. The blocks created before will become more difficult to tamper, if the new blocks are added to it. The newly added blocks data will be copied to all the nodes in the blockchain network according to the predefined rules.

In the centralized system, the data can be altered or modified by the central authority. But in the blockchain network, none of the blocks could be modified (tamper resistant) once it is created, because of the distributed fashion [2]. In 2008, the combination of few of the technologies and the computation concepts came out to create the cryptocurrencies as Electronic cash and are protected through the puzzle solving mechanism.

The first modern cryptocurrency in the blockchain network was widely spread in the year 2009 namely the Bitcoin. In the distributed system, the transfer of information in a digital form represented as Electronic cash takes place in the bitcoin systems. The nodes in the bitcoin network can transfer and digitally sign their rights in that data to all the nodes and the blockchain record the transaction publicly [3]. So that all the nodes in the network are allowed to verify the validity of the transactions. These details were maintained independently in the distributed group of nodes. Along with this mechanism, the cryptographic values are made tamper proof.

The blockchain technology enabled many of the cryptocurrencies like and the Ethereum and Bitcoin. Due to this, the blockchain is seen as bound to the cryptocurrency solution. It is being investigated for many of the sectors and also available for many of the applications. It is a challenge for the investigators to understand the distributed systems and the cryptographic primitives.

2 Overview of Blockchain

The tamper-proof digital ledgers were implemented in the distributed fashion without the centralized authority which are the blockchains. Initially, it is to enable the nodes to record the transactions in the network. In this way no such transactions in the network could be modified. This idea was implemented in the concept of cryptocurrencies of electronic cash such as bitcoin.

Bitcoin was the first blockchain-based cryptocurrency. The information stored in the bitcoin blockchain represents the electronic cash and are attached to the digital address [4]. The nodes in the bitcoin blockchain use the digital signatures and transfer the data to the other nodes in the blockchain network. The data recorded are available transparently for all the nodes in the network to validate the transactions. Along with the cryptographic mechanism, the blockchain is made tamper proof for the nodes which try to modify or forge the data.

The blockchain creates the trust in the environment where all the nodes could not easily be identified. This trust is typically delivered to all the nodes in the blockchain network. The characteristics that are needed for the trust for blockchain technology is described in Fig. 1 as follows,

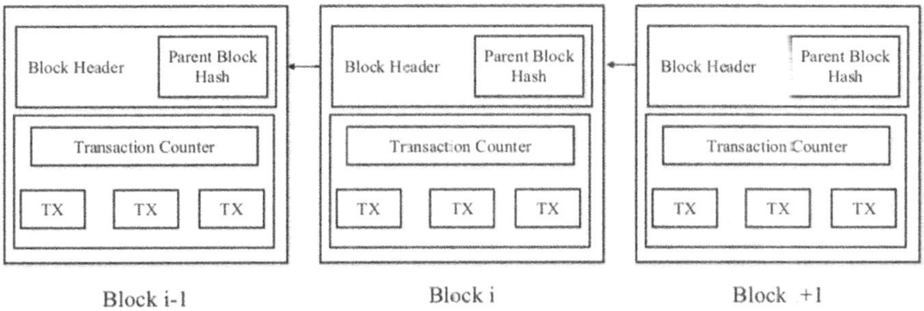

Fig. 1 Block chain architecture

Ledger—like the traditional ledgers for storing the transactions and the data, the blockchain uses the digital ledgers to store the transactions in the network that could not be overwritten.
Distributed—the blockchain is distributed. This allows the nodes in the blockchain to join the blockchain network and prevents from the attackers [5]. The possibility of involving the attacker may happen if the number of nodes in the blockchain increases.
Secure—the blockchain is secured by a cryptographical puzzle, so that the transactions are tamper proof.
Shared—the digital ledgers are shared among all the nodes in the blockchain network. It provides the nodes to transparently available for all the nodes.

3 Hash Functions

Cryptographic hash is the most important component in the blockchain technology [6]. The method for applying the cryptographic hash function which gives the relative outputs for the inputs given. The nodes are allowed to check the data, hash of that data, and finding the results of that data for proving that the data is immutable. If there is any change in the input, there will be different output digest. This is shown in Table 1.

Table 1 Hash function

	SHA-1	SHA-224	SHA-256	SHA-384	SHA-512
Message digest size	160	224	256	384	512
Message size	$<2^{64}$	$<2^{64}$	$<2^{64}$	$<2^{128}$	$<2^{128}$
Block size	512	512	512	1024	1024
Word size	32	32	32	64	64
Number of step/round	80	64	64	80	80

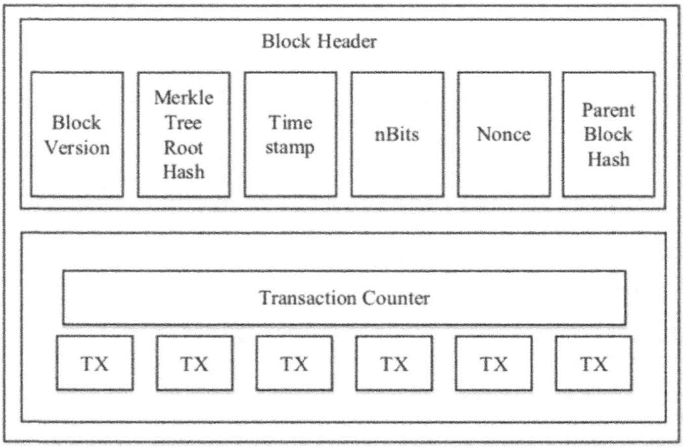

Fig. 2 Block header

The more specifically used algorithm for securing the data in the blockchain technology is the Secured Hash Function (SHA) which is of 256-bits [7]. Most of the computers nowadays use this algorithm in hardware for more high computation.

Few of the important security properties of the cryptographic hash functions are:

- **Preimage resistant**: it is also called the one-way function, where the input hashes cannot be found for the inputs.
- **Collision resistant**: it means that one could not find the two of the input hash to the same outputs.

A. **NONCE**

Nonce is cryptographically generated random number by using the algorithms. The different hash values were hash digest [8] which were produced by combining the data and the nonce as shown in Fig. 2.

$$\textbf{DIGEST} = \textbf{HASH}(\textbf{DATA} + \textbf{NONCE})$$

B. **Ledger**

The data about the transactions that are stored in a place is named as ledger. In the traditional ledger, the papers and pens were used for keeping the records of all types of transactions. By using the computers, the ledgers were updated to digital ledgers that are operated by centralized organizations. These were under centralized control. It is now updated to decentralized system, especially in blockchain.

C. **Block**

Each and every transaction in the blockchain network is stored in the digital ledger called the blocks in the blockchain network. The block consists of data about all the transactions that occurred in the distributed network. The copy of a single block data is shared with every other node in the network by the consensus protocols. The block contains the nonce value (hash value), the timestamp of every transaction.

D. **Block header**

The block header in the blockchain holds the following data,

a. Consists of the previous block hash value.
b. The hash value representation shows the root hash for that block.
c. The Timestamp.
d. Block size.
e. Value of nonce (generated by solving the mathematical crypto puzzle).

E. **Data in block**

Like block header, the block data consist of all the data or the transactions that occurred in a block in the blockchain network.

F. **Merkle tree**

The hash tree is formerly known as Merkle tree or Merkle root. For the data synchronization and the verification of the data structure is used. The tree data structure is holding the values of hash of non-leaf nodes and that represents the values of its child nodes. The leaf nodes in the Merkle tree maintains the same depth as possible. The hash functions are maintained for its integrity.

G. **Digital signatures**

The digital signatures are the keys for the security of data in the blockchain technology, where it works under the algorithm of SHA-256. In the blockchain network, the signed and verified transaction is shared among all the nodes in the network [9]. It has two phases, they are Verification Phase and Signing Phase. For example, two nodes in the blockchain network agree under the consensus protocol and send the data to each other. That data is signed and encrypted by the user who sends. This is the Signing phase. Then that data is being Verified by the user who receives it. This is the Verification phase. So that the data is highly secured and tamper proof.

H. **Forking**

In the blockchain technology, the changes and updating of data will become difficult for some time. In the permission-less blockchain networks, due to heavy competition and increase in the nodes agreed by the consensus protocols, it becomes much difficult. Few of the changes to the blockchain network and to the data are called Forking as shown in Fig. 3. It is divided into two types, they are soft forking and hard forking. It is compatible for the nodes to change backwards, the need not be updated in the soft forks. But in hard forking, it is not compatible for the nodes to go back and the blocks will not be updated and also be rejected. In the permissioned blockchain network, as all the nodes come under the consensus protocol and all nodes were known to each other, there will not be any chances for forking.

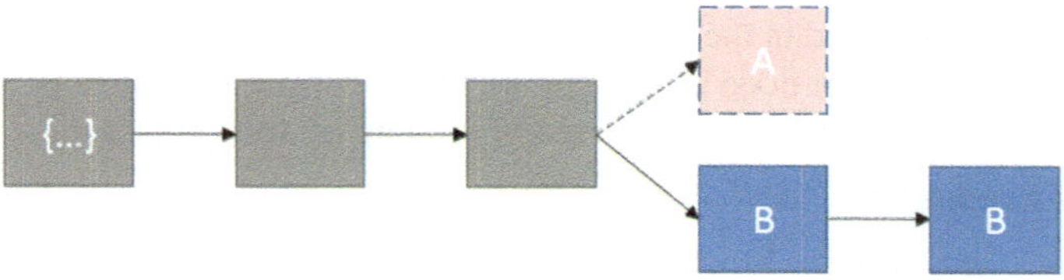

Fig. 3 Forking

4 Proof of Work Consensus Model

The Proof of Work consensus protocol is to solve the crypto puzzle by using all the computational powers by the nodes. This is a proof that the node has performed the work by solving the puzzle. It is designed in such a way that the nodes find it difficult to verify the block. It proposes the nodes to proceed to create a new block and the proposed block will be rejected if it is not satisfied.

A. **Crypto Puzzle**

The crypto puzzle is a method of solving a mathematical problem for creating a new block by verifying the previously created block's transaction. The puzzle is a hash value that is generated as the nonce (the block header) by the block. For each and every attempt tried for creating a block by the node to compute the hash value to that block header, it takes a minimum of 10 min for creating a block by solving the crypto puzzle. For increasing the difficulty, the target value may be modified.

B. **Computation Power**

In the proof of work consensus protocol, the node uses its computational power for solving the mathematical crypto puzzle to create a new block. As mentioned earlier, it takes a minimum of 10 min to solve the crypto puzzle by using the computational power. By using this method, the nodes in the network can influence others by having high computational powers. This type of attack is called the Sybil attack.

C. **Double-Spending Attack**

The digital assets are transacted more than once at the same set are known to be double spending. The group of nodes in the blockchain network do double spending for verifying a transaction to create a new block is called 51% attack. This attack can be prevented by using the timestamp for all the nodes in the bitcoin network.

5 Distributed Applications in Blockchain

There are three types of blockchain technology [10]. They are,

1. Public
2. Private
3. Consortium.

A. **Public Blockchain**

The platform where every node in the blockchain network are able to access the data and have to submit the proof of work for accessing the same data. Bitcoin is the best example for the public blockchain, where it is a decentralized network as the number of nodes uses this technology.

Few of the other example for the public blockchain are:

- **Ethereum**: helps running the smart contracts and also allows the users to build the dapps.
- **Factom**: the record used to store the details for the use of governance and business.
- **Blockstream**: provides the focus on extending the bitcoins for the sidechain technology.

B. **Private Blockchain**

In the private blockchain network, only the authorized nodes are allowed to access the data. Similar to the centralized servers, only the owner of the block has permission to grant access to the other nodes in the private blockchain network. Many of the use cases were built using this network. A shared software is provided by Eris industries using this blockchain technology. The operations based on the financial institution including the settlement and the clearance were provided by this private blockchain. For the financial transactions, the distributed opensource database is provided by Multichain.

C. **Consortium Blockchain**

The consortium blockchain is a combination of both public and private blockchain. But it usually follows the private blockchain consensus protocols. By collaborating with other networks, the users create different decentralized applications for many of the use cases. Even though they restrict access to users, it is maintained that the solutions and the property rights within the consortium is maintained.

D. **Smart Contract**

The collection of data and the codes that are deployed by using the algorithms (SHA-256) signed for the transactions in the blockchain networks [11]. The Ethereum uses the programming language named Solidity. It is more or less similar to JavaScript and C++. Few other smart contracts are Vyper and Bamboo. The languages like

Serpent and Mutan were used before Solidity language. For example, Ethereum smart contract and the Hyperledger fabric chain codes. The blockchain technology is maintained and leveraged to an extent that is done by the Smart contracts [12]. Within the blockchain network [13], the smart contracts are executed and the nodes that execute the smart contracts should proceed for the same results and results are recorded in the blockchain.

6 Ethereum

Ethereum is an open-source software developed by Vitalik Buterin. It is executed with the help of smart contracts for computing the decentralization based blockchain technology. Vitalik has extended the version of Ethereum in the blockchain technology and the bitcoin protocols to enable the Ethereum applications most of the cryptocurrencies [14, 15]. Ethereum has the inbuilt cryptocurrency as Ether and the language used here is Solidity. It runs on the network named as Ethereum Virtual Network (EVM) and it can deploy and write the Ethereum smart contracts in a short period of time [16]. The solidity cannot be used directly to connect with Ethereum Virtual Machine, so the opcodes were used to connect with Ethereum.

A. **Opcodes**

As mentioned above, Ethereum Virtual Machine uses the set of rules to execute the specified tasks that are named as Opcodes. While writing the opcodes, 140 unique opcodes were used. It means that the Ethereum virtual machine can compute any type of task with enough resources. It is because of 1 byte opcodes. There is a maximum of 256 (16^2) opcodes. Few of the opcodes are as follows;

- Opcodes for stack manipulation (push, pop, dup, swap)
- Opcodes for comparison/arithmetic/bitwise (add, sub, gt, lt, and or)
- Opcodes for environment (caller, call value, number)
- Opcodes for memory manipulation (mload, mstore, mstore8, msize)
- Opcodes for storage manipulation (sload, sstore)
- Opcodes for program counters (jump, jumpi, pc, jumpdset)
- Opcodes for halting (stop, return revert, invalid, selfdestruct).

B. **Deploying the Smart Contract**

A transaction is created while deploying a smart contract without the to address. Few of the bytecodes were into the input data named as constructor, which are needed to write the starting variables to store before copying the runtime bytecode for the contract's code. During the runtime it will run on every call for contract but runs only once when the bytecode creation is deployed.

```
.
├── contracts
|   ├── Migrations.sol
|   └── Storage.sol
├── migrations
|   ├── 1_initial_migration.js
|   └── 2_deploy_contracts.js
├── test
├── truffle-config.js
└── truffle.js
```

Fig. 4 Truffle migration

C. **Migrating and Compiling the Smart Contract**

Guiding the way for the developers to automate deployment data and their supporting structures are named as migrations. For managing the new version of software, it is very much useful and are not exclusive for blockchain development.

a. **Truffle Migration**

Migrating the truffle enables the smart contracts to push it into Ethereum blockchain and to set the prerequisites for linking the contracts to the other nodes with the initial data. Where the migrations are the management for the contract address on the blockchain network. It all happens in the truffle for entire migration as shown in Fig. 4. For this truffle migrations, the prerequisites should be verified that truffle framework and the ganache cli has been installed.

With the migrations, the smart contracts were deployed, for this the artifacts needed to be accessed. The files to describe the migrations are Contact addresses, the networks (where the contracts to be deployed), and the functions.

The description for the smart contracts in the .json files

- Name of the contract
- Application Binary Interface of the contract
- Bytecode of the contract
- Deployed bytecode of the contract
- The last compiled version of the contract data
- Each of the contract addresses and the networks on which the contracts deployed.

With these files, the communication with the smart contract is enabled by the JavaScript to the Truffle. For an illustration, when a contract. Address in the JavaScript code called the .json file is read by the Truffle framework and the networks and the contracts versions were enabled.

7 Hyperledger

The Linux foundations have developed an open-source platform named Hyperledger. The Hyperledger comes under the permissioned blockchain technology [17]. It is also a customizable application for few of the organizations. According to the needs and the necessity of the organizations, it is customizable as plug and play. Scalability, resilience, and confidentiality were delivered by Hyperledger. It uses the container technology in the blockchain to host the smart contract. In comparison with the Ethereum, it is more secure by authorized nodes [18].

A. Components in Hyperledger Fabric

1. **Ledger**: the fabric ledger consists of the state about the blockchain and the world state. In blockchain, the history of all the state changes were stored. Where the world state holds the status of all assets which are tracked. All the nodes in the channel or network has the ledgers and are stored on the blockchain network.
2. **Peers**: peers are participants in the organizations who make the physical structure of the blockchain networks. It is maintained by the particular organization that are known well to the blockchain network. The peers in the network can endorse the ledger and maintains the multiple ledgers in the network.
3. **Channels**: the peers in the blockchain network want to get connected or to communicate with the other peers in the network privately. Where the ledgers are channel specific, the peers in the network willonly be able to access and see the transactions of their ledgers.

B. Hyperledger Composer

Similar to Hyperledger, the Hyperledger composer is an open-source software development tool and the framework for building the blockchain technology. The composer provides a comfortable layer and the business abstraction for implementing the smart contracts by using the fabric instead of using the smart contracts. It is easier for the users to connect with the business networks through mobile applications or web services. This network consists of all the assets, transactions, and the participants. It provides the access granting control to all the peers in the network.

C. Hyperledger Composer Deployment

The development of a business network is provided with multiple tools by the Hyperledger composer. It permits to develop the tools both local and online. The deployment, configuration, and the manual testing of the network were provided for the user interface by the composer playground. It is not intended for the development of the automated testing for the use of source code control of the versions. The following diagram explains the business network in Hyperledger composer.

Hyperledger composer is the fastest way for creating new business networks [19]. After creating all the above files like model file, script file, access control file, and query file the structure will be formed and generated to compose the blockchain network as shown in Fig. 5.

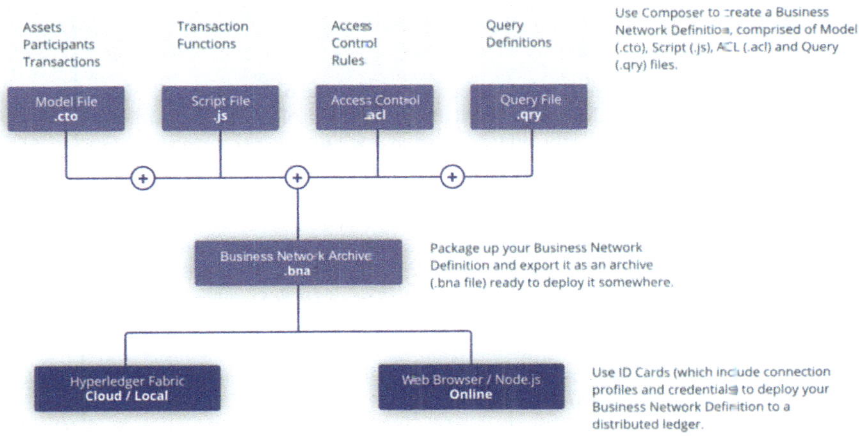

Fig. 5 Hyperledger composer

8 Hyperledger and Ethereum Comparison

Characteristics	Ethereum	Hyperledger
Platform description	Platform with generic blockchain	Platform with modular blockchain
Operations types	Ethereum for developers	By the Linux foundation
Consensus	Ledger levels by the proof of work (PoW)	Approaches with multiple level transactions
Smart Contracts	Smart contract (solidity)	Chain code (Go, java)
Currency	Ether	No currency needed

The comparison of the Ethereum and Hyperledger [20] are as follows.

9 Conclusion

In this paper, the high-level overview of the blockchain technology was provided. The many ways of implementing the blockchain have been discussed and the components and few diagrams were provided along with the mechanism how consensus protocols were working. The Smart contracts for deploying the network in the blockchain and the forking techniques were also explained. Some of the comparisons between the Ethereum and the Hyperledger was also shown. This paper also presents some of the areas where the blockchain technology can be implemented. To conclude, blockchain is a high-level secured technology for sharing the transaction both in public and private networks.

References

1. S. Nakamoto, Bitcoin: a peer-to-peer electronic cash system (2008)
2. N. Harish, S. Krause, H. Gradstein, Distributed ledger technology (DLT) and blockchain (2017)
3. Hash per Second. Bitcoin Wiki, http://en.bitcoin.it/wiki/Hash_per_second
4. W. Akins, J.L. Chapman, J.M. Gordon, A whole new world: Income tax considerations of the bitcoin economy (2013). Online Available: https://ssrn.com/abstract=2394738
5. M. Sharplesand, J. Domingue, The blockchain and kudos: a distributed system for educational record, reputation and reward, in *Proceedings of 11th European Conference on Technology Enhanced Learning (EC-TEL 2015)* (Lyon, France, 2015), pp. 490–496
6. National Institute of Standards and Technology, *Secure Hash Standard (SHS)*, (Federal Information Processing Standards (FIPS) Publication 180–4, August 2015). https://doi.org/10.6028/NIST.FIPS.180-4
7. National Institute of Standards and Technology (NIST), Secure Hashing website. https://csrc.nist.gov/projects/hash-functions
8. National Institute of Standards and Technology, *SHA-3 Standard: Permutation-Based Hash and Extendable-Output Functions* (Federal Information Processing Standards (FIPS) Publication 202, 2015). https://doi.org/10.6028/NIST.FIPS.202
9. National Institute of Standards and Technology (NIST), *Digital Signature Standard* (Federal Information Processing Standards (FIPS) Publication 186–4, July 2013. https://doi.org/10.6028/NIST.FIPS.186-4
10. Hertig, What is a decentralized application? Coindesk (2017). Online available: https://www.coindesk.com/information/what-is-a-decentralized-application-dapp/. Accessed on 14 May 2018
11. G. Foroglou, A.-L. Tsilidou, Further applications of the blockchain (2015)
12. A. Kosba, A. Miller, E. Shi, Z. Wen, C. Papamanthou, Hawk: the blockchain model of cryptography and privacy-preserving smart contracts, in *Proceedings of IEEE Symposium on Security and Privacy (SP)* (San Jose, CA, USA, 2016), pp. 839–858
13. N. Singh, Ethereum or hyperledger fabric? Medium (2018). Online available: https://medium.com/quillhash/ethereum-or-hyperledger-fabric-259f3c9b8da6. Accessed on 02 May 2018
14. Exchange Ethereum Stack https://ethereum.stackexchange.com/questions/760/how-is-the-address-of-an-ethereum-contract-computed
15. Y. Zhang, J. Wen, An iot electric business model based on the protocol of bitcoin, in *Proceedings of 18th International Conference on Intelligence in Next Generation Networks (ICIN)* (Paris, France, 2015), pp. 184–191
16. H.-H. Buerger, Ethereum white paper, Github.Com (2016). Online available: https://github.com/ethereum/wiki/wiki/White-Paper. Accessed on 18 Feb 2018
17. P. Sajana, M. Sindhu, M. Sethumadhavan, On blockchain applications: hyperledger fabric and ethereum. Int. J. Pure Appl. Math. **118**, 2965–2970 (2018)
18. LDAP.com. https://www.ldap.com
19. S. Pongnumkul, C. Siripanpornchana, S. Thajchayapong, Performance analysis of private blockchain platforms in varying workloads, in *2017 26th International Conference on Computer Communication and Networks (ICCCN)* (2017), pp. 1–6
20. M. Valenta, P. Sandner, Comparison of ethereum, hyperledger fabric and corda (2017)

Blockchain Technology: Consensus Protocol Proof of Work and Proof of Stake

B. Sriman, S. Ganesh Kumar, and P. Shamili

Abstract The core technology used for many of the cryptocurrencies is blockchain. The extensive research attention has been received by the blockchain, a distributed ledger technology. The peer to peer network and the cryptography are playing a major role in blockchain technology. In addition to that, the consensus protocols are also playing an important role in the fundamentals of the blockchain technology. In blockchain systems, the security and the fault tolerance were guaranteed by the consensus protocols. There are two broad categories of the blockchain consensus protocols. They are Probabilistic finality consensus and the absolute finality consensus protocols. In this paper, the consensus protocols of the above two categories and the strengths and the weakness of blockchain types that could be applicable are presented.

Keywords Blockchain · Consensus protocols [POW, POS] · Cryptocurrency · Cryptography

1 Introduction

A new decentralized cryptocurrency was described by Nakamoto in the white paper for the first time in blockchain. The digital cryptocurrency [1] like bitcoin took the blockchain technology attention among all the people. After that, most of the ideas came out with the blockchain technology. With the combination of all the techniques like cryptography, peer to peer network [1] and distributed system technology, the

B. Sriman (✉) · S. Ganesh Kumar · P. Shamili
Computer Science and Engineering, SRM Institute of Science and Technology, Chennai 603203, India
e-mail: srimanb@srmist.edu.in

S. Ganesh Kumar
e-mail: ganeshk1@srmist.edu.in

P. Shamili
e-mail: shamilip@srmist.edu.in

S. S. Dash et al. (eds.), *Intelligent Computing and Applications*, Advances in Intelligent Systems and Computing 1172,
https://doi.org/10.1007/978-981-15-5566-4_34

blockchain uses it. Blockchain provides a secure framework to the cryptocurrencies, where no one could tamper the data of the transactions. Due to this, the wide use of blockchain technology [5] is used in the fields of Supply chain, Internet of Things (IoT), medical, and the financial fields.

There is a big issue based on applying the blockchain technology and many of the challenges; how to design a suitable consensus protocol. Here the consensus is the distributed ledger, [6] that is maintained by all the nodes in the network. In the existing system (Traditional software), the main problem is maintaining the consensus because of the centralized servers [7]. But, in the blockchain distributed system, each and every node itself is a host and a server. Moreover, it needs to share the information to all the nodes in order to complete the consensus protocols.

This is an excellent consensus protocol that could tolerate the phenomena and minimize the dangers which are not affected by the consensus results. The consensus protocols should be adopted by the systems that are suitable for the system based on blockchain. The three types of blockchain are public, private, and consortium blockchain [19]. Each one has its own application scenarios. In this paper, the different consensus protocols of the blockchain are analyzed and their applications and the performances based on the scenarios.

2 Consensus Protocols Proof of Work [Pow]

Satoshi Nakamoto [1] applied the bitcoin white paper with the concept of Proof of Work (PoW) in 2008, which was first published by Bonneau et al. [7] Cynthia Dwork and Moni Naor in 1993. The most popular consensus mechanism for cryptocurrency is the Proof of Work. In 1999, the term "Proof of Work" was initially used by Markus Jakobsson and Ari Juels [3]. The principle followed here is "Easy to verify but difficult to find".

Principle: A solution that is easy to verify but difficult to find

A. **Miners**

The process of solving the crypto puzzle is called mining [12]. The mining activity performed by the node in the network is named as minor. Then the new block is added to the blockchain as shown in Fig. 1. Here comes the proof of work as an answer to the mathematical problem (crypto puzzle) which is considered as the solution that is difficult to find but it is easily verified.

B. **Mining Pool**

For finding the new blocks, the amount of work done by the miners over a network with the shared processing power equally in the pooling of resources are the mining pool [14]. The members in the mining pool are awarded with the amount of "Share," who provides the partial verified Proof of Work. The difficulty level for the minors is increased for mining in the pools where it takes more time for the slower miners

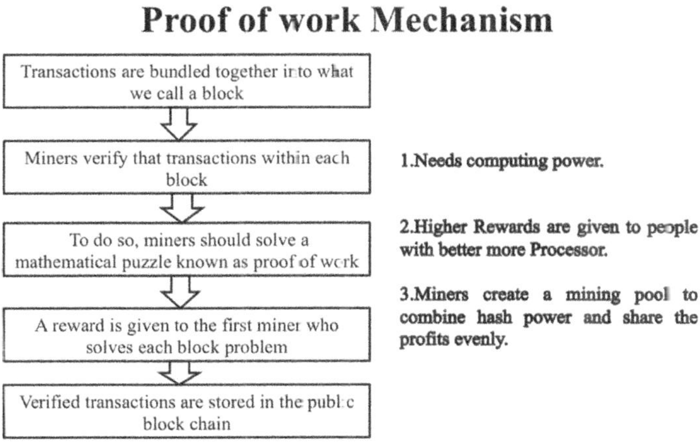

Fig. 1 Flow of POW

to create a block as shown in Fig. 1. Rather than randomly finding once few years, it is better to get rewards for every newly generated block. This is the solution for the problem to the miners to pool in their resources.

C. **Solving the Crypto Puzzle**

The puzzle is solved by guessing at random hash value [1]. Predicting the output is made impossible by using the hash function. For that difficulty, the minors choose the value randomly and apply it to the hash function and the data in a particular block. The number of zeros that are preestablished will be the starting for the resulting hash. There may be many results with the combination of two consecutive integers and it leads to numbers that are impossible to find. Now, what to do if there is more than one nonce produced? (all minors trying for the same block).

If a particular miner as shown in Table 2, in the peer to peer network solves the crypto puzzle [10] or finds the hash within the time period and the announcement takes place. Then the other miners will stop their work and go for the next nodes to solve the crypto puzzle. As a result, the miners will be rewarded as they get new bitcoins for solving the puzzle.

3 Cryptography SHA-256

A. **Cryptography Major Role in Proof of Work**

SHA 256 is the most popularly used Proof of Work consensus and was announced as the portion of the Bitcoin. [1] The others were SCRYPT, SHA-3, SCRYPT JANE, SCRYPT-N, etc.

Table 1 SHA HASH calculation

	SHA-1	SHA-224	SHA-256	SHA-384	SHA-512
Message digest size	160	224	256	384	512
Message size	$<2^{64}$	$<2^{64}$	$<2^{64}$	$<2^{128}$	$<2^{128}$
Block size	512	512	512	1024	1024
Word size	32	32	32	64	64
Number of step/round	80	64	64	80	80

With the help of the hash value, the miners try to find the random nonce [14] (random data of small size) and find the block that holds the hash (of binary values) as shown in Table 1, with particular 0's. A hash with number of zeros that has the rare hashes to find. If it's a good hash, the data is not found and have to try many times to find a perfect nonce [16].

The number of zeros were based on the difficulty [17] faced by the miners to search for a perfect block. For every 10 min, a new block is created by an average of how many blocks that are previously added to it.

B. **Nonce**

The cryptocurrencies and the blockchain are working under the mining algorithm like Proof of Work with the concept of the central part as nonce [9]. A nonce, which produces the hash which is lower than or equal to the hash value is fixed by the network as difficult as shown in Fig. 2. For finding these values the miners do mining and compete with each other. In that network, the miner who finds a nonce is named as the golden nonce. Then the miner gets the reward and adds that block to the

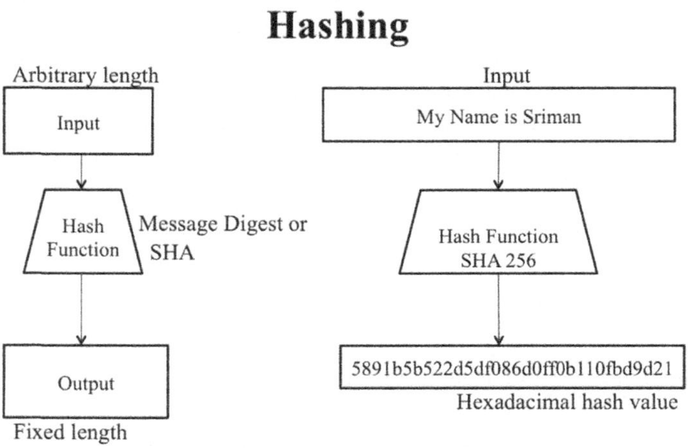

Fig. 2 Hashing techniques

Summary	
Number Of Transactions	3124
Output Total	9,692.38953688 BTC
Estimated Transaction Volume	495.63664992 BTC
Transaction Fees	0.31843823 BTC
Height	596977 (Main Chain)
Timestamp	2019-09-28 12:37:51
Received Time	2019-09-28 12:37:51
Relayed By	Unknown
Difficulty	12,759,819,404,408.79
Bits	387321636
Size	1303.714 kB
Weight	3992.803 kWU
Version	0x3FFFE000
Nonce	854393374
Block Reward	12.5 BTC

Hashes	
Hash	0000000000000000009cdfc7cd3a28c4c6deab-4a[illegible]b1ceb188a8520cce7755
Previous Block	0000000000000000000a0f0b85b1279adbf3f3c8[illegible]84[illegible]740e70e2424db7ac81ef
Next Block(s)	
Merkle Root	a25d22ef9ab25e361ab14801407d814fab80e933[illegible]71[illegible]eb9d2c265315443b5065

Fig. 3 Block summary

blockchain network. It all happens during the Proof of Work mining. Now let us see how the process of nonce is done in mining.

C. **Building Blocks with Nonce**

Like timestamp and the difficulty target, the block header stays with its key data value. A key with 32-bit in block is the nonce as shown in Fig. 3. For creating a new block, the miners choose a random nonce and add it to the block header while building a block. As stated before in nonce, if the miners are not having the number of zeros, then they avoid the hash and go for the new nonce. It will be repeated until a nonce is discovered by the miners.

D. **Block Reward**

Every block that is mined successfully by the miners in the blockchain network are rewarded with the Bitcoin block.

- The number of bitcoins a miner gets, is the number of rewards that miner mines a block.
- The reward becomes half for every four years or every 210,000 blocks.
- The expected reward that hit be zero to around 2140

E. **Concept of Block Reward**

The size of every bitcoin block is 1 MB and holds the data of transaction information. Illustration, when a transaction occurs between two nodes in the network and information about transaction are stored in a block [16].

At the initial stage (in 2009), for each and every bitcoin block was rewarded as a worth of 50 BTC (bitcoin). In February 2019, its reward was 12.5 BTC for one block and bitcoin price is \$3500, which is of $12.5 \times 3500 = \$42{,}000$. The block rewards were given to the miners those who do mining by using their computation powers to find a new block as shown in Fig. 3 Similarly, the other cryptocurrencies in the blockchain have the same mechanism for rewarding the miners. The miner who wins the block reward adds the first transaction on the block.

4 Main Issues with Proof of Work Consensus

The following are some of the issues for the Proof of Work consensus mechanism.

A. **The 51% Attack**

In the blockchain network, if any node gains 51% or more than 51%, [8] the nodes could influence the blockchain by gathering most of the network as shown in Table 2.

Time consuming: for solving the crypto puzzle, the miners have to check for the nonce which must be solved to mine the block. This helps in time consumption.

- **Resource consumption**: in order to solve the mathematical crypto puzzle, the miners consume high computational power. This results in wastage of resources like hardware, space, money, and energy. It is estimated that the world's electricity spent for verifying transactions in 2018 was 0.3%.
- It takes a minimum of 10–60 min for the confirmation of any of the transactions in the blockchain network. This is due to the time taken for mining the transactions and adding it to the blockchain.

B. **Computation Power**

- **Carbon footprint: 34.73 Mt CO_2** In comparison with the carbon footprint of Denmark over 723,140 VISA transactions with 48,872 h of time watching YouTube.
- **Electrical Energy: 73.12 TWh** In comparison with the power consumption of Austria, the equivalent power consumption has an average of over 20.61 days in US as shown in Fig. 4.

Table 2 Analysis and comparison

System function	Proof of work [POW]	Proof of stake [POS]
Mining power	The work (solving the crypto puzzle) done by the miner	Depends on the stake of the miners
51% attack	Level incentive to avoid 51% attack	Highly expensive in 51% attack
Energy consumption	Higher	Lesser
Decentralized versus centralized	Becomes powerful when nodes which tend to be centralized over a period of time	Mining communities will become decentralized based upon their stake
Target time for a block	For every 10 min	For every 15 s

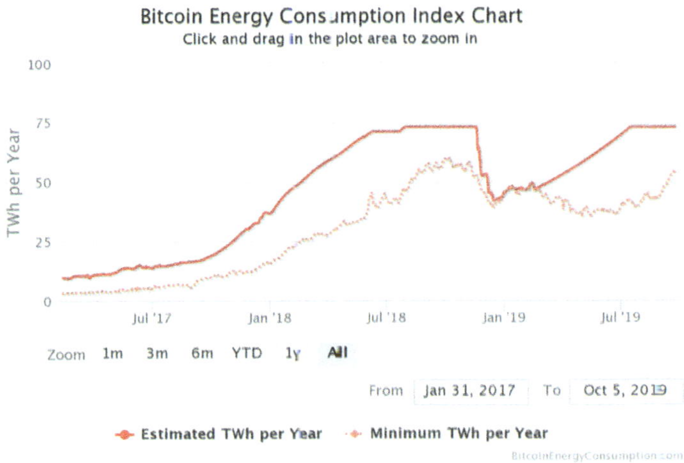

Fig. 4 Energy consumption chart

- **Electronic Waste: 11.49 kt** In comparison with the e-waste generation of Luxembourg, as shown in Fig. 4, the equivalent weight of 1.48 "C"-size batteries or golf balls around 2.09

C. **Definition of 51% Attack**

This attack on blockchain is the 51% attack. [15] Usually the bitcoins, the attacks which are mainly by a group of miners who control more than 50% of the mining hash in the network or by the computational power. The ability as shown in Table 2, of the attackers is to prevent the transactions that gain confirmations and can halt the payments between the users [19]. They can also reverse the completed transactions of any user in the network and this leads to Double Spend coins. The attackers could not alter old blocks or create new coins in the blockchain-based cryptocurrency.

D. **Double–Spending**

The risk at which the spending of digital currency is twice is called Double spending. The savvy individuals, who understand about the blockchain network and the computational power needed were reproduced to obtain the digital currencies that are unique potential problem [15]. This type of issue will not happen during the cash transactions, the parties involving in transactions can easily verify the authenticity and the ownership of the physical currency. But in digital currency, the digital token can be copied by the holder and it might be sent to any of the parties or the merchants with the original one.

E. **Proof of Work System Features**

The two features that could contribute to the wide spread of the consensus protocol are

- It is difficult to do the mathematical crypto puzzle problem.
- The correctness of the solution could be easily verified.

5 Consensus Protocols Proof of Stake [Pos]

The complete virtual consensus mechanism was made by the Proof of Stake. The way of achieving the goal differs and the process remains the same as proof of work. While in PoW, the miners solve the mathematical crypto puzzle with the help of the high computing resources.

A. **Validators**

The validators will be there in PoS instead of miners in PoW [13]. In the ecosystem, the validators lock some of their Ether [2] as a stake. Like betting, the validators as shown in Fig. 5, will bet on those blocks that are likely to be added in the chain. In a proportion to the stake, the validators will be rewarded when a new block is added.

1. One who becomes a validator may hold the cryptocurrency and also sometimes locked up deposit were required.
2. It is also done based on how much stake or cryptocurrency the validators having are the chances to mine a new block.
3. In the PoS protocol, it will randomly assign the validator and give the right to create a new block in between as shown in Fig. 5, the validators and that is based on the stake value.
4. The reward as shown in Fig. 5, will be provided for the chosen validator.

The Proof of Stake is mainly suitable for resolving the BFT (Byzantine Fault Tolerance) [4] as the validators were tracked in the network and the known identities.

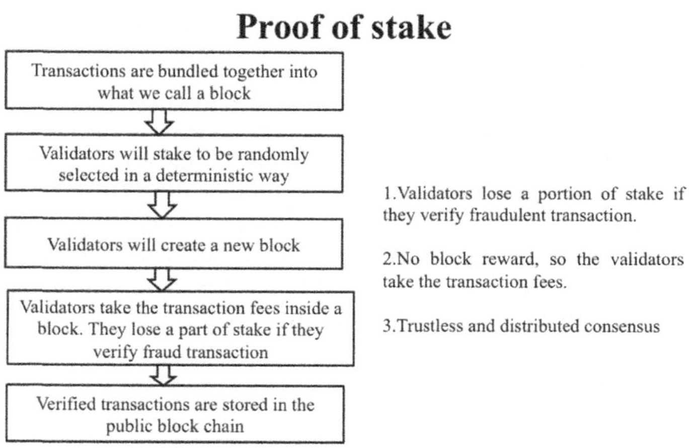

Fig. 5 Flow of POS

For example, the list wallet address. In Byzantine Fault Tolerance, it requires 2/3 of the validators to be honest and keeping all these individuals helps to maintain the status.

B. **Advantage**

- **Efficiency in energy**

The PoS algorithms were efficient in energy in comparison with PoW [13]. The mining process in PoS makes it a greener option in cutting out the energy intensively.

- **Security**

For security in PoS, the attackers should proceed with their stakes and the assets in order to attempt 51% attacks. In comparison with PoW, the attackers will not lose their hardware when attempting the 51% attacks.

- **Decentralization**

Leading to the real threat of centralization, the group of miners mining their resources (mining pool), could control over 51% [15] of the networks that are running in PoW systems. As a result, there is an exponential increase in the reward for a single investment in PoW system because of opposition in the linear increase in the PoS system. For an example, if the validator in PoS network invests twice as much as any other validator, they will be granted the control.

C. **Conclusion**

The PoS made the validators not to have their own computing power as the factors that may influence the validators to win number of their own coins and the complexity of the network.

The benefits that switched from PoW to PoS are as follows.

- Savings in energy
- If it becomes highly expensive, there will be safer network.

D. Casper Blockchain

The consensus mechanism in which both the POW/POS works together is named as Casper Blockchain. Where the Proof of Stake is overlaying on the ethash of the Proof of Work protocols. In this way, the Casper is a security deposit protocol that are based on the economic consensus system. The new consensus will be thanked for making the validators to deposit the security amount to be a part of consensus protocol. The Casper protocol is a control over security deposit that determines the specified amount of rewards the validators receive. The validators security deposit will be deleted and the privilege to be a part of the consensus network will be removed, if the validator creates the "invalid" block. The Casper security can also be defined as doing something like Bets. In the PoS protocol, the validators who won the bet will be rewarded with the money prize based on the transactions.

6 Analysis and Comparison

A. Fault Tolerance

In PoW and PoS protocols [10], the attacker needs a large amount of computing power or the stake [18] for the creation of a long chain instead of the valid chain. In bitcoin, the double-spending attack is completed by creating the longer private chain with a fraction of 50% by using the high computing power. The blockchain network becomes undetermined, if the attackers fraction becomes 50% or more than that. The stakeholder who has less than 50% stake only be allowed in the PoW and PoS.

B. Limitation

The PoW consumes most of the computational power in comparison with the other consensus protocol and the output of PoW per second in the bitcoin, which limits the applicable transactions of PoW in the payments [20]. The PoW and PoS have the same features, even though the consumption of the computation power is reduced and the stakeholders get the reward for the block. Each node communicating in the PoS requires the message transfer in every round of the consensus, thus it has a high performance for the network. There will be no anonymity due to the nodes participating in the consensus. In the PoS, the process in the consensus is completed in a few seconds, that is suitable for the scenario in payments.

C. Scalability

The PoW and PoS are having good scalability. But there are some ways to improve the scalability. For example, in Bitcoin it adopts the network of lightning to give the Off-Chain system to improve the scalability. The plasma layers of layer 1 and layer 2 are the scaling solution and the sharding technology were proposed by the Ethereum [8], respectively. With the small number of nodes, the PoS is limited to the high-performance networks.

D. Scenarios

The current blockchain system is of three parts. They are public, private, and consortium blockchain as mentioned above. Everyone in the public blockchain will participate in the distributed ledger and the consensus process. The PoW and PoS [11] are applicable to the public blockchain. In the private and consortium blockchain, the nodes in the network participates in the consensus process that belongs to the permissioned blockchain. Even though, the public and the consortium are not like a public blockchain, due to the high efficiency and their strong consistency of the consensus, it will be applicable in the medical and the commercial scenarios.

7 Conclusion

In the blockchain system, the consensus protocols are guaranteed for the perfect operations in the blockchain systems. The nodes in the blockchain network agree [12] to the transaction or the values through the consensus protocols. In this paper, the blockchain consensus protocols and the strengths and the weaknesses were discussed through the application scenarios through the analysis and the comparisons. It is concluded that the consensus protocols were of best use and applicable in the appropriate application scenarios.

References

1. S. Nakamoto et al., Bitcoin: a peer-to-peer electronic cash system
2. V. Buterin, On public and private blockchains (2015). https://blog.ethereum.org/2015/08/07/on-public-and-private-blockchains/
3. M. Swan, *Blockchain: Blueprint for a New Economy* (O'Reilly Media, Newton, MA, USA, 2015); A.M. Antonopoulos, *Mastering Bitcoin: Unlocking Digital Cryptocurrencies* (O'Reilly Media, Newton, MA, USA, 2014)
4. J. Fan, L.T. Yi, J.W. Shu, Research on the technologies of Byzantine system. J. Softw. **24**(6), 1346–1360 (2013). Online available: http://www.jos.org.cn/1000-9825/4395.htm
5. T.T.A. Dinh, R. Liu, M. Zhang, G. Chen, B.C. Ooi, J. Wang, 'Untangling blockchain: a data processing view of blockchain systems'. IEEE Trans. Knowl. Data Eng. **30**(7), 1366–1385 (2018)
6. F. Tschorsch, B. Scheuermann, Bitcoin and beyond: A technical survey on decentralized digital currencies, IEEE Commun. Surv. Tuts. **18**(3), 2084–2123 (2016). 3rd Quart
7. J. Bonneau, A. Miller, J. Clark, A. Narayanan, J.A. Kroll, E.W. Felten, SoK: research perspectives and challenges for bitcoin and cryptocurrencies, in *Proceedings of IEEE Symposium on Security and Privacy* (San Jose, CA, USA, 2015), pp. 104–121
8. G. Wood, Ethereum: a secure decentralised generalised transaction ledger, Yellow paper (2015)
9. C. Natoli, V. Gramoli, The blockchain anomaly, in *Proceedings of the 15th IEEE International Symposium on Network Computing and Applications, NCA'16* (2016), pp. 310–317
10. C. Natoli, V. Gramoli, The balance attack against proof-of-work blockchains: the R3 testbed as an example, Tech. Rep. 1765133 (2016)
11. A. Black, Hashcash-Adenial of service counter-measure, Tech. Rep. Cypherspace, (2002). URL http://www.hashcash.org/papers/hashcash.pdf
12. Alyssa Hertig, What Is Ether? Coin Desk, 21 Apr (2017). https://www.coindesk.com/information/what-is-ether-ethereum-cryptocurrency
13. F.B. Crane, Proof of work, proof of stake and the consensus debate. Cointelegraph, 20 Dec. 2014. https://www.cointelegraph.com/news/proof-of-work-proof-of-stake-and-the-consensus-debate
14. M.A Alahmad, I. Al-shaikhli et al., Joux multicollisions attack in sponge construction, in *The 6th International Conference on Security of Information and Networks (SIN), 2013 6th International Conference on, ACM* (2013)
15. M.A Al-Ahmad, I.F. Alshaikhli, Broadview of cryptographic hash functions. Int. J. Comput. Sci. Issues **10**(4), 239–246(2013)
16. S. Wu, D. Feng, W. Wu, J. Guo, L. Dong, J. Zou, (Pseudo) preimage attack on round-reduced Grøstl hash function and others, in *Canteaut,* vol. 9, pp. 127–145
17. X. Wang, H. Yu, Y.L. Yin, Efficient collision search attacks on SHA-0, in *Advances in Cryptology–CRYPTO 2005* (Springer, Berlin, 2005)

18. Merkle tree URL https://en.wikipedia.org/wiki/merkletree
19. J.R. Douceur, The Sybil Attack, in: International Workshop on Peer-to-Peer Systems (Springer, Berlin, 2002), pp. 251–260
20. Total cryptocurrency market cap hits new all-time high (2018). URL https://www.coindesk.com/150-billion-total-cryptocurrency-market-cap-hits-new-time-high. Accessed on 14 May 2018

Non-Invasive Techniques of Nutrient Detection in Plants

Amit Singh and Suneeta V. Budihal

Abstract Plants are really important for the earth and for all living beings, as they absorb carbon dioxide released by the human being and release oxygen for their survival. Plants require major nutrients such as Nitrogen, Phosphorus and Potassium (NPK) in a proper ratio to keep them healthy and more resistant to pests. The deficiency of nutrients in plants makes it impossible for the plants to complete the vegetative and reproductive stage of their life cycle. It is needed to monitor the amount of nutrients present in plants, so that if there is a deficiency of nutrients, manure and fertilizers can be added to the desired extent. There can be many methods to monitor the amount of nutrients in plants such as invasive and non-invasive methods which include testing the sample in the laboratory, spectroradiometers, reflectometers and image processing techniques. This paper proposes a survey work carried out and guidelines for researchers over the various non-invasive methods for the detection and estimation of macronutrients.

Keywords NPK concentration · Non-invasive · Machine learning · Neural networks

1 Introduction

The crop yield relies on the continuous interaction amongst the soil and plant features. A healthier plant provides good crop yield. Hence fertile soil is the main source of nutrients' supplier from the plants. There is an increase in demand for food production. In order to have plants and crops healthy, proper proportions of nutrients in the plants are needed. In reality, majority of macronutrients are acquired from the soil

A. Singh (✉) · S. V. Budihal
School of ECE, KLE Technological University, Hubli 580031, India
e-mail: amitsingh11097@gmail.com

S. V. Budihal
e-mail: suneeta_vb@kletech.ac.in

S. S. Dash et al. (eds.), *Intelligent Computing and Applications*,
Advances in Intelligent Systems and Computing 1172,
https://doi.org/10.1007/978-981-15-5566-4_35

such as NPK. Any kind of deficiency in micronutrients can be managed but a deficiency of macronutrients results in a major damage to the growth of the plants and yield of the crops. Any improper use of fertilizers will affect the plants' growth and fertility leading to reduced crop yield, reduced fruit and vegetable size, changed taste of fruits. Many times, improper utilization of the fertilizers will affect the quality of the fertile soil also. Excessive use of external fertilizers will contaminate the surface and groundwater resources. Hence the discussions by the agricultural researchers and engineers realized that the following topics are to be investigated further.

1. Soil-based agriculture.
2. Timely testing of the soil.
3. Need for pesticides and fertilizers.
4. Early detection of plant diseases and its remedies.
5. Estimation of the macro- and micronutrients.

Nitrogen is abundantly available in the soil but all other nutrients are affected by the pH value, temperature, organic materials and moisture of the soil. We can monitor the amount of nutrients in many ways like visual deficiency, soil test, plant tissue test and the behavior of the plant's response to the fertilizers or organic manure. The symptoms of nutrients' deficiency can be visible in the leaves. Some symptoms include interveinal chlorosis, marginal chlorosis, edge distortion, changes in the shape of the leaf, burning and holes in the leaf and change in the colour of the leaf. Therefore, there is a requirement of nutrient analysis on a timely basis which should be non-harmful to plants.

2 State-of-the-Art Techniques

A study over the existing literature demonstrated that the NPK nutrients are identified not only through the health of the leaves, but also through the quality of the soil. This is carried out by processing the soil and leaves chemically to detect the pH values. Based on the values, the nutrient concentrations are detected and further analyzed for actions. Here the leaves of the plants need to be crushed and dried and processed chemically for further initiatives. Such traditional modes of nutrient deficiency detection and estimation are termed as an invasive mode in agricultural language. As these are completely dependent on the amount of the chemical to be used, type of the chemical to be used, the temperature and lighting conditions change in the colour of the ingredients after the chemical reaction, the expertise in handling the laboratory experimentation, human observation errors, etc.

It led to the development of non-invasive mode of nutrient concentration detection without harming the plants and to provide results that are close to actual values. It is needed to detect NPK concentrations using sensors, micro-controllers, IoT-based techniques, Image processing techniques, spectrometers, optical sensors, Machine Learning techniques, etc.

Currently, large groups of researchers are evaluating the nitrogen content through non-destructive techniques. Here the leaf chlorophyll content is mapped to nitrogen concentration as in [1], such as the Soil Plant Analysis Development (SPAD), which was used to investigate N concentration [2]. In order to monitor the plant nutrition concentration [3] in large scale, spectral remote sensing analysis technology was developed. It is very difficult to capture and quantify micronutrients with hyper-spectrometer [4]. Added to it is the difficulty with the data processing and computation processes are challenging [5], which restrict the application. However, the high cost of [6] fluorescence detection sensors for N nutrition detection limits its usage. Another technique called as phenotyping based on imaging processing was employed to evaluate N nutrition. Most of the research was focused on colour traits [7] and [8]. It is indeed needed to help in identification [9], classification [10] and prediction [11] in plants. ML approaches are applied when a large data set is available, relating inputs to output quantities. The ML algorithms Support Vector Regression (SVR), Random Forest (RF) and Neural Network (NN) that supports a non-destructive, easy, low-cost realization.

3 Non-Invasive Techniques

There are many studies done by the researchers for the nutrient's analysis. Electrochemical sensor has been developed to determine NPK ratio. Multispectral imaging which is for a few discrete spectral channel or wavebands, Hyperspectral imaging which is used with a sequence of continuous waveband covering a specific region, digital image processing, colour histogram were reviewed by many of the researchers and they all are used as a non-invasive technique for the analysis of nutrients. Many methods have been proposed by the researchers for diagnosing the nutrient deficiencies in plants and crops. In these papers, the various methods and algorithms for monitoring the nutrient in plants are discussed. Some of the techniques have been discussed below.

3.1 Detection Using Sensor

In [12], the author proposed the system for monitoring nutrient and moisture level of the soil using antimony electrode for pH measurement, and the temperature of soil is measured using DS18B20 sensor; the temperature sensor works on a Dallas one wire protocol architecture. In [13], the author has made the project which includes various devices like GPS-based remote controlled device, moisture and temperature sensing, leaf wetness and security facilities. A wireless sensor is used for reading the soil properties. In the control system, data received is compared with data set. According to that, the values are generated in the android application and farmers get the information. In [14], the authors take important features like temperature, sunlight, humidity,

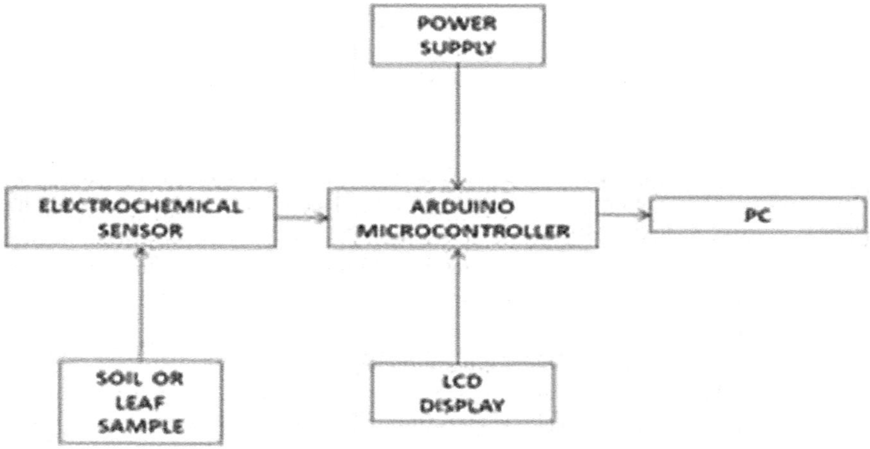

Fig. 1 A block diagram of NPK detection using sensors and microcontrollers

soil moisture, pH and design a plant monitoring and smart gardening system with the help of IoT. All the features are measured using sensors and interface with the raspberry pi microcontroller. All the information of the garden can be directly monitored by the owner of the farm/garden in his/her mobile phone using IoT. In [15], the author has proposed a system to determine NPK using electrochemical sensor dipped under soil. Electrochemical sensor works under the principle of absorption of ions from the soil. Sensors are connected with Arduino for monitoring nutrients deficiency.

In [16], the author proposed a system which has a hand held device which will give the pH value and estimate NPK from the pH of that soil. A classification algorithm to predict the crops based on the values got from the device is used. In [17], a plant watering system is proposed where the sensor is used to check soil moisture content, and then Arduino switches ON water pump if the level is less and the information is sent to the users using IoT modules. In [18], the author proposed a system by using features like pH level and temperature (Fig. 1).

3.2 Detection Using Image Processing

To analyze the nutrients using image processing technique required the following:

1. The area of the leaf to be processed.
2. Integration of edges and veins of the leaf.
3. Processing of the area of the leaf taken.
4. Extraction of colour feature in the leaf.
5. Identification of deficient nutrients.

A rapid, non-destructive, repetitive and relatively inexpensive imaging technology is better compared to optical techniques to provide an approach to non-invasive sensing of N in plants. In [19], a system to measure the pH value, temperature, various spots and patterns is developed using image processing technique and using sensors like soil sensor and temperature sensor. The input pixel of the colour images in three different planes RGB and stored into three variables. After nutrient level is detected all the data is sent to the cloud server.

In [20], the author discussed the detection of the leaf diseases using the leaf images and process the image. Nutrient detection is using a web application which is using image processing and CNNs for finding diseases in leaves. The author s objective was to introduce algorithms which can be operated on leaf images include RGB to HIS. After feature extraction, clustering methods and CNN are used to detect the diseases in plants. Colour, Hu's moment and flood fill algorithm are the three features vector used for the feature extraction purpose. In [21], the author proposes a method using colour image analysis for monitoring the nitrogen contained in the grapes. The first step is image acquisition, then the sample is processed in the lab the method used in the lab is Kjeldahl digestion method.

In [22], the authors propose the accurate detection of diseases by using plant leaf images. The features like spots in the leaf, various leaves pattern, edge detection, colour detection are some of the features that were taken by the author for detecting the diseases present in the leaves. The technique is inexpensive compared to other techniques. In [23] the rice leaf features were extracted and stepwise discriminate analysis for the cross-validation is used to identify NPK. In [24], the author proposed a method using colour image analysis for the estimation of nitrogen for the cotton plants. The image processing technique is used for the accurate and faster result and the technique is inexpensive also. Instead, DIP technique is better than the hyperspectral imaging because it is inexpensive and it is easy to operate. DIP can extract spectral image, can give morphological, colour and texture information. Image processing is a non-destructive technique in agriculture field that can be used for greater accuracy and for faster results. The first step is image acquisition where images are captured. The second step is pre-processing, where removal of noise takes place. For this purpose, various pre-processing techniques are used like image clipping, etc. The third step is image segmentation where the image is separated into various parts having same features. *K*-means clustering is used or segmentation, converting RGB to HSI model. The fourth step is feature extraction, used for identification purpose. The last step is the classification stage; for classification stage SVM is used (Fig. 2).

3.3 Detection Using Spectral Imaging

Concentration of elements in plants is determined using advanced spectroscopy. The commonly employed methods are Ultra Violet, Visible and Infrared reflectance spectroscopy. In [25], the author reviewed existing methods and techniques used for plants' nitrogen detection. Estimation of nitrogen in plants is carried out through

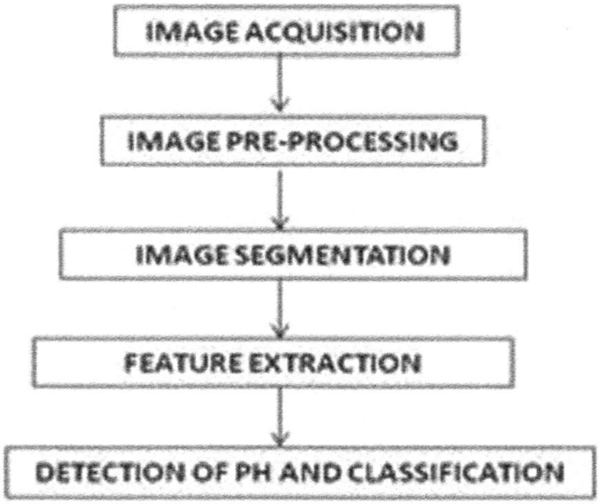

Fig. 2 A block diagram of NPK detection using image processing technique

many techniques such as spectroradiometer, reflectometer, digital camera to measure optical properties such as chlorophyll and polyphenol fluorescence, crop canopy reflectance, leaf transmittance, etc. The information of the nutrients is easily transferred to the farmer in android application using IoT module. In [26], the normal plant data is collected and compared with the data collected by the sensor. When illuminated with a source of appropriate wavelength, based on diffused reflectance, soil sample is obtained. In the paper [27], the author discussed estimation using visible near-infrared source and reflectance value was observed for NPK. In [28], the author proposed a system for the estimation of NPK nutrients depending on various factors like soil type, soil microbes and soil pH value. The samples were analyzed using Kjeldahl method, UV Spectrophotometer and Atomic Absorption Spectroscopy (AAS) for the estimation of NPK, respectively. The spectral data is collected using the assembled handheld spectroradiometer, which consists of a transmitter/receiver setup (Fig. 3).

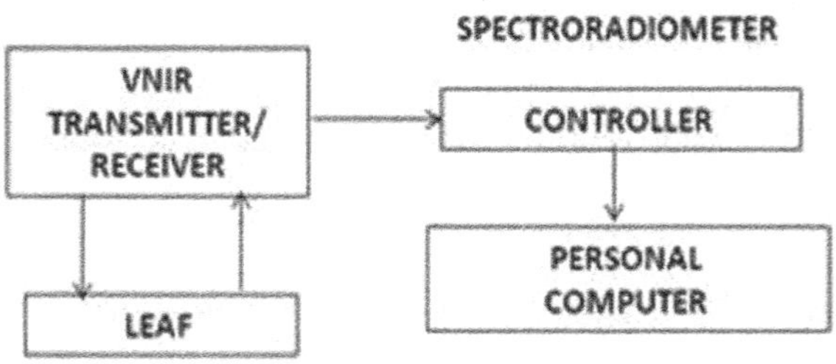

Fig. 3 A block diagram for spectral imagining for nutrient detection

3.4 Detection Using Sensing Techniques

The nutrient concentration detection and estimation using various sensing methods are discussed in [29]. Most of the soil nutrient sensing techniques described in literature involve either an optical sensing using reflectance spectroscopy or an electrochemical sensing using ion-selective electrodes. However, this review focuses mainly on optical sensing methods and principles for measuring the macronutrients and other properties in soil. The advantages associated with the optical sensing over electrochemical technology are non-destructive measurement and no need to take the sample [30] which has attracted the researchers. In order to evaluate the properties of the material, the reflectance, absorption/transmittance are used [31]. Optical sensors are frequently affected by the different soil properties and exhibit different responses in different regions of the spectral field.

In [32], author used Near Infrared Reflectance (NIR) spectroscopy to detect the nitrogen concentration through Nitrates and soil moisture content, respectively. The authors in paper [33] demonstrated usage of admittance spectroscopy to detect the variation in the nutrient content in plants. The soil organic carbon, nitrate and moisture content is detected using VIS-NIR MIR spectroscopy by Bogrekci and Lee [34]. The phosphorous is detected through NIR absorbance spectroscopy by authors in paper [35]. The VIS and NIS methods are used to detect phosphorous and pH values in [36]. The Raman spectroscopy is used to detect phosphorous in [37]. Lee et al. [38] discussed and analyzed the Raman scattering, Reflectance spectroscopy to detect nitrogen and phosphorous. The paper [39] is considered to discuss the optical sensor for nutrient detection. A MEMS-based technique is explored in paper [40] for nitrogen detection. Diffuse reflectance spectroscopy is elaborated in [41] for the detection of nitrogen and organic matter. The authors in [42] explained about the concentrations of potassium and phosphorous using diffuse reflectance spectroscopy. It includes optical, acoustic, electrical and electromagnetic, electrochemical and mechanical [43] techniques. Its sensitivity and quick response [44], the optical detection method has higher potential towards real-time scenarios. Optical method was used in several reports on NPK soil detection [45] and [48]. Most of the devices used additional optical components such as fibre optic to drive light to soil [46] and [47].

In Fig. 4, NPK detection of soil with optical transducer is overviewed. Light transmission, detection integrates to form optical transducer. An Arduino was used to operate the light source in a transmission system as data acquisition from the light detection system and provides LCD display.

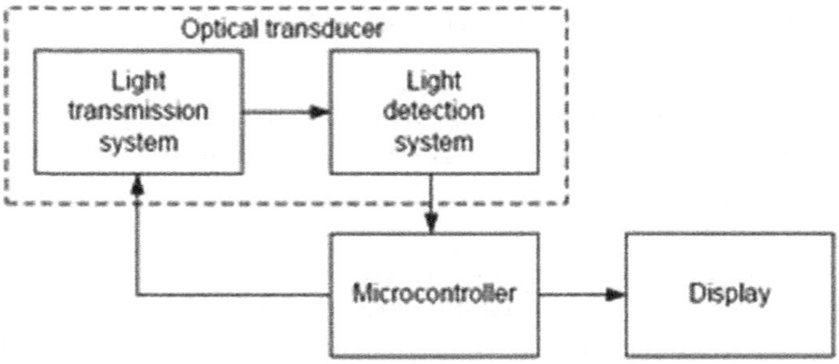

Fig. 4 A block diagram for integrated microcontroller with optical transducer

3.5 *Future Research Directions*

A survey over the state-of-the-art techniques provides great insights for the NPK detection using non-invasive method. The following research problems are identified as outcomes of the work.

1. The deep learning techniques can be applied when the data set is available for the plant leaves.
2. A prediction framework may be proposed to detect the value of one nutrient having known with the other two.
3. Advanced controllers may be used to process data and speedy operations.
4. A common framework for nutrient detection and disease detection may be proposed using plant leaf as reference data.
5. A strong data set may be generated for nutrient detection in future use.
6. Using MEMS, a sensor may be developed to detect the NPK nutrients in one stretch.

4 Conclusion

Plants are an important part of every living being on the earth. It is the responsibility of humans to take care of plants and get good yield by proper usage of pesticides and fertilizers. The plants need proper proportions of macro- and micronutrients for their proper growth. The study in this paper concentrates on the detection of NPK the macronutrients at the early stage for proper handling. The paper provides state-of-the-art techniques for the NPK concentration detection in plants. Deficiency of nutrients in plants makes it impossible for the plants to complete the vegetative and reproductive stage of their life cycle. It is needed to monitor the amount of nutrients present in plants, so that if there is a deficiency, manure and fertilizers can be added to the desired extent. There can be many methods to monitor the amount of nutrients in plants such as invasive and non-invasive methods. This paper proposed a survey work carried out and guidelines for researchers over the various non-invasive methods for the detection of macronutrients. The paper concludes by suggesting a few research directions in non-invasive nutrient detection techniques.

References

1. R.H. Fox, C.L. Walthall, Crop monitoring technologies to assess nitrogen status. *Nitrogen Agriculture Systems Nitrogeninagric* (2008) pp. 647–674
2. S.T. Ata-Ul-karim et al., *Non-Destructive Assessment of Plant Nitrogen Parameters Using Leaf Chlorophyll Measurements in Rice* (Frontiers in Plant Science, 2016), pp. 1829–1836
3. T.J. Nigon et al., Hyperspectral aerial imagery for detecting nitrogen stress in two potato cultivars. Comput. Electron. Agri. **112**, 36–46 (2015)
4. L. Chen, L. Lin, G. Cai, Y. Sun, T. Huang, K. Wang, J. Deng, Identification of nitrogen, phosphorus, and potassium deficiencies in rice based on static scanning technology and hierarchical identification method. PLoS ONE **9**(11), 113–200 (2014)
5. F.N. Gehan et al., Lights, camera, action: High-throughput plant phenotyping is ready for a close-up. Curr. Opin. Plant Biol. **24**, 93–99 (2015)
6. G. Agati, L. Tuccio, B. Kusznierewicz, T. Chmiel, A. Bartoszek, A. Kowalski, A. Grzegorzewska, R. Kosson, S. Kaniszewski, Nondestructive optical sensing of flavonols and chlorophyll in white head cabbage grown under different nitrogen regimens. J. Agriculture. Food Chem. **64**, 85–94 (2015)
7. S.B. Sulistyo et al., Building a globally optimized computational intelligent image processing algorithm for on-site nitrogen status analysis in plants. IEEE Int. Syst. **33**(3), 15–26 (2018)
8. Sulistyo et al., Regularized neural networks fusion and genetic algorithm based on-field nitrogen status estimation of wheat plants. IEEE Trans. Ind. Inform. **13**(1), 103–114 (2017)
9. F. Alvaro, Y. Sook, K. Sang, P. Dong, A robust deep-learning-based detector for real-time tomato plant diseases and pests recognition. Sensors **17**(9), 20–22 (2017)
10. A. Picon et al., Deep Convolutional neural networks for mobile capture device- based crop disease classification in the wild. Comput. Electron. Agric. **161**, 280–290 (2019)
11. X. Yu et al., Prediction of regulatory interactions in Arabidopsis using gene- expression data and support vector machines. Plant Phys. Biochem. **49**, 280–283 (2011)
12. N.A. Abdullah et al., An IoT based system for remote monitoring of soil characteristics, in *International Conference on Information Technology (InCITe) the Next Generation IT Summit on the Theme-Internet of Things: Connect your Worlds* (2016), pp. 316–320
13. N. Suma et al., IOT based smart agriculture monitoring system. Int. J. Recent Innov. Trends Comput. Commun. **5**(2), 177–181 (2017)
14. R. Elangovan, R. Rozario, A Plant monitoring and smart gardening system using IoT (2018)
15. R. Sindhuja, B. Krithiga, Soil nutrient identification using Arduino. Asian J. Appl. Sci. Technol. (AJAST) **1**(4), 40–42 (2017)
16. K. Abhang, S. Chaughule, P. Chavan, S. Ganjave, in *Soil Analysis and Crop Fertility Prediction* (2018)
17. S.R. Pratibha et al., IOT based monitoring system in smart agriculture. Int. Conf. Recent. Adv. Electron. Commun. Technol. **5**(2), 81–84 (2017)
18. P. Sihombing, N.A. Karina, J.T. Tarigan, M.I. Syarif, Automated hydro-ponics nutrition plants systems using arduino uno microcontroller based on android. J. Phys: Conf. Ser. **978**(1), 12–14 (2018)
19. P.R. Harshani, M. Newlin Rajkumar, T. Maheshwari, in *Monitoring the Nutrient Level for Efficient Crop Productivity Using IOT* (2017)
20. J.K. Patil et al., Advances in image processing for detection of plant diseases. J. Adv. Bioinform. Appl. Res. **2**(2), 135–141 (2011)

21. A. Vibhute et al., Colour image processing approach for nitrogen estimation of vineyard. Int. J. Agric. Sci. Res. **3**(3), 189–196 (2013)
22. B. Patil, H. Panchal, S. Yadav, A. Singh, in *Plant Monitoring Using Image Processing, Raspberry Pi IoT* (2017)
23. Y. Sun et al., Identification of nitrogen, phosphorus, and potassium deficiencies based on temporal dynamics of leaf morphology and colour. Sustainability **10**(3), 762–769 (2018)
24. A.P. Janwale, *A survey on Nitrogen estimation techniques for plant using Image Processing Techniques* (2016)
25. R.F. Muñoz-Huerta, R.G. Guevara-Gonzalez, in *A Review of Methods for Sensing the Nitrogen Status in Plants* (2013)
26. S. Mukherjee, S. Laskar, VIS-NIR-based optical sensor system for estimation of primary nutrients in soil. J. Opt. **48**(1), 87–103 (2019)
27. Mani et al., Estimating plant macronutrients using VNIR spectroradi-ometry. Pol. J. Environ. Stud. **28**(3), 1831–1837 (2019)
28. F.A. Azizan et al., Soil NPK variability mapping for harumanis mango grown in greenhouse at perlis. Int. J. Adv. Sci., Eng. Inf. Technol. **9**(2), 495–501 (2019)
29. Joseph V. Sinfield, Daniel Fagerman, Oliver colic, evaluation of sensing technologies for on-the-go detection of macro-nutrients in cultivated soils. Comput. Electron. Agric. **70**(1), 1–15 (2010)
30. H.J. Kim et al., Soil Macro nutrient sensing for precision agriculture. J. Environ. Monit. 1810–1824 (2009)
31. V.I. Adamchuk et al., On-the-go soil sensors for precision agriculture. Comput. Electron. Agric. **44**(1), 71–91 (2004)
32. A. Ehsani et. al., A NIR technique for rapid determination of soil mineral nitrogen. Precis. Agric. 1385–2256 (1999)
33. J. Jeosadaque et al., A sensor for monitoring the volume of nutrient in a solid substrate-based growth media by using electrochemical admittance spectroscopy. Sens. Actuators **87**, 268–273 (2002)
34. B.R. Jahn et al., Mid-infrared spectroscopic determination of soil nitrate content. Bio Syst. Eng. **94**(4), 505–515 (2006)
35. I. Bogrekci, W.S. Lee, Effects of soil moisture content on absorbance spectra of sandy soils in sensing phosphorus concentrations using UV-VIS-NIR spectroscopy. Trans. ASABE **49**(4), 1175–1180 (2006)
36. A.M. Mouazen et al., On-line measurement of some selected soil properties using a VIS-NIR. Sensor **93**, 13–27 (2007)
37. W.S. Lee, I. Bogrekci, *Portable Raman Sensor for Soil Nutrient Detection*. US Patent: US20070013908 (2007)
38. W.S. Lee et al., Sensing technologies for precision speciality crop production. Comput. Electron. Agric. **74**, 02–33 (2010)
39. G. Dooly et al., Highly selective optical fiber ammonia sensor for use in agriculture. Proc. Eur. Sensors **XXV**, 1113–1116 (2011)
40. Y.S. Xiaoyu et al, MEMS Near-infrared reflectance absorption spectrometer for detection of composition in soil. Int. J. Digit. Content Technol. Its Appl. (JDCTA) **7**(1), 2013
41. V.B. Suneeta, et al., Facial Expression Recognition Using Supervised Learning. *International Conference On Computational Vision and Bio Inspired Computing*. Springer, Cham (2019)
42. Yubing Wang et al., Soil pH value, organic matter and macronutrients contents prediction using optical diffuse reflectance spectroscopy. Comput. Electron. Agric. **111**, 69–77 (2015)
43. Y. Kulkarni et al., Primary nutrients determination in the soil using UV spectroscopy. Int. J. Emerg. Eng. Res. Technol. **2**, 198–204 (2014)
44. D.Q. Huo et al., Recent advances on optical detection methods and techniques for cell-based micro fluidic systems. Chin. J. Anal. Chem. **38**(9), 1357–1365 (2010)
45. N. Singh et al., D NPK measurement in soil and automatic soil fertilizer dispensing robot. Int. J. Eng. Res. Technol. **3**(7), 635–637 (2014)

46. L.C. Gavade, Detection of N, P, K using fiber optic sensor and PIC controller. Inter Natl. J. Eng. Sci. **13787** (2017)
47. D.V. Ramane et al., Detection of NPK nutrients of soil using fiber optic sensor. Int. J. Res. Advent Technol. pp. 13–14, 2015
48. A. Rashid, LED based soil spectroscopy. Bul. Opt. **2016**(3), 1–7 (2016)

Implementation of Cryptographic Approaches in Proposed Secure Framework in Cloud Environment

Manoj Tyagi, Manish Manoria, and Bharat Mishra

Abstract Through Internet, cloud computing shows extraordinary potential to give on-demand services to its clients with higher suppleness in an economical manner. Security is the primary need for this leading approach of computing ability. This paper presents the analysis and implementation of some symmetric cryptographic algorithms. It finds the best symmetric method and combines it with asymmetric approach and shows the importance of hybrid encryption model for cloud security. Moreover, this paper presents a framework by integrating CAPTCHA, Two Factor Authentication (TFA), hybrid encryption approach, and separate server concepts like authentication server with cryptographic server. It also suggests the use of optimization for attaining availability and security strength.

Keywords Cloud computing · Authentication · Symmetric encryption · Asymmetric encryption · Data confidentiality · Integrity · TFA · CAPTCHA · Hybrid encryption

1 Introduction

In the last few years, online users and data demand have enormously grown, the conventional computing infrastructure getting more expensive and more challenging to maintain. Conventional computing is inappropriate for accessing data at anytime and anyplace. To achieve this, it requires keeping the data on outside storage. Internet usage is rapidly increasing which introduces new model for handling the quantity,

---

M. Tyagi (✉) · B. Mishra
Mahatma Gandhi Chitrakoot Gramodaya Vishwavidyalaya, Chitrakoot, India
e-mail: manojtyagi80.bhopal@gmail.com

B. Mishra
e-mail: bharat.mgcgv@gmail.com

M. Manoria
Sagar Institute of Research Technology, Bhopal, India
e-mail: manishmanoria@gmail.com

S. S. Dash et al. (eds.), *Intelligent Computing and Applications*,
Advances in Intelligent Systems and Computing 1172,
https://doi.org/10.1007/978-981-15-5566-4_36

availability, and variety of data. This new model is cloud computing allowing resource pooling, ubiquitous, convenient, on-demand access which are delivered by various service provider interfaces [1]. Cloud computing allows its client to pay for only the utilized time of resources. In this model, the clients request for the required resources, then the cloud fulfills the clients' request by providing such resources to the clients. It gives enormous benefits to the enterprise as well as individual clients. By virtualization technique, cloud computing allocates resources to its clients efficiently. Cloud computing has characteristics like availability, manageability, and scalability. Additionally, it has various features such as multitenant, ubiquitous, expedient, elasticity, on-demand service, stability and also it is economical [2].

2 Security Schemes

Security is an essential factor for cloud computing. End-users keep their data in the cloud, remotely, so that users have no control over its data and don't know about the location of its stored data. Authentication and cryptographic schemes are used to fulfill the security objectives. CAPTCHA can prevent Internet services from bots through a simple test. The basic idea to design the CAPTCHA test is that a Human can get success in it, but the computer program can fail [3]. Single-Factor-Authentication (SFA) is simple to implement, and security is to depend on password strength. SFA suffered if a hacker stole/break the password and access to the cloud through an identity check process. To overcome this problem, Multi-Factor-Authentication (MFA) is an excellent option to protect the cloud from unauthorized access [4].

2.1 Symmetric Encryption

National Institute of Standards and Technology (NIST) has published DES which suffered from weakness in both cipher-design and key-size where small key-size invites brute-force attacks [5]. 3-DES is secure and the replacement of DES but it is slow [6]. NIST announced the AES scheme which is widely accepted and replaced the DES and 3-DES [7]. Famous cryptologists Bruce Schneier designed a secure approach named Blowfish but it does not suit for large data due to its small block size. Twofish is the successor of blowfish [8]. Lai and Massey introduced a block cipher IDEA, which also provide good security [9]. Ross-Anderson, Lar-Knudsen, and Eli-Biham designed block cipher named as serpent which is a very secure approach but slow than AES [10]. Stafford Tavares and Carlisle Adams also developed symmetric encryption approach CAST-128 and CAST-256 [11]. Ron Rivest designed RC4 which is a renowned stream cipher known for its simplicity and execution speed [12]. Ronald L. Rivest invented RC5 and RC6 where RC6 is providing higher security, less round, and increased throughput [13]. Here symmetric encryption is implemented in java on Intel core i5-8250U CPU 1.60–1.80 GHz with 8 GB RAM.

2.2 *Asymmetric Encryption and Hashing*

Rivest, Shamir, and Adleman designed an asymmetric scheme RSA which is based on prime number factorization. RSA is secure when using large prime number factorization because of its hardness [14]. Koblitz and Miller independently developed an elliptic curve based cryptosystem which is faster and needs less computing power. A popular elliptic curve scheme is known as Elliptic-Curve-Integrated-Encryption-Scheme (ECIES) [15]. The hash function generates a message digest to detect any change in message to maintain integrity. Many hashing approaches are available like SHA-3, SHA-2, SHA-1, MD5, where SHA approaches are better than MD5 due to less secure behavior of MD5 [16] (Figs. 1 and 2).

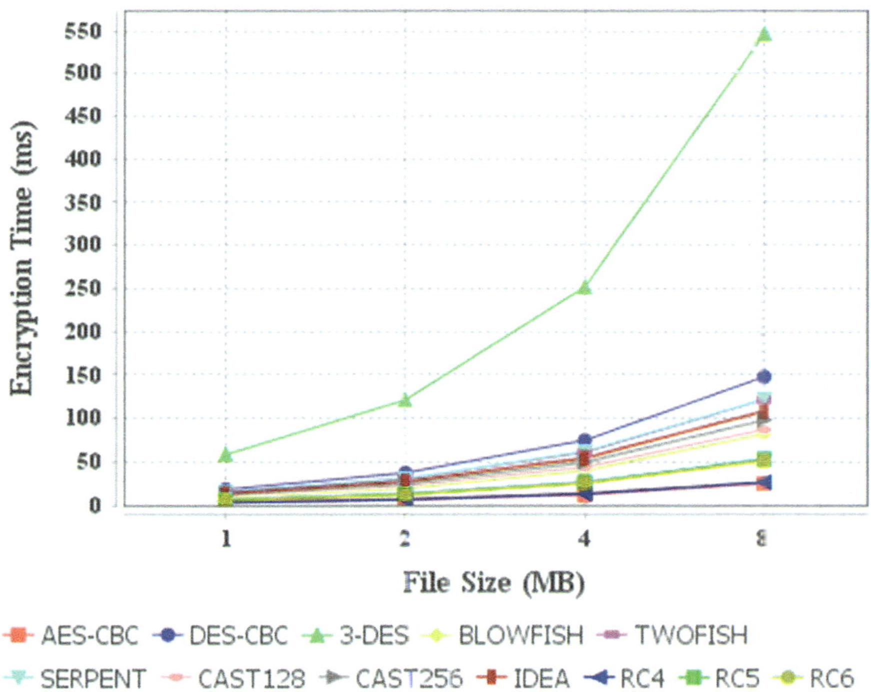

Fig. 1 Encryption time of symmetric algorithms except read/write time of file

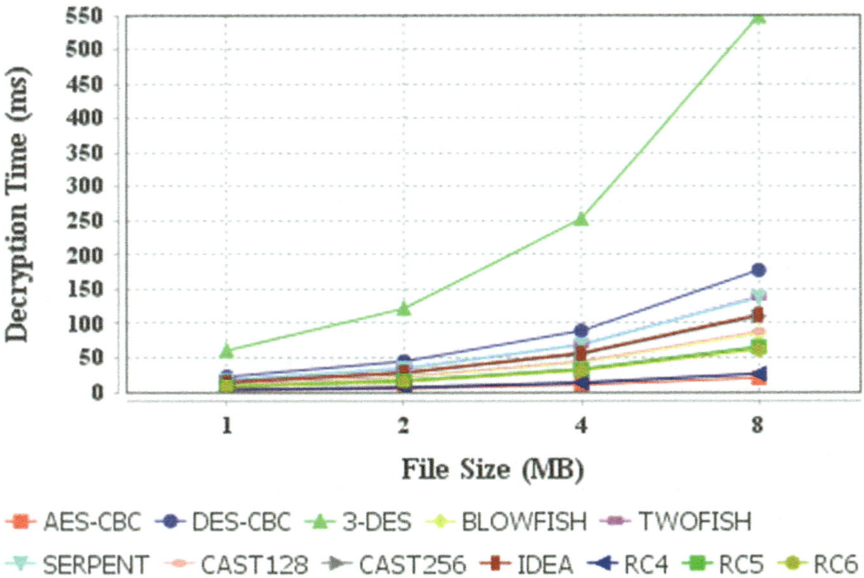

Fig. 2 Decryption time of symmetric algorithms except read/write time of file

3 Problem Definition

Lack of protection affects the acceptability of cloud. Practically, the computation burden in cloud computing reduces its performance, and a single computation server is not preferable in terms of security. Then, use two separate servers, one for authentication operation and other for cryptography operation to reduce the burden and improve security aspects [17]. CAPTCHA is accessible in a web-based application to protect the system from bots and MFA is commonly accepted to verify the genuine users. Generally, two factors are in practice for user verification in clouds which is called TFA. By analysis of various symmetric key-methods, it is clear that AES is better than other methods. It is fast as well as it offers excellent security.

Whereas, public-key cryptography like RSA is not practical to encrypt extensive data due to its slow execution. This issue introduces an idea of hybrid encryption which combines the RSA and AES. The hybrid approach applies a symmetric method AES to encrypt/decrypt the original message, and asymmetric method RSA encrypts/decrypt the symmetric key of AES [18]. ECC is also another option of asymmetric approach, and ECIES gives the complete integrated approach to secure the cloud data. So the need for framework is required to achieve confidentiality, authentication, and integrity as well as also it reduces the computation burden.

4 Proposed Framework

- Effective framework averts the data from various attacks in the cloud. Performance of framework depends on techniques that are applied in a particular framework for preventing the data. It shows that the selection of security techniques for framework is very important. Various approaches are being applied for solving the security issues like CAPTCHA, OTP, authentication, encryption, and integrity schemes. Various techniques and methods are available for each approach, so selection of effective techniques or methods is directly proportional to the security of the system. Separate server enhances the security by reducing the chances of server compromise situations and also reduces the overhead which improves the speed. Proper resource utilization improves the availability and reduces the time complexity of all running processes (Fig. 3).
- The user has data and wants to keep it on the cloud. First, it sends the verification request which is verified by Authentication Server (AS) and allows the genuine user to store the data. AS applied two factors, namely username–password and OTP, to authenticate the user. CAPTCHA is also used to prevent the cloud from bots attack.
- At the user side, encipher the original data by utilizing AES approach then use an asymmetric approach like RSA or ECC to encipher the AES key. This process is handled by the Cryptographic Server (CS). The user also generates the tag of enciphered data using SHA-256.
- The user sends the data in an enciphered form through the Internet to the cloud server, and keeps their data in the cloud.
- When the user wants to access the data from the cloud, then first it is verified by the authentication server through CAPTCHA and the above two verification factors. After successful authentication, the user accesses the enciphered data and generates the tag again using SHA-256, and matches both tags. If tags are same, then accept the data otherwise reject.
- At the user side, first, decipher the AES key by applying the asymmetric approach and with the help of AES key interpret the original data.
- Strong password prevents dictionary attacks. OTP secures the system from Keylogger and replay attacks. CAPTCHA averts from brute-force attack.
- Security algorithms are having two major issues in cloud, where one is time complexity and the other is security. Time complexities are depending on server and allotted resources at cloud, so efficient utilization of resources at cloud directly improves the other issues like availability, time complexity of all running algorithms including security algorithm, and access time. Nowadays, optimization methods also play an imperative role in recourse allocation, server selection at the cloud side. Bio-inspired optimization techniques like Cuckoo search, Honey bee, Particle swarm, Bat, Flower pollination algorithms are widely researched in the last few years for resource scheduling. Optimization algorithm like CMA-ES is also giving good results to find the quality cipher in case of DES symmetric encryption. Avalanche effect is applying to calculate the avalanche value which

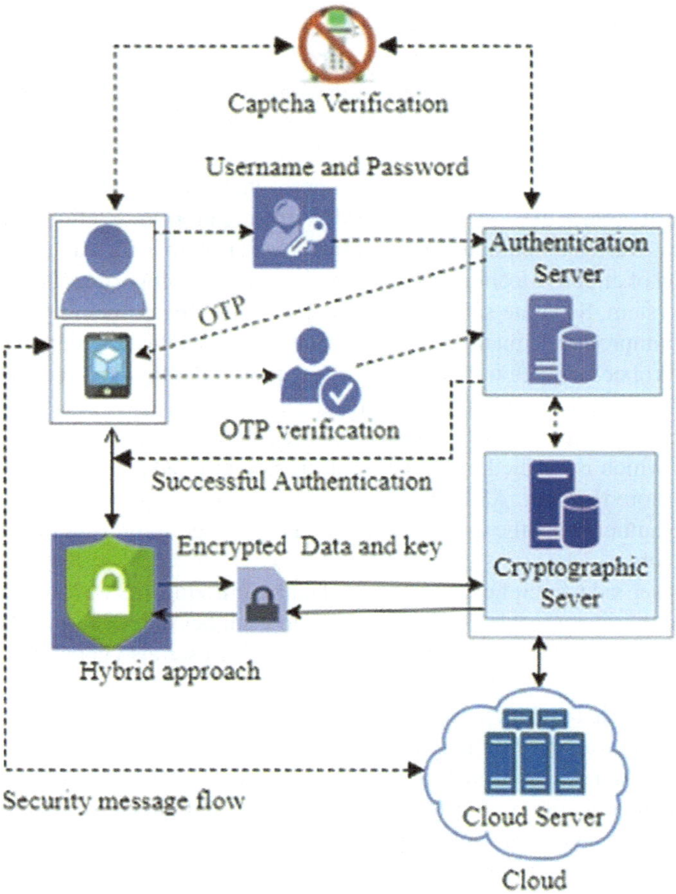

Fig. 3 Proposed framework with hybrid approach

measures the quality of cipher-text, where higher value presents the higher strength of the cipher. Any change in plaintext or key generates a lot of change in cipher-text. CMA-ES produces the many cipher-texts for the same data by applying a little bit change in key, then find avalanche value. In this way, find the symmetric key which provides the quality cipher with high avalanche value by CMA-ES optimization [19]. So optimization improves efficiency by resource utilization as well as enhances the security by providing an excellent key.

5 Conclusion

In the modern era, cloud computing services are top-rated and used by individuals as well as organizations. A lot of users keep their data on the cloud which are situated at a remote location. It creates issues regarding trust and security, which are an essential requirement for cloud acceptability. This paper implements various symmetric approaches in the cloud environment and proposes a framework including CAPTCHA, TFA, separate computing server, and hybrid encryption. CAPTCHA protects from bots, TFA applying two factors username-password and OTP to confirm the genuine user. Two different computing-servers help to reduce the overhead of computing at cloud as well as enhancing security aspects, hybrid encryption combined symmetric approach AES with asymmetric approach RSA or ECC for better security and speed. Bio-inspired optimization Cuckoo search approaches are suggested for better resource utilization and safe server selection. CMA-ES is a good option to optimize the ciphertext in case of DES symmetric encryption so, in this way, this paper gives better implementation of various cryptographic methods in secure framework for achieving privacy, confidentiality, and integrity in cloud. Moreover, it protects from bots attack and reduces the burden of the server.

References

1. D. Zissis, D. Lekkas, Addressing Cloud Computing Security issues. J. Future Generation Comput. Syst. Elsevier **28**, 583–592 (2012)
2. S. Singh, Y.S. Jeong, J.H. Park, A survey on cloud computing security: issues, threats, and solutions. J. Netw. Comput. Appl. (2016)
3. N. Roshanbin, J. Miller, A survey and analysis of current captcha approaches. J. Web Eng. **12**(1&2), 001–040 (2013)
4. A. Ometov, S. Bezzateev, N. Makitalo, S. Andreev, T. Mikkonen, Y. Koucheryavy, Multi-factor authentication: a survey. J. Cryptography, MDPI 1–31 (2018)
5. D. Coppersmith, The data encryption standard and its strength against attacks. J. IBM J. Res. Develop. **38**(3) (May, 1994)
6. D. Coppersmith, D.B. Johnson, S.M. Matyas, A proposed mode for triple-DES encryption. J. IBM J. Res. Develop. **40**(2) (March, 1996)
7. S. Almuhammadi, I. Al-Hejri, A comparative analysis of AES common modes of operation, in *30th Canadian Conference on Electrical and Computer Engineering (CCECE)*. IEEE (2017)
8. B. Schneier, J. Kelsey, D. Whiting, D. Wagner, C. Hall, N. Ferguson, Twofish: A 128-Bit Block Cipher (1998)
9. E. Biham, O. Dunkelman, N. Keller, A. Shamir, New attacks on idea with at least 6 rounds. J. Cryptology (2013)
10. H.M. Elkamchouchi, A.E. Takieldeen, M A. Shawky, A modified serpent based algorithm for image encryption, in *35th National Radio Science Conference, IEEE* (2018)
11. F. Alaa Kadhim, G.H. Abdul-Majeed, Enhancement CAST block algorithm to encrypt big data. in *Annual Conference on New Trends in Information and Communications Technology Applications-(NTICT'2017)* (2017)
12. A. Mousa, A. Hamad, Evaluation of the RC4 algorithm for data encryption. Int. J. Comput. Sci. Appl. **3**(2) (2006)
13. R.L. Rivest, M.J.B. Robshaw, R. Sidney, Y.L. Yin, The RC6 Block Cipher, Version 1.1 (1998)

14. A. Karakra, A. Alsadeh, A-RSA: augmented RSA, in *SAI Computing Conference 2016, IEEE* (2016)
15. V.G. Martinez, L. Hernandez, A.Q. Dios, Security and practical considerations when implementing the elliptic curve integrated encryption scheme. Taylor Francis **39**, 244–269 (2015)
16. A.A. Alkandari, I.F. Al-Shaikhli, M.A. Alahmad, Cryptographic hash function: A high level view, in *International Conference on Informatics and Creative Multimedia, IEEE* (2013)
17. F.F. Moghaddam, S.G. Moghaddam, S. Rouzbeh, S.K. Araghi, N.M. Alibeigi, S.D. Varnosfaderani, A Scalable and Efficient User Authentication Scheme for Cloud Computing Environments, IEEE (2014)
18. X. Li, J. Chen, D. Qin, W. Wan, *Research and Realization based on hybrid encryption algorithm of improved AES and ECC* (ICALIP, IEEE, 2010)
19. M. Tyagi, M. Manoria, B. Mishra, A framework for data storage security in cloud, in *Advances in Data and Information Sciences, Lecture Notes in Networks and Systems* (2017)

Home Automation With NoSQL and Node-RED Through Message Queuing Telemetry Transport

Naman Chauhan and Medhavi Malik

Abstract In the last decades, home automation technology has evolved and achieved popularity. This paper discusses the current and emerging home automation technologies that will make life easier and swifter. Most automation systems use a smartphone and a microcontroller. Current systems are mostly dependent on local Wi-Fi or Bluetooth connections for the transfer of data from the user to the microcontroller. In this paper, wireless communication through mobile data or the Internet and controlling the data through a centralized and secure NoSQL database on Google Firebase are studied. The user will be connected to the system using a mobile application built for both Android and iOS. Moreover, the advantages and drawbacks of different home automation systems are discussed.

Keywords Home automation systems · Firebase · NoSQL · MQTT · Wi-Fi · Raspberry pi

1 Introduction

Today the technology is progressing rapidly to make the life of mankind easier, efficient, and secure. The home automation system not just provides automated access to the devices and appliances of your home to make your day easy but it also saves you valuable time. Automation of home appliances can also be taken into consideration for timed control of appliances for saving electricity. But with the advantages come the drawbacks. Many systems of home automation are localized and thus cannot be remotely accessed. Those systems which are able to provide you remote access

N. Chauhan · M. Malik (✉)
Department of Computer Science and Engineering, SRMIST, Delhi-NCR Campus, Ghaziabad, Uttar Pradesh, India
e-mail: medhavimalik28@gmail.com

N. Chauhan
e-mail: chauhannaman98@gmail.com

S. S. Dash et al. (eds.), *Intelligent Computing and Applications*,
Advances in Intelligent Systems and Computing 1172,
https://doi.org/10.1007/978-981-15-5566-4_37

come with a proprietary, dedicated device which act as the control center and often they have low-level security for the database or high latency in toggling the devices remotely.

This paper will discuss the various technologies which have been used to overcome the snags in the previous systems. This home automation system has a cloud-based NoSQL (Not only SQL) service which is being provided by Google Firebase. The database can be accessed through the mobile application signed-in with the Google account. This provides a personalized and secure way to store and access. Each home has a dedicated server that shall handle all the requests and updates. This can be a Raspberry Pi or any other minicomputer supporting IEEE 802.11 connectivity [1]. Most of the automation uses the HTTPS protocol for communication but this paper will discuss integrating the lightweight Message Queuing Telemetry Transport (MQTT) for data transfer. The microcontroller being used is the ESP8266 by Espressif Systems running on NodeMCU firmware. This further controls the relays to switch on or off the appliances.

Different home automation systems are developed by different authors to automatically on and off the appliances. We will discuss the design and development of activation and control of the home automation system via the Android or iOS application using a microcontroller. It predominantly concentrates on the control of home appliances remotely when the person is far from the home.

2 Previous Technologies

2.1 Infrared

The most primitive and cheap method of automating a device is using the IR radiation. In the electromagnetic spectrum, IR radiation lies between microwaves and visible light. Consumer electronics like TV, security system, and short-range communications use this technology. It is simple and cheap but it comes with downsides. It can only be used for line-of-sight communication and cannot penetrate walls. It can only be used for short-range [2].

2.2 Radio Frequency

A pair of radio frequency modules (transmitter and receiver) can be used to automate the home. Common modules used to work on the frequency of 433 MHz or modules like ZigBee which is a more powerful and long-range device operates on radio bands of 868, 915, and 2.4 MHz [2]. It creates a wireless mesh network but it has a disadvantage of having radio interference with other radio services in the surroundings.

2.3 Bluetooth

Many consumer electronics are equipped with Bluetooth. At the speed of 1 Mbps, Bluetooth can transfer data to about 10 m wirelessly [2]. As mobiles have Bluetooth, this can be used for controlling appliances through mobile easily using an application. But it has a snag as its operating range is not more than 10 m.

2.4 Global System of Mobile communication (GSM)

By using the cellular network GSM/GPRS module delivers 850/900/1800/1900 MHz performance for voice, SMS, and data with low power consumption [2]. As cellular networks have a wide range of coverage; it is possible to control home appliances from very far off. Offering operations from remote regions and vast coverage, these are very much reliable technologies for home automation. These modules require a Subscriber Identification Module (SIM) to connect to a telecommunication service provider. Also, the telecommunication provider charges rental fees and data charges for SMS, voice, and data services. Thus, these are comparatively costlier as they incur charges while operating.

2.5 Wireless Fidelity

Most Wi-Fi devices use 2.4 GHz frequency and implement frequency division multiplexing technology. Either personal computer or smartphone can connect with the home Wi-Fi router which gives access to the local router's network through which we can communicate with other devices connected with the network [2]. These systems have a server that handles the operations which provide better-centralized control as compared to the above-discussed technologies. Most of the systems made on Wi-Fi technology are usually localized to the router's network only. If Internet access is provided to the network, we can spread the reach of the system. This paper is all about using this technology for expanding the reach of the automation system.

3 Proposed System

3.1 Overview

The proposed home automation uses the IEEE 802.11 Wireless Fidelity connectivity at 2.4 GHz frequency to establish a connection between the ESP8266 and the server which is actually a Raspberry Pi 3B running on Raspbian OS [1]. Debian-based Raspbian operating system works mostly like an Ubuntu desktop but unlike a computer, it is as small as a credit card [1, 3]. For the server's backend, Node-RED developed by IBM is used which has a graphical flow editor. Node-RED makes a Nodejs server without writing any JavaScript code. Wi-Fi router creates a network in the home which is to be automated and server and the ESP8266 hubs connect to it which further controls the relays.

3.2 Architecture

ESP8266 is connected to the server through the Wi-Fi network and communicates using the Message Queuing Telemetry Transport protocol. It is a machine to machine extremely lightweight connectivity protocol and has a very small code footprint that requires less bandwidth and does decreases latency in the device toggling. The client application is logged into the Google account to which the firebase is connected. Now, the client application changes the state of an appliance in the app and the update is sent to the firebase's real-time database over the mobile Internet or any Wi-Fi connection. On the server, Node-RED tracks the updates in the database on firebase [4]. When any update takes place in the database, the payload is injected into the flow and published to the specific topic via MQTT. The microcontroller which has subscribed to that specific topic takes the command and changes the relay state of the one having the required appliance attached to it [5].

Figure 1 explains the whole architecture of the proposed system with a sequence diagram. Server, ESP8266, and relays can be collectively known as a home hub. There can be many home hubs in a single system and there can be many ESP8266 within a home hub which depends upon the requirements like the number of rooms or number of appliances. Also, there can be many relays connected to a single ESP8266 depending upon the number of appliances to be controlled.

3.3 Working and Testing

When a button is pressed, the application sends an update to the firebase over the Internet, be it mobile data, Wi-Fi, or any other means of connecting to the Internet. Firebase authenticates the update request and then the update to the database. As soon as the real-time database is updated, Node-RED running on the home server takes the updates into the flow of the server and injects the payload to the other node where the payload is published to the topic corresponding to the appliance state changed via MQTT protocol [5].

When the subscribed ESP8266 gets the payload, it checks for the case which needs to be executed and then changes the state of the pin required. Pin state defines the state of the relay attached to it. Relays are the best way to control the appliances which

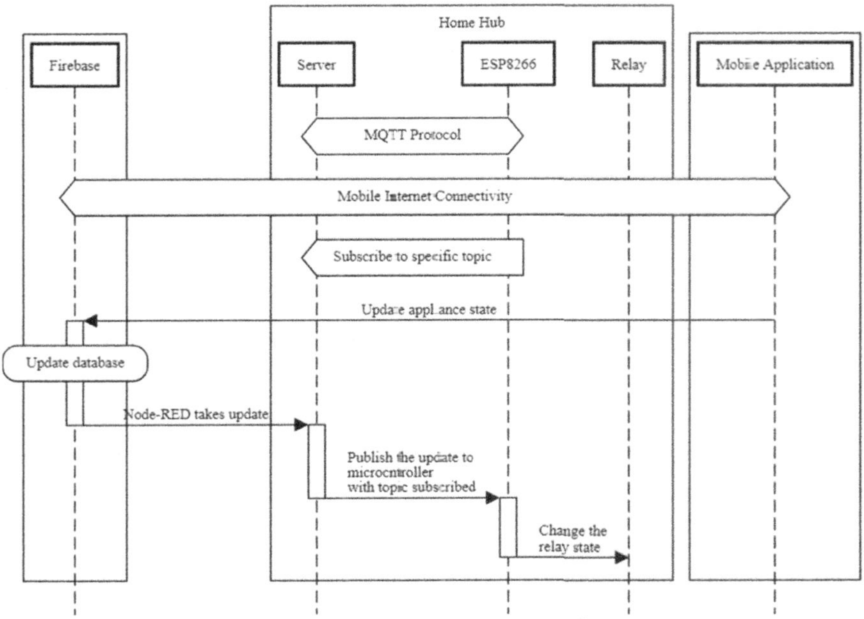

Fig. 1 Sequence diagram of the proposed automation system

make a bridge between the high potential alternating current and low potential direct current. Figure 1 having the sequence diagram explains this working in a graphical mean.

A minimal working example has been tested by the black box testing technique. Black box testing ensured that the system gave correct and desired output, given various possible inputs. The average time taken by the system to switch on or off the appliance when a button is clicked from the client application was calculated which was further compared with the average time the system took to do the same in reality while testing. It can be more or less depending upon the speed of the Internet to which the mobile and server are connected.

4 Conclusion

4.1 Testing and Issue Fixing

While testing it was observed that when the server goes offline due to some power failure or connectivity issue, database updates from the firebase do not reach the server. If the buttons are pressed repeatedly from the application while the server is offline, the firebase updates are not sent to the server but they are stored in a queue. When the server comes online, all the updates are sent to the server from the queue and the server publishes them further to the microcontroller. This leads to the toggling of appliances when the server comes online. Repetitive switching on or off appliances can damage them which is not desirable.

This issue can be solved by using the firebase on disconnect node which writes data when Node-RED disconnects from the firebase servers. When the Node-RED server goes offline, the value of server status on firebase is modified to false which can be read by the mobile application. The mobile application can now get to know that the server is offline and thus, it disables the buttons so that no repetitive toggling is done. When the server comes online again, Node-RED sets the value of server status on firebase to true. When the application reads the change, it enables the button for the operations. Figure 2 explains this feature through a sequence diagram.

Also, if the Internet of the system is down for a longer period due to some unforeseen reason, the server will no longer be able to get updates from the database. Without the Internet, the system will not work as no updates can be made to the database. This issue can be solved by manually overriding the automation system. Figure 3 shows the circuit diagram for the manual overriding for this issue fix.

This can be done by connecting the relays in parallel to the manual switches in the home. If the relays keep the circuit open, the other parallel connection by switchboard can close the circuit manually.

4.2 Advantages and Disadvantages

This system provides access to the home automation system from any corner of the world. No matter how far the client application is sending updates to the firebase, as long as the mobile phone is connected to the Internet, the home automation system can be operated. The system maintains a NoSQL database of the home automation system, unlike other previous systems that are secure with Google authentication. The communication of data over the Internet is encrypted using Secure Socket Layer (SSL) encryption. Moreover, the home server to microcontroller communication is done using the Message Queuing Telemetry Transport which has a very lightweight and small footprint.

Even though two major issues of this system have been addressed, the system still has some snags which cannot be fixed. This system is totally dependent on the Internet which means that if the Internet goes down for some reason, the system fails to operate and manual overriding is done. Even though it works on Wi-Fi like previously discussed systems that work on a local network only and does not depend on the Internet connection, it cannot work in a similar manner. It requires some major changes to be done on the home server to do so.

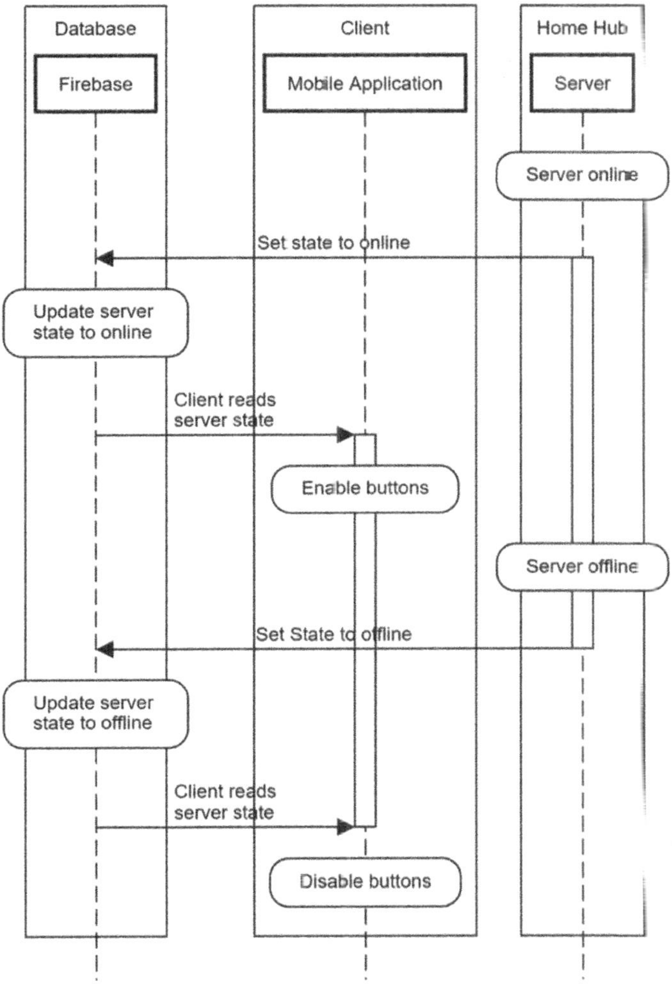

Fig. 2 Sequence diagram of server status updating to fix switch repetition issue

4.3 *Future Prospects*

The system has been experimentally proven to work with sample appliances attached to it and the devices were controlled successfully through a wireless mobile phone. The system still has a few disadvantages like every other system and these can be fixed in the future.

Features like voice-controlled commands from the application or Google Assistant's integration through Google actions can also be done to make a personalized

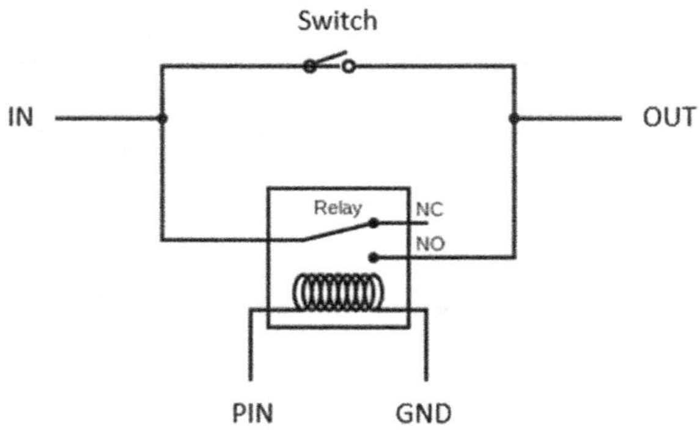

Fig. 3 Circuit for manual overriding

chatbot in the Google Assistant. Also, the home automation system can not just be available for android or iOS devices but it also is accessed through a web client by hosting a web application that updates the database just like a mobile application does.

Home automation systems have the scope of several technological advancements as new technologies are being introduced with time. This will take this system to homes and it can be commercially available to common masses in the near future.

References

1. R. Prashant, S. Khizaruddin, R. Kotian, S. Lal, Raspberry Pi based home automation using Wi-Fi, IoT, and android for live monitoring. Int. J. Comput. Sci. Trends and Technol. **5**(2), 363–386 (April, 2017)
2. Types of wireless communication technology used in home automation. Electronics For You (19, April 2018). Available: https://electronicsforu.com/resources/learn-electronics/wireless-technology-types-home-automation. Accessed on 12 October 2019
3. M.P. Sathish, Dr. S.A.K. Jilani, D. Girish Kumar, Home automation through e-mail using Raspberry Pi. Int. J. Adv. Res. Electron. Commun. Eng. **4**(9), 2475–2480 (September, 2015)
4. Node-RED-contrib-firebase, Node-RED (February, 2018). Available: https://flows.nodered.org/node/node-red-contrib-firebase. Accessed on June 2019
5. Documentation|MQTT v5.0, MQTT Version 5.0 (03, April 2019). Available: https://docs.oasis-open.org/mqtt/mqtt/v5.0/os/mqtt-v5.0-os.html. Accessed on June 2019

Naïve Bayes Algorithm Based Match Winner Prediction Model for T20 Cricket

Praffulla Kumar Dubey, Harshit Suri, and Saurabh Gupta

Abstract In the previous statistical analysis of sports, it is very difficult to predict the winning team of the game/sports before the game starts. Nowadays, it has become a challenge for all. There are a lot of indoor and outdoor games but Cricket is a very popular game in every country over the world which is played between two teams. Each team has 15 players in which only 11 selected players play a match. The objective of this paper is to predict the winner of a cricket match through various machine learning algorithms like logistic regression, support vector machines, k-NN, Naive Bayes, and random forest classifiers applied on the past match performance records and improve the accuracy of the model.

Keywords Logistic regression · Naïve Bayes · SVM · K-NN · Random forest

1 Introduction

In the last few years, we have contributed to the success in the field of sports with the help of statistical modeling which has been used in sports for decades, and has contributed to the success of sports. There are two most popular outdoor games in the world, i.e., Cricket and football. Data analytics is easy in cricket as compared to other games such as baseball, football, and basketball [1]. Various natural factors influencing the game of cricket are huge media coverage, and a large betting market has given sturdy incentives to model the game from different views [2]. However, the complex rules governing a game like cricket, the ability of the team's players on

P. K. Dubey (✉) · H. Suri · S. Gupta
Computer Science and Engineering, SRM Institute of Science and Technology, Delhi NCR Campus, Ghaziabad, India
e-mail: praffullakrdubey@gmail.com

H. Suri
e-mail: suriharshit007@gmail.com

S. Gupta
e-mail: saurabh256837@gmail.com

S. S. Dash et al. (eds.), *Intelligent Computing and Applications*, Advances in Intelligent Systems and Computing 1172,
https://doi.org/10.1007/978-981-15-5566-4_38

a given day of the match, and their performance play an important role in influencing the final outcome of a cricket match. This presents several challenges in predicting game outcomes with accuracy. The game of cricket is played in all three formats—Test match, ODI, and T20, but our research focuses on limited over matches, i.e., T20 (IPL) and ODI— the most popular format of cricket. In this paper, we use Naive Bayes, random forest, and k-NN classifier algorithms for predicting the match's winner. The algorithms used in this research are explained in the model below.

1.1 Supervised Learning

The task of machine learning to teach a computer's function that maps input to an output based on input–output examples is known as supervised learning. It deduces a function from the marked training data consisting of various training examples sets. In this technique, each example is a pair which consists of an object as input object (generally a vector) and a value of output as desired (also known as supervised signal). The observed technique analyzes the training data and creates a deduced function that can be used to map new instances. This scenario allows the learning algorithm to uniquely identify class labels for unseen instances. The various supervised learning algorithms are

1.1.1 Linear Regression

If the class and attribute data are numerals, a linear regression classifier is used for classification. This is the primary statistical method. The approach in linear regression is to derive the class expressions in terms of a linear combination of pre-established weights and attributes [3]. It is of two types:

1. **Simple Linear Regression**

$$Q = cP + f$$

where

f is the y-intercept
c is the slope
Q is the dependent variable
P is the independent variable

2. **Multilinear Regression**

$$Y = q_0 + q_1 r_1 + \cdots + q_n + r_n$$

where

q_1 is the regression coefficient
r_1 is the independent variable
Y is the dependent variable

1.1.2 K-Nearest Neighbor

In machine learning, the k-nearest neighbor (k-NN) algorithm is a non-parametric technique that is applied for regression and classification. In these cases, k-closest training example acts as input. The output depends on whether k-NN is applied for regression or classification analysis. Distance measurements are used to determine which examples of K in the training data set are similar to the new inputs. For real-valued input variables, Euclidean distance is the most common distance measure [4]. The distance of two data points can also be evaluated using Manhattan, Hamming, Jaccard, Minkowski, Mahalanobis, and some other distances [1]. The k-NN algorithm applies the closest neighboring proposal for classification. It takes a set of examples from the training data that is divided into several groups or categories; training data set is taken as an input and it is classified into unlabeled test data set examples into one of those groups or categories. K-NN can also be termed as a sluggish learner technique. K-NN models are never explicitly formed. This contributes to a faster development training phase but a slower classification phase [5].

Algorithm

Classify (T, U, g)
for e from 1 to g;

Compute the distance (T_i, g)

End
Calculate set H containing indices for the k smallest distance (T_i, g).
return majority label {U_i where I belongs to H}
where training data is specified by T. class labels of T are specified by U, and an unknown sample is specified by g.

1.1.3 Naive Bayes Classifier

Naive Bayes is one of the classifiers of the Bayesian type that uses predicting as the likelihood of class membership and assigning a target class to investigate the data instances. A Bayesian classifier is a statistical classifier that approximates the probability of a given tuple belonging to a class. Naive Bayes is a curious learner [1]. Various trends show that Naive Bayes is more accurate in comparison to other more sophisticated and complex techniques. The Naive Bayes classifier makes an assumption that each attribute possesses a distinct effect on class label, irrespective of the other attribute values. Hence, this independency is called class-conditional independence. These classifiers are based on Bayes' theorem [4].

Bayes Theorem:
Let E be a data tuple and F be a class label
Let E belong to class F, then $P(F|E) = P(E|F)P(F)/P(E)$
where

$P(F|E)$ is the posterior probability of class F given predictor E
$P(F)$ is the prior probability of class
$P(E|F)$ is the posterior probability of E given the class F
$P(E)$ is the prior probability of the predictor

The classifier evaluates $P(F|E)$ for each class Fi for a given tuple E. Then it will predict that E belongs to the highest posterior probability class, conditioned on E. It means E is affiliated to the class Fi if and only if $P(Fi|E) > P(Fj|E)$ for $1 \leq j \leq x, j \neq i$.

1.1.4 Random Forest Classifier

Random forest is an extremely powerful algorithm in machine learning and is an ensemble method. Random forest classifiers create a set of discordant or inconsistent trees that are disjoint and that choose the best class for creating a random forest [6]. Each of these trees has the same chance of sampling and is drawn randomly from a set of possible trees. The random trees can be effectively created, by combining with a collection of trees that leads to accurate models. The algorithm used is

1. **Bagging**

The random forest training algorithm applies a general technique of bootstrap aggregation or bagging to tree learners.

With the given training set and response set bagging repeatedly (B times) selects a random sample with replacement of the training set and fits into the samples:
For $b = 1, \ldots, B$
Sample, with replacement of the training examples
Train the regression or obtained classification tree
After training, predictions for unseen samples can be made by taking the average of all the predictions from all the individual regression trees or by taking the majority when classification trees are used.

1.1.5 Support Vector Machines

Isabelle Guyon, Bernhard Boser, and Vladimír Vapnik introduced the thorough idea of support vectors in their work. SVMs are less prone to overfitting and are highly accurate. Using non-linear mapping, SVM converts real data into a data which contains data with high dimension. SVM is used for both classification and numerical prediction. Then it looks for a hyperplane that is linearly optimal in this new dimension that separates the tuples of different classes with each other. The tuples belonging to the two classes can always be differentiated by a hyperplane when properly mapped to a sufficiently high dimension. Using the support vectors and the margins defined by the support vectors, the algorithm obtains this hyperplane. The support vectors found by the algorithm gives a useful description of the learned prediction model. The separating hyperplane is expressed in the following form: $Q \cdot A + b = 0$, where Q denotes the weight vector, $Q = \{Q_1, Q_2, Q_3, \ldots, Q_n\}$, the number of attributes is specified by n and the scalar, often referred to as bias is specified by b [4].

If we provide input with two attributes $P1$ and $P2$, and the training tuples are two dimensional, (e.g., $A = (A_1, A_2)$), where A_1 and A_2 are the values of attributes $P1$ and $P2$, respectively. Then, the points occurring above the separated hyperplane are classified to be belonging to the

Class $P1$:

$$Q \cdot A + b > 0$$

And any points occurring bottom to this separating hyperplane can be classified to be belonging to the

Class $P2$:

$$Q \cdot A + b < 0$$

1.1.6 Logistic Regression

It is widely applied for problems in classification. It can be used to manage different types of relationships as it uses a non-linear logarithmic transformation for the odd ratio prediction. It is not necessary for logistic regression to have a linear relationship between variables that are dependent and independent. In order to avoid under-fitting and overfitting, we must not fail to include all the variables that are important. A better technique for ensuring such practice is to estimate logistic regression by using methods like gradual method. This leads to the requirement of big sized samples as maximum possible estimates are least efficient than normal at least square at low sized samples. There is no correlation between independent variables, i.e., no multiplicity. However, the interaction effects of categorical variables can be included in the analysis and model. If the value of the dependent variable is ordinary, it is referred to as ordinal logical regression. If a dependent variable relies on multiple classes, it is termed as multinomial logical regression [7].

The simple form is

$$L(q) = 1/1 + e^{-q}$$

where

$L(q)$ is the variable that increases with time and is a fraction between 0 and 1.

Variable q stands for time.

1.2 Unsupervised Learning

Unsupervised learning can be defined as a technique of machine learning where you do not have to oversee the model. In fact, we must allow the model to process on its own to predict information. It mainly deals with unlabeled data. These algorithms allow us to execute more sophisticated processing tasks in comparison with supervised learning. Compared to other natural learning methods, unsupervised learning can be highly unpredictable [8].

1.2.1 Partition Algorithms

The partition clustering algorithm gets one partition of data in place of structures as clustering structures, such as a dendrogram created by techniques of hierarchy. The problem accompanying in using a partial algorithm is the selection of the number of output clusters that are required. These techniques have benefits in implementations involving various big sets of the data where dendrogram construction is computationally prohibited. Divided or partitioned techniques typically create clusters by optimizing the criterion function defined either globally (defined overall significant patterns) or locally (on a small subset of patterns). The combinatorial search for a set of possible labeling for the ideal criterion value is plainly prohibited from the calculation point of view. Therefore, in use, the partition algorithms are usually implemented by running them multiple times with various initial states that are different from each other, and after various runs, the most significant configuration that is obtained will be final output clustering [9].

1.2.2 K-Means Algorithm

The K-means algorithm, possibly one of the firstly proposed cluster algorithms, which has its base on a very basic and simple notion, provided the initial set of clusters, assigning each point to one of them available, and then each cluster center is replaced by a midpoint on the corresponding cluster. These 2 basic steps are repeated until convergence. The cluster is assigned a point that is close in the Euclidean distance to the point. Even though K-means are preferred as they have the benefit of being easily put into various operations, it has 2 major disadvantages. The first one is that it can be proved to be one of the extremely slow algorithms because at each step the distance among each point and each cluster must be evaluated, which can be highly expensive in the existence of a large data set. Secondly, the technique is indeed sensitive to the initial clusters provided. But in recent years this trouble has been solved with a certain degree of success [9]. At last, the algorithm is aimed at optimizing the objective function which is known as the squared error function, and is given by

$$J(O) = \sum (i = 1 \; to \; c) \sum (j = 1 \; to \; c_i)(\|r_i - o_j\|)^2$$

here,

'$\|r_i - o_j\|$'	is the Euclidean distance between x_i and v_j.
'c_i'	is the number of data points in ith cluster.
'c'	is the number of cluster centers.

Steps for k-means clustering

- Let $R = \{r_1, r_2, \ldots, r_n\}$ be the sets of data points and $O = \{o_1, o_2, \ldots, o_n\}$ be the set of centers
- Randomly select the cluster centers 'c'
- Calculate the distance among each point of data and centers of clusters and assign data points to cluster center whose distance is minimum and recalculate the new center for the cluster by the equation:

$$O_i = (1/c_i) \sum (j = 1 \; to \; c_i) O_i$$

- Now calculate the new distance between the data points and new center
- If no data point was reassigned the halt, else repeat the procedure [10].

1.2.3 Hierarchical Clustering Algorithms

An efficient algorithm of such kind is hierarchical clustering; it is applied in the famous arithmetic-based software like MATLAB. It is an algorithm which belongs to agglomerative algorithm having few changes based on the metric used to determine the distances of clusters. For single points Euclidean distance is used. No criteria are known as to which clustering distance should be used, and it seems to be sturdily reliant on the data set [9].

1.3 Reinforcement Learning Algorithm

In reinforcement learning algorithm, the evaluation of learner's views is used to set the standards of correctness in the service of behavioral goals [11].

2 Methodology

Data science is a technique of processing and analyzing data, i.e., applied for a precise intention. This technique is also beneficial in different project performance predictions, i.e., useful in business analytics [12]. It contains six steps that are

- **Data Import**: we have obtained the relevant data set for our project, i.e., 'matches.csv'. It is a csv file and it contains six hundred and thirty-six rows and eighteen columns. This data set has been imported and used for the analysis as well as prediction.
- **Data Cleaning**: Data cleaning is a very time-consuming process. Most of the data comes with missing parameters or duplicate values so it becomes important to clean our data so that our data set contains only the required data. In our data set, we found many columns that were useless as they were not required for prediction so we dropped them and our data set also contained many columns that had null values in them so we dropped them also. Many values were redundant and some of the columns in the data set had no significance in the prediction process so those columns were dropped during the cleaning process.
- **Data Munging/Wrangling**: It is the process of transforming and mapping data from raw format into other forms of data that is more appropriate and valuable for a variety of purposes such as analysis and visualization of the data present in the data set [13].
- **Data Visualization**: In this, we try to show the various analysis of the data present in the data set through various graphs or patterns. The analysis is shown with the help of various graphs present in mat plot library of python.
- **Data Modeling**: The process of creating a descriptive diagram illustrating the relationship between different types of information stored in a database is called data models. These diagrams can be graphs which include joint plot, histogram, scatter plot, bar-plot, etc.
- **Cross-Validation**: The process of training one set of data and testing it using other data set is called cross-validation. We have used Naïve Bayes to train our model and test our model to do prediction.

3 Proposed Algorithm

Step 1: Collect match Statistics of various countries T20 cricket (RD_1, RD_2, …. RD_n) where RD = Resource Data
Step 2: Data Preprocessing

1. Cleaning of data, i.e., clean of data
2. Normalization, i.e., normalize the values
3. Transformation
4. Integration of the data

Step 3: The data is to be split into Training data and Test data for training the model
Step 4: Create a model using Naïve Bayes Algorithm
model = sc.Naive Bayes(x_train)
Step 5: Apply Test data on model
test = sc.fit_transform(test)
model.fit(x_train,y_train)
prediction = model.predict(x_test)
Step 6: Predictive outcome

1. Knowledge Discovery in terms of various graphs like histogram, scatter plot, line graph, etc.
2. Performance accuracy of the model
3. Score prediction in terms of overs and score of the team.

4 Result

In our project, we have applied support vector machine, random forest, logistic regression, k-NN, and Naïve Bayes algorithm and after training our model on the data set we found that the accuracy of our model is approximate 71% for Naïve Bayes Algorithm, 66% for k-NN, 64% for logistic regression, 63% for random forest, and 59% for Support Vector Machine. We found that using Naïve Bayes Algorithm we get the maximum accuracy for our model therefore we can conclude that using Naïve Bayes Algorithm we get more accurate results as compared to other algorithms used. Figure 2 shows the most succesful IPL team till the year 2017 (Figs. 2, 3, and 4).

5 Conclusion and Future Scope

The important limitation in carrying out this project was the limited data set, which I had at my end. The next step in the direction to improve the accuracy of prediction problems at hand would be to test out the approaches and various techniques proposed in this paper using a larger and more representative data set. Also, I would like to extend the features like venue, pitch condition, and weather condition. If Deep Neural network (Tensorflow, Thaeno, and Keras) comes into the implementation, the accuracy will be even higher. Also, a similar model can be developed for other sports like Football, Kabaddi, Olympics, Asian Games, and so on. We can further increase the accuracy of our model by using unsupervised learning and reinforcement learning.

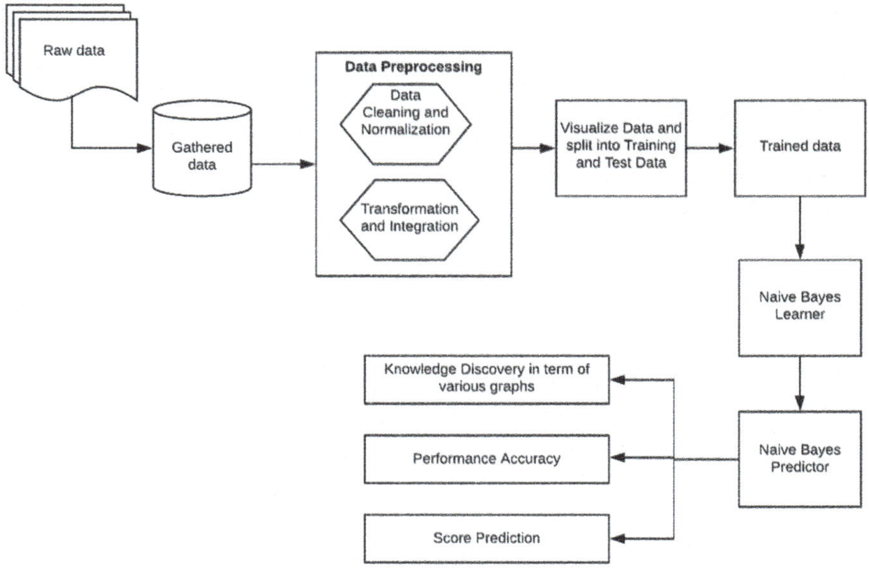

Fig. 1 Methodology

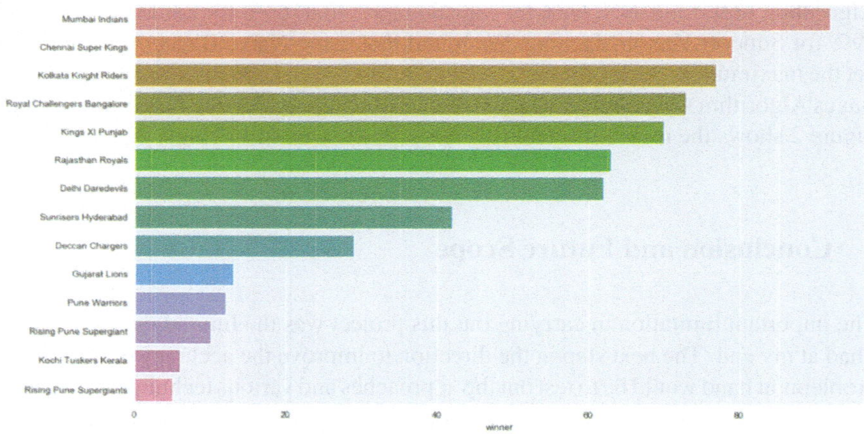

Fig. 2 The most successful IPL team till 2017

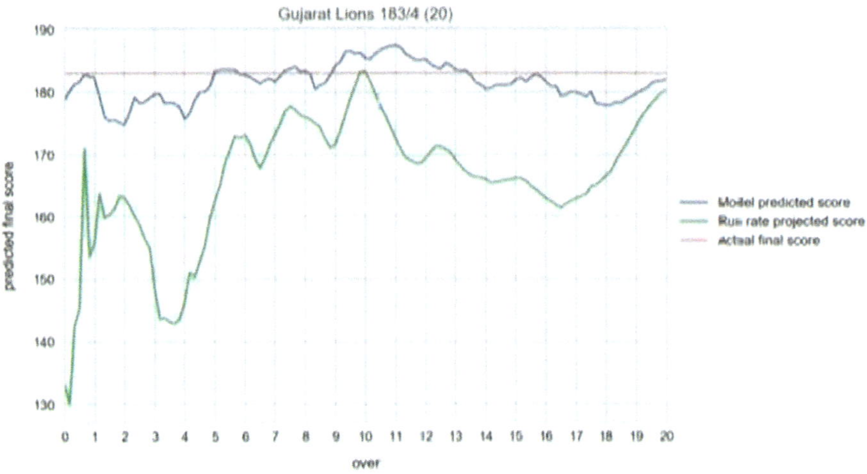

Fig. 3 Performance of the model trained

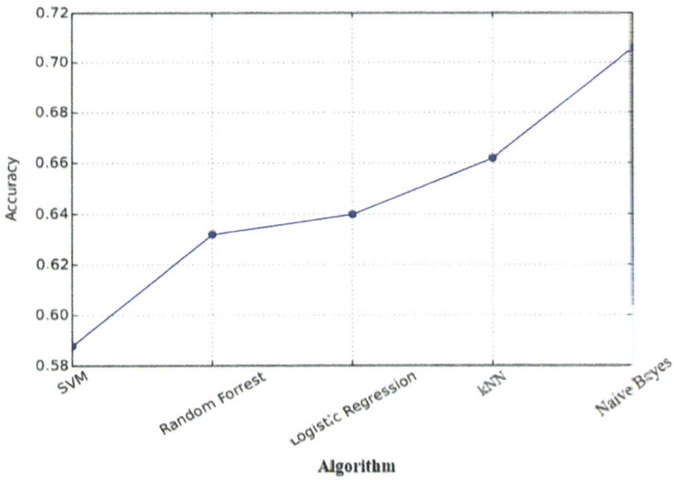

Fig. 4 Accuracy of various classifiers applied

References

1. V. Purkayastha, Predicting outcome of cricket matches using classification learners. MSc Research Project, 2017
2. M.G. Jhawar, V. Pudi, Predicting the outcome of ODI cricket matches: a team composition based approach, in *European Conference on Machine Learning and Practice of Knowledge Discovery in Databases (ECML-PKDD 2016)*. Report No: IIIT/TR/2016-1

3. T. Singh, V. Singla, P. Bhatia, Score and wining prediction in cricket through data mining, in *International Conference on Soft Computing Techniques and Implementation (ICSCTI)* (2015)
4. S. Kumar, S. Roy, Score prediction and player classification model in the game of cricket using machine learning. Int. J. Sci. Eng. Res. **9**(8) (2018). ISSN 2229-5518
5. https://cse.buffalo.edu/~jing/cse601/fa12/materials/classification_methods.pdf
6. H.A. Sulaiman, M.A. Othman, M.F.I. Othman, Y.A. Rahim, N.C. Pee, *Advanced Computer and Communication Engineering Technology: Proceedings of ICOCOE 2015*, vol. 362 (Springer, 2015)
7. A. Nimmagadda, N.V. Kalyan, M. Venkatesh, N.N. Sai Teja, C.G. Raju, Cricket score and wining prediction using data mining. Int. J. Adv. Res. Dev. **3**(3) (2018)
8. https://www.guru99.com/unsupervised-machine-learning.html
9. S. Mythili, E. Madhiya, An analysis on clustering algorithm in data minings. Int. J. Comput. Sci. Mob. Comput. (IJCSMC) **3**(1) (2014)
10. https://sites.google.com/site/dataclusteringalgorithms/k-means-clustering-algorithm
11. R. Lokhande, P.M. Chwan, Live cricket score and wining prediction. Int. J. Trend Res. Dev. **5**(1) (2018). ISSN: 2394-933
12. https://medium.com/@praffullakrdubey/understanding-data-science-pipeline-and-its-general-terms-5898c6927707
13. Wikipedia, https://en.wikipedia.org/wiki/Data_wrangling

Design of an Efficient Deep Neural Network for Multi-level Classification of Breast Cancer Histology Images

H. S. Laxmisagar and M. C. Hanumantharaju

Abstract Breast cancer is the second most prominent cause of death among women, and it is a major health problem across the world over the past many years. In this work, four hundred different biopsy images were collected from the Bio-imaging 2018 breast histology classification challenge by participating Grand Challenge on breast cancer histology images (BACH) for multiclass breast cancer histology image classification using deep learning technique. Computer-aided detection or diagnosis (CAD) system plays an important role in the detection and to increase death survival rate of women suffering from breast cancer, which can decrease death rate among women. The main purpose of this paper is to develop a CAD system which can be used to detect whether the input biopsy image fits one of the four different classes like benign, in situ, Invasive and Malignant. We have incorporated an efficient lightweight neural network such as MobileNet2.10ex instead of normal convolution neural network for feature extraction and at top a fully connected deep neural network is designed to do classification among different classes. The confusion matrix is developed in order to draw the model accuracy of both techniques. It is observed that MobleNet2.10ex is giving about 88.92 % training accuracy at 257 epochs and the corresponding loss value is 0.2249. In order to progress the performance during pre-processing, a staining technique was used to remove staining from a digital scan of histology microscopic image of fine-needle aspirate (FNA) slides and feature invariant approach such as data augmentation technique was used to rotate each image in pipeline about the axis 45° apart three times in order to increase dataset and prediction accuracy. The model was trained till 280 epochs to evaluate training and validation accuracy using Google Colab having strong GPU 2496 CUDA cores, 12 GB GDDR5 VRAM and 12.6 GB RAM.

Keywords Breast cancer · Biopsy · CAD · Benign · In situ · Invasive · FNA · MobileNet · Google Colab · GPU

H. S. Laxmisagar (✉) · M. C. Hanumantharaju
BMS Institute of Technology and Management, Yelahanka, Bangalore 560064, India
e-mail: sagar8.hs@bmsit.in

M. C. Hanumantharaju
e-mail: mchanumantharaju@bmsit.in

S. S. Dash et al. (eds.), *Intelligent Computing and Applications*,
Advances in Intelligent Systems and Computing 1172,
https://doi.org/10.1007/978-981-15-5566-4_40

1 Introduction

Cancer continues to be one of the second [1] biggest health risks in all over the world. There is no doubt that cancer today is a major threat to the society. Breast cancer is a foremost basis of death and touches millions of lives every year. In primary finding could benefit to intensification the endurance of several lives in [2] addition to saving billions of dollars. Hence this area of research has been largely unexplored.

Breast cancer is the greatest vastly destructive cancer disease and a foremost health [3] problem in women. Breast cancer for women is very serious and common cancer. Digital Mammography is the preliminary method used to find breast cancer [4, 5].In filtering all the important features to support the clinical disease identification is a challenging and time-consuming task.

Some of the authors have done a lot of research in the advance of computer-aided detection (CAD) system. In 2009, an expert [6] system was developed for finding breast cancer using association rules (AR) and neural network (NN). In 2017, Breast lump [7] detection and classification in ultrasound images using surface analysis and super-resolution approaches improves the performance of the system. Computer-aided detection or diagnosis (CAD) systems [8] play an important role in the earlier identification of breast cancer which can decrease the death rate among women suffering from breast cancer In [9], machine learning techniques used to improve the diagnostic accuracy of breast cancer from nuclear features of fine-needle aspirates. To speed up the execution process [10] make use of high-performance hardware such as GPU, CPU and RAM. One of the major problems in the analysis of breast cancer microscopic image because of H&E stained so its appearance is variability [11]. It is due to notable differences in light detectors, Optics and light sources used in the scanners.

2 Breast Cancer Dataset

The microscopic image dataset contains 400 high-resolution (HR) uncompressed images and each image has 2040 $\times$ 1536 pixels. It is ordered of hematoxylin and eosin (H&E) stained histology microscopic images from the Bio-imaging breast histology classification challenge 2018. We restrict the scope of the project to just the classification of histology microscopy [12, 13] images. Each image class is labelled with one of four categories such as benign, normal, in situ and invasive. The image class that is labelled was accomplished by two expert pathologists to provide a diagnostic from the image contents. The desired goal of the Bio-imaging challenge is to perform classification between different classes of data.

The dataset has 400 different microscopic images, which are distributed as follows:

1. Normal images: 100
2. Benign images: 100

3. In situ carcinoma images: 100
4. Invasive carcinoma images: 100
5. Histological images have RGB colour model with .tiff format
6. Microscopic image size: 2048 × 1536 pixels
7. Size of the pixel (Height × width): 0.42 μm × 0.42 μm
8. Type of label: image-wise.

3 Data Pre-processing

After obtaining the dataset, we pass the images through two steps of pre-processing namely

1. Normalization
2. Data Augmentation.

Normalization is a pre-processing technique that uses staining technique to remove staining for the biopsy image slides preparation. The HR images are transformed to optical density (OD) using logarithmic [14, 15] transformation. The singular value decomposition (SVD) is then applied to the optical density tuples and the obtained colour space transform is used to the unusual biopsy image. Then the histogram image [16] is stretched so that it covers the lower 90 % of the data. Figure 1 shows some image samples before and after normalization. The staining allows the pathologist to identify the type of cancer and its severity in the tissue under the microscope. But computer does not require staining because using algorithm it can operate on pixel wise. Computer does not require different colour for understanding.

Data augmentation is nothing but adding similar data to the existing dataset to further substantiate the model. Further, a dataset is created from the normalized images to rotate the image in pipeline. To avoid under fit due to low number of samples, data augmentation technique using image rotation and flipping [16] is incorporated without altering the diagnosis and image quality to further increase [17] the dataset. Once the data has been normalized and augmented, it is fed as input to the CNN model. The model proceeds to learn the underlying features in the four classes of images following which, it will be able to make predictions on images without predefined labels.

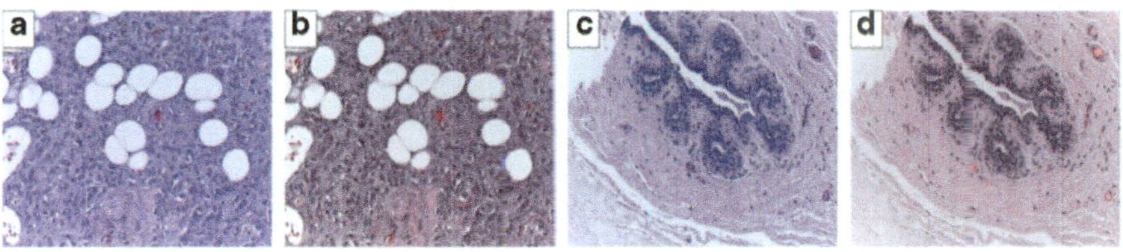

Fig. 1 Normalization of images (before *v/s* after)

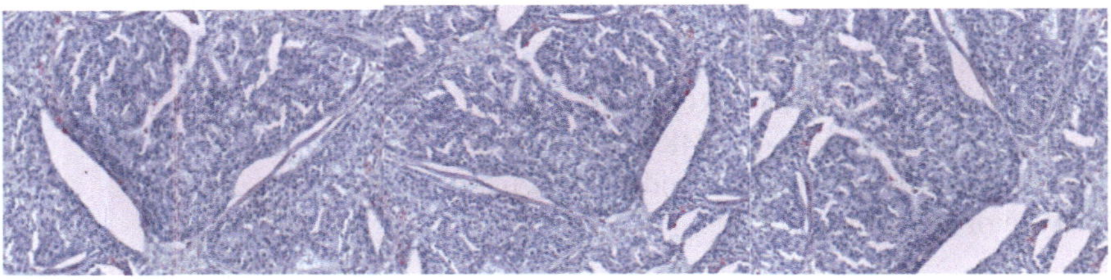

(a) Original Image (b) Data augmentation-1 (c) Data augmentation-2

Fig. 2 Image augmentation (rotation about the axis)

Keras has this ImageDataGenerator class which lets the handlers to achieve image augmentation in a very simple way. The ImageDataGenerator class has three different approaches flow_from_directory(), flow_from_dataframe() and flow() to read the images from folder having images and a big NumPy array.

The dataset whatever we are using is very small for our deep neural network for training the model. To avoid it, we used different data augmentations on pre-processed histology images to rotate about the axis, cropping and flipping of the image in order to develop robust model and to prevent overfitting problem. Figure 2 depicts data augmentation of original image with respect to axis.

4 Training and Testing Phase of Proposed CAD System

Figures 3 and 4 represent the two phases such as training and testing phase of the proposed CAD system. During pre-processing about 400 h (High-resolution) images were converted to low-resolution (LR) images using image normalization. In that 352 LR images were used for training and remaining 48 LR images were used for testing.

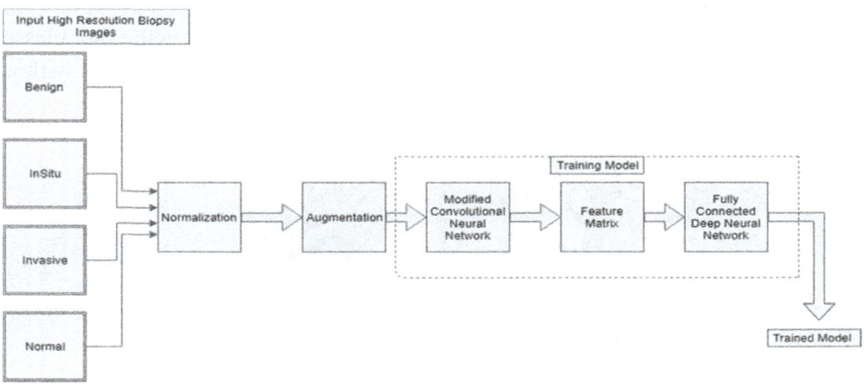

Fig. 3 The training stage of the proposed CAD system

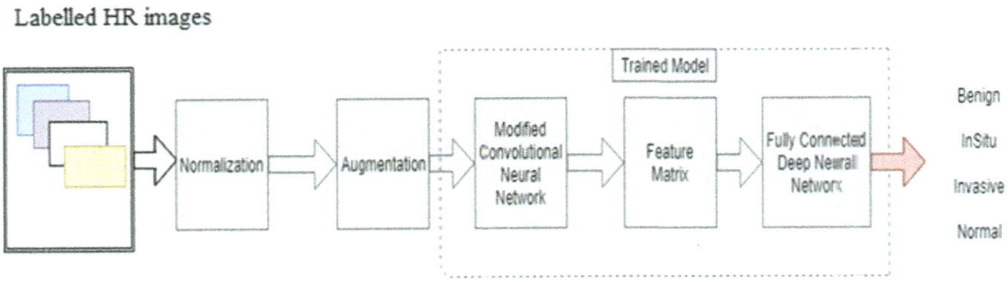

Fig. 4 The testing stage of the proposed CAD system

During training phase about 88 LR images were allocated for each class such as benign, in situ, Invasive and Normal. During testing phase about 12 LR images were reserved for each of the above class of data. Observe that during training phase the class labels (benign, in situ, Invasive and Normal) of the input LR image sequences are known. The difference between testing and training phase is that during testing the known class label of the LR input image is fed into the trained model after pre-processing to detect whether the patient belongs to one of the four categories.

4.1 Framework of Histology Image Classification

To perform convolution operation a MobileNet2.10ex convolutional neural network approach is used, which is depth-wise separable convolution that includes two layers one layer is used to perform depth-wise convolutions and another layer is used for point-wise convolutions. At the output of each neuron, a ReLU activation function is used to avoid dying of neuron. We can use depth-wise convolution layer to apply a single filter for each input depth channel and a 1×1 convolution for [18] combining. This process is continued and feature matrix is extracted which contains labelled data. The feature data is given as input to deep neural network which contains four hidden layers to perform classification between four different class of data (Fig. 5).

4.2 Architecture of MobileNet

The architecture is shown in Fig. 6. The depthwise separable convolutions divide standard convolution into a depthwise convolution and a 1×1 pointwise convolution [19]. The standard convolutional filters (Fig. 6a) are replaced by two layers first one is depthwise convolution (Fig. 6b) and second one is pointwise convolution (Fig. 6c). The depth-wise separable convolution uses two layers in that one layer is used to perform filtering and another layer is used for combining. Computation cost and the size of model can be reduced by using this factorization technique. The

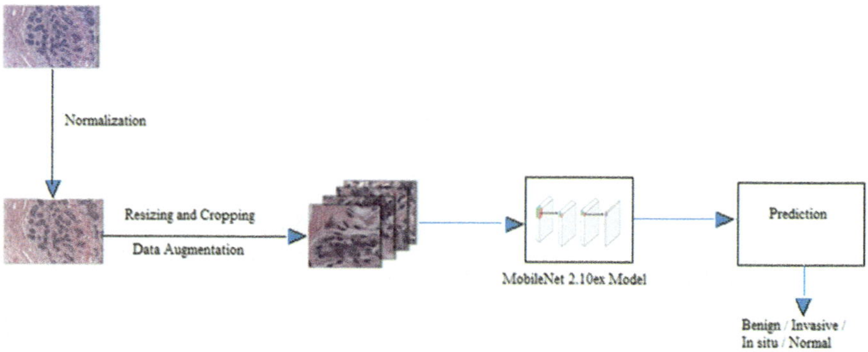

Fig. 5 High-level operational overview

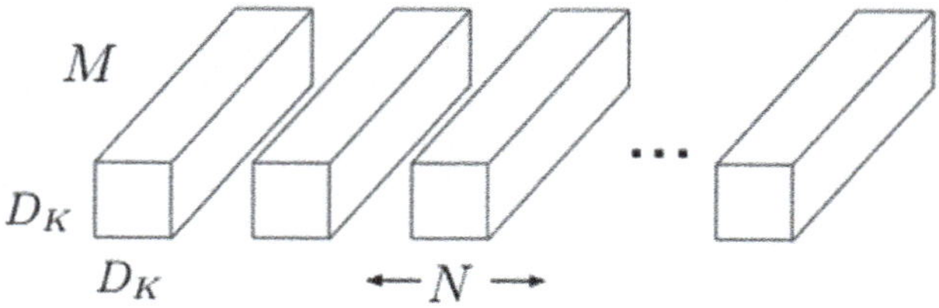

(a) Standard Convolution Filters

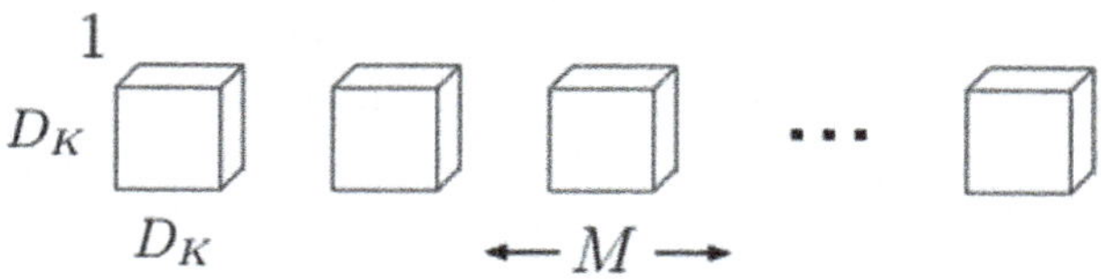

(b) Depthwise Convolutional Filters

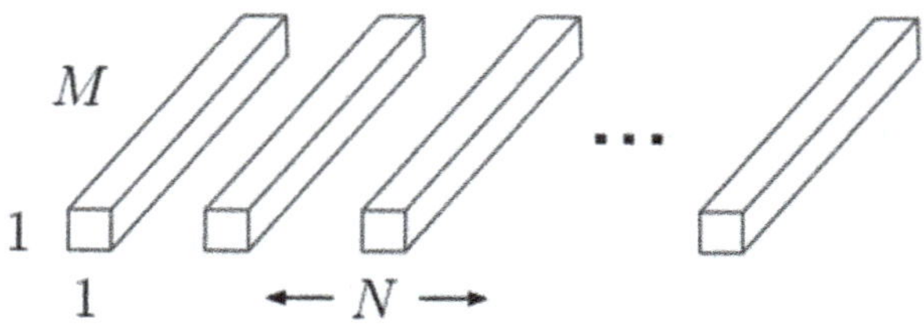

(c) 1×1 Convolutional Filters called Pointwise Convolution in the context of Depthwise Separable Convolution

Fig. 6 Comparison between Standard, depth-wise and point-wise convolutions presented in [19]

MobileNet architecture is a lightweight convolutional neural network. It uses depth-wise separable convolutions as it does a single convolution on each colour channel instead of combining all three channels such as RGB and flattening it. MobileNet [20] is a stack of the separable convolution modules which are composed of depth-wise convolution and a 1×1 point-wise convolution.

Suppose in Fig. 6, we have five channels, then it is almost five $D_K \times D_K$ spatial convolution. Point-wise convolution is the 1×1 convolution to change the dimension.

Operation cost = Depth-wise Separable Convolution Cost: Depth-wise + Point-wise Convolution Cost

$$D_K \cdot D_K \cdot M \cdot D_F \cdot D_F + M \cdot N \cdot D_F \cdot D_F$$

where 'N' represents the output channels, 'D_K' and 'D_F' indicates Kernel size and Feature map size, 'M' represents the input channels.

The Standard Convolution Cost is:

$$D_K \cdot D_K \cdot M \cdot N \cdot D_F \cdot D_F$$

Thus, the computation reduction is:

Depth-wise Separable Convolution Cost/Standard Convolution Cost

When $D_K \times D_K$ is 3×3, 8–9 times less computation can be achieved, but with only small reduction in accuracy.

Depth-wise convolution can be preferred to decrease the complexity and size of the model. The number of multiplications required is very large using conventional convolution method. Suppose there are 512 $5 \times 5 \times 3$ kernels that change 8×8 times, it results in $512 \times 3 \times 5 \times 5 \times 8 \times 8 = 2{,}457{,}600$ multiplications.

In the depth-wise convolution, we have three $5 \times 5 \times 1$ kernels that move 8×8 times. That's $3 \times 5 \times 5 \times 8 \times 8 = 4800$ multiplications. In the point-wise convolution, suppose we have 512 $1 \times 1 \times 3$ kernels that transfer 8×8 times, i.e $512 \times 1 \times 1 \times 3 \times 8 \times 8 = 98{,}304$ multiplications. Now we need to add them together, that's about 103,104 multiplications. It is observed that 103,104 is very less than 2,457,600. With less multiplications, the deep neural network is able to process more data in a shorter amount of time. In deep neural network, the speed and power consumption depend on number of multiplications and accumulates unit, which directly measures the number of multiplication and addition operations.

4.3 *Model Training and Validation*

4.3.1 Model Specifications

- Architecture—MobileNet2.10ex
- Epochs trained for—280
- Classifier used—Fully connected Neural network

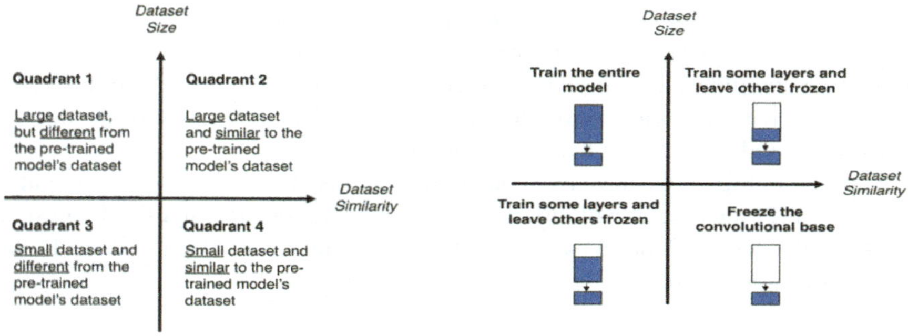

Fig. 7 Transfer learning strategies [21]

- Activation function—Soft Max
- Other layers—ReLU
- Loss function—Categorical Cross Entropy
- Batch Size—32
- Optimizer—Adam.

4.3.2 Transfer Learning

During the process of training the model, the concept of transfer learning is implemented. In computer vision, transfer learning can be expressed using pre-trained models. Consequently, due to this less multiplication is required for training such models, it is very good practice to use models such as VGG, Inception, MobileNet (Fig. 7).

4.4 Various Architectures of Convolution Neural Networks

4.4.1 MobileNetV2 (Apr. 2018)

MobileNetV2 architecture was developed in April 2018 [21] appropriate for embedded and mobile-based vision applications in which there is deficiency of computational power.

4.4.2 AlexNet (2012)

In 2012, AlexNet was developed. It is similar to MobileNet [22] architecture but has more filters in each layer with stacked convolutional layers.

4.4.3 VGGNet (2014)

VGGNet was developed in the year 2014, and it is having sixteen convolutional layers and is very analogous to AlexNet but has lots of filters [23]. Presently, it is one of the best algorithm choices in the community to do research for extracting different features from each image.

4.4.4 ResNet (2015)

At last, a new architecture called residual [24] neural network (ResNet) has lower complexity compared to VGGNet. Using this technique, we can able to train a neural network model with 152 layers.

5 Experiments and Results

Our dataset consists of total 400 h hematoxylin and eosin (H&E) stained images categorized as benign, in situ, invasive and normal (total 100 images of each class). Figure 8 depicts the sample of four different classes of images. The required histology microscopy image dataset is available at http://iciar2018-challenge.grand-challenge.org/dataset/.

The dataset whatever we are using is very small for our deep neural network for training the model. To avoid it, we used different data augmentations on pre-processed histology images to rotate about the axis, cropping and flipping of the image in order to develop robust model and to prevent overfitting problem.

In total, we have conducted 280 iterations of the model training procedure, with the objective of obtaining the best possible results with the given technical constraints. In order to progress the performance during pre-processing a staining technique was used to remove staining from a digital scan of histology microscopic image of fine-needle aspirate (FNA) slides and feature invariant approach such as data augmentation technique was used to rotate each image in pipeline about the axis 45° apart three times in order to increase dataset and prediction accuracy. Below is

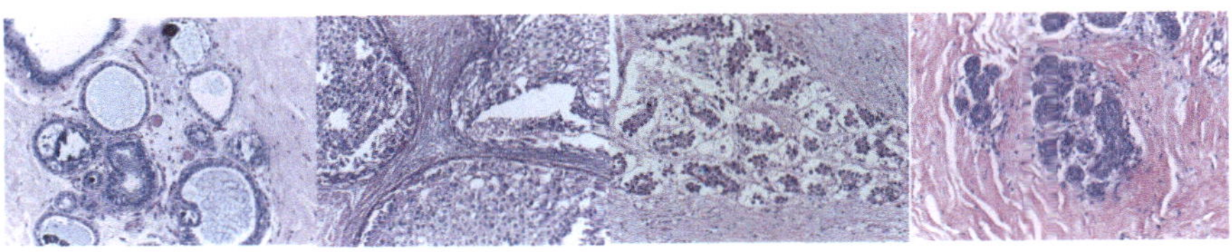

(a) Benign (b) In situ (C) Invasive (d) Normal

Fig. 8 Sample histology images of ICIAR 2018 challenge on Breast Cancer Histology [12] (BACH) a graphical summary of the results obtained over the course of the model training process.

5.1 *Model 2.10ex*

The model was trained till 280 epochs to evaluate training and validation accuracy using Google Colab having strong GPU 2496 CUDA cores, 12 GB GDDR5 VRAM and 12.6 GB RAM. Once the model is trained it is required to plot training and validation percentage accuracy related to the number of epochs used for deep learning. Figure 9 depicts the graphical illustration of training and validation accuracy which is moving very close to each other and the gap between the two is minimum (Fig. 10).

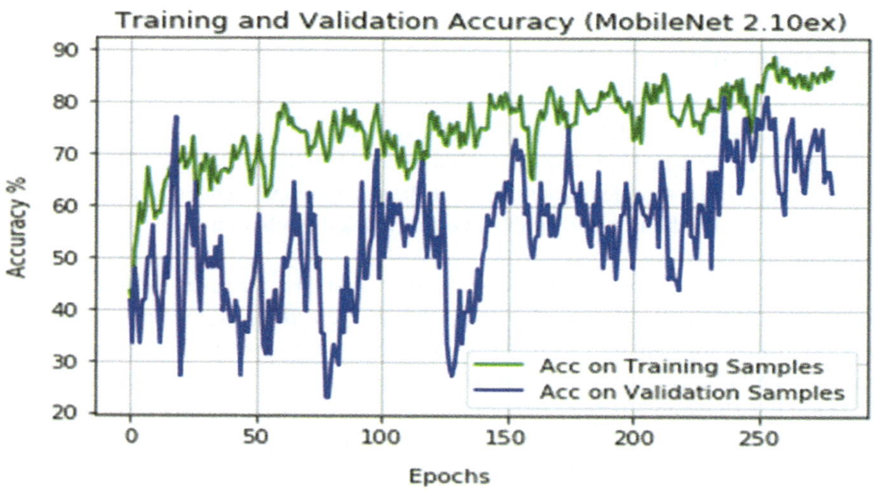

Fig. 9 Depicts training and validation accuracy using MobileNet2.10ex

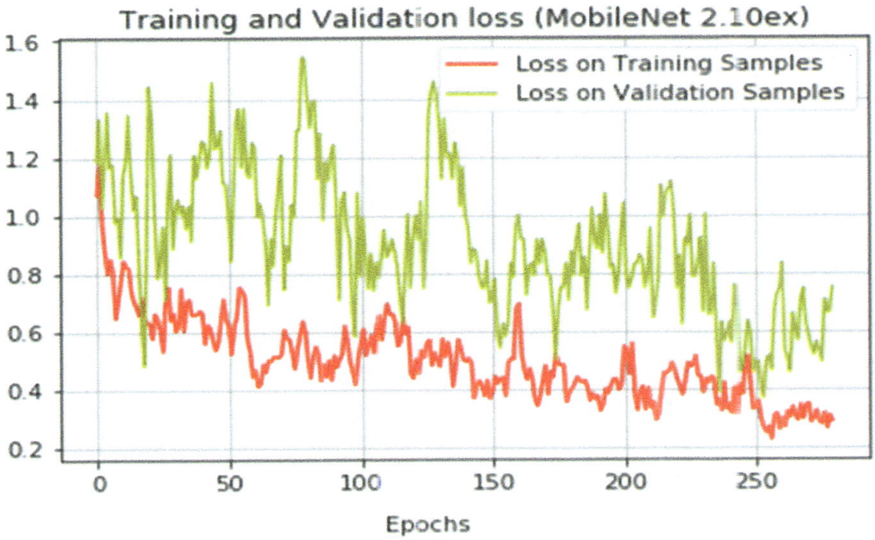

Fig. 10 Depicts training and validation loss using MobileNet2.10ex

Sample Code:

```
def plot_training(hist):
   accr = hist.hist['accr']
   accr=[ele*100 for ele in accr] # Precentage Value of Accuracy
   value_accr = hist.hist['value_accr']
   value_accr=[ele*100 for ele in value_accr] # Precentage Value of Acc
uracy
   loss = hist.hist['loss']
   value_loss = hist.hist['value_loss']
   epochs = mention range (len(accr))

   plt.grid(visible=True)
   plt.plot(iteration, accr, 'g',label='Accr on Training Samples')
   plt.plot(iteration, value_accr, 'b',label='Accr on Validation Samples')
   plt.legend(loc='best')
   plt.xlabel('Epochs')
   plt.ylabel('Accuracy %')
   plt.title('AccuracyTraining and Validation (MobileNet 2.10ex)')
   plt.savefig('Training_and_ValidationAccuracy (MobileNet 2.10ex)
.png')
plt.figure()
   plt.grid(visible=True)
   plt.plot(iteration, loss, 'r',label='Loss on Training Samples')
   plt.plot(iteration, val_loss, 'y',label='Loss on Validation Samples')
   plt.xlabel('Iteration')
   plt.legend(loc='best')
   plt.title('Training and Validation loss (MobileNet 2.10ex)')
   plt.savefig('Training_and_ValidationLoss (MobileNet 2.10ex).png'
)

plot_training(ModelEvolution)
```

6 Conclusion

The major work done in this paper is to do the classification between four different classes of data such as normal, benign, in situ and invasive using his to pathological images. The dataset, obtained from the ICIAR-BACH challenge 2018, was augmented to create a more versatile dataset to feed into the convolutional neural network. In the feature extraction step, the deep CNN architecture of MobileNet is used. The MobileNet was trained to differentiate among four classes of data and its parameters were altered to classify microscopic medical images. The creation and exploration of various architecture versions produced a high performing architecture with a calculated accuracy of 87.5 %.

This version of the MobileNet architecture is very promising not only due to the high accuracy obtained but also due to its resource-efficient architecture. This low-cost model works out to be an exciting opportunity for industry-related projects in the arena of breast cancer detection. In future, we can implement this lightweight model using FPGA. For future scope, other architecture of deep neural networks will be recommended to evaluate performance characteristics which include visual geometry group (VggNet) and the residual network (ResNet) architecture.

References

1. D. West, P. Mangiameli, R. Rampal, V. West, Ensemble strategies for a medical diagnosis decision support system: a breast cancer diagnosis application. Eur. J. Open. Res. **162**, 532–551 (2005)
2. Deep learning for cancer diagnosis. https://software.intel.com/en-us/articles/deep-learning-for-cancer-diagnosis-a-bright-future
3. D. Sun, M. Wang, A. Li, A multimodal deep neural network for human breast cancer prognosis prediction by integrating multi-dimensional data (2018), pp. 1–10
4. S.-M. Choua, T.-S. Leeb, Y.E. Shaoc, I.-F. Chenb, Mining the breast cancer pattern using artificial neural networks and multivariate adaptive regression splines. Exp. Syst. Appl. **27**, 133–142 (2004)
5. B. Zheng, S.W. Yoon, S.S. Lam, Breast cancer diagnosis based on feature extraction using a hybrid of *K*-means and support vector machine algorithms. Exp. Syst. Appl. **41**, 1476–1482 (2013)
6. M. Karabatak, M.C. Ince, An expert system for detection of breast cancer based on association rules and neural network. Exp. Syst. Appl. **36**(2 Part 2), 3465–3469 (2009)
7. M. Abdel-Nasser, J. Melendez, A. Moreno, O.A. Omer, D. Puig, Breast tumor classification in ultrasound images using texture analysis and super-resolution methods, in *Engineering Applications of Artificial Intelligence*, Dec 2016 (2017), pp. 84–92
8. J. Tang, R.M. Rangayyan, J. Xu, I.E. El Naqa, Y. Yang, Computer-aided detection and diagnosis of breast cancer with mammography: recent advances. IEEE Trans. Inf Technol. Biomed. **13**(2), 236–251 (2009)
9. W.H. Wolberg, W.N. Street, O.L. Mangasarian, Machine learning techniques to diagnose breast cancer from fine needle aspirates. Cancer Lett **77**(2–3), 163–171 (1994)
10. P. Rajendra, A. Subbarao, G. Ramu, V. Brahmajirao, Prediction of drug solubility on parallel computing architecture by support vector machines. Network Model Anal Health Inf Bioinform **7**(1), 130 (2018). https://doi.org/10.1007/s13721-018-0174-0
11. M. Veta et al., Breast cancer histopathology image analysis: a review. IEEE Trans. Biomed. Eng. **61**(5), 1400–1411 (2014)
12. ICIAR Grand Challenge 2018 on Breast Cancer Histology Images. https://iciar2018-challenge.grand-challenge.org/dataset/
13. P. Athology, A. Sethi, Fast GPU-enabled color normalization for digital pathology (2019)
14. X. Li, S. Member, K.N. Plataniotis. A complete color normalization approach to histo-pathology images using color cues computed from saturation-weighted statistics, Feb 2015.https://doi.org/10.1109/TBME.2015.2405791

15. M. Macenko, M. Niethammer, J.S. Marron, D. Borland, J.T. Woosley, X. Guan,, C Schmitt, N.E. Thomas, A method for normalizing histology slides for quantitative analysis. 3 Statistics and Operations/Research, 4 Lineberger Comprehensive Cancer Center, 5 Renaissance Computing Institute, 6 Pathology and Laboratory Medicine, 7 Dermatology University of Nor. 1107–1110 (2009)
16. Y. Roohani, E. Kiss, Improving accuracy of nuclei segmentation by reducing histological image variability (2018)
17. M.D. Bloice, C. Stocker, A. Holzinger, Augmentor: an image augmentation library for machine learning (2017). arXiv preprint arXiv:1708.04680
18. D. Scherer, M. Andreas, S. Behnke, Evaluation of pooling operations in convolutional architectures for object recognition, Sept 2010
19. Transfer learning using MobileNet and Keras. https://towardsdatascience.com/transfer-learning-using-mobilenet-and-keras-c75daf7ff299. Last access 2018
20. A.G. Howard, M. Zhu, B. Chen, D. Kalenichenko, W. Wang, T. Weyand, M. Andreetto, H. Adam, MobileNets: efficient convolutional neural networks for mobile vision applications (2017). arXiv:1704.04861
21. M. Sandler, M. Zhu, A. Zhmoginov, C.V. Mar, MobileNetV2: inverted residuals and linear bottlenecks. Transfer learning from pre-trained models: https://towardsdatascience.com/transfer-learning-from-pre-trained-models-f2393f124751.
22. Z. Alom, et al., The history began from AlexNet : a comprehensive survey or deep learnin approaches
23. S. Liu, W. Deng, Very deep convolutional neural network based image classification using small training sample size (2015)
24. Y. Chen, Q. Zhang, Y. Wu, B. Liu, M. Wang, Y. Lin, *Fine-Tuning ResNet for Breast Cancer Classification from Mammography* (Springer, Singapore, 2019)

Autonomous and Adaptive Learning Architecture Framework for Smart Cities

Saravanan Muthaiyah and Thein Oak Kyaw Zaw

Abstract The context of smart cities should really be anchored onto two key attributes. First is the ability for a city to learn adaptively with the aid of machine learning (ML) or artificial intelligence (AI) and second is the ability for a city to sustain operations autonomously without any human intervention. While Internet of Things (IoT) is seen as an enabler by making all devices connect to a network that communicates with one another with minimal human interference; however, critical problems such as sewer management, health, parking woes, traffic congestion, pollution, waste management, and noise are not fully being addressed. In this paper, we discussed the most recent literature for smart city initiatives across the globe including the comprehensive Alcatel–Lucent study in 2011 and proposed an overarching autonomous learning city baseline and target architecture with specific functionalities for each layer.

Keywords IoT · AI · Machine learning · Adaptive · Autonomous · Architecture · Smart city

1 Introduction

IoT has been drawing in enormous volumes of research as the world progresses toward making everything connected. It is a future vision that everyday devices will be equipped with microcontrollers, transceivers, and protocol stacks that will communicate with each other making it integrated with the Internet [1]. Connected devices will enable easier access to the sensors, actuators, and so on, which will create many applications that will make life for the people easier. Currently, there are a wide range of applications of IoT in the market for different areas such as medical aids,

S. Muthaiyah (✉) · T. O. K. Zaw
Faculty of Management, Multimedia University, Cyberjaya, Malaysia
e-mail: saravanan.muthaiyah@mmu.edu.my

T. O. K. Zaw
e-mail: rahman93.iu@gmail.com

S. S. Dash et al. (eds.), *Intelligent Computing and Applications*,
Advances in Intelligent Systems and Computing 1172,
https://doi.org/10.1007/978-981-15-5566-4_41

traffic management, home automation, and many more [2]. However, each of the IoT systems has its own mechanism and methods in realizing its objectives. This makes integration for IoT applications into a system or control almost impossible.

Given this complexity, the driving force comes from users, technology partners, and government to truly realize the Smart City vision [3]. Dameri [4] stated that the main problem for the vague definition of a smart city is due to a bottom-up problem where the ideas arise from the technology application itself toward urban problems as shown by Fig. 1. The author stated that the main driver for the smart city was the birth and development of the technology, especially ICT, which connects different digital services. IoT, ML, and AI have the potential for solving urban problems such as crime, pollution (e.g., carbon footprint), and congestion. Nevertheless, with stakeholders having their own objectives without a unified vision or direction efforts for Smart City initiatives seemed to be misaligned. Focus, vision, policy, and rules toward Smart City that are vague is the main reason for the lack of a comprehensive framework. The moment we can obtain the unified vision, a comprehensive definition can be established and urban problems can be solved in a more effective manner toward realizing a true Smart City. As no comprehensive definition exists for a Smart City, one could argue that a city may not reach the status of smart even though most of the devices if not all, are connected. One could also argue that IoT may not be the answer in realizing it. Therefore, elements to achieve in a Smart City must have solid guidelines or statements so that directions toward it can be realized. From literature review and scenario stated above, a clear gap exists for a unified vision and definition on what a Smart City truly should be and how should it be realized. In this paper, we intend to narrow the gap by proposing a baseline and target architecture for an autonomous and adaptive learning Smart City framework.

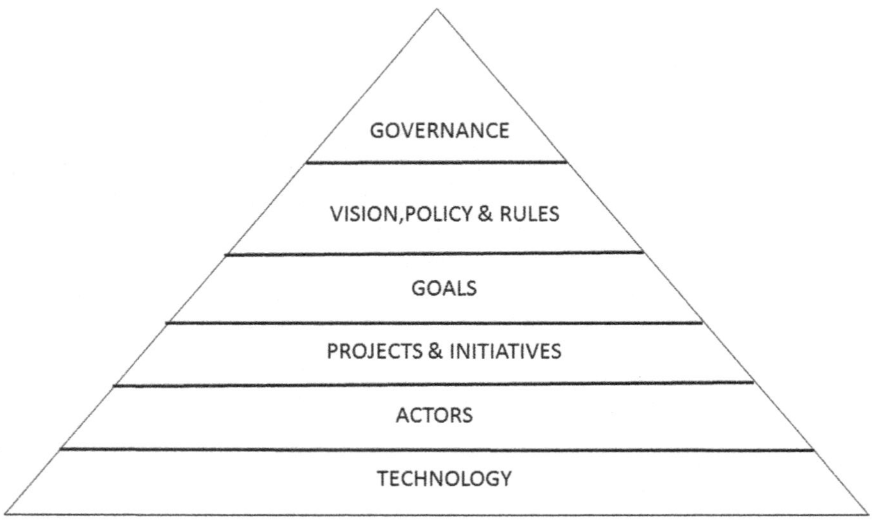

Fig. 1 Bottom-up Smart City development path

2 Smart City Definitions

Giffinger et al. [5] stated that "A Smart City as a well-performing city built on the 'smart' combination of endowments and activities of self-decisive, independent and aware citizens." Caragliu et al. [6] highlighted that a city is smart "when investments in human and social capital and traditional (transport) and modern (ICT) communication infrastructure fuel sustainable economic growth and a high quality of life, with a wise management of natural resources, through participatory governance." Both the definitions differ from each other with the former discusses results and the latter highlights components of achieving it. Rouse [7] on the other hand, stated that "a smart city is a municipality that uses information and communication technologies to increase operational efficiency, share information with the public and improve both the quality of government services and citizen welfare." It means information, communication, and technology should be used by the city to increase the operational efficiency. When the situation happens, a city will be regarded as a smart city. Again, the definition is obscure without any specific methods, technology, or measurement to be used to achieve it. Techopedia [8, 9] bears the same idea on the Smart city. Both authors mentioned the usage of ICT in making a city smart to enhance quality, performance, and address urbanization problems. Perhaps, the most realistic definition is "a smart city is said to be a well-defined geographical area, in which technologies such as ICT, logistics, energy and production cooperate to create benefits for citizens in terms of their well-being" [4]. IoT offers a wide range of capabilities for a Smart City in the context of solving core urban challenges and also being very cost-effective at the same time [10]. Listed below are some of the main points that advocate the need for smart solutions according to a study by Alcatel–Lucent in 2011.[1] The study done across 52 cities around the globe indicate common problems and issues of cities that require smart solutions, they are (1) energy, (2) telecommunications, (3) traffic congestion, (4) intelligent community, (5) public utilities and lastly, (6) industry. In energy adoption, one of the key areas that is being addressed is the Smart Grid [11]. The major contribution is for an efficient power grid to support energy requirements for the city in terms of energy conservation monitoring and quality of service. Telecommunications is needed to support home automation smart buildings and public infrastructure for data acquisition using smart sensors. Key areas also include home surveillance, lighting, panic button, and energy conservation. Although there are research developments, implementation is still low [12]. Traffic congestion is a norm that happens in cities and one of the proposed solutions is public transportation. This problem has persisted after the emergence of cheaper cars. Smart cities should offer a solution by providing monitoring of the traffic and even controlling signal lights when needed. The monitoring also can be possible by having vehicles equipped with GPS sensors [13].

Information from monitoring can greatly help authorities to build better and bigger roads for places that have high traffic. Also, provide an early warning for potential

[1] https://www.tmcnet.com/tmc/whitepapers/documents/whitepapers/2013/7943-alcatel-lucent-getting-smart-smart-cities-recommendations-smart.pdf.

traffic burst that could happen so that prevention can be taken earlier. Another aspect is parking management. Limited parking spaces for cars and motorcycles is a norm in the city (Table 1).

Smart parking, smart meters, and gazetted parking bays can utilize sensors and intelligent displays [14]. Therefore, smart parking will be made available for cars and motorcycles alike as well as reservation of parking bays ahead of time. Intelligent community framework focuses on adding green spaces, recreation, which includes noise as a part of a community.

Although, there are laws enforcing in making noise pollution lesser, it is still rampant especially in city areas where there are many businesses located. Sensors

Table 1 Alcatel–Lucent 2011 Smart City survey

No.	City	No.	City
1	Amsterdam (The Netherlands)	27	Malmö (Sweden)
2	Ballarat (Australia)	28	Masdar (UAE)
3	Besançon (France)	29	Moncton (Canada)
4	Birmingham (U.K.)	30	Ottawa (Canada)
5	Bottrop (Germany)	31	Paredes (PlanIT Valley, Portugal)
6	Bristol (U.S.A.)	32	Pedra Branca (Brazil)
7	Cape Town (South Africa)	33	Porto Alegre (Brazil)
8	Chattanooga (U.S.A.)	34	Quebec City (Canada)
9	Cleveland (U.S.A.)	35	Recife (Brazil)
10	Copenhagen (Denmark)	36	Riverside (U.S.A.)
11	Curitiba (Brazil)	37	Rotterdam (The Netherlands)
12	Dakota County (U.S.A.)	38	Shanghai (China)
13	Dongtan (China)	39	Shenyang (China)
14	Dublin (Ireland)	40	Songdo (South Korea)
15	Dublin (U.S.A.)	41	Sopron (Hungary)
16	Eindhoven (The Netherlands)	42	Suwon (South Korea)
17	Gdansk (Poland)	43	Tallinn (Estonia)
18	Gold Coast City (Australia)	44	Taoyuan (Taiwan)
19	Gujarat international financial tech-city (GIFT, India)	45	Tianjin Binhai (China)
20	Ipswich (Australia)	46	Toronto (Canada)
21	Issy-les-Moulineaux (France)	47	Trikala (Greece)
22	Jubail (Saudi Arabia)	48	Trondheim (Norway)
23	Kalundborg (Denmark)	49	Urumqi (China)
24	Lavasa (India)	50	Windsor-Essex (Canada)
25	Lyon (France)	51	Winnipeg (Canada)
26	Malaga (Spain)	52	Wuxi (China)

Table 2 ML/AI techniques with use cases

Machine learning (ML) techniques	Use cases
Neural network	Traffic woes and weather
Regression and clustering	Healthcare
K-means nearest neighbor	Waste management and home surveillance
Anomaly detection	Weather, pollution, carbon footprint, and traffic
Support vector machine	Weather and pollution

can be deployed for monitoring noise pollution [15]. The information gathered will be provided to authorities so that timely action can be taken. Residents can report or request for noise scanning (dB) smart devices if they feel that their area is getting too noisy. In this area, the community plays a role as well. Public utilities include waste management and carbon footprint per square foot calculator. Cities create problem for waste storage, shortage of garbage landfills. Using intelligent containers that are able to detect the level of waste, collectors can plan their routes more effectively. This in return will reduce cost for collecting waste and improving the quality of recycling [16] and also enable household waste tracking. the industry sector focuses on environment management, which includes air quality. Air in urban areas is heavily polluted. Smart technologies will be able to help to monitor air quality in crowded places such as parks and fitness trails [17].

It will be able to monitor factory concentrated areas as well. Companies residing in the polluted area can be imposed with carbon taxes so that efforts will be undertaken in making the quality of air better in those places. Public can also be warned of affected areas beforehand to live an environmentally friendly lifestyle. The technology can also be used in buildings for rainwater harvesting and creating smart building index as well. Although a city may differ in problems that it faced in its city, smart solutions will be able to intelligently adjust to its requirements. However, IoT alone can only be an enabler, what is needed is an overarching framework which is proposed in the next section (Table 2).

3 Overarching Framework for Smart Cities

While we agree with the author, our contribution is the overarching framework that includes specific layers, i.e., (1) Acquisition, (2) Data, (3) Business, and (4) Application architectures that make a city smart. Therefore, after connecting all the dots, we propose the framework (see Fig. 2) with emphasis on the framework being autonomous and self-learning at the Business Architecture layer. These two items were proposed because Smart should consist of the ability to reason and with very little or no human input (i.e., independent or autonomous). Ability to think on its own

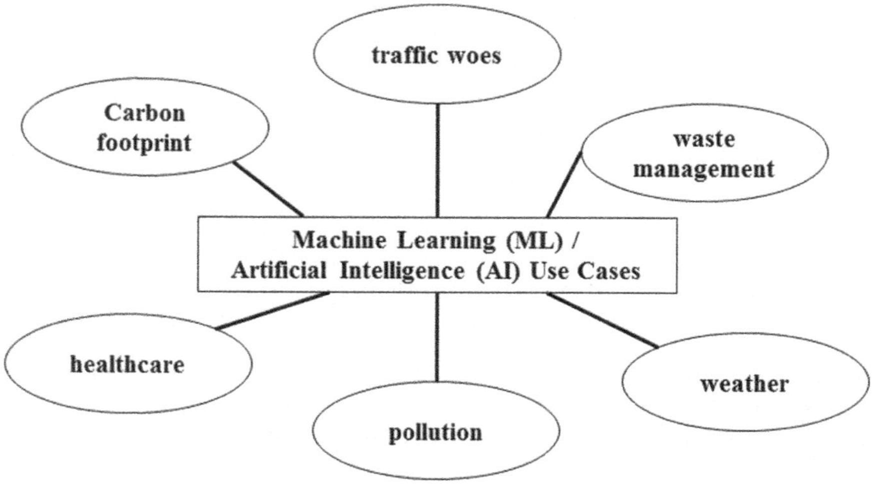

Fig. 2 Smart City ML/AI use cases for smart solution

will require learning algorithms and the actions must be autonomous. Therefore, it is clear that in order to be called smart, a system or device needs to be able to achieve two things: (1) ability to think independently in difficult situations and (2) ability of independent actions. With current IoT as the solution, both of the elements will not be able to be fulfilled, as IoT does not possess learning algorithms for it to think independently in difficult situations. IoT only possesses the first main component, which is completely autonomous with minimum human interaction. Even so, it still requires human intervention even though the interactions are minuscule. It will be made possible by the combination of two technologies, which are IoT and artificial intelligence (AI). Smart technology implies automatic computing principles like self-configuration, self-healing, self-protection, and self-optimization [18]. Therefore, in order to reach a Smart City status, IoT alone is not adequate and a combination of IoT and AI is needed (Fig. 3).

Table 3 illustrates each layer's capabilities of the proposed framework. AI for self-learning must perform tasks that will require intelligence to perform task with machine learning reasoning capabilities. Douglas [19] stated that AI has four main abilities. They are the ability to sense, converse, act, and learn. Although AI and IoT as a solution is not the mainstream solution, the ideas and initiatives with both technologies combined in realizing a Smart City are catching up.

Huawei in 2018 unveiled its artificial intelligence Smart City platform, which utilizes IoT capabilities [20]. Roven [21] and Bailey and Coleman [22] highlighted solutions with both the elements mentioned earlier.

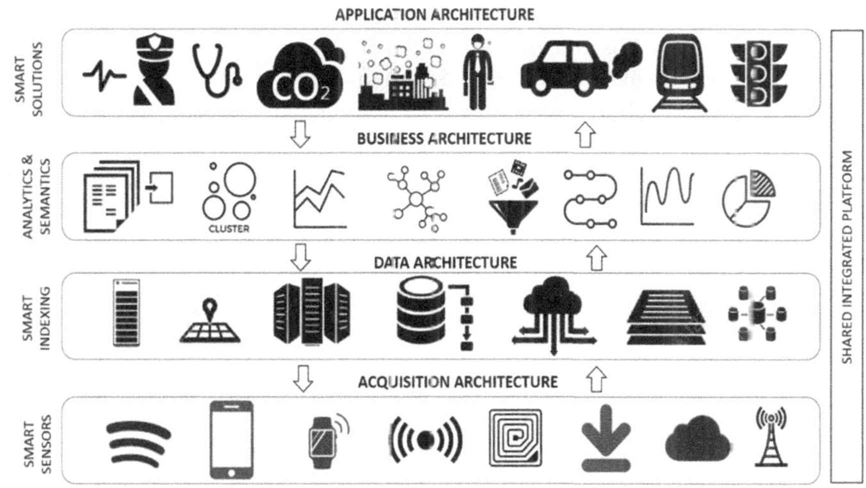

Fig. 3 Overarching autonomous learning framework

Table 3 Layer functionality

Layers	Functionality
Application architecture	i. Present data for service allocation ii. Respond to waste traffic, carbon emission, healthcare, and surveillance iii. Mitigation responses and decision-making
Business architecture	i. Data analytics—prescriptive and predictive ii. Semantic analysis with aided reasoning capabilities iii. Agent learning capabilities supported by autonomous features
Data architecture	i. Data acquisition—multiple sources ii. Data optimization and gleaning—from structured as well as unstructured data iii. Learning and training capabilities
Acquisition architecture	i. Smart readers—feeder for data architecture ii. Smart sensors—feeder for data architecture iii. IoT sensors—feeder for data architecture iv. Smart contracts for participative user network

4 Baseline to Target Architecture

Understanding baseline architecture ("As-Is") is necessary for us to determine the target architecture ("To-Be") for Smart Cities. In order to be fully autonomous with learning capabilities, our Business Architecture layer should be capable of execution for decision without human intervention. Müller and Bostrom [23] estimated that AI will have equal cognitive ability with humans by 2050 and the ability that will exceed human's capability by 2075. A combination of different types of machine learning, can be used to achieve this and they are (1) Supervised Learning, (2) Unsupervised Learning (3) Reinforcement Learning, and (4) Semi-supervised learning. The transition strategy for Acquisition Architecture layer must include full deployment of smart readers and IoT sensors coupled with smart contracts for decision-making. As for the Data Architecture layer, data optimization and inclusion of learning algorithms supported by machine learning will be the core priority. Metadata that are structured and unstructured should be also considered and normalized. This will ensure that the analytics done at the Business Architecture layer will provide accurate reasoning for actions to be executed on-the-fly. For this, a proper taxonomy for definitions must be in place with objects defined for each use case, e.g., surveillance, healthcare, CO_2 emission, and traffic routing, sewer, and waste management. We recommend the Java Agent Development Framework (JADE) which is based on the Foundation of Intelligent Physical Agents (FIPA) to be used as reference (see Fig. 4). A specification of conceptualization for different stakeholders can be developed with ontologies and semantic processing will enable mediation among these ontologies. Agent softbots will then fire services needed to be presented at the Application Architecture layer for each use case discussed earlier (Fig. 5).

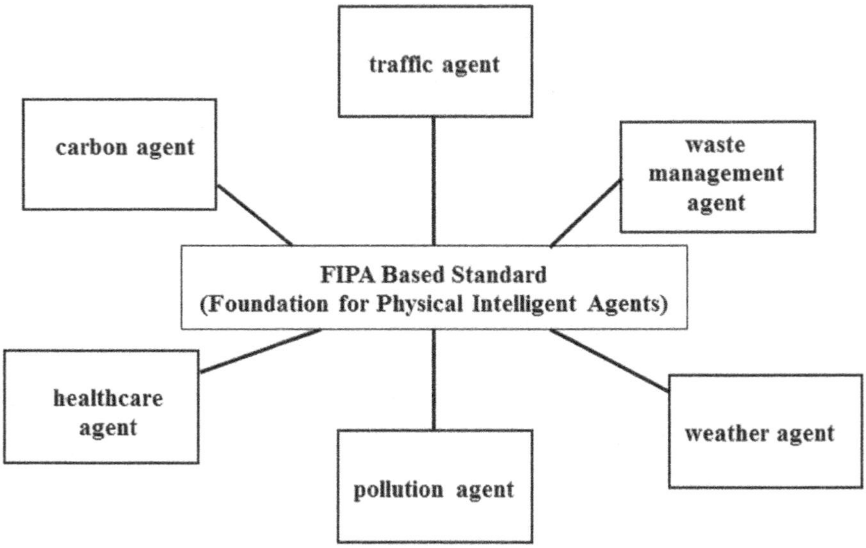

Fig. 4 Multi-agent system (MAS) for use cases

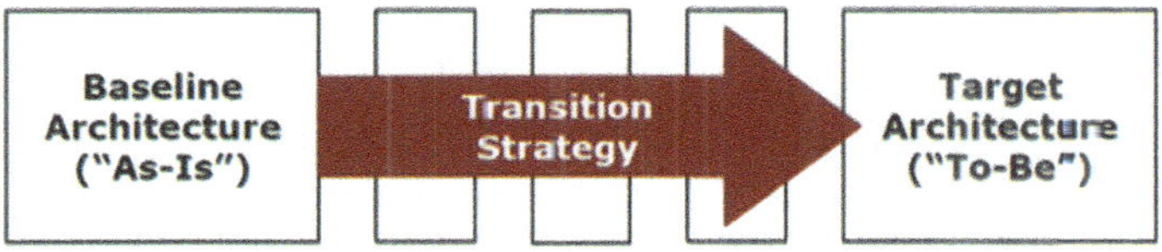

Layers	Functionality
Application Architecture	i. *Present data for service allocation* ii. *Respond to waste, traffic, carbon emission, healthcare and surveillance* iii. *Mitigation responses and decision making*
Business Architecture	i. *Data analytics - prescriptive and predictive* ii. *Semantic analysis with aided reasoning capabilities* iii. *Agent learning capabilities supported by autonomous features*
Data Architecture	i. *Data acquisition – multiple source* ii. *Data optimization and gleaning – from structured as well as unstructured data* iii. *Learning and training capabilities*
Acquisition Architecture	i. *Smart readers – feeder for data architecture* ii. *Smart sensors – feeder for data architecture* iii. *IOT sensors – feeder for data architecture* iv. *Smart contracts for participative user network*

Fig. 5 Baseline and target architecture

5 Conclusion

Due to population expansion our world will soon realize a total population of 10 billion by 2050. The enormous pressure on clean water, sanitation, waste management, carbon footprint, noise, urban healthcare, and crime management will push forward the need for smart solutions in the context of smart cities. The World Urban Forum no. 9 (WUF 9) declaration summarized sustainability as the core driving force for smart inclusion for cities. Urban governance, however, cannot be a one size fits all approach. Core needs of a city must be understood, carefully planned out, and deployed using technology. For this, a use case for cities must be addressed first in their own context. After that a baseline and target architecture must be developed and lastly proper maintenance is required. In this paper we have (1) proposed Smart City use cases based on the Alcatel–Lucent study in 2011, (2) introduced the overarching autonomous learning framework (OALF), (3) transition strategies for OALF, and (4) a multi-agent approach as to how to deploy the use cases keeping in mind autonomous and learning attributes for city sustainability and governance.

References

1. L. Atzori, A. Iera, G. Morabito, The Internet of Things: a survey. Comput. Netw. **54**(15), 2787–2805 (2010). https://doi.org/10.1016/j.comnet.2010.05.010
2. P. Bellavista, G. Cardone, A. Corradi, L. Foschini, Convergence of MANET and WSN in IoT urban scenarios. IEEE Sens. J. **13**(10), 3558–3567 (2013). https://doi.org/10.1109/jsen.2013.2272099
3. H. Schaffers, N. Komninos, M. Pallot, B. Trousse, M. Nilsson, A. Oliveira, Smart cities and the future internet: towards cooperation frameworks for open innovation. Future Internet **6656**, 431–446 (2011). https://doi.org/10.1007/978-3-642-20898-0_31
4. R.P. Dameri, Searching for Smart City definition: a comprehensive proposal. Int. J. Comput. Technol. **11**(5), 2544–2551 (2012). https://doi.org/10.24297/ijct.v11i5.1142
5. R. Giffinger, C. Fertner, H. Kramar, E. Meijers, N. Plichler-Milhanovic, *Smart Cities: Ranking of European Medium-sized Cities* (2007)
6. A. Caragliu, C. de Bo, P. Nijkamp, Smart cities in Europe, in *3rd Central European Conference in Regional Science* (2009)
7. M. Rouse, What is Smart City?—Definition from WhatIs.com (2017). Retrieved from: https://internetofthingsagenda.techtarget.com/definition/smart-city. 24 Jul 2017
8. Techopedia, What is a Smart City?—Definition from Techopedia (2015). Retrieved from: https://www.techopedia.com/definition/31494/smart-city. 30 Nov 2015
9. Gemalto, What is a Smart City? (2018). Retrieved from: https://www.gemalto.com/iot/inspired/smart-cities. 18 Jan 2018
10. M. Dohler, I. Vilajosana, X. Vilajosana, J. Llosa, Smart Cities: an action plan, in *Proceedings of Barcelona Smart Cities Congress* (2011), pp. 1–6
11. M. Yun, B. Yuxin, Research on the architecture and key technology of Internet of Things (IoT) applied on smart grid, in *2010 International Conference on Advances in Energy Engineering* (2010). https://doi.org/10.1109/icaee.2010.5557611
12. A. Monacchi, D. Egarter, W. Elmenreich, Integrating households into the smart grid, in *2013 Workshop on Modeling and Simulation of Cyber-physical Energy Systems (MSCPES)* (IEEE, 2013), pp. 1–6
13. X. Li, W. Shu, M. Li, H.Y. Huang, R.E. Luo, M.Y. Wu, Performance evaluation of vehicle-based mobile sensor networks for traffic monitoring. IEEE Trans. Veh. Technol. **58**(4), 1647–1653 (2009). https://doi.org/10.1109/tvt.2008.2005775
14. S. Lee, D. Yoon, A. Ghosh, Intelligent parking lot application using wireless sensor networks, in *2008 International Symposium on Collaborative Technologies and Systems* (2008). https://doi.org/10.1109/cts.2008.4543911
15. N. Maisonneuve, M. Stevens, M.E. Niessen, P. Hanappe, L. Steels, Citizen noise pollution monitoring, in *Proceedings of the 10th Annual International Conference on Digital Government Research: Social Networks: Making Connections between Citizens* (2009), pp. 96–103
16. T. Nuortio, J. Kytöjoki, H. Niska, O. Bräysy, Improved route planning and scheduling of waste collection and transport. Expert Syst. Appl. **30**(2), 223–232 (2006). https://doi.org/10.1016/j.eswa.2005.07.009
17. A.R. Al-Ali, I. Zualkernan, F. Aloul, A mobile GPRS-sensors array for air pollution monitoring. IEEE Sens. J. **10**(10), 1666–1671 (2010). https://doi.org/10.1109/jsen.2010.2045890
18. J. Sairamesh, A. Lee, L. Anania, Information cities. Commun. ACM **47**(2), 28–31 (2004)
19. E. Douglas, Artificial Intelligence: Abilities & Expectations (2018). Retrieved from: https://orzota.com/2018/02/08/artificial-intelligence-abilities/. 8 Feb 2018
20. C. Reichert, Huawei Unveils Artificial Intelligence Smart Cities Platform (2018). Retrieved from: https://www.zdnet.com/article/huawei-unveils-artificial-intelligence-smart-cities-platform/. 14 Nov 2018
21. A. Roven, AI in Smart Cities (2018). Retrieved from: https://medium.com/neuromation-blog/ai-in-smart-cities-dfe2fa7d2829. 27 Nov 2018

22. D.J. Bailey, Y. Coleman, Urban IoT and AI: How Can Cities Successfully Leverage This Synergy? (2019). Retrieved from: https://aibusiness.com/future-cities-iot-aio/. 8 Jan 2019
23. V.C. Müller, N. Bostrom, Future progress in artificial intelligence: a survey of expert opinion. *Fundamental Issues of Artificial Intelligence*, pp. 555–572 (Springer, Cham, 2016). https://doi.org/10.1007/978-3-319-26485-1_33

A Predictive Analysis for Heart Disease Using Machine Learning

V. Rajalakshmi, D. Sasikala, and A. Kala

Abstract Predictive analysis plays a major role in healthcare industry where forecasting the disease will reduce the risk that happen to patients. Statistics show that cardiovascular diseases have increased the mortality rate in India. Machine learning which is used in developing a predictive model for various domains is nowadays applied in the field of medical diagnostics. Machine learning is playing an integral role in predicting the presence or absence of heart diseases. Such predictions, if done well in advance, can help the doctors to carry out the treatment for the patients and mitigate their health risk. Biological samples such as blood or tissues are collected from the human body to predict cardiovascular diseases. The proposed work is focused on developing various machine learning predictive models using support vector machine, decision tree, neural network and K-nearest neighbour for prediction of heart disease. For this work Cleveland heart disease dataset is used which consists of 14 attributes and 294 records. A comparative analysis on the prediction models were carried out. From the results, it was found that support vector machine, decision tree and KNN ($k = 15$) classifiers yield better accuracy to predict heart disease than the other models.

Keywords Cardiovascular disease · Machine learning · K-nearest neighbour · Decision tree · Neural network

V. Rajalakshmi (✉) · D. Sasikala · A. Kala
Sri Venkateswara College of Engineering, Chennai, India
e-mail: vraji@svce.ac.in

D. Sasikala
e-mail: dsasikala@svce.ac.in

A. Kala
e-mail: akala@svce.ac.in

S. S. Dash et al. (eds.), *Intelligent Computing and Applications*,
Advances in Intelligent Systems and Computing 1172,
https://doi.org/10.1007/978-981-15-5566-4_42

1 Introduction

There is a major shift in traditional healthcare system after the boom of Internet and telecommunication. Today almost majority of the common people have Internet in some devices which they use in their day-to-day life. In earlier days, healthcare system means hospitals with doctors, caretakers, skilled lab technicians, etc. In traditional systems, patient's health is examined with the lab test or with devices. These devices are operated with the help of doctors or people who are skilled in that field. The results from the health records are reported to doctors and they assess these records with their knowledge of expertise. These records are analyzed by the medical professionals and they assess the patient's health. In the nineteenth century, the industry started using Internet and this made a big revolution in industrial sectors. Slowly this technology has entered into the medical industry and aided the medical professionals in many ways. In some countries, machine-assisted operations are performed on patients by the doctors. Internet technologies are playing a major role in medical sectors. For example, medical professional assistance is given through video conference; these are due to technology evolution.

1.1 Motivation and Background

The mortality rate due to cardiovascular disease is high when compared to other diseases which is pointed out by the statistical inferences. If these diseases are predicted much earlier then it can save human life. Prediction accuracy is very vital to prevent any judgmental errors. Machine learning technology is applied in many scientific and medical fields wherein it has brought a new lease of life. Building a good machine learning model to predict diseases can save human life. There are many machine learning algorithms and each of them produces the results based on the training and hypothesis it has built from the data. This paper is focused on finding the suitable machine learning model for accurate prediction of heart disease from the Cleveland dataset.

Cardiovascular or heart disease generally refers to a narrow/blocked vessel causing cardiac arrest. There are many other conditions which may affect the heart muscle which also results in blocking the blood flow from the heart to other parts of the body. Cardiovascular disease can be prevented by diagnosing them earlier. A simple and cost-effective procedure should help mankind rather than costlier ones. In conventional procedure diagnosis of a patient to find a normal or abnormal condition is assessed through the blood sample or tissue collected from them. Many works related to this research were done using image processing by other researchers. Our aim is to build a machine learning model using numerical clinical dataset. In the first phase, Cleveland dataset is used in the proposed system to build a machine learning model using support vector machine where the prediction accuracy was good. Later other machine learning models like decision tree, neural network and K-NN classifier were built. These models' accuracy of prediction for the Cleveland dataset was analyzed to find the best model.

In many parts of the world, healthcare systems built by machine learning model and artificial intelligence are being widely used and successfully accepted by the medical professionals. But there is a need for an accurate model which would further help the mankind. There are still researches going on in this field for the betterment of patient-centric care. Many researches had been done by the scientific and research community and their few works are presented in the next section.

2 Literature Survey

Alba et al. [1] proposed a model for segmenting highly abnormal hearts using statistical shape model. Their focus was to accurately identify the abnormal hearts from the cardiac image even though there is an abnormal reshaping due to other conditions like pulmonary hypertension. They employed virtual remodelling transformation from the image generated from the patient and the reference modelling which was obtained from the boundary analysis. A reference statistical shape model from a sample of normal subjects was built and then the original and pathological statistical model was compared to find any abnormality.

This work [2] finds the heart failure by employing Complex Event Processing along with the statistical approach. Here the patient health record is monitored and the history of data is stored and compared with the current data, based on the reference, prediction was done. But patient-centric threshold is set and the values are dynamically changed based on the health generated data. The project which Karan Bhanot [3] presented uses machine learning algorithms like support vector machine, random forest, decision tree and K nearest neighbour. Among these algorithms, K-nearest neighbour algorithm has higher accuracy prediction rate when compared to other classifiers.

Improving the accuracy of heart disease prediction at an early stage and reducing the error rate in prediction [4] was the target of their work. Ensemble techniques such as bagging and boosting reduce the error rate of weak classifiers. Another similar work was done [5], in this work they tested their heart disease dataset across multiple classifiers like support vector machine, Naïve Bayes classifier, logistic regression, gradient boosting and random forest. They built a model using the classifier for prediction.

For diagnosing [6] a heart disease, a hybrid model was built with seven classifiers, three feature selection algorithms and a cross-validation method. They improved their model accuracy by removing the irrelevant feature using feature selection method and improved the accuracy of prediction. This model has improved the execution time as well. In this work [7], they developed machine learning models using decision tree, random forest, multivariate adaptive regression splines, tree model from genetic

Table 1 Dataset description and attributes

Clinical features (attributes)	Description
Age	Age in years
sex	Gender (1 = male; 0 = female)
cp	Chest pain type
trestbps	Resting blood pressure
chol	Serum cholesterol in mg/dl
fbs	Fasting blood sugar > 120 mg/dl (1 = true; 0 = false)
restecg	Resting electrocardiographic results
thalach	Maximum heart rate achieved
exang	Exercise-induced angina (1 = yes; 0 = no)
olpeak	ST depression induced by exercise relative to rest
slope	The slope of the peak exercise ST segment
ca	Number of major vessels (0–3) coloured by fluoroscopy
thal	3 = normal; 6 = fixed defect; 7 = reversible defect
num	Diagnosis of heart disease

algorithm and used Cleveland dataset [8] to predict the heart disease. Their study proved that decision tree outperforms the other machine learning algorithms.

3 Dataset Description

The Cleveland database is the heart disease dataset which is popularly used by the machine learning researchers in heart disease prediction. The dataset consists of 14 attributes, namely (Table 1):

The target field is 'num' which refers to the presence of heart disease in the patient. '0' refers to absence of heart disease and '1' refers to presence of heart disease. The dataset consists of 294 records.

4 Proposed System

The objective of the work is to build a machine learning model to predict the heart disease with high accuracy. Cleveland dataset is used in this work. In the first step,

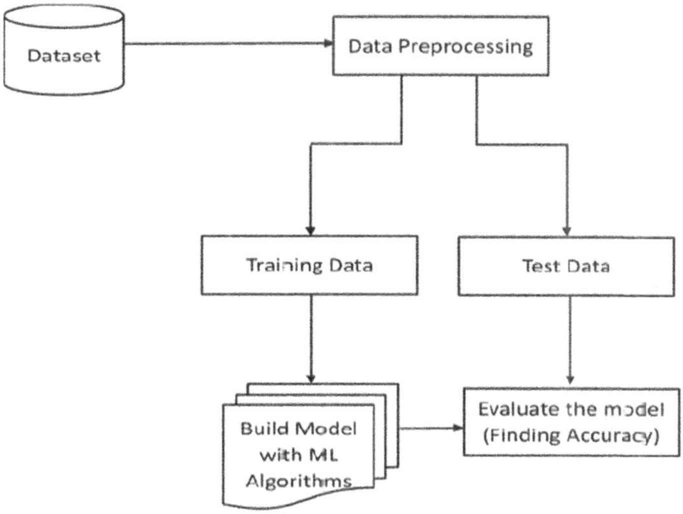

Fig. 1 Heart disease prediction architecture using machine learning

data cleaning is done where the missing data and the duplicate records are identified and removed. The data is divided into training and test set for further model building. The ratio of the training and test data split is 70:30. The model for prediction of heart disease is built with support vector machine, decision tree, neural network and K-nearest neighbour. The model is evaluated with the test data to identify the best model that predicts the disease with least errors. Figure 1 describes the architecture for heart disease prediction.

4.1 Machine Learning Models

- **Artificial Neural network (Perceptron)**
  Perceptron is a supervised linear classifier that classifies the data into two classes.
- **Decision Tree**
  Decision Tree is a supervised classifier where the data is repeatedly split based on a certain parameter.
- **Support Vector Machine**
  Given a labelled training data, the support vector machine defines a separating hyperplane, that categorizes the new data.
- **K-Nearest Neighbour**
  Given an input, which consists of *k* closest training data, KNN assigns the new data to the class that exists in maximum among the K-nearest neighbours.

5 Implementation and Results

The prediction of the heart disease is done by implementing four models: the perceptron, the decision tree, support vector machine and K-nearest neighbour with *k* values 1, 5, 10 and 15. The models are implemented in R Studio.

Figure 2 shows the decision tree classification and perceptron model built for the given training set. The models are evaluated by computing various metrics. Let '*a*' be true positive, '*b*' be true negative, '*c*' be false positive, '*d*' be false negative and '*n*' be the total number of samples. The metrics computed are as follows (Table 2):

$$\text{Accuracy} = (a + d)/n \tag{1}$$

$$\text{Precision} = a/(a + c) \tag{2}$$

$$\text{Recall} = a/(a + d) \tag{3}$$

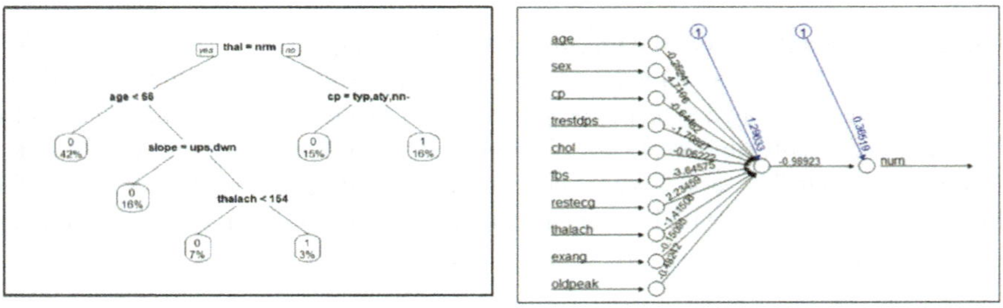

Fig. 2 Decision tree and perceptron model

Table 2 Comparison of results from ML models

ML model	Accuracy	Precision	Recall
Perceptron	78.5	83.3	60.6
Decision tree	84.5	87.8	60.6
SVM	84.5	86.3	61.9
K-NN ($k = 1$)	77.3	82.9	60
K-NN ($k = 5$)	79.7	83.6	61.2
K-NN ($k = 10$)	78.5	84.7	59.1
K-NN ($k = 15$)	**85.71**	**88**	**61.1**

6 Conclusion

From the results, it was found that support vector machine, decision tree and KNN ($k = 15$) classifiers yield better accuracy to predict heart disease than the other models. The work is planned to extend further by implementing the Committee of Machines and deep learning models.

References

1. X. Alba, M. Pereanez, C. Hoogendoorn, A.J. Swift, J.M. Wild, A.F. Frangi, et al. An algorithm for the segmentation of highly abnormal hearts using a generic statistical shape model. IEEE Trans. Med. Imaging **35** (2015)
2. A. Mdhaffar, I.B. Rodriguez, K. Charfi, L. Abid, B. Freisleben, Complex event processing for heart failure prediction. IEEE Trans. NanoBiosci. **16**(8), 708–717 (2017)
3. https://towardsdatascience.com/predicting-presence-of-heart-diseases-using-machine-learning-36f00f3edb2c
4. C. Beulah Chirstalin Latha, S. Carolin Jeeva, Improving the accuracy of prediction of heart disease risk based on ensemble classification techniques. Inform. Med. (2019)
5. K.G. Dinesh, K. Arumugaraj, D. Santhosh, V. Mareeswar, Prediction of cardiovascular disease using machine learning algorithms, in *International Conference on Current Trends towards Converging Technologies*, pp. 1–7 (2018)
6. P. Li, M.H. Memon, S. Nazir, R. Shah, A hybrid intelligent system framework for the prediction of heart disease using machine learning algorithms. Mob. Inf. Syst. (2018)
7. T. Sharma, S. Verma, Kavita, Prediction of heart disease using cleveland dataset. A machine learning approach. Int. J. Rec. Res. Aspects **4**, 17–21 (2017)
8. https://archive.ics.uci.edu/ml/machine-learning-databases/heart-disease/

Application of Data Mining Algorithms for Tourism Industry

Promila Sharma, Uma Meena, and Girish Kumar Sharma

Abstract This paper presents different Data Mining Algorithms to be used to access data patterns in Tourism Industry and their applications using Data Mining Tool like WEKA which is a popular software, and useful patterns can be found efficiently. Data Mining, as the name implies is the science by which hidden and interesting information can be found from large datasets analogous to Gold Miners who extract precious metal sieving through a lot of raw material. Normal day to day business operations create huge voluminous data which is needed to be utilized properly to find information to make efficient, effective, and prompt decisions otherwise businesses can not sustain in competitive environment. There are some techniques and algorithms used in data mining for determining interesting hidden patterns. data mining algorithms are becoming popular day by day in various applications like Travel and Tourism. Nowadays, a number of data mining applications are available for Public use. Therefore, we have to judge which particular algorithm or application suits a problem domain. The idea behind applying data mining is very straight forward, it is similar to a human being becoming intelligent from learning by examples and experience. Similarly, in data mining the behavior of given system (training dataset) is learned by application, and certain rules are developed. Then these rules are used to evaluate the behavior/result for the given situation (test dataset). The overall goal of this paper is to classify some data mining algorithms and applications, draw a comparative study that which one is better for tourism industry and under what circumstances.

P. Sharma (✉)
Department of Computer Science, Mewar University, Chittorgarh, Rajasthan, India
e-mail: prom_sharma@yahoo.com

U. Meena
CSE Department, SRMIST, Delhi-NCR Campus, Modinagar, UP, India
e-mail: uma.b18@gmail.com

G. K. Sharma
Department of Computer Applications, Bhai Parmanand Institute of PG Studies (Under DTTE, GNCT of Delhi), Delhi, India
e-mail: Gkps123@gmail.com

S. S. Dash et al. (eds.), *Intelligent Computing and Applications*,
Advances in Intelligent Systems and Computing 1172,
https://doi.org/10.1007/978-981-15-5566-4_43

Keywords Travel and tourism industry · Travel pattern · Apriori algorithm · Weka tool · Data mining · Decision trees · Association rules · Classification · Clustering

1 Introduction

Data mining is the latest technology and is becoming popular because of the increasing use by companies who have a strong customer focus like retail, marketing, tourism, etc. It provides great help to companies who can collect data about the behavior of their potential customers and can discover hidden information within the data that queries and reports cannot reveal properly.

Data mining best known as analysis part of the "Knowledge Discovery in Databases" process, or KDD. It is an interdisciplinary part of computer science, is a data-driven research which applies algorithms on computer and discovers patterns in huge datasets. It sits at the intersection of statistics database systems, artificial intelligence, and machine learning but with few differences. The databases provide transactional and query reports as output whereas data mining provides analytical output, Statistics makes generalizations from observations whereas data mining provides specific information, Machine intelligence deals with small-sized data and data mining with huge datasets. The main objective to use data mining is that it can extract information from large datasets and transform to readable form for further uses. Data mining uses some techniques to extract interesting patterns from large datasets [1].

1.1 Steps of Data Mining

Data mining or Knowledge Discovery in Databases (KDD) comprises the following processing steps.

1. Data cleaning step or Preprocessing—This is the first step to remove noisy data that is missing, duplicate inconsistent data from large datasets.
2. Data integration step—This step combines data from multiple sources.
3. Data selection step—All attributes are not related or needed to be included for analysis task. So in this step only data relevant to a particular problem domain is selected.
4. Data transformation—Data mining techniques require that data should be transformed to consolidated or some standard normalized form before beginning data mining.
5. Data mining application—It is the important phase in which various data mining algorithms are applied on selected datasets to extract unknown useful patterns.
6. Pattern recognition—This is the phase where we see benefits of data mining. It is used to find unknown and interesting patterns from datasets.

7. Knowledge interpretation or Presentation—Various visualization techniques can be used to present mined patterns to the user and managers.

1.2 What are Data Mining Applications

A Human analyst manually cannot evaluate and analyze huge volume of data so computer-based efficient and speedy software applications are required. Therefore, data mining techniques are becoming popular for businesses and other fields as medical, engineering, sports, marketing, etc. With the help of data mining techniques, businesses can analyze their marketing and sales patterns, develop latest digital marketing campaigns, and accurately predict customer behavior and profile. Some of the applications of data mining are [1]:

1. Market Identification—Common features/characteristics of customers can be found who bought similar products from company.
2. Customer churning—One can predict which customers had left your company and had gone to the competitor.
3. Fraud detection—Through this fraudulent transactions/hacking can be detected.
4. Direct sales and marketing—In order to obtain higher response rate of customers, it accesses which prospects of mailing lists can be changed.
5. Interactive/Online/Digital marketing—Through the use of website mining, potential customers behavior and buying patterns can be known as who have done online shopping.
6. Market/Product basket analysis—Through association rule mining algorithms one can make judgment as which products or services customers had purchased together, i.e., Bread and milk.
7. Trend/Growth rate analysis—Sales differences between previous year and current year can be known to analyze trends or growth rate of products.

1.3 Data Mining Uses for Travel and Tourism Industry

Travel is known as the movement of people for different purposes between remote geographical locations. There can be different modes of travel by foot, by bicycle, auto, railway, ship, bus, aeroplane, or other means one way or round trip. Information Technology has found great uses in travel and tourism industry for information processing [2]. Advancements made in information technology have influenced the services and facilities offered and these are efficiently delivered and promoted. It also has affected the way how organizations are structured and communication channel between customers and services providers has improved [3]. Travelers too are benefited from Internet and communication technology. They can easily search places of their interests and avail services accordingly.

Tourism in India is considered the largest service industry as both inbound and outbound tourists have contributed to 6.23% to the national GDP and 8.78% of the total employment. According to a survey report travel industry in India has generated about $100 bn USD in 2018 and it is expected to increase to $275.5 bn USD in near future with an annual growth rate of 9.4% (India Tourism Statistics) [4].

Because of increasing business competition, the situation of sustainability for Indian travel and tourism industry has declined. Therefore, the industry needs to increase their position in the market and keep control of their businesses. In order to develop market and to predict potential customers, to plan best tour packages and services, they should use data mining tools and techniques efficiently [5]. Modern technology has had a great impact on tourism (Domestic and International Both) when Internet and communication technologies became the most dominant channel. Tourists' expectations are related not only to tourism services, but also to technology. For this to be successful, one must have domain knowledge of tourism as well as how it can be implemented using data mining [6].

In the tourism industry, tour operator companies can formulate better marketing strategies and maximize their profits by knowing guests' behavior—where they have come from, the money they have spent, and when and on what they have spent. In day to day businesses, travel and tour companies accumulate large amounts of customer profile data, which they can organize in proper format and integrate in databases. Advanced technology as data mining can be used to identify important variables and analyze the information hidden in datasets which guides marketing decision [7]. Nowadays businesses specially related to customer focus are raising information and services system by using data mining techniques, for example, uses in hospitality management and service industries like hotel management, tourism, ticket booking, railways, Information technology has been used to assist and expedite the delivery services [8]. Improved capacity management and operational efficiency, inventory control of items, check availability of room information, increasing management control and capability, creating consolidated or summary marketing, sales and operational reports, keeping track of customers and hotel guests who visit frequently, transforming transactions from internal operations of management to human resources are some of the results of IT technology uses [9]. Most of these items apply only to hotels and accommodation managers and providers. The latest technologies like data mining helps Hotel Industry to predict future trends and access customer transactions which are useful in sales and marketing by making high-quality market research and planning [10].

2 Introduction to Data Mining Algorithms/Strategies

A data mining algorithm can be defined as a set of some rules and heuristics that can be used to create Data mining model from dataset. In order to develop a model, the algorithm first analyzes datasets (Training dataset) and views specific unknown patterns. Then it defines optimal parameters for the model by using output/result of the training set. These parameters are applied in the remaining dataset (Test dataset) to extract hidden information, i.e., actionable rules and statistics.

2.1 *Popular Data Mining Algorithms*

There are different types of techniques/algorithms used in data mining which depend on the application domain and type of data recovery process that provide useful answers to problems. These are listed below:

1. **Classification algorithms** also known as discrimination or pattern recognition which makes prediction of one or more discrete or class variables, based on the other independent variables/attributes in the dataset. These come under supervised learning.
2. **Regression or Estimation algorithms** also come under supervised learning. This method produces numerical attributes as output based on other input variables/attributes. For example, prediction of profit or loss, depending on other attributes in the dataset.
3. **Segmentation or Clustering algorithms come under unsupervised learning.** These divide datasets into unlabelled groups or clusters of objects that have similar properties. Examples are image segmentation, market segmentation, outlier detection, etc.
4. **Association Rule Mining algorithms** are also known as Market basket analysis or Affinity analysis which is unsupervised learning. In this, relationship between different attributes in a dataset is found whose most common application is to identify buying patterns of customers and accordingly design store layouts.
5. **Sequence Detection algorithms.** A time series is a set of observations collected as sequences. This algorithm summarizes frequent sequences or time series in data, such as a website path flow. These are important in bioinformatics, sales forecasting, stock market analysis, etc.

3 Significance of the Problem

3.1 *Problem Statement*

To Access Data Patterns for Indian Tourism Industry.

This research work can provide the solutions to the following questions pertaining to tourism sector as following:

1. How many and what age group Domestic/Indians travelers travel in Delhi or other places of their choices per year? How much per year they spend on tours?
2. How many foreign tourists come to Delhi and what are the places of their interest, purposes of travel, what age group people usually come and from which countries. The tour packages they are offered and their likings.
3. What are the lean and peak seasons/timings during the year to travel?
4. What is the probability that certain tourists will respond to the upcoming promotion for a holiday tour package?

5. What places and packages are expected to be in the priority list of customers whether Domestic Travelers/NRI (Non-Resident Indians)/FTA (Foreign Travel Arrivals)?

4 Research Methodology

Research methods may be defined as all those methods/techniques that are applied by researchers for conduction of research. Research methods or techniques, thus, refer to the processes and methods and tools, the researchers use in performing research operations. For solving the above problems some research methods and techniques are used for obtaining the desired result. Some tools and algorithms are required for obtaining results. Main steps followed under the research methodology are listed below.

4.1 Literature Review of Research Papers

In this paper, all previous work from journal literature and research papers were studied. It gave insight into what type of work has been done by previous scholars/researchers in the particular domain and by what methodologies/techniques of research were followed, and what tools were used to carry out research. Then more knowledge was gained about the problem.

4.2 Identify Tools

Then a tool was identified for carrying various data mining applications on tourism dataset and the popular tool "WEKA" (Waikato Environment For Knowledge Analysis) was selected from all. It was found very useful and reliable. There are algorithms built in the software for Classification, Regression Analysis, Clustering, Association Rule Mining, and Visualization.

4.3 Defining Required Database Structure and Attributes

The structure and attributes were prepared for collecting Tourists Information, their profile, places they visited, and others about their tour packages. Questionnaire method was used for primary data collection.

4.4 Data Collection

The questionnaire was filled by visiting domestic tour operators in Delhi and a few data were collected by emailing questionnaires to friends group who visited Delhi and its excursions, nearby places of interest.

4.5 Data Pretreatment

The Database has 1050 records and 14 useful attributes are selected. First the Data is prepared in Excel sheets and missing values are dealt with. Then raw data file is saved with CSV extension. It was converted to arff file format because this is Weka's native file format used for data analysis.

4.6 Data Mining

After determining nature and workflow of Research Problem Database is organized and data mining algorithms are applied using WEKA in order to generate results. The following techniques are applied.

5 Analysis Using Arm (Association Rule Mining)

5.1 Conceptual Framework

ARM is a type of unsupervised learning algorithm. Its main objective is in transactional database to describe the rules of data relationship. The set of items which are purchased in a transaction is called its market basket such as shopping cart. Through this one can predict the buying patterns of customers like which products are bought together [11]. The knowledge gained from ARM analysis can be used in maximizing sales and accordingly store layouts can be managed. For this data mining discipline offers a useful strategy known as Association Rule Mining. It is a popular algorithm credited to Agarwal, Imielinski and Swami in 1993 [12].

Let us assume that a transaction contains data items A and B. If any transaction that contains "A" is also likely to contain "B" is expressed as the rule "A" => "B" ($A \cap B = \varphi$). This is an expression of Association rule. A is denoted as antecedent or Hypothesis and B is called a consequent or result which follows B. They are related as "If A then B." In ARM two estimates Support S and Confidence C from transaction datasets are calculated. Support is the percentage of items that contain both "A" and "B". It is described as $P(A \cap B)$ where P is probability. Suppose there are 100 transactions in which 80 transactions contain Item "A", 70 contain item "B" and 60 contain both "A" and "B". Then Support S is 60%. The rule has Confidence C as the percentage of items "B" is present in transaction containing "A". It is denoted y Conditional probability P('B'|'A'). In the above example C is 60/80 that is 75%. Support S is used to measure the frequency of an Association Rule and Confidence C is used to measure the strength of the Association Rule. The Three important phases of ARM are

1. First the set of Items that has significant impact on businesses are found.
2. In the second step, information from numerous transactions on dataset items from many sources are combined.
3. Then Association Rules are generated from counts of significant items in Transactions.

The Apriori algorithm (Bodon, Zaki, Agarwal, and Srikant) is the most widely used tool for ARM. This algorithm uses previous knowledge of properties of data items that appear frequently. Firstly, Apriori algorithm generates candidate itemsets of a given size and then analyzes the dataset items if their counts are large.

In the first pass all singleton itemsets are generated and those are selected whose support value is above the minimum specified. Remaining singleton items are combined to form two member candidate itemsets. Support values of this candidate itemsets are again calculated and the next phase creates three candidate itemsets. The process is iterative and it stops when all frequent itemsets are accounted for [13].

5.2 Discovering Travel Pattern Through ARM Using Weka

An ARM Algorithm cannot handle numeric attributes. Therefore, all the numeric attributes have to be converted to discrete or categorical attributes. This is called discretization and this can be accomplished by using Weka's Internal discretization tools.

The complete dataset of Indian tourists is refined/preprocessed and loaded in Weka. 110 instances are selected as training set for analysis. Numeric attributes as age, stay_in_days, no_persons are discretized. All other attributes except 8 are removed which are useless for the analysis through ARM. The best 10 rules were found supporting each other according to their dominant category. These are generated by taking default settings of Apriori algorithm in Weka. These are listed in Fig. 1.

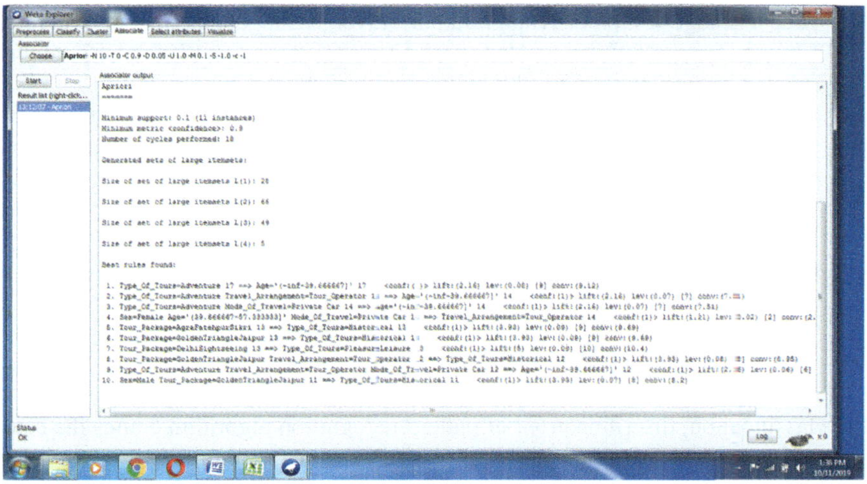

Fig. 1 Result of ARM in Weka

5.3 Best Rules Found Are

1. Type_Of_Tours = Adventure 17 ==> Age = '(-inf-39.666667]' 17 means there are 17 adventure types of tours and they are taken by age group of people up to 40 yrs of age.
2. Type_Of_Tours = Adventure Travel_Arrangement = Tour_Operator 14 ==> Age = '(-inf- 39.666667]' 14 means it follows above rule 1. In addition to that all 14 people have traveled through tour operator.
3. Type_Of_Tours = Adventure Mode_Of_Travel = Private Car 14 ==> Age = '(-inf- 39.666667]' 14 means there are 14 records of adventure types of tour and have traveled through Private car taken by 14 people of age up to 40 yrs.
4. Sex = Female Age = '(39.666667-57.333333]' Mode_Of_Travel = Private Car 14 ==> Travel_Arrangement = Tour_Operator 14 means There are 14 females who are in the age group 40–58 and have traveled by private car and tour is arranged by tour operator.
5. Tour_Package = AgraFatehpurSikri 13 ==> Type_Of_Tours = Historical 13 Means there are 13 records of tour package as AgraFatehpurSikri is type of Historical tour.
6. Occupation = Service Mode_Of_Travel = Private Car 21 ==> Stay_In_Days = '(-inf-3]' means there are 21 persons in service who have traveled by private car and stayed for 3 days.
7. Type_Of_Tours = Religious 20 ==> Religion = Hindu 19 means the type of Religious tour taken by Hindu religious people.
8. Tour_Package = DelhiSightseeing 13 ==> Type_Of_Tours = PleasureLeisure 13 means there are 13 records in which tour package was Delhisightseeing and type of tour is PleasureLeisure.

9. Tour_Package = GoldenTriangleJaipur Travel_Arrangement = Tour_Operator 12 ==> Type_Of_Tours = Historical 12 means there are 12 GoldenTriangle-Jaipur tour package arranged by tour operator which is historical.
10. Sex = Male Tour_Package = GoldenTriangleJaipur 11 ==> Type_Of_Tours = Historical 11 means 11 male persons have taken GoldenTriangleJaipur tour package which is historical.

6 Analysis Using Classification

6.1 Conceptual Framework

The classification, discrimination, or pattern recognition algorithm comes the category of supervised learning. The process evaluates a function and a relationship between a dependent variable (Output/Discrete) and contingent independent variables (Input parameters) is established. In the classification problem an object is identified and its belongingness to a given class is analyzed [14].

In Classification algorithms rules are found to organize given datasets into predefined classes. A part of the historical data is chosen (Training data), algorithm rules are generated, and then are tested on the remaining data (Test data). During Training phase the model is trained or learned and the model is developed. Then the developed model is run on Test data. The Traditional and well-accepted method of classification is the induction of decision trees, which partitions the dataset to develop rules. Another approach uses probability theory for classification and is known as Naïve Bayes's algorithm.

The other popular algorithms that are used for classification are Artificial Neural Networks (ANN) and Super Vector Machines (SVM) [13].

6.2 Decision Tree for Classification Using Weka

Decision trees are used for classification and prediction tasks of data mining, and they are the most popular techniques. A Decision Tree is a flowchart hierarchical tree-like structure in which the condition is tested on each internal node attribute, each next level branch node represents an outcome of the test, and finally leaf node holds a class label. During the late 1970s and 1980s, J. Ross Quinlan, a researcher in machine learning had developed Tree Algorithm known as ID3 (Iterative Dichotomiser). He later on modified it and presented C4.5 which has become a benchmark to newer supervised learning algorithms. It is the advanced version of ID3. In Weka it is called J48. It has some additional features like missing values are taken care of, continuous attributes are converted, pruning of decision trees is done internally to avoid overfitting and the rule can be derived. A heuristic approach is used for pruning the Tree which is based on the statistical significance of splits. Many ways are defined

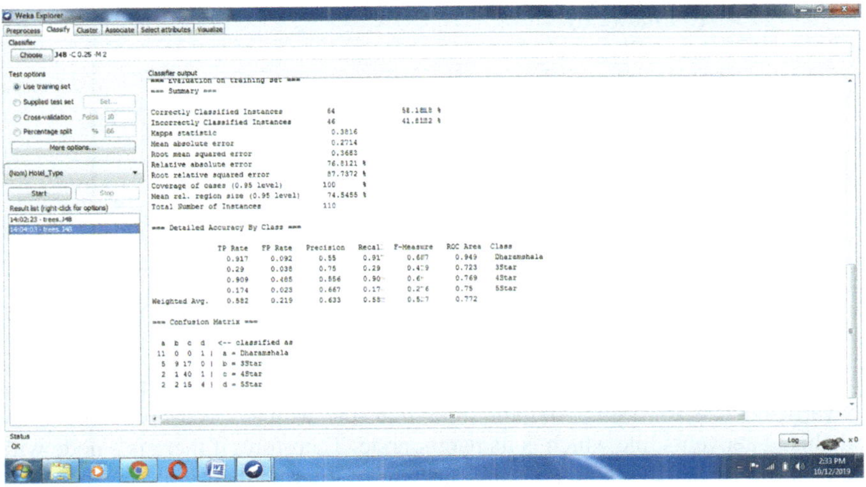

Fig. 2 Decision tree algorithm run in Weka

to select the attributes in the nodes of decision tree and to arrange these in order. Information gain or entropy measurement is one of the simpler ways [14]. Information gain is the measurement of much information/answer to a specific question that is provided. It is the measure of how much uncertainty there is in the information. For n number of possibilities with P_i the probability of the event i, the Entropy $H(X)$ is given by

$$H(X) = -\Sigma^n = 1 p_i \log 2 p_i$$

Figure 2 is the decision tree shown by C4.5. The database is selected and loaded in Weka. Tree algorithms J48 with default setting was run on 110 records as Training set of Indian_tourist database which classifies travelers as they have stayed at which hotel type as per their attributes as Age, Types Of Tours, Sex, etc. Similarly, other decision trees can also be built according to the type of attributes one wants to classify.

6.3 *Analysis of Result*

The generated tree is as shown below:
J48 pruned tree:
Type_Of_Tours = Religious: Dharamshala (20.0/9.0)
Type_Of_Tours = Historical
| Sex = Male: 4Star (21.0/11.0)
| Sex = Female
| | Age <= 51: 3Star (5.0/1.0)

```
| | Age > 51: 4Star (2.0)
Type_Of_Tours = PleasureLeisure
| Age <= 51: 4Star (19.0/10.0)
| Age > 51: 5Star (3.0/1.0)
Type_Of_Tours = Adventure
| Sex = Male
| | Age <= 26: 5Star (3.0/1.0)
| | Age > 26: 4Star (7.0/2.0)
| Sex = Female: 3Star (7.0/2.0)
Type_Of_Tours = Wildlife: 4Star (13.0/5.0)
Type_Of_Tours = HeritageCultural: 4Star (10.0/4.0)
```

Each line represents a node in the Tree. The Root node is the Attribute without any vertical bar '|' to its left. All Other lines that start with a '|' is a child node of the node listed above its line which is its parent node. Therefore, if there is a node with one or more '|' characters before the statement of the rule then it is the child node of the node immediately preceding that line. Leaf nodes are denoted by a colon (:) and the class is identified by text after colon (:). Node that generates a classification such as Age <= 51: 3Star (5.0/1.0) are followed by two numbers. First number 5 tells how many instances set are correctly classified by this node where age <= 51 and 1 tells that this is incorrectly classified by algorithm.

7 Analysis Using Clustering

7.1 Conceptual Framework

Clustering comes under the category of unsupervised way of learning. Clustering algorithm is different but the idea is similar to classification. The difference lies that classes are not predefined. We can only imagine the benefits of grouping customers and merchandise, giving us insight into which customer is likely to buy which product. There are two approaches to performing clustering tasks—Hierarchical and Non-Hierarchical. Based on implementation technique, hierarchical has either agglomerative method or divisive method to build clusters. Non-hierarchical clustering is used to perform data segmentation/market segmentation. Some measurement of dissimilarity between segments and the degree of segmentation is required to be done. In this, three kings of algorithms are there partitioning methods, density-based, and probability-based methods.

Simple K-Means is the most popular and commonly used partitioning method of non-hierarchical clustering. The centroid of a cluster is defined as the mean value of points within the cluster. First of all, it randomly selects k value of the objects in D, which represents a cluster mean or center value. Then for the remaining objects, Euclidean distance between the object and the cluster mean is calculated and the object is assigned to the cluster if it is most similar or the distance is less. The K-means algorithm then iteratively assigns objects and improves the within cluster variation [15].

7.2 Clustering Using Weka

In Weka, the SimpleKMeans algorithm is used for clustering which can automatically handle mix types of attributes whether they are categorical or numerical attributes. The algorithm does distance computations and also normalizes the numerical attributes. It uses Euclidean distance measurement to calculate distances between instances and clusters [13].

7.3 Analysis

Following is the result of SimpleKMeans Clustering on Indian_tourist database.

Training data of 110 records and 7 useful attributes were selected. The number of clusters initially is taken as 2 so it formed 2 clusters with following statistics (Fig. 3).

This is very useful for the future planning of tour packages which are more saleable. There is another group of 3 clusters which was taken by reducing the number of attributes as 4. From the demonstration, it is clear that the SimpleKMeans algorithm is a straightforward algorithm for finding clusters in datasets (Fig. 4).

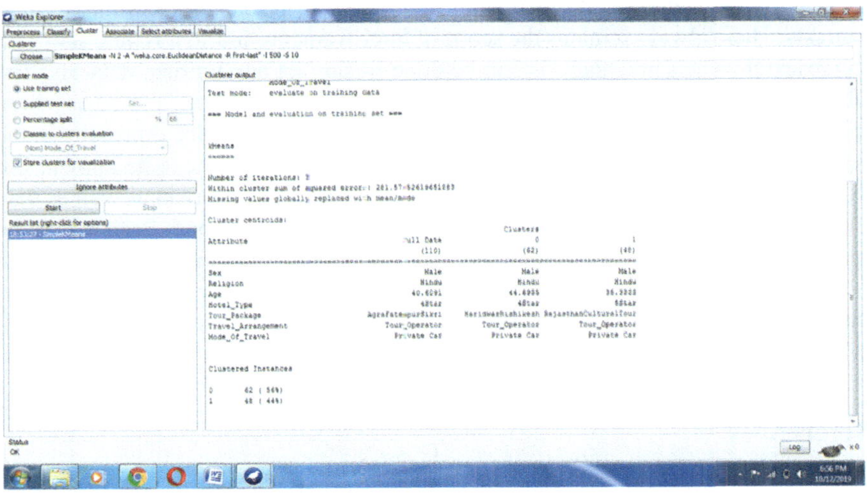

Fig. 3 Clustering result in Weka

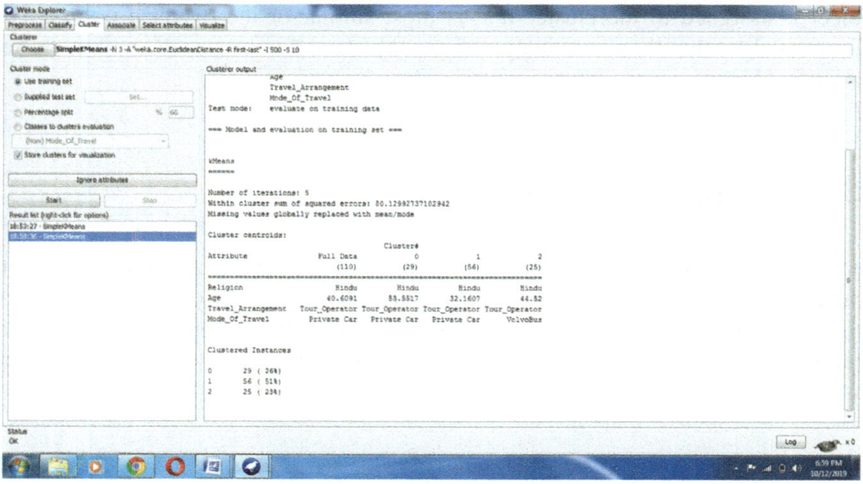

Fig. 4 Clustering result in Weka

8 Conclusion and Future Work

The Paper has presented applications of different data mining algorithms using popular tool for analysis WEKA. Each algorithm has different objectives to classify and analyze datasets. The overall objective of the paper is to evaluate the use of data mining algorithms for the use of the tourism industry. This work used data of Indian travelers who come to the capital of India, Delhi from different parts of the country to visit Delhi and its excursions and avail tour packages according to their interests. There always remains the scope of improvements and so it is here also with this research. The same analysis can be done for in fact any tourism datasets. The inbound tour operators who cater to NRIs and Foreigner Travelers can also use these applications to analyze different datasets. Using data mining tools, especially in the field of tourism can improve quality and customer preferences can be known. For this person should have domain knowledge of both sectors: Tourism as well as data mining technology. This certainly would bring in more tourists both from within the country and from abroad. In future we can evaluate more datasets of the tourism industry and draw a comparative study like which one is more efficient to access patterns or behavior of tourists/travelers and under which circumstances.

References

1. J. Han, M. Kamber, *Data Mining and Techniques,* 2nd ed. (Simon Fraser University, 2006)
2. P.J. Sheldon, *The Tourism Information Technology* (CAB International, Wallingford, 1997)
3. M. Olsen, D. Connolly, Tour. Anal. **4**(1), 29–46 (1999)

4. D. Junping, Z. Min, T. Xuyana, The realization of distributed sampling association rule mining algorithm in tourism, in *7th World Congress on Intelligent Control and Automation. WCICA 2008* (2008), pp. 183–187. https://doi.org/10.1109/wcica.2008.4592921
5. R. Omar, A.O. Md Tap, Z.S. Abdullah, Web usage mining: a review of recent works, in *2014 the 5th International IEEE Conference on Information and Communication Technology for the Muslim World (ICT4M)* (2014). https://doi.org/10.1109/ict4m.2014.7020638
6. M.B. Karathiya, R.S. Sakshi, D.S. Sakshi, D.R. Kathiriya, Data mining for travel and tourism. J. Inf. Oper. Manag. **3**(1) (2012). ISSN: 0976-7754 & E-ISSN: 0976-7762
7. M. Danubianu, V. Hapenciuc, Improving customer relationship management in hotel industry by data mining techniques, in *Proceeding of Competitiveness and Stability in the Knowledge-Based Economy*, CD, 30–31 May 2008, Craiova, Romania (2008), pp. 2444–2452. ISSN/ISBN: 978-606-510-162-3
8. Bartwal, *Incredible India Loosing Sheen* (McClatchy—Tribune Business News, Washington)
9. D. Buhalis, *eTourism: Information Technologies for Strategic Tourism Management* (Financial Times Prentice Hall, 2002)
10. N. Padhy, P. Mishra, R. Panigrahi, The survey of data mining applications & future scope. Int. J. Comput. Sci. Eng. Inf. Technol. (IJCSEIT) **2**(3) (2012)
11. G.G. Emel, Akat, Prolifing a domestic tourism market by means of association rule mining. Anatolia (2007)
12. P. Rani et al., Association rule mining in discovering travel pattern in passport data analysis. (IJCSIT) Int. J. Comput. Sci. Inf. Technol. **5**(4), 5015–5019 (2014)
13. A.B.M. Shawkat Ali, S.A. Wasimi, C. Learning, *Data Mining Methods and Techniques* (2009)
14. E. Irawan et al., Application of classification algorithm C4.5 in recommendations for natural tourism development in district Simalungun. J. Phys.: Conf. Ser. **1255**, 012078 (2019)
15. N. Iswandhani, M. Muhajir, K-means cluster analysis of tourist destination in special region of Yogyakarta using spatial approach and social network analysis. J. Phys. Conf. Ser. **974**, 012033 (2018), *International Conference on Mathematics*

Real-Time Safety and Surveillance System Using Facial Recognition Mechanism

Sachi Pandey, Vikas Chouhan, Rajendra Prasad Mahapatra, Devansh Chhettri, and Himanshu Sharma

Abstract Face recognition mechanisms can be used in many diverse applications such as Security, Retail, Healthcare, and Marketing. The webcams are often used as a security standard with the integration of the facial recognition technology which allows authorized users to unlock objects such as Computer, Laptop, and Mobile. Cameras have a major role in recognition technology to track human faces and keep a track of the number of people in a shot or a certain location such as in an entrance or public places. This technology can be further narrowed down to the recognition and tracking of eyes. In the proposed work, we used an existing Viola–Jones algorithm as a basis for face detection to create a surveillance mechanism by analyzing captured data. Besides, the e-complaint registration feature has been achieved by sending security breach email alerts to the nearby police station or registered users. The proposed system uses cloud storage along with cryptographic techniques to achieve confidentiality, availability and provides surveillance against any uncertain incidents.

Keywords Face recognition · Surveillance · Availability · Home security · Intruder detection · Cloud storage

S. Pandey (✉) · V. Chouhan
Department of Computer Science and Engineering, Indian Institute of Technology, Roorkee, India
e-mail: sachipandey.231@gmail.com

V. Chouhan
e-mail: vikaschouhan.iitr@gmail.com

S. Pandey · R. P. Mahapatra · D. Chhettri · H. Sharma
Department of Computer Science and Engineering, SRM Institute of Science and Technology, Delhi-NCR, India
e-mail: Mahapatra.rp@gmail.com

D. Chhettri
e-mail: chhettridevansh@gmail.com

H. Sharma
e-mail: himanshu.sharma0756@gmail.com

S. S. Dash et al. (eds.), *Intelligent Computing and Applications*, Advances in Intelligent Systems and Computing 1172,
https://doi.org/10.1007/978-981-15-5566-4_44

1 Introduction

Face Recognition Security is an important biometry technology that captured the digital images and evaluated the patterns to identify a particular person in a digital snapshot. Biometric facial recognition technology is a highly effective mechanism which mainly anchors on the same recognizer that a human primarily uses to differentiate between one person's face respective to another. The primary goals are to distinguish the humans' faces to individualize the identification with high accuracy and to interpret the byzantine human-based optic approach. The very first semi-automated facial recognition system involves the system user or administrator to discover facial features that are related in a pattern such as eyes, nose, mouth, and ear. On the digital snapshot, first, we evaluate the difference in ratios based on a common remark point, subsequently which were analyzed the remarked information.

The proposed system also deploys as a home security system to detect an intruder and securely control the door access. In the home environment, the intruder's motions are identified and recognition is accomplished by capturing the live video feed from the web camera and framing. The web camera captures the current frame when the intruder motion is detected where the framing process implements on the captured video frames. The captured image is sent automatically to the registered email users and the nearby police station control rooms for further immediate necessary actions. The proposed technique is implemented in multiple phases. The first phase recognizes the intruders based on human face from the live video streaming for intruder detection. In this phase, we exploit the Paul Viola and Michael Jones Algorithm. In the second phase, it recognizes the human face, and then it transforms into a digital image snapshot. In the third phase, the recognized image is automatically sent to the respective registered user email ID. The automatic email sending process is executed persistently until the user closes the program manually. The performance of the current system is more desirable as compared to existing methods [1, 7, 11]. The proposed mechanism provides a concealed and continuous time covert security recording feature, which is triggered when the camera is activated. The hidden recording will be initiated along with image processing that runs in the background and will be stored in the client's local storage, such that it can be viewed later by the user for surveillance purposes. We stored the captured footage in 'avi' extension form. The footage will be recorded within the time frame of the image snapshot. Thus the recording turns out to be in faster motion than being normal. Better accuracy in recognition is realized with the existing proposed method [3]. In the face recognition system, a major challenge is encountered inside the multidimensional visual model and a trending extent of research. This system utilizes for accessing the door. The facial recognition incorporates several features extracted from the face image, recognition, and classification process [8]. In our work, we utilize the on-demand resource availability of cloud computing, especially data storage. We store and retrieve the captured video anytime, anywhere in ubiquitously using any computing devices such as Mobile, Laptop, and Desktop [9]. Therefore, we store the recorded data on the cloud data centers in encrypted form to achieve data confidentiality along with high availability.

2 Related Work

The authors in [7] discuss that several approaches have been proposed by different individuals where some of them are an emphasis on high-resolution image surveillance and automation. They utilize the single pan-tilt-zoom camera that enables the zoom property, which helps in-depth analysis of the face images. The authors proposed a system for traffic surveillance in [6] based on the background subtraction method, which is implemented in the MATLAB environment. Mishra et al. [10] proposed the detection of moving objects with the help of Eigen-object from the captured frames. They perform the frame pre-processing and processing for better detection accuracy and construct the video frames, respectively. They subtract the background images from the present images pixel by pixel. The multiple video camera surveillance systems are discussed in [5]; they emphasized on how to handle the complicated challenges based on real-world constraints. However, detection and tracking, along with recognition features, are achieved by the adequate use of geometric complexities. There are several algorithms that exist, out of them Viola–Jones algorithm [2] is more vigorous and speedy for the detection of the faces. Haar feature is used by the authors in [12] for detecting regions such as the nose, mouth, and eye by adapting the Adaboost algorithm.

The aforementioned literature provides image detection, surveillance, and concentrates on the accuracy, but they did not consider the confidentiality, availability, and accessibility of the stored information. Hence there is a need to develop a mechanism that provides the surveillance features along with the confidentiality, availability, and accessibility.

3 Problem Analysis

A highly effective and secure facial recognition mechanism is required in a real-time scenario to capture the digital images and evaluating the patterns. Therefore, the proposed system provides secure facial recognition mechanisms and surveillance against any uncertain incidents using cloud storage along with cryptographic techniques to achieve confidentiality and availability.

The preservation of the data or information is becoming essential and difficult. Nowadays, the surveillance cameras are commonly present in Offices, Airports, ATMs, Universities, Banks, and in any locality with a security system. The facial detection system recognizes or verifies an individual from a digital image. However, this approach is typically useful for security and surveillance purpose. Even though the facial detection process identifies a face automatically in an image and extracting its features, then it identifies regardless of expressive behavior, a sign of aging, poor lighting, transitions, and different poses. The extraction of these features is a strenuous task [4]. The semi-automated recognition started in the late 90s, and it has recently caught the attention of the scientific association. Many facial analysis techniques with facial modeling methods have extremely advanced in the last decade. However, the loyalty of the facial detection system and various constraints structure still present a great challenge to the scientific association [13]. Face recognition grasps various preferences over other biometric methods. It is natural, non-obstructive, and quite easy to use. In a research study of the face recognition security system, we consider the rapport of six biometric techniques (eye, finger, signature, hand, face, and voice) with machine-readable travel documents. Figure 1 shows that the facial feature attains the highest level of percentage compatibility. In this study, parameters like the enrolment, renewal, machine specifications, and public perception were considered.

Figure 1 illustrates the basic comparison of machine-readable travel documents (MRTD) compatibility in percentage unit scale along with six different biometric elements. The enhanced attention of automated facial detection systems has been obtained from a domain other than the scientific community which is widely widespread due to enhancing public concerns for security and safety, especially due to the many occurrences of terrorists around the world after September 11, 2001. However, automated face detection or identification system can be utilized in widespread areas other than typically security-based applications (for instance, access-control/verification systems, surveillance systems) and such as computer customized computer–human based interaction.

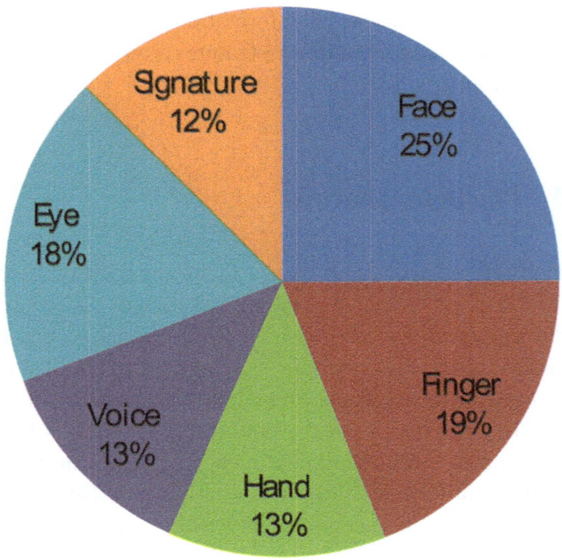

Fig. 1 Illustrates the basic comparison of machine-readable travel documents compatibility in percentage unit scale along with six different biometric elements

We exploit the widely used Viola–Jones Algorithm for our proposed face recognition system. The algorithm performs the following tasks:

Haar Detection: In this phase, the algorithm is to perform the Haar detection. However, every human face has some common structural integrity. This integrity can be identical and identified using feature detection recognized as Haar Detection. Haar detection is applied to detect the eye region, nose bridge, location of eye, mouth, etc. Each feature detection outcome is an isolated value which is determined by the difference between black pixels and white pixels.

Creating Integral Image: The image representation is also known as the integral image that assesses rectangular characteristics in constant time, which provides them a significant speed benefit over more complex alternative features [6]. An integral image at a particular location (x, y) contains the sum of all left pixels (x, y) and above pixels.

AdaBoost Training: In this phase, the machine learning system algorithm is applied to many other diverse kinds of learning languages to improve the rendering of the system. AdaBoost serves for the feature detection and trains the classifier.

Cascading Classifier: The concept is constructed over the recognition process where the detector works every time through the same frame image. The classifier is worked to be fast in demand to be implemented on low-power CPUs, like cameras and mobiles. Major features are used and operated in a repository, and the rest of them are aligned in a cascade container to save some storage space.

4 Proposed Mechanism

The proposed security scheme (i.e., Human Facial Recognition Security) constructs up on the frame subtraction method. The main tactic is to build a static frame series (background frame), and then compare every upcoming frame of the series to the static frame in sequence to identify the motion detection.

The web camera continuously streaming the video and then applies the corresponding operations to determine the intruder as shown in Fig. 2. First, we captured the frame, and then it accomplishes the sequencing of the frames. In the intruder investigation module, we considered two cases. In the first case, if there are no changes in further subframes, then the frame sequence is compared in the video stream on the basis of its first or background frame. In the second case, if there are any changes detected in subframes, then we can determine and handle the variation of subframes of the captured streaming video. Further, the safety operations are initiated, the intruder detection is thoroughgcing as rapidly as frame changes and auto alert system forward the intruder frame to the registered user via email.

The face detection steps are classified and illustrated in Fig. 3a and the surveillance and alert strategy steps are illustrated in Fig. 3b.

In Fig. 4 we present the methodology of the proposed facial recognition system. An input device such as webcam, external cam, CCTV, etc. can be used to capture the video. After providing input from the camera, the system will automatically discover

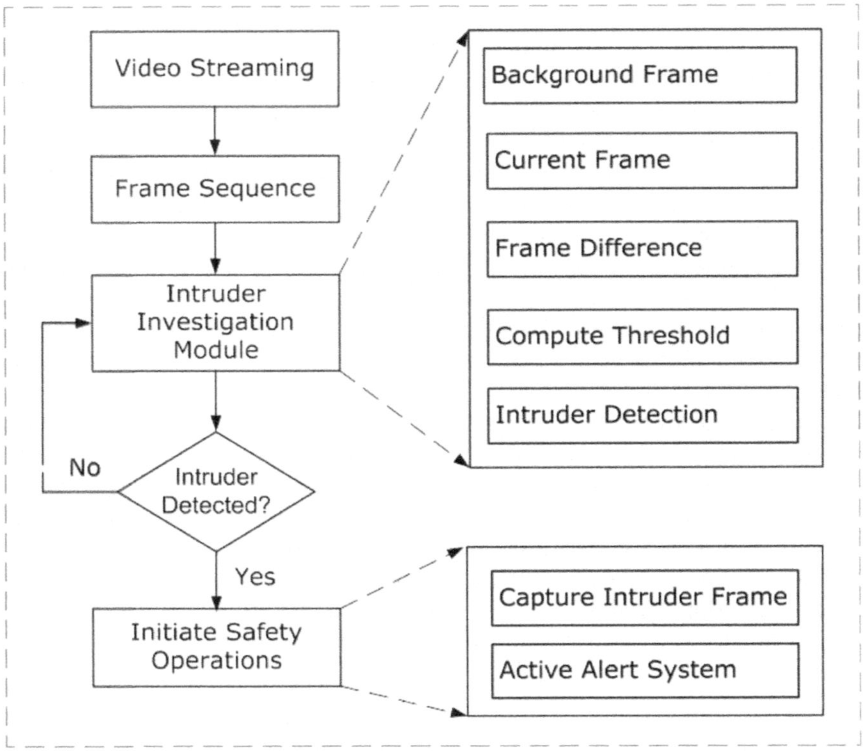

Fig. 2 Investigation of intruder

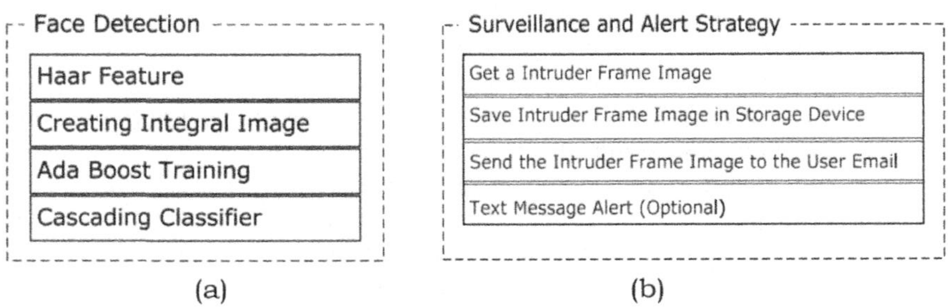

Fig. 3 **a** Face detection steps. **b** Surveillance and alert strategy

faces in real time which is based on the Viola–Jones Algorithm. Haar uses structural integrity to detect identical features, an integral image that assesses rectangular characteristics in constant time, which provides them a significant speed benefit over more complex alternative features.

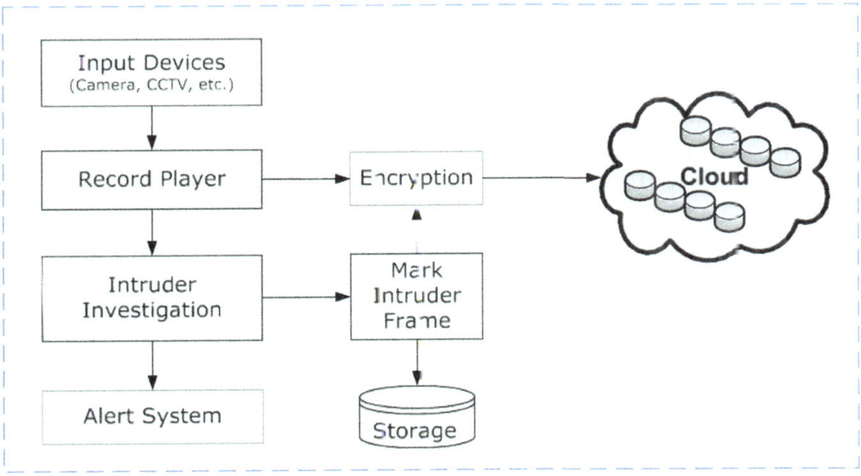

Fig. 4 The methodology of the proposed recognition system

Adaboost training is used for feature detection and to train the classifier lastly cascading the classifier. Recorded video is forwarded to the cloud storage in encrypted form to ensure confidentiality. We used the Secure Hash Algorithm to perform the Encryption and Decryption operations. Consequently, the intruder investigation process marks the intruder frame. After detecting faces in real time, it then captures the frame and converts it into the desired image in encrypted form; then the image is saved in the desired folder of local storage and also uploads this frame to the cloud storage. Further, the alert system is activated, and it would be sent to registered user email through Simple Mail Transfer Protocol (SMTP) protocol. Meanwhile, a hidden recording will be initiated with image processing in a background area and will be stored in the local client storage, such that it can be viewed later by the registered user for surveillance purposes. The AVI encoder or decoder is used to convert camera input into video output in 'avi' extension. The AVI record player will capture whole image frames entirely until the process is stopped manually. The security footage will be recorded within the time frame of the image snapshot. There would be an alert of intrusion recognition on the given User email ID. The whole process is automated and would start again until the user stops it manually. Hence the mail transfer is executed through the SMTP Server, where the SMTP is an Internet standard for electronic mail transmission. In the SMTP protocol process, data (i.e., a document, image, and file) would be sent to the mail server by using open ports of that server. Every server has default communication open ports. These open ports are the way of communication to the desired server.

5 Result Analysis

In this section, we discuss the implementation that incorporates a real-time safety and surveillance system. Our implementation setup uses MATLAB computing environment, which combines the visualization with computer intended vision, to process the captured frames. We set the input device module priority to one that enables the camera as an input device.

Basically, in the case of the video in constant frame sequence mode (constant background frame rate), the algorithm is executed and attempts to search for human face's structural integrity.

The algorithm marks the face boundary and captures the frame immediately when the human visage or face is detected or the background frame rate changes which are shown in Fig. 5.

Figure 6a shows that the camera detects the multiple faces precisely, therefore capable of detecting various faces simultaneously. Thus it captures the frame immediately and processes it to the next phase.

Figure 6b demonstrates the captured frame in the form of the email notification. The captured frame is transformed into an image and is automatically mailed to the registered user email. This sending process will be continuous in every 10 s of a time interval or frame delay.

The recording operation is run in background (hidden security recording) that stored the captured video in local client storage, which can be accessed and played

Fig. 5 Face detected

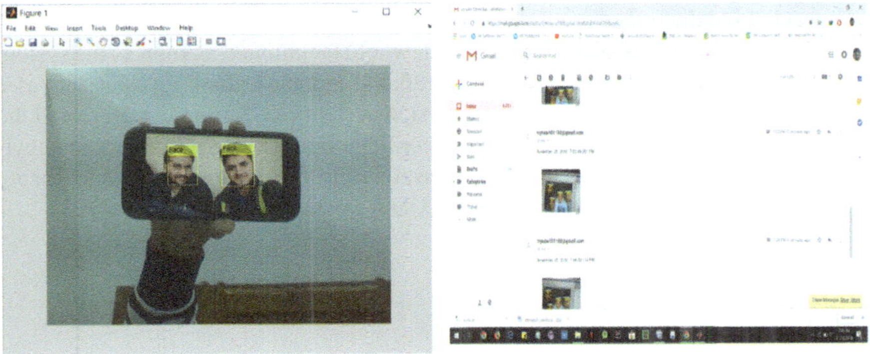

(a) (b)

Fig. 6 **a** Multiple face detection. **b** User Email inbox

anytime. This security recording can be played with any media player which supports 'avi' extension. Moreover, cloud storage allows the accessibility and availability of the stored captured video ubiquitously.

6 Conclusions and Future Work

In highly restricted security areas, the face identification security structure is very beneficial in the detection of intruders. This system helps in minimizing human need and error detection. The security scheme comprises two primary modules. One is the hardware module (tangible) and the other is the software module (Intangible).

The Tangible module entails of any camera, whereas the Intangible module comprises face detection and recognition algorithms. When an individual arrives at the deployed zone, a sequence of snapshots is taken via the installed camera and directed to the registered user as an alarm indication. This process goes on every 10 s of the time interval. The proposed procedure is entirely automated and does not involve any manual labor to tackle the situation.

Robust automated facial or expression identification system can be applicable or be beneficial as individual identification arrangements at the examination, travel documentation, grocery stores, and safety, and also criminal tracking. Nowadays, face detection systems work very fast under restricted circumstances, although all working methods nowadays are much better with frontal mug-shot snapshots and incessant illumination.

Near in future generations, the person identification system will be more secure, reliable, fast, and easy to identify people in real time and in extremely less restricted conditions. It can be done possibly with the help of auxiliary support of the video recording behind the auto capturing image frame. A hidden recording process executes behind the image processing which leads to a covert recording that simultaneously works with image and video, which would be initiated as soon as the system process starts. Hence, as a result, it is a reliable hidden technology that acts as a clandestine for the human identification system. Additionally, our system stored the user data in an encrypted form to achieve confidentiality. Moreover, the cloud environment increases accessibility and data availability in a ubiquitous manner using computing devices such as Mobile, Laptop, and Desktop.

References

1. K. Bonsor, R. Johnson, How facial recognition systems work. HowStuffWorks. Com Np **4** (2001)
2. M.K. Dabhi, B.K. Pancholi, Face detection system based on Viola-Jones algorithm. Int. J. Sci. Res. (IJSR) **5**(4), 62–64 (2016)
3. N.T. Deshpande, S. Ravishankar, Face detection and recognition using Viola-Jones algorithm and fusion of PCA and ANN. Adv. Comput. Sci. Technol. **10**(5), 1173–1189 (2017)
4. P. Divyarajsinh, B. Mehta, Face recognition methods and applications. Int. J. Comput. Technol. Appl. **4**(1), 84–86 (2013)
5. J. Doring, H. Sprave, Multiple video camera surveillance system, U.S. Patent 10/371,192, 16 Oct 2003
6. C.H. Low, M.K. Lee, S.W. Khor, Frame based object detection—an application for traffic monitoring, in *2010 Second International Conference on Computer Research and Development* (IEEE, 2010), pp. 322–325
7. N.A. Manap, G. Di Caterina, J. Soraghan, V. Sidharth, H. Yao, Smart surveillance system based on stereo matching algorithms with IP and PTZ cameras, in *2010 3DTV-Conference: The True Vision-Capture, Transmission and Display of 3D Video* (IEEE, 2010), pp. 1–4
8. R. Manjunatha, R. Nagaraja, Home security system and door access control based on face recognition. Int. Res. J. Eng. Technol. **4**(3), 437–442 (2017)
9. P. Mell, T. Grance, The NIST definition of cloud computing (2019), http://dx.doi.org/10.6028/NIST.SP.800-145. Accessed 12 May 2019
10. S.K. Mishra, K.S. Bhagat, Human motion detection and video surveillance using matlab. Int. J. Emerg. Res. Manag. Technol. **5**(5) (2016)
11. J. Pagliery, FBI launches a face recognition system. CNN.com (2014)
12. H.M. Phuong, L. Dung, T. de Souza-Daw, N.T. Dzung, T.M. Hoang, Extraction of human facial features based on haar feature with Adaboost and image recognition techniques, in *2012 Fourth International Conference on Communications and Electronics (ICCE)* (IEEE, 2012), pp. 302–305
13. S. Tathe, A.S. Narote, S.P. Narote, Human face detection and recognition in videos, in *2016 International Conference on Advances in Computing, Communications and Informatics (ICACCI)* (IEEE, 2016), pp. 2200–2205

Liver Disease Prediction Using an Ensemble Based Approach

B. Muruganantham, R. P. Mahapatra, Kriti Taparia, and Mukul Kumar

Abstract Lately, the usage of Information system and strategic tools in the domain of medical science is constantly growing. The liver is an Exocrine Gland that helps in the digestion process, especially for fats along with altering the pH of food due to its alkaline nature. One of the most common signs for the majority of liver disease is hyperbilirubinemia which is very difficult to identify at an early stage. While certain diseases such as obstructive Jaundice and acute viral hepatitis present themselves with an early rise of bilirubin along with yellowish discoloration of the skin, many other diseases don't usually present themselves with an early rise of serum bilirubin or skin discoloration because of which sometimes liver disease are overlooked or misdiagnosed in primary level. However, serum bilirubin is not the only way to diagnose liver disease because it is not specific. The most specific way to diagnose liver disease is by liver function test. With the help of the detection of the enzyme level, we can identify and confirm the presence of liver disease and intensity of liver damage when coupled with suitable imaging modalities like Ultrasound, CT scan, or MRI scan. The dataset contains patient parameters such as Age, Sex, Total Proteins, Alkaline Phosphatase, Alanine Phosphatase, Total Proteins, Total Bilirubin, Albumin, Albumin and Globulin Ratio, Direct Bilirubin, and the Result. We are using Binary Classification which is basically classifying the element of given set into two given sets i.e., Patient suffering from Liver disease or not. We will try to use an Ensemble Based Approach to find the best prediction accuracy.

B. Muruganantham (✉) · K. Taparia · M. Kumar
Department of Computer Science and Engineering, SRM Institute of Science and Technology, Kattankulathur, Chennai, Tamil Nadu 603203, India
e-mail: muruganb@srmist.edu.in

K. Taparia
e-mail: krititaparia@gmail.com

M. Kumar
e-mail: mukulsharma1998@gmail.com

R. P. Mahapatra
Department of Computer Science and Engineering, SRM Institute of Science and Technology, Ghaziabad, India

S. S. Dash et al. (eds.), *Intelligent Computing and Applications*,
Advances in Intelligent Systems and Computing 1172,
https://doi.org/10.1007/978-981-15-5566-4_45

Keywords Liver disease · Prediction · Classification · Accuracy

1 Introduction

The liver is the largest gland of the body. Liver Disease is a very tricky disease to diagnose. It is very hard to diagnose it at an early stage based on its symptoms. It is an Exocrine Gland. This organ is present in the right hypochondriac region of the abdomen. It serves many important functions such as digestion, storing glucose, metabolizing toxin. Fatty Liver Disease which occurs due to fat build-up in the liver is also one of the major liver diseases. It is a very prevalent disease accounting for more than 10 million cases per year in India. It usually causes no symptoms and needs a test to determine if a person is suffering from the disease.

Artificial Intelligence (AI) is the ability of computer programs to think learn and apply the knowledge it has gained from training in various fields. Nowadays artificial intelligence is applied in almost every field of life and it consists of various parts such as Machine Learning and Deep Learning which further divides into many parts. Artificial intelligence proceeds in its course of action in such a way as to obtain goal maximization at each stage.

Machine Learning, is a sub-domain of artificial Intelligence basically involves techniques which help the machine experience to learn to take decisions and respond in certain scenarios based on the prior information it has gained. It consists of two types: Supervised Machine Learning and Unsupervised Machine Learning.

Supervised Machine Learning is a learning in which we train our machine by providing labeled datasets that means data is already well labeled with correct answers. This data is known as training data and the data which we use to test if our model has learned and giving expected result is known as testing data. In this type of learning, we first train our model using training data and after training, we test our model using test data. Some examples of Supervised Machine Learning are Linear regression, Random Forest, and Support Vector Machine

Unsupervised Machine Learning is self-supervised learning which does not contain training dataset it finds its solution using previous unknown patterns available in the dataset. It can also be further grouped into clustering and association.

Semi-Supervised Learning—Some data is labeled and most of the data is unlabeled, i.e., combination of testing and training data. It is a mixture of both Supervised and Unsupervised Machine Learning Techniques in the sense that testing and training datasets are not explicitly specified but are together.

2 Related Work

Nahar and Ara [1] have used various algorithms related to decision tree such as Random Forest, Random Tree, Decision Stump, J48, LMT, and Hoeffding tree for the classification process and it has been compared on the performance measures such as Mean Absolute Error, Precision, Accuracy, Recall, and Runtime and shows that Decision Stump has high accuracy. Jacob et al. [2] have used SVM, Logistic Regression, K-NN, and ANN with 10 input neurons showing that ANN is highly efficient. Sontakke et al. [3] have proposed the use of genetic microarrays along with neural networks for the prediction showing that the molecular biology approach can strengthen the predictive power. Kefelegn et al. [4] have shown that Random Forest algorithm is not suitable due to the problem of overfitting and has suggested the utilization of oversampling methods to address this problem. Thirunavukkarasu et al. [5] have shown that both Logistic Regression and K-NN generated equal accuracy but in medical biology, sensitivity plays a key role in differentiating the performance. Hamid et al. [6] have proposed the use of the abstention paradigm to ensure that no result is generated when the model is not confident about its prediction so as to avoid the wrong classification. Ayelkeen et al. [7] have shown that biochemical markers have a huge impact on the positive prediction of different stages of liver fibrosis and HA level increases with increase in the grade of fibrosis. Belavigi et al. [8] have proposed that in comparison to Rprop and SAG on text dataset and CNN on image dataset, CNN gives high accuracy. Kumar et al. [9] have used different classification algorithms such as Random Forest, Naive Bayes, K-NN, K-Means, and C 5.0 proposing that C.5.0 works best after adaptive boosting. Hashem et al. [10] have used various inclusion and exclusion parameters to determine which data can be used for the training of the model to ensure that irregular data patterns do not affect the classification. Sontakke et al. [3] have proposed the usage of unsupervised learning methods such as K-Means, Affinity Propagation, and DBSCAN rendering Silhouette coefficient as the primary factor for judging the performance (Table 1).

3 Proposed Methodology and Overview of System Architecture

This proposed work is to process the datasets into the training and testing part and apply various techniques to find the best approach. The architecture of the proposed work as shown in Fig. 1, shows the various modules of our proposed system.

3.1 Data Cleaning

Data Collection contains a collection of datasets of various patient data. The data collection contains a collection of patients various parameters and details. It is basically the first step to collect as much data as possible. The model will benefit proportionally to the size of dataset, therefore the larger the size of the dataset the more the model learns from it.

Table 1 Related work

Ref. No.	Proposed objective of work	Algorithms used/architecture developed	Dataset/input parameters used	Applicability	Performance measures used	Limitations found
[1]	Prediction of Liver Disease with the help of different techniques of Decision Tree	1. J48 2. REP Tree 3. Random Tree 4. Decision Stump 5. LMT 6. Random forest 7. Hoeffding Tree	UCI Machine Learning Repository	For the given parameters, Decision stump has proved to provide the highest accuracy as compared to other algorithms	1. Accuracy 2. MAE 3. PRE 4. REC	Does not use advanced decision trees such as CART
[2]	Machine learning algorithms for the detection of liver diseases	1. Logistic Regression 2. K-NN 3. ANN 4. SVM	Indian Liver Patient Dataset (ILPD) (UCI Machine Learning Repository)	A back propagation neural network with 10 input neuron layers serves as the model with highest accuracy	1. Accuracy 2. Sensitivity 3. Precision 4. Specificity	The results obtained can be varied based on the number of neurons in the input layer
[3]	Prediction of Liver Disease using various approaches of Machine Learning	1. Back Propagation 2. Support Vector Machines Algorithm	Indian Liver Patient Dataset (ILPD) (UCI Machine Learning Repository)	Microarray analysis and comparison of two algorithms	1. Accuracy 2. Sensitivity 3. Precision 4. Specificity	Natural factors, for example, age, sexual orientation, ethnicity and diet keep on influencing the consistency of the outcomes
[4]	Usage of data mining algorithms on imbalanced data for liver disease prediction	1. Logistic Regression 2. Random Forest 3. Auto Neural Network 4. K-NN	Datasets of liver patients from Andhra Pradesh region of India	Imbalanced data handled with the help of sampling method	1. Accuracy 2. Train rate 3. Train-ROC	Problem of overfitting in random forest algorithm due to lack of data set

(continued)

Table 1 (continued)

Ref. No.	Proposed objective of work	Algorithms used/architecture developed	Dataset/input parameters used	Applicability	Performance measures used	Limitations found
[5]	Usage of different classification algorithms for liver disease prediction	1. Logistic Regression 2. Support Vector Machines 3. K-Nearest Neighbor	Indian Liver Patient Dataset (ILPD) (UCI Machine Learning Repository)	Given the set of input parameters, Logistic Regression and K-Nearest Neighbor have been shown to have the highest accuracy whereas logistic regression gives the highest sensitivity	1. Accuracy 2. Sensitivity 3. Specificity	Does not consider the precision of results as a performance measure for the algorithms
[6]	Automated detection of liver abnormalities with the use of ultrasound images	1. Nearest Neighbor (NN) 2. Support Vector Machine (SVM) 3. Learning with Abstention (LWA)	Dataset involves a total of 99 liver ultrasound images comprising of 56 images of liver abnormalities example FLD and 43 images of healthy individuals	Compares usage of LWA with conventional classification	1. Specificity 2. Sensitivity	Needs to incorporate working on large independent test set with elastography data

(continued)

Table 1 (continued)

Ref. No.	Proposed objective of work	Algorithms used/architecture developed	Dataset/input parameters used	Applicability	Performance measures used	Limitations found
[7]	Decision Tree Approach for the prediction of different stages of Liver Fibrosis	1. Decision Tree	dataset includes laboratory tests conducted and fibrosis markers collected from the Department of Medical Biochemistry and Molecular Biology, Faculty of Medicine, Cairo University	The study shows that HA level increases with the advancement in the stage of liver fibrosis and is a crucial marker for prediction	Accuracy	Other biomarkers need to be considered for significance of calculation and also increase the sample population
[8]	Comparison of Prediction of Liver Disease using Rprop, SAG on text dataset and CNN on image dataset	1. Resilient Back propagation Neural Network (Rprop) 2. Stochastic Average Gradient (SAG) 3. Convolutional Neural Network (CNN)	Indian Liver Patient Dataset (ILPD) (UCI Machine Learning Repository)	Shows that the image dataset used for CNN, including the ultrasound images of the liver gives the best accuracy	1. Accuracy 2. Sensitivity 3. Precision 4. Specificity 5. Misclassification rate	CNN get issues of low accuracy

(continued)

Table 1 (continued)

Ref. No.	Proposed objective of work	Algorithms used/architecture developed	Dataset/input parameters used	Applicability	Performance measures used	Limitations found
[9]	Data mining techniques for liver disorder analysis and diagnosis	1. K-Nearest Neighbor Algorithm 2. K-Means 3. Naive Bayes 4. C5.0 Algorithm 5. Random Forest	UCI repository	Shows that Random Forest Algorithm performs best when no changes are applied but after adaptive boosting the C5.0 algorithm provides an increasingly accurate result	1. Accuracy 2. Recall 3. Precision	Different algorithms work best with respect to different performance measures used
[10]	Prediction of advanced liver fibrosis in chronic hepatitis C patients using various machine learning approaches	1. Decision Tree Learning Algorithms 2. Genetic Algorithms 3. Particle Swarm Optimization	Cohort data of 39,567 chronic hepatitis C patients enrolled in Egyptian National Committee for the Control of Viral Hepatitis database in National Treatment Program of HCV patients in Egypt	Approach to develop the classification models by combining the clinical information and serum biomarkers	1. Accuracy 2. Sensitivity 3. Specificity 4. ROC analysis	Low sensitivity of the ADT model

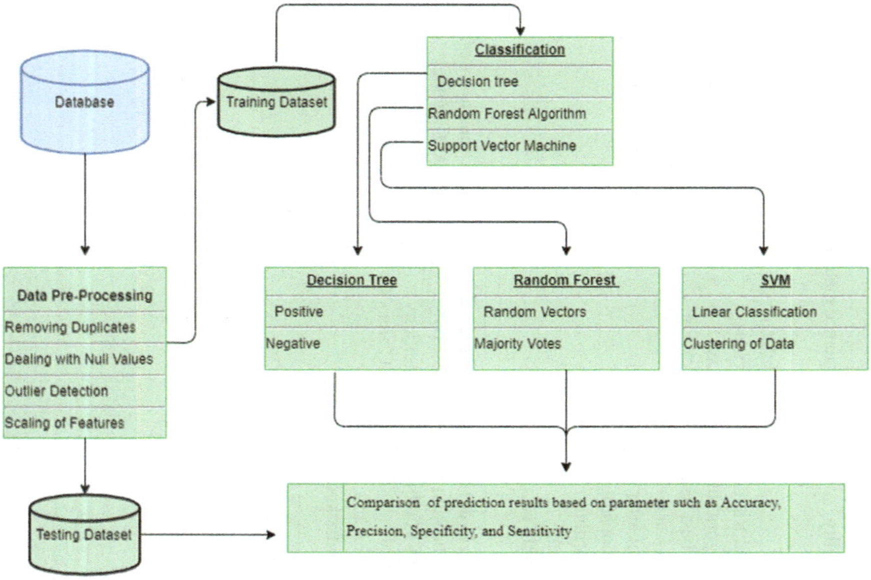

Fig. 1 Architecture diagram of the proposed system

3.2 Data Preprocessing

Data Preprocessing is the next crucial step after Data collection. The real-world data available is not fit for direct use as it contains a lot of noisy, incomplete and inconsistent data so we must clean it before we can use it. Data preprocessing is a method to convert the raw data into useful data for further processing. Data preprocessing help us get better results when we apply classification algorithms to the data. The various stages involved in Data Preprocessing are Data cleaning, Data integration, and Data transformation.

3.2.1 Data Cleaning and Data Integration

Data Cleaning is the process that involves the removal of incomplete, corrupt, or inaccurate data from the dataset for example if a patient does not contain some important parameters so data cleaning is used. Data integration basically involves combining the various data residing at different sources into a single unified view.

3.2.2 Data Transformation

Data Transformation is used to convert the image from one form or structure to another.

3.3 Classification

As the name suggests Classification is the process of classifying. The main goal is to predict the result accurately or predict the target goal accurately. A classification model can be used in many real-life applications. It is the initial phase that uses algorithms like Decision Tree, Random forest etc.

3.3.1 Decision Tree Algorithm

Decision Tree is one of the examples of supervised machine learning which is used for classification problem. Ayeldeen et al [7]. In this technique, we split the sample into two or more homogeneous sets which are based on input variables which are most significant and it also includes a node, branches, and leaf nodes. As shown in Fig. 2, each depicts different things such as internal node shows test on attributes, branch shows the outcome and leaf node holds a class label. The nodes expand on a branch having Higher Information Gain.

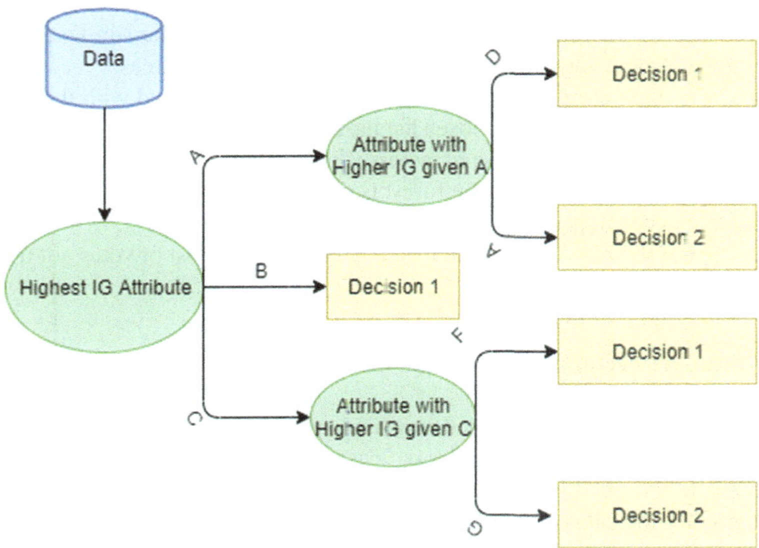

Fig. 2 Decision tree algorithm

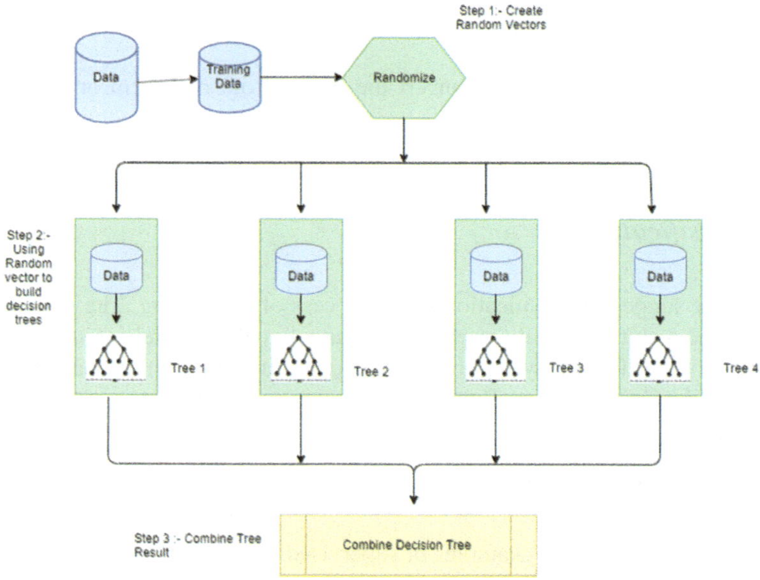

Fig. 3 Random forest algorithm

3.3.2 Random Forest Algorithm

The algorithm works in the following manner as shown in Fig. 3.

1. Given there are n cases in the preparation dataset, from these n cases, sub-tests are picked indiscriminately with substitution. These irregular sub-tests browsed the preparation dataset are utilized to assemble individual trees.
2. Expecting there are *k* factors for input, a number *m* is picked with the end goal that $m < k$ [1]. *m* factors are chosen haphazardly out of *k* factors at every hub. The split which is the best of these *m* factors is picked to part the hub. The estimation of *m* is kept unaltered while the forest is developed.
3. Each tree is developed as huge as conceivable without pruning.
4. The class of the new item is anticipated dependent on most of votes got from the blend of all the choice trees.

3.3.3 Support Vector Machine Algorithm

SVM is an example of supervised machine learning algorithm that is used for analyzing data that is required for classification and regression analysis [2]. These models intently look like neural systems. Consider an *X* dimensional dataset. Utilizing SVM, the data can be plotted into an *X* dimensional space. The preparation information focuses are then plotted into *k* various areas depending on their names as characterized by hyper-planes of *X* various dimensions [3]. After the testing stage is finished, the test focuses are plotted in a similar N dimensioned plane. Contingent upon which area the focuses are situated in, they are fittingly ordered in that region SVM. The fundamental objective is to arrange all testing vectors into two classes so it can recognize them and give its very own predictions.

4 Performance Evaluation

The comparison of the results has been done based on the performance measures of prediction such as Accuracy, Sensitivity, Specificity, Precision, and Recall. We will perform various evaluations to differentiate the techniques among each other. The performance and its measures are compared statistically using different types of plots to view their performance against all performance measures.

5 Conclusion

The objective of this paper is to compare different classification algorithms on the dataset and compare their results based on different performance measures specified. This comparison will help in identifying which algorithm works best with this type of text dataset with the specified set of input parameters. This model can aid medical experts in accurate liver disease prediction as it is not easily identifiable in its early stages. A GUI will be incorporated for this model to provide a user-friendly interface for all medical professionals to be able to use this tool for running predictions on the patients. In future usage, this proposed system can be used for dynamic and more complex datasets of a much larger volume and with more number of parameters taken into consideration to make the model perform more accurately.

References

1. N. Nahar, F. Ara, Liver disease prediction by using different decision tree techniques. Int. J. Data Min. Knowl. Manag. Process (IJDKP) **8**(2) (2018)
2. J. Jacob, J. Chakkalakal Mathew, J. Mathew, E. Issac, Diagnosis of liver disease using machine learning techniques. Int. Res. J. Eng. Technol. (2018)
3. S. Sontakke, J. Lohokare, R. Dani, Diagnosis of liver diseases using machine learning, in *International Conference on Emerging Trends & Innovation in ICT (ICEI)* (2017)
4. S. Kefelegn, P. Kamat, Prediction and analysis of liver disorder diseases by using data mining technique: survey. Int. J. Pure Appl. Math. (2018)
5. K. Thirunavukkarasu, A.S. Singh, Md. Irfan, A. Chowdhury, Prediction of liver disease using classification algorithms, in *4th International Conference on Computing Communication and Automation (ICCCA)* (2018)

6. K. Hamid, A. Asif, W.A. Abbasi, D. Sabih, Machine learning with abstention for automated liver disease diagnosis, in *International Conference on Frontiers of Information Technology* (2017)
7. H. Ayeldeen, O. Shaker, G. Ayeldeen, K.M. Anwar, Prediction of liver fibrosis stages by machine learning model: a decision tree approach, in *Third World Conference on Complex Systems (WCCS)* (2015)
8. D.H. Belavigi, G.S. Veena, D. Harekal, Prediction of liver disease using Rprop, SAG and CNN. Int. J. Innov. Technol. Expl. Eng. (IJITEE) **8**(8) (2019). ISSN: 2278-3075
9. S. Kumar, S. Katyal, Effective analysis and diagnosis of liver disorder by data mining, in *Proceedings of the International Conference on Inventive Research in Computing Applications (ICIRCA 2018)*
10. S. Hashem, G. Esmat, W. Elakel, S. Habashy, S.A. Raouf, M. Elhefnawi, M. El-Adawy, M. El Hefnawi1, Comparison of machine learning approaches for prediction of advanced liver fibrosis in chronic hepatitis C patients, IEEE/ACM Trans. Comput. Biol. Bioinform. **15**(3) (2018)

Significance of Route Discovery Protocols in Wireless Sensor Networks

Guntupalli Gangaprasad, Kottnana Janakiram, and B. Seetha Ramanjaneyulu

Abstract Wireless Sensor Networks (WSN) have numerous applications in various fields like military, navigation, agriculture, health care sector, etc., due to their data sensing and transmission abilities. In spite of their wide range of applications in many fields, finding appropriate methods for data gathering, routing, enhancing network lifetime, and efficient communication between the sensor nodes have become the issues of continuing research during the last two decades. Among them, finding the best route from source to destination is one of the prime issues. Many routing mechanisms are proposed in the literature and implemented, but there is no single routing mechanism that addressed all the issues like throughput, latency, reliability, and energy efficiency. This article provides a review of the routing techniques in wireless sensor networks, along with the discussion of the issues they addressed.

Keywords Wireless sensor networks · Routing protocols · Network life time · LEACH · PEGASIS

1 Introduction

A wireless sensor network comprises a large number of sensor nodes placed in the field. The sensor nodes are of small size, limited battery, less computing power, and small memory for data storing. These sensors sense the physical information like temperature, pressure, humidity, moisture, etc., from the surroundings, and forward it to the base station or sink node, through many intermediate nodes. At the base

G. Gangaprasad (✉) · K. Janakiram (✉) · B. S. Ramanjaneyulu (✉)
Department of ECE, VFSTR University, Guntur, AP, India
e-mail: Ganga.prasad96@gmail.com

K. Janakiram
e-mail: Janakiram006@gmail.com

B. S. Ramanjaneyulu
e-mail: ramanbs@gmail.com

S. S. Dash et al. (eds.), *Intelligent Computing and Applications*,
Advances in Intelligent Systems and Computing 1172,
https://doi.org/10.1007/978-981-15-5566-4_46

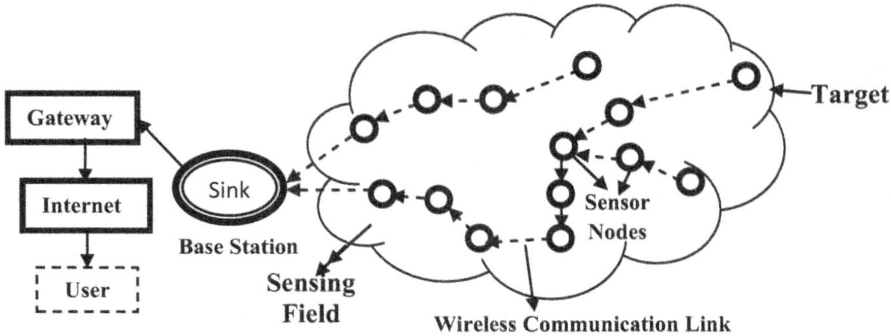

Fig. 1 Wireless sensor network architecture

station, the processing is performed on the data and appropriate action will be taken like forwarding it to remote locations through devices like gateways [1] (Fig. 1).

The WSNs have gained a lot of importance in many fields of applications because they can be deployed for many useful purposes in the real world. The applications of these networks are mainly categorized into two types, namely monitoring applications and tracking applications [2, 3]. The first category is used for inhouse and outside environment monitoring, Seismic and structural monitoring, physical condition monitoring, and control monitoring, whereas the second category is used for tracking the vehicles, human beings, animals, etc.

From the application point of view, WSNs are used in the broad categories of the following:

- Water quality testing
- Environmental monitoring
- Military applications
- Disaster relief management
- Soil moisture monitoring
- Health care monitoring
- Body area networks
- Home applications, etc.

Though WSNs have many applications in different fields, they are still facing many issues. Due to limited power available with the sensor nodes, they are unable to stay alive for longer durations. There are a number of design issues and challenges which degrade the performance of the WSNs. The most important are:

- There is no prior topology to sensor networks; once placed into the field they setup themselves only. Human involvement to place them in a specific topology may not be possible in many cases.
- These networks are infrastructure-less and hence distributed routing mechanism is required.

- Sensor nodes depend on their battery power and there is no chance to recharge the battery frequently. So energy efficiency is important for long-time survival.
- Designing of sensor node hardware that minimizes energy consumption is needed.
- Providing communication among all the nodes in the field is essential. Otherwise, it will form isolated nodes, which is an indication of non-reachable nodes in WSN.

In sensor networks, the best routing method is expected to minimize energy consumption at all nodes to improve the network lifetime. Along with energy savings, the best routing protocol should also offer the best throughput, minimum latency, and guaranteed delivery of data. Extensive research is going on to find efficient routing techniques. So many routing protocols are already proposed in the literature, but no single protocol solves all the issues. Designing efficient routing protocols for WSN is a challenging task for researchers from the last two decades.

The main scope of this paper is to provide a survey on energy-efficient routing protocols and their classification existed at present. Any route-finding protocol in sensor networks falls under one of the categories like Network Structure, Communication model, Topology based, and Reliable Routing Scheme [4]. Among all the above-stated methods Network Structure protocols offers the best energy efficiency in the network because they implement cluster-based routing. They are also more scalable compared to all traditional routing protocols.

The remaining part of the paper is framed in the following way. Challenging issues of wireless sensor networks are presented in section-2. Information about existing routing protocols of WSNs is presented in Sect. 3. In Sect. 4, a detailed discussion of various routing protocols is carried out. The conclusion of the paper is presented in Sect. 5.

2 Challenges of Designing Wireless Sensor Networks

WSNs have a wide range of applications in various fields, but at the same time, they are plagued with many problems also. In this section, some important challenges of WSN are presented. These are listed below.

Quality of service Reliability, availability, throughput, and latency are some of the quality attributes that a network can offer. They are usually described in terms of the metrics associated with them. Appropriate routing, clustering, and channel access mechanisms will help in achieving the quality of service expected from the given WSN.

Fault Tolerance When the nodes in WSN are damaged or run out of energy that leads to permanent failure of communication between the nodes. So it is important that such faults should be addressed immediately or precaution taken for their avoidance. Also, the network needs to be employed with some pre-tolerant methods.

Lifetime In WSNs, once nodes are placed, it is highly impossible to replace the drained batteries. So, the nodes are considered with limited energy. But, in addition to its own data transmissions, every sensor node in WSN also acts as a relay node to forward the data of neighbor nodes to the destination or base station. So the possibility of energy drain is high and hence achieving enhancement of lifetime in the network is a challenging task.

Scalability Due to the requirement of WSN with a varied number of nodes for each application, network scalability is important. In spite of it, effective communication has to be maintained between the nodes.

Programmability The nodes employed in WSN are not only for sending the pre-decided information to the sink. In addition to that, they may also need to perform the tasks for changing requirements of a later time. To facilitate this, the nodes should be programmable on-the-fly, after their deployment also. They also need to reconfigure as per changing topologies so that the network will perform in an optimum manner. So, the nodes in the network should be programmed to fulfill the above.

Maintainability In WSN, some nodes may get drained quickly due to their higher quantum of work. It may lead to the failure of ongoing communication because the route that passes through this drained node may disappear. In that scenario, the network should automatically find new routes. It must also monitor its health and respond to the disturbances in the ambiance.

Multi-hop wireless communication The sensors forward the sensed content to the destination via multiple nodes. It makes energy consumption less compared to the direct communication to sink from all the nodes.

Energy-Efficient operation Designing of each protocol in an energy-efficient manner is the desired approach. For that, it needs to address many constraints.

Collaboration and in-network processing The sensed data from all sensors in the network need not be sent in the same for the sink, because some redundant data may exist. So network should ensure that data aggregation is done prior to transmission of data to the sink, thereby duplicate or redundant data is reduced. To do this collaboration among all the nodes is required.

3 Related Work

The energy-efficient routing protocols are implemented based on the network model and application needs. Different factors are taken into consideration while designing routing protocols like power, memory size, communication, routing, and network lifetime. Among all, routing is an important issue for energy-efficient sensor networks to improve the network lifetime. There are many proposals available in the literature for energy minimization protocols in WSN. In [1], the authors explained some protocols belonging to the quality of service and their metrics in detail. In [2], the authors provide details about the networking of wireless sensors and their issues in detail. In [3], the authors describe various applications of WSNs and their issues while designing real networks. In [5], energy-aware protocols belonging to the QoS concept and their techniques in WMSNs are explained. In [4], the authors provide well-defined methods for energy efficiency and internal classification with their applications and limitations in detail. In [6], the authors made an extensive survey of design issues and techniques for WSN. They explained the constraints of sensor nodes and networks stack. Then, they discussed various applications of WSNs. In [7], the authors presented various routing concepts in WSN. They describe the routing concepts based on network structure into three types i.e., flat, hierarchical, and position-based routing methods. And more ever, these protocols are categorized into energy, multipath, negotiation, and QoS-type routing protocols. Different energy minimization methods for sensor networks are provided in [8]. These methods are designed based on data-centric, location-based, and hierarchical.

In [9], the authors present a classification of protocols based on broadcast/multicast problem is in WSN. The proposed minimum energy broadcast/multicast is aimed at reducing the total transmission energy usage at all sensor devices. In [2], the authors present multiple applications and different aspects of WSNs. These aspects are divided into the following types: underlying operating system, a communication protocol stack, internal platform, services of network, and placement issues. In [10], the authors presented details about the quantum of battery energies used for various parts of the node. The hardware of the wireless sensor device is divided into four parts: a subsystem for sensing, a computing subsystem including microcontrollers and data storing, a unit for communication, and a battery unit. This article explained the characteristics and advantages of energy consumption schemes in detail. In [11] the author presented a categorization of route-finding methods and issues in WSN. This work explained a number of route-finding methods briefly. In [12] the important challenging design parameters of medium access control protocols in WSN are presented. It describes many MAC protocols with their strengths and weaknesses. It didn't provide any energy-efficient routing protocols for network lifetime improvement in WSN. In [13], the authors presented routing protocols for WMSNs. They explained some design parameter issues of routing protocols for WMSN and their issues.

4 Classification of Routing Methods in Sensor Networks

There are a number of routes discovering methods in wireless sensor networks. Generally, these routing protocols are classified into the following types [4].

(A) Network structure.

(B) Communication model.
(C) Topology type.
(D) Reliable routing.
(E) Nature of nodes.

(A) Network Structure

The network structure can be classified based on how the nodes are placed in the network and connected to each other and to the sink. The protocols under this classification describe the way nodes are interconnected to provide the routing according to the network structure. The models in this type are (a) Flat protocols, and (b) Hierarchical protocols.

(a) Flat Protocols

In this category of protocols, all the nodes have the same level of connectivity in the network. This type of network uses to minimize the overhead for communicating nodes [4]. The protocols under this category are described below.

Proactive Protocols Here complete route-finding information is provided with the help of routing tables at each node. Through the periodic exchange of data between nodes, it builds a routing table at each node. In these protocols, the bandwidth requirement is high and extra battery power is required to maintain routing tables. Here a good number of protocols have existed and one of them is explained below.

Wireless Routing Protocol (WRP) is a table-based method, which is implemented by using the Bellman-ford algorithm [14]. In this method, each node focuses on checking the route content received repeatedly from all neighbor nodes. These protocols are not useful for very large scale WSN. These proactive protocols require more bandwidth.

Reactive protocols This type of protocol works based on on-demand requests only. Whenever a particular request is raised, that time only, the node will initiate the route-finding process by sending/receiving route request/route reply information, respectively. The below-mentioned protocol comes under this category.

The temporary ordered routing algorithm is distributed in nature and utilizes the principle of link reversal with the loop-free mechanism. This protocol is implemented with the help of topologies of the network to the minimization of communication overhead for decreasing power usage. In this flat type of protocols, some more are proposed for broadcasting, flooding, rumor routing, and energy-aware temporary ordered routing algorithm. So all are used to minimize the energy and to improve the network lifetime.

(b) Hierarchical Based Routing

In this routing protocol, the entire network is divided into different clusters, each cluster maintaining one cluster head which is responsible for sending data to the base station from all sensors in that cluster [4]. For energy efficiency purposes, hierarchical routing is used. It uses two stages approach. The first stage is to make clusters and elect cluster leaders based on some threshold value whereas the second stage is about transferring data from cluster members to the sink through various cluster leaders. The communication in hierarchical routing networks is through single-hop communication and multi-hop communication. In the former case, the communication is between cluster members and cluster head whereas in the latter case the communication is between cluster heads and the base station. Some of the advantages of Hierarchical based routing are [15]:

- Reduction in transmission cost and energy saving.
- Improvement in network Scalability.
- Reduction in routing table size at each sensor node.

The protocols of this type are discussed below.

Low Energy Adaptive Clustering Hierarchy (LEACH) is implemented to minimize energy consumption. In LEACH, cluster heads are selected based on threshold value in different random rounds. For communicating data to the cluster leader, respective TDMA time slots are provided. The aggregated data is sent to the sink node in CDMA format to avoid the interference in the medium. This protocol is implemented in two stages, namely the Setup stage, and Data transmission stage. In the setup stage, cluster creation, and cluster head election are carried out. In data transmission stage aggregation of data is performed. Aggregated data is transferred to the destination through various cluster heads. Each sensor node provides various random values between 0 and 1. If a sensor node gets a value less than the threshold value, that node becomes a cluster leader for that round. If any node acted as cluster head, it is not allowed to become a cluster head one more time for that round. The formula for selection of cluster head is based on threshold value given by,

$$T(n) = \frac{P}{1 - P\left(r * \operatorname{mod}\left(\frac{1}{p}\right)\right)} \text{If } n \varepsilon G \tag{1}$$

where P is the probability of selection of Cluster head, r is the existing round and G is the group of sensor nodes that have not been Cluster head in the current round. By using this protocol maximum energy efficiency is achieved and network lifetime is increased.

Power Efficient Gathering in Sensor Information Systems (PEGASIS) protocol is a Hierarchical type. In this method, the near-optimal chain concept is used for minimizing energy consumption in WSNs. It offers better performance than the previous protocol. Here communication between the nodes in the network takes place by communication between closest neighbor's nodes only. The same nodes participate in sending the information to the sink. This process is carried out in different rounds and in each round, one node sends the data to the destination. After completion of the round, a new round will begin. In PEGASIS protocol nodes are formed in chain structure to share the data per each round. The aim of this method is to minimize the battery power while forwarding the sensing information in each round and to maintain uniform energy consumption at all nodes in the network. To get all these advantages in PEGASIS, the following considerations need to be implemented in the network.

- To enhance the lifetime of each sensor node in WSN, collaborative methods are used.
- To optimize the bandwidth, the closest nodes should communicate with each other (Figs. 2 and 3).

In addition to the above, the following are some of the advanced routing protocols which are aimed at improving the network lifetime in WSN.

- **AZ-SEP**: Advanced Zonal Stable Election Protocol is hybrid in nature. Here sensor nodes directly communicate with the base station or sink node and remaining sensor nodes participate in creating clustering. In this protocol, the energy is distributed equally and improves network stability [16].
- **Cross technology method**: This protocol is implemented based on the usage of various communication technologies between the sensor nodes. Generally in WSN, ZIGBEE technology is preferred, but here both BLUETOOTH and

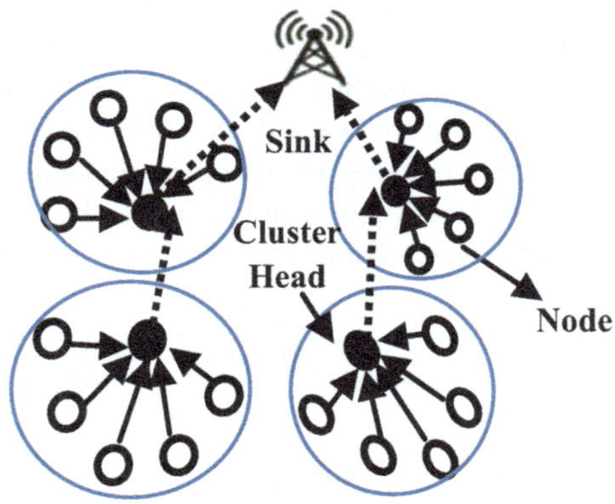

Fig. 2 LEACH protocol

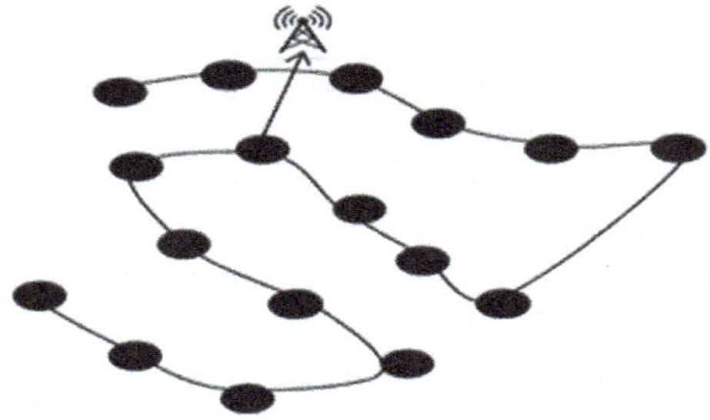

Fig. 3 PEGASIS protocol

ZIGBEE are used together for providing the best routing paths, thereby improving the network performance [17].

- **Based on Heterogeneous energy levels**: By using energy levels in the heterogeneous network the following methods are proposed. These are [18]
  - **Circle Zone stable election protocol**: A circle region is created in the center of the network and provides random distribution around this circle. It is a two-level energy protocol.
  - **Circle Zone stable election protocol with Helping nodes**: It is similar to the above but some helping nodes are placed in between the circle. This mechanism improves the reduction of energy usage in the WSN. It is a three-level energy protocol.

(B) **Communication Model**

This category is based on the working principle of the protocol to send the routing information packets in the network. The classifications under this section are (a) Query-type protocols and (b) Coherent and non-coherent type protocols.

- **Query-Type Protocols**: Queries are generated in the network from the base station for access to some data and upon receiving a query for a particular node send the data to the sink node.
- **Coherent and Non-Coherent Protocols**: All sensed data in the network need not be sent to the destination, because it contains some redundant information. Redundancy needs to be eliminated from the data by the mechanism of data aggregators. Incoherent type, the nodes perform some small processing and then transfer it to the data aggregators whereas in non-coherent type the sensed information is processed by some employed intermediate node for aggregation.

(C) **Topology type Protocols**: The name itself indicates that protocols in this category have some structural or topology information present at sensor nodes in the network. It contains effective route details of all nodes. These are divided into two types

- **Mobile Agent Type**: In this type of protocol one node is completely allocated to collect the sensed data from all the nodes in the network and it is responsible to send the same data to the base station for further processing.
- **Location Type**: In the network, useful information resides in some regions only for our requirement satisfaction. In this case, getting data to the destination from that region needs to be done via some intermediate nodes.

(D) **Reliable Routing type protocols**: The routes in the WSNs are not always available due to node failures and battery drain etc. It leads to communication failure in the network. These protocols provide alternative routes in those situations. These are divided into

- **Multipath-Based Protocols**: They find alternate paths when a link failure occurs or network is loaded with more packets.

Table 1 A basic comparison of various routing mechanisms

Parameter	Network structure	Communication model	Topology type	Reliable routing
Classification	1. Flat routing 2. Hierarchical routing	1. Query type or coherent based 2. Non-coherent or negotiation type	1. Location-based 2. Mobile agent-based	Quality of service (or) Multipath based
Importance	They find the path details from the way the sensor nodes are structured	The main operation is based on the manner the protocol forwards the packets in the network	Network maintains topology information	It maintains QoS metrics like delay, energy, bandwidth, services
Advantages	Good for a small network Routing can be done against sink mobility also	Suitable for all networks Bandwidth can be reduced	Effective routing is possible	Quality of service is provided to ensure network lifetime

- **QoS-Based Protocols**: It offers the quality of data at the base station.

(E) **Nature of nodes**. This classification is mainly based on the type of nodes used in the network i.e., Homogeneous or Heterogeneous. If the sensor nodes in the network are similar type, equal power supply levels, communication bandwidth, and same storage capacities they belong to Homogeneous sensor networks whereas counter to the above properties comes under Heterogeneous sensor networks [19] (Table 1).

5 Conclusion

From the past two decades, WSNs are proposed for many application areas. In WSNs, energy conservation is one of the main factors to describe the network efficiency and performance, because nodes are equipped with low battery capacities. To suit a wide range of applications using WSN, appropriate routing mechanisms are essential. In this article, various routing techniques and their classification are provided along with their energy consumptions. Here we mainly focused on Hierarchical routing protocols, because among all the techniques they provide maximum reduction of energy usage for many operations in the network.

References

1. M. Asif, S. Khan, R. Ahmad, M. Sohail, D. Singh, Quality of service of routing protocols in wireless sensor networks: a review. IEEE Access **5**, 1846–1871 (2017)
2. J. Yick, B. Mukherjee, D. Ghosal, Wireless sensor network survey. Comput. Netw. **52**(12), 2292–2330 (2008)
3. Th. Arampatzis, J. Lygeros, S. Manesis, A survey of applications of wireless sensors and wireless sensor networks, in *Proceedings of the 2005 IEEE International Symposium on, Mediterranean Conference on Control and Automation Intelligent Control, 2005* (IEEE, 2005)
4. N.A. Pantazis, S.A. Nikolidakis, D.D. Vergados, Energy-efficient routing protocols in WSN: a survey com. IEEE Commun. Surv. Tutor. **15**(2) (2013)
5. Samina Ehsan, Bechir Hamdaoui, A survey on energy-efficient routing techniques with QoS assurances for wireless multimedia sensor networks. IEEE Commun. Surv. Tutor. **14**(2), 265–278 (2011)
6. I.F. Akyildiz, W. Su, Y. Sankarasubramaniam, E. Cayirci, Wireless sensor networks: a survey. Comput. Netw. **38**(4), 393–422 (2002)
7. J.N. Al-Karaki, A.E. Kamal, Routing techniques in wireless sensor networks: a survey. IEEE Wirel. Commun. **11**(6), 6–28 (2004)
8. K. Akkaya, M. Younis, A survey on routing protocols for wireless sensor networks. Ad Hoc Netw. **3**(3), 325–349 (2005)
9. S. Guo, O. Yang, Energy-aware multicasting in wireless ad hoc networks: a survey and discussion. Comput. Commun. **30**(9), 2129–2148 (2007)
10. G. Anastasi, M. Conti, M. Francesco, A. Passarella, Energy conservation in wireless sensor networks: a survey. Ad Hoc Netw. **7**(3), 537–568 (2009)
11. R.V. Biradar, V.C. Patil, S.R. Sawant, R.R. Mudholkar, Classification and comparison of routing protocols in wireless sensor networks. Spec. Issue Ubiquitous Comput. Secur. Syst. **4**(2), 704–711 (209)
12. R. Yadav, S. Varma, N. Malaviya, A survey of MAC protocols for wireless sensor networks. UbiCC J. **4**(3), 827–833 (2009)
13. S. Ehsan, B. Hamdaoui, A survey on energy-efficient routing techniques with QoS assurances for wireless multimedia sensor networks. IEEE Commun. Surv. Tutor. **14**(2), 265–278 (2011)
14. A.S. Toor, A.K. Jain, A survey of routing protocols in Wireless Sensor Networks: IEEE Hierarchical routing, in *2016 International Conference on Recent Advances and Innovations in Engineering (INCREASE)* (2016), pp. 1–6
15. S. Kellner, M. Pink, D. Meier, E. Blass, Towards a realistic energy model for wireless sensor networks, in *Proceedings of 5th Annual Conference on Wireless on Demand Network Systems and Services, Garmisch* (2008), pp. 97–100
16. F.A. Khan et al., Hybrid and multi-hop advanced zonal-stable election protocol for wireless sensor networks. IEEE Access **7**, 25334–25346 (2019)
17. D. Gao et al., RowBee: a routing protocol based on cross-technology communication for energy-harvesting wireless sensor networks. IEEE Access **7**, 40663–40673 (2019)
18. F.A. Mohamed et al. Energy-efficient circle zones stable election protocol with helper nodes for heterogeneous WSNs. IET Wirel. Sensor Syst. **9**(5), 313–322 (2019)
19. J. Yan, M. Zhou, Z. Ding, Recent advances in energy-efficient routing protocols for wireless networks: a review. IEEE Access **4**, 5673–5686 (2016)

In-Memory Computation for Real-Time Face Recognition

Nikhil Kumar Gupta and Girijesh Singh

Abstract In the modern era, a facial recognition-based authentication system is quite demanding. Facial recognition systems are widely used in many disciplines like branches of biometric verification, video-based surveillance systems, interaction between human and computers, access control systems and network security. The traditional face recognition systems work well for detecting and recognizing faces but when it comes to real-time operation, the system usually faces a problem of drop in frame rate with an increase in the number of faces which might affect the performance of the whole system. Due to the dropping frame rate, we might miss some frames during the operation which in turn makes our whole system unreliable to use in a real-time environment. In this paper, we propose a real-time face recognition system that uses in-memory computation to keep frame rate constant during the entire process even with an increase in the number of faces and hence makes the system consistent and precise during its operation.

Keywords Face detection and recognition · Attendance system

1 Introduction

Accuracy is one of the important aspects of any facial recognition system. Any fault, error or inaccuracy can make the whole system inconsistent to use in a real environment. Face recognition is comprised of various steps. From capturing a frame to detecting faces and then recognizing it. A recognized face is matched against the stored images in the database, and consequently, actions are performed. With

N. K. Gupta (✉) · G. Singh
Primus Software Development India Pvt. Ltd, Noida, Uttar Pradesh, India
e-mail: activenikhilg@gmail.com

G. Singh
e-mail: singhgirijesh1996@gmail.com

S. S. Dash et al. (eds.), *Intelligent Computing and Applications*,
Advances in Intelligent Systems and Computing 1172,
https://doi.org/10.1007/978-981-15-5566-4_47

increase in computation capabilities of hardware, the algorithm based on neural network are trained extensively to achieve human level accuracy in face recognition. These algorithms are getting smarter day by day and being widely used in daily life application.

There are enough algorithms available for face detection and recognition that work very well for a variety of faces and are widely used in the face recognition-based attendance system. These algorithms give outstanding results even for multiple faces in a frame. But when it comes to video stream processing, it was seen that these algorithms show a significant downfall in frame rate which might be a serious drawback of the whole system as any drop in a frame can make whole system unsuitable to use in real-time application.

Processing frames in real time in series created a problem of frame rate drop so to address the issue, the proposed system uses in-memory computation to process the frames. The execution flow of processing frames has also been changed, rather than performing operation in series, a method of side by side operation is used, in which, the system detect faces from an image, write into memory and at the same time a recognition module read images from the memory and recognize the faces.

2 Related Work

In this section, different methodologies which are used to detect faces in the image processing are described. The framework is majorly split into two parts: first being face detection and the second being face alignment and recognition. The literature review of both parts is described below.

2.1 Face Detection and Face Alignment

An image frame may consist of more than one face in it. To determine the count, size and location of the faces, we have to find a definite common structure of a human face consisting of major parts like nose, forehead, eyes, mouth and chin.

For face detection, Viola and Jones [1] proposed an algorithm that involves the scanning of an image with a sub-window that can detect faces in an image. Three major components of the Viola-Jones face detector are image integral, AdaBoost classifier learning and the attentional cascading for efficient computational resource allocation. In this method, the image is rescaled to different sizes and then the algorithm with a fixed size detector is executed on the rescaled images. It's a time-consuming process since it operates on multiple sizes of images.

To detect objects in the image, a Histogram of Oriented Gradients (HOG) feature descriptor [2] is widely used. The picture is first split into sub-images or "cells", then a histogram of edge orientation is accumulated inside the cell. The algorithm then counts the number of gradient orientations in a localized segment of an image. At last, the HOG descriptor uses the classification algorithm, Support Vector Machine (SVM), to find out whether there is a face or not.

Several research results have shown a significant improvement in the performance of face detection systems with the help of a neural network. A Deep Convolutional Neural Network based model known as Deep Dense Face Detector (DDFD) (by Farfade et al. [3]) which, irrespective of a wide range of orientations could detect faces easily. There is no need for bounding-box regression, segmentation or even Support Vector Machine classifier when using this method. It was found using the study that the method has the capability of detecting faces from various angles and an assumption of having correlation among the distribution of positive examples and negative examples in the training data and score of the suggested face detector.

Zhang et al. [4] introduced an algorithm that had the capability of detecting face and alignment by making use of unified cascaded Convolutional Neural Networks (CNN) with the help of multi-task learning. There are 3 stages in this framework: firstly, it acquires the regression of bounding box, employing a CNN named Proposal Network (P-Net). After calculating the estimated bounding box regression, non-maximum suppression (NMS) is utilized which helps in ignoring the bounding boxes which remarkably converge on each other. The qualifying applicants are further passed through the Refine Network (R-Net) which results in rejecting a plethora of erroneous candidates. It also performs NMS candidates merge and calibration with the bounding box regression. The third and the final step involves describing precise facial characteristics and outputting 5 facial landmark positions using the Output Network (O-Net). If using a 2.60 GHz CPU, an approximate frame rate of 16 frames per second (fps) is achieved. The Nvidia Titan Black GPU has the capability of providing a frame rate of up to 99 fps. The above-mentioned strategy delivered a 95% average precision in each stage complying against major benchmarks and had a stellar real-time outcome.

2.2 *Face Recognition*

The utmost integral component of a face recognition system is to be able to distinguish distinct faces by identifying the unique attributes of a face. The design of a face recognition algorithm which stays robust even under ill-illuminated conditions is highly significant. The conventional face recognition approaches which worked on the basic principles of geometric features analysis [5], principal component analysis (PCA) [6] and linear discriminant analysis [7] were able to attain a precise recognition but only in certain conditions. CNN based facial recognition algorithms were able to recognize faces with greater efficiency. A brief review of some of the state-of-art methods is given below.

DeepFace [8], an algorithm for facial recognition which was proposed by Taigman et al. combined 3D alignment and similarity metric with a deep neural network. The inspiration for the network architecture came from the raw pixel RGB values. Support Vector Regressor (SVR) extract 67 fiducial points, which was trained to predict point configurations from an image descriptor based on Local Binary Pattern (LBP) histogram. DeepFace uses a single-core Intel 2.2 GHz CPU which takes 0.33 s per image resulting in the system attaining 97.35% in precision. DeepID3 is another method suggested by Sun et al. [9]. DeepID3 is composed of 2 deep network architectures, one is, stacked convolution and the other one is, which makes it more accurate to perform facial recognition. DeepFace is an improvement of DeepID2+ method. DeepFace utilizes Joint identification-verification supervisory signals to fully connected layers branching out from the pooling layer along with the last fully connected layers. DeepID3 also uses the same method which helps the architecture in better supervision of the early feature extraction process. DeepID3 uses a 10–15 non-linear feature extraction layer network which is significantly deeper compared to DeepID2+. These layered networks helped DeepID3 achieve 99.53% accuracy for face verification and 96% for face identification.

FaceNet [10] which is a state-of-art face recognition based on Deep convolutional network was suggested by Schroff et al. This network is so well trained that it has the capability to directly optimize the embedding of the input image. This network learns directly from the pixel of a face image. Triplet loss which helps in minimizing the distance of the embedding in the feature space between an anchor (reference face) and a positive (the same person) and maximize the anchor and negative (a different person) is used by FaceNet. It is believed that the faces of distinct people have large distances whereas the face of the same person has small distances. The network is trained in such a way that the squared L2-norm distances in the embedding space directly correspond to face similarity. FaceNet is based on CNN which is trained with the help of Stochastic Gradient Descent (SGD) with standard backpropagation and AdaGrad. An accuracy of 99.63% was achieved using the above approach. The system only uses 128 bytes per face and make it more suitable for running in the embedded computer vision.

3 Proposed Framework

This part of the paper describes the architecture and framework used to accomplish the task of real-time face recognition. There are two major parts: face detection and face alignment and face recognition.

3.1 Face Detection and Face Alignment Module

Face detection which works as an input to the recognition system is the primary and initial step in the framework. Face detection is performed with the help of Max-Margin object detector (MMOD) [11]. Rather than performing sub-sampling this method Optimizes overall sub-windows. Object Detection methods can be enhanced in performance by MMOD if the learned parameters are linear. The model used here uses CNN based feature extractor that can detect faces in different variations and illumination condition. A bounding box is generated around the detected face in the input image, and is given as output.

Face alignment is performed using Ensemble of Regression Trees [12]. Out of other available options like DeepFace by Facebook and Constrained Local Model (CLM), Ensemble of Regression Trees was selected as it was trained by using MMOD face detector and hence had the best accuracy among others. From a sporadic of pixel intensities, this model directly estimates the positions of face's landmark. An Ensemble of Regression trees based 68-point face landmarking model was used and affine transforms were applied on the obtained output to get a vertically aligned face without changing its shape. The vertically aligned face image is required as this is the type of input taken by the face recognition model to work.

3.2 Face Recognition and Classification Module

Face recognition is a critical part of the attendance process. In this paper, a ResNet network was used for mapping the face to a 128D feature vector. For Image Recognition we used a modified version of ResNet-34 network which is based on Deep Residual Learning [13], with only 29 convolutional layers and the filters decreased to half as compared to the original model in each of the layer. This network takes 150 * 150 px image of a face and embeds its features into a 128D vector. Pair-wise hinge loss was used to train this network. It operates over a pair of matching set in a mini-batch and comprise hard-negative mining at the mini-batch level. Any identities fall in non-overlapping radius of 0.6 are projected.

For classifying the faces, KNN classifier was used. KNN classifies the new 128D face feature vector to predict the identical face from previously learned feature vectors. If no feature vector lies within the euclidean distance threshold of 0.6, the obtained vector is marked as unidentified.

3.3 The Framework

The complete abstract of the framework is shown in Fig. 1. The process starts with input captured from the camera. The feed from the camera was captured in the form of video stream and was stored in a buffer memory from where the most recent frame was kept in the memory and rest are dumped.

In the initial stage, the system picks a frame from the buffer memory to process. The frame was given to the face detector model as input and the model returned an array of coordinates of bounding boxes. The array gives the number of faces detected in the image along with the location of each detected faces in the image. The data received from the face detector model and the original image frame was encapsulated and was saved into a secondary memory.

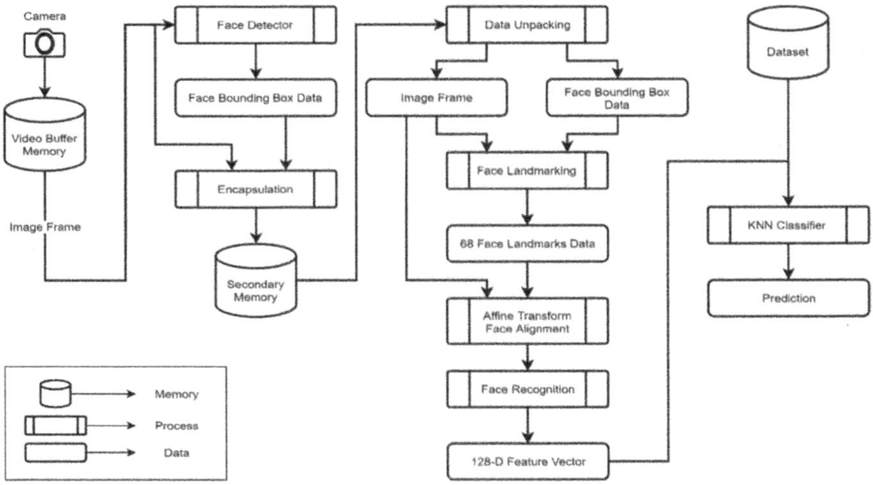

Fig. 1 Process framework

The second stage, which runs parallelly with the first, picks the encapsulated data from the secondary memory continuously in series with time and unpacks it to restore the original image frame and the face detector data array. Each detected face in the frame was given as an input to the face landmarking model which gives 68 coordinate points, landmarking the facial features like eyes, nose, mouth, face boundary, and certain other features. These coordinate points describe the alignment of the face and hence used to perform affine transforms to align the face vertically without affecting the shape of the face. The output obtained by this step was a vertically aligned face.

The vertically aligned image of the face is then given as input to the face recognition model. This model returns a 128-dimensional feature vector for the input face image. This 128-D feature vector is specific to a person and hence this data was employed to classify the frame and determine the identity of an individual. To classify, a pre-trained KNN classifier was used. This takes the 128-D feature vector and determines the person in the image.

To classify an unknown person, the euclidean distance threshold of 0.6 was considered. If there are no other learned (labelled) vector within the threshold of 0.6, then the current vector was marked as unidentified. This step returns the best match of the current vector from the pre-learned vectors, which defines the identity of the person present in the image.

Fig. 2 Working of recognition process

4 Experimental Setup and Results

4.1 Experimental Setup and Dataset

To accomplish the task of face detection and recognition, a Linux based system with Intel i7-7700T processor with 4 cores and 8 threads and 16 GB of DDR4 System RAM along with Nvidia 1060-Ti with 6 GB RAM and with 1280 Cuda cores was used. The camera used to capture the frames was Logitech C920 webcam. To validate the efficiency of the algorithm a custom dataset of a video stream with people passing by the camera and number of people varying from 0 to 8 was created.

4.2 Model and Classifier

The models used in the system are mentioned in this part of the paper. For face detection, a CNN feature-based Max-Margin Object Detector model was used with pre-trained weights provided with the Dlib toolkit by Davis E. King.

The face landmarking task for face alignment was done by Ensemble of regression trees model which was trained on ibug 300-W dataset [14], with pre-trained weights provided by Davis E. King in Dlib toolkit.

The face recognition was done by a 29 layered ResNet network, a modified version of ResNet-34 network [13], which was trained on 3 million faces (most of them from face scrub dataset and VGG dataset). On LFW benchmark, the pre-trained model available with Dlib toolkit shows 0.993833 as a mean error and 0.002727 as standard deviation.

A K-nearest neighbour classifier (KNN) was used to classify the input face vector on the basis of previously learned vectors from the database of known faces (Fig. 2).

4.3 Results

From Table 1 and Fig. 3, it can be seen that there was a significant downfall in frame rate with the increase of number of faces when we are running the algorithm without

Table 1 Framerate in CPU and GPU with/without in-memory computation

Number of faces in frame	Frame rate			
	Without in-memory computation		With in-memory computation	
	CPU	GPU	CPU	GPU
1	5.24	20.05	5.03	20.26
2	3.25	17.39	5.02	20.20
3	2.31	15.07	5.01	20.12
4	1.82	12.28	4.99	20.01
5	1.39	10.01	4.97	19.96
6	0.97	8.12	4.94	19.89
7	0.76	7.05	4.92	19.82
8	0.62	5.16	4.88	19.79

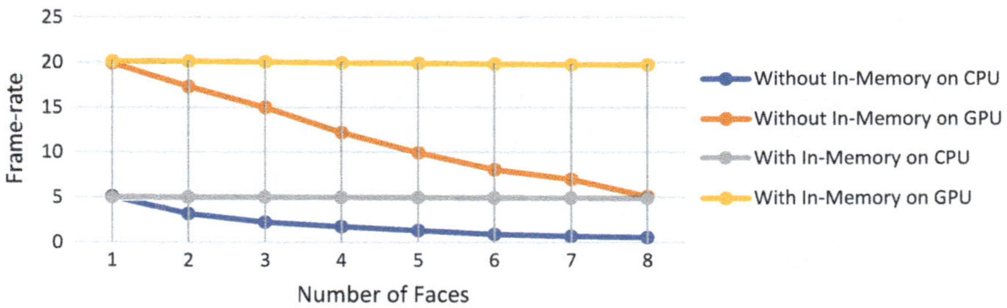

Fig. 3 Framerate versus number of faces graph

in-memory computation on CPU and GPU. On the other hand, in the case of in-memory computation, it was observed that there was a very minor fall in frame rate on both CPU and GPU. This result encourages, to use in-memory computation on the same hardware specification to achieve a better performance along with a better result.

5 Conclusion

In this paper, a framework for real-time multiple face recognition systems was proposed which uses in-memory processing to handle the drop of frame rate during the processing of frames with an increase in the number of faces. This framework shows a very consistent behaviour with no or very minor loss in frame rate during the increase of the number of faces in a frame. This helps to achieve results even with limited resources. This framework suggests that better and efficient computation capabilities could be obtained with the same algorithms with no extra cost just by changing the method of computation.

References

1. P. Viola, M.J. Jones, Robust real-time face detection. Int. J. Comput. Vis. **57**(2), 137–154 (2004)
2. N. Dalal, B. Triggs, Histograms of oriented gradients for human detection, in *IEEE Computer Society Conference on Computer Vision and Pattern Recognition, 2005. CVPR 2005*, vol. 1 (IEEE, 2005), pp. 886–893
3. S.S. Farfade, M.J. Saberian, L.-J. Li, Multi-view face detection using deep convolutional neural networks, in *Proceedings of the 5th ACM on International Conference on Multimedia Retrieval* (ACM, 2015), pp. 643–650
4. K. Zhang, Z. Zhang, Z. Li, Y. Qiao, Joint face detection and alignment using multitask cascaded convolutional networks. IEEE Signal Process. Lett. **23**(10), 1499–1503 (2016)
5. L. Wiskott, N. Krüger, N. Kuiger, C. Von Der Malsburg, Face recognition by elastic bunch graph matching. IEEE Trans. Pattern Anal. Mach. Intell. **19**(7), 775–779 (1997)
6. J. Yang, D. Zhang, A.F. Frangi, J.-Y. Yang, Twodimensional PCA: a new approach to appearance-based face representation and recognition. IEEE Trans. Pattern Anal. Mach. Intell. **26**(1), 131–137 (2004)
7. W. Zhao, A. Krishnaswamy, R. Chellappa, D.L. Swets, J. Weng, Discriminant analysis of principal components for face recognition, in *Face Recognition* (Springer, Berlin, Heidelberg, 1998), pp. 73–85
8. Y. Taigman, M. Yang, M.A. Ranzato, L. Wolf, Deepface: closing the gap to human-level performance in face verification, in *Proceedings of the IEEE Conference on Computer Vision and Pattern Recognition* (2014), pp. 1701–1708
9. Y. Sun, D. Liang, X. Wang, X. Tang, Deepid3: face recognition with very deep neural networks, arXiv preprint arXiv:1502.00873 (2015)
10. F. Schroff, D. Kalenichenko, J. Philbin, Facenet: a unified embedding for face recognition and clustering, in *Proceedings of the IEEE Conference on Computer Vision and Pattern Recognition* (2015), pp. 815–823
11. D.E. King, Max-margin object detection, arXiv preprint arXiv:1502.00046 (2015)
12. V. Kazemi, J. Sullivan, One millisecond face alignment with an ensemble of regression trees, in *Proceedings of the IEEE Conference on Computer Vision and Pattern Recognition* (2014), pp. 1867–1874
13. K. He, X. Zhang, S. Ren, J. Sun, Deepresidual learning for image recognition, in *Conference on Computer Vision and Pattern Recognition (CVPR)* (2016), pp. 770–778
14. C. Sagonas, G. Tzimiropoulos, S. Zafeiriou, M. Pantic, 300 Faces in-the-Wild Challenge: the first facial landmark localization challenge, in *Proceedings of IEEE International Conference on Computer Vision (ICCV-W), Workshop on 300 Faces in-the-Wild Challenge (300-W)*, Sydney, Australia, 2013, pp. 397–403

A Novel Approach for Detection of Basketball Using CFD Method

G. Simi Margarat, S. SivaSubramanian, and K. Ravikumar

Abstract Video analysis plays a vital role in detecting and tracking an object. In a dynamic environment, such as sports or military area, it is difficult to detect and track specified object. In this paper, we have presented a novel algorithm to detect a basketball in the court. The objects are detected from video frames using different types of realms like HSV space, time, frequency, histogram, etc. For ball detection, Color Based and Frame Differencing (CFD) method is proposed and experimented. The complexities arise in each stage such as preprocessing, segmentation, feature extraction, background subtraction, and hole filling. Color quantization is applied to minimize the number of distinct colors in the video. Frame differencing method utilizes the difference in components between frames for ball detection. Morphological operations on connected components are applied to increase the performance and reduce the complexities.

Keywords Basketball tracking · Ball detection · Video capture · Frame difference · Histograms

1 Introduction

The elemental steps are to detect, identify, and classify the object. The machine based vision detection method segments the pixels of moving objects and thereafter

G. S. Margarat (✉)
Department of Computer Science and Engineering, Bharath Institute of Higher Education and Research, Chennai, Tamilnadu, India
e-mail: simimargaratphd@gmail.com

S. SivaSubramanian (✉)
Dhanalakshmi College of Engineering, Chennai, Tamilnadu, India
e-mail: drsivatbm@gmail.com

K. Ravikumar (✉)
Department of Computer Science and Engineering, Manonmanium Sundaranar University, Thirunelveli, Tamilnadu, India
e-mail: ravikumarcsephd@gmail.com

S. S. Dash et al. (eds.), *Intelligent Computing and Applications*, Advances in Intelligent Systems and Computing 1172,
https://doi.org/10.1007/978-981-15-5566-4_48

tracks it. During these processes unnecessary interventions such as noise (the shadow which is a connected set of background points are detected as relevant object) and dynamic background curb the work. The video is read frame by frame. The frames that contain the ball and ball similar object are retrieved based on the histogram of the frames. The color palates run to around 2^{24} combinations of RGB, the color quantization algorithm is experimented to minimize the color palates and improve the image quality. Further, implementation of edge detection methodology exhibits robustness against the shadows, illuminations, repeated motions, etc. The Background Subtraction is appropriate for the moving object detection system to extort the moving basketball from the static background. However, dynamic environment changes such as shadows, illumination changes, camera noise, etc. make the ball detection less accurate.

2 Literature Survey

Motion based recognition is the method which had been used to detect the object in the array of an image. A huge number of images could be grouped to form a video and by this method motion information had been extracted. Problem of holes had occurred by employing two frame differencing. In order to overcome this problem, three frame differencing and background subtraction had been used. Background subtraction had been used in immobile background where dynamic background subtraction had been used in mobile background for detecting objects. This would help to solve problem on holes and could detect the objects in the mobile background. This literature had presented an idea for solving the problem on holes in the mobile background [1]. A motion based recognition method [1] solved the problem due to holes in two frames. This would be overcome with the help of three frame differencing and background subtraction. The dynamic background subtraction had been used to detect the object in the video which has background changes. This could reduce the problem of holes and had given the better detection of objects.

To track the objects, point tracking, kernel tracking, and silhouette tracking had been used. The proposed work [2] tracks and detects the moving objects as their existing system had failed to give the noiseless results in the narrow variations. To extract the feature color, size, and motion had been extracted. Efficient motion estimation algorithm had been used to extract the motion of feature. The same had been used for other feature extractions.

This literature had presented an algorithm for motion detection and segmentation which was simple and robust [3]. This algorithm had worked on the basis of background subtraction. A Differencing and Summing Technique (DST) had been employed for motion detection and segmentation. This had reduced the difficulties in computation. In addition to this, the method had detected and segmented the moving objects in both indoor and outdoor environments with immobile background.

Reference [4] developed an algorithm for highlighting the levers to broadcast the table tennis. To track the ball, the color, size, and position of the ball candidates were attained using Bayesian decision framework. And from this framework, the position of the ball can be identified by Kalman filter.

An efficient approach for detection of moving object under a static background [5] is presented in the literature survey. Selection of maximum pixel intensity value of the two frames had been calculated by segmenting the frames into non-overlapping block. Reflections of objects captured in certain cases create haziness and hinder the detection of targeted object. The joint probability density function (PDF) or the scatter plot of two observed or mixed images is used for separation [6].

The object reconstruction algorithm for moving vehicle detection is based on three frame differencing [7]. The algorithm implements Preset-Threshold-based Region Filling (PTRF) method to reconstruct the moving object. The ORTFD has high-accuracy in detecting dark moving vehicle object, and accelerates the detection speed.

To derive a good color model, a general optimization technique such as simulated annealing is used. Color quantization [8] is regarded as a clustering problem and is solved using fuzzy set [9]. The color palette is seen each as clusters of color from which the suitable color with regard to the video frame can be derived. ViBE (Visual Background Extractor) is foreground object segmentation in a dynamic background. As an extension to this ViBE algorithm [10], the paper has included additional parameter 'blinking pixel' to analyse the pixels of the inner border of the background area.

3 Examination of Methodologies

3.1 Color Based Method

In this method, the features of video frames are extracted by implementing histogram on it. The colors of histogram are compared against the frames. There are 16, 777, 21 color in Red, Green, Blue (RGB) slices, which increases the computational complexities. The color quantization technique produces minimum number of color palettes along with quality, is implemented to retrieve the best quality of the object frame. The color based techniques have more provisions or features compared to geometric cues and grayscale intensity. Generally, the color pixels are represented by three 8 bits each for Red, Green, and Blue. During the process of object segmentation, the values are converted to an equivalent color model parameters. A comparison of these color palette parameters is made to the video frame.

The most popular color models used are Hue, Lightness, Saturation (RGB, HSV, HLS), Hue, Saturation, Intensity (HIS), and Y is the luma component and CB and CR are the blue-difference and red-difference chroma components (YCbCr). To retrieve intensity variations HSV and YCbCr color spaces are used [11]. The histogram gives

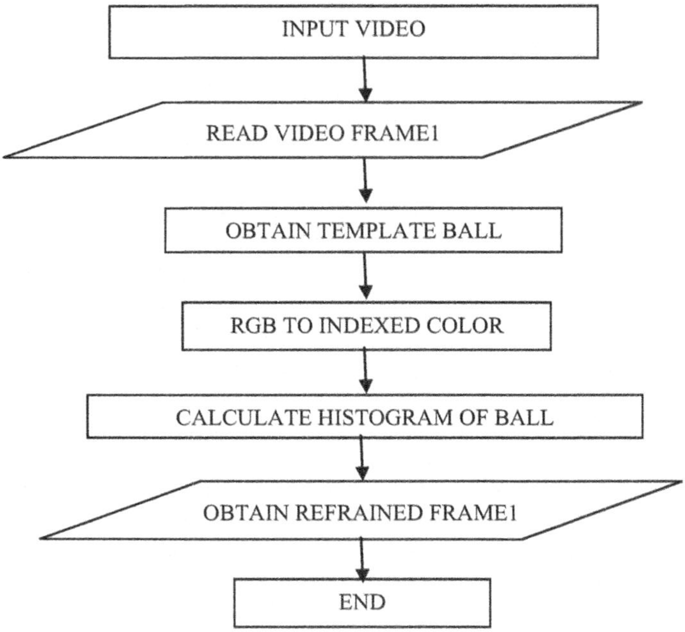

Fig. 1 Block diagram showing the working of the color based algorithm

the summary of the data allocation in the video or the image. In this paper, it is with regard to the video frames.

The resultant framework proceeds for detecting the ball automatically. Algorithm has been designed to retrieve the group of ball objects from each frame and test their locations in the consecutive frames (Fig. 1). Tracking operations are executed on the ball objects by data associated algorithms. The foreground and background moving objects are segmented by implementing the background subtraction technique.

3.2 Frame Differencing Method

This method is used for segmenting moving object in the frames. Two or more successive frames are worked out, to get the desired result. The frame difference method uses a very small time interval between successive frames and thresholding, extracts the image region of the movement.

In a sequence, a morphological operation has been applied in order to bridge the gaps and remove the interventions such as noise. After segmenting, morphological operations [12] had been executed in order to bridge the gaps and eliminate the noise. In order to differentiate the objects in motion in the background, an edge detection technique had been exploited. The Prewitt technique [13] reduces the blurring effect

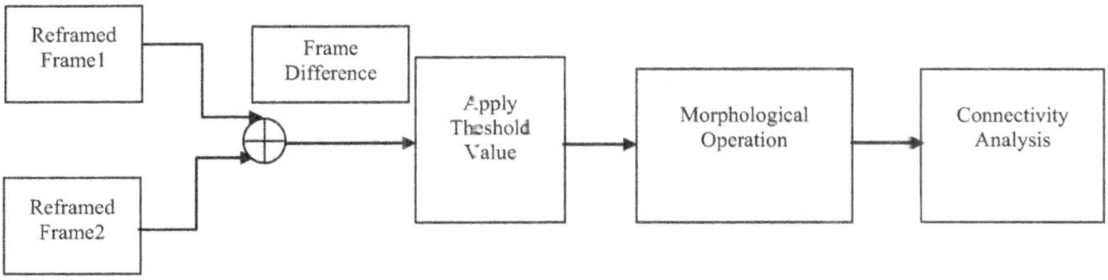

Fig. 2 Block diagram illustrates working of frame difference method

while detecting edges [14]. This technique uses a 3 × 3 gradient operator. A group of ball candidates and the objects which had looked like balls were obtained in the segmentation results and from that the ball candidates had been strained according to their color, shape, and compaction trait of the ball range.

While pointing the ball candidates' centroid location over time, the generation of 2D candidate generation plot had been done. The computational morphological operation includes erosion and dilation (Fig. 2). These take data as an image to be eroded and a structuring element. The combined operations are applied to include or exclude pixels from the boundaries of features to get a smoothened. Further, these operations separate portions of features or touching features and to remove isolated pixel noise from the image.

3.3 *Proposed Method*

The Color Based and Frame Differencing (CFD) algorithm is employed to detect the basketball, tracks all the visual involved. The algorithm gives a clear map to track the place and color of the high-density area which would be around the detectable ball. The implementation of color based technique enables to track the object and object similar identities. The color models which are around 2^{24} that is 16 million color are available for comparison. However to minimize these color models and to retrieve the best quality object frame, the color quantization technique is implemented.

In CFD algorithm, Hue, Saturation, and intensity level of the frame is computed, these diminish the noises (illuminations, shadows of the ball) from the frame. The Hue, Saturation, Intensity Value (HSV) of each frame gives the hue and saturation level. The object's intensity level is analyzed and connectivity between the frames is established. This establishment endorse the detection of the ball.

The edge detection algorithm using fuzzy sets requires threshold values to remove the noisy edges from and around the ball in the frame. There can be strong and weak edges, which needs different levels of focus by the edge detection [15, 16]. Thus in some cases, without the threshold value the noises can be edged out.

Further, the morphological operation for segmentation and the threshold value for the detection of ball removes the unnecessary intervenes and noises. The morphological operation is a filtering measure. This blends the narrow breaks and cut-offs, and bridges the gap in the contour. This operation results, the elimination of small gaps between cut off segments and bridge the region. The morphological operation is employed by erosion after dilation. The dilation process, in the proposed algorithm, converts the ball pixel in one frame to other ball pixel, thus summing to find the boundaries of the ball and closes the isolated background pixels. This is followed by the erosion method. This gives the position of the basketball in the first frame.

Algorithm

Step 1: Input the Video and read the frames in order
Step 2: For every frame
Step 3: Convert RGB to HSV image
Step 4: Consecutive two frames are subtracted using Frame differencing
Step 5: Set threshold value and obtain binary frame.
Step 6: Apply morphological operation
Step 7: Display detected ball
Step 8: End

4 Conclusion

In this paper, the detection of basketball in a vision based domain is done. A range of existing algorithms were also reviewed. Their performance and complexities were discussed. The proposed method is based on color and frame difference algorithm. In support of the proposed technique, color quantization, edge detection, and morphological operation are implemented. These reduce the illuminations and other basketball similar objects from detection. On comparison, the proposed has a better performance. The enhancement of this proposal is to design and experiment an algorithm to identify the trajectory of the basketball in the video. The trajectory of the basketball will be used to envisage the make or miss of the ball.

References

1. N. Srivastav, S.L. Agrwal, S.K. Gupta, S.R. Srivastava, B. Chacko, H. Sharma, Hybrid object detection using improved three frame differencing and background subtraction, in *7th International Conference on Cloud Computing, Data Science & Engineering—Confluence*, 2017, pp. 613–617
2. G. Sharmila Sujatha, V. Valli Kumari, An efficient motion based video object detection and tracking system, in *International Conference on Innovative Trends in Electronics Communications and Applications (ICIECA15)*, vol. 1, no. 13, 2015, pp. 85–97

3. G. Thapa, K. Sharma, M.K. Ghose, Moving object detection and segmentation using frame differencing and summing technique. Int. J. Comput. Appl. **102**(7), 20–25 (2014)
4. W. Chen, Y.J. Zhang, Tracking ball and players with applications to highlight ranking of broadcasting table tennis video, in *Proceedings of IMACS Multi Conference on Computational Engineering in Systems Applications*, 2006, pp. 1896–1903
5. S.S. Sengari, S. Mukhopadhyal, A novel method for moving object detection based on block based frame differencing, in *3rd International Conference on Recent Advances in Information Technology*, 2016, pp. 1–6
6. M. Wajid, M. Sharma, Digital image separation algorithm based on joint PDF of mixed images. Image Process. Commun. **20**(1), 5–12 (2015)
7. Y. Lin, M. Fang, D. Shihong, An object reconstruction algorithm for moving vehicle detection based on three-frame differencing, in *2015 IEEE 12th International Conference on Ubiquitous Intelligence and Computing and 2015 IEEE 12th International Conference on Autonomic and Trusted Computing and 2015 IEEE 15th International Conference on Scalable Computing and Communications and Its Associated Workshops (UIC-ATC-ScalCom)*, 2015, pp. 1864–1868
8. G. Schaefer, Soft Computing-based colour quantization. EURASIP J. Image Video Process. **2014**, 1–9 (2014)
9. P.A. Khaire, N.V. Thakur, A fuzzy set approach for edge detection. Int. Jo. Image Process. (IJIP) **6**(6), 403–412 (2012)
10. Tomasz Kryjak, Foreground object segmentation in dynamic background scenarios. Image Process. Commun. **19**(2–3), 25–36 (2015)
11. Giuliani Donatella, Color image segmentation based on principal component analysis with application of firefly algorithm and Gaussian mixture model. Int. J. Image Process. **12**(4), 101–112 (2018)
12. S. Behera, M.N. Mohanty, Land boundary detection of an island using improved morphological operation. Int. J. Image Process. (IJIP) **6**(6), 413–421 (2012)
13. R. Maini, J.S. Sohal, Performance evaluation of Prewitt edge detector for noisy images. GVIP J. **6**, 39–46 (2006)
14. E. Kermani, D. Asemani, A robust adaptive algorithm of moving object detection for video surveillance. EURASIP J. Image Video Process. **2014**, 1–9 (2014)
15. W. Yang, Analysis of sports image detection technology based on machine learning. EURASIP J. Image Video Process. **2019**, 1–8 (2019)
16. K. Aggarwal, M.S. Bhamrah, H.S. Ryait, Detection of cirrhosis through ultrasound imaging by intensity difference technique. EURASIP J. Image Video Process. **2019**, 1–10 (2019)
17. Z. Weng, Y. Guan, Action recognition using length-variable edge trajectory and spatio-temporal motion skeleton descriptor. EURASIP J. Image Video Process. **2018**, 1–15 (2018)

Ensemble Similarity Clustering Frame work for Categorical Dataset Clustering Using Swarm Intelligence

S. Karthick, N. Yuvaraj, P. Anitha Rajakumari, and R. Arshath Raja

Abstract Many attempts are made to resolve the clustering problem over categorical dataset through the process of clustering ensemble. However, the conventional techniques failed to generate proper data portioning to cluster the data due to the presence of incomplete information. Hence, this paper uses a modified cluster ensemble similarity framework to cluster the relevant categorical datasets, and a weight-based similarity approach has been proposed to work with type II full-space and subspace ensembles that involve random-*k* and fixed-*k* techniques. The centroid selection is initiated using swarm-based bio-inspired optimization method or swarm-based similarity ensemble clustering (WSE). Finally, Division Reminder Hashing (DRH) technique is used to determine the weights of the cluster using similarity measurement. This technique helped in eliminating the null and duplicate datasets. The measurement for testing includes Adjusted Rand (AR), Normalized Mutual Information (NMI), and Classification Accuracy (CA). The performance of the proposed technique is tested over University of California Irvine (UCI) dataset, and the results proved that the proposed technique provides better clustering using Type-2 and Type-3 ensemble.

S. Karthick (✉) · P. A. Rajakumari
Department of Computer Science and Engineering, SRM Institute of Science and Technology, DelhiNCR Campus, Ghaziabad, India
e-mail: karthick.usilai@gmail.com

P. A. Rajakumari
e-mail: sweetanip@gmail.com

N. Yuvaraj
Department of Computer Science and Engineering, St. Peter's Institute of Higher Education and Research, Chennai, Tamil Nadu, India
e-mail: yraj1989@gmail.com

R. A. Raja
Department of Electronics and Communication Engineering, B.S. Abdur Rahman Crescent Institute of Science and Technology, Chennai, Tamil Nadu, India
e-mail: arshathraja.ru@gmail.com

S. S. Dash et al. (eds.), *Intelligent Computing and Applications*, Advances in Intelligent Systems and Computing 1172,
https://doi.org/10.1007/978-981-15-5566-4_49

Keywords Full-space ensemble · Subspace ensembles · Weight-based clustering · UCI data

1 Introduction

Data clustering is a learning method with inherent practical applications. Analysis of clustered data is important for data analysis and data processing. Most clustering methods divide the dataset into classes as per the data similarity and enable the class element to have similar character and keep distance of the dataset cluster of different classes as large as possible.

There are several popularly used clustering algorithms, which include k-means [1] and PAM [2]. This facilitates the numerical data with its intrinsic properties in its natural state. This measures the distance between vectors through distance metrics [3, 4]. The direct application is unfeasible to cluster categorical data with distinct values and possess no order.

Several other clustering algorithms for the categorical data include uncertainty clustering [6], Context-based Distance Learning [6], which is employed for real and synthetic datasets. Other algorithms include concept-drifting categorical data [7].

The varied parameters in specific dataset, diverse, or same algorithms provide distinct solutions. Consequently, it is difficult to determine the precise algorithm for a specified dataset. This issue motivates to form cluster ensembles. In [8], there exists an application that comprises reusing possibility of conventional knowledge concerning the data and mining, wherein data partitions exist based on its locations. Lately, clustering ensembles [9, 10] gained popularity to overcome the issues of clustering algorithms. It is apparent that clustering finds the patterns of any dataset. No cross-validation is executed for tuning the parameters with inherent clustering. If the user gets equipped without any guidelines, the clustering technique for specific dataset is selected.

In [11], the relationships between clustering categorical data and clustering ensemble are carried out. Categorical data clustering through maximal K-partite cliques [12] determines the clusters in categorical datasets using K-partite maximal cliques. The clustering ensemble process possesses two stages: population generation with clustering partitions through resampling [13] attribute subspace [14] homogenous algorithm [15] and others. Second stage integrates clustering results into a final solution. Clustering ensemble uses supervised bagging and boosting algorithms. In [16], an adaptive approach, ensemble individual partitions are generated sequentially by subsamples of dataset. Also, clustering categorical dataset related issue is analyzed and investigated in the proposed study. A suggested generic framework is effectively applicable in all the other data types.

The paper proposes Weight-based Similarity Clustering Ensemble (WSE) method. The weights of the each dataset point change during each iteration. A strategy is adopted while weights are updated using similarity matrix. This is the base concept for weight-based similarity method. The centroid values of the clusters are chosen based on swarm behavior with respect to Fixed and Random-k clustering. The cluster ensemble is generated through randomly chosen centroid values. The reduction is maintained minimal for removing the irrelevant dataset from the clusters. This is done using DRH technique with weight-based similarity function. This makes the probability of Consensus Function to be applied with respect to RM. The new swarm algorithm with weight-based ensemble framework has suggested deliberately for engendering the precise and accurate measures in an economical manner.

Section 2 provides the proposed method for clustering ensemble. Section 3 evaluates the proposed clustering ensemble with conventional techniques. Section 4 concludes the paper.

2 Proposed Method

1. *Weight-based Cluster Ensemble*

A similarity method acquires λ (a), $a \in N_r = 1,2, \ldots,r\}$ from r clustering algorithms $\pounds(a)$. This forms a similarity matrix $(m \times m)$ M (a) and m—dataset size. Once ensemble is applied, consensus similarity matrix is finally produced. This gets transformed to the result for final clustering.

Based on the results of $\pounds(a)$, similarity matrix M (a) is applied. The clustering algorithms used here are crisp clustering, and the data point share goes to a single class. The clustering results are expressed in two forms: λ (a) and M (c).

$$\lambda(a) = 1 \times m \text{ vector},$$

where λ (a, i)—class label belonging to ith data point.

When the similarity matrix M (a) is generated, and if λ $(a, m) = \lambda$ (a, n) or the mth and nth data lie in similar classes, then $M_{(m,n)}(a) = 1$, or else it is 0. Similarly, if M (a) = binary matrix, then each data point pair occurs inside similar clusters.

Hence, for all similar matrices, r average matrix a_m is calculated after the generation of consensus similarity matrix s_m. Here, if s_m is binary, $a_m = [0, 1]$. If M $(m, n) = 1$, the possibility of mth and nth dataset lies in similar classes. A threshold of 0.5 is set. If M $(m, n) >$ threshold, s_m is 1 or else s_m is 0. Then, binary consensus definite similarity matrix is generated and consensus similarity matrix is generated, and hence the final clustering result is obtained.

The weights are added with the similarity method, and clustering ensemble method is proposed. Finally, the Weighted Similarity Clustering Ensemble (WSE) is formed with weighted information which is shown in algorithm.

Swarm Algorithm

Input: Dataset $D = \{x_1, x_2, \ldots, x_n\}$;
Weight of individual data
Weight = $\{w_1, w_2, \ldots, w_n\}$
Base Cluster $\alpha(a)$, $a \in N_\alpha$.
Process:
For $a = 1, \ldots, m$;

$\lambda(a) = \alpha(a)(D)$;
Derive $m \times m$ similarity $M(a)$ using $\lambda(a)$;

End
Evaluate and record λ;
$M_1 = \sum_{a=1}^{r} M^{(a)}$
$a = 1$
$M(a)$;
For $i = 1, \ldots, m$

For $j = 1, \ldots, m$

If $M_1(i, j) == r$
$M_1(i, j) = 1$.
Else $M_1(i, j) = 0$

End

End
Reduce w of each cluster points of same class over r process
For $i = 1, \ldots, m$

$$w = \frac{w(i)}{\sum_{i-1}^{m} w(i)}$$

End
If final condition is satisfied

Clustering result is obtained;

else
Clustering is repeated with updated w.
Output: λ-clustering result

3 Results and Discussion

This section evaluates the proposed ensemble algorithm using weight cluster ensemble or WSE framework. This employs several validity indices on real datasets.

Table 1 Datasets

Dataset	D	N	K	AV
Zoo	16	101	7	36
Lymphography	18	148	4	59
Primary tumor	17	339	22	42

Data partition quality is created and evaluated based on categorical data ensemble technique (Table 1).

The five ensemble types for evaluation are:

- Type-1,
- Type-2 Fixed-*k*,
- Type-2 Random-*k*,
- Type-3 Fixed-*k*, and
- Type-3 Random-*k*.

The *k*-modes clustering generates the clustering with proposed algorithm with

- Weighted Similarity Ensemble framework (WSE),
- Weight-based Cluster Ensemble (LCE),
- Cluster-based Similarity Partitioning Algorithm (CSPA),
- Hypergraph Partitioning Algorithm (HGPA).

Table 2 shows the clustering accuracy for entire dataset in the Table 2.

Table 2 Clustering accuracy

Dataset	Ensemble type	LCE	HGPA	WSE
Lymphography	I	0.743	0.66	0.7569
	II-Fixed *k*	0.796	0.709	0.812
	II-Random *k*	0.784	0.633	0.823
	III-Fixed *k*	0.782	0.692	0.795
	III-Random *k*	0.777	0.633	0.7925
Zoo	I	0.894	0.594	0.932
	II-Fixed *k*	0.921	0.836	0.941
	II-Random *k*	0.931	0.826	0.942
	III-Fixed *k*	0.941	0.831	0.954
	III-Random *k*	0.931	0.835	0.946
Primary tumor	I	0.445	0.293	0.589
	II-Fixed *k*	0.478	0.438	0.545
	II-Random *k*	0.496	0.428	0.594
	III-Fixed *k*	0.463	0.423	0.6945
	III-Random *k*	0.484	0.408	0.6936

Table 3 Performance comparison

Ensemble type	Methods	CA		NMI		AR	
		B	W	B	W	B	W
1	WSE	187	28	146	53	164	53
	LCE	170	35	137	65	149	61
	HGPA	21	193	19	208	24	201
2–fixed *k*	WSE	225	3	228	5	212	9
	LCE	208	4	204	9	201	13
	HGPA	64	80	93	92	93	91
2–random *k*	WSE	235	2	214	8	219	8
	LCE	209	4	203	12	201	11
	HGPA	67	121	41	139	49	132
3–fixed *k*	WSE	219	5	203	8	197	9
	LCE	197	7	191	13	182	16
	HGPA	71	82	56	101	65	101
3–random *k*	WSE	227	3	206	9	196	9
	LCE	203	6	191	13	181	11
	HGPA	51	117	38	134	63	131

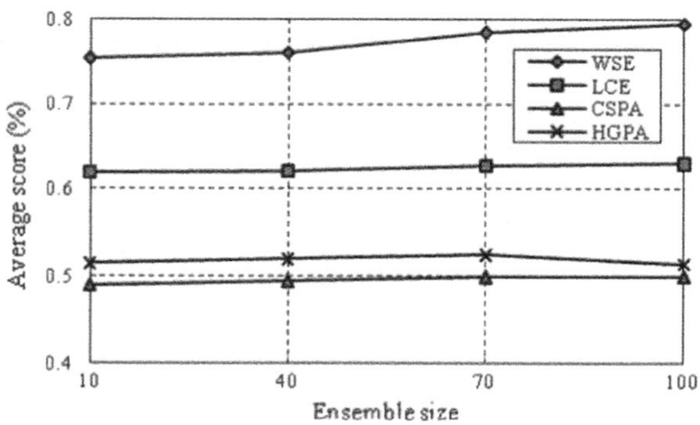

Fig. 1 Type 2 ensemble: fixed data clustering

Table 3 shows significant performance of various ensemble types as per NMI, AR, and CA evaluation indices. An enhanced effectiveness is proved in terms of suggested weighted techniques in comparison with other conventional clustering method. The *B* and *W* indicate *better* and *worse* performance than the others.

The parameter analysis presents a practical means and that allows the users for employing the WSE at fullest range. Performance of the technique depends entirely on decay factor, where DC $\in$ [0, 1] and that deploys the ascertained clustering similarity. The relationship between DC $\in$ {0.1, 0.2, …, 0.9} and the clustering performance values is presented in *XY*-axis.

The average score WSE value results of ensembles depend on DC. The Figs. 1, 2, 3, and 4 show the weighted average scores of proposed technique with the conventional techniques. The WSE approach for the users ascertains consistent, high quality, and best results as of DC values between 0.7 and 0.9. This is due to centroid calculation through swarm intelligence and the clustering results for all DC.

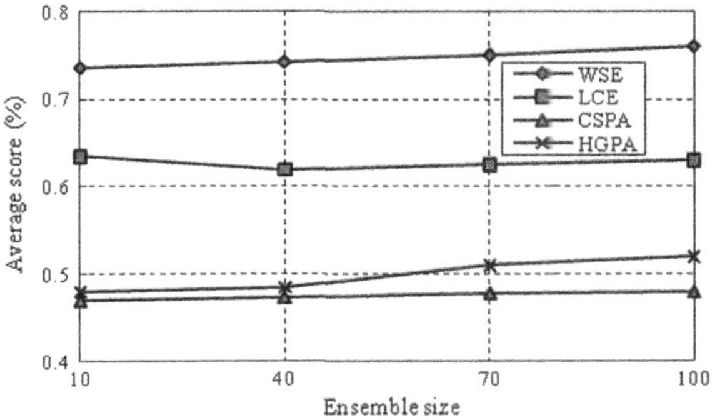

Fig. 2 Type 2 ensemble: random data clustering

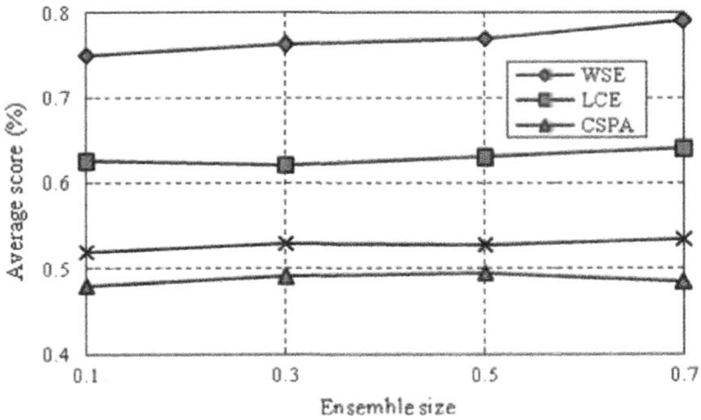

Fig. 3 Type 3 ensemble: fixed data clustering

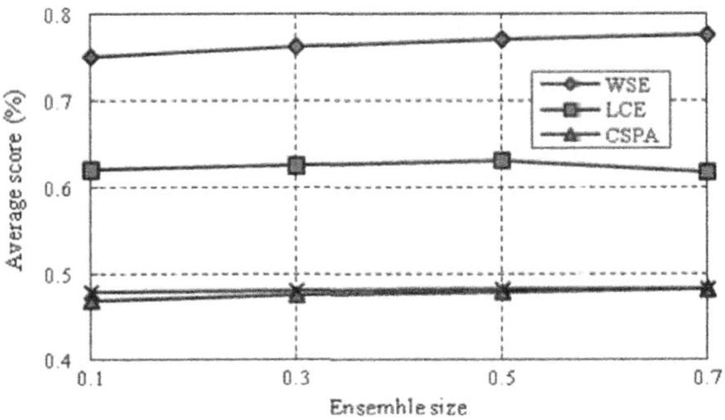

Fig. 4 Type 3 ensemble: random data clustering

4 Conclusion and Future Work

The weight-based ensemble clustering algorithm turns out to be an effective method that enhances the performance of the categorical data clustering. The use of weight-based similarity ensemble technique clusters the categorical data without empty datasets. Here, the cluster ensemble calculates the centroid value using swarm intelligence process. Then, the matrix transformation is carried out to transform the original data matrix to RM. The duplicate entries are finally removed using DH method and that uses graph partitioning method. The weight-based similarity is used to resolve the issue while constructing a RM. The weight information is often changed due to limitations in K-means and other conventional algorithms. This information is applied to the forthcoming iteration, and the final score is calculated through similarity matrix. The framework is substantiated and results prove its validity and performance of clustering algorithm on different categorical datasets. The proposed framework can be incorporated in other clustering algorithms, and the evidences prove the practical possibility of ensemble technique.

References

1. J. Wu, Advances in K-means Clustering: A Data Mining Thinking (Springer, 2012)
2. L. Kaufman, P.J. Rousseeuw, Finding Groups in Data: An Introduction to Cluster Analysis (Wiley, 2004)
3. A.S. Shirkhorshidi, S. Aghabozorgi, T.Y. Wah, A comparison study on similarity and dissimilarity measures in clustering continuous data. Phys. Rev. Lett. **10**(12) (2015)
4. L.O. Mafteiu-Scai, A new dissimilarity measure between feature-vectors. Int. J. Comput. Appl. **64**(17), 39–44 (2013)
5. D. Parmar, T. Wu, J. Blackhurst, MMR: an algorithm for clustering categorical data using rough set theory, in 25th International Conference on Conceptual Modeling, vol. 63, Issue No. 3, pp. 879–893 (2007)
6. D. Ienco, R.G. Pensa, R. Meo, Context-based distance learning for categorical data clustering, in 8th International Symposium on Intelligent Data Analysis, pp. 83–94 (2009)
7. H.-L. Chen, M.-S. Chen, S.-C. Lin, Catching the trend: a framework for clustering concept-drifting categorical data. IEEE Trans. Knowl. Data Eng. **21**(5), 652–665 (2009)
8. N. Iam-On, T. Boongoen, S. Garrett, C. Price, A weight-based approach to the cluster ensemble problem. IEEE Trans. Pattern Anal. Mach. Intell. **33**(12), 2396–2409 (2011)
9. Sandro Vega-Pons, Jose Ruiz-Shulcloper, A survey of clustering ensemble algorithms. Int. J. Pattern Recognit. Artif Intell. **25**(3), 337–372 (2011)
10. T.Y. Li, Y. Chen, Fuzzy clustering ensemble with selection of number of clusters. J. Comput. **5**(7), 112–1118 (2010)
11. Z.Y. He, X.F. Xu, S.C. Deng, A cluster ensemble method for clustering categorical data. Inf. Fusion **6**(2), 143–151 (2005)
12. M.J. Zaki, M. Peters, Clicks: mining subspace clusters in categorical data via k-partite maximal cliques, in International Conference on Data Engineering, pp. 355–356 (2005)
13. J. Jia, X. Xiao, B. Liu, L. Jiao, Bagging-based spectral clustering ensemble selection. Pattern Recogn. Lett. **32**(10), 1456–1467 (2011)
14. M. Al-Razgan, C. Domeniconi, D. Barbara, Random subspace ensembles for clustering categorical data. Supervised Unsupervised Ensemble Methods Appl. 31–48 (2008)

15. H.G. Ayad, M.S. Kamel, Cumulative voting consensus method for partitions with variable number of clusters. IEEE Trans. Pattern Anal. Mach. Intell. **30**(1), 160–173 (2008)
16. B. Minaei-Bidgoli, H.P.H. Alinejad-Rokny, H. Alizadeh, W.F. Punch, Effects of resampling method and adaptation on clustering ensemble efficacy. Artif. Intell. Rev. **41**(1), 27–48 (2014)

Multi-Focus Image Fusion Using Conditional Generative Adversarial Networks

A. Murugan, G. Arumugam, and D. Gobinath

Abstract Multi-focus image fusion is a task of generating a composite image from the numerous limited depth of focus images of the identical clips. The aim of multi-image fusion is to achieve a better quality image for better human visual perception or better machine interpretation. Calculating the focus map to distinguish the focused and unfocused pixels is the challenging task in image fusion algorithms. Numerous ways of calculating focus map have been proposed in the last two decades. Very recently, researchers show more interests on artificial intelligence and deep learning based algorithms to identify the focused regions from the limited depth of field image. Image fusion involves the following three subproblems: (1) Focus region detection, (2) Selecting the focus region based on confidence level, and (3) Final fusion. Motivated by Conditional Generative Adversarial Network (cGAN), a novel way of detecting the focused regions same as image-to-image translation has been proposed here. GAN requires more number of images to train the network. Due to lacking of limited depth of field images dataset, we synthesize a dataset based on PASCAL VOC 2012. We compare the results of the proposed deep learning based fusion methods with the traditional fusion methods. Results prove that the proposed unsupervised neural network based approach outperforms the other traditional methods.

Keywords Image fusion · Deep learning and generative adversarial network

A. Murugan (✉) · G. Arumugam · D. Gobinath
Department of CSE, SRMIST, Kattankulathur Campus, Chennai, India
e-mail: murugana@srmist.edu.in

G. Arumugam
e-mail: arumugam@gmail.com

D. Gobinath
e-mail: gobinatd@srmist.edu.in

S. S. Dash et al. (eds.), *Intelligent Computing and Applications*,
Advances in Intelligent Systems and Computing 1172,
https://doi.org/10.1007/978-981-15-5566-4_50

1 Introduction

Acquiring an image with all the regions in focus is a confronting problem in the current camera systems. Due to the shallow Depth Of Field (DOF), the partial regions are in good focus and remaining area is in out of focus/blurred. An algorithm to combine several incompletely focused pictures/images into a composite image is called image fusion. This improves the visual perception of the image to human and information to the machine. So far, many of the researchers have proposed several algorithms to accomplish this task.

Shallow depth of focus makes images to blur in the region outside DOF. Very small range which falls inside DOF gets fully focused. The algorithm to extract all the focused pixels in the multiple shallow depth images and create a composite image is multi-image fusion. It uses various camera parameters to combine the images.

Basic multi-focus fusion algorithms are classified into three levels based on the complexity of the algorithm as low, mid, and high. Some authors have referred this as pixel, feature, and characteristic level. But most of the pixel level algorithms work either on spatial or on the transformed domain [1].

In the transform-based steerable image pyramid, the images are decomposed into sub-bands, and the sub-bands are having the advantages of rotation and translation invariant. Basic steerable image pyramid is a multiple orientation, self-inverting, and multi-scale image decomposition [2]. In [3], Jing and Li apply wavelet transform to all the input images. Then, for all the input images, the coefficient (wavelet approximation) along with detail coefficient is combined by applying different rules (fusion). The final composite image is generated by applying inverse transform (wavelet). Region-based wavelet coefficient is also used in [4]. The authors have implemented two stages, where the textured and other regions are handled in the initial stage. Later, in the second stage, spectral clustering method groups together in segmented primitive regions. The DT-CWT coefficient was used in the texture process.

As compared against the transform-based methods, spatial methods do the fusion task easily by processing the source images. Spatial methods are classified into three subgroups, region, block, and pixel-based methods. Block-based approach has been used in [5] as to decompose the input images into spatial blocks, and then the spatial frequency is computed for each blocks in the source images.

Even though all of these handed craft recent algorithms performing better and producing better fused image, choosing the right parameters and filter size manually is a tedious task, and always producing some artifacts on the fused image and also designing an ideal design is practically impossible due to some limitations or lack of relevant knowledge. In [5], an already-trained Convolutional Neural Network (CNN) based model is used to calculate the focus measure score map which basically has all the focused pixels and non-focused pixels in the input image. And, a focus map is calculated from the focus score map by taking average of the overlapping patches. Then, the binary mask image is generated by applying the binarization to the focus map; the pixel-wise weighted mean method is used to generate fused image from the decision map.

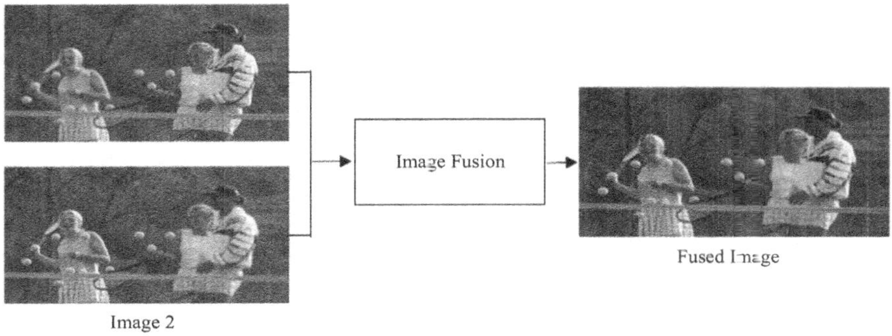

Fig. 1 Basic image fusion

Inspired by generative models, Goodfellow [6] came up with the Generative Adversarial Network (GAN). GAN has achieved success, and many algorithms have been proposed on top of the generative modeling. Few of the algorithms are given below: Image-to-Image translation [9], In-painting [7], and Text-to-Image translation [8]. As mentioned in [9], image fusion may be viewed as multiple images to single image translation problem. We propose this N to 1 image translation using conditional GAN [10].

The rest of the paper is organized as follows: related works in Sect. 2, proposed method in Sect. 3, results and comparison in Sect. 4, and conclusion and future works in Sect. 5 (Fig. 1).

2 Related Works

(A) **Generative Adversarial Network**

Generative Adversarial Network (GAN) [6] is an effective generative model which has two major components, Generator (G) and Discriminator (D). Generator takes in random numbers and outputs an image. Discriminator accepts samples from training data and G and outputs a scalar between 0 and 1. The D's scalar output is an indicator (probability) that the sample comes from training data. During adversarial training, G is trained in such a way that it can generate a fake data and fool D. D's responsibility of distinguishing real and fake data is getting stronger, while the G's output is more close to real. Whenever D finds the differences between real and G's output, G's parameter gets adjusted to generate a data more close to real. Eventually, G will have the capability to generate a sample which cannot be distinguished by D.

(B) Conditional Generative Adversarial Network

Just feeding random noise and training the model in GAN is instable, and sometimes it introduces artifacts in result image. To reduce the instability, Mirza et al. [10]

proposed an effective way of training GAN by adding an auxiliary input to both Generator and Discriminator. This auxiliary input can be a label or data or information from other model. This forms the GAN network with conditional input. By comparing the MINIST results, authors prove that training GAN with this auxiliary (conditional) input improves the Fused Image stability than vanilla GAN. The loss function with auxiliary input is given below in (1):

$$\text{minmax} C(G, D) = E, \sim P(x, y)[\log D(x, y)]GD + E \sim P(x), z \sim (z)[\log 1 - D(x, G(x, z))] \quad (1)$$

(C) **Image-to-Image Translation**

Image-to-image translation may be explained as the task of generating an image that is translated from other image using generative models. The objective is to learn the spatial map between a source and output grid. They do map a full resolution input image into output grid. And also both input and output differ in surface appearance.

3 The Proposed Method

In this method, we construct a cGAN-based network model to generate a focus measure map to distinguish focused pixels and non-focused pixels from the shallow depth of field images. The focus map is generated with different labels for the focused and non-focused regions. In the second step of focus map selection, we are selecting the exact pixels and making the final selective focus map. General fusion map is applied on the selective focus map to generate fusion image. The architecture of the proposed method is shown in Fig. 2. The remaining part explains about architecture and design.

Same like the vanilla GAN, this model also has two sub-networks/components, a Generator (G) and a Discriminator (D). The input to the generator and discriminator in this model is having multiple images and single image output. The multiple images are being sent to network to extract the features, and in the feature concatenation layer, the input features are concatenated and sent to discriminator. The network is being forced to learn/train the same features for the multiple input images. The multiple input features weights are shared between networks [11]. Since the tied weights, the features are same for all the input images. And it is easy for concatenating the different features. The proposed method detects the focused regions and generates the focus measure map.

Accepts multiple input images and returns a focus measure map.

In Generator, the network accepts multiple images, and each image passes through Encoder which has Convolution, Batch normalization, and Relu. To extract the high-level features, Encoder output goes through the Resnet Blocks [12]. The Decoder has deconvolution, batch normalization, and Relu to resize the concatenated feature

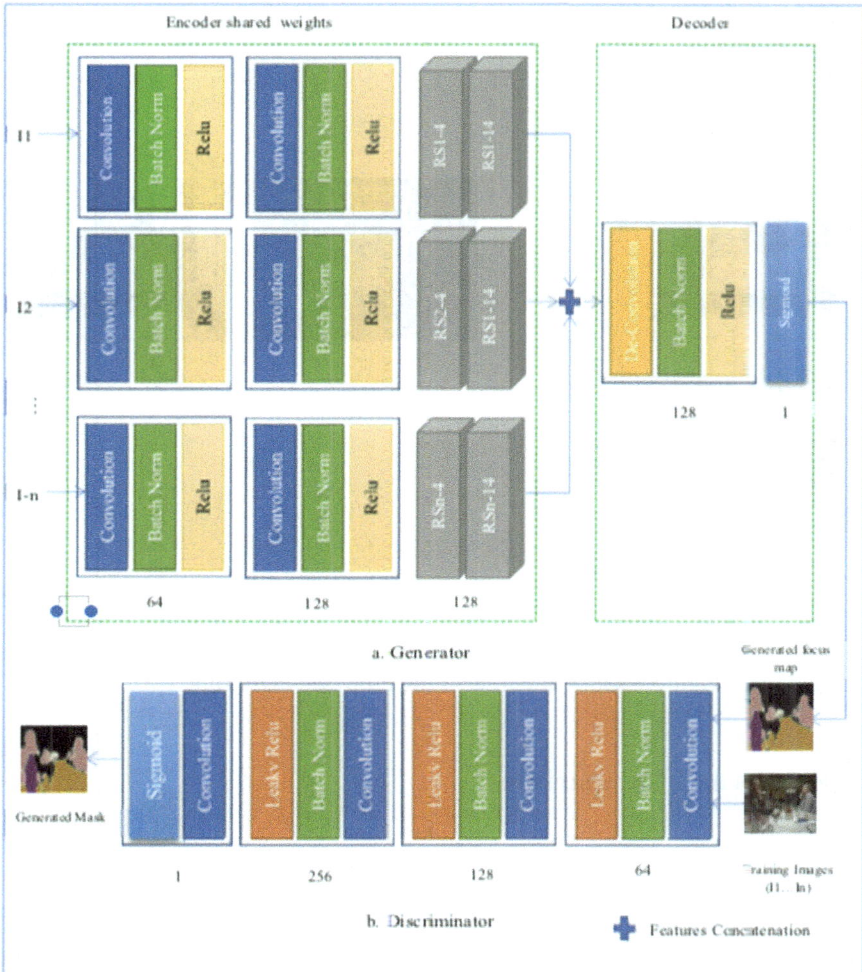

Fig. 2 Architecture of the proposed method. It has two sub-networks, "Generator (G) and Discriminator (D)"

to the input images. Then the focus map is mapped through sigmoid operation. The discriminator accepts both generated focus map and input images using cGAN's objective function to train the model.

The proposed algorithm is as follows:

1. The multi-focus images are passing through the GAN network and focus map is generated. Each pixel is mapping to a focused region in the focus map image.
2. Selecting the focused pixel from the map and weighted average neighborhood operation is applied to select the focused pixel.

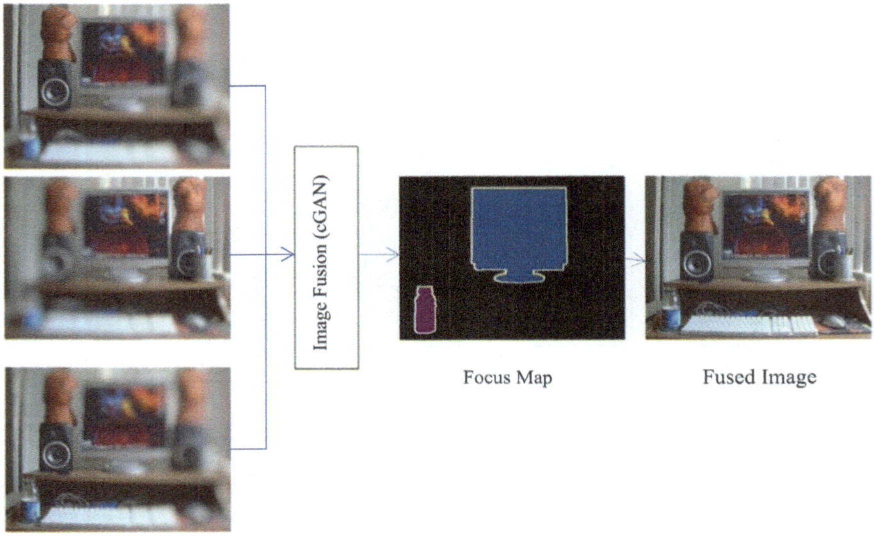

Fig. 3 Proposed algorithm flow

3. Final fusion rule is applied to the selected focus map, and fused image is generated. Figure 3 shows the flow of the proposed algorithm.

4 Results and Comparison

(A) Data Preparation

We have used the PASCAL Visual Object Classes 2012 (PASCAL VOC 2012) [13]. This dataset is having 2913 images with segmented labels with ground truth. For the de-focused images, there is no large dataset available. To train the network, we would need a large number of images with segmented labels. We synthetically created a de-focused dataset using VOC 2012 images by applying Gaussian blur with different sigma.

(B) Comparison

We compared the proposed method with the following three best conventional image fusion techniques: (1) PCA-based method [14], (2) Spatial frequency [15], and (3) Gray-level local variance [16]. The fused image generated by PCA method is using weighted mean. The weights are calculated from the Eigen vector of the corresponding Eigen values.

This method produces weak fused image as compared against other methods. The spatial frequency is measuring the activity level of the each pixel in the source image.

Fig. 4 Subjective results, left to right: PCA, SF, all in focus, and proposed method

It is affecting the fusion quality due to blocking effect. The high level of weak features in this method lacks in un-artifact fused image. The third method mainly relies on the neighborhood size choosen for focus measurement. This leads to artifacts if the proper neighborhood size is not chosen. The result of all the methods is given in Fig. 4.

C. **Quantitative Comparison**

The quantitative comparison of stacked images is a tedious task as it involves ground truth comparison. There are no proper quantitative measures proposed to measure the fused image results. Comparing the mutual information may give us the proper comparison. We used Normalized Mutual Information [17]; it is more powerful than Mutual Information (MI). The definition of NMI is given below (2):

$$Q = 2M(A, F) + H(A) + H(F)$$
$$M(B, F)(2)H(B) + H(F) \tag{2}$$

5 Conclusion and Future Works

It is determined from the discussions above that there exist many common challenges in typical activity level measuring techniques. Calculating the correct filter size is one among the key challenges. Deep learning based image fusion research reduced the hand crafting works and improved the quality of fusion for any real-time multi-focus images. However, designing of neural network model for a specific fusion task is a challenging task in deep learning based techniques. Another important drawback is generating a dataset for model training (a sensible problem in existing deep learning based image fusion is that the lack of large available dataset), lack of the applying of domain information or lack of hardware, etc. In practice, these problems are having close relations to one another and may be collectively thought-about. In this proposed method, the features are extracted using the cGAN and there is no manual hand craft involved in image fusion. And, the result shows that this method produces quality fused image without manual work involved. The future work is to exploit the usage of Generative Adversarial Network (GAN) to generate the large set of dataset without manual labeling of segmented images.

References

1. G.A. Ratt, J. Aacute, A. V1ega, J.E. Murari, Image fusion: advances in the state of the art. Inf. Fusion **8**(2), 114–118 (2007)
2. Z. Liu, K. Tsukada, K. Hanasaki, Y.K. Ho, Y.P. Dai, Image fusion by using steerable pyramid. Pattern Recogn. Lett. **22**(9), 929–939 (2001)
3. J. Tian, L. Chen, Adaptive multi-focus image fusion using a wavelet based statistical sharpness measure. Sig. Process. **92**(9), 2137–2146 (2012)
4. J.J. Lewis, R.J. O'Callaghan, S.G. Nikolov, D.R. Bull, N. Canagarajah, Pixel- and region-based image fusion with complex wavelets. Inf. Fusion **8**(2), 119–130 (2007)
5. Y. Liu, X. Chen, H. Peng, Z. Wang, Multi- focus image fusion with a deep convolutional neural network. Inf. Fusion **36**, 191–207 (2017)
6. I. Goodfellow, NIPS 2016 tutorial: Generative adversarial networks. arXiv preprint arXiv: 1701.00160, 2016
7. R.A. Yeh, C. Chen, T.Y. Lim, A.G. Schwing, M. Hasegawajohnson, M.N. Do, Semantic image inpainting with deep generative models, in P*roceedings of IEEE Conference on Computer Vision Pattern Recognition*, pp. 6882–6890
8. S. Reed, Z. Akata, X. Yan, L. Logeswaran, B. Schiele, H. Lee, Generative adversarial text to image synthesis. arXiv preprint arXiv:1605.05396 (2016)
9. P. Isola, J.Y. Zhu, T. Zhou, A.A. Efros, Image-to-image translation with conditional adversarial networks, in *Proceedings of IEEE Conference on Computer Vision And Pattern Recognition*, pp. 5967–5976 (2017)
10. M. Mirza, S. Osindero, Conditional generative adversarial nets. Comput. Sci. 2672–2680 (2014)
11. K.R. Prabhakar, V.S. Srikar, R.V. Babu, *DeepFuse: A Deep Unsupervised Approach for Exposure Fusion with Extreme Exposure Image Pairs*, pp. 4724–4732
12. K. He, X. Zhang, S. Ren, J. Sun, Deep residual learning for image recognition, in *Proceedings of IEEE Conference on Computer Vision And Pattern Recognition*, pp. 770–778 (2016)
13. M. Everingham, S.M.A. Eslami, L.V. Gool, C.K.I. Williams, J. Winn, A. Zisserman, The pascal visual object classes challenge: a retrospective. Int. J. Comput. Vision **111**(1), 98–136 (2015)
14. O. Rockinger, T. Fechner, Pixel-level image fusion: the case of image sequences. Proc SPIE **3374**, 378–38829 (1998)
15. S. Li, T. James Kwok, Y. Wang, Combination of Images with diverse focuses using the spatial frequency. Inf. Fusion **2**(3), 169–176 (September 2001)
16. S. Pertuz et al., Generation of all-in-focus images by noise-robust selective fusion of limited depth-of- field images. IEEE Trans. Image Process. **22**(3), 1242–1251 (2013)
17. M. Hossny, S. Nahavandi, D. Creighton, Comments on 'Information measure for performance of image fusion. Electron. Lett. **44**(18), 1066–1067 (2008)

Privacy Preservation Between Privacy and Utility Using ECC-based PSO Algorithm

N. Yuvaraj, R. Arshath Raja, and N. V. Kousik

Abstract In this paper, we propose a new privacy utility method, where privacy is maintained by lightweight Elliptical Curve Cryptography (ECC) and utility is maintained through Particle Swarm Optimization (PSO) clustering. Initially, the datasets are clustered using PSO, and then the privacy of clustered datasets is maintained using ECC. The proposed method is experimented over medical datasets, and it is compared with existing methods through several performance metrics like clustering accuracy, F-measure, and data utility and privacy metrics. The evaluation shows that proposed method obtains improved privacy preservation using clustering algorithm than existing methods.

Keywords PSO clustering · ECC · Privacy and utility

1 Introduction

Data anonymization [1] is a major technique for preserving the data with privacy in the field of data publishing or data mining. It involves anonymizing the data records from linkage attacks [2] or probabilistic inference attack [3]. The process of anonymization is achieved either through suppression of QID attributes or through generalization. The former one leads to increased loss of data and the latter one is an NP-hard problem. Hence, privacy preservation using data anonymization needs

N. Yuvaraj
Deputy Manager, Research And Development, ICT Academy, Chennai, Tamil Nadu, India
e-mail: yraj1989@gmail.com

R. Arshath Raja (✉)
Senior Associate, Research And Development, ICT Academy, Chennai, Tamil Nadu, India
e-mail: arshathraja.ru@gmail.com

N. V. Kousik
Department of Computer Science and Engineering, Galgotias University, Greater Noida, Uttar Pradesh, India
e-mail: nvkousik@galgotiasuniversity.edu.in; nvkousik@gmail.com

S. S. Dash et al. (eds.), *Intelligent Computing and Applications*,
Advances in Intelligent Systems and Computing 1172,
https://doi.org/10.1007/978-981-15-5565-4_51

proper improvement to avoid the constraints like increased data loss and NP-hard problem.

On the other hand, privacy preservation process involves measuring the utility of dataset and provides information on how the privacy-preserved data is useful to the user. The sensitive information in the dataset is diminished such that the data is removed, distorted, or transformed to obtain confidentiality [4]. The process of finding the equilibrium between the utility and privacy of dataset remains obstinate that leads to trade-off between them [5].

Specifically, in large datasets, the privacy and utility is a contradicting factor, where either one should be sacrificed to achieve the other. If perfect privacy is achieved without publishing the data, then perfect utility cannot be achieved since original data is tend to be published without privacy. Hence, maintaining equilibrium between privacy and utility of datasets in e-health record datasets is of prime importance.

The utilization of data anonymization in data publishing would be preferred to obtain privacy [6], and clustering improves the data utility [2] in a fully distributed database. Hence, in this paper, we propose an anonymization technique to preserve the privacy of data in a fully distributed environment. Further, a clustering algorithm is used to achieve the anonymization and provide resilient to probabilistic inference attack and linking attacks. Later, Hadoop distributed file system [7] is used for distributing the privacy-preserved anonymized datasets.

The main contributions of the paper are given below:

1. A new privacy utility method is proposed using lightweight Elliptical Curve Cryptography with PSO clustering.
2. The privacy is attained using lightweight ECC, and privatized data tends to get published after the process of clustering.
3. The machine learning classification is carried out using PSO multi-label clustering that reduces the errors in forming proper cluster.

2 Proposed Method

Initially, the privacy preservation is applied using lightweight ECC algorithm. After the generation of privatized datasets, it is clustered using PSO clustering algorithm. The clustering error obtained from PSO algorithm is used for the selection of cluster with lowest clustering error. If the clustering error is lesser than the threshold level, the data is published directly. On the other hand, the privacy parameters are adjusted, and then the clustered instances are reclustered to obtain accurate clustered results. Once the criteria are satisfied, the results are published.

The proposed method can perform data anonymization in order to provide resilience against most common attacks, and then the data is sent to third party. The third party access ensures privacy in distributed environment since the data is anonymized in prior. Also, the data can be utilized to maintain utility without privacy violation on a distributed environment.

A. ***PSO Clustering Process:***

Various techniques are used for achieving diversity by setting threshold values for private attributes. On the other hand, it is very essential to categorize the private attributes into different levels of confidentiality. It should further provide suitable background information to assign threshold level for the private attributes. However, skewing of data does not provide desired anonymization level and that satisfies *t*-closeness constraints. The PSO data clustering [8] is carried out in this work to avoid similarity attack. The similarity attack is handled using S-diversity principle.

The PSO clustering aims at dividing the entire class into various subclasses. The algorithm is further divided into two major steps (i) partitioning and (ii) adjusting step. The first step divides the data into multiple clusters, and the second step adjusts the cluster size.

In order to further improve the uniformity of distributed clusters, we employ J48 and Naïve Bayes (NB) over PSO algorithm. This algorithm helps to generate the clusters with uniformity such that the sensitive information is distributed uniformly. The combination of J48 and NB with PSO finds the best neighbor in each cluster group, and it assigns the values one at a time to the previous cluster The proposed method modifies the PSO algorithm to avoid the problem of skew in distributing clusters of sensitive information using J48 and NB techniques. The nearest neighbors are found using the J48 and NB algorithms over each cluster group with sensitive information. Hence, the distribution of cluster is same in both the formed clusters and original dataset. This avoids the probability of probabilistic inference attack in data publishing.

The pseudocode of proposed J48 and NB with PSO is given in Algorithm 3.

Input: Privacy datasets

Output: Unequally clustered datasets with uniform distributions of sensitive information

Step 1. The private dataset is sorted based on the values of sensitive attributes
Step 2. Then, the sorted input dataset is split into subgroups based on *K*, *S*—diversity and D_{min}—minimum distance between each itemset
Step 3. Check if the subgroup has the identical sensitive attributes values
Step 4. Else go to step 1, End
Step 5. Repeat the process for other subgroups or clusters
Step 6. Depending on the value of *K* and *S*, we use two cases
Step 7. Case one, the value of *K* is greater than *S*

(a) Cluster size = D_{min}
(b) Distribute the records to all clusters, and it creates single element cluster
(c) Add nearest neighbor from the remaining cluster to the existing cluster
(d) Thus, the cluster formed has *S* instances with *S*-diverse private information or sensitive information
(e) Go to 7(c)
(f) Remove the instances from clusters that are existing in each cluster
(g) Repeat the process

Step 8. Find the centroid using ACO
Step 9. Find the instances in nearest neighbor from the clusters using Euclidean distance estimation

Merge the single dataset with other subgroups.

3 Results and Discussion

The proposed method is implemented in Java environment on Eclipse IDE. The Hadoop 1.2.1 is used to distribute the privacy-preserved dataset. Different classifiers are executed to find the data utility of proposed algorithms using Weka 3.7.12. The experiments are conducted to show the cluster size and total number of cluster formed using the proposed method for measuring the data utility, privacy degree, execution time, and scalability in HDFS. The experiments are conducted to compare the data anonymization between the proposed and existing methods.

(A) ***Dataset***

The proposed method is compared with conventional methods based on several experiments conducted on the datasets. It includes adult dataset available [9], and its description is given in Table 1. The experiments are further conducted on synthetic data to determine the scalability proposed method. Two synthetic datasets are used that consist of 200,000 and 500,000 records, respectively, each with ten attributes. The experiments carried out with the given dataset are tested between the proposed and existing Mondrian [11], Datafly [10], and Incognito [11] algorithms.

The experiments are conducted using Adult dataset. The proposed method is evaluated in terms of discernibility cost and average equivalence class size. The performance is compared in terms of F-measure and percentage of accurately classified instances. Finally, the proposed method is compared with existing methods in terms of execution time.

The proposed system is tested in terms of two metrics, namely discernibility cost and equivalence class size for determining the data utility loss of privacy-preserved dataset. The lesser value of above metrics denotes the reduced information loss.

Table 1 Discernibility cost ($\times 10^8$)

k	Incognito	Mondrian	Datafly	PSO
50	0.6	0.83	2.35	0.21
100	6.9	0.93	2.43	0.23
150	6.8	1.23	1.33	0.23
200	6.83	0.6	1.32	0.24
250	3.25	0.32	1.2	0.22
300	3.33	0.21	0.9	0.25

Table 2 Average values of equivalence class size

k	Incognito	Mondrian	Datafly	PSO
50	60	7	75	2
100	60	6	38	2
150	40	5	50	2
200	30	5	38	2
250	30	5	18	2
300	25	5	15	2

The Table 1 provides discernibility cost based on the experiments conducted on case one and case two datasets, respectively. The result shows that proposed method obtains better performance in both the cases. The proposed method obtains reduced discernibility cost than other algorithms, and it outperforms existing metrics in terms of reduced discernibility cost.

The Table 2 shows the results of average values of equivalence class size between the proposed and existing algorithms, where the experiments are conducted using anonymization toolbox. The average values of equivalence class size of proposed method is lesser than existing algorithm in case of case one, and in case two, the value is found higher for low k values, and it reduces gradually with increasing k values. However, Mondrian algorithm obtains reduced equivalence class size than proposed method.

Tables 3 and 4 show the percentage of correctly classified data instances and *F*-measure for varying *k* values, say 5, 10, 25, and 30, where the experimentation is carried out over case one. The result shows that performance of PSO algorithm attains better performance than PSO algorithm and exhibits closer performance with original datasets than other algorithm with Naïve Bayes and J48 classifier.

Further, the degree of privacy (Table 5) is experimented on both the cases using the proposed algorithm. It is seen that attack links the value of individual QID attribute to

Table 3 Accuracy of PSO clustering

k	Algorithm	Incognito	Mondrian	Datafly	PSO
5	**J48**	84.2	84	84.24	84.05
	NB	81.64	81.82	81.84	82.35
10	**J48**	84.2	83.96	84.24	84.05
	NB	81.33	81.82	81.84	82.35
25	**J48**	84.2	83.99	84.28	84.05
	NB	81.6	81.83	81.84	81.01
30	**J48**	84.2	84.04	84.26	84.24
	NB	81.6	81.84	82	82.33

NB—Naïve Bayes; J48—J48 Classifier

Table 4 *F*-measure of clustered results

k	Algorithm	Incognito	Mondrian	Datafly	ACO
5	**J48**	0.818	0.818	0.821	0.831
	NB	0.809	0.809	0.81	0.818
10	**J48**	0.819	0.818	0.821	0.83
	NB	0.81	0.809	0.81	0.818
25	**J48**	0.819	0.819	0.821	0.832
	NB	0.809	0.809	0.81	0.805
30	**J48**	0.819	0.818	0.821	0.83
	NB	0.809	0.809	0.81	0.805

NB—Naïve Bayes; J48—J48 Classifier

Table 5 Results of privacy degree

Case	*k*	NB-ACO	J48-ACO
One	2	3227	1423
	5	3227	1423
	10	3227	1423
	15	3234	1365
	25	3875	2476
Two	50	5	3
	100	5	3
	150	5	3
	200	5	3
	250	5	3
	300	5	3

the value sensitive attribute with confidence level of P^{-1}. In case one, the probability of record linking is higher in ACO clustering algorithm than in KNN-ACO clustering algorithm. The probability of record linking using case two is set to exhibit higher linkage in ACO clustering algorithm than KNN-ACO clustered dataset. There exists a 50% chance of record linking in case two using ACO clustering algorithm than KNN-ACO clustering algorithm where it shows only 25% record linkage.

From the above three measures, we found that there exists a trade-off between the data utility and data privacy. This is determined in terms of total number of clusters formed using the proposed method, and this exhibits the trade-off. As the total number of clusters increases, the data privacy tends to reduce with increased data utility. Such reduction of data privacy with increased data utility is due to reduced number of instances with increased clusters from the given datasets. This sets to form reduced information loss; however, the level of privacy is low.

4 Conclusions

In this paper, we propose data anonymization technique to provide equilibrium between data utility and data privacy. The proposed data anonymization technique using PSO clustering and ECC privacy-preserved datasets is experimented over adult datasets. The evaluation metrics like discernibility cost and average equivalence class size shows reduced value with improved performance in terms of data utility. Additionally, the data utility is measured in terms of accuracy and F-measure and the result shows that proposed method obtains improved performance than existing methods.

References

1. L. Sweeney, k-anonymity: a model for protecting privacy. Int. J. Uncertainty Fuzziness Knowl. Based Syst. **10**(05), 557–570 (2002)
2. W.Y. Lin, D.C. Yang, J.T. Wang, Privacy preserving data anonymization of spontaneous ADE reporting system dataset. BMC Med. Inform. Decis. Mak. **16**(1), 58 (2016)
3. J. Soria-Comas, J. Domingo-Ferrer, D. Sánchez, S. Martinez, t-closeness through microaggregation: strict privacy with enhanced utility preservation. IEEE Trans. Knowl. Data Eng. **27**(11), 3098–3110 (2015)
4. K. Mivule, C. Turner, S.Y. Ji, Towards a differential privacy and utility preserving machine learning classifier. Procedia Comput. Sci. **12**, 176–181 (2012)
5. R.C.W. Wong, A.W.C. Fu, K. Wang, J. Pei, Minimality attack in privacy preserving data publishing, in *Proceedings of the 33rd International Conference on Very Large Data Bases* (VLDB Endowment, 2007), pp. 543–554
6. K. Chen, L. Liu, Privacy-preserving multiparty collaborative mining with geometric data perturbation. IEEE Trans. Parallel Distrib. Syst. **20**(12), 1764–1776 (2009)
7. I. Polato, R. Ré, A. Goldman, F. Kon, A comprehensive view of Hadoop research—a systematic literature review. J. Netw. Comput. Appl. **46**, 1–25 (2014)
8. Y., Kao, K. Cheng, An PSO-based clustering algorithm, in International Workshop on PSO and Swarm Intelligence (Springer, Berlin, Heidelberg, 2006), pp. 340–347
9. M. Lichman, UCI Machine Learning Repository, 2013. http://archive.ics.uci.edu/ml. Accessed 28 Aug 2018
10. L. Sweeney, Datafly: a system for providing anonymity in medical data, in *Database Security XI* (Springer, Boston, MA, 1998), pp. 356–381
11. K. LeFevre, D.J. DeWitt, R. Ramakrishnan, Incognito: efficient full-domain k-anonymity, in *Proceedings of the 2005 ACM SIGMOD International Conference on Management of Data* (ACM, 2005), pp. 49–60

Predicting Energy Demands Constructed on Ensemble of Classifiers

A. Daniel, B. Bharathi Kannan, N. Yuvaraj, and N. V. Kousik

Abstract Analysis of time series data and perfect future prediction are the most stimulating tasks that the data analysts face in countless fields. Forecasting of energy demands is very essential because both excess energy cost and delay lead to a significant reduction storing costs. In order to discover the uniformities in dynamic, non-stationary data and time series prediction needs the use of models to be integrated with multiple forecast models. The Ensemble learning model discovers the dynamic patterns in energy time series data. The performance of two different Ensemble learning techniques is compared against Bagging and stacking in forecasting energy time series data. Stacking technique used in this paper combines different classifiers like Multilayer Perceptron (MLP), Radial Basis Function (RBF), and Support Vector Machine (SVM).

Keywords Predicting · Ensemble · Neural networks · Stacking techniques · Dynamic pattern

1 Introduction

Time series analysis and perfect prediction play a significant role in various fields of science like electricity demand predicting, financial, weather predicting, etc. Short-term forecasting is the process of predicting events in short time periods such as days,

---

A. Daniel (✉) · B. Bharathi Kannan · N. V. Kousik
School of Computing Science and Engineering, Galgotias University, Greater Noida, Uttar Pradesh, India
e-mail: a.daniel@galgotiasuniversity.edu.in

B. Bharathi Kannan
e-mail: bharathi.kannan@galgotiasuniversity.edu.in

N. V. Kousik
e-mail: nvkousik@galgotiasuniversity.edu.in; nvkousik@gmail.com

N. Yuvaraj
ICT Academy, Chennai, Tamil Nadu, India
e-mail: yraj1989@gmail.com

S. S. Dash et al. (eds.), *Intelligent Computing and Applications*,
Advances in Intelligent Systems and Computing 1172,
https://doi.org/10.1007/978-981-15-5566-4_52

weeks, and months. Medium-term forecasting involves prediction of future values which can extend up to two years, whereas long-term forecasting problems can be extended for many years [3].

Energy demand planning is an important factor for any country's economic growth. Predicting allows decision makers to make accurate conclusions and to get competitive advantage. Improving the accuracy of the forecasting technique is the major issue in time series analysis. Well-organized energy scheduling requires accurate forecast of future energy demands. There are quite a lot of statistical methods for time series prediction such as ARX, ARIMA, ARMA, GARCH, Box and Jenkins, and Smoothing techniques. ML Techniques such as Artificial Neural Networks (ANN), SVM, and Decision Trees are also widely used for time series prediction. The forecasting accuracy of each model is different from each other.

The selection of a particular model is based on the accuracy needed the computational overhead incurred. In [2], it is reported that ANN provides better forecasting accuracy than statistical techniques such as ARIMA and Multiple Linear Regression (MLR). But the computational complexity of ANN was higher over other two techniques. Each forecasting method has its own advantage as well as disadvantage in time series prediction. Ensemble methods improve the forecasting accuracy by combining the hypothesis generated from individual base learners for the same training data set [7]. Ensemble learning objective is to improve the performance of classifier or predictor. Model Selection is an important task in ML. Accuracy depends on type of classifier used and which realization of the classifier is to be chosen. For example, in ANN, different initialization of parameters gives different outcomes.

An Ensemble model which consists of a group of learners is usually weaker than an Ensemble. Ensemble learning is more advantageous because it can boost weak learners to become strong learners which makes highly accurate forecast. Most Ensemble methods such as Bagging and Boosting generate a homogeneous base learner by making use of a single learning algorithm. There are also some other methods such as Stacking and Voting which produce a heterogeneous learner using multiple learning algorithms [7].

Two different steps can be used to build an Ensemble. The base learner generation can be sequential or parallel while the base learner generation has an effect on the future generation of learners. The base learners should be as accurate as possible to get a more precise Ensemble. The accuracy of the learners is assessed by many effective steps; for example, cross-validation, random sampling, stop testing, etc.

The advantages of various individual techniques are combined in Ensemble learning, thereby improving the accuracy of predicted values. Accuracy is also based on the Meta-learner to be chosen in the process of creating an Ensemble. This paper compares the performance of the Ensemble with that of individual methods in the prediction of energy demand.

2 Literature Survey

From literature, a number of pattern recognition techniques are used for time series forecasting. Those techniques include ARIMA and GARCH models, artificial neural networks, fuzzy logic, genetic algorithms, and Support Vector Machines (SVM). Ensemble of multiple classifiers can improve the prediction power rather than using a single classifier. Combining the predicted results of a number of classifiers will significantly improve the accuracy of the prediction algorithm. Martínez-Álvarez proposed the Pattern Sequence-based Forecasting (PSF) algorithm [3], which predicts the future values of a time series based on pattern sequence similarity.

PSF produced a better prediction of energy time series compared to other well-known techniques. Karin Kandananond [6] compared the performance of ANN approach with that of ARIMA and Multiple Linear Regression (MLR) models. ANN model outperforms MLR and ARMIA models in terms of accuracy. They reported that even though accuracy of ANN is higher the computational overhead of ANN is also higher compared to other models.

Wen and Vahan [1] proposed an algorithm Pattern Forecasting Ensemble Model (PFEM) for day-ahead Energy Demand Forecasting. They combined various clustering techniques like K-Means, SOM, Hierarchical Clustering, Fuzzy C Means, K-Mediods, and applied forecasting models for the clustered output prediction of the day ahead energy demand. The performance of PFEM was proved to be better than other models.

Pasapitch and Nittaya [4] compared the performance of ARIMA and ARMA models in forecasting household power consumption patterns. ARIMA model is proved to be better for monthly and quarterly analysis, and ARMA model performs better for daily and weekly analysis.

Reinaldo and Javier [8] forecasted day ahead electricity prices of Spain and California Electrcity Markets with GARCH model.

Chitra [5] used an Ensemble of multiple classifiers such as Self-organizing Map, K-Nearest Neighbors, and Radial Basis Function for time series prediction. The performance of individual learners with that of Ensemble model is compared using three different data sets. In this paper, the performances of two different Ensemble learning techniques were compared by combining various classifiers such as RBF, SVM, and MLP.

3 Methodology

Numerous types of classifiers are existing for pattern prediction. RBF, MLP, and SVM are considered as the best methods for non-linear time series analysis and forecast. Ensemble learning is one of the machine learning techniques in which multiple learners are trained to solve a single problem. There are various Ensemble methods: Bagging, Boosting, Stacking, and Voting.

Model Selection is a vital factor in supervised learning. Which model is best suited for the given problem? There are two different ways to clarify this question: i) what type of model is to be chosen among many competing models, such as MLP, SVM, etc.; ii) given a particular prediction algorithm, which realization of this algorithm is to be chosen, i.e., different initializations of SVMs give rise to different decision boundaries, even if all other parameters are kept constant.

The most commonly used technique is to choose the algorithm which gives the smallest error and high accuracy. But it is a flawed one; most of the techniques evaluating the accuracy such as cross-validation may be misleading because it provides less accuracy for previously unseen data. So, Ensemble learning is the best alternative for this model selection problem. The following section describes two different Ensemble learning techniques, Stacking and Bagging, respectively.

3.1 Stacking

The block diagram of Stacking is shown in the following Fig. 1. The Ensemble process is done with the following four steps: Data Processing, Applying Individual Learner, Classification of the new data set with the Meta-learner, and Evaluation of predicted output. It consists of three different classifiers: Radial Basis Function Neural networks, MLP, and SVM. In the Stacking algorithm, a number of first-level individual learners are generated from the training data set D by employing different first-level learning algorithms L1, L2, …, LT.

Those individual learners are then combined by a second-level learner "L" which is called as meta-learner. Ensemble learners are superior to single learners because

Fig. 1 Stacking of classifiers

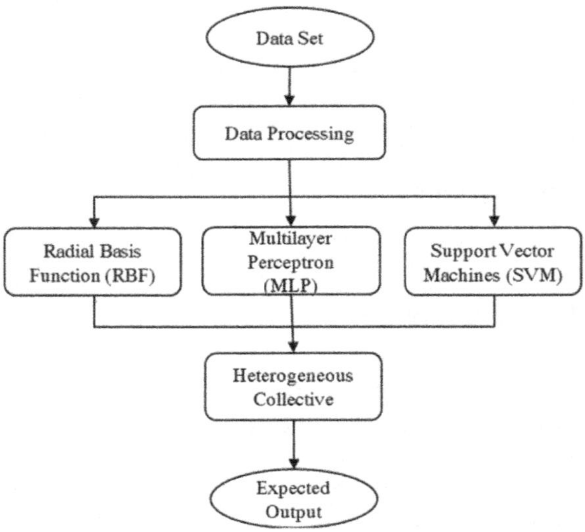

of various reasons. The training data might not contain sufficient inputs for choosing a single learner. The search process of the Ensemble learners is better than individual learners. The hypothesis space might not have actual target function. Ensemble learners are used whereever machine learning techniques can be applied.

Algorithm:

Input: Data Set D
First-level Learning algorithms RBF, MLP, and SVM.
Second-level Learning algorithm L.

1. Train the first-level individual learners h1, h2, and h3 by applying the first-level learning algorithms RBF, MLP, and SVM to the original data set D.
2. Generate a new data set D'.
3. Apply individual learners h1, h2, and h3 to train the data set D.
4. Assign the result of first-level learning to the new data set D′.
5. Train the second-level learner h' by applying second-level learning algorithm to the new data set D′.

Output

Output of Stacking is the final predicted output of two-level Ensemble. The above algorithm gives steps of the two-level Ensemble approaches with the set of base learners and the Meta-learner. As a first step, the input data is preprocessed and given to the set of base learners. Finally, the resultant data set is trained with the Meta-learner. Hence, it uses the two-level heterogeneous Ensemble approach; the generalization ability of the Ensemble is much higher than that of individual learners.

3.2 Bagging

Figure 2 trains various base learners with the help of bootstrap samples obtained from the training data set. Bootstrap sample is obtained from creating subsamples from the original data set with replacement. The size of the bootstrap sample is same as that of the training set. Bootstrap samples are generated from the given training set as random samples. So there is a possibility of choosing the same tuples again. For instance, the given data contains d tuples, and the data set is sampled d times, thereby generating d bootstrap samples. Therefore, each tuple has a probability of 1/d to be chosen and has the probability of 1–1/d not to be chosen.

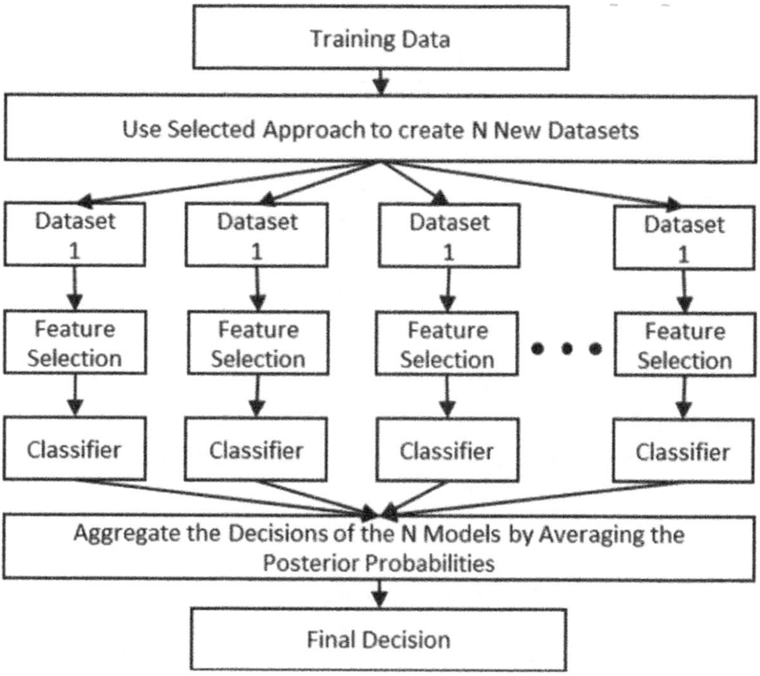

Fig. 2 Bagging of classifiers

Algorithm:

Input: Data Set D

Base Learning Algorithm L Number of rounds I

Process

For i = 1,2, … I

1. Create Bootstrap sample Di from the data set D.
2. Train the base learner hi with the bootstrap

sample Di End

Output

Output of bagging is the average of predictions from the base learner. Like in Voting Ensemble Learning, for numeric prediction, predictions of base learners are averaged. For classification problems, Predictions of base learners are taken majority voting.

4 Results and Discussions

The data set was obtained from US department of Energy. The data set contains the monthly energy consumption data. The data set is available at the following site: http://www.eia.gov/totalenergy/data/monthly/index.cfm#consumption. The performance of the above algorithms is evaluated with the following metrics.

Performance Metrics

The following are various metrics that are used for evaluating the performance of the classifiers with that of Ensemble Model: Mean Absolute Error (MAE), Root Mean Squared Error (RMSE), Mean Squared Error (MSE), and Mean Absolute Percentage Error (MAPE).

A comparison of Prediction Accuracy, MAE, and RMSE of the algorithms RBF, MLP, SVM, and Stacking and bagging is given below.

The data set is evaluated with hold-out evaluation technique, where the given data is randomly partitioned in training and testing sets. Therefore, one-third of the data is used as the test set. The data set consists of 490 records in which 327 records are chosen as training set and remaining 163 records are used as test set (Table 1).

RBF Ensemble is the Ensemble obtained as a result of applying bagging to RBF neural network. Similarly, SVM Ensemble and MLP Ensemble are the results of bagging of SVM and MLP, respectively (Table 2).

The following chart shows the accuracy of the different algorithms in forecasting of energy time series data (Figs. 3, 4).

Table 1 Performance comparison of classifiers with ensemble

Methods	MAE	RMSE	MSE	MAPE
RBF	0.2854	0.3783	0.1431	3.3788
MLP	0.395	0.499	0.249	4.8151
SVM	0.2153	0.2792	0.0779	2.6585
Stacking	0.2861	0.3477	0.1209	3.4222

Table 2 Performance comparison of bagging of different classifiers

Classifiers	MAE	RMSE	MSE	MAPE
RBF	0.2989	0.3956	0.1565	3.5394
MLP	0.2643	0.3563	0.1269	3.1497
SVM	0.1864	0.2482	0.0616	2.2767

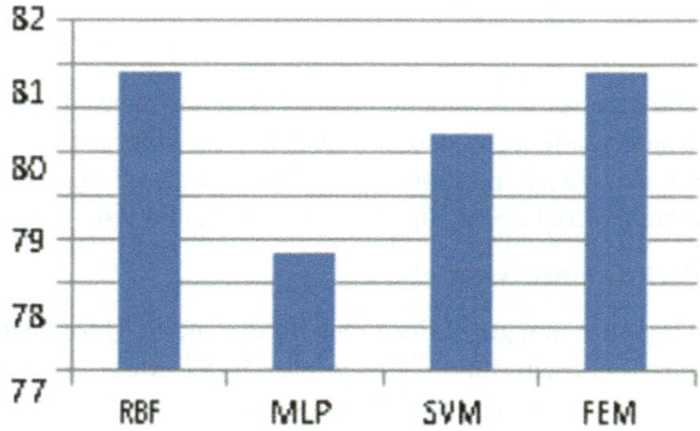

Fig. 3 Prediction accuracy of classifiers in time series forecasting

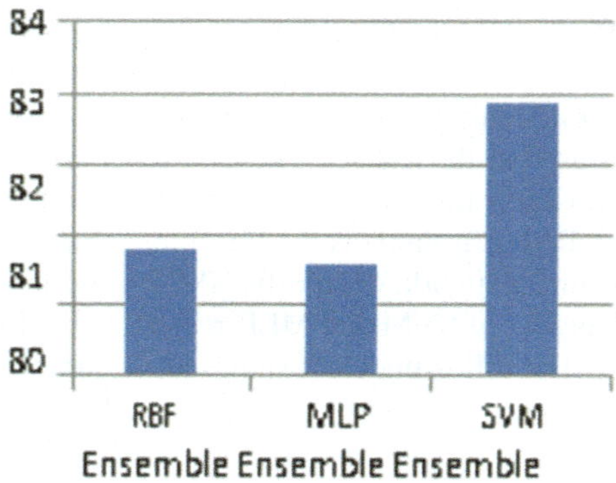

Fig. 4 Accuracy of classifiers after bagging

5 Conclusion

The energy demand forecasting is very essential for optimization of energy resources and application of green trends. The proposed Ensemble model can be applied for prediction of both linear and non-linear time series data and can be used in wide range of applications in time series forecasting. The experimental results show that the Ensemble model outperforms all other individual learners.

References

1. S. Wen, B. Vahan, An Ensemble Model for Day-ahead Electricity Demand Time Series Forecasting, e- Energy'13 (2013)
2. L. Felix, K. Wolfgang, Classification and clustering of electricity demand patterns in industrial parks. Energies **5**, 5215–5228 (2012)
3. A. Francisco, T. Alicia, Energy time series forecasting based on pattern sequence similarity. IEEE Trans. Knowl. Data Eng. **23**(8), 1230–1243 (2011)

4. C. Pasapitch, K. Nittaya, Time series analysis of household electric consumption with ARIMA and ARMA models, in *Proceedings of the International Multi Conference of Engineers and Computer Scientists* (2013), pp. 295–300
5. A. Chitra, S. Uma, An ensemble model of multiple classifiers for time series prediction. Int. J. Comput. Theor. Eng. **2**(3) (2010)
6. K. Karin, Forecasting electricity demand in thailand with an artificial neural network approach. Energies **4**, 1246–125 (2011)
7. Z.-H. Zhou, *Ensemble Methods: Foundations and Algorithms*, 1st edn (CRC Press, 2012)
8. C.G. Reinaldo, C. Javier, A GARCH forecasting model to predict day-a head electricity prices. IEEE Trans. Power Syst. **20**(2), 867–874 (2005)

Use of RNN in Devangari Script

Madhuri Sharma and Medhavi Malik

Abstract In India, Devanagari script is used most prevalently. But recognizing the handwritten documents of Devanagari Script is a challenging task due to variations in writing, rendering, and cursive nature of handwriting. However, there has been a lot of improvement in research in the field of recognizing Handwritten Documents. In this, we will implement a framework that inputted the number of images gives an output for recognized characters.

Keywords Devanagari script · Natural language processing · RNN

1 Introduction

Devangari script is used in Hindi and Sanskrit. Devanagari script contains 47 primary characters; 14 are vowels and 33 are consonants. In HWR (Handwritten Recognition), the model is trained for correct classification. For this, Supervised Learning, Unsupervised Learning, and Reinforcement Algorithm are applied to train the model. Supervised learning is learned from Label data, while for unsupervised algorithm there is no labeled data; so Clustering algorithm and K-Means algorithm are used. Reinforcement algorithm provides learning from previous experience. They can depict syntactic data (e.g., grammatical form labeling, piecing, and parsing) or semantic data (e.g., word-sense disambiguation, semantic job naming, named element extraction, and anaphora goals) [1]. There are no sufficient number of work on Indian language character recognition although there are 12 major scripts in India [2].

M. Sharma (✉) · M. Malik
SRMIST, Delhi-NCR Campus, Modinagar, India
e-mail: madhurisharma44@gmail.com

M. Malik
e-mail: medhavimalik28@gmail.com

S. S. Dash et al. (eds.), *Intelligent Computing and Applications*, Advances in Intelligent Systems and Computing 1172,
https://doi.org/10.1007/978-981-15-5566-4_53

2 Identification of Handwritten Recognition (HWR)

After analyzing various word samples in Devanagari script, it has been found that, in this script, generally words are written in non-cursive fashion which has prompted us to use Recurrent Neural Network (RNN) models, as RNN models require pre-segmented data [3]. To recognize handwritten words, it usually undergoes number of stages. Following are the steps that can recognize a word:

- **Process a word**: This is the first step to process the handwritten words. First, convert the image into Gray Scale image having white background and black pixels as text. Then, noise has to be removed since it will create a problem in the process of segmentation. Various types of noise removal techniques are used: Smoothing, Filtering, and Extraction. After this, normalization procedure is performed to ensure the values lie in between −1 and 1. There are a number of different ways for handwritten recognition style.
- **Segmentation**: Segmentation is in which text is divided into number of meaningful units: Graphemes represent the lowest level characters; one or more characters represent words, and sentences are made up of number of words (Fig. 1).

Word Segmentation is the process of dividing a text into number of tokens. Sentence Segmentation is the process of breaking down into number of words separated by boundaries in Devanagari Script (puran viram (।)). Both word and sentence segmentations cannot be applied simultaneously and are independent of each other. The inputted text is tokenized, and all the inconsistencies like spelling checking, punctuation marking, grammar checking, etc., are removed. If there is any mistake in grammar and spelling, it automatically corrects and gives the suggestions to the user. The special words like numbers, date, time, abbreviations, and special symbols (like: ॐ) are identified [4].

But, there are some limitations that must be addressed in developing algorithms for text segmentation which are described below:

Language Dependency: Written adaptations of language have a diverse set of features.

Character Dependency: All texts are encoded in seven-bit ASCII-encoded set which allowed only 128 characters and essential characters for writing English. Extended character sets allow eight-bit encodings, but an eight-bit encoding allows just 256 distinct characters. Still, there are tens of thousands of distinct characters in all the writing systems. So, the main reason behind these characters in different languages is currently encoded in a large number of overlapping character sets. The

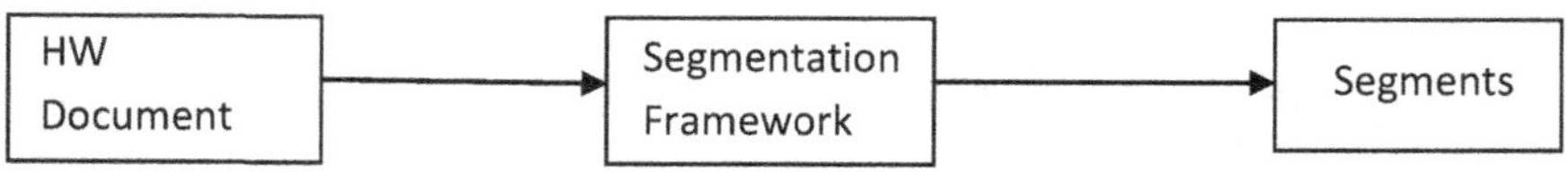

Fig. 1 Segmentation

Unicode Standard, Version 12.0 specifies a single two-byte encoding system that includes a range from U+090x to U+ 097x.

Application Dependency: These may include Morphological Analyzers, Part-of-Speech taggers, Lexical Look-up Routines, or Parsers which are being used in its process.

Corpus Dependency: This is due to the availability of large amount of corpus, misspellings, erratic punctuation, and irregular white spaces.

- **Feature Extraction**: Feature extraction is the main component of the text recognition. Various techniques are used in this process: LDA, PCA, and other various algorithms to train the model.

Statistical Features: Statistical features include size, shape, and intensity. Various statistical features are Area, Eccentricity, Orientation, Centroid, BoudingBox, MajorAxisLength, MinorAxisLength, ConvexHull, ConvexImage, ConvexArea, FilledImage, and FilledArea. Various methods used are Moment, Zoning, Projection Histogram, and Crossing and Distances.

Structural Features: Structural features are the geometrical features of the image of the character. Some of the features are Vertical Line, Horizontal Line, End Points, Number of Crossing Points, and Number of Contours. Various methods used in the process are Geometrical and Topological Features, Coding, Graphs, and Trees.

Raw Features: Pixel values and deep learning like Convolution Neural Network are used to learn features by themselves.

- **Recognition Method**: This step will assign a label from the given set of classes or predict the labels using RNN. Various recognition methods are used in this process: HMM (Hidden Markov Model), SVM (Support Vector Machine), CNN (Convolution Neural Network), and MLP (Multilayer Perceptron) (Fig. 2).

3 Recurrent Neural Network (RNN)

RNN is a feedforward neural network which requires an internal memory for storing the output at any stage since it is inputted back to the previous stage. RNN for every input data performs the same function since it is recurrent in nature; however, input is independent of each other. For decision, it depends on the previous input, current input, and output from a node at a stage.

Recurrent nature of this helps in performing various tasks: Speech Recognition, Content Handwritten Recognition, and Unsegmented task.

Current state is described as in (Fig. 3).

$$c_t = f(c_{t-1}, i_t)$$

Applying Activation Function:

$$c_t = \text{activation\_ function}\left(w_{\text{ps}} c_{t-1}, w_{\text{ic}} i_t\right)$$

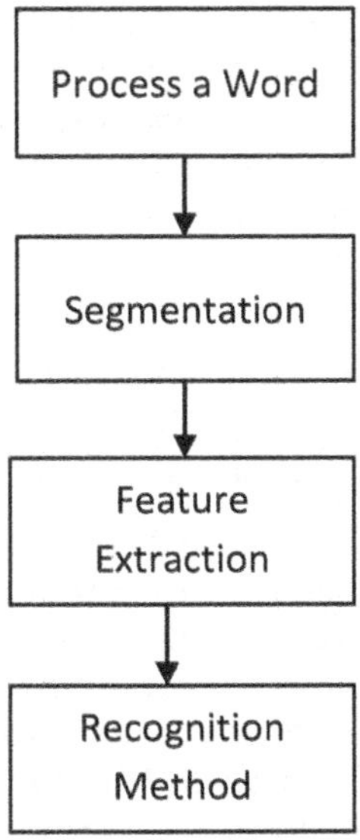

Fig. 2 Steps

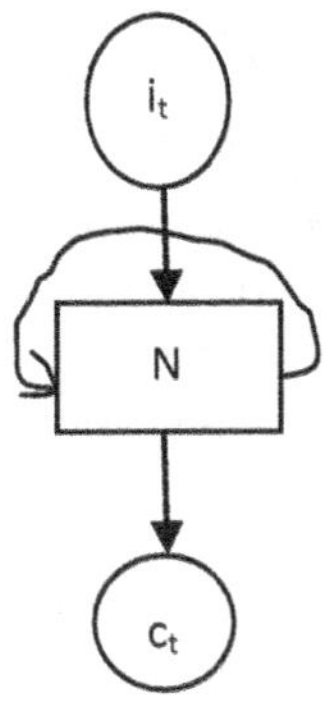

Fig. 3 RNN network

Output:

$$o_t = w_{\mathrm{co}} c_t$$

4 Types of RNN

Different types of RNN are as follows:

One to One: It maps the fixed size of input to the fixed size of output, which does not depend on the previous output. Example: Classifying an image, depending upon the visual.

One to Many: It maps the fixed size as an input of data and provides the output as a sequence of data. Example: Gives the textual description of an image.

Many to One: It gets the sequence of information as an input and provides a fixed size output. Example: Automatically processes the data to detect positive and negative opinions from a text.

Many to Many: It takes the sequence of information and process it recurrently, produced an output a sequence of data Example: In Machine Translation, to convert the one language to another.

5 Construct a Next Word Predictor

Main goal is to predict the next word; which can be used in many applications.

Pre-processing: First step is to make a corpus of unique words. But, it has to be remembered that as the number of unique words in the corpus increased, complexity also increased a lot.

It starts with the collection of how frequently each word co-occurs with one other in a corpus.

Probabilistic Model: Each word is replaced with a frequently used word which makes a vector of it. This vector makes a probabilistic model of it. The result of this provides useful information for the next phase.

Model RNN Architecture: A Recurrent Neural Network (RNNs) is a model which contains a self-connected hidden layer. This gives an efficient method for the connections between neurons and makes a directed cycle.

It makes sequences of data to predict the favorable outcome. Suppose, we are taking follwoing as an input: आपका, दिनand it will predict the output. First, input is provided as x_0, and produces an output h_0. For the next round with input x_1, it produces an output h_1. At last, the output is got from h_t, which demonstrates the next word prediction.

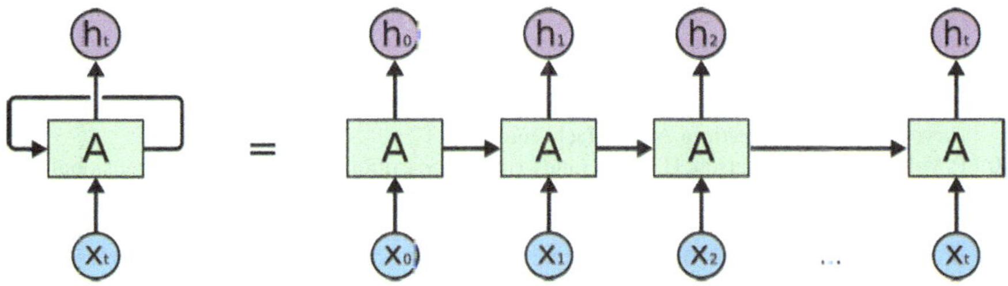

The output particularly depends on the percentage of favorable words and predicts the output:

आपका दिन मंगलमय हो or आपका दिन शुभ हो

ये वक़्त भी गुजर जायेगा or ये वक़्त भी बीत जायेगा

6 Conclusion

This paper demonstrates the development of a structure of Devangari Script using RNN concept. This work will be extended toward online handwritten recognition that will be more beneficial to the user. In the future, we would like to relax the assumption of having pre-segmented images. We also plan to extend this work to include other Indic scripts.

References

1. Medhavi Malik, Bersha Kumari, Madhuri Sharma, "Steps: To Make NLP More Enhance by New Techniques". Int. J. Eng. Adv. Technol. (IJEAT) **8**(2S2), ISSN: 2249 – 8958 (2019)
2. U. Pal, B.B. Chaudhuri, Indian script character recognition: a survey. Patt. Recogn. **37**(9), 1887–1899 (2004)
3. P. Keshri, P. Kumar, R. Ghosh, RNN based online handwritten word recognition in devanagari script, in 2018 16th International Conference on Frontiers in Handwriting Recognition (ICFHR) (2018)
4. M. Sharma, M. Malik, N. Gupta, An efficient & learning approach of POS tagging using rule-based for Devanagari script. Int. J. Innov. Technol. Exploring Eng. (IJITEE) **8**(7C2). ISSN 2278-3075 (2019)
5. R.J. Milewski, V. Govindaraju, and A. Bhardwaj, Automatic recognition of handwritten medical forms for search engines. IJDAR (2009)
6. N. Sankaran, A. Neelappa, C.V. Jawahar, Devanagari text recognition: a transcription based formulation. ICDAR (2013)
7. J. Li, W. Monroe, A. Ritter, M. Galley, J. Gao, D. Jurafsky, Deep reinforcement learning for dialogue generation. arXiv preprint arXiv: 1606.01541 (2016)
8. B. Shaw, S.K. Parui, M. Shridhar, Offline handwritten devanagari word recognition: a holistic approach based on directional chain code feature and hmm, in International Conference on Information Technology (ICIT) (2008)
9. R. Plamondon, S.N. Srihari, Online and off-line handwriting recognition: a comprehensive survey. IEEE Trans. Pattern Anal. Mach. Intell. **22**(1), 63–84 (2000)
10. P. Voigtlaender, P. Doetsch, H. Ney, Handwriting recognition with large multidimensional long short-term memory recurrent neural networks, in 15th International Conference on Frontiers in Handwriting Recognition (ICFHR) (2016)
11. B. Shaw, U. Bhattacharya, S.K. Parui, Combination of features for efficient recognition of offline handwritten devanagari words, in 14th International Conference on Frontiers in Handwriting Recognition (ICFHR) (2014)
12. R. Ishida, An introduction to indic scripts, in *Proceedings of the 22nd International Unicode Conference* (2002)

13. U. Consortium et al., *The Unicode Standard, Version 2.0* (Addison-Wesley Longman Publishing Co., Inc., 1997)
14. N. Sankaran, C. Jawahar, Recognition of printed devanagari text using BLSTM neural network, in 21st International Conference on Pattern Recognition (ICPR) (2012)
15. V. Bansal, R. Sinha, A complete OCR for printed hindi text in devanagari script, in 6th International Conference on Document Analysis and Recognition (DAS) (2001)
16. R.J. Milewski, V. Govindaraju, A. Bhardwaj, Automatic recognition of handwritten medical forms for search engines. Int. J. Doc. Anal. Recog. (IJDAR) (2009)
17. P. Liu, X. Qiu, X. Chen, S. Wu, X. Huang, Multi-timescale long short-term memory neural network for modelling sentences and documents, in *Proceedings of the Conference on Empirical Methods in Natural Language Processing* (2015)
18. A. Maas, R.E. Daly, P.T. Pham, D. Huang, A.Y. Ng, C. Potts, Learning word vectors for sentiment analysis, in *Proceedings of the ACL* (2011), pp. 142–150
19. B. Pangand, L. Lee, Asentimental education: sentiment analysis using subjectivity summarization based on minimum cuts, in *Proceedings of ACL* (2004)
20. L.-C. Yu, J. Wang, K.R. Lai, X. Zhang, Refining word embeddings for sentiment analysis, in *Proceedings of the 2017 Conference on Empirical Methods in Natural Language Processing* (2017), pp. 545–550
21. L. He, K. Lee, M. Lewis, L. Zettlemoyer, Deep semantic role labeling: what works and what's next, in *Proceedings of the Annual Meeting of the Association for Computational Linguistics* (2017)
22. D. Bahdanau, K. Cho, Y. Bengio, Neural machine translation by jointly learning to align and translate. arXiv preprint arXiv:1409.0473 (2014)
23. A. Graves, Generating sequences with recurrent neural networks. arXiv preprint arXiv: 1308.0850 (2013)
24. Y. Kim, Convolutional neural networks for sentence classification, in *Proceedings of the 2014 Conference on Empirical Methods in Natural Language Processing (EMNLP 2014)*, pp. 1746–1751 (2014)
25. C. Dyer, M. Ballesteros, W. Ling, A. Matthews, N. A. Smith, Transition-based dependency parsing with stack long short-term memory. arXiv preprint arXiv: 1505.08075 (2015)

Role of Data Science for Combating the Problem of Loan Defaults Using Tranquil-ART1NN Hybrid Deep Learning Approach

Chandra Shaardha and Anna Alphy

Abstract Banking sector is facing a watershed with the problem of huge NPAs being tackled with various measures such as IPB, Recapitalization to name a few. Credit monitoring is indispensable to protect the lenders' investment. An effective monitoring system will protect and identify the loans which are to be red-flagged. The resources of the bank manager to monitor the loans are limited. The parameters for such surveillance are many. Some of them are available to the banker through systems in place, while some others have to be instituted with the networking capabilities and oversight of transactions of borrowers. With thousands of loans at various stages of fulfillment, it would be a herculean and cumbersome task for the conventional RDBMS and SQL systems to segregate the loans into categories that require better attention and follow-up. This paper aims to present the possibilities with the banking managers to make use of the advancements and also proposes Tranquil-ART1NN approach using Artificial Intelligence (AI) to give advanced signals to bank officers regarding NPAs/Special Mention Accounts. Tranquil algorithm is meant for the identification of an initial better neighborhood and the Intelligent Huddle Creation algorithm is meant for creating clusters of borrowers based on their performance. The clusters of borrowers are trained using ART1NN approach to identify the optimized clusters of borrowers for early identification of NPAs. The quality of the proposed system is compared with various traditional algorithms like Genetic clustering (GC), K-means, Ant Colony Optimization (ACO), etc.,and the *experimental* results show that our proposed system has a better performance in terms of purity and DB Index, when compared to K-Means, ACO, and GC. The early detection of default will enhance the profitability, liquidity of the banks. Technology will help bankers to eliminate the gap between resources and workload available and will lead to improved risk management and efficiency of the banks.

C. Shaardha (✉)
Department of Management Studies, SRMIST, Delhi-NCR Campus, Ghaziabad, UP, India
e-mail: chandras@srmist.edu.in

A. Alphy
Department of Information Technology, SRMIST, Delhi-NCR Campus, Ghaziabad, UP, India

S. S. Dash et al. (eds.), *Intelligent Computing and Applications*,
Advances in Intelligent Systems and Computing 1172,
https://doi.org/10.1007/978-981-15-5566-4_54

Keywords Lending · Credit monitoring · Defaults · NPAs · Borrower profiles · Pheromones · Neural network · Adaptive resonance theory · Ant colony

1 Introduction

"Money," is the vitamin source for human life. "Fund," another term of Money is the essential source for any business development activities. In the banking scenario, lending funds to an individual, organizations, and institutions is the major role to play, to earn profit for the investors. When the borrower defaults in repayments, it affects the profitability and liquidity of the banks.

Indian banking sector is under pressure to control the defaults of repayments and dues from the borrowers. As per RBI guidelines, when the borrower fails to repay the principal/interest for 90 days and when the loan account ceases to generate income, the loan has to be classified/termed as "Non-Performing Assets" (NPAs). As on date, the rising NPAs percentage gives increasing pressure to the banking sector. Gross Non-Performing Assets (GNPAs) of Public Sector Banks (PSBs) have reduced by Rs. 89,189 crores from Rs. 8,95,601 crores in March 2018 to Rs. 8,06,412 crores in March 2019. The GNPAs as a percentage to overall loans remained at 9.3% as of 31 March 2019.

One of the major reasons for the increase in the NPAs percentage is the lending design adapted to the banking sectors. The Nirav Modi's Letter of Guarantee fraud is the latest case in point to illustrate the faulty methods of lending by the banking institutions.

RBI has taken various measures to curtail the NPAs percentage over years. But, still it is very difficult to control as well to monitor the stressed assets. Credit rating agencies have also initiated different methods to analyze the creditworthiness of the borrower and submit the report to banking institutions before the disbursement of loans. Even after the compliance of the credit rating agencies report, the borrower makes default in repayment after availing the loans. The bankers are only giving importance to analyze the borrower's background before the disbursal of loans. Once loans get disbursed, due to the workload/work pressure/manpower shortage, bankers do not have much time to see or have an eye on the repayment of dues on time in the loan accounts. Hence, it is important to report the issue of non-performance much in advance before the loans convert to a stressed or Special Mention Account.

It is important to make certain the early recognition of loan repayments in the borrower's account. The bankers should be trained to identify the borrower's accounts with the help of symptoms they could see during the transactions. It becomes necessary to have a proper mechanism to identify the factors that could avoid the loan turning into NPAs. When the symptoms are seen in the borrower's loan accounts, it is the duty of the officer to make an honest assessment of that account and give a proper alarm or follow-up to initiate the steps to identify the reason behind the cause.

Generally, the steps are not taken on time due to various issues. It becomes mandatory for the bankers to incorporate the monitoring procedures with the risk management process to ensure that the early warning symptoms are duly recognized and followed up with due diligence to recover the dues without any default

To identify the preventive measures of loans turning to be NPAs, RBI have given the guidelines to monitor the loan accounts. The checkpoints are given below, where bankers should be cautious about

1. Non-proper submission of (a) Stock statement, (b) Balance sheet, (c) P&L accounts
2. Cheque Returns
3. Default in deferred payment guarantee
4. Default in payment of Letter of Credit
5. Continuous default in payment of Bank Guarantee
6. Return of Cheque/Bills discounted
7. Default in payment of Bills/Cheque discounted
8. Decline in Sales/Profit
9. Cash/Fund loss
10. Decline in Net worth
11. Incomplete documentation
12. Not fulfilling the terms and conditions
13. Special Mention Accounts I, II.

This paper tries to identify the possibilities of utilizing the latest technology of the information systems with the help of Machine learning/Artificial Intelligence to prompt/give a signal to bank officers in advance about the capabilities/capacity of the borrower to make the repayment of loan due before it turns into NPAs/Special Mention accounts.

2 Related Works

Author Naveenan [1] in his paper titled "Warning signals—A Tool to control NPAs in banks" has strived to analyze various opinions of bank managers about the warning signals identified by the Reserve Bank of India in predicting slippages and the resultant NPAs. He has analyzed four types of warning signals, namely Financial, Operational, Banking, and Managerial. Under each category, the author has chosen different parameters. For example, under the managerial category, he has narrowed on diversion of funds, laxity in financial regime, poor interpersonal teamwork, miscalculated risks, etc. Using non-parametric Friedman test, he concludes that ranking based on means of warning signals in each category would be a valid approach.

HackerEarth [2] hosted a hypothetical problem of a bank facing huge bank defaults. It challenged hackers to develop machine learning solutions to help the bank. The challenge was attempted by more than 3000 ML enthusiasts across the world. The solution of top performers with their statements is described in the site.

The participants have used feature engineering to create new features such as to estimate the number of periods when the loan was paid, ratio of insurance paid to the total insurance due, etc., and used ML algorithms such as XGB.

Dai et al. [3] introduced particle swarm chaos optimization mining algorithm (PSCOMA). It used the strong global search ability of PSO and the strong local search ability of chaos optimization for the process of web usage mining. It offered a balancing between exploration and exploitation. Kennedy et al. [4] presented particle swarm optimization (PSO) which is an evolutionary swarm intelligence-based computational model PSO inspired by bird flocks. Here, each swarm represents a solution set. The swarms or particles fly through the solution space. Each position of the particle in the problem space represents a solution. At each move, a fitness function is evaluated to identify the closeness of particle solution to the global optimal solution. Cui et al. [5] introduced a bio-inspired clustering model called the Multiple Species Flocking clustering model (MSF). The flocking behavior consisted of making decisions on movements based on certain rules that depended on an action in response to the neighboring members and the environment it wit. In this algorithm, each document vector was anticipated as a boid in 2D virtual space. Azzag et al. [6] introduced an AntTree algorithm directly inspired by the ants' self-assembly behavior and its application to the unsupervised learning problem. This biologically inspired clustering algorithm proceeded in a distributed way to build a tree-structured organization of the data. Ramos et al. [7] introduced a clustering algorithm based on corpse clustering. In certain species of ants (Messor Sancta) for cleaning up their nest, they sort their larvae or form piles of corpses to form cemeteries.

3 Proposed Tranquil-ART1NN System

Data mining is the method to recognize patterns and to find liaisons to elucidate problems via data analysis. It allows predicting future behavior. In the proposed Tranquil-ART1NN system, we use simulated annealing approach to identify the initial clusters of similar borrowers. The clusters obtained are optimized through the ART1NN approach. It includes the following steps.

1. Preprocessing the banking data to identify borrower periods.
2. Pattern discovery based on data mining methods [8–12]
3. Generate borrower profiles from clusters
4. Trajectory of budding Borrower Profiles.

Figure 1 describes the steps involved in web usage mining.

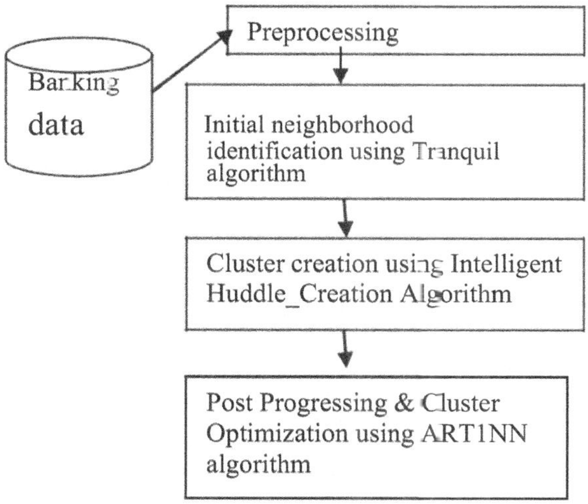

Fig. 1 Steps involved in early identification of NPA using Tranquil-ART1NN approach

3.1 Preprocessing the Banking Data to Extract Borrower Periods

Data mining is the automatic discovery of borrower activity.

The first step in data mining processing is the preprocessing of input data available.

After availing of the loans, some borrowers make defaults in repaying loans installments/interest. When they default in repayments for 90 days, then the accounts become NPAs. To avoid or to identify or to alert, the bank managers have to follow the parameters given below.

These are the guidelines by RBI which are mandatory to adhere to in the system of banking. When the bank manager fails to identify the non-receipts of the statements mentioned below, these are considered for creating binary borrower vector 'ƛi'

Non-proper submission of Stock statement. Balance sheet
P&L accounts Cheque Return
Default in Deferred Payment Guarantee—Default in payment of Letter of Credit
Continuous default in payments of Bank Guarantee Return of cheque discounted
Return of Bills discounted
Default in payment of Bills Discounted Default in payment of Cheque discounted
Decline in Sales and profit
Cash Loss Fund loss
Decline in Net Worth
Incomplete documentation
not full filling the terms and conditions Special Mention Account 1
Special Mention Account 2

$ƛi^{j} = \{1$, if borrower did a valid submission 0, otherwise
For the ith *borrower and* jth *parameter.*

This binary input vector is given as input for the proposed data mining process to identify the borrower who can fall into NPA category. This binary vector is given as an input to the proposed tranquil algorithm.

3.2 Tranquil Algorithm for Identifying Initial Borrower Segregation for Early Detection of NPA Assets

Initially, each borrower ƛ is placed randomly on the visualization panel. Visualization panel is a two-dimensional plane represented by *x–y* coordinates. The *x*-axis and *y*-axis values range from 0 to 1. To convey the similar borrowers nearer and the dissimilar borrowers far apart in the visualization plane, the proposed Tranquil algorithm based on simulated annealing concepts used in metals is used.

In annealing, a metal is heated to its melting point and then cooled back to solid state. The eminence of the annealing process depends upon in what way the cooling is performed. Slow cooling fallouts in low energy state with stable crystals, whereas hasty cooling results in high energy state with imperfect crystals. The parameters of thermodynamic annealing can be mapped into simulated annealing approach, in which a state of the system epitomizes viable solutions, energy epitomizes outlays, transformation of state epitomizes neighboring function, temperature epitomizes control parameters, and the final state is represented by the frozen state.

The proposed Tranquil algorithm (Algorithm 1) uses a greedy heuristics allowing the μs to travel from recent positions to the superlative neighboring solution.

$\mu = \mu 1, \mu 2, \mu n$ represents the activities of a borrower

Algorithm 1: Tranquil algorithm

Notations used:
Ʊ= borrower ƛ position on the visualization plane
ł = temperature length,
Ʊ *= borrowers updated locus on the visualization plane
U_i = ith borrower on the visualization panel
Pos (U_i) = position of borrower on the visualization plane
Ѝ= total number of borrowers on the x-y visualization plane
Input: U_i where *i*=1 to Ѝ, Ʊ, ł, Ѝ
Output: Ʊ*

Engender a primary solution Ʊ $_0$ ϵ Ʊ *= Ʊ $_0$ // *ϵ is the element of operator*
Engender primary temperature ł >0

Repeat

For 1<= *temp*<= *t*

Identify λ_j neighboring Ʊ* *using Euclidean distance* $ED(\lambda_i, \lambda_j) = \sqrt[2]{\sum_{x=1}^{n} |\lambda_i^k - \lambda_j^k|^2}$

Ḓ= *Sim* (λ_i, λ_j) - Simλ_i, λ_{i+1}) // where *Sim* (λ_i, λ_j) = $\frac{\lambda_i \cdot \lambda_j}{\|\lambda_i\| \|\lambda_j\|}$,

Where $\lambda_i \cdot \lambda_j = \sum_{k=1}^{n} \lambda_i^k \lambda_j^k$ $\|\lambda_i\| = \sqrt{\lambda_i \cdot \lambda}$

If Ḓ <=0 //downhill move

Ʊ *= {λ_j, Pos (λ_j)}

Else

IF random (0, 1) < e^ (Ḓ/*T*) then Ω*= {λ_j, Pos (λ_j)}

End if

Increment i by one.

End for

temp=Ÿ**temp*//reduce temperature

Until bottom value of temperature or looping doesn't accept a new solution.

Return Ʊ*

The inputs to the algorithm Tranquil are borrower vector μ, their locus on the visualization panel, and the temperature span. Initially, the primary solution ƛ0 is generated randomly and this is allocated as the final solution ƛ*. Now, the primary temperature value is created. At this juncture, a new solution is bent by selecting neighboring ƛ*i* that is analogous to the solution ƛ*. A budget function (Ḓ) is calculated with a cosine similarity of input vectors with an idea of maximizing the cosine similarity (*Sim*) of ƛ. Cosine similarity, which is able to handle the qualitative and quantitative data between any two ƛs, signifies the similarity between borrowers that are mapped to that ƛ. It can also deal with high-dimensional sparse data. Then the budget function is calculated and checked for the decrease in the same. The budget functions are considered as the verve for simulated annealing. If the verve is dwindled, the newfangled is accepted. Otherwise, we accept the newfangled state with probability *e*^ (Δ/*T*). A linear temperature lessening with $0.8 \leq \acute{R} \leq 0.99$ is used here. The T is adjusted in such a way that the number of iterations in higher temperature are small and the larger number of iterations at the lower temperatures. To make the final solution sovereign of the starting one, the initial temperature should be high enough. There will be no uphill moves when the temperature is low.

The Tranquil algorithm returns the ƛs with their new position on the visualization panel. Now the alike ƛs lie nearer in the visualization panel. As the aloofness between ƛs on the visualization panel increases, their likeness decreases. The uphill moves are supported to avoid local minima. As a result of the Tranquil algorithm, ƛs converge to a freezing profound state, where similar ƛs are located nearby.

In this proposed method, when the temperature is high, bad moves are accepted due to the difficulty in escaping from the neighborhood. When the temperature is low, bad moves are rejected and the finest results are kept in final solutions.

The usage of Tranquil algorithm reduces the number of iterations thereby decreasing the time to identify the default behaviors. The Tranquil algorithm identifies the borrowers with similar loan payment behavior and keeps them as neighbors. The 'doubtful category' borrowers will be grouped into the same cluster thereby increasing the easy identification of malfeasance by the borrowers.

The borrowers with the updated positions are given as input to the Intelligent_Huddle Creation algorithm. In this algorithm, each borrower is mapped to an intelligent agent. The intelligent agents have the self-learning capability to identify clusters of borrowers.

Algorithm 2: Intelligent_Huddle Creation Algorithm ***Notations used***:

Notations used:

Pos(ƛ) = *locus* of agent on the visualization plane
d_{th}=distance threshold.

Input: agent_$ƛ_i$, Pos (agent_$ƛ_i$)
Output: Clusters *C1, C2*... C_N

Read the agent_$ƛ_i$
Assign the ƛ to cluster C_k
For all $ƛ_j$ do
If (Distance (agent_$ƛ_i$, .agent_$ƛ_j$)) < $d_{th}$and σ_k^2 <σ_{th}^2 // σ_k^2 *is calculated using equation (1)*
Assign agent_$ƛ_j$ to C_k
Else
Assign agent_$ƛ_j$ to C_{k+1}
End if
End for

In Intelligent_Huddle Creation Algorithm (algorithm 2), Distance (agent_*ƛi*, agent_.*ƛj*) represents the distance between agent_*ƛi* and agent_*ƛj* on the visualization panel, σk^2 represents the mean squared error or average dissimilarity between the cluster prototype and the data records. Mean squared error is calculated using Eq. (1).

$$\sigma_k^2 = \frac{\Sigma_{S(i) e \chi x_k} d_{ki}^2}{d_{ki}^2} \tag{1}$$

where $S^{(i)}$ represents the *i*th borrower, *k* represents the set of borrowers assigned to *k*th cluster, *dki* is the distance from $S^{(i)}$ to k. Initial clusters of borrowers are formed by grouping the agent_ƛs that lie within a distance threshold and whose mean squared error lies within σth^2 into a cluster.

Thus, the ƛs within a cluster represent similar borrowers. These clusters are given as input for the ART1NN approach.

In the intelligent_HuddleCreation algorithm, borrowers having similar loan repayment behavior will be grouped into the same cluster. The intelligent agents can identify borderline behavior such as 'special mention' category. It can also identify the borrowers with genuine interest and they are grouped into the same cluster. The self-learning, decision-making, and dynamic behavior of intelligent agents allow identifying the borrowers who take multiple loans and regularly repaying all the loans but not concentering on a single loan so that the borrower cannot be categorized as NPA.

To improve the quality of identification, the clusters obtained as the out of the Intelligent_HuddleCreation algorithm are given as input to the ART1NN.

3.3 Postprogressing Using ART1 Neural Network Approach

In the postprogressing phase, the input is the clusters generated by Intelligent_HuddleCreation algorithm. It gives better refined and optimized clusters as output so that the slippage of NPA assets can be avoided.

ART1NN helps to cluster binary vectors. It is an unsupervised learning. The algorithm finds a pattern and proceeds to examine the members to be included in the cluster or for the creation of another cluster based on a distance threshold. This is iterated to cover the entire dataset.

The ART1NN approach operates on three levels. The first level is the identification phase. In this phase, the existing output classification is matched with the input vector. The output of the neuron becomes '1' for the best match and '0' otherwise. This level is called the recognition phase. The second phase is the Collation phase, in which the input vector is compared with the collation layer vector. A reset function will be applied if the degree of likeness is lower than the vigilance parameter. The third level is the Exploration level. Here, the neural network will examine for the no reset and good match. If this is obtained, then the network will consider as the classification is over. Otherwise, the same process is repeated with the other stored patterns to identify the best match.

Algorithm 3: ART1NN algorithm

Notations Used:

α= Number of attributes in the input vector
β=Maximum No. of Clusters that can be formed
μ=Vigilance Parameter
Wij=bottom-up weights.
Xij=top-down weights

Input: Clusters C1,C2,…CN
Output: Optimized Clusters C1,C2,…CM, M<N

Initialize the learning rate r as r>1, 0<=μ>1,

$0<W_{ij}(0)<\frac{r}{r-1+\alpha}$ and Xij(0)=1

Repeat
  For every training input
    Set activations of F1(a)=input vectors and F2=0
    Sent input signal from F1(a) to F1(b) as $sig_i = y_i$
      For each subdued F2 node
        $X_j = \sum_i b_{ij} y_i$ // $X_j \neq -1$
    End for
      Repeat
        Identify J for $X_J \geq X_j$
        Calculate activation on F1(b) as $y_i = sitJ_i$
      Until($\|v\|/\|sig_i\| < \mu$)
      $W_{ij}(new) = \frac{r y_i}{r-1+\|y\|}$
      $X_{ij} = y_i$
  End For
Until stopping condition is not true

The stopping condition of the above algorithm includes no weight change, maximum number of iterations reached, or no reset is performed.

4 Performance Analysis

Figure 2 shows the 2D visualization of the Visualization plane. Initially, borrowers are placed randomly on the visualization plane.

Figure 3 shows the output clusters generated by the Intelligent_HuddleCreation algorithm: borrowers with similar behavior are grouped into the same cluster.

In Fig. 4, the refined and optimized clusters obtained after applying the ART1NN approach is shown.

Here, quality is evaluated in terms of purity and DB Index.

Davies–Bouldin(DB) index [13] calculates the average similarity between each cluster and its most similar one, in a clustering result. Let $Clus_i$ and $Clus_j$ be two

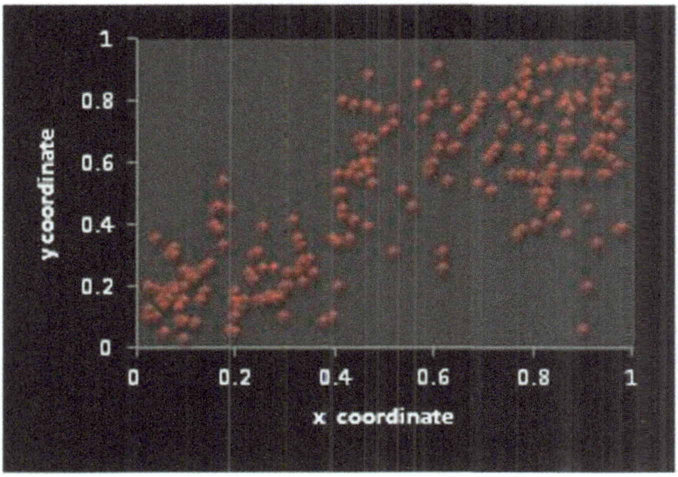

Fig. 2 Random placements of borrowers in visualization panel

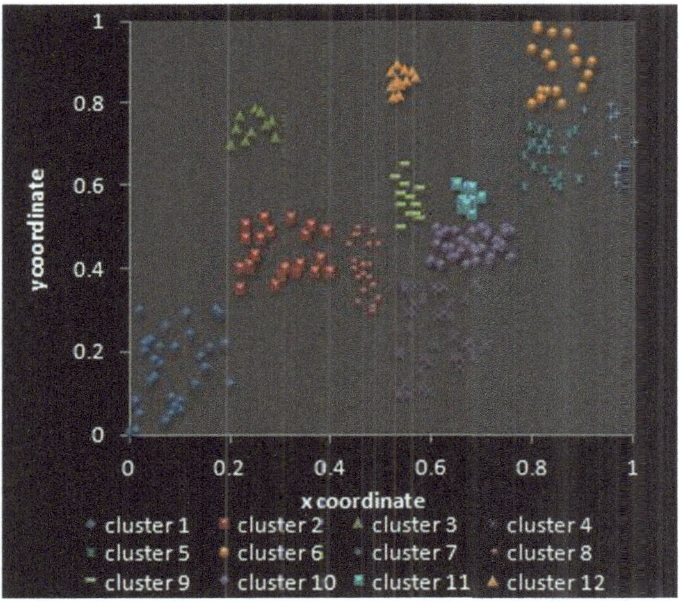

Fig. 3 Intial clusters of borrowers using Intelligent Huddle_Creation algorithm

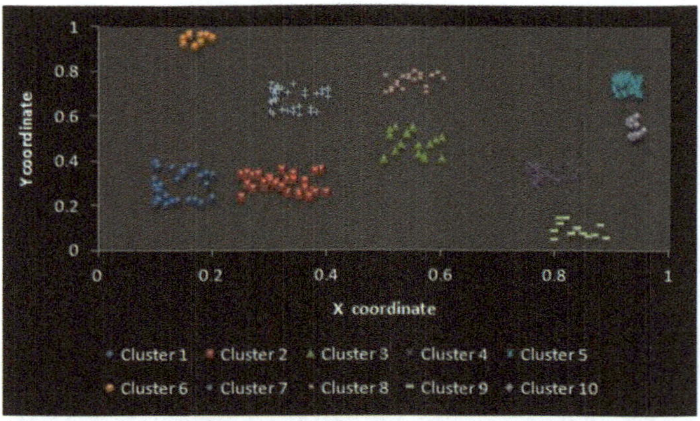

Fig. 4 Optimized clusters of borrowers using ART1NN approach

clusters in a clustering result. i represents the measure of dispersion in a cluster Clus_i and dij is the cluster dissimilarity measure. To calculate the similarity measures between the clusters (Sij), it has to satisfy the following conditions.

$$S_{ij} > 0$$
$$S_{it} = S_{jt}$$

If $i = 0$ and $j = 0$, then $S_{ij} = 0$

If $j > k$ and $d_{ij} = d_{kj}$ $d_{ij} = d_{kj}$, then $S_{ij} > S_{ik}$ If $j = k$ and $d_{ij} < d_{kj}$, then $S_{ij} > S_{ik}$

$$S_{ij} = \frac{x_i + x_j}{d_{ij}} \tag{2}$$

DB index is defined by

$$\text{DB}_{n_c} = \frac{1}{n_c} \Sigma_{i=1}^{n_c} s_i, \tag{3}$$

where $S_i = \max\limits_{i=1-n_o i \neq j} S_{ij}, i = 1, \ldots n_c$. n_c is the number of clusters.

Since DB index represents the average similarity between each cluster and its most similar one and it is desirable for the clusters to have the minimum possible similarity to each other, a clustering that minimizes DB is more desirable (Fig. 5).

From the above figure, it is observed that DB Index is higher for Tranquil-ART1NN approach, when compared to K-means, Genetic clustering, and Ant Colony Optimization (Fig. 6).

Purity [14] is calculated using the equation given below

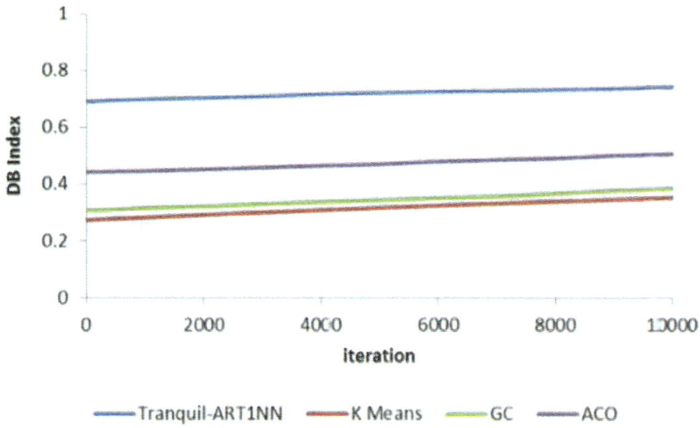

Fig. 5 Quality comparison in terms of DB index

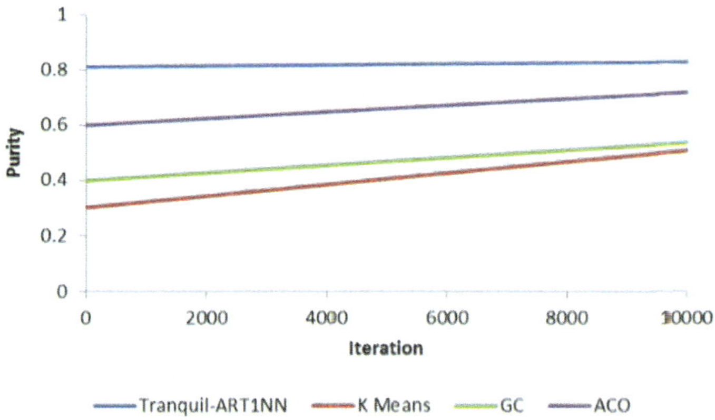

Fig. 6 Quality measure in terms of purity

$$\text{purity}_j = \max p_{ij} \tag{4}$$

The overall purity of a clustering approach is given by

$$\text{purity} = \sum_{i=1}^{k} \frac{m_i}{m} \text{purity}_j \tag{5}$$

It is noticed from the above figure that purity is higher for Tranquil-ART1NN approach when compared to K-means, Genetic clustering, and Ant Colony Optimization

5 Conclusion

The complex problem of NPA demands superior techniques for resolution. With myriad parameters affecting the issue on the one hand and the rich experience of the regulators and the industry to combat the NPA menace on the other, the meeting ground can be none other than the ground of AI which throws up various tools to apply the experience to the problem domain. An attempt has been made in this paper, to apply the rigorous AI algorithms to this problem domain of predicting defaults.

References

1. Naveenan, R.V., Warning signals—a tool to control NPA in banks. Int. J. Adv. Res. Comput. Sci. Manage. Stud. 4(7) (2016). Available at www.ijarcsms.com
2. 'Bank Fears Loanliness', Available at https://www.hackerearth.com/problem/machine-learning/bank-fears-loanliness/
3. L. Dai, W. Wang, W. Shu, An efficient web usage mining approach using chaos optimization and particle swarm optimization algorithm based on optimal feedback model. Math. Prob. Eng. (2013). Article ID: 340480
4. J. Kennedy, R. Eberhart, Particle swarm optimization, n Proceedings of IEEE International Conference on Neural Networks, vol. 4, 1995, pp. 1942—1948
5. X. Cui, T.E. Potok, A distributed agent implementation of multiple species flocking model for document partitioning clustering model for document partitioning clustering. Coop. Inf. Agents X **4149**, 124–137 (2006)
6. H. Azzag, N. Monmarche, M. Slimane, G. Venturini, Anttree: a new model for clustering with artificial ants, in 2003 Congress on Evolutionary Computation, 4 Dec 2003, pp. 2642–2647
7. V. Ramos, F. Muge, P. Pina, Self-organized data and image retrieval as a consequence of inter-dynamic synergistic relationships in artificial ant colonies. Soft Comput. Syst. Des. Manag. Appl. Frontiers Artif. Intell. Appl. **87**, 500–509 (2002)
8. Cooley, R., Mobasher, B., Srivastava, J., Web mining: information and pattern discovery on the world wide web, in Proceedings of Ninth IEEE International Conference on Tools with AI (ICTAI '97), 1997, pp. 558–567
9. Nasraoui, O., Krishnapuram, R., Joshi, A., Mining web access logs using a relational clustering algorithm based on a robust estimator, in Proceedings of Eighth International World Wide Web Conference (WWW '99), 1999, pp. 40–41
10. O. Nasraoui, R. Krishnapuram, H. Frigui, A. Joshi, Extracting web borrower profiles using relational competitive fuzzy clustering. Int. J. Artif. Intell. Tools **9**(4), 509–526 (2000)
11. J. Srivastava, R. Cooley, M. Deshpande, P.N. Tan, Web usage mining: discovery and applications of usage patterns from web data. SIGKDD Explor. **1**(2), 1–12 (2000)
12. O. Nasraoui, M. Soliman, E. Saka, A. Badia, R. Germain, A web usage mining framework for mining evolving borrower profiles in dynamic web sites. IEEE Trans. Knowl. Data Eng. **20**(2), 202–215 (2008)
13. D.L. Davies, D.W. Bouldin, Cluster separation measure. IEEE Trans. Pattern Anal. Mach. Intell. **1**(2), 95–104 (1979)
14. P.H. Choua, P.H. Lib, K.K. Chenc, M.J. Wua, Integrating web mining and neural network for personalized e-commerce automatic service. Expert Syst. Appl. **37**, 2898–2910 (2010)

Optimized Multi-Walk Algorithm for Test Case Reduction

U. Geetha, Sharmila Sankar, and M. Sandhya

Abstract Software testing is the most expensive process of the development and maintenance phase of the software product. Whenever changes are made in the code, it becomes very difficult to execute all the test cases in regression testing. Therefore, that testing process needs some test case reduction and prioritization techniques to improve the regression testing process. Although there are many existing techniques for test case reduction as well as for prioritizations, this paper will take its focus on the multi-level random walk algorithm which has been used for test case reduction. In a multi-level random walk selection algorithm, test case selection for further reduction is done randomly on every iteration that degrades the performance of the testing process in terms of coverage and will also generate a situation for random test case tie. To overcome this situation of random test case selection and handling test case tie, a solution is being proposed in this paper which includes a combination of optimized multi-level random walk and optimized algorithm (Genetic algorithm). This proposed approach proves that the combination of both the techniques improves the overall performance better than the existing multi-level random walk.

Keywords Test data generation · Test reduction techniques · Test case optimization

U. Geetha (✉)
Department of Information Technology, B.S. Abdur Rahman crescent Institute of Science and Technology, Chennai, India
e-mail: geetha_it_phd_17@crescent.education

S. Sankar · M. Sandhya
Department of Computer Science and Engineering, B.S. Abdur Rahman crescent Institute of Science and Technology, Chennai, India
e-mail: sharmilasankar@crescent.education

M. Sandhya
e-mail: sandhya@crescent.education

S. S. Dash et al. (eds.), *Intelligent Computing and Applications*,
Advances in Intelligent Systems and Computing 1172,
https://doi.org/10.1007/978-981-15-5566-4_55

1 Introduction

Software testing is the process by which bugs and faults can be found in a software product. It can also be done through the verification and validation process in the early and late stages of software development. This can be done in functional and non-functional development of the product. Regression testing is more important in the maintenance phase of the software product that affects the cost of the testing process. Still many issues are faced by regression testing in terms of resources, time, and cost. So, to reduce the cost of regression testing, software testers introduced the concept of prioritization of their test cases. Prioritization is done to find the useful and representative set of test cases, by some measures, which are made to run on earlier phases of the regression testing process. Another important goal of prioritization is to increase a test suite's rate of fault detection. An extensive review has been carried out to find faults in the existing literature which says that the existing reduction technique used in this paper faces some problems with test case tie due to the random selection of test cases on every iteration, thus making the overall test suite complex. Therefore, to improve the test suite and reduce the complexity by maintaining the coverage ratio, some optimization techniques have been introduced. It considers two optimization techniques which includes the combination of optimized multi-level random walk and optimized algorithm (genetic algorithm). One of the most common algorithms for test case selection is the random walk algorithm that uses local optima and backbone test cases to simplify the original problem into small problems by removing the shielded test cases that means test case reduction problem. On the other hand, to improve the ordering of test cases and to reduce the prioritized test cases, an optimization algorithm (genetic algorithm) is used which is very powerful and is a widely used stochastic search process. Genetic algorithm (GA) is an evolutionary algorithm based on natural selection. It is used to find approximate solutions for optimization and search problems [4]. The aim of GA is to achieve better results through selection, crossover, and mutation. Genetic processes are most effective for optimizing test cases. Due to this, the proposed approach uses genetic and multi-walk random algorithms for reduction problems.

2 Problem Description

Multi-level random walk is a software test case reduction technique that is taken as a focus area in this research. It tries to find an optimal and refined solution for the original problem instance [2, 3]. But this algorithm still has some problems associated with it. On each level, a search is being performed, selection of random test cases is made, and this selection of random test cases increases the complexity of the entire test suite. Moreover, re-execution of test cases will affect the regression testing which is a very time-consuming and expensive process. Also, there will be a situation when the random test cases will meet a test case tie at some point in time due to which the statement coverage ratio will also get affected. To overcome all these scenarios and to make the overall test suite more effective, a solution that is preferred is the incorporation of optimization technique (genetic algorithm) with the existing reduction and prioritization technique.

2.1 *Existing Methodology*

Test case reduction is more important for regression testing because the number of test cases affects the cost of the regression testing process. In this situation, the system needs effective test cases from the original test suite to check whether existing product is getting affected by the modified one.

Multi-level random Walk algorithm is used in this paper for test case reduction [5].

One of the most common algorithms for test case selection is the random walk algorithm that uses local optima and backbone test cases to simplify the original problem into small problems by removing the shielded test cases which means test case reduction problem.

In each level, a random walk is made and an intersection or the common part is locked and discarding those test cases which are not locked or unshielded. However, this algorithm reduces the problem through random selection during the selection process that removes some effective test cases. To overcome the problem, the proposed approach uses genetic and multi-walk algorithms for optimizing test cases in spite of random selection [7, 8].

This time, the solution obtained by the multi-level random walk is not much optimized as the statement coverage ratio is not maintained properly. So some optimized algorithm can be thought of invocation with this.

Let us take a look at how it works:

Initially, we will take a coverage matrix with some assigned weights and test cases satisfying some of the statements at each level of the matrix.

(i) ***Initial Coverage Matrix***

The steps given below are followed in the multi-walk algorithm for the reduction process (Table 1).

Step 1: Selection of two random test cases, i.e., {T3, T7}
Step 2: Selection of these test cases is covering statements {S2, S6}
Step 3: These test cases and the statements that are taken into consideration will become 0 and
will contribute to reduced level 1 matrix.
Step 4: Now, these covered statements and test cases will get a locally optimal solution by finding
shielded test cases.

Table 1 Initial Coverage Matrix

Weights	0.5	0.7	0.8	0.2	0.4	0.6	0.8
Test cases	S1	S2	S3	S4	S5	S6	S7
T1	1	0	0	1	1	0	0
T2	0	0	1	0	1	1	1
T3	0	1	0	0	1	1	0
T4	1	1	0	1	0	0	1
T5	1	1	1	0	1	0	0
T6	0	0	0	1	1	1	1
T7	0	1	0	0	0	0	0

Step 5: Here, the shielded test case taken is {T6} which is shown in the reduction level 1 matrix.

Step 6: Further,

the next random test cases selected are {T4, T5} covering statements {S1, S3, S4, S5, S7}.

Step 7: Reductive level 1 will give a reduced and refined matrix but the solution so obtained is

Not the expected optimal solution according to the statement weightage covered and

Remaining test cases will not even contribute to any statements and test case coverage (Table 2).

(ii) ***Reductive Level 1 Matrix***

Backbone Test cases for initial coverage matrix: {T3, T7}

Shielded Test case: {T6}

Backbone test cases after reductive level 1 matrix: {T4, T5}

Therefore, after covering most of the test cases and statements, the entire matrix is on the verge of

becoming 0 throughout and the optimal solution found by test cases **{T3, T4, T5, T7}**.

Table 2 Selection of Shielded Test Case

Test cases	S1	S2	S3	S4	S5	S6	S7
T1	1	0	0	1	1	0	0
T2	0	0	1	0	1	0	1
T3	0	0	0	0	0	0	0
T4	1	0	0	1	0	0	1
T5	1	0	1	0	1	0	0
* T6	0	0	0	1	1	0	1
T7	0	0	0	0	0	0	0

(iii) ***Test Case Reduction Percentage***: After performing the multi-level random walk, the test case reduction percentage comes out to be **57%** which has more probability of getting enhanced and improvised.

(iv) ***Effectiveness of test cases in terms of statement weightage***:

If we find the statement coverage and its effectiveness in terms of statement weightage, it comes out to be: **4.7**

T3 = 1 * 0.6 = 0.6
T4 = 1 * 0.5 + 1 * 0.3 + 1 * 0.2 + 1 * 0.8 = 1.8
T5 = 0.5 + 0.3 + 0.8 + 0.4 = 2
T7 = 0.3,
Total = 0.6 + 1.8 + 2 + 0.3 = 4.7

2.2 Proposed Approach

A comparison being performed proves that the GA is more agile and profound in terms of coverage weightage w.r.t multi-level random walk. But to get the optimal solution, a proposed approach is thought of which describes for a combination of multi-level random walk and optimized algorithm (genetic algorithm). The proposed approach of the combined algorithm is more effective and improvised than the existing methodology.

GAs or Hereditary Algorithms are effective and broadly relevant stochastic pursuit and advancement strategies in light of the thoughts of normal choice and characteristic assessment. It chips away at a populace to the advancement issue These issues which either can't be defined in correct or in an exact numerical shape may contain uproarious or sporadic information or they basically can't be settled by the customary computational techniques [7, 9].

(i) Algorithm # Combination of optimized multi walk algorithm and genetic algorithm:

Input:
T = < T1.......Tn > set of test cases
S = < S1........Sn > set of statements

R1—Reduction level 1

Output:

S1—Reduced test-suite

for i = 1 to R1
F = call genetic ()
bi = bi U F

Collect Refined and reduced test cases covered by F

```
        and reduce instance
      Find shielded test cases (SHi)
        Discard SHi in T
      End
        Return (T = F)
  Int[] genetic ()
       //initial population &return value
     For every Test case Ti in Tn and for every statement Si in Sn
      If (Ti covering Si)
        Mark = 1;
        A[i] = A[i] +w;
      Else
        Mark = 0;
```

1. Select two test cases which has the highest weight and find the target
   While (target not reached)
2. If (! target)
   Crossover (T1 (2), T2 (6))
3. If (! target)
   Mutation (Crossover (1))
4. if (! target)
   Select highest fitness one and one from the existing suite

```
     Repeat step 1
      Else
        return (Selected test cases)
  End algo
```

(ii) **Configuration setting for GA** (Table 3).
(iii) **Reduction process**

1. **Initial Population**: The bits stored in the form of '0' or '1' in the coverage matrix are the chromosomes here for the reduction techniques.
2. **Fitness Values**: The fitness values can be calculated using the associated test cases by the formula **F (t1)** $\sum_{s1}^{sn}$ **wi$*$ coverage statement(Si)** (Table 4).

For further manipulation, we will select those test cases which have the highest weighted fitness value.

Table 3 Experimental Setup

Genetic steps	Parameters
Initial population	Statement coverage information is used with respective chromosomes as '0' (or) '1'
Fitness value	Calculated by fitness value F(t1) $\sum_{s1}^{sn}$ wi* coverage statement(Si)
Selection	Tournament-based selection
Crossover	1st and 2nd of First fitness value and 4th and 5th of 2nd fitness value of the test cases
Mutation	5th bit of crossover result (discrepant bit)
Termination	New four test cases to be tested that should have 75% coverage value

Table 4 Fitness Value

T1	T2	T3	T4	T5	T6	T7
1.1	2.6	0.6	1.8	2	2	0.3

3. **Selection**: Selection of the test cases is done based on Tournament-Based Selection. Select any two test cases and perform XOR operation (Table 5).
4. **Crossover**:

Crossover can be performed on the test case 2 and 6 with some selected bits and again the XOR operation is performed to get crossover value satisfying 75% coverage of statements (Table 6).

5. **Mutation**:

Most bit of the output will get changed into 1 and, after performing mutation, it will be on the verge of achieving 100% coverage on statements which will become 1 throughout after crossover. Now, suppose if 100% coverage on the statement is not achieved, then it is needed to continue for all possible combinations. All combinations will involve the fitness value with the highest weightage and a moment will come

Table 5 Selection Output

T2	0	0	1	0	1	1	1
T6	0	0	0	1	1	1	1
XOR	0	0	1	1	1	1	1

Table 6 Crossover Output

T2	0	1	1	0	1	1	1
T6	0	0	0	1	1	0	1
XOR	0	1	1	1	1	1	1

where all these combinations will contribute to achieve the 100% target. In this example, we can take {T2, T6} as the backbone test case for reduction problem.

As we have performed multi-level random walk in the initial coverage matrix, it is difficult to get the optimal solution by reduced test cases, so invocation of GA is performed to compare the effectiveness genetic algorithm with multi-level random walk.

We will take {T2, T6} as backbone test cases for reduced level 1 matrix which will cover the respective coverage statements as depicted (Table 7).

Here, two shielded test cases possible for test cases T2 and T6, i.e., T1 and T4, but we will go for T4 as it is covering more number of 1; moreover, T1 will ultimately become optimal after further coverage with T4 as a shielded test case (Table 8).

Only one backbone is needed to be selected so no need to go for optimization (Table 9).

Optimal solution:

As 1 is left uncovered in T5 test case, therefore, it is taken as the optimal solution: {T5}

Reduction Test case in terms of weightage:

Backbone test case 1 = {T2, t6}

Backbone test case 2 = {t4}

Optimal solution = {t5}

Total = 2.6 + 2 + 1.8 + 2 = 8.4

Table 7 Reduction Level 1

Test cases	S1	S2	S3	S4	S5	S6	S7
T1	1	0	0	1	1	0	0
T2	0	0	1	0	1	1	1
T3	0	0	0	0	0	1	0
T4	1	1	0	1	0	0	1
T5	1	1	1	0	1	0	0
T6	0	0	0	1	1	1	1
T7	0	1	0	0	0	0	0

Table 8 Reduction Level 2

Test cases	S1	S2	S3	S4	S5	S6	S7
T1	1	0	0	1	0	0	0
T2	0	0	0	0	0	0	0
T3	0	0	0	0	0	0	0
T4	1	1	0	1	0	0	0
T5	1	1	1	0	0	0	0
T6	0	0	0	0	0	0	0
T7	0	1	0	0	0	0	0

Table 9 Reduction Level 3

Test cases	S1	S2	S3	S4	S5	S6	S7
T1	0	0	0	0	0	0	0
T2	0	0	0	0	0	0	0
T3	0	0	0	0	0	0	0
T4	0	0	0	0	0	0	0
T5	0	0	1	0	0	0	0
T6	0	0	0	0	0	0	0
T7	0	0	0	0	0	0	0

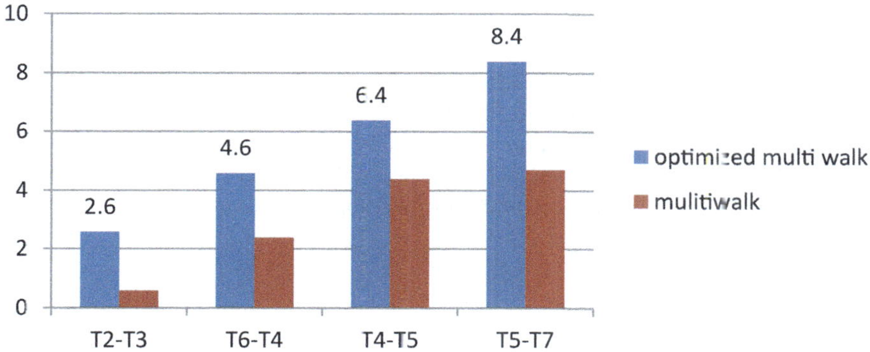

Graph 1. Multi-walk versus optimized multi-walk

Therefore, after covering most of the test cases and statements the entire matrix is on the verge of becoming 0 throughout and the optimal solution found by test cases **{T2, T6, T4, T5}**.

2.3 *Performance Analysis*

The following graph shows that the overall performance of multi-walk and optimized multi-walk algorithms and optimized approach is better than multi-walk approach (Graph 1).

2.4 *Conclusion and Future Work*

On taking a tour through this paper, we experienced that regression testing is a very expensive form of testing which includes re-execution of all test cases, thus making the overall test suite more cumbersome and leading to an increase in test cases. So to enhance the regression testing, initially, a multi-level random walk is performed and a common intersection point is found, but that walk being performed is not optimized. Therefore, an optimized random walk is made which tries to decrease the redundancy in the test suite and improves the overall test suite. When it comes to redundancy, test case prioritization also plays a very important role. To get a more profound test suite, one of the optimization techniques (genetic algorithm) is incorporated with the existing multi-level random walk reduction technique. Thus, the combination of both reduction and prioritization techniques with optimization technique (genetic algorithm) is a good recommendation that is tried to be concluded in this paper.

In this paper, we have tried to overcome the random test case tie situation and to decrease the redundancy in the test suite by using a combination of multi-level random walk and optimization algorithms (genetic algorithm). But prioritization can also work as a very good remedy for reducing the redundancy of test cases in the test suite as it will provide a sequence and order to test cases, which can help in simplifying the entire test suite. In future, a good prioritization technique can be combined with an optimization technique. Moreover, some faults will also be generated at the time of acceptance testing phase that can also be worked upon for optimization using some reduced and prioritized optimization techniques.

References

1. M. Gligoric, L. Eloussi, D. Marinov, Practical regression test selection with dynamic file dependencies, in *Proceedings of International Symposium on Software Testing and Analysis*, Baltimore, MD, 12–17 July 2015, pp. 211–222
2. Hadoop, Apache Hadoop [Online] (2015). Available http://github.com/apache/hadoop/graphs/commit-activity
3. S. Yoo, M. Harman, Regression testing minimization, selection and prioritization: a survey. Softw. Test. Verif. Reliab. **22**(2), 67–120 (2012)
4. M. Harman, S. A. Mansouri, Y. Zhang, Search-based software engineering: trends, techniques and applications. ACM Comput. Surv. **45**(1), 1–61 (2012)
5. A. Caprara, P. Toth, M. Fischetti, Algorithms for the set covering problem. Ann. Oper. Res. **98**(1–4), 353–371 (2000)
6. R.M. Karp, Reducibility among combinatorial problems, in *Complexity of Computer Computations, The IBM Research Symposia Series*, ed. by R. Miller, J. Thatcher, J. Bohlinger (Springer, New York, 1972), pp. 85–103
7. H. Hsu, A. Orso, MINTS: a general framework and tool for supporting test-suite minimization, in *Proceedings of 31st International Conference on Software Engineering*, Vancouver, Canada, 16–24 May May 2009, pp. 419–429
8. J. Black, E. Melachrinoudis, D. R. Kaeli, Bi-criteria models for all-uses test suite reduction, in *Proceedings of 26th International Conference on Software Engineering*, Edinburgh, Scotland, 23–28 May 2004, pp. 106–115
9. A. Gotlieb, D. Marijan, FLOWER: optimal test suite reduction as a network maximum flow, in *Proceedings of International Symposium on Software Testing and Analysis*, San Jose, CA, 21–26 July 2014, pp. 171–180
10. D. Hao, L. Zhang, X. Wu, H. Mei, G. Rothermel, On-demand test suite reduction. in *Proceedings of 34th International Conference on Software Engineering*, Zurich, Switzerland, 2–9 June 2012, pp. 738–748

Chennai Water Crisis—Data Analysis

Deepak Shankar, N. Aaftab Rehman, Sharmila Sankar, Aisha Banu, and M. Sandhya

Abstract Chennai is a coastal city and is the capital city of the Indian state of Tamil Nadu. It is densely populated and currently facing a severe drought as of June 2019. Chennai mainly depends on reservoir water, i.e., metro water and its groundwater resources to meet its water needs. The four main reservoirs of the city are Red Hills, Chembarambakkam, Cholavaram, and Poondi. With a combined total capacity of 11,057 Mcft, these act as the major source of water for the inhabitants of the city. With the ever-increasing population and the added effects of climate change which induces a decreasing rate of total rainfall, the situation of drought is likely to increase over the upcoming years. This project aims to explain the severity of drought in Chennai and how it can get worse in the upcoming years with the help of data analysis to find the per capita water that is currently present for a civilian, and how this has changed over the years due to various factors. This project also aims to create awareness and provide scope for possible solutions and measures that can be taken to mitigate the situation of drought in upcoming years.

Keywords Water crisis · Reservoir · Dataset · Rainfall

1 Introduction

Chennai, one of the four major metropolitan cities of India, is hit with a never-experienced before tremendous impact of drought, which is the highest so far recorded in terms of water availability levels of the reservoirs: Red Hills, Cholavaram,

D. Shankar · N. A. Rehman · S. Sankar (✉) · A. Banu · M. Sandhya
Department of Computer Science & Engineering, B.S. A Crescent Institute of Science & Technology, Chennai, India
e-mail: sharmilasankar@crescent.education

D. Shankar
e-mail: 01deepakshankar@gmail.com

N. A. Rehman
e-mail: aaftabrehman@gmail.com

S. S. Dash et al. (eds.), *Intelligent Computing and Applications*,
Advances in Intelligent Systems and Computing 1172,
https://doi.org/10.1007/978-981-15-5566-4_56

Poondi, and Chembarambakkam which act as the potable source for the whole city other than groundwater. The reason that can be accredited for this worse condition of Chennai is its low monsoon rainfall record in the months between March and July in the year 2019. Naturally, the city that is close to the equator might fall under the list of places that experiences harsh climate and harsh temperature. Analysis and records show that the rainfall record after 2016 has drastically lowered down and almost less than 10 cm rainfall was totally recorded in the monsoon of 2019 which is a catastrophic condition for any city. Chennai also witnessed two consecutive failures of monsoons in 2018 and 2019. As a result, the city had to import water from other states and had to circulate it to its residents to meet its water requirements. Again, the climatic change could be reasoned as an effect of global warming. We have limited ourselves to the city of Chennai which is now on a verge to turn into next Somalia as its water resources are depleting. In this paper, we tried to cover up with all those factors that have resulted in this condition of Chennai and we have calculated the per capita water based on the details and data available only on the four major reservoirs of the city.

2 Related Work

Many algorithms have been proposed for dealing with drought. We discuss some of these in this section.

Agana et al. [1] discussed basically on drought conditions in California and how big data analysis is used to build a real-time model for drought modeling and prediction. Various predictions are made using diverse data sets such as weather data, drought condition, climate sensor and satellite data, and water usage report's dataset. So, it is much more accurate in predicting the drought and its severity. It provides a more realistic visualization and offers a real-time model for drought prediction which describes the need for water conservation and drought awareness. Further, Anumalla et al. [2] designed a method for groundwater monitoring systems based on a network of wirelessly linked pressure sensors to measure the groundwater level. Gu et al. [3] used the extended empirical orthogonal function along with a standardized precipitation index to develop a dataset on spatial–temporal data. Jokhio et al. [4] presented with an idea to innovate the wireless sensor networks and employ them in forecasting systems to predict the drought and floods. Agana and Homaifar [5] have proposed a method to predict the droughts in the future by using deep learning technology. However, drought patterns are fluctuating in nature. Therefore, neural networks are also employed to predict the patterns accurately and thereby on further application of Standardized streamflow index (SSI) as an input in predicting this nonlinear time series. Nivedha Deve et al. [6] talks about the agricultural drought and how it varies change in temperature and rainfall levels with help of remote sensing which is used to monitor the temperature and vegetation variation over the Trichy District over the span of four years 2013–2016. The dataset used for calculating this factor is Surface Temperature (LST) and Normalized Difference Vegetation Index (NDVI) and the correlation between these two factors can help describing the severity of drought and how it changed over the years. Rao and Chaudhari [7] discusses how soil moisture data can be used to calculate and predict the drought with minimum accuracy. A comparison is done with estimated soil moisture data and rainfall data to find the correlation between the two factors. It can be used to find out the drought and flooding conditions and it is more suitable for bare fields with uniform conditions due to a poor spatial resolution of AMSR-E. Kollukuduru Sravanthi and Rajan [8] talk about identifying the spatial co-occurrence of droughts across various subdivisions of India. The results show that similarities in the climate forcing factors have a role in spatiotemporal correlation of events. De Azevedo et al. [9] analyzed the general survey about the drought patterns across India using satellite remote sensing. The project uses the Moderate Resolution Imaging Spectroradiometer (MODIS) to study the stress on vegetation over India. Dynamics of monthly Normalized Difference Vegetation Index (NDVI) and its relation to the climate conditions over a period of 10 years have been studied and presented in this paper. Liao et al. [10] discusses monitoring the dryness or drought scale in grassland. The GDI was constructed by integrating the canopy water content (CWC), soil moisture content, and precipitation together with different weights. A detailed analysis is made on GDI regarding temporal stats.

3 Algorithm

import and read the dataset into project.
import numpy, pandas, matplotlib libraries.

The dataset initially acquired is in the raw format. So, the collected data are first cleaned using Pythonic Data Cleaning with Pandas and NumPy. Then the data are filtered by choosing only the attributes that are necessary for plotting the graph. Finally, the various attributes like 'date' are modified and converted into an appropriate format that is recognizable by Python compiler and can be meaningfully plotted using Python Matplotlib and Seaborn Library (Fig. 1).

for each reservoir:
plot the graphs for various reservoir on a scale: Mcft every year.
plot the graph for rainfall at reservoirs on a scale: Mcft every year.

Water levels in each of the major reservoirs (y-axis) are plotted simultaneously against the date (x-axis). This graph helps to visualize the water holding capacity of all the reservoirs at each date of the year for a span of 15 years. These attributes help to tell how the reservoir levels have changed over the years and whether any effort was made to increase the capacity over the years. Also, the rainfall received at the major reservoirs (y-axis) is plotted simultaneously against the date (x-axis). This graph helps to visualize the total rainfall received at all the reservoirs on each date

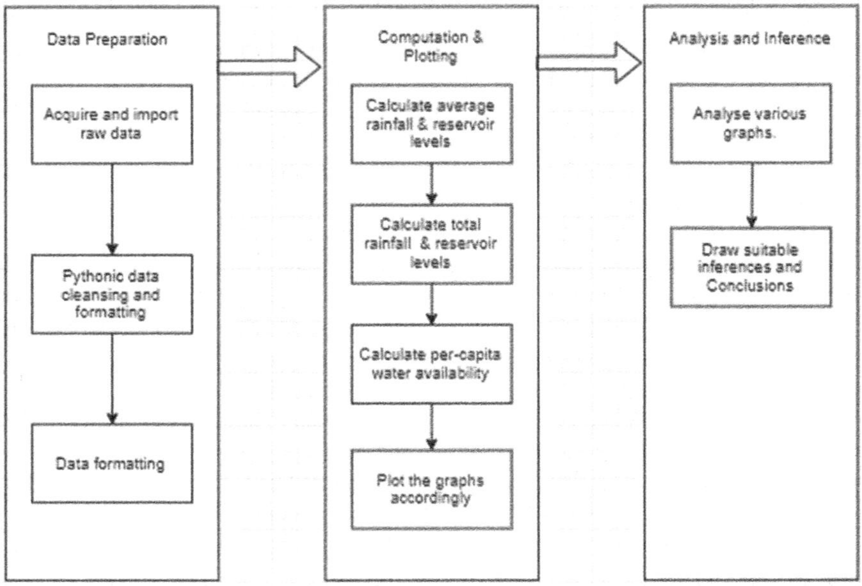

Fig. 1 Architecture diagram

of the year for a span of 15 years. These attributes help in detecting fluctuations in rainfall and analyze extreme scenarios such as floods.

total_water_level =Σ (water level at each reservoir)
total_rainfall =Σ (rainfall at each reservoir)
plot the graph for total_water_level and total_rainfall with scale in Mcft.

The total reservoir water level is calculated by finding the sum of water level at all the reservoirs and the average rainfall level is also calculated. Finally, each factor is individually plotted against the date. This helps to analyze the overall reservoir levels for a given rainfall on any given day. This also helps to analyze the total water that is actually made available to the public despite heavy rainfall.

read the population dataset
plot the graph of population growth taking a scale that measures years in x-axis and population in y-axis.

The graph of population growth over a period of 15 years is plotted. This can help analyzing the growth in population and how this factor is increasing at an accelerated rate. This attribute is chosen because drought mainly occurs when water resource is not able to meet the demand which is caused by increasing population.

(1) $\text{Per Capita Water} = \frac{\text{Total Reservoir Level}}{\text{Population}}$

plot the graph for per Capita Water taking years in x-plot and liters in y-plot.

The per capita water describes the water that is actually available to each person through the reservoirs as metro water. This factor is calculated as it shows the severity of drought over the given years. Thus, this factor is calculated and plotted against the year. Analysis can give possible insights into remedies and solutions that can be taken to mitigate the situation of drought in the upcoming years and thus safeguard the future generation.

Finally, analyze the various graphs that are generated to arrive at possible conclusions.

4 Analysis

As inferred from Figs. 2 and 3, The Redhills Reservoir has the highest capacity out of holding the water and Cholavaram has the least capacity. The average total holding capacity of all the reservoirs is around 4195.66 Mcft.

The graphs in Figs. 4 and 5 show how the water at individual reservoirs rise and fall according to the rainfall pattern. Cholavaram seems to contribute the least with hater holding capacity. So, it can meet only the needs of a small portion of the Chennai public.

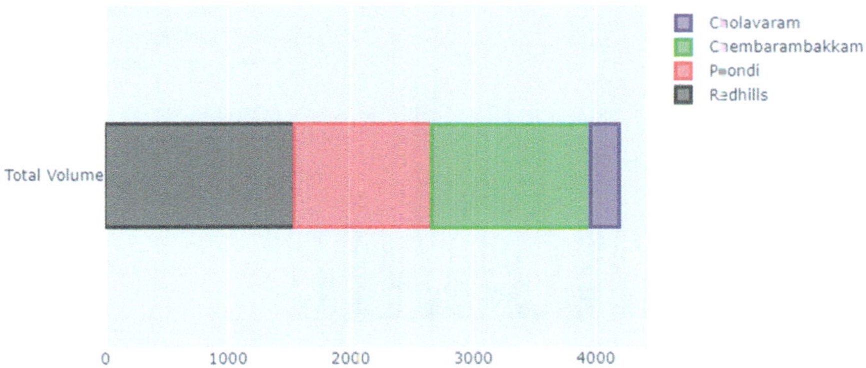

Fig. 2 Average water levels in reservoirs graph

Fig. 3 Overall Average water availability in a year (in terms of MCFT)

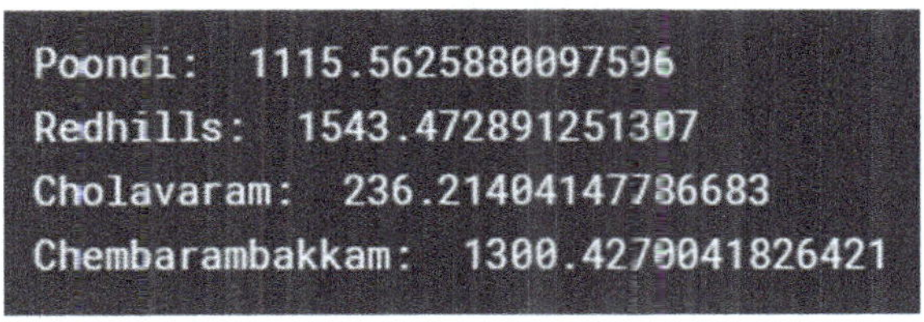

Poondi: 1115.5625880097596
Redhills: 1543.472891251307
Cholavaram: 236.21404147786683
Chembarambakkam: 1300.4270041826421

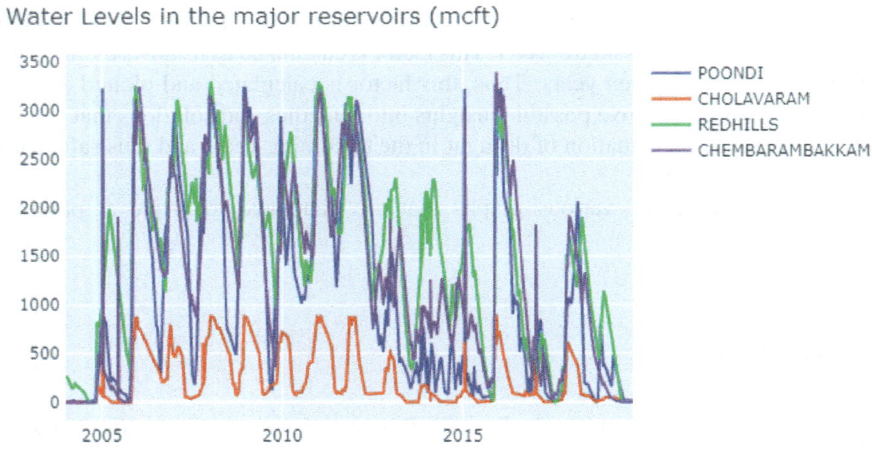

X-Axis: Year, Y-Axis: Water level

Fig. 4 Graph on water levels at various reservoirs

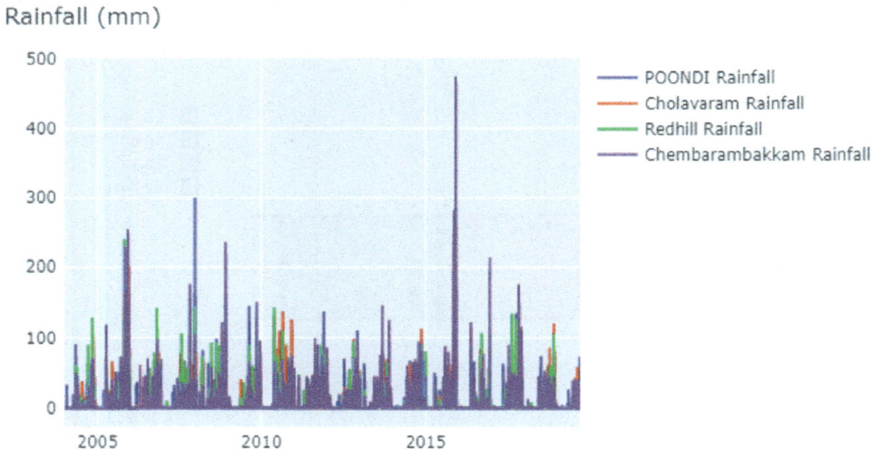

X-Axis: Year, Y-Axis: Water level

Fig. 5 Graph on rainfall levels at various reservoirs

During every year, rainfall peaks during November–January season which is seen as the maxima at both the graphs. In Fig. 6, there's a significant peak in the year 2005 and 2015 with a very steep rise and fall which marks two of the biggest floods of Chennai history. A broader peak describes the gradual rise and fall of the water level. This scenario is observed when there's a low to medium rainfall. A steep and sharp peak describes the scenario of a flood (Fig. 7).

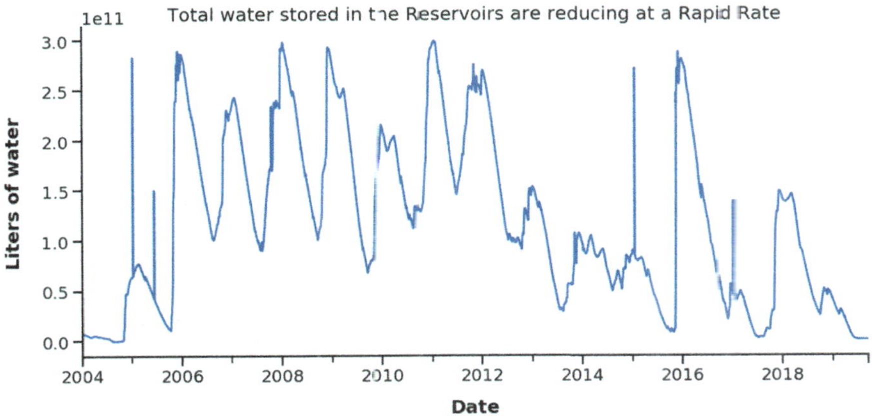

Fig. 6 Graph on total water level in all reservoirs

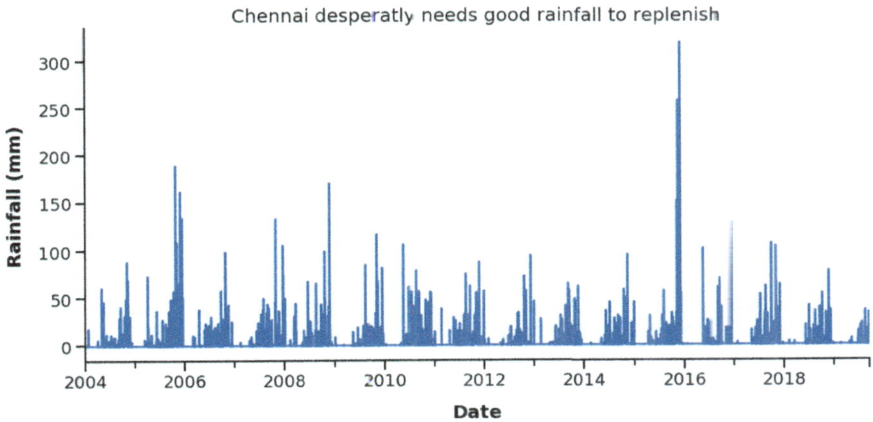

Fig. 7 Graph on total rainfall received overall

Also, the steepness in these two regions indicates that the water holding capacity of the reservoir is extremely low during excessive rainfall. So, a major amount of water is wasted into the sea (Fig. 6).

On average, the groundwater table has gone down by 100% from earlier levels. So, in locations where people got water at 150 feet, now they have to bore for 300 feet to even get a glimpse of water (Fig. 8). This is due to a significant reduction in the number of water bodies, an increase in construction, and insufficient and defective percolation pits (Rainwater harvesting).

From Fig. 9a, b, we infer that there is a steady rise in the population level. But the capacity of the reservoirs remains the same. Hence, there's an exponential decrease in the per capita water from the reservoir. And as of June 2019, all four reservoirs ran out of water, putting a large section of Chennai population underwater crisis. Water

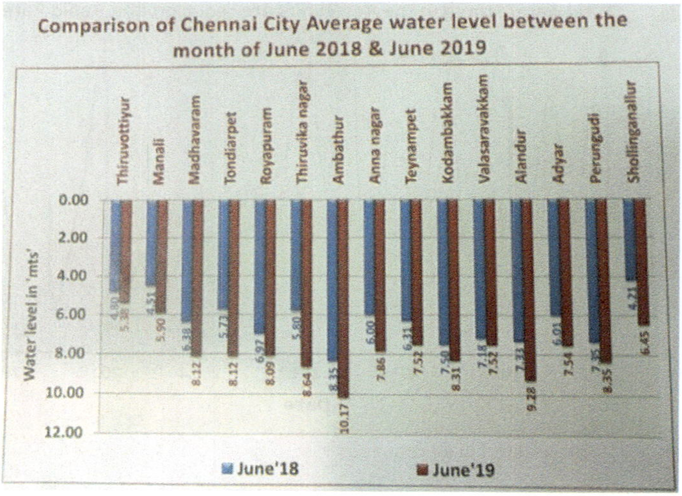

Fig. 8 Article on average water level in Chennai [11]

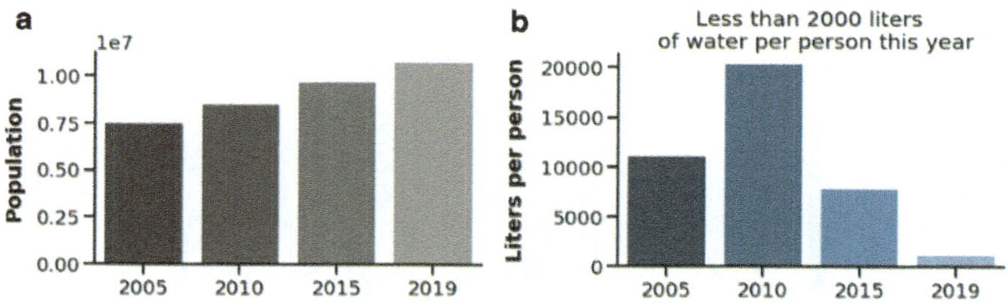

Fig. 9 **a** Graph on population dynamics over the years. **b** Graph on per capita water over the years

crisis is dependent on multiple factors like rainfall, climate, population, changes in land topography, and also groundwater replenishment level. But, many other projects tend to miss out on many of these factors and solely blame on climate change. But this project aims to relate and bring in all the possible factors influencing a drought condition in a metropolitan city like Chennai.

5 Inference

As we can see, there is a variable increase and decrease of rainfall and water levels in the reservoirs which makes it totally unfit to be completely dependent upon. As inferred from the graphs plotted above, there are significant bars and plots above representing the data of overall rainfall and water availability which were abundant in the years 2004 and 2015 and has rapidly vanished in no time, which portrays the irregular and careless behavior of its citizens and governing body to conserve water and use it ideally. The study has also proved that the water resources in Chennai are very scarce and the conventional source of water may no longer support its residents to meet their needs on their basic amenities. Another graph also shows the study on the population dynamics of the city and is a city of tedious growth the city has invited millions of immigrants in the subsequent years after 2005.

The various factors that have resulted in drought condition as analyzed are:

- Insufficient rainfall due to climate change.
- Increase of population inducing an increase in the demand for the water resource.
- Reduction in the number of water bodies [12] by at least 50% due to illegal encroachment and construction over existing water bodies.
- Water holding capacity during floods is extremely low. This can be attributed to two factors:
  - Insufficient de-silting done by the government. As we see, the reservoirs must be properly de-silted every 5 years or so. But it's not the case.
  - The structure of the reservoirs is not that good. So, the officials fear for the worst and release most of the water. If the structure was strong enough, this wouldn't be the case.
- Reckless mushrooming of bore-wells which has caused a reduction in groundwater table by more than 100%
- The inefficient rainwater system:
  - More than 40% of the houses don't have a rainwater harvesting system present [13].
  - The existing rainwater harvesting pits are extremely old and not maintained well. The rainwater percolation pits have to be cleaned at least once every five years to avoid clogging. But not everyone does this. There are many modern and much cost-effective methods present in recent times to implement this model effectively. So if people step up and follow it, the problem could be reduced several folds.
- Illegal sand mining:
  - It is a very popular crime in southern states where the top layer of river sand is mined and extracted illegally for the construction of sand. Due to this, the land becomes arid, and water-retaining and holding capacity of the river basin decreases severely. This also affects the groundwater table.
- Reckless irrigation methods:
  - It is estimated that 80% of the freshwater in India is used for agriculture. Due to improper implementation by the government, the farmers tend to over irrigate the land. This is an extremely bad practice as it pollutes the excess water with pesticides and fertilizers.

  - Drip irrigation will prove to be a better alternative saving a major portion of water.

- No system to Recycle and reuse wastewater:
  - Almost all the water is let out into the sea even without being treated. If we treat the water from households or from industries, more than 40% of pure water can be recovered. But no proper infrastructure is present in the state due to prejudice and carelessness of the government when it comes to conserving essential resources.

6 Conclusion and Future Scope of Work

The project was based on the domain—Data Analysis under which we have analyzed the water scarcity of Chennai that is on a verge to turn into a dry land sooner if not acted upon. The study has detailed graphs on the water resources of the city mainly concentrating on its water reservoirs and also has graphs on rainfall levels in those areas and data on the population dynamics. The graphs on rainfall and water level are far more terrorizing as they are decreasing at an alarming rate, if this goes on there is no stopping of this land turning into barren. Different plots above show that Chennai received adequate rainfalls before it fell short of rainfall in the years preceding 2015, but due to the irregular management of water the city has been struck with a drought-like condition to a level which was never experienced before.

However, the government has taken few steps to turn this city, the government made it mandatory for all the houses which are built to install a rainwater harvesting system and have also approved to provide small funding to build it. The government is now taking up steps to build factories on the outskirts of Chennai.

The water resources of this city are drying up, the city with no rivers as a source could be helped by its neighboring states by simply allowing the water rather stopping it through dams. The condition of the city went to worse and if not supported by the neighboring state the residents may face severe problems.

As observed, we have only provided this paper with the data on water level and rainfall level of the major reservoirs of the city but this alone won't suffice to completely analyze the pattern of drought in Chennai. Additionally, people who engage in this project in the future may take up additional parameters which are naturally occurring like temperature, humidity, and many more. Also, we have not covered up with all the sources of water to the city, the residents alternatively use water from the ground, as these factors are mainly important and participate in the overall water cycle.

References

1. N.A. Agana, A. Homaifar, A Deep Learning Based Approach for Long—Term Drought Prediction, in SoutheastCon (2017)
2. S. Anumalla, B. Ramamurthy, D.C. Gosselin, M. Burbach, Ground water monitoring using smart sensors, in IEEE International Conference on Electro Information Technology (2005)
3. X. Gu, N. Li, Study of droughts and floods predicting system based on spatial-temporal data mining, in 6th International Conference on New Trends in Information Science, Service Science and Data Mining (ISSDM2012) (2012)
4. S.H. Jokhio, I.A. Jokhio, F. Khan, S. Memon, Wireless Sensor Network Based Flood/DroughtForecasting System, in IEEE SENSORS (2015)
5. N.A. Agana, A. Homaifar, A deep learning based approach for long-term drought prediction, in Southeastcon-IEEE (2017)
6. S. Nivedha Deve, M. Jasmineniketha, P. Geetha, K.P. Soman, Agricultural drought analysis for Thuraiyur Taluk of Tiruchirappali District using NDVI and land surface temperature data, in 11 th International Conference on Intelligent Systems and Control (ISCO) (2017)
7. Y.S. Rao, A.A. Chaudhari, Analysis of 7 years Aqua AMSR-E derived soil moisture data over India, in 2009 IEEE International Geoscience and Remote Sensing Symposium
8. K. Sravanthi, K.S. Rajan, Mining spatial co-occurrence of drought events from climate data of India, in 2010 IEEE International Conference on Data Mining Workshops
9. S.C. de Azevedo, R.P. Singh, E.A. da Silva, Assessing 2016 drought progression over India using remote sensing data for the period 2006–2015, in 2017 IEEE International Geoscience and Remote Sensing Symposium (IGARSS)
10. Z. Liao, B. He, X. Quan, X. Bai, C. Yin, X. Li, S. Qiu, Constructing a global grassland drought index (GDI) product based on MODIS and ancillary data, in 2015 IEEE International Geoscience and Remote Sensing Symposium (IGARSS)
11. Chennai Metro Water—Twitter Handle (@CHN_Metro_Water)
12. https://timesofindia.indiatimes.com/city/chennai/chennais-two-third-water-area-vanishes-in-a-year/articleshow/68857592.cms
13. https://timesofindia.indiatimes.com/city/chennai/only-50-households-in-chennai-have-rain-water-harvesting-structures-finds-corporation/articleshow/70205838.cms

Generalized Canonical Correlation Based Bagging Ensembled Relevance Vector Machine Classifier for Software Quality Analysis

Noor Ayesha and N. G. Yethiraj

Abstract Early detection of defects helps to save cost and efforts. Few research works have been designed in existing works for analyzing the quality of software program using various machine learning techniques. However, the classification performance of existing work was lower which reduces the defect detection accuracy. In order to overcome the above existing issues, the Generalized Canonical Correlation Analysis based Bagging Ensembled Relevance Vector Machine Classification (GCCA-BERVMC) model is proposed. The GCCA-BERVMC model considers the number of source code lines from software program dataset as input. After that, the generalized Canonical Correlation Analysis (gCCA) algorithm designed in GCCA-BERVMC model selects the relevant features (i.e., code metrics) in order to improve the quality of software program. By applying Bagging Ensembled Relevance Vector Machine Classification (BERVMC) algorithm, the classification of program files is performed in GCCA-BERVMC model. The boosting algorithm creates '*n*' number of weak learners to classify the input source code lines as normal or defected by analyzing the source codes and chosen metrics. After that, the weak learner's results are combined into strong classifier by using the majority votes. This helps for GCCA-BERVMC model to enhance the accuracy of defect prediction for software quality analysis with a lower amount of time. Experimental evaluation of GCCA-BERVMC model is conducted using metrics such as defect detection accuracy, false positive rate, and time complexity with respect to various software code sizes. The experimental result shows that the GCCA-BERVMC model is able to increase the defect detection accuracy and also minimizes the amount of time required for software quality analysis when compared to state-of-the-art works.

N. Ayesha (✉)
Department of Computer Science, Bharathiar University, Coimbatore, India
e-mail: razia.ayesha.bi@gmail.com

N. G. Yethiraj
Department of Computer Science, Maharani's Science College for Women, Bengaluru, India
e-mail: rajsbmjce@gmail.com

S. S. Dash et al. (eds.), *Intelligent Computing and Applications*,
Advances in Intelligent Systems and Computing 1172,
https://doi.org/10.1007/978-981-15-5566-4_57

Keywords Bagging Ensembled Relevance Vector Machine Classification · Generalized canonical correlation · Majority vote · Software defect · Strong classifier and weak learner

1 Introduction

Software testing is the procedure of discovering faults in software while executing it. The results of the testing are employed to identify and correct faults. Software defect prediction finds where faults occur in source code. The results from the defect prediction can be utilized to optimize testing and get better software quality. Software Defect Prediction impacts quality and has attained significant popularity in the last few years [1, 2]. Therefore, a novel GCCA-BERVMC model is introduced in this research work by using the generalized Canonical Correlation Analysis (gCCA) algorithm and Bagging Ensembled Relevance Vector Machine Classification (BERVMC) algorithm.

GCCA-BERVMC model is developed in this research work. The main contributions of GCCA-BERVMC model are described in below

- To enhance the software defect detection performance through classification when compared to state-of-the-art works, GCCA-BERVMC model is introduced using generalized Canonical Correlation Analysis (gCCA) algorithm and Bagging Ensembled Relevance Vector Machine Classification (BERVMC) algorithm.
- To improve the software metric selection performance with a minimal amount of time when compared to conventional works, generalized Canonical Correlation Analysis (gCCA) algorithm is employed in GCCA-BERVMC model. The gCCA algorithm determines the associations between each input software metric 'μ_i' and objective function to significantly choose the related metrics.
- To increase the defect prediction accuracy when compared to traditional works, BERVMC algorithm is applied in GCCA-BERVMC model. BERVMC algorithm formulates a strong classifier with aiming at classifying each source code line as normal or defect with lower error for effective software quality analysis.

The remaining article structure is formulated as follows. In Sect. 2, the proposed GCCA-BERVMC model is explained with the aid of architecture diagram. In Sect. 3, experimental settings are presented, and the comparative result of GCCA-BERVMC model is discussed in Sect. 4. Section 5 portrays the conclusion of the paper.

2 Generalized Canonical Correlation Analysis Based Bagging Ensembled Relevance Vector Machine Classification Model

Software metrics (features or attributes) are collected during the software development cycle. Metric selection is one of the most important preprocessing steps in the process of building defect prediction models and may improve the final prediction result. To improve the effectiveness and quality of software development and to predict defects in software, various machine learning techniques were designed in conventional works [3, 4]. But, the classification performance of conventional work was not sufficient to achieve better defect detection accuracy [5]. Therefore, Generalized Canonical Correlation Analysis based Bagging Ensembled Relevance Vector Machine Classification (GCCA-BERVMC) model is designed in this research work. The architecture diagram of GCCA-BERVMC model is presented in Fig. 1.

Figure 1 depicts the overall processes of GCCA-BERVMC model. As demonstrated in the below figure, GCCA-BERVMC model contains two main processes namely metrics selection and classification [6]. Initially, GCCA-BERVMC model gets a number of source code lines from an input software program dataset. During the metrics selection process, GCCA-BERVMC model discovers the most related software metrics to test the quality of input software program. During the classification process, GCCA-BERVMC model predicts the error in a given program by classifying each source code line as normal or defect [7]. From that, GCCA-BERVMC model improves the software quality analysis performance when compared to state-of-the-art works. The detailed processes of GCCA-BERVMC model are explained in the below subsections.

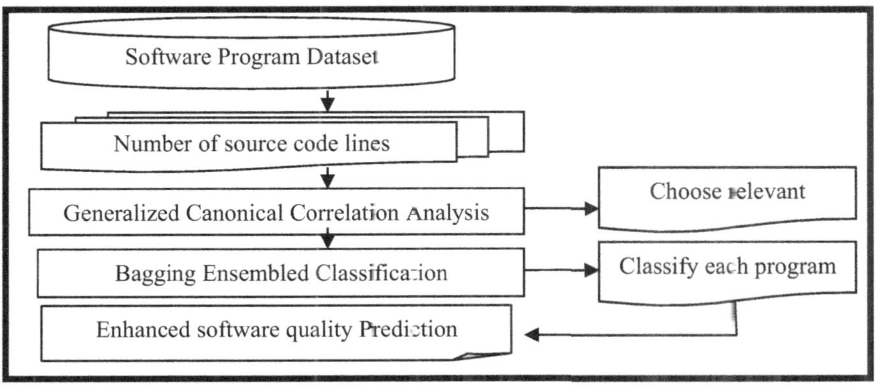

Fig. 1 Architecture diagram of GCCA-BERVMC model for software quality analysis

2.1 Generalized Canonical Correlation Analysis

The selection of software metrics is significant for predicting the software quality [8]. Selecting such metrics is not feasible because of limited project resources, particularly if the number of available metrics is large. Therefore, generalized Canonical Correlation Analysis (gCCA) algorithm is designed in GCCA-BERVMC model in order to choose only a more significant metrics to enhance the quality of software program. The gCCA is designed to improve the feature selection (i.e., software metrics) performance with minimal amount of time in order to efficiently improve the software quality analysis. The process involved in gCCA is depicted in Fig. 2.

Figure 2 shows the flow process of gCCA algorithm to perform metrics selection during the software quality analysis process. As depicted in figure, generalized canonical correlation analysis is used in gCCA algorithm in order to compute the associations between each input software metric 'μ_i' and objective function 'o_i', i.e., software quality analysis. Let us consider two features 'μ_i' and 'o_i', and there are correlations among the information. The gCCA algorithm at first finds linear combinations between software metric 'μ_i' and objective function 'o_i' which have a maximum correlation with each other. From that, generalized canonical correlation analysis is mathematically estimated as below

$$\mathrm{Cr} \rightarrow \sum \mu_i o_i = \mathrm{Cov}(\mu_i, o_i) \tag{1}$$

From the above mathematical Eq. (1), the covariance matrix is calculated by using information of software metric 'μ_i' and objective function'o_i'. In gCCA algorithm, the covariance between input software metric 'μ_i' and objective function 'o_i' is mathematically determined as below

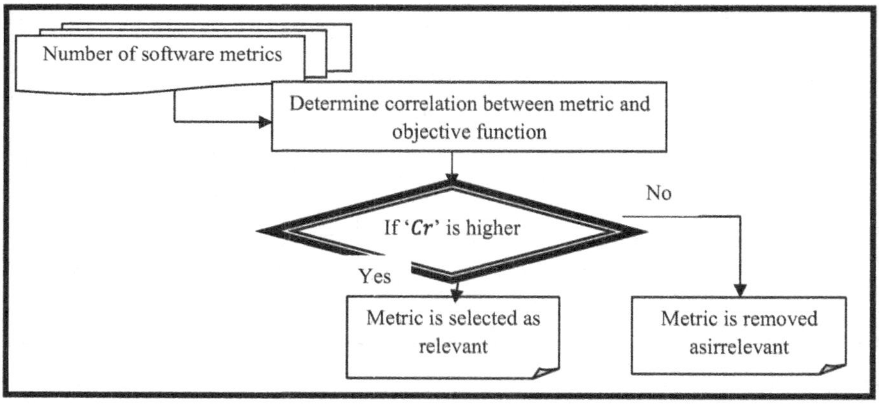

Fig. 2 Flow processes of gCCA algorithm

$$\mathrm{Cv}(\mu_i, o_i) = \frac{\sum(\mu_i - \bar{\mu})(o_i - \bar{o})}{n} \tag{2}$$

From the above mathematical formula (2), 'μ_i' signifies the values of software metric and 'o_i' denotes the values of objective function. Here, '$\bar{\mu}$' indicates the mean of the software metric, '$\bar{o}$' refers the mean of objective function, and 'n' represents a total number of the software metrics considered for testing input software program. With assistance of the Eqs. (1) and (2), gCCA algorithm estimates the relationship between the software metric 'μ_i' and objective function 'o_i'. Thus, gCCA algorithm selects the software metrics with higher correlation value as best in order to enhance the performance of software quality prediction with lower time complexity.

The algorithmic processes of gCCA algorithm are explained below.

Algorithm 1 Generalized Canonical Correlation Analysis

//Generalized Canonical Correlation Analysis Algorithm
Input: large number of software metrics '$\{\mu_1, \mu_2, \ldots, \mu_n\}$';
Output: Select relevant code metrics for improving the software quality
Step 1: Begin
Step 2: For each input software metric 'μ_i'
Step 3: Determine covariance '$\mathrm{Cv}(\mu_i, o_i)$' using (1)
Step 4: Evaluate the canonical correlation 'Cr using (2)
Step 5: **If (Cr is higher) then**
Step 6: Select software metric 'μ_i' as relevant
Step 7: **Else**
Step 8: Remove software metric 'μ_i' as irrelevant
Step 9: **End If**
Step 10: **End For**
Step 11: End

Algorithm 1 describes the step by step processes of generalized canonical correlation analysis. By using the above algorithmic process, gCCA algorithm efficiently chooses the features that are more significant for increasing software quality with a lower amount of time consumption. Hence, gCCA algorithm attains enhanced feature selection performance for effective software quality management as compared to existing works. As a result, GCCA-BERVMC model provides better metrics selection performance with a lower time complexity as compared to state-of-the-art works.

2.2 Bagging Ensembled Relevance Vector Machine Classification

Bagging Ensembled Relevance Vector Machine Classification (BERVMC) algorithm is designed in GCCA-BERVMC model with aiming at improving the performance

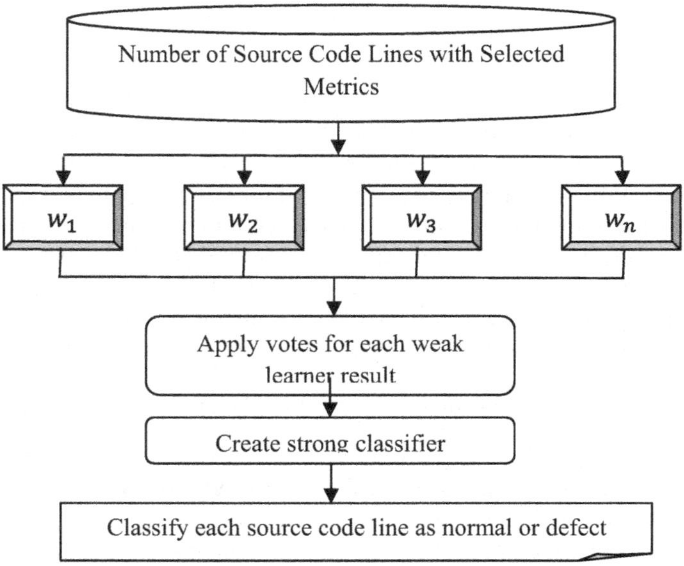

Fig. 3 Block diagram of bagging ensembled relevance vector machine classification algorithm

of software quality analysis. On the contrary to conventional works, BERVMC algorithm is proposed by using the Relevance Vector Machine and bagging ensemble algorithm. The designed BERVMC algorithm takes a Relevance Vector Machine classification as a weak learner to classify software code as normal or defect using chosen software metrics. The flow processes of BERVMC algorithm are presented in Fig. 3.

Figure 3 shows the flow processes of BERVMC algorithm. As demonstrated in the above figure, BERVMC algorithm at first takes number of source code lines with selected metrics as input. Then, BERVMC algorithm designs 'n' number of weak learner result for each source code. Let us consider an input software dataset that includes of many source code lines denoted as '$\{Sc_1, Sc_2, Sc_3, \ldots, Sc_n\}$'. Here, '$n$' point outs the total number of source code lines in the input dataset. In the BERVMC algorithm, each source code is trained using weak learner, i.e., relevance vector machine classification model to classify the source code as normal or defect. The weak learner is represented as '$\{(Sc_1, \gamma_1), (Sc_2, \gamma_2), \ldots, (Sc_n, \gamma_n)\}$' where '$Sc_i$' indicates a set of training samples (i.e., input source code lines) and 'γ_i' refers to the output (prediction result). From that, weak relevance vector machine classification result is mathematically obtained using below

$$z(Sc_i, w) = \sum_{i=1}^{n} w_i K(Sc, Sc_i) + w_0 \tag{3}$$

From the mathematical Eqs. (3). '$K(\mathrm{Sc}, \mathrm{Sc}_i)$' symbolizes the kernel function, 'w_i' refers to the weight of the 'ith' kernel function '$w = [w_0, w_1, \ldots, w_n]^T$', and '$w_0$' denotes the bias. In weak learner, Gaussian kernel function is applied in order to map '$z(\mathrm{Sc}_i, w)$' to (0, 1) for two class classification, i.e., normal class or defect class because, the weak learner returns output as '0' or '1'. For minimizing the over-fitting of weak learner due to excessive support vectors utilized, all weight vectors fulfill a zero-mean Gaussian prior distribution with help of below expression,

$$P(w, a) = \prod_{i=0}^{n} n(w_i|0, a_i^{-1}) = \prod_{i=0}^{n} \frac{a_i}{\sqrt{z\pi}} \exp\left(-\frac{a_i w_i^2}{2}\right) \tag{4}$$

From the above mathematical formula (4), '$a = [a_0, a_1, a_2, \ldots, a_n]^T$' signifies the hyper-parameter vector which determines the prior distribution of weight vector 'w' and controls the degree of weight diverges from its zero-mean. For a given prior probability distribution and the likelihood distribution, Bayes' rule is applied in weak learner to find out the posterior probability of models 'w' and 'a' with aid of below equation,

$$P(w, a|t) = P(w|t, a)P(a|t) \tag{5}$$

Subsequently, approximation procedure is employed in weak learner with aid of Laplace's method which identifies maximum hyper-parameter vector 'a' using below,

$$w_{\mathrm{MP}}^{\mathrm{new}} = w_i + \Delta w, \tag{6}$$

$$a_{\mathrm{MP}}^{\mathrm{new}} = \frac{1 - a_i \sum_{ij}}{w_{\mathrm{MP}}^2} \tag{7}$$

From the above mathematical representation (6) and (7), optimal hyper-parameter vector 'a' is identified in weak learner to classify each input source code line as normal or defect. In weak learner, the result '$\gamma_i = 0$' represents that a source code line is normal, whereas '$\gamma_i = 1$' symbolizes that the source code line is defect. To further enhance the accuracy of weak learner when compared to conventional works, bagging algorithm called bootstrap aggregation is utilized in GCCA-BEC model. Let consider a BERVMC algorithm constructs 'n' number of weak learner results for each source code in bootstrap samples. Subsequently, BERVMC algorithm all weak learner results into a strong classifier with help of below equation,

$$w(\mathrm{Sc}_i) = w_1(\mathrm{Sc}_i) + w_2(\mathrm{Sc}_i) + \cdots + w_n(\mathrm{Sc}_i) \tag{8}$$

Next, BERVMC algorithm applies vote 'δ_i' for each weak learner results '$w(\mathrm{Sc}_i)$' using below

$$\delta_i \rightarrow \sum_{i=1}^{n} w(\mathrm{Sc}_i) \tag{9}$$

Consequently, the majority vote of all weak learner results is employed to formulate a strong classifier for efficiently categorizing each source code line as normal or defect. Accordingly, strong classifier result is mathematically performed as follows:

$$A = \arg\max_{n} \delta(w(\mathrm{Sc}_i)) \tag{10}$$

From the above mathematical formulation (10), 'A' designates the final strong classifier result, whereas '$\arg\max_{n} \delta$' signifies majority votes of weak learner results. The strong classifier helps for BERVMC algorithm to correctly classify each source code line as normal or defect based on selected software metrics with a minimal time when compared to state-of-the-art works. The algorithmic steps of BERVMC are explained below.

Algorithm 2 Bagging Ensembled Relevance Vector Machine Classification

//Bagging Ensembled Relevance Vector Machine Classification Algorithm
Input: Software Program Dataset; Number of Program Files '$p_{f1}, p_{f2}, p_{f3}, \ldots, p_{fn}$'; Number of Source Code '$\mathrm{Sc}_1, \mathrm{Sc}_2, \mathrm{Sc}_3, \ldots, \mathrm{Sc}_n$'
Output: accurately classify source codes as normal or defect
Step 1: Begin
Step 2: **For** each input program files 'p_{fi}'
Step 3: **For** each source code line 'Sc_i'
Step 4: Design bootstrap samples
Step 5: Obtain 'n' number of weak learner results
Step 6: Aggregate all weak learner results using (8)
Step 7: Apply voting scheme using (9)
Step 8: Design strong classifier by considering majority voting results using (10)
Step 9: Exactly Classify source code line as normal or defect
Step 10: **End For**
Step 11: **End For**
Step 12: End

BERVMC algorithm as presented in the above algorithmic process, initially gets a number of source code lines as input and then produces a number of bootstrap training samples. For each source code line in bootstrap training samples, next BERVMC algorithm generates 'n' number of weak learner results.

Then, BERVMC algorithm aggregates the results of all weak learners together and then applies voting scheme. Finally, BERVMC algorithm considers majority votes of all weak learners to design strong classifier that correctly classifies each source code line as normal or defect according to selected software metrics with a lower amount of time consumption when compared to conventional works. Thus,

GCCA-BERVMC model attains better software quality analysis performance when compared to conventional works.

3 Experimental Settings

In order to determine the performance of the proposed, GCCA-BERVMC model is implemented in Java Language using schoolmate dataset. The data set using open source projects is obtained from https://sourceforge.net/. This dataset contains 66 PHP programs for elementary, middle, and high schools. The GCCA-BERVMC model takes different sizes of software code as input in the range of 10–100 KB. Then, GCCA-BERVMC model classifies each source code line as normal or defect in order to increase software quality. The performance of GCCA-BERVMC model is estimated in terms of defect detection accuracy, time complexity, and false positive rate with respect to different software code sizes. The experimental result of GCCA-BERVMC model is compared with weighted majority voting techniques [9] and neural network based classification technique [10].

4 Results

In this section, the comparative result of GCCA-BERVMC model is discussed. The performance of GCCA-BERVMC model is compared with weighted majority voting techniques [9] and neural network based classification technique [10], respectively. The performance of KPFS-SBPRVM technique is determined along with the following metrics with the help of tables and graphs.

4.1 Performance Measure of Defect Detection Accuracy

In GCCA-BERVMC model, Defect Detection Accuracy 'DDA' is evaluated as the ratio of number of source code lines that are accurately identified as normal or defect through classification to the total number of source code lines taken as input. The defect detection accuracy is calculated using below

$$\text{DDA} = \frac{n_{\text{CD}}}{\text{SC}_n} * 100 \tag{11}$$

From the above mathematical representation (11), accuracy of software defect prediction is determined. Here, 'SC_n' point outs the total number of source code lines and 'n_{CD}' indicates number of source code lines that are properly detected. The defect detection accuracy is evaluated in terms of percentage (%).

Sample Calculations

- **Proposed GCCA-BERVMC model**: DDA $= \frac{45}{50} * 100 = 90\%$
- **Existing weighted majority voting techniques**: DA $= \frac{40}{50} * 100 = 80\%$
- **Existing neural network based classification technique**: DDA $= \frac{38}{50} * 100 = 76\%$.

The experimental result analysis of defect detection accuracy obtained during the processes of software quality analysis using three methods namely the proposed GCCA-BERVMC model and conventional weighted majority voting techniques [9] and neural network based classification technique [10] is demonstrated in Table 1.

Table 1 Tabulation result of defect detection accuracy

Software code size (KB)	Defect detection accuracy (%)		
	GCCA-BERVMC model	Weighted majority voting techniques	Neural network based classification technique
10	90	80	76
20	91	82	78
30	89	82	77
40	89	81	79
50	88	80	79
60	90	81	77
70	91	79	77
80	92	78	75
90	95	77	74
100	91	77	73

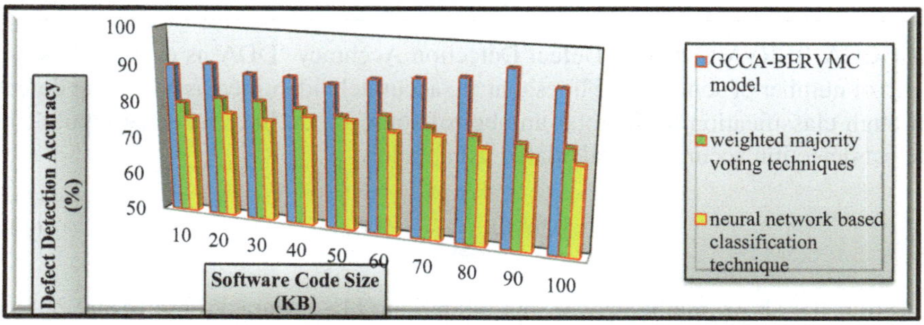

Fig. 4 Graphical result analysis of defect detection accuracy versus different software code sizes

Figure 4 presents the impact of software defect detection accuracy along with diverse number of software code size in the range of 10–100 KB using the proposed GCCA-BERVMC model and traditional Ensemble weighted majority voting techniques [9] and neural network based classification technique [10]. As demonstrated in the above graphical representation, the proposed GCCA-BERVMC model achieves higher defect detection accuracy with increasing size of software code as input when compared to existing weighted majority voting techniques [9] and neural network based classification technique [10]. This is because of application of generalized Canonical Correlation Analysis (gCCA) algorithm and Bagging Ensembled Relevance Vector Machine Classification (BERVMC) algorithm in the proposed GCCA-BERVMC model on the contrary to state-of-the-art works.

By using the concepts of gCCA algorithm, the proposed GCCA-BERVMC model selects the features that are more relevant in order to enhance software quality. Further with application of BERVMC algorithm, the proposed GCCA-BERVMC model designs strong classifier that accurately classifies each source code line as normal or defect based on selected software metrics with higher accuracy when compared to conventional works. Hence, the proposed GCCA-BERVMC model enhances the ratio of number of source code lines that are accurately identified as normal or defect when compared to other works. As a result, the proposed GCCA-BERVMC model increases the software defect detection accuracy by 14 and 19 % when compared to weighted majority voting techniques [9] and neural network based classification technique [10].

4.2 Performance Measure of Time Complexity

In GCCA-BERVMC model, time complexity 'TC' determines the amount of time needed to identify the defects in given software program. The time complexity is mathematically estimated using below

$$TC' = n * \text{Time(DDSSC)} \tag{12}$$

From the above mathematical Eq. (12), the time complexity of software defect prediction is calculated. Here, 'n' refers to the number of source code lines and 'T(DDSSC)' symbolizes the time utilized for detecting the defect in a single source code line. The time complexity is evaluated in terms of milliseconds (ms).

Sample calculations

- **Proposed GCCA-BERVMC model**: TC $= 50 * 0.2 = 10$ ms
- **Existing weighted majority voting techniques**: TC $= 50 * 0.3 = 15$ ms
- **Existing neural network based classification technique**: TC $= 50 * 0.33 = 17$ ms.

Table 2 Tabulation result of time complexity

Software code size (KB)	Time complexity (ms)		
	GCCA-BERVMC model	Weighted majority voting techniques	Neural network based classification technique
10	10	15	17
20	12	16	19
30	14	18	20
40	19	26	26
50	25	31	32
60	29	40	41
70	33	44	46
80	35	46	48
90	41	51	54
100	45	57	59

The tabulation result analysis of time complexity involved during the processes of software quality analysis using three methods namely the proposed GCCA-BERVMC model and traditional weighted majority voting techniques [9] and neural network based classification technique [10] is depicted in Table 2.

Figure 5 illustrates the impact of time complexity of software quality analysis based on different number of software code sizes in the range of 10 KB to 100 KB using the proposed GCCA-BERVMC model and state-of-the-art weighted majority voting techniques [9] and neural network based classification technique [10]. As presented in the above graphical figure, proposed GCCA-BERVMC model attains lower time complexity to predict the software defect with increasing size of software code as input when compared to existing weighted majority voting techniques [9] and neural network based classification technique [10]. This is owing to application of generalized Canonical Correlation Analysis (gCCA) algorithm and Bagging Ensembled Relevance Vector Machine Classification (BERVMC) algorithm in the proposed GCCA-BERVMC model on the contrary to existing works.

With the help of gCCA algorithmic process, the proposed GCCA-BERVMC model picks the key software metrics with minimal amount of time consumption to

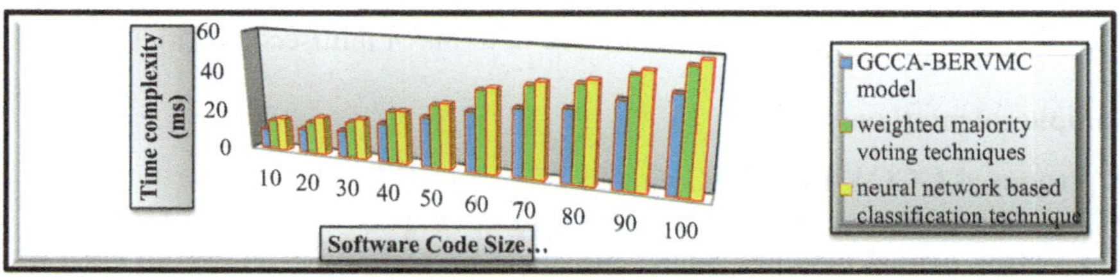

Fig. 5 Graphical result analysis of time complexity versus different software code sizes

improve software quality. In addition with application of BERVMC algorithm, the proposed GCCA-BERVMC model generates strong classification result and thereby perfectly classifies each source code line as normal or defect using chosen software metrics with lower amount of time consumption when compared to existing works. Therefore, the proposed GCCA-BERVMC model minimizes the amount of time needed to discover the defects in given software program when compared to other works. Accordingly, the proposed GCCA-BERVMC model reduces the time complexity by 24 and 29 % when compared to conventional weighted majority voting techniques [9] and neural network based classification technique [10].

4.3 Performance Measure of False Positive Rate

In GCCA-BERVMC model, False Positive Rate 'FPR' measures the ratio of number of source code lines that are wrongly detected as normal or defect to the total number of source code lines taken as input. The false positive rate is determined as follows.

$$\mathrm{FPR} = \frac{n_{\mathrm{WD}}}{\mathrm{SC}_n} * 100 \tag{13}$$

From the above mathematical expression (13), false positive rate of software defect prediction is calculated. Here, 'SC_n' indicates the total number of source code lines, whereas 'n_{WD}' denotes number of source code lines that are mistakenly detected. The false positive rate is estimated in terms of percentage (%).

Sample Calculations

- **Proposed GCCA-BERVMC model**: $\mathrm{FPR} = \frac{5}{50} * 100 = 10\%$
- **Existing weighted majority voting techniques**: $\mathrm{FPR} = \frac{10}{50} * 100 = 20\mathrm{FPR}\%$
- **Existing neural network based classification technique**: $\mathrm{FPR} = \frac{12}{50} * 100 = 24\%$.

The performance result analysis of false positive rate of software defect prediction during the processes of software quality analysis using three methods namely the proposed GCCA-BERVMC model and state-of-the-art weighted majority voting techniques [9] and neural network based classification technique [10] is shown in Table 3.

Figure 6 demonstrates the impact of false positive rate of software defect prediction according to diverse number of software code size in the range of 10–100 KB using the proposed GCCA-BERVMC model and traditional weighted majority voting techniques [9] and neural network based classification technique [10]. As shown in the above graphical figure, the proposed GCCA-BERVMC model obtains minimal false positive rate to correctly detect the software defects with increasing size of software code as input when compared to conventional weighted majority voting techniques [9] and neural network based classification technique [10]. This is due

Table 3 Tabulation result of false positive rate

Software code size (KB)	False positive rate (%)		
	GCCA-BERVMC model	Weighted majority voting techniques	Neural network based classification technique
10	10	20	24
20	9	18	22
30	11	18	23
40	12	19	21
50	11	18	21
60	10	19	23
70	9	21	23
80	8	22	25
90	5	23	26
100	9	23	27

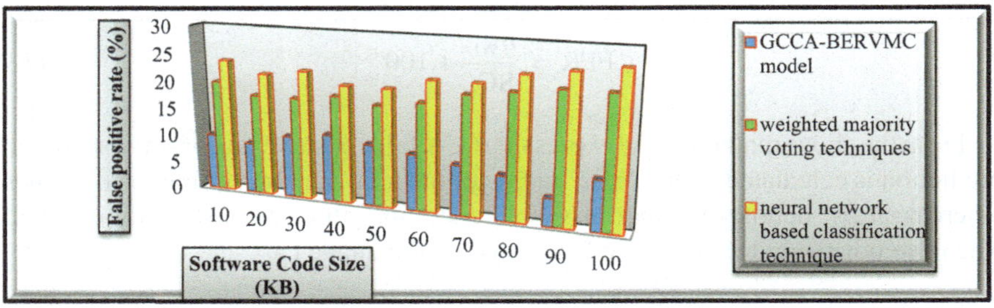

Fig. 6 Graphical result analysis of false positive rate versus different software code sizes

to application of generalized Canonical Correlation Analysis (gCCA) algorithm and Bagging Ensembled Relevance Vector Machine Classification (BERVMC) algorithm in the proposed GCCA-BERVMC model on the contrary to state-of-the-art works.

With the support of gCCA algorithmic steps, the proposed GCCA-BERVMC model finds the most important software metrics to accurately perform software quality analysis process. As well with application of BERVMC algorithm, the proposed GCCA-BERVMC model categorizes each source code line as normal or defect considering selected software metrics with minimal misclassification error when compared to conventional works. For that reason, the proposed GCCA-BERVMC model reduces the ratio of number of source code lines that are imperfectly detected as normal or defect when compared to other works. Thus, the proposed GCCA-BERVMC model decreases the false positive rate of software defect prediction by 52 % and 59 % when compared to conventional weighted majority voting techniques [9] and neural network based classification technique [10].

5 Conclusion

The GCCA-BERVMC model is proposed with the intention of increasing the performance of software defect prediction via performing classification with a minimal error rate. The objective of GCCA-BERVMC model is obtained with the application of generalized Canonical Correlation Analysis (gCCA) algorithm and Bagging Ensembled Relevance Vector Machine Classification (BERVMC) algorithm the contrary to traditional works. The proposed GCCA-BERVMC model increases the ratio of number of source code lines that are correctly predicted as normal or defect by using BERVMC algorithm when compared to conventional works. As well, the proposed GCCA-BERVMC model minimizes the time utilized to predict software defect with the support of gCCA and BERVMC algorithmic process when compared to other state-of-the-art works.

In addition, the proposed GCCA-BERVMC model minimizes the ratio of number of source code lines wrongly predicted for software quality analysis when compared to other existing works. Therefore, proposed GCCA-BERVMC model gives better software quality analysis performance in terms of defect detection accuracy, time complexity, and false positive rate when compared to conventional works. The experimental result shows that the proposed GCCA-BERVMC model presents better software defect prediction performance with an enhancement of accuracy and minimization of time when compared to state-of-the-art works.

References

1. H.B. Yadav, D.K. Yadav, Early software reliability analysis using reliability relevant software metrics. Int. J. Syst. Assur. Eng. Manag. **8**, 2097–2108 (2017)
2. S. Chatterjee, B. Maji, A new fuzzy rule based algorithm for estimating software faults in early phase of development. Soft. Comput. **20**(10), 4023–4035 (2016)
3. A. Okutan, O.T. Yıldız, Software defect prediction using Bayesian networks. Empir. Softw. Eng. **19**(1), 154–181 (2014)
4. S. Huda, K. Liu, M. Abdelrazek, A. Ibrahim, S. Alyahya, H. Al-Dossari, An ensemble oversampling model for class imbalance problem in software defect prediction. IEEE Access **6**, 24184–24195 (2018)
5. T. Wang, Z. Zhang, X. Jing, L. Zhang, Multiple kernel ensemble learning for software defect prediction. Autom. Softw. Eng. **23**(4), 569–590 (2016)
6. H. Tong, B. Liu, S. Wang, Software defect prediction using stacked denoising autoencoders and two-stage ensemble learning. Inf. Softw. Technol. **96**, 94–111 (2018)
7. Z. Xu, S. Li, J. Xu, J. Liu, X. Luo, Y. Zhang, T. Zhang, J. Keung, Y. Tang, LDFR: learning deep feature representation for software defect prediction. J. Syst. Softw. **158** (2019)
8. C.-Y. Huang, T.-Y. Kuo, Queuing-theory-based models for software reliability analysis and management. IEEE Trans. Emerg. Top. Comput. **5**(4), 540–550 (2017)
9. S. Moustafa, M.Y. El Nainay, N. El Makky, M.S. Abougabal, Software bug prediction using weighted majority voting techniques. Alexandria Eng. J. **57**(4), 2763–2774 (2018)
10. R. Jayanthi, L. Florence, Software defect prediction techniques using metrics based on neural network classifier. Cluster Comput. **22**, 77–88 (2019)

A Deep Learning Approach Against Botnet Attacks to Reduce the Interference Problem of IoT

Pramathesh Majumdar, Archana Singh, Ayushi Pandey, and Pratibha Chaudhary

Abstract Today we are witnessing a world where hacking into a user's computer using tiny bots or intercepting a group of interconnected devices is no more impossible. These tiny bots are called botnets which are a group of malicious codes that can hamper the whole security system without the knowledge of the user. As Internet of Things (IoT) is emerging rapidly, the interconnected devices are susceptible to breach as one affected device can hamper the whole network. The security threat remains as botnet attacks increase their presence to the interconnected devices. In this work, we are implementing Restricted Boltzmann Machine (RBM) algorithm of deep learning approach on the CTU-13 dataset to train the algorithm about the botnet attack patterns in IoT and to prevent the botnet attacks on IoT devices, thus reducing the interference problem in the network.

Keywords Bitnets · DeBot · Interference problem · Internet of Things (IOT) · Security

1 Introduction

Today we are living in an era of advanced technology where sneaking into a remote system is not impossible anymore Malwares have become so advanced today that they even don't need to write anything into the system files. The concept of dropping a malicious file into a target system is long gone now. Malwares are now able to use legitimate inbuilt applications to access the mainframe of the system. These

P. Majumdar · A. Singh · P. Chaudhary
Amity University, Noida, India
e-mail: Pramathesh@outlook.com

A. Singh
e-mail: asingh27@amity.edu

A. Pandey (✉)
SRM IST, Ghaziabad, India
e-mail: Pandeyayushi185@gmail.com

S. S. Dash et al. (eds.), *Intelligent Computing and Applications*,
Advances in Intelligent Systems and Computing 1172,
https://doi.org/10.1007/978-981-15-5566-4_58

advanced class of malwares are called file-less malwares. The concept of botnets is somewhat like file-less malwares. Botnets don't need to drop a malicious file into the target system. Their main goal is to hamper as much devices in the network as possible without leaving any trace of the source. In IoT, where every device is connected, botnet attacks have proven themselves to be an alarming challenge. Until the IoT concerns improve their security mechanisms, botnet attacks have the potential to become the new weapon of future cyberattacks. Since the last decade, the statistics of the botnet attacks have increased rapidly. The more technology is becoming advanced, the more advanced botnets are becoming.

Present technology of botnet detection mechanisms like signature or flow-based error detection is providing sound results when there is a small-scale botnet attack. In larger scale, they are not meeting the desired output. Nowadays we are witnessing the rise of artificial intelligence-based botnets which can learn the pattern of the users' usage mechanism and deploy themselves accordingly. Therefore, there is an unavoidable need of learning-based mechanisms to be implemented in the field of botnet attack prevention.

We are implementing a deep learning approach (Restricted Boltzmann Machine) to train the algorithm more accurately with a higher accuracy level about the attack patterns of the botnets. The dataset used in this work is the CTU-13 dataset of botnet attacks which shows several botnet attacks patterns and their feature-set and parameters.

This paper proposes a solution to the detection of botnet activity within consumer IoT devices and networks. A novel detection algorithm was developed based on deep learning mechanism. Detection was performed at the packet level with Wireshark by creating a fake network using ApateDNS.

2 Literature Review

Many researches have been done on providing an efficient solution against botnet attacks. Torres et al. proposed a methodology to compare several Recurrent Neural Network (RNN) models and the efficiency is analyzed to behavior of the traffic by designing it as a sequence of states that changes [1]. The focus of this work is to analyze the behavioral characteristics of the botnets. This work analyses the RNN methodologies because of the two main reasons—unbalanced network traffic and length of the data sequence. To perform the proposed concept, a K-fold cross-validation and test was conducted on unseen data extracted from different botnet samples. From the results, it was found that the RNN model is capable of efficiently differentiating the botnet attacks with a high detection rate. But it is not as much efficient when using imbalanced network traffic.

The ongoing development of the Internet of Things (IoT) has brought about an ascent in IoT-based DDoS assaults. McDermott et al. exhibit an answer for the identification of botnet action inside purchaser IoT gadgets and systems [2]. A tale use of deep learning is utilized to build up an identification model dependent on a Bidirectional Long Short-Term Memory based Recurrent Neural Network (BLSTM-RNN). This paper shows that although the bidirectional methodology adds overhead to every age and expands preparing time, it demonstrates to be a superior dynamic model after some time. A named dataset was created as a component of this examination and is accessible upon solicitation.

Yan et al. worked on the new patterns and attributes of DDoS assaults in distributed computing and gave a far-reaching overview of characterizes systems against DDoS assaults utilizing SDN [3]. Moreover, the examinations about propelling DDoS assaults on SDN, just as the techniques against DDoS assaults in SDN were analyzed. Apparently, the repudiate or connection among SDN and DDoS assaults has not been all around tended to in past works. This work shows how to utilize SDN's focal points to crush DDoS assaults in distributed computing conditions and how to avert SDN itself from turning into a casualty of DDoS assaults, which are significant for the smooth advancement of SDN-based cloud without the diversion of DDoS assaults.

Botnets comprise an essential risk to Internet security. The capacity to precisely recognize botnet traffic from non-botnet traffic can help essentially in moderating pernicious botnets. Roosmalen et al. presented a novel way to deal with botnet discovery that applies profound learning on streams of TCP/UDP/IP-parcels [4].

As of late, botnets have turned out to be one of the significant dangers to data security since they have been continually developing in both size and advancement. A few botnet identifications measures, for example, honeynet-based and Intrusion Detection System (IDS)-based, have been proposed. Nonetheless, IDS-based arrangements that utilization marks appear to be insufficient on the grounds that ongoing botnets are outfitted with advanced code update and avoidance strategies. A few investigations have appeared unusual as botnet recognition strategies are more successful than mark-based techniques since oddity-based botnet discovery strategies don't require pre-constructed botnet marks and henceforth they have the capacity to recognize new or obscure botnets. Toward this path, Hoang X. and Nguyen Q. proposed a botnet discovery model dependent on AI utilizing Domain Name Service question information and assessed its viability utilizing famous AI strategies [5]. Exploratory outcomes demonstrate that AI calculations can be utilized successfully in botnet recognition and the irregular timberland calculation creates the best by and large identification precision of over 90%.

Since the past decade, botnet has risen as an intense risk to digital security by giving its ability of bargaining billions of PCs and making them do the unlawful work. There are various existing ways by which botnet can be recognized. A thorough review of the current strategies is likewise expressed in this paper. Because of the inclusion of immense measure of information, recognition of botnet utilizing AI calculations is in tremendous pattern. Mathur et al. proposed a method to utilize AI to prepare classifiers by a system stream dataset. From that point, the prepared classifiers were connected on the gathered information to assess the outcomes [6]. Investigation of system stream information is utilized as a technique for location since it doesn't rely on the bundle content henceforth giving resistance toward the most recent type of encryption and confusion utilized by assailants to shroud their bots.

Results are unmistakably demonstrating that the proposed technique can separate the typical traffic and the bot-traffic with a high precision and low false positive rate. Also, pretty much every sort of botnet can be distinguished utilizing the proposed model.

In the age of the Internet of Things (IoT), a gigantic measure of detecting gadgets gathers as well as produces different tangible information after some time for a wide scope of fields and applications. Considering the idea of the application, these gadgets will result in huge or quick/constant information streams. Applying investigation over such information streams to find new data, anticipate future experiences, and settle on control choices is a significant procedure that makes IoT a commendable worldview for organizations and a personal satisfaction improving innovation. Mohammadi et al. proposed an analysis outline on utilizing a class of cutting-edge AI strategies, Deep Learning (DL), to encourage the examination and learning in the IoT area [7]. The procedure is performed by articulating IoT information qualities and recognizing two noteworthy medicines for IoT information from an AI viewpoint, IoT huge information examination, and IoT spilling information investigation. Additionally, why profound learning is a promising way to deal with accomplish the ideal examination in these sorts of information and applications were talked about.

Kudugunta and Ferrara proposed a system utilizing streak against 2D parodying assault to be utilized as a face liveness identification component [8]. Alongside the upgrade of separation among genuine just as ill-conceived clients, there is minimization of the impacts caused because of the natural components present because of the nearness of glimmer. A picture that incorporates streak and the one that does exclude streak is taken from the subject here. To catch the data of two pictures, present inside the model, four surfaces just as 2D structure descriptors are used that have low computational unpredictability.

Wireless devices are the largest part of the Internet of Things (IoT) units. Radiated, such as specific absorption rate, and spurious related issues are regulated by the standards bodies. The functionality performance will greatly influence IoT system stability, user experience, and cost. Radio receiver desensitization and coexistence aspects of IoT devices are discussed in work done by Yihong Qi et al. [9].

Measurement technologies are briefly reviewed for transmit/receive mode only of single-input and single-output, and multiple-input multiple-output devices, to provide a means of device performance evaluation and electromagnetic interference troubleshooting. An understanding of the electromagnetic compatibility aspects of the IoT and its measurement methods can help ensure the best user experience for ubiquitous wireless systems.

Bakshi et al. presented central thoughts of EMIT by portraying the worldwide impedance insights as far as single-gadget task and creates control rate allotment techniques to ensure low-postpone high-unwavering quality execution [10]. The future Internet of Things (IoT) systems are required to be made from a huge populace of ease gadgets discussing progressively with passages or neighboring gadgets to convey little packages of postponing delicate information. To help the high-force and brief requests of these developing systems, we propose an Efficient MAC worldview for IoT (EMIT). The worldview sidesteps the high overhead and coordination expenses of existing MAC arrangements by utilizing an impedance averaging methodology that enables clients to share their assets all the while. As opposed to the prevalent obstruction stifling methodologies, EMIT abuses the thick and dynamic nature of IoT systems to diminish the spatio-fleeting fluctuation of impedance to accomplish low-postponement and high-unwavering quality in administration.

The Internet of Things (IoT) imagines inescapable, associated, and shrewd hubs cooperating self-rulingly while offering a wide range of administrations. Wide appropriation, transparency, and generally high preparing intensity of IoT objects made them a perfect focus for digital assaults. Additionally, the same number of IoT hubs is gathering and preparing private data, they are turning into a goldmine of information for pernicious entertainers. Along these lines, security and explicitly the capacity to recognize traded off hubs, together with gathering and protecting confirmations of an assault or malevolent exercises rise as a need in effective arrangement of IoT systems. In this paper, we initially present existing significant security and legal sciences challenges inside IoT area and after that quickly talk about papers distributed in this exceptional issue focusing on distinguished difficulties.

Deep learning is a promising methodology for separating exact data from crude sensor information from IoT gadgets sent in complex situations. Due to its multilayer structure, profound learning is additionally fitting for the edge processing condition [11]. Consequently, in this article, we initially bring profound learning for IoTs into the edge figuring condition. Since existing edge hubs have restricted handling ability, we likewise structure a novel offloading technique to advance the presentation of IoT profound learning applications with edge processing. In the presentation assessment, we test the exhibition of executing various profound learning undertakings in an edge registering condition with our technique. The assessment results demonstrate that our strategy outflanks other advancement arrangements on profound learning for IoT.

Kasprzyk et al. performed an investigation of digital dangers, with specific accentuation on the dangers coming about because of botnet movement [12]. Botnets are the most widely recognized sorts of dangers and frequently saw as essential regarding national security. Their grouping and strategies for spreading are the reason for making the internet model including the nearness of various kinds of digital dangers. A well-structured internet model empowers to build an exploratory situation that takes into consideration the investigation of botnet attributes, testing its protection from different occasions and recreation of the spread and advancement. For this reason, committed stages with capacities and practical attributes to meet these prerequisites have been proposed.

The expansion of IoT gadgets that can be more effectively bargained than workstations has prompted an expansion in IoT-based botnet assaults. To alleviate this danger, there is a requirement for new techniques that recognize assaults propelled from traded off IoT gadgets and that separate among hours-and milliseconds-long IoT-based assaults. Meidan et al. proposed a novel system based on abnormality discovery strategy for the IoT called N-BaIoT that extricates conduct previews of the system and uses profound autoencoders to recognize bizarre system traffic from traded off IoT gadgets [].

3 Problem Statement

Many technologies have been developed over the years to cope with the increasing threat of the cybersecurity, but the ration of cybercrime is increasing instead of decreasing. Heuristic-based tools use rules to examine suspicious codes and classify them as malware. This approach is limited, however, due to the fact that it relies on the sequence of repeated code that is indicative of malicious intent. Hence, in this work, we are presenting a view on the combined approach of static and dynamic analyses with tools based on real-time extraction (Fig. 1).

3.1 Botnets in Internet of Things

A botnet is a network of interconnected small group of computers which runs on one or more bots. Their main target is to deploy botnet framework into the target system's mainframe to gain access to the system's administrative privileges. Once this step is done, they use their framework to manipulate the users' behavior and intervention to the system processes. For example, the Torii botnet attack (September 20, 2016) was one of the most severe botnet attacks which took over the IoT network. The "Bot-master" uses the C&C server to gain access to the victim system and transfer the information via IP trafficking through the botnet's framework. Botnet is an abbreviation which was taken from "Robot" and "Network" (Bot + Net). The main motive behind a botnet attack is to collapse as much IoT devices as possible and to gain access to the privileged information of a target system. The effects of botnets are

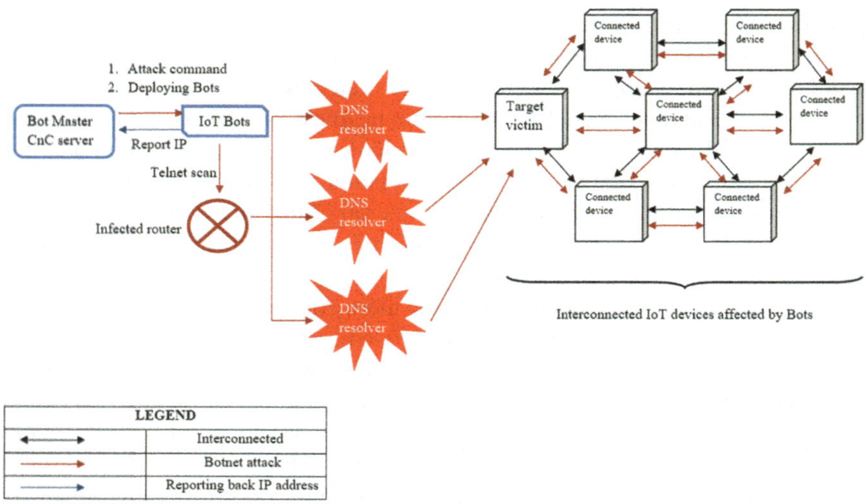

Fig. 1 Existing problem of botnet attack on IoT network

increasing in a rapid manner which was anticipated in the early ages. Statistics show that most of cyberattacks somehow involve in botnet implementations.

3.2 *Work-Flow Model*

The above work-flow model shows the step by step process of the proposed concept. They are as follows (Fig. 2):

i. At first, we need to collect the required botnet sample to update the dataset repository. Here, we used the CTU-13 dataset.
ii. A secure lab environment was created to ensure the safety of the public DNS and internal resources. Oracle Virtual Machine was installed using Windows 10 operating system to run the botnet sample. ApateDNS tool was used for this purpose, which creates a fake network to restrict the botnet from realizing that it is running inside a virtual machine.
iii. Analyze the botnet sample and record its severity level to determine the number of epochs to run.

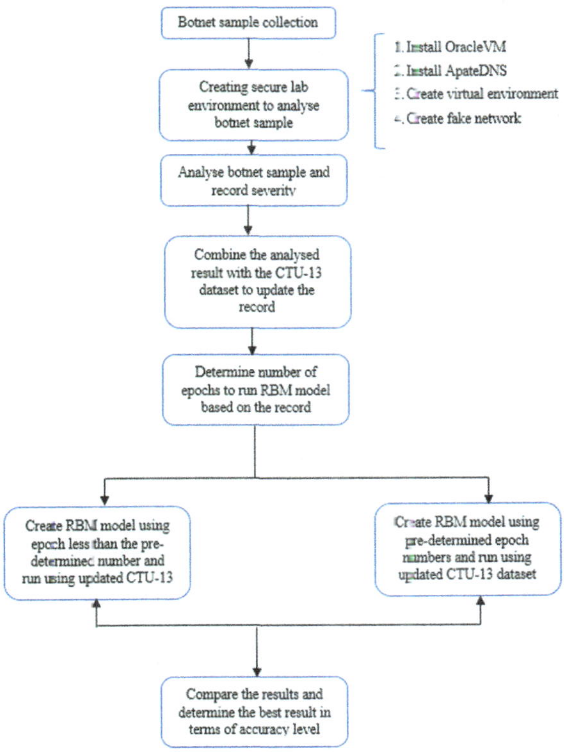

Fig. 2 Work-flow model of the proposed concept

iv. In the next step, the analyzed result was combined using the CTU-13 dataset to update the data repository.
v. Based on the updated record, the number of epochs was determined.
vi. Now, in this step, two experiments were performed using different epoch numbers. The first one used epoch number less than the pre-determined number and the second one used the pre-determined epoch number. This step was performed to cross-verify the accuracy of the epoch selection and accuracy level.
vii. Compare both the results from the previous step and determine the best result among them. The model is then ready and trained to detect botnet samples using the updated repository.

3.3 Deep Learning Against Botnets

Now we are leaving in a world of IoT where almost every device is interconnected to ease our lives. But the same mechanisms can and are being used against the security of the cyberspace. As we know, botnets target the most vulnerable devices and connections in an IoT network to spread into other devices, the question remains that how we are going to stop this attempt cybersecurity invasion. Malwares, ransomwares and botnets are becoming advanced faster than we can improve our defense against them. AI-enabled malwares, ransomwares, botnets are not new today. This threatens cybersecurity and raises the question that are we secure enough and advanced to defend ourselves in case in future any more devastating security threats raise concern.

Deep Learning (DL) mechanism is an advanced extension of the Machine Learning (ML) mechanism which includes the following main types—Boltzmann Machine (BM), Neural Network (NN), and Stacked Auto Encoders. In this work, we are comparing machine learning approaches with the deep learning Neural Network Sequential algorithm to find out whether machine learning or deep learning does more accurate job in learning the attack patterns accurately.

Figure 3 shows the implementation of Restricted Boltzmann Machine (RBM) on the CTU-13 dataset in our secure lab environment by creating a imitated network using ApateDNS tool to restrict the botnet sample from gathering information about the virtual environment. ApateDNS starts a Virtual Private Network (VPN) inside the Virtual Machine to limit the botnet's access to the public DNS.

As we can see from the result, the accuracy of the algorithm is 64.88% when we are using epoch = 5. Epoch is the count of iterations that are implied on the dataset to train the algorithm in a recursive manner. The more the count of epoch, the better the accuracy level is.

Count of epoch $\propto$ Accuracy level

```
Successfully downloaded train-images-idx3-ubyte.gz 9912422 bytes.
Extracting /tmp/data/train-images-idx3-ubyte.gz
Successfully downloaded train-labels-idx1-ubyte.gz 28881 bytes.
Extracting /tmp/data/train-labels-idx1-ubyte.gz
Successfully downloaded t10k-images-idx3-ubyte.gz 1648877 bytes.
Extracting /tmp/data/t10k-images-idx3-ubyte.gz
Successfully downloaded t10k-labels-idx1-ubyte.gz 4542 bytes.
Extracting /tmp/data/t10k-labels-idx1-ubyte.gz
Epoch 0 completed out of 10 loss: 1736390.3468322754
Epoch 1 completed out of 10 loss: 410037.0322418213
Epoch 2 completed out of 10 loss: 223206.3439064026
Epoch 3 completed out of 10 loss: 131157.39508372545
Epoch 4 completed out of 10 loss: 86741.48155975342
Accuracy: 0.6488
```

Fig. 3 Implementation of Restricted Boltzmann Machine (RBM) using epoch = 5

Using the CTU-13 dataset, we extracted the features from the dataset and applied Restricted Boltzmann Machine (RBM) algorithm on it.

We adjusted the number of epochs in different phases (e.g., epoch = 1, 2, 3 and 4) and found that the model gives underfitting results till epoch = 9. In epoch = 10, the model provides optimal result with an accuracy of 95.13%. Again, in epoch = 11, the model becomes overfitted. Hence, we selected number of iterations of epoch = 10 (Fig. 4).

4 Conclusion and Future Scope

In this paper, we worked on a deep learning-based botnet detection algorithm, which trains the IoT devices few ways for future research have been distinguished. To show the capacity of our created model to identify new varieties of botnets, a changed adaptation of the Torii sample will be utilized in the next phase of the work, to produce a second combined dataset and will be looked at against existing mark and stream-based oddity location strategies.

This work will be extended using a sample analysis of Torii botnet and combining the parameter set extracted with the CTU-13 dataset. A novel comparison will be done in the next phase on the dataset between machine learning approach and deep learning approach to find out which algorithm provides the best result and creating a model which can predict the attacks.

The following figure shows the implementation result of the RBM model on the CTU-13 dataset with epoch = 10 (0–9). This model provides an optimal result with a minimum processing time. Also, this model analyses the dataset and its patterns more accurately than other phases performed.

```
Successfully downloaded train-images-idx3-ubyte.gz 9912422 bytes.
Extracting /tmp/data/train-images-idx3-ubyte.gz
Successfully downloaded train-labels-idx1-ubyte.gz 28881 bytes.
Extracting /tmp/data/train-labels-idx1-ubyte.gz
Successfully downloaded t10k-images-idx3-ubyte.gz 1648877 bytes.
Extracting /tmp/data/t10k-images-idx3-ubyte.gz
Successfully downloaded t10k-labels-idx1-ubyte.gz 4542 bytes.
Extracting /tmp/data/t10k-labels-idx1-ubyte.gz
Epoch 0 completed out of 10 loss: 1736390.3468322754
Epoch 1 completed out of 10 loss: 410037.0322418213
Epoch 2 completed out of 10 loss: 223206.3439064026
Epoch 3 completed out of 10 loss: 131157.39508372545
Epoch 4 completed out of 10 loss: 86741.48155975342
Epoch 5 completed out of 10 loss: 51217.02963626385
Epoch 6 completed out of 10 loss: 38083.232104536146
Epoch 7 completed out of 10 loss: 28213.538931260107
Epoch 8 completed out of 10 loss: 22449.897190473974
Epoch 9 completed out of 10 loss: 21965.627538987996
Accuracy: 0.9513
```

Fig. 4 Implementation of Restricted Boltzmann Machine (RBM) using epoch = 1

References

1. C.D. McDermott, F. Majdani, A.V. Petrovski, Botnet detection in the internet of things using deep learning approaches, in *2018 International Joint Conference on Neural Networks (IJCNN)* (IEEE, 2018), pp. 1–8
2. Q. Yan, F.R. Yu, Q. Gong, J. Li, Software-defined networking (SDN) and distributed denial of service (DDoS) attacks in cloud computing environments: a survey, some research issues, and challenges. IEEE Commun. Surv. Tutor. **18**(1), 602–622 (2018)
3. P. Torres, C. Catania, S. Garcia, C.G. Garino, An analysis of Recurrent Neural Networks for Botnet detection behavior, in *2016 IEEE biennial congress of Argentina (ARGENCON)* (IEEE, 2016), pp. 1–6
4. J.V. Roosmalen, H.P.E. Vranken, M.C.J.D. van Eekelen, H.H. Haddad, Applying deep learning on packet flows for botnet detection, in *SAC18: The 17th edition of the Computer Security Track at the 33rd ACM Symposium on Applied Computing, 9–13 April 2018, Pau, France*, ed. by H.H. Haddad (ACM, New York, 2018), pp. 1629–1637
5. H. Zin, C. Kim, M. Wu, S. Kim, Avoidance of channel interference in polygonal IoT networks. Concurr. Comput.: Pract. Exp. **29**(11), e4060 (2017)
6. X. Hoang, Q. Nguyen, Botnet detection based on machine learning techniques using DNS query data. Future Internet **10**(5), 43 (2018)
7. L. Mathur, M. Raheja, P. Ahlawat, Botnet detection via mining of network traffic flow. Procedia Comput. Sci. **132**, 1668–1677 (2018)
8. S. Homayoun, M. Ahmadzadeh, S. Hashemi, A. Dehghantanha, R. Khayami, BoTShark: a deep learning approach for botnet traffic detection. Cyber Threat Intelligence (2018), pp. 137–153
9. M. Mohammadi, A. Al-Fuqaha, S. Sorour, M. Guizani, Deep learning for IoT big data and streaming analytics: a survey. IEEE Commun. Surv. Tutor. **20**(4), 2923–2960 (2018)
10. S. Kudugunta, E. Ferrara, Deep neural networks for bot detection. Inf. Sci. **467**, 312–322 (2018)

11. A. Bakshi, L. Chen, K. Srinivasan, C.E. Koksal, A. Eryilmaz, EMIT: an efficient MAC paradigm for the Internet of Things, in *IEEE INFOCOM 2016—The 35th Annual IEEE International Conference on Computer Communications*, April (IEEE, 2016), pp. 1–9
12. M. Conti, A. Dehghantanha, K. Franke, S. Watson, Internet of Things Security and Forensics: Challenges and Opportunities (2018)
13. H. Li, K. Ota, M. Dong, Learning IoT in edge: deep learning for the internet of things with edge computing. IEEE Netw. **32**(1), 96–101 (2018)

Predicting Movie Success Using Regression Techniques

Faaez Razeen, Sharmila Sankar, W. Aisha Banu, and Sandhya Magesh

Abstract Hollywood is the largest and most profitable movie industry in the world. In 2018 alone, it generated a massive global box office of over $42 billion. A single production company with multiple movies may benefit greatly from knowing which movies are likely to succeed—it would help them focus their resources on the required advertisement and promotion campaigns. Furthermore, theaters would get a preference on which movies to run for a longer duration based on its success rate. Large-scale investments come with large risks. Using machine learning to predict revenues may help investors mitigate these risks. The algorithms in this paper aim to recognize historical patterns in the movie industry to try and predict the success of upcoming movies using a variety of machine learning algorithms. The success metric used is the box office, i.e., the commercial success of a film in terms of overall money earned. The results show that it is indeed possible to predict revenue with a considerable amount of accuracy, with better results than a majority of the papers that were reviewed.

Keywords Regression · Machine learning · Movie revenue

1 Introduction

The movie industry is one of the most lucrative industries in the world. Including the global box office and home entertainment revenue, the global film industry is worth $136 billion as of 2018. Marketing and advertising costs take up to 50% of the entire budget allotted for a movie. For a big chunk of the budget, spending needs to be done wisely. Here is where machine learning comes to play. A movie production company often has many movies and TV shows under production at the

F. Razeen (✉) · S. Sankar · W. A. Banu · S. Magesh
B.S. Abdur Rahman Crescent Institute of Science and Technology, Chennai, India
e-mail: razeenfaaez@gmail.com

S. Sankar
e-mail: sharmilasankar@crescent.education

S. S. Dash et al. (eds.), *Intelligent Computing and Applications*,
Advances in Intelligent Systems and Computing 1172,
https://doi.org/10.1007/978-981-15-5566-4_59

same time. Since advertising is a huge expense, it would help the company to know which movie to put their money in. This is better supported by the fact that not all movies become blockbusters, i.e., movies that are of great commercial success. In this paper, the movies used are of only U.S. release in hopes of better performance due to homogeneity.

Successful prediction of a movie's success is complicated, as it depends on multiple factors. Success here can be defined in two ways—the revenue it earns or a quantification of how well it is received, i.e., user or critic score. Revenue is a much better definition of success in this case, as the amount of money a movie earns compared to its budget is a more rewardable measure of success.

The data used in this paper was a combination of data taken from IMDb (Internet Movie Database), TMDb (The Movie Database), and Facebook. Due to the fact that this paper uses metadata about a movie, predicting the revenue before it releases was not possible. Other papers which had done this used data like Twitter tweet rate, number of theaters the movie was going to be released in, number of Wikipedia edits at the time before release, etc. These were possible to retrieve before release. This paper uses Facebook data. The disadvantage here is that Twitter is more representative of the hype surrounding a movie. Someone excited about a movie who uses both Facebook and Twitter will obviously be using Twitter to talk about it due to the fact that Twitter tweets are meant to be real-time and to share trending information, while Facebook is more about connecting with friends, sharing photos and videos, etc.

The approach presented in the paper mainly focuses on regression algorithms. This paper aims to see if using machine learning algorithms can help predict movie revenue accurately. It also aims to find the features that are most important in determining a movie's revenue.

2 Literature Review

Quader et al. [1] used multi-class classification with five classes to predict whether a movie would be a "flop" or a "blockbuster," along with three other categories in the middle. Sentiment analysis was done by calculating the sentiment score for each review of a movie using Microsoft Azure's Text Analytics API and multiplying it by the total number of reviews. Using a variety of classification techniques, they found that IMDb vote count and number of screens were the best predictors. They have used two types of predictions: one, an exact match which rewards correct classifications, and the other being a "one-away" prediction, where the predicted class is considered correct if it is one away from the actual class. Considering the fact that there are only five classes, this method of classification is not indicative of actual accuracy. The inaccuracy is evident by the fact that using one-away prediction gave accuracies in the high 80s, while using normal prediction gave accuracies hovering around 50%. Such a difference is significant and should not be considered for real usage scenarios.

Yoo et al. [2] found that budget had the most correlation (~0.63) with revenue. They have used numerical and textual features and a combination of both, "sentiment" features, which are taken from movie reviews and given a positive or negative weight based on a subjectivity lexicon obtained from the University of Pittsburgh Opinion-Finder project, although it only leads to a minuscule increase in accuracy. They have theorized that this was due to two factors: the sentiment is already captured by the "rating" feature, and the number of people who watched the movie matter more than the number of people who think the movie is good. When using linear regression, they managed to achieve a correlation of 0.7479 between the predicted revenue and the actual revenue, which they consider not accurate enough to be used in practice.

Vr et al. [3] had used movies released in the USA and in the English language, in hopes of getting higher accuracies due to the fact that reviews are written in English. They implemented Linear Regression, Logistic Regression, and SVM Regression, and had accuracies of 51, 42, and 39%. The supposed cause for this was the small size of their dataset (1050 samples). The overall low accuracies would not prove useful in practice.

Liu et al. [4] had used purchase intention based on tweets to find out movie revenue. They found that the best correlated factor to box office revenue was how many people are willing to see the movie. They took tweets from Twitter and split them based on purchase intention. They used this to predict a movie's first week income and eventual revenue. They used Linear Regression and Support Vector Machine as their models. Their best results are an adjusted ^{2} value of 0.95 for predicting first week revenue, and 0.74 for predicting eventual gross. Their experiment verified their assumption that purchase intention is a better indicator than both popularity of the movie and sentiment analysis.

Jernbäcker et al. [5] have used classification to predict success of movies, using both revenue and rating as metrics. They have taken inflation into account in order to reduce the number of possible outcomes from millions to a little more than a hundred. They have used three classification algorithms: decision trees, support vector machines, and variants of nearest neighbor classifiers. Although the accuracies for predicting rating were on par with other papers, the highest accuracy they achieved on predicting the gross revenue was 15%.

Asur and Huberman [6] have shown how social media can be used to predict future outcomes. They have used tweets referring to movies prior to their release to predict the box office revenue generated by a movie in its opening weekend. Their research shows that the tweet rate (the average number of tweets per hour) of any movie is highly indicative of the success of a movie's opening week revenue. Using only the average tweet rate, they got an ^{2} value of 0.80. Additionally, they have used seven variables, created using time series values of the tweet rate for the seven days before release, each corresponding to a particular day, along with another variable which contains the number of theaters the movies were released in. Using sentiment analysis to calculate the ratio of positive tweets to negative tweets yielded a slight improvement in performance; however, these sentiments were not as important as the tweet rates themselves. Maximum prediction accuracy was achieved using both tweet rates time series and theater count, with an ^{2} value of 0.97 in predicting opening weekend gross. They have demonstrated that a simple linear regression model with only tweet rate as a feature can perform better than artificial money markets like Hollywood Stock Exchange (HSX).

Afzal [7] used various classification algorithms and managed to achieve an accuracy of 84% using logistic regression. However, they have not mentioned the number of output classes, which is important in gauging the efficacy of the algorithm. Classification would not prove useful in practice if a low number of classes were used.

Pal et al. [8] have used a combination of movie metadata and sentiment analysis to predict movie revenue. They extracted sentiment scores from movie reviews and applied various regression techniques with and without the sentiment scores to see if they actually helped improve regression accuracy. The sentiment scores were calculated using Naive Bayes and VADER (Valence Aware Dictionary for Sentiment Reasoning). Using the RMSE (Root Mean Square Error) scoring metric did not seem appropriate for this problem due to the fact that RMSE is very sensitive to outliers. Thus, they decided to convert their regression problem into a classification problem. A leeway of 15% was given, which means that if the prediction lies in this range, it is counted as an accurate prediction. The highest accuracies were achieved using Decision Tree Regression and Random Forest Regression. Although the inclusion of sentiment scores improved accuracy and decreased RMSE, the improvement was minor and deemed low enough to be not significant. Additionally, because they have converted their regression problem into a classification one, there is bound to be a loss in accuracy when used in real-time applications.

Mestyán et al. [9] have used data from Wikipedia to predict movie box office success. To estimate the popularity of a given movie, they measured four types of activity on a page: number of views, number of users, number of edits, and collaborative rigor (quality of thoroughness of the article). They also included another feature which they considered essential: number of theaters. A multivariate linear regression model was used to predict revenue. Using all five features, the R^2 value achieved was 0.77.

3 Data Preprocessing and Feature Engineering

3.1 Preprocessing

Two separate datasets were merged using an inner join on the IMDB ID column. Since they were merged, there were columns with the same attributes. Columns with either lesser information or which contained a lesser degree of relevancy were discarded. Columns which played no statistically significant role in prediction were also removed.

The merged dataset had two of the same columns showing movie revenue. However, the difference between them was found to be 43%. The cause of this was because one dataset contained revenue for the region of USA, while another had global revenue. The column with the global revenue was dropped in favor of USA revenue in order to keep the scope of the project homogeneous. The samples used were only chosen if the country of production was USA and the language was English. Other samples were dropped. Samples containing at least one null value were also dropped. The table containing all the features and their descriptions are shown below (Table 1).

Table 1 Dataset features and their description

Feature name	Description
vote_average	TMDB (The Movie DataBase) average vote rating
vote_count	TMDB total vote count
num_critic_for_reviews	Number of critic reviews on IMDb (Internet Movie Database)
num_user_for_reviews	Number of user reviews on IMDb
director_facebook_likes	Number of likes on Facebook for the director of the movie
actor_1_facebook_likes	Number of likes on Facebook for actor 1 of the movie
actor_2_facebook_likes	Number of likes on Facebook for actor 2 of the movie
actor_3_facebook_likes	Number of likes on Facebook for actor 3 of the movie
cast_total_facebook_likes	Number of likes on Facebook for the entire cast
movie_facebook_likes	Number of likes on Facebook for the page of the movie
revenue	Box office earnings of the movie inside U.S.A. (in dollars)
num_voted_users	Number of people who voted for the movie
facenumber_in_poster	The number of actors featured on the movie poster
duration	Duration of the movie in minutes
budget	Total budget allocated for the movie in dollars
imdb_score	IMDb score of the movie
release_year	Year of release

3.2 Feature Engineering

Feature Engineering is a process of creating new features by utilizing available features and domain knowledge in order to improve the performance of machine learning models. Three features were engineered.

Genre. The revenue across different genres was also studied on. Upon visual analysis, it was found that Action and Adventure movies on an average earned more revenue when compared to other genres. Thus, a new column was created, which contains a 1 if the genre of the movie is either action/adventure, else contains a 0.

Top Director. The top ten directors sorted in descending order by average revenue per movie were used to create a new column, called "top_director," which contains a 1 if the director of that particular movie is a top director, else contains a 0.

Collection. If a movie was part of a series, this column contains a value of 1. If the movie was a standalone movie, this column contains a 0.

4 Feature Selection

Feature selection is the process of selecting only a set of useful features from the available lot. It is used to prevent overfitting, a condition where an algorithm captures too much noise of the data, rendering it inefficient on data it has not seen before. The removed features usually have little to no effect on prediction accuracy.

4.1 Numerical Attributes

Since only regression techniques are used, all nominal features were either dropped or converted to categorical features.

4.2 Backward Elimination

Backward Elimination is an iterative process starting with all candidate variables, and in each iteration, deleting the variable whose loss gives the most statistically insignificant deterioration of the model fit, and repeating this process until no further variables can be deleted without a statistically significant loss of fit.

In this paper, the probability value (p-value) was used. According to Wasserstein and Lazar [10], for a given statistical model, the p-value represents the probability that, when the null hypothesis is true, the statistical summary (such as the sample mean difference between two groups) would be equal to or more extreme than the actual observed results. The smaller the p-value, the greater the statistical incompatibility of the data with the null hypothesis.

In simpler terms, if the p-value of a certain feature is 0.05, then it means that there is a 5% chance that the results obtained were due to pure chance rather than due to the statistical features of the data. The industry standard for p-value is 0.05 or 5%, and the same is used here. In each iteration of calculating the p-values, the feature with the highest p-value is removed until all features have p-values <0.05.

5 Regression Models

All algorithms were carried out using the Scikit-learn library [11] in Python. The efficacy of each model used would be judged based on four criteria.

- ^{2}, a value which measures goodness of fit of the model. $^2 = 1$ indicates a perfect fit.
- Mean Absolute Error (MAE) measures the difference between two variables: the actual value (y_i) and the predicted value (y_i). For easy interpretation of MAE, the features are normalized between 0 and 1 before applying algorithms.

$$\text{MAE} = \frac{1}{n}\sum_{i=1}^{n}\left|y_i - \hat{y}_i\right|$$

- Root Mean Squared Error (RMSE) is the square root of the Mean Squared Error (MSE). RMSE is a better metric compared to MAE, as it is sensitive to outliers.

$$\text{RMSE} = \sqrt{\frac{1}{n}\sum_{i=1}^{n}\left(\left|y_i - \hat{y}_i\right|\right)^2}$$

- Correlation between the predicted revenue and actual revenue. From the final subset of features, vote_count was found to be the highest correlated feature with revenue, with a Pearson correlation coefficient of 0.7515. This is used as a baseline—all models used in this paper should be able to achieve a correlation equal to or greater than 0.7515 (Fig. 1).

All metrics except correlation were cross validated over ten sets. Cross validation is a resampling procedure where each subset of the data is used both as a training set and as a testing set. Doing this helps prevent the model from overfitting on the data.

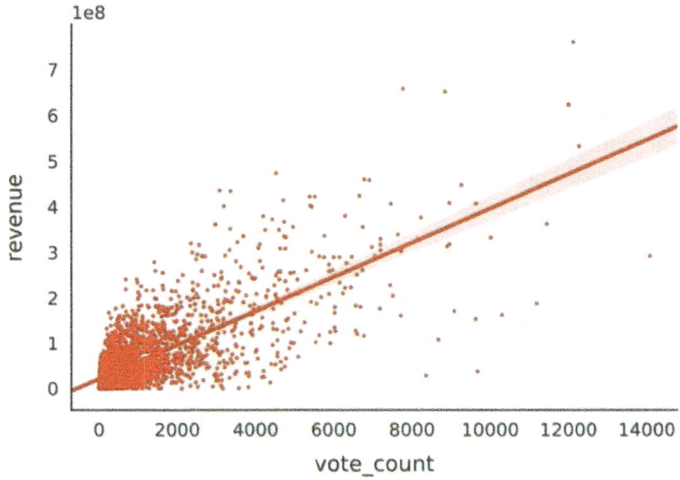

Fig. 1 Correlation between vote_count and revenue

5.1 *Linear Regression*

Multiple linear regression is a technique that uses multiple explanatory variables to predict the outcome of a single response variable through modeling the linear relationship between them. It is represented by the equation below:

$$y = \beta_0 + \beta_1 x_1 + \cdots + \beta_n x_n$$

where y_i = dependent variable, x_i = independent variables, β_0 = y-intercept, and β_n = slope coefficients.

5.2 *Support Vector Regression*

A Support Vector Machine is a classifier that aims to find the optimal hyperplane (the separation line between the data classes with the error threshold value epsilon) by maximizing the margin (the boundary between classes and that which has the most distance between the nearest data point and the hyperplane). In this paper, a linear kernel was used.

5.3 *Decision Tree Regression*

A decision tree is a supervised classification model that predicts by learning decision rules from features. It breaks down data into smaller subsets by making a decision based on asking a series of questions (the answers are either True or False), until the model gets confident enough to make a prediction. The end result is a tree, where the leaf nodes are the decisions.

The questions asked at each node to determine the split are different for classification and regression. For regression, the algorithm will first pick a value, and split the data into two subsets. For each subset, it calculates the MSE (Mean Squared Error). The tree chooses the value with the smallest MSE value. After training, the algorithm runs it through the tree until it reaches a leaf node. The final prediction is the average of the value of the dependent variable in that leaf node. The usage of a single decision tree gave the worst results. While decision trees are supposed to be robust against collinearity, they did not perform better than linear regression.

5.4 *Random Forest Regression*

Random forest is an ensemble method, which means that it combines predictions from multiple machine learning algorithms, in this case, decision trees. The problem with decision trees is that they are very sensitive to training data and carry a big risk of overfitting. They also tend to find the local optima, as once they have made a decision, they cannot go back. This was evident by the fact that the ^{2} value and correlation varied in each iteration of running the algorithm. Random forest contains multiple decision trees running in parallel, and in the end, averages the results of multiple predictions. Random forests with 100 trees were found to have the best results among all algorithms used in this paper.

5.5 *Ridge Regression*

Ridge regression L2 uses regularization, which is a method used to avoid overfitting by penalizing high-valued regression coefficients through shrinkage, where extreme values are shrunk toward a certain value. Particularly, in L2 regularization, the coefficients are penalized toward the square of the magnitude of the coefficients.

Ridge regression is a technique used to mitigate multicollinearity in linear regression. Multicollinearity is where one feature in a multiple regression model can be predicted from other features due to a positive linear relationship. This causes inaccurate estimates of the regression coefficients. While the fulfillment of the multicollinearity assumption is not necessary for prediction and is only necessary for inference, using Ridge Regression decreased performance. The cause of this was not

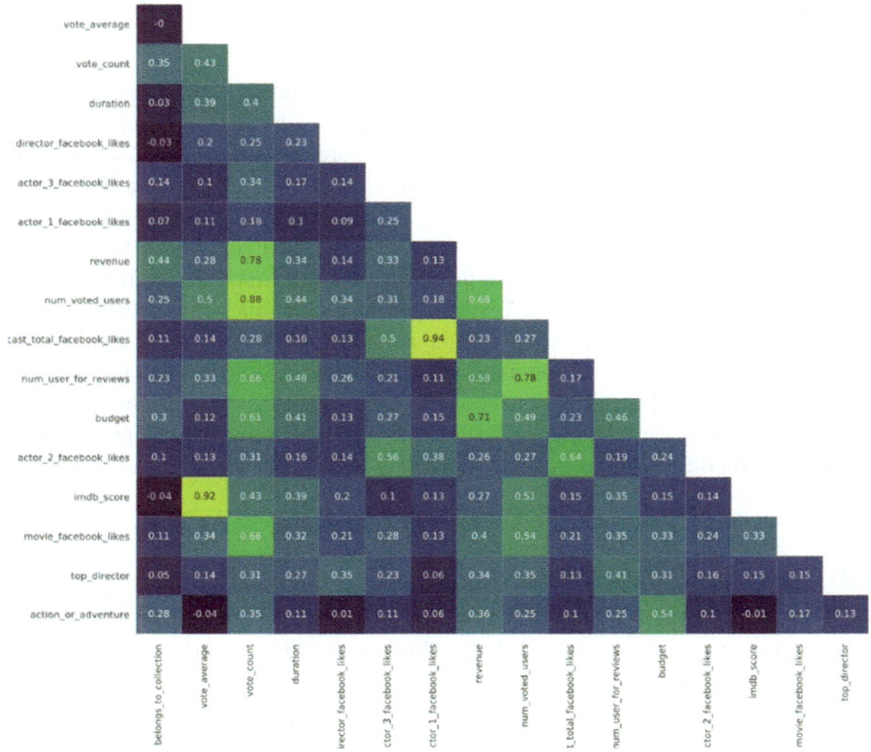

Fig. 2 Correlation heatmap of features indicating the presence of multicollinearity

clear. The features themselves were confirmed to contain multicollinearity, indicated by the correlation heatmap which is shown in Fig. 2.

5.6 *Lasso Regression*

Similar to ridge regression, lasso regression shrinks all coefficients toward a value, in this case, the absolute value of the magnitude of coefficients. This is called L1 regularization, and can sometimes lead to elimination of some coefficients. Lasso regression had similar performance to Ridge regression.

6 Results

6.1 Comparison of Models

Table 2 compares the performance of different regression algorithms based on certain metrics. The usage of a single decision tree was found to be the least effective algorithm for this scenario. The best performing algorithm was Random Forest, while Linear, Ridge, and Lasso fell shortly behind. While the former three algorithms seem to be similarly performing in terms of correlation and [2], the error metrics are much higher in Ridge and Lasso compared to Random Forest and Linear Regression. Overall, Random Forest is the most suitable algorithm, followed by Linear Regression (Figs. 3 and 4).

Table 2 Aggregation of results

Regression model	R^2	MAE	RMSE	Correlation
Linear	0.7299	0.0331	0.0509	0.8584
Support vector	0.6893	0.0408	0.2337	0.8532
Decision tree	0.6893	0.0408	0.2337	0.7626
Random forest	0.7356	0.0326	0.051	0.8941
Ridge	0.6893	0.0408	0.2337	0.8614
Lasso	0.6893	0.0408	0.2337	0.8614

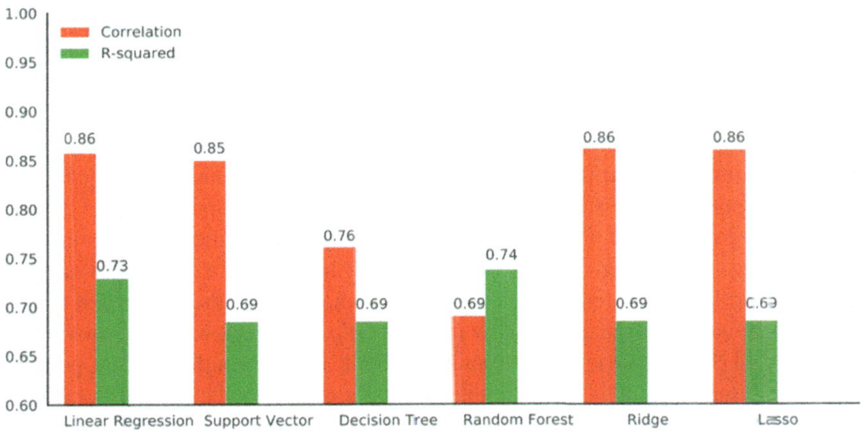

Fig. 3 Comparison of correlation and R^2 between algorithms

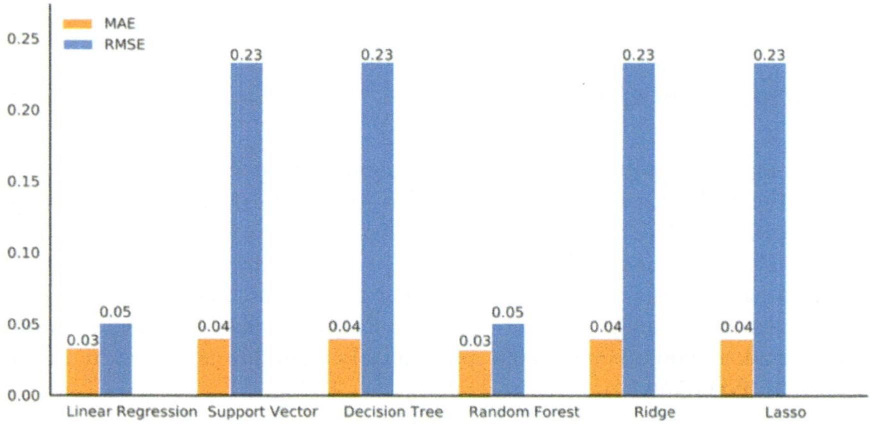

Fig. 4 Comparison of error metrics between algorithms

6.2 *The Most Important Features*

To find out the most important features in determining the success of a movie, random forest regressor was used. In each individual tree, the decision is made based on the MSE (Mean Squared Error). When training individual trees, the degree of how each feature decreases the MSE can be averaged. The features are then ranked accordingly in ascending order. The following bar plot shows that *vote_count* and budget are the most important in determining the revenue of a movie. Similarly, *vote_count* was the best predictor in [1] and budget was the best predictor in [2]. This shows that people's opinion of a movie is more important in determining the success of a movie than the budget it was allotted (Fig. 5).

The bar plot also shows that, out of the feature-engineered columns, the *belongs_to_collection* column had more importance compared to the other engineered columns. The other two columns (*action_or_adventure* and *top_director*) had very less importance but still contributed to a slight increase performance, proven when they were removed, and the performance metrics were measured again. Next to *vote_count*, the budget of a movie is the most indicative of a movie's success. The next most important features are all social media type features. They tell us how many people voted for the movie, how many people left a review for the movie, and how many liked the movie on Facebook. Overall, it can be inferred that people's opinion of a movie matters most in determining a movie's success. Furthermore, a movie's genre does not have any role in predicting the revenue, nor does the success of the director.

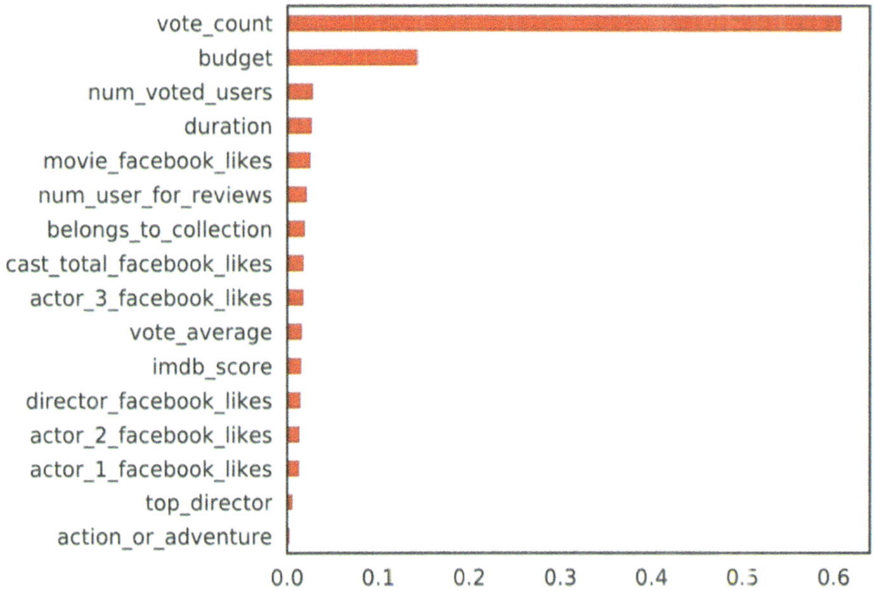

Fig. 5 Feature importances using random forest regressor

7 Conclusion and Future Work

The aim of this paper was to use regression techniques in order to predict movie revenue using movie metadata. Compared to other papers that were reviewed, the R^2 values were higher, except in one case [6], where Twitter data was used. While it is well researched that using real-time trend-based social media data like Twitter tweets can have excellent results, most other papers using movie metadata and non-hype based social media data like data from Facebook had lower performing algorithms when compared to the results found in this paper.

Error metrics could not be compared as other papers did not mention them. The best algorithm used was Random Forest, which had an R^2 value of 0.89. Most papers had implemented classification algorithms, which are unsuitable for predicting numerical values, as they have a window within which the actual value lies. If the size of this window is too big, results obtained may vary from actual results.

Overall, the methods presented in this paper show that movie revenue can in fact be predicted successfully, although some might argue that the accuracy is not enough for real life use. People's opinion of a movie matters the most in determining the success of the movie. The measure of opinion here is based on the number of likes, votes, and reviews on social media websites.

For future work, an aggregation of trend-based social media data like Twitter, social media data from Facebook, and data from IMDb could be used to improve the prediction results vastly, to the point of deploying these algorithms for real-time use.

References

1. N. Quader, M. Gani, D. Chaki, M. Ali, A machine learning approach to predict movie box-office success, in *2017 20th International Conference of Computer and Information Technology (ICCIT)* (2018) (online). https://doi.org/10.1109/iccitechn.2017.8281839
2. S. Yoo, R.K. Kanter, D.C. Cummings, A. Maas, *Predicting Movie Revenue from IMDb Data* (2011)
3. N. Vr, M. Pranav, P.B.S. Babu, A. Lijiya, Predicting movie success based on IMDB data. Int. J. Bus. Intell. **03**, 365–368 (2014) (online). https://doi.org/10.20894/ijbi.105.003.002.004
4. T. Liu, X. Ding, Y. Chen, C. Haochen, M. Guo, Predicting movie Box-office revenues by exploiting large-scale social media content. Multimed. Tools Appl. **75**, 1–20 (2014) (online). https://doi.org/10.1007/s11042-014-2270-1
5. C. Jernbäcker, P. Shahrivar, *Predicting Movie Success Using Machine Learning Techniques* (Stockholm, Sweden, 2017)
6. S. Asur, B. Huberman, Predicting the future with social media, in *Proceedings—2010 IEEE/WIC/ACM International Conference on Web Intelligence, WI 2010*, vol. 1 (2010) (online). https://doi.org/10.1109/WI-IAT.2010.63
7. H. Afzal, Prediction of movies popularity using machine learning techniques. Int. J. Comput. Sci. Netw. Secur. **16**, 127–131 (2016)
8. L. Pal, L. H. Bui, R. Mody, Predicting Box Office Success: Do Critical Reviews Really Matter?
9. M. Mestyán, T Yasseri, J. Kertész, Early prediction of movie box office success based on Wikipedia activity big data. PLoS ONE **8**, e71226 (2013) (online). https://doi.org/10.1371/journal.pone.0071226
10. R.L. Wasserstein, N.A. Lazar, The ASA statement on p-values: context, process, and purpose. Am. Stat. **70**(2), 129–133 (2016) (online). https://doi.org/10.1080/00031305.2016.1154108
11. F. Pedregosa, G. Varoquaux, A. Gramfort, V. Michel, B. Thirion, O. Grisel, M. Blondel, P. Prettenhofer, R. Weiss, V. Dubourg, J. Vanderplas, A. Passos, D. Cournapeau, M. Brucher, M. Perrot, E. Duchesnay, Scikit-learn: machine learning in Python. J. Mach. Learn. Res. **12**, 2825–2830 (2011)

Vehicle Recognition Using CNN

V. K. Divyavarshini, Nithyasri Govind, Amrita Vasudevan, G. Chamundeeswari, and S. Prasanna Bharathi

Abstract We are designing the software with the help of Convolution Neural Network (CNN) with 25 layers of max pooling. The proposed work is achieved by python or MATLAB with Support Vehicle Machine (SVM) with HOG. The SVM is used first to recognize the vehicle and then CNN is used to split or classify them into domains and keyframes of a specific video. This automation can be used in toll gates, highways, heavy traffic surveillance, etc.

Keywords Convolution neural networks · Max pooling · Support vector machine · Deep learning · Machine learning

1 Introduction

In modern years, we have observed fascinating growth in fields related to artificial intelligence like automatic vehicles, vision of computers, machine learning and deep learning. As a newcomer, it is very difficult to remain updated due to the drastic changes that occur in this field. Till date there are no journal and conference

V. K. Divyavarshini (✉) · N. Govind · A. Vasudevan (✉) · S. Prasanna Bharathi
Department of ECE, SRM Institute of Science and Technology, Vaadapalani, Chennai 600026, India
e-mail: divyavarshini13@gmail.com

A. Vasudevan
e-mail: vasudevanritu@gmail.com

N. Govind
e-mail: nithyasrigovind23@gmail.com

S. Prasanna Bharathi
e-mail: prasanns@srmist.edu.in

G. Chamundeeswari
Saveetha School of Engineering, SIMATS, Chennai, India
e-mail: easwarig@gmail.com

S. S. Dash et al. (eds.), *Intelligent Computing and Applications*,
Advances in Intelligent Systems and Computing 1172,
https://doi.org/10.1007/978-981-15-5566-4_60

papers which deal with generalized study of automatic vehicles using large amount of datasets, but for specific problem related papers are available.

This project provides the software which can be used accurately to solve such problems. Here as it turns out, deep neural networks are outperforming the approach. Categorization of vehicles manually is strenuous; hence by automating this work will enhance the efficiency to the user. So, for the proposed problem, we have designed a system using machine learning by developing a software for identifying each vehicle separately in an image which is captured in a high definition camera. In real time, taking photographs are a tedious process, in order to overcome we are using a video which splits into images which we normally call that as 'keyframes'.

Each keyframes can be splitted from minutes to nanoseconds, for example, if we captured a video of one minute of moving vehicles in a pedestrian, we can split the image from 1 s to 1 ns. Here, we are using three main algorithms for developing this software which is CNN 'Convolution Neural Network', SVM 'Support Vector Machine', HOG 'histogram of oriented gradients'. Earlier, the classification of vehicles was done by observing and analyzing a single vehicle at a time. This in turn reduces the accuracy of the classification.

Every year, new algorithms/models keep on outperforming the previous ones. The latest state of the art software system for object detection was just released last week by Facebook AI team. The software is called Detectron that incorporates numerous research projects for object detection and is powered by the Caffe2 deep learning framework. Now, this is made easier through data science and the people who work with these tools are called data scientist. In order to figure out the complicated analytical problems, data science can be used which is a unique combination of both datasets and algorithms. Datasets is a collection of information that has been accumulated for a very long time in a specific location; this information can be used for research purposes.

This data can be much learned in the mining process [1]. We can build with the data available through advanced algorithms. The main purpose of this project is to detect and classify the objects in an image using image processing of machine learning for example vehicles. The sole purpose of this project is to help the government in regulating the traffic issues that is being by the commoners. This can be achieved through MATLAB, though there are different tools available in the market. The most popular tool used for image processing recognition is python 3.7.0 (latest version). The method by which this classification has to be done is simplified in this project to make it user friendly and easy for execution. Mainly, with the renaissance of deep learning, convolutional neural networks have automated this task while significantly boosting performance.

The basic requirements to develop this software are

Hardware:

- Laptop
- High definition camera.

1.1 Software Used

MATLAB—The tool which we are executing is MATLAB. It merges a desktop environment which is adaptive for the repetitive analysis and design processes with a programming language that disclose matrix and array mathematics directly. Within few lines of MATLAB code, we can frame deep learning models without being an experienced person. MATLAB can run models up to 7× faster than tensor flow and up to 4.5× faster than caffe2. It uses the functions and tools to envision the common results and debug deep learning models.

1.2 Proposed Work

Here is a short introduction into traffic surveillance cameras for vehicle detection. Vision-based algorithms for vehicular detection can be divided into three categories: movement-based approaches, hand-made practical approaches and CNN-based approaches. Motional solutions include subtracting frames and subtracting the history. Motion-based methods include subtraction of frames and subtraction of history. Image subtraction computers for the identification of the movement target the variations in two or three series images. Frame subtraction has a simple calculation and a dynamic background, but is not ideal to move too quickly or too slowly. Handcrafted feature-based approaches include Histogram of Oriented Gradients. CNN methods have shown a rich power of representation and achieved promising results. CNN uses object proposal to prepare CNN for detection tasks by selective searching. In the framework of CNN, ALEX-Net and Fast CNN speed up the generation of regional proposal on the map. computer only needs these approaches once. Instead of selective search, Faster CNN uses the region proposal network, then it can train end to end and improve speed and accuracy.

2 Algorithms

The breakthrough for this project is done by three main algorithms:

- CNN—Convolution Neural Networks
- SVM—Support Vector Machine
- HOG—Histogram of Oriented Gradients.

2.1 Convolution Neural Network

The convolution neural networks have tiny fragments as their input images and these are called s receptive fields (Fig. 1). This type of feature extraction was inspired by the visual mechanism in living organisms, where cells in the visual cortex are perceptive to small regions of the visual field. There are two specific types of layers in CNNs: the convolution layer and the pooling layer. In the convolution layer, the images are convoluted by divergent convolution filters via shifting the receptive field gradually (Fig. 1). The convolution filters contribute the identical parameters in every small portion of the image, largely condensing the representation of hyperparameters in the model. The pooling layer, taking leverage of the 'stationary' property of the images, takes the mean, the max or other data of the countenance at discrete positions in the feature maps, thus reducing the variance and capturing necessary features. In the above table, the convolution layer fields are represented by coloured blocks.

The input patch is expressed by the letter 'a', and it is actually multiplied with the matrices of the pooling layer. The outcome of this matrix is represented by a_{ij}, c_{ij} and k_{ij} at line I and column J. Convolution layers operate by passing on the input images to the next layer by applying a convolution [2]. The function of the convolution is to imitate a similar response of a respective neuron to a visual stimuli. Each field is taken concentrated individually and the data from each neuron is processed. It is not practical to use can architecture on images which learn the features as well as classify the images which is done by a fully connected neural network.

As the number of neurons that are required is very high, even in a shallow architecture, because the pixels in a very large input image which is relevant are very high.

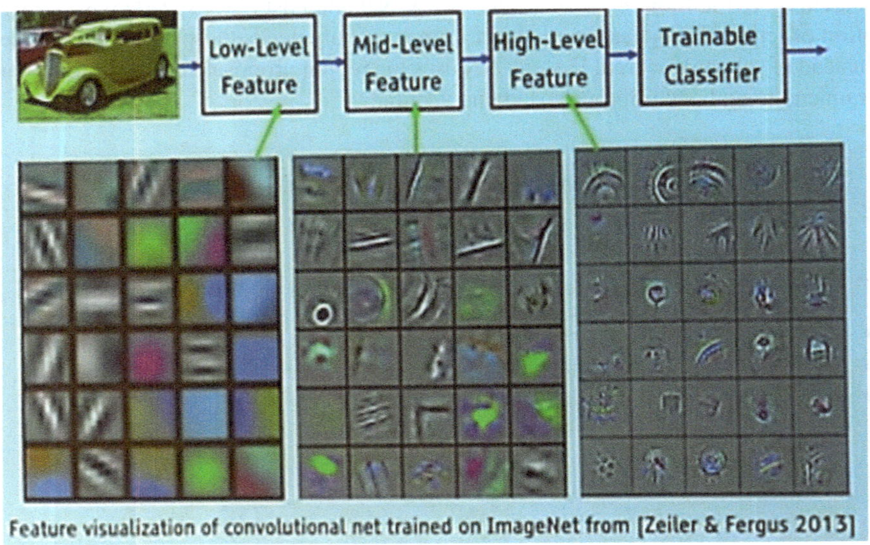

Fig. 1 CNN

For example, a small image which is actually connected with the layers will eventually have a size of 100 × 100 and its weight will be almost equal to 1000 neurons which are present in the second layer. A solution is put forward by the convolution operation by decreasing the free parameters. This allows the network to function deeper with the help of fewer parameters. For example, regardless of the size of the image, tiling regions present in an image is of size 5 × 5. They share their weights equally and the learnable parameters required are only. In this method, the problem of vanishing or exploding of gradients is resolved by training traditional multi-layer neural networks with many layers. This method is called sing back propagation.

3 Model-Based on Vehicle Recognition Tensor

3.1 Tensor Flow

Today, Google's TensorFlow is the world's best-known deep learning library. Google products use machine learning to enhance search engine, translation, underline picture and suggestions in all of their products.

Google users may experience a more rapid and streamlined search with AI to give a specific example. Google makes a recommendation on what the next word might be when the user types the keyword in a search bar. Google would like to use machine education, so that users get the best experience from their massive datasets. Machine learning is used by three different groups: researchers, data scientists, programmers.

They can all work together and improve their effectiveness using the same toolset.

Google has not only data; it's the largest machine in the world, so Tensor Flow has been designed for scale. TensorFlow is a Google brain group library built to expedite machine learning and groundbreaking research on neural networks. It has been designed for running on several CPUs or GPUs and handheld devices, and is bundled in several languages such as Python, C++ and Java.

3.1.1 Pooling

A pooling layer is another section of Knits main action to successively decrease the structural size of the representation which will thereby decrease the computation and the number of parameters which is necessary in the network. The pooling layer operates only on a single feature map independently [3]. Both global and local pooling layers can be included in the convolutional networks, by this the outputs of a neuron cluster are combined in one layer with a neuron in another layer. One example is average pooling, which takes the average value from each of a cluster of neurons at the prior layer. And another example is max pooling which makes use of the maximum value from each of a cluster of neurons at the prior layer. The most common approach used in pooling is max pooling.

3.2 *Support Vector Machine*

In the support vector machine the machine learning has a very good role in classifying the images into different domains. It is a classifier defined by a hyperplane. The algorithm learns to classify the new images when it is trained by a labelled dataset. For instance, if an image is allowed for classification, this algorithm will divide itself into a set of hyperplanes or a single hyperplane in an infinite or high-dimensional space, which is then used to perform various tasks like outliers detection.

The tuning parameters in SVM classifier are

- Kernel
- Regularization parameter
- Gamma
- Margin.

By varying these parameters, we can achieve considerable accuracy in reasonable amount of time.

3.2.1 SVM Classifier

Figuring out the (soft-margin) SVM classifier in minimum expression is

$$\left[\frac{1}{n}\sum_{i=L}^{n}\max(0.1-y_i(w,x_i-b))\right]+\lambda\|\omega\|$$

The problem of quadratic programme involves by selecting small value of λ yields (Fig. 2) and that's the reason behind the full focus to the classifier first.

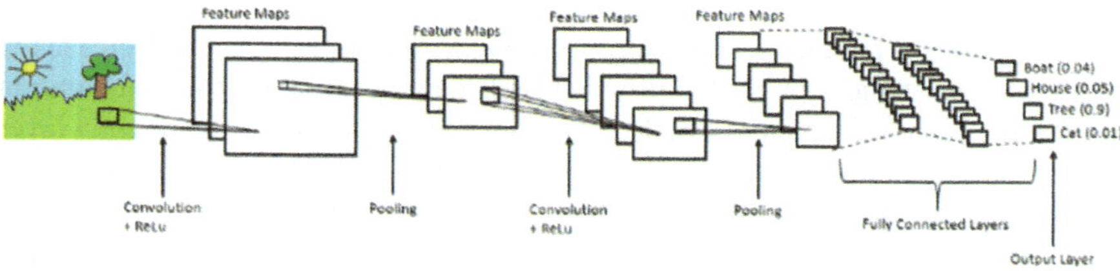

Fig. 2 CNN flow diagram

3.2.2 Kernel Classifier

The nonlinear distribution rule which coincides with linear distribution rule for the transformed data point $\varphi(\overline{x_i})$ is given by kernel function k which satisfies $k(\overline{x_i}, \overline{x_i}) = \varphi(\overline{x_i}), \varphi(\overline{x_i})$.

The classification vector $\bar{\omega}$ in transformed space gives

$$\bar{\omega} = \sum_{i=L}^{n} C_i y_i \varphi(\overline{x_i})$$

where the C_i is obtained by explaining the problem.

3.2.3 Linear SVM

The training dataset of n points are taken into consideration

$$(\overline{x_i}, y_i), \ldots, (\overline{x_n}, y_n)$$

where y_i can be -1 or 1, that indicates the group the point where $\overline{x_i}$ belongs. Every ρ dimensional real vector has $\overline{x_i}$. The 'maximum-margin hyperplane' has to be calculated that splits the group of points $\overline{x_i}$ for which $y_i = 1$ and from the group $y_i = -1$ is taken and the distance between the nearest point $\overline{x_i}$ and the hyperplane from any one of the group is zoomed.

3.2.4 Nonlinear Classification

In Fig. 3 authentic maximum-margin hyperplane algorithm known as linear classifier is shown in a graph with x axis as 'x1' and y axis as 'x2'. It is distributed by employing the kernel trick to maximum-margin hyperplanes. The emerging algorithm is generally similar, but the dot product is restored by a nonlinear kernel function. Hence for a particular area with the help of the algorithm, the maximum margin of the hyperplane is estimated. Even though a hyperplane is the classifier and charged feature area, it might be the nonlinear original input, but the changed area will have higher dimensions. Engaging in higher dimensional area can be an advantage as it reduces rationalization error in SVM without effecting the performance of the algorithm (Fig 4).

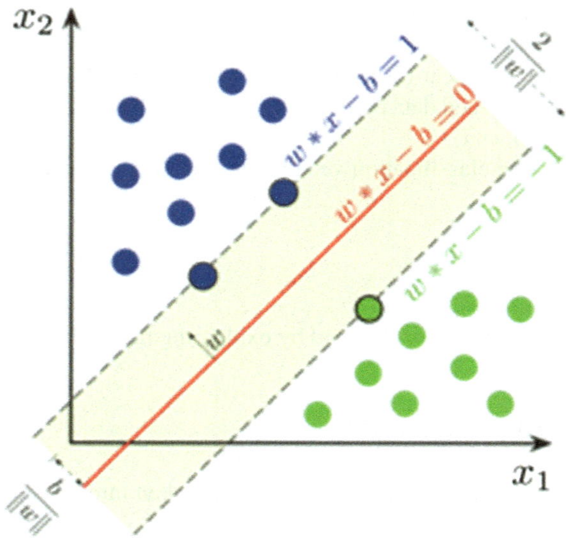

Fig. 3 Maximum-margin hyperplane and margins for an SVM trained with samples from two classes. Samples on the margin are called the support vectors

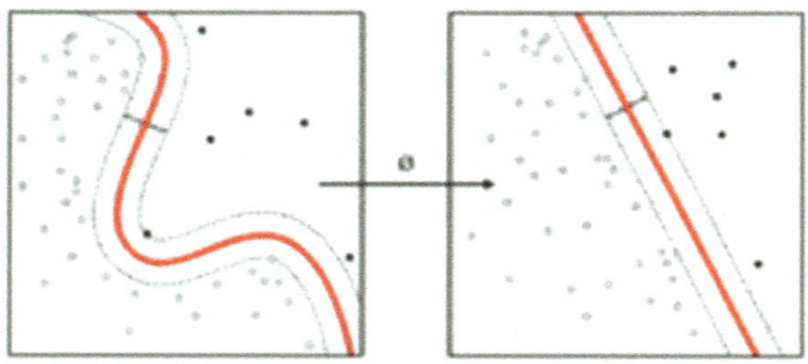

Fig. 4 Nonlinear classifier

3.3 Histogram of Oriented Gradients

In image processing and computer vision, a component called Histogram of Oriented Gradients (HOG) is used which is a feature descriptor used for detection of object. The function of a feature descriptor is to delete the extraneous information and obtain the relevant content by compressing the image or image patch and provide an indication. The method calculates the localized occurrences of gradient orientation in an image. This method is same as that of changes descriptors, and shape contexts, but varies in the fact that it is evaluated on a thick grid which is of equally spaced cells and makes use of overlapping [4] local contrast generalization to achieve better accuracy.

To calculate this the following are used:

- Pre-processing
- Calculate the Gradient Images
- Calculation of Histogram of Gradients in 8 × 8 cells

Fig. 5 Pre-processing

- 16 × 16 Block Normalization
- Calculate the HOG feature vector.

3.3.1 Pre-processing

For pedestrian detection HOG feature descriptor is used and 64 × 128 image is applied for the estimation (Fig. 5). The size of the image can vary. Generally, each patch is monitored at different image locations at multiple scales. The only limitation is that there are fixed aspect ratios for the patches that are being monitored. To explain this point, a patch of size 100 × 200 is selected for determining our HOG feature descriptor. The image is then cropped and resized to 64 × 128; the image is retrieved from the patch. The gamma correction is also done as a pre-processing step, but the results obtained are minimum thus we are avoiding this step.

3.3.2 Calculate the Gradient Images

The HOG descriptor can be calculated using the vertical and horizontal gradients. Figure 6 shows the gradient images. This is quickly accomplished with the filtered images of the following kernels

$$g = \sqrt{g_x^2 + g_y^2}$$

$$\theta = \arctan \frac{g_y}{g_x}$$

The vertical lines represent the x-gradient and the horizontal for y-gradient lines. When there is change in the intensity, the magnitude of the gradient changes. At

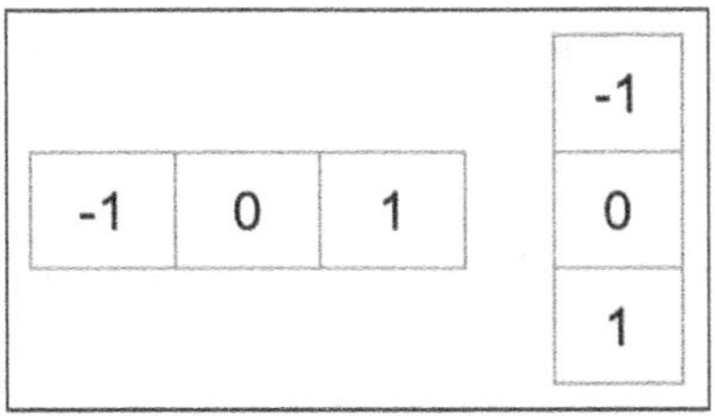

Fig. 6 Gradient image

each pixel, the gradient has vector and magnitude. For coloured images, it helps to identify the gradients of the three channels (as shown in Fig. 6). The magnitude of the gradient depends upon the magnitude of each pixel inside the gradient.

3.3.3 Calculate Histogram of Gradients in 8 × 8 Cells

Here, the picture is segregated as 8 × 8 cells and the histogram is found in each cell. The main reason to use the feature descriptor is to get a patch of image which in turn gives a compact representation of the gradients. Each 8 × 8 image patch has 8 × 8 × 3 which is equal to 192-pixel range.

The gradient contains vector and magnitude per pixel in which the dimensions have 8 × 8 × 2 which is 128 in number, these 128 pixels of magnitude and vector representation are denoted by 9-bin histogram gradient which is kept at an array of 9 numbers. Now, by determining a histogram in an 8 × 8 patch will make the gradient into less noise. Each gradient will have some noise, but this type of representation will make the gradient into less disturbance (Figs. 7 and 8).

The histogram has 9 bins parallel to angles 0°, 20°, 40°, …, 160°. Now, the gradients look like 8 × 8 in an image.

The image mentioned above as Fig. 8 shows the gradient and the direction of the gradient with its length showing the magnitude. The direction of arrows directs to the change in intensity and the magnitude shows how big the direction change is.

From Fig. 8, we are able to observe some raw numbers displaying with one minimum difference inside an 8 × 8 cells—the angles are between 0 and 180° than 0 and 360°.

They are called 'unsigned' gradients. Else, the same category is obtained by the 180 degree opposite and gradient arrow. Now, the final step is to generate 8 × 8 cells of histogram of gradients. The histogram has 9 bins parallel to angles 0°, 20°, 40°, … 160°.

3.3.4 16 × 16 Block Normalization

Previously, image was generated using histogram based on the gradient. The lighting is the main part of the gradient orientation (Fig. 9). The darker image is segregated by pixels first, all the pixel values are divided by 2 so that the gradient magnitude

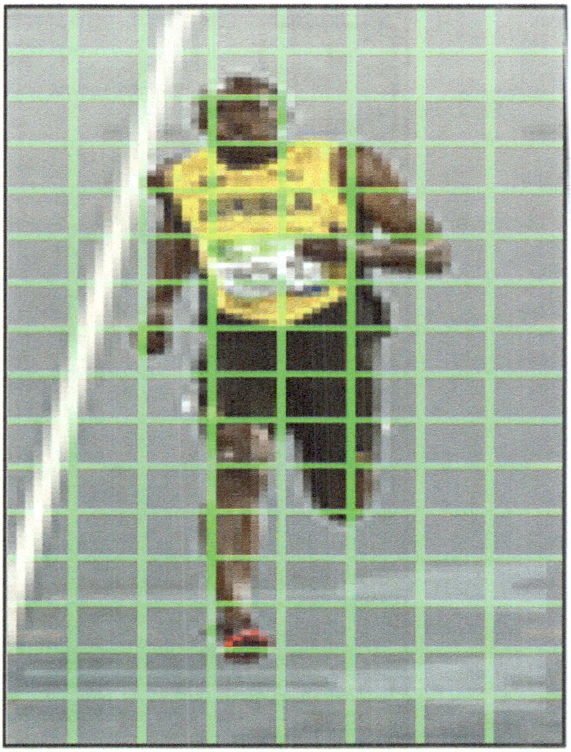

Fig. 7 Fully layered image

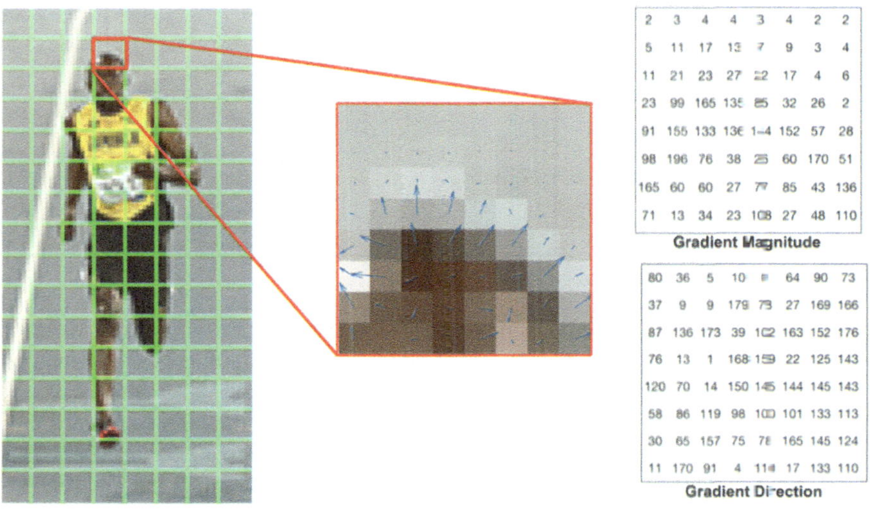

Fig. 8 8 × 8 cell image

Fig. 9 16 × 16 block normalization

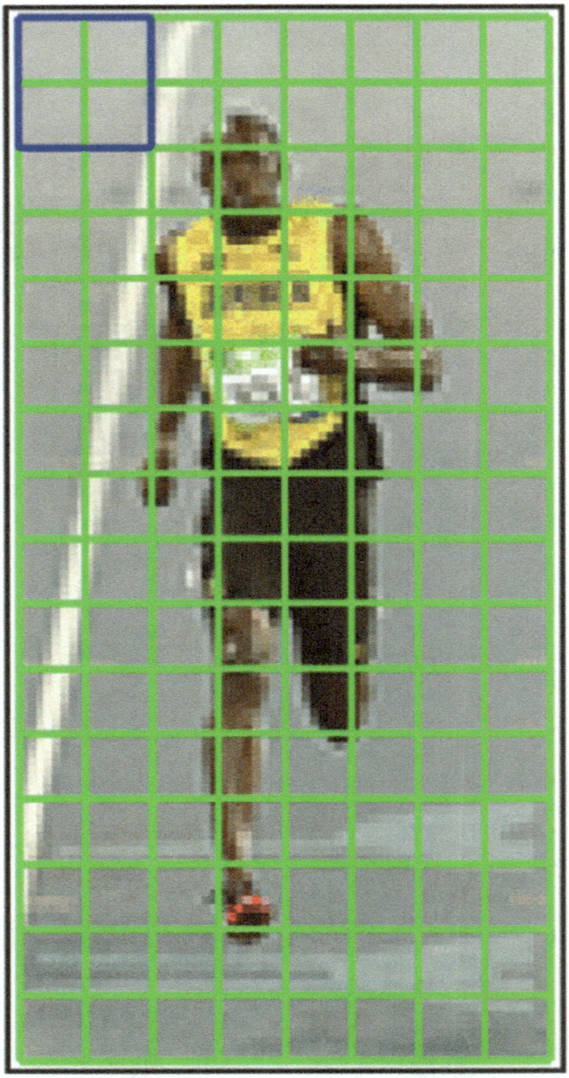

becomes half. Now, automatically, histogram value becomes half. Basically, we don't want any lightning differences in the captured image, to rectify that we can normalize our descriptor of the histogram.

Colour vector is of RGB type [128, 64, 32]. L2 norm is the length of the vector. By dividing each element by 146.64 and the resultant normalized vector [0.87, 0.43, 0.22]. Now, consider another vector which has twice the value than the first one. So, the equation becomes 2x [128, 64, 32] = [256, 128, 64]. The normalized value of the second vector is [0.87, 0.43, 0.22], which is exactly same as the first vector.

3.3.5 To Calculate the HOG Feature Vector

For calculating the complete image patch, we have to multiply 36 × 1 vector with one giant vector [5] which forms a final feature vector. And to calculate that

1. Totally, there are 105 positions which 7 in horizontal and 15 in vertical positions.
2. So, in each 16 × 16 block, we have 36 × 1 vector (Fig. 10) which internally calculated, we get 36 horizontals and 105 vertical vectors of totally 3780-dimensional vector.

4 Data Science

To expand the software, the primary concept of data science strategies is needed. It is known as 'data product'. It is an advantage for every data scientist. More precisely, a 'data product' is a biggest credit which makes use of the input and processes each and every data to get logically and algorithmically generated outputs. The most common example is an engine, which analyze the user data and personalize itself based on the data which has been received previously. There are examples of these data products:

- The recommendation engines of the Amazon website which shows the suggested items in one corner of the page which the engine suggests based on the interest of the user by some algorithms.
- The next is Netflix which suggests movies for us.

4.1 Data Scientist

The interest and the training of data is the capital contribution for the scientist. The people who deal with data with science are known as 'Data Scientist'. A common personality character of data scientists, they are with high curiosity and very deep thinkers. Being analytical is the part of Data science like exploring new things, different methods of solution for a single problem, asking questions for every single problem thrown at them. Data science is multidisciplinary.

Fig. 10 36 × 1 patch image

4.2 Analytics

If a person thinks quantitatively, then that thinker is analytical thinker by nature. Materialistically, 'science of analysis' defines analytics. A person who connects both science and statistics together with some math, technology and business awareness is said to be Data Scientist. They basically work with raw data and analyze it and finally build a product out of it. Though data scientist analyses the data but not as deep as analyst. Their job is to interact with the data scientist and develop a new database for further research purpose. They mainly try to get the insights of the raw data which they acquire from data scientist.

5 Detailed Description Using a Flow Chart

Step 1: A road swamped with vehicles is selected and a video of few seconds is recorded. The video can be recorded for about few nanoseconds or milliseconds.

Step 2: Now the images in the video are splitted into few keyframes for few seconds or milliseconds. As the number of keyframes increases the accuracy increases.

Step 3: From the keyframes, the object is recognized as vehicle by using SVM then the data is forwarded to CNN.

Step 4: The data forwarded to CNN was nothing but the different keyframes. These keyframes are recognized as vehicle by SVM and categorized with the help of CNN. The classification can be done based upon the user need.

Step 5: These keyframes which were recognized and categorized are now attached back and the original video is retrieved back.

Step 6: Now the final video is played in which the vehicles are categorized properly.

The following task of training a software with an input data set as well as the expected output data is a method that is followed in machine learning. In order to complete this, the training process has to undergo various stages.

- Construction of software.
- Training the datasets.
- Testing the images collected from the datasets.
- Evaluation of the processed image.

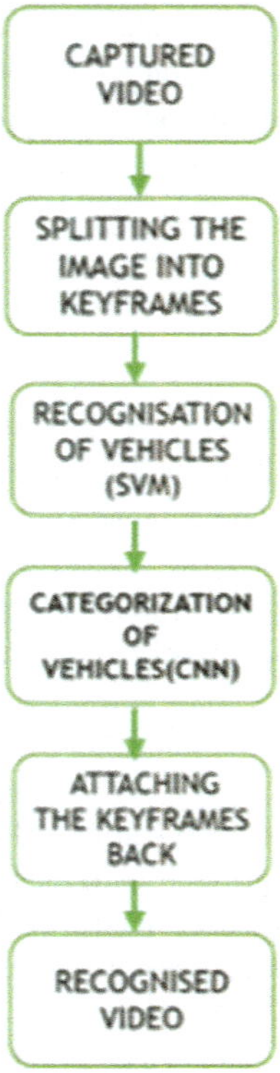

6 Training Model

Training data is provided by images of vehicles (Fig. 11). In order to detect a vehicle on the image, we need to identify feature(s) which uniquely represent a car. We could try using simple template matching or relaying on colour features but these methods are not robust enough when it comes to changing perspectives and shapes of the object. Here are two samples.

In order to have a robust feature set and increase our accuracy rate, we will be using Histogram of Oriented Gradients (HOG). This feature description is much more resilient to the dynamics of the traffic. MATLAB library provides us with the necessary API for calculating HOG feature. Here Y Cr Cb colour space and all its channels are used as inputs for HOG features extraction (Fig. 12). Here's a sample of vehicle and nonvehicle image with HOG features from the same images as above:

In order to detect the vehicle based on our feature set, we would need a prediction model. Linear Support Vector Machines (Linear SVMs) is an administered learning model that will be able to classify whether something is a car or not after we train it.

In order to improve the accuracy of the final output, we will be trying to find multiple hits for the object of interest in the similar area. This approach is equivalent

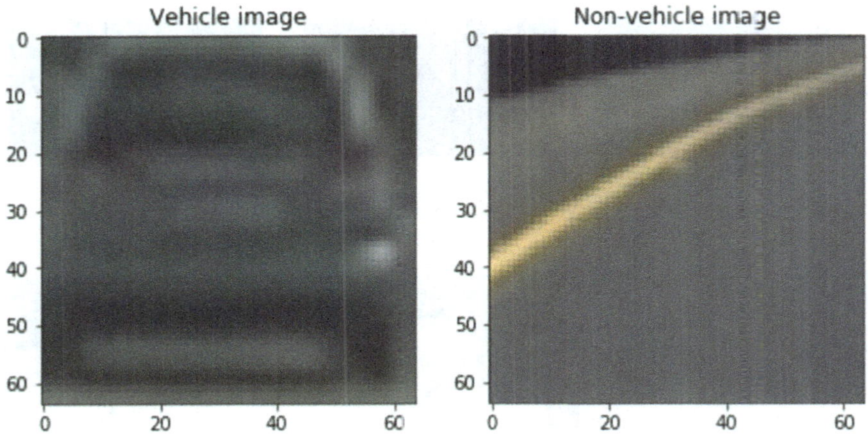

Fig. 11 Training model from the dataset

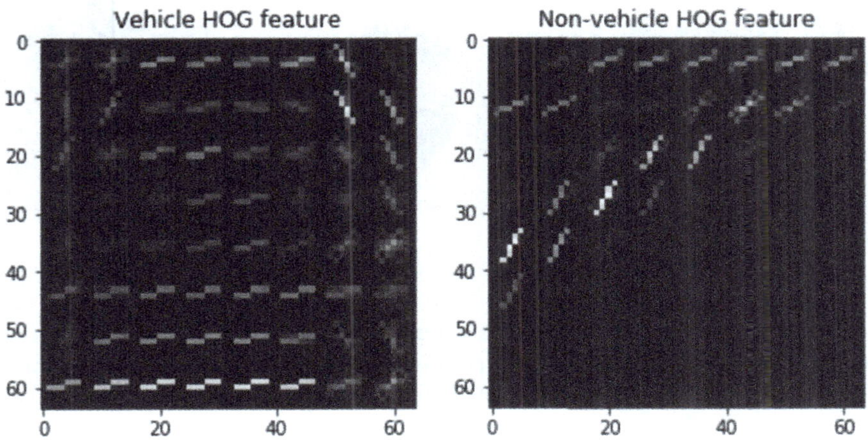

Fig. 12 Differentiation in HOG identified vehicle and nonvehicle

to creating a heat map. The next step is threshold which needs to be met in order for a specific hit count from the heat map to be accepted as a detected vehicle (Fig. 13).

7 Testing Model and Output

Here, the CNN process is tested with a real time video in the swamped roads of vehicles in the busy morning time. The file here is saved as 'test.mp4' (Fig. 14).

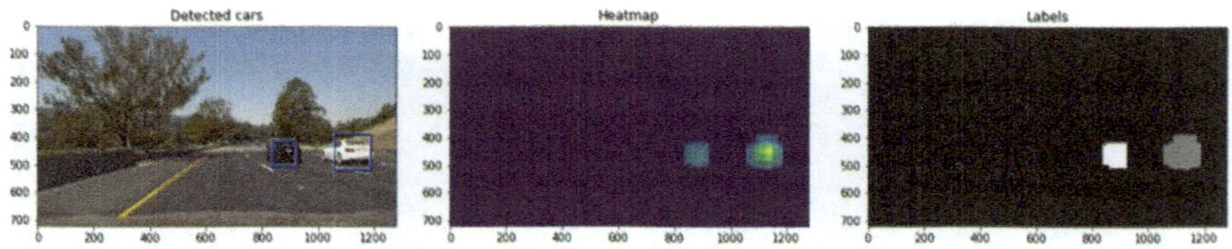

Fig. 13 Stereotypic identification by any machine learning

Fig. 14 Output of test.mp4

The performance of the pipeline is not great and can be improved. Deep Neural Network approach would have better performance numbers. Since this car detection approach is based on camera it's prone to usual challenges with this kind of sensor (bad visibility, reflections, etc.).

8 Advantage and Disadvantage

The coding is less complicated when compared to other software. The software is effectively used in areas such as toll gates in classifying vehicles without human intervention.

It can also be used for statistical analysis of the vehicles.

The software is smart enough to find the vehicle in less distorted image or keyframe of specific video. The CNN is basically slow in every software regardless of whether its python or MATLAB. The training process is slow in CPU when compared to GPU.

9 Acknowledgement

This software can be effectively implemented not only in the field of automobiles but also in various other fields like health care, entertainment, electronic, security, merchandise, etc.

9.1 Smart Parking

This system can be used for smart vehicle parking where the number of vehicles enters the parking plot. If the number of parking slots in a particular level is stored then when the vehicles entered exceeds that number, an indication will be shown.

9.2 Health Care

This software has been able to provide a new method for generating genetic data. This recent advancement in technology will open new doors in the field of biotechnology and help us to solve the mysteries of human genetic structure. Genomic sequence analysis structure can be achieved using deep learning algorithm [6].

- Colorization of b/w frames.
- Adding some background music to no sound movies.
- Classification of objects in a picture.
- Generation of caption in an image.
- Speech recognition like Google assistant and Amazon echo.
- Face recognition like Google photos, Microsoft how-old.

10 Conclusion

The vehicle categorization is a very promising access since it incorporates the advantage of noticeable types of information. The efficiency that is acquired so far in this project is 67 % (Fig. 15), but the efficiency can be further increased by using more data sets and training the software with them.

Vehicles	Expected results (%)	Arrived result (%)
Car	100	60
Auto	100	10
Motorbike	100	30

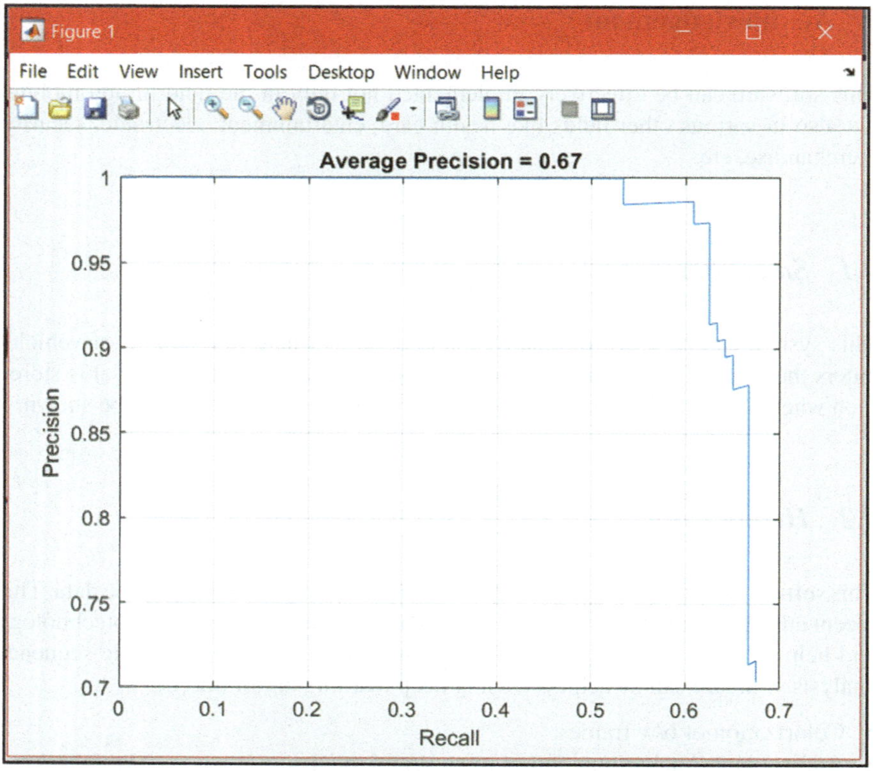

Fig. 15 Accuracy of the output

References

1. B. Selbes, M. Sert, Multimodal vehicle type classification using convolutional neural network and statistical representations of MFCC, in *2017 14th IEEE International Conference on Advanced Video and Signal Based Surveillance (AVSS)* (2017)
2. C. Cao, F. Liu, H. Tan, D. Song, W. Shu, W. Li, Y. Zhou, X. Bo, Z. Xie, Deep learning and its applications in biomedicine. Genomics Proteomics Bioinform. (2018)
3. *Proceedings of the International Conference on ISMAC in Computational Vision and Bio-Engineering 2018 (ISMAC-CVB)*. (Springer Nature, 2019)
4. Z.-S. Huang, C.-C. Chuang, C.W. Tao, M.-Y. Hsieh, C.-X. Zhang, C.-W. Chang, IOS-based people detection of multiobject detection system, in *2016 Joint 8th International Conference on Soft Computing and Intelligent Systems (SCIS) and 17th International Symposium on Advanced Intelligent Systems (ISIS)* (2016)
5. D. Prasanna, M. Prabhakar, An efficient human tracking system using Haar-like and hog feature extraction. Cluster Comput. (2018)
6. J. Ravikumar, A.C. Ramachandra, K.B. Raja, K.R. Venugopal, Convolution based face recognition using DWT and HOG, in *2018 International Conference on Intelligent Informatics and Biomedical Sciences (ICIIBMS)* (2018)

GLCM and GLRLM Based Texture Analysis: Application to Brain Cancer Diagnosis Using Histopathology Images

Vaishali Durgamahanthi, J. Anita Christaline, and A. Shirly Edward

Abstract *Objective* This work focuses on design and development of an automated diagnostic system using brain cancer histopathology images. *Background* Detection and classification of cancerous tissues are one of the challenging tasks in histopathology image analysis. The features based on cell morphology, cell distribution, and randomness in growth and placements have been considered as important aspects of the cancer diagnosis. Results of the conventional methods are subjective and dependent on the skills of histopathologists. The computer-assisted diagnosis (CAD) offers the diagnostic results based on standard algorithms and standard test database. *Material and Methods* This work shows the contribution of gray-level co-occurrence matrix (GLCM) based Haralick features and gray-level run length matrix (GLRLM) features in analysis and classification of Brain histopathology images. The primary focus of this study is to analyze the complex random field of cancer images with co-occurrence matrix features, run length matrix features, and also the fusion of both features. These features are used for classifying healthy and malignant tissue. The SVM classifier with RBF kernel has been used for classification. The Confusion matrix is formed showing true positive (TP), true negative (TN), false positive (FP), false negative (FN) information. The classifier performance has been described with sensitivity, specificity, precision, accuracy, and F-score measures. *Results* Outcome of this work shows that GLCM-based and GLRLM-based features offer excellent discriminating features for statistical study of histopathology images and can be useful for cancer detection. Overall accuracy improvement is seen by fusing the GLCM and GLRLM based texture features. *Conclusion* The work describes an innovative way of using GLCM and GLRLM based textural features to extract underlying information in brain cancer imagery.

V. Durgamahanthi (✉) · J. Anita Christaline · A. Shirly Edward
Department of Electronics & Communication Engineering, SRM University, Vadapalani, Chennai, Tamil Nadu, India
e-mail: vaishalb@srmist.edu.in

J. Anita Christaline
e-mail: anitaj@srmist.edu.in

A. Shirly Edward
e-mail: edwards@srmist.edu.in

S. S. Dash et al. (eds.), *Intelligent Computing and Applications*, Advances in Intelligent Systems and Computing 1172,
https://doi.org/10.1007/978-981-15-5566-4_61

Keywords Computer-assisted diagnostics (CAD) · Gray-level co-occurrence matrix (GLCM) · Gray-level run length matrix (GLRLM) · Support vector machine (SVM)

1 Introduction

1.1 Background and Motivation

For many decades, researchers have been working to understand cancer, its causes, and its various development stages so that they can offer the best solution and arrest its development. The main cause of cancer is uncontrolled mitosis of human cells caused by a mutation in a cell's DNA. This reaction triggers production of more and more damaged cells. The accumulation of these cells forms a malignant tumor in later stages [1, 2]. As the entire mitosis process has become exceptionally random, cancer detection is a challenging task. Histopathology is the study of microscopic cells and tissue structures, cells growth and its placement forming which is identified as a perticular textured pattern in a diseased cases. The texture is a main feature of cancer histopathological image [3]. According to WHO reports, 80% of biopsy samples studied for cancer diagnosis have been found to be benign. The processing of biopsy samples has become quicker and efficient with the development of whole slide imaging (WSI) or digital pathology. The introduction of fast scanning devices and the advancement in computer technology have replaced the subjective judgments of a histologist with an accurate objective judgment with computers' assistance. The vital aim of cancer research is to study its spreading activity (metastasis) and to design the best treatment to confine it to a small area of the human body. In a computer-assisted diagnosis (CAD), researchers have developed many algorithms for detection of symptoms as prognosis tool and detection of disease-specific features as a diagnosis tool. Early detection of the disease leads to accurate treatment and also increases survival rate.

1.2 Our Contribution

This research is aimed to study brain histology images and classify these images into normal and malignant, based on textural attributes in cancer biopsy samples [4, 5]. Our aim is to use the Haralick texture features which are based on gray-level co-occurrence matrix (GLCM) and gray-level run length matrix (GLRLM) features to classify histopathology images into the normal and malignant sample using support vector machine (SVM) classifier [5, 6]. This work also describes the fusion of extracted features to achieve better accuracy in diagnosis. The performance of SVM classifier has been checked by describing confusion matrix. Comparison of

GLCM and GLRLM based classification with model-based approach has also been discussed.

The article is organized as follows. Section 2 gives an overview of related work done in the past. Section 3 describes materials and methods; it describes basics of GLCM and Haralick texture features and GLRLM features. The brief description on TMAD is also given. It also describes classification with performance evaluation measures. Section 4 describes experimentation and results. In Sect. 5, we draw a conclusion.

2 Histopathology Image Analysis: An Overview

Advancement in imaging modalities and computer resources has opened the new doors for the research to analyze disease-specific image features. Several researchers in the field of clinical pathology and image processing have acknowledged the importance of digital pathology and histopathology images. The work has been performed to improve algorithms involved in computer-assisted diagnoses (CAD), mainly image enhancement, underlying information extraction in terms of model parameters, and classification for disease prognosis and diagnosis [7–9]. Histopathology image analysis has been classified into two parts. object-level and spatial-level feature analysis. Object-level features are further characterized into (i) Size and shape, (ii) Radiometric and densitometric, (iii) Textural features, and (iv) Chromatin-specific features. In these image analysis techniques, irregularities in cell distribution and morphologies have been used to detect the malignancy and also the level of malignancy [10–13]. To avoid interobserver and interobserver variation in results, researchers have suggested quantitative analysis of tissue and cell structures. Gunduz et al. have considered cell graph and have suggested how to quantify organizational characteristics using cell graph for cancer diagnosis. O. Sertel et al. have proposed the method to grade tissue abnormalities based on the amount of cytological components present in a tissue sample. He has performed texture analysis based on gray-level co-occurrence matrix [14, 15]. Our previous work describes the use of autoregressive parameters for classification of brain cancer histopathology images considering model-based image analysis [16, 17]. The lymphocytic infiltration has been considered as a key component in cancer prognosis and diagnosis through which one can study the interactions of the tumor with immune systems. This study has further been used for setting therapeutic targets [18]. Multiresolution analysis in wavelet sub-band has been proved as a revolutionary tool to analyze complex and heterogeneous random field of histology images. Many researchers have addressed the heterogeneity problem of cancer tissue in wavelet transform domain [19, 20]. The selection of correct classifier is one of the key steps while handling histology image database. Several researchers have worked on different classification approaches to achieve better classification accuracy. Several machine learning methods have been established for classification; they are neural network based multilayer perceptron (MLP), k-nearest neighborhood (KNN), a logistic regression method, fuzzy systems, support vector machine (SVM), etc. Considering the amount of database in pathology imagery, it is suggested by many researchers to use multiclassifier systems for a better result [20–29]. This work has been focused on texture-based feature analysis. The texture is a connected set of a pixel which shows repetitive occurrence. Texture provides information about the intensities in the form of smoothness, coarseness, and regularity. Textural feature analysis primarily considers GLCM features, GLRLM features, fractal features, multiresolutional features in wavelet sub-bands, entropy-based features, and chromatin-specific features. These features form a base for analysis of textural region seen in cancer images [11, 12]. Graph run length matrix (GRLM) is the major tool to study tissue organization. The analysis of random field image texture for diagnosis purpose has been presented by AkifBurak et al. [30]. They have proposed a new homogeneity measure to describe the tissue components. The study of arrangement and cell structure through GRLM has been undertaken by Kannan et al. [31] Morphological approach and GLCM approach in texture feature analysis of histology images have been well described by Omar Al-kadi [32]. Even though the researchers have been working with the histopathology image analysis for past few decades, we found that there is a very limited work on brain cancer histopathology. This work proposes a novel concept of using statistics-based texture features to understand abnormalities in histopathology samples of brain cancer.

3 Material (Data) and Methods

3.1 Image Database

This work has been focused on brain tissues for brain cancer detection using histopathology samples. The tissue database has been taken from the Stanford tissue microarray database (TMAD) and tissue microarray virtual slides available on Biomax.us. These are the public resources providing reference images for researchers and histopathologists. This database consists of stained tissue images from Tissue Microarray experiments as shown in Fig. 1 [33, 34]. The tissue samples have been stained to enhance the visibility of the structures on the slide. The tissue sections have been dyed with standard stains such as Hematoxylin and Eosin (H & E) and Immunohistochemical (IHC). H & E stained slides enhances the spatial nuclei features like size, shape, texture, spatial arrangements, tubules, and stromal, etc. [35]. The color related and intensity changes were found enhanced in IHC-stained tissue slides. [36] Once the tissues are stained, the imaging of stained samples has been performed by whole slide scanner with adjustable resolution based on histological and cytological requirements. For the study of nuclei deformities in cytology, resolution is set to 400X, and for histological analysis, it is below 40X, as entire tissue structure has been examined in histology image.

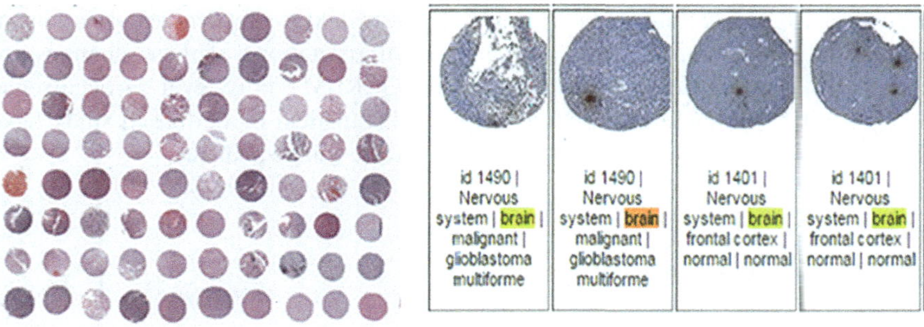

(a) (b)

Fig. 1 **a** Tissue microarray core. **b** TMA dataset (TMAD)

4 Methodology

This work has been carried out to quantify deformities in cell, regional lymph node metastasis, distant metastasis, and growth factor for cancer diagnosis using GLCM-based Haralick features and GLRLM-based textural features. It consists of three major stages: image enhancement, feature extraction, and analyzing extracted features using SVM classifier.

4.1 Image Enhancement

The preprocessing and segmentation are the two major stages in enhancement. To reduce noise component arising from image acquisition and staining process, it is necessary to preprocess an image prior to feature extraction. The noise removal and improving overall quality of an image is the main objective of preprocessing. This step also involves segmentation of key areas for analysis and diagnosis. Depending upon cell-based and tissue-based studies, segmentation of cell has been performed from a segmented tissue sample. Cell-based study considers region-based and boundary-based segmentations. But when there is a need to consider global features alone, such as first order, second-order statistical features like mean, variance, autocorrelation, etc., segmentation is not required. Researchers have proved that calculating global features can help to diagnose presence and absence of the disease [8].

4.2 Feature Extraction

In this step, the entire image has been represented by extracted features. The spatial independency or spatial dependency can be seen depending on the cell-based features or the tissue-based features, respectively [7]. Tissue texture determines variations in tissue/cell organization. In cancer diagnosis, interpretations of these changes are significant. The gray-level run length matrix (GLRLM), gray-level co-occurrence matrix (GLCM), transform-based methods, and statistical model based study have been considered as widely used methods for texture image analysis.

4.3 Haralick Feature

Textural features extracted in this work have been based on the hypothesis that textural information on an image 'I' represents overall spatial association among the different gray levels of the image. In other terms, texture information is effectively given by a set of gray-level spatial co-occurrence matrix (GLCM) [5]. GLCM is computed for various angular relationships and distances between neighboring pixels carrying aspecific set of gray levels. So while constructing GLCM for texture representation, there have been three main parameters: the quantification of the image gray levels, the displacement, and angular orientation. Sample structure of GLCM is given in Fig. 2 where N is the total number of gray levels existing in an image and probability $P(i, j)$ is the probability of finding a pixel with gray-level value 'i' together with a pixel gray-level value 'j'. An example of the gray-level image and corresponding calculated GLCM is given in Fig. 3.

Haralick defined textural features using GLCM. These features are related to specific textural property such as contrast, correlation, homogeneity, absence or presence of aspecific pattern in an image, and also quantify complexities and gray-level transitions. Features have been defined from F1 to F14 and shown in APPENDIX A. They are F1: Angular second moment (ASM): It represents energy contents in the image. Indirectly, it gives the measure of uniformity. F2: Contrast: This is the measure of local intensity variation in the histological sample. F3: Correlation: It measures linear dependencies of gray levels of two neighboring pixels. In other words, it is a measure of deformities in the textural image. F4: Variance: The variance is a measure of how far the gray levels have been spread from their mean value

$$\begin{bmatrix} P(1,1)\ P(1,2)\ P(1,3) \ldots\ldots\ldots P(1,N) \\ P(2,1)\ P(2,2)\ P(2,3) \ldots\ldots\ldots P(1,N) \\ P(3,1)\ P(3,2)\ P(3,3) \ldots\ldots\ldots P(3,N) \\ P(4,1)\ P(4,2)\ P(4,3) \ldots\ldots\ldots P(4,N) \\ \cdot \\ \\ \\ \cdot \\ \\ P(N,1)\ P(N,2)\ P(N,3) \ldots\ldots\ldots P(N,N) \end{bmatrix}$$

Fig. 2 GLCM matrix format

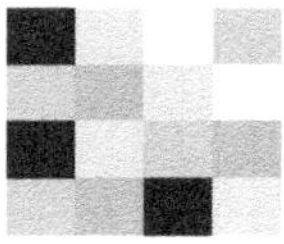

A

	0	1	2	3
0	0	3	4	2
1	2	1	3	4
2	0	3	2	1
3	2	1	0	3

B

	0	1	2	3	4
0	0	1	0	3	0
1	1	0	3	1	0
2	0	3	0	1	1
3	3	1	1	0	2
4	0	0	1	2	0

C

Fig. 3 **a** Sample image. **b** Intensity values. **c** GLCM at angle 0°

in the given sample. It is one of the measures of the probability distribution. F5: Inverse difference moment (IDM): It is the degree of homogeneity. In heterogeneous image, this measure will be small. This is one of the important measures for cancer diagnosis. F6, F7, and F8: Sum of Average, Sum of Variance, and Sum of Entropy are the average, variance, and entropy of the normalized gray tone image in the spatial domain. F9: Sum of Entropy is the measure of randomness in the image. F10: Measure of redundancy or information lying in the image which is further used for coding while compression. F11: Difference variance is another measure of randomness. F12, F13, and F14: a measure of the correlation in a more precise form [5]. This work considers thirteen Haralick features as they can be made invariant to translations and rotations. As mentioned above, they have the ability to understand something instinctively in given images of biopsy samples, for example, coarse versus smooth, directionality of the pattern, image complexity, etc. Gray-level run length matrix (GLRLM), $P(i, j/\theta)$ gives information about pixel gray level 'i' present 'j' times consecutively in direction 'θ.' GLRLM can be generated at 0°, 45°, 90°, and 135° like GLCM. The rows of GLRLM show gray levels in the image, and the columns signify length of runs, with the entries corresponding to the number of runs of the given length in the given image. Thus, if we observed a large number of neighboring pixels of the same gray level, the texture is a coarse texture, and if we find a small number of neighboring pixels of the same gray level, the texture is a fine texture. GLRLM calculated for 'θ' = 0° is shown in Fig. 4. We have considered seven statistical features derived by Galloway [37] and Chu et al. [38]. They are the short-run emphasis (SRE), long-run emphasis (LRE), gray-level non-uniformity (GLN), run length non-uniformity (RLN), run percentage (RP), low gray-level run emphasis (LGLRE), and high gray-level run emphasis (HGLRE). The detailed definitions are given in APPENDIX B.

4.3.1 Classification

In this phase, we allocate one among the two classes (Malignant, Normal) to the sample after extracting GLCM and GLRLM based texture features. This work uses support vector machine (SVM) classifier to separate malignant tissues from a healthy

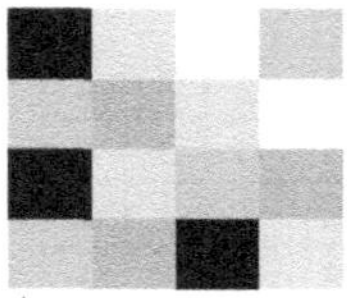

A

	0	1	2	3
0	0	3	4	2
1	2	1	3	4
2	0	3	2	1
3	2	1	0	3

B

0^0		Run Lengths			
		1	2	3	4
Gray Level	0	3	0	0	0
	1	3	0	0	0
	2	3	0	0	0
	3	4	0	0	0
	4	2	0	0	0

C

Fig. 4 **a** Sample image. **b** Intensity values. **c** GLRLM at angle 0°

one. In support vector machine (SVM), several hyperplanes are constructed to separate dataset into two classes. We need to find the optimal hyperplane which works as the linear decision boundary with maximum margins between the vectors of two classes. This margin has been optimally computed by critical components of the training data called support vectors. This optimal hyperplane can be described as

$$W^{\mathrm{T}}x + b = 0 \quad x\varepsilon R^{\mathrm{d}} \tag{1}$$

SVM classifier is able to classify miscellaneous data by varying kernel functions. These kernels are linear, polynomial function, radial basis function (RBF), and sigmoid functions. This work considers the RBF kernel function.

4.4 Testing Classifier Performance

A confusion matrix as shown in Fig. 5 reveals the information about actual and predicted outcomes. The performance of classifier is commonly evaluated using the information described in this matrix. Sensitivity (Recall), Specificity, Accuracy,

TP	True Positive	Sick person correctly identified as sick.
TN	True Negative	Healthy person correctly identified as healthy
FP	False Positive	Healthy person incorrectly identified as sick
FN	False Negative	Sick person incorrectly identified as healthy

(a) Definitions

		Predicted	
		Negative	Positive
Actual	Negative	TN	FP
	Positive	FN	TP

(b)Confusion Matrix Standard form

		Predicted	
		Negative	Positive
Actual	Negative	1.00	0.00
	Positive	0.14	0.86

(c)Observed Confusion Matrix with a fusion of GLCM and GLRLM Data Set

Fig. 5 Confusion matrix

Table 1 Statistical measures

Sensitivity/recall	TP/(TP + FN)
Specificity	TN/(TN + FP)
Accuracy	(TN + TP)/(TN + TP + FN + FP)
Precision	TP/(TP + FP)
F-Scores	2 * (Precision * Recall)/(Precision + Recall)

Precision, F-Score, etc., are most chosen statistical measures to evaluate the performance of diagnostics test, and in this work, it is higher order statistical features (GLCM and GLRLM features) with SVM Classifier. Each measure is used to describe how good and reliable the test is. Sensitivity/Recall identifies how much worthy the test is at detecting a positive disease, while specificity estimates how likely patients without disease to be correctly ruled out. Accuracy gives how good the test is to identify as well as ruling out disease. The disease prevalence is a statistical term that depends on population of the disease and always affects the accuracy measure. Precision or confidence is the measure of predicted positive cases that are truly positives. F-Score is another measure of tests accuracies based on precision and recall. Sometimes, "F-score" is referred as balanced mean or weighted average mean. The confusion matrix consists of terms such as true positive (TP), true negative (TN), false positive (FP), and false negative (FN) which are associated with diagnostic tests as shown in Fig. 5a, 5b, and 5c. The statistical measures calculated with confusion matrix have been listed in Table 1. In this study, fusion of GLCM and GLRLM features has been considered to evaluate the performance of SVM classifier.

5 Results and Discussions

5.1 Experimentation

In this study, we have considered 78 image samples of astrocytomas, 118 image samples of glioblastomas, and 54 image samples of normal (54) tissue. This work has been carried out in three phases: preprocessing, features extraction, and classification. In the first part, we cropped the image of size 51 × 512 pixels before preprocessing. After preprocessing, we have calculated thirteen Haralick Texture features based on gray-level co-occurrence matrix and seven features based on gray-level run length matrix. Table 2 and Table 3 show Haralick and GLRLM based features calculated for malignant and healthy brain tissues. We have formed three feature datasets using GLCM, GLRLM, and fusion of GLCM and GLRLM features. The SVM classifier using the k-fold technique, with '*k*' value set to 5 and number of iteration set to 100 has been considered for classification (Diagnosis). Initially, individual GLCM and GLRLM feature sets have been considered for classification. In the next step, we

Table 2 Haralick features for normal and malignant samples

Haralick texture feature	Normal sample	Malignant sample
F1: Angular second moment	0.7594	0.11875
F2: Contrast	0.1385	0.63417
F3: Correlation	347.0757	1415.6
F4: Variance	35.2915	31.747
F5: Inverse difference moment	0.9453	0.75925
F6: Sum of average	11.8893	11.119
F7: Sum of variance	128.2208	86.779
F8: Sum of entropy	0.5841	1.9885
F9: Entropy	0.6874	2.4946
F10: Difference variance	0.0984	0.048007
F11: Difference entropy	0.3681	0.87048
F12: Information measure of correlation	− 0.2079	− 0.23777
F13: Information measure of correlation	0.3840	0.69993

Table 3 GLRLM features

GLRLM texture feature	Normal sample	Malignant sample
SRE	0.7199	0.8268
LRE	4.0671	2.1502
GLN	2193.4157	1627.6164
RP	0.6149	0.7688
RLN	4840.3810	8034.9178
LGRE	0.0108	0.0170
HGRE	132.8836	92.9851

have considered the fusion of both the datasets to see the improvement in classification accuracy. To test the performance of diagnostic system, i.e., SVM classifier, we constructed the confusion matrix as shown in Fig. 5a. It consists of true detection cancerous patients(TP), true elimination of non cancerous patients (TN), and false detection of non-cancerous patients (FP), false elimination cancerous patients (FN) terms. Using confusion matrix, we have calculated statistical measures like sensitivity, specificity, accuracy, precision, and F-score.

5.2 Results

In this proposed work on brain cancer diagnosis, in the first part, GLCM feature and GLRLM feature datasets have been separately given to classifier, and classification

Table 4 SVM classifier

Dataset	% classification accuracy
GLCM	87.33
GLRLM	91.14
GLCM + GLRLM	93.00

Table 5 Classifier performance based on confusion matrix

Test parameters	GLCM + GLRLM	AR (4) model features
TP	86.00	87.50
TN	01.00	81.25
FP	00.00	18.75
FN	14.00	13.50
Sensitivity/recall	100.00	87.50
Specificity	100.00	81.25
Accuracy	93.00	84.50
Precision	87.79	82.35
F-Score	92.66	84.84

accuracies have been calculated. In the next step, fusion of thirteen GLCM and seven GLRLM features has been performed, and this fused dataset has been tested with SVM classifier. As shown in Table 4, the classification accuracy with GLCM-based Haralick features has been observed as 87.33% and with GLRLM as 91.14%. The improvement in classification accuracy has been observed with the fusion of GLCM and GLRLM features which shows 93.00% classification accuracy as described in Table 4. As described in Table 5, the numerical value of sensitivity is 100%, which represents the probability of identifying the person who is suffering from the disease; similarly, specificity value is 100%, which represents ability of the classifier to detect the person who is not suffering from the disease. The classification accuracy is 93% which represents the overall effectiveness of the classifier for true positive results in the total population. Similarly, F-score (92.66%) value reveals the harmonic mean of precision and recall and gives a better evaluation of performance test.

5.3 *Discussions*

In brain cancer detection, this work reveals the use of GLCM-based Haralick Textural features and GLRLM features. In textural feature analysis for cancer diagnosis, several methods have been described opting for various methods [39]. We believe that in the texture-based study of brain cancer histopathology imagery this work presents a novel approach where the fusion of thirteen Haralick Textures features and seven run length features have been considered for cancer diagnosis. As shown in Table 2, Haralick features such as ASM, correlation, entropy, sum of entropy, a difference of entropy, and information measures of correlation have proved to be good biomarkers to significantly differentiate between malignant and normal tissue. Similarly, as shown in Table 3, GLRLM features such as long run emphasis (LRE), gray-level non-uniformity (GLN), run length non-uniformity (RLN), and high gray-level run emphasis (HGLRE) terms have been proved as biomarkers. This work also gives performance evaluation of SVM classifier with a fusion of GLCM and GLRLM features. Table 5 displays major statistical measures for checking classifier performance with fusion of dataset. We have compared this result with model-based study [17]. It shows that textural analysis of histopathology images with GLCM and GLRLM features offers better results than fourth-order autoregressive model based features. The performance evaluation of SVM classifier checks two major requirements; first is the confirmation about the correct predictions of positive and negative labels, and second is the ability of the classifier to avoid failures. There are a few limitations to the proposed method. The issues related to heterogeneity and long-range pixel dependency found in cancer pathology images have not been resolved in this work. We propose some future tasks such as Markov random field modeling (MRF) and radial basis function (RBF) modeling which can be considered while interacting with nonlinear heterogeneous textures in histopathology imaging [40]. Similarly, long-range dependency issues can be addressed if the same work is attempted in transformed domain by taking Fourier or Wavelet transforms. There have been three more performance measures which can be considered in evaluating diagnostic tests performance; these are Youden's Index (γ), Likelihoods (ρ_+ and ρ_-), and discriminant power (DP). The Youden's index evaluates the ability of the diagnostic test to avoid failure. While likelihood evaluates classifier performance on the finer scale, high value of positive likelihood and low value of negative likelihood represent better classification algorithm. The discriminant power (DP) is the power of classifier to distinguish between positive and negative labels [41]. Analyzing the diagnostic performance of classifier algorithms on above measures will be considered as our future tasks.

6 Conclusion

This work presents a textural analysis of brain cancer histopathology images using GLCM-based Haralick features and Gray-level run length matrix based features. Survey of texture analysis for histopathology images and its applications in cancer diagnostics has also been carried out. The estimated texture-based features are exceptional discriminating capacity that provides the basis for statistical analysis. These features can be used to quantify and classify tissue deformations in cancer diagnostics. The work also shows the SVM classifier performance based on standard statistical measures. It also reveals fusion of GLCM and GLRLM features forms excellent dataset which provides better diagnosis. As a future work, one can extract more complex features from histopathology image using transform domain approach. Similarly, one can also consider model-based approach for analyzing heterogeneity, and generate vector data of textural features so as to capture more deformities in architecture and hence key regions in histology samples.

Acknowledgements The authors would like to thank S.R.M. University, Vadapalani, Chennai for their continued support and encouragement during this research work.

Conflicts of Interest The authors declare that there are no conflicts of interest regarding the publication of this article.

Appendix 1: GLCM Haralick Features

Haralick texture feature	Standard formulae
F1: Angular second moment	$F1 = \sum_{i-0}^{Ng-1} 1 \sum_{j-0}^{Ng-1} \{p(i, j)\}^2$
F2: Contrast	$F2 = \sum_{n=0}^{Ng-1} n * n \sum_{i-0}^{Ng-1} 1 \sum_{j-0}^{Ng-1} \{p(i, j)\} \quad \{i - j\} = n$
F3: Correlation	$F3 = \sum_{i=1}^{Ng} 1 \sum_{j=1}^{Ng} \{(i - \mu x)(j - \mu y) p(i, j) /\sigma x \sigma y$
F4: Variance	$F4 = \sum_{i=1}^{Ng} 1 \sum_{j=1}^{Ng} (i - \mu)^2 p(i, j)$
F5: Inverse difference moment	$F5 = \sum_{i=1}^{Ng} 1 \sum_{j=1}^{Ng} \{p(i, j)\}/1 + (i - j)^2$
F6: Sum of AVERAGE	$F6 = \sum_{k=2}^{2Ng} k p_{x+y}(k)$
F7: Sum of Variance	$F7 = \sum_{k=2}^{2Ng} (k - F6)^2 p_{x+y}(k)$
F8: Sum of entropy	$F8 = \sum_{k=2}^{2Ng} p_{x+y}(k) \log\{p_{x+y}(k)\}$
F9: Entropy	$F9 = \sum_{i=1}^{Ng} 1 \sum_{j=1}^{Ng} \{p(i, j)\} \log\{p(i, j)\}$
F10: Difference variance	$F10 = \sum_{k=0}^{Ng-1} k - \left\{\sum_{l=0}^{Ng-1} l p_{x-y}(l) l p_{x-y}(l)\right\}^2 p_{x-y}(k)$
F11: Difference entropy	$F11 = \sum_{l=0}^{Ng-1} p_{x-y}(k) \log\{p_{x-y}(k)\}$
F12: Information measure of correlation	$F12 = \frac{F9-HXY!}{\max(HX,HY)}$
F13: Information measure of correlation	$F13 = (1 - \exp(-2(HXY2 - F9)))^{1/2}$
F14: Maximal correlation coefficient	$F14 = Q(i, j) = \sum_k 1 p(i, k) p(j, k)/pi(i)pk(k)$

Appendix 2: GLRLM Features

GLRLM features	Standard formulae
SRE	$\text{SRE} = \frac{1}{n_r} \sum_{i=1}^{M} 1 \sum_{j=1}^{N} \{p(i, j)\} \frac{1}{j^2} = \frac{1}{n_r} \sum_{j=1}^{N} \{p(j)\} \frac{1}{j^2}$
LRE	$\text{LRE} = \frac{1}{n_r} \sum_{i=1}^{M} 1 \sum_{j=1}^{N} \{p(i, j)\} * j^2 = \frac{1}{n_r} \sum_{j=1}^{M} \{p(j)\} * j^2$
GLN	$\text{GLN} = \frac{1}{n_r} \sum_{i=1}^{M} 1 \left(\sum_{j=1}^{N} \{p(i, j)\} \right) *^2 = \frac{1}{n_r} \sum_{j=1}^{M} \left\{ p_g(j) \right\} *^2$
RLN	$\text{RLN} = \frac{1}{n_r} \sum_{i=1}^{N} 1 \left(\sum_{j=1}^{M} \{p(i, j)\} \right) *^2 = \frac{1}{n_r} \sum_{j=1}^{N} \{p_r(i)\} *^2$
RP	$\text{RP} = \frac{n_r}{n_p}$
LGLRE	$\text{LGRE} = \frac{1}{n_r} \sum_{i=1}^{M} 1 \sum_{j=1}^{N} \{p(i, j)\} * \frac{1}{j^2} = \frac{1}{n_r} \sum_{j=1}^{N} \left\{ p_g(i) \right\} * \frac{1}{j^2}$
HGLRE	$\text{HGRE} = \frac{1}{n_r} \sum_{i=1}^{M} 1 \sum_{j=1}^{N} \{p(i, j)\} * j^2 = \frac{1}{n_r} \sum_{j=1}^{N} \left\{ p_g(j) \right\} * j^2$

References

1. D.C. Cireşan, A. Giusti, L.M. Gambardella, J. Schmidhuber, Mitosis detection in breast cancer histology images with deep neural networks, medical image computing and computer-assisted intervention, in *MICCAI 2013. 16th International Conference* (Nagoya, Japan)
2. Glioblastomas and Malignant Astrocytomas, American Brain Tumor Association, Publications and Services (2014)
3. C.H. Chen, L.F. Pau, P.S.P. Wang, *The Handbook of Pattern Recognition and Computer Vision*, 2nd edn. (World Scientific Publishing Co., 1998), pp. 207–248
4. R. Chellappa, Stochastic models in image analysis and processing, Ph.D. thesis, Purdue University, 1981
5. R.M. Haralick, K. Shanmugam, I. Dinstein Texture features for image classification. IEEE Trans. Syst. Man Cybern. **SMC-3**(6), 610–621
6. A. Baraldi, F. Parmiggiani, An investigation of textural characteristics associated with gray level co-occurrence matrix statistical parameters. IEEE Trans. Geosci. Remote Sens. **33**(2), 293–304 (1995)
7. C. Demir, B. Yener, Automated cancer diagnosis based on histopathological images: a systematic survey, Technical Report, Rensselaer Polytechnic Institute, Department of Computer Science (TR-05–09) (2009)
8. M.N. Gurcan, L. E. Boucheron, A. Madabhushi, N.M. Rajpoot, B. Yener, Histopathological image analysis: a review. IEEE Rev. Biomed. Eng. **2**, 147–171 (2009)
9. S. Bhattacherjee, J. Mukherjee, S. Nag, I.K. Maitra, S.K. Bandopadhyay, Review on histopathological slide analysis using digital microscopy. Int. J. Adv. Sci. Technol. **62**, 65–96 (2014)
10. O. Sertel, J. Kong, U. Catayurek, G. Lozansski, J. Saltz, M. Gurcan, Histopathological image analysis using a model based intermediate representation of color texture, follicular lymphoma grading. J. Sig. Process. Syst. **55**, 169–183 (2009)
11. L.E. Boucheron, Object and spatial level quantitive analysis of multispectral histopathology images for detection and characterization of cancer. Ph.D. thesis, 2008
12. J.C. Caicedo, Features for histology images. Image Segmentation, pp. 1–9 (2009)

13. A.D. Belsare, P.O. Brown, C.A. Ball, M.M. Mushrif, Histopathological image analysis using image processing techniques: an overview. Sig. Image Process. Int. J. (SIPIJ) **3**(4), 22–31 (2012)
14. M.K.B. Sims, A. Murray, Image analysis can be used to detect spatial changes in the histopathology of pancreatic tumors. Phys. Mach. Biol. **48**, 183–191 (2003)
15. C. Gunduz, B. Yener, S. Gultekin, The cell graphs of cancer. Bioinformatics **20**, 145–151 (2004)
16. D. Vaishali, R. Ramesh, J. Anita Christine, Histopathology image analysis and classification for cancer detection using 2D Autoregressive model. Int. Rev. Comput. Softw. **10**(2), 182–188 (2015)
17. D. Vaishali, R. Ramesh, J. Anita Christaline, Autoregressive modeling: application to mitosis detection in brain cancer histopathology. Int. J. Biomed. Eng. Technol. **20**(2), 179–194 (2016)
18. A.N. Basavanthy, S. Ganesan, S. Agner, J.P. Monaco, M.D. Feldman, J.E. Tomaszewski, G. Bhanot, A. Madabhushi, Computerized image- based detection and grading of lymphocytic infiltration in breast cancer histopathology. IEEE Trans. Biomed. Eng. **57**(3), 642–653 (2010)
19. S.G. Mallat, A theory for multiresolution signal decomposition: the wavelet representation. IEEE Trans. Pattern Anal. **11**, 674–693 (1989)
20. B. Weyn, G. Van de Wouwer, A. Van Daele, P. Scheunders, D. Van Dyck, E. Van Marck, W. Jacob, Automated breast tumor diagnosis and grading based on wavelet chromatin texture description. Cytometry **33**, 32–40 (1998)
21. M.N. Gurcan, T. Pan, H. Shimada, J. Saltz, Image analysis for neuroblastoma classification: segmentation of cell nuclei. In: *Proceedings of the 28th IEEE EMBS Annual International Conference New York City, USA*, 30 Aug–Sept 2006
22. P. Spyridonos, P. Ravazoula, D. Cavouras, K. Berberidis, G. Nikiforidis, Computer-based grading of haematoxylin-eosin stained tissue sections of urinary bladder carcinomas. Med. Inform. Internet Med. **26**, 179–190 (2001)
23. A.N. Esgiar, R.N.G. Naguib, B.S. Sharif, M.K. Bennett, A. Murray, Microscopic image analysis for quantitative measurement and feature identification of normal and cancerous colonic mucosa. IEEE Trans. Inf. Technol. B. **2**, 197–203 (1998)
24. P.W. Hamilton, P.H. Bartels, D. Thompson, N.H. Anderson, R. Montironi, Automated location of dysplastic fields in colorectal histology using texture image analysis. J. Pathol. **182**, 68–75 (1997)
25. M. Wiltgen, A. Gerger, J. Smolle, Tissue counter analysis of benign common nevi and malignant melanoma. Int. J. Med. Inform. **69**, 17–28 (2003)
26. S. Rajeev Srivastava, S. Sharma, S.K. Singh, Design, analysis and classifier evaluation for CAD tool for breast cancer detection from digital mammograms. Int. J. Biomed. Eng. Technol. **13**(1), 270–300 (2013)
27. M. Muthu Rama Krishnan, M. Pal, S. Bomminayuni, C. Chakraborty, R.R. Paul, J. Chatterjee, A. Roy, Automated classification of cells in the sub-epithelial connective tissue of oral submucous fibrosis—an SVM-based approach. J. Comput. Biol. Med. **39**, 1096–1104 (2009)
28. S. Doyle, M. Feldman, J. Tomaszewski, A. Madabhushi, A Boosted Bayesian Multi-Resolution Classifier for prostate cancer Detection from digitized needle biopsies. Trans. Biomed. Eng. **59**(5), 2005–2018 (2005)
29. M. Murat Dundar, S. Badve, G. Bilgin, V. Raykar, R. Jain, O. Sertel, M.N. Gurcan, Computerized classification of intraductal breast lesions using histopathological images IEEE Trans. Biomed. Eng. **58**(7), 1977–1984 (2011)
30. A. Baurak, M. Kanemir, C. Sokmensure, C. Demir, Object oriented texture analysis for unsupervised segmentation of biopsy images for cancer detection. Pattern Recogn. **42**, 1104–1112 (2009)
31. K.P. Kannan, A. Ananthakumarari, Texture analysis of histopathology images to identify anomalous region. Int. J. Manag. IT Eng. **2**(8), 1–10 (2012)
32. O. Al-Kadi, Texture measures combination for improved meningioma classification of histopathology images. J. Pattern Recogn. **43**, 2043–2053 (2010)
33. Stanford Tissue Microarray Database. https://tma.im/
34. www.biomax.us/tissue-arrays/Brain/

35. M. Macenko, M. Niethammer, J.S. Marron, D. Borland, J.T. Woosley, X. Guan, C. Schmitt, N.E. Thomas, A method for normalizing histology slides for quantitative analyses. In: *IEEE International Symposium on Biomedical Image Processing* (2009), pp. 1107–1110
36. R.A. Walker, Quantification of immunohistochemistry Issues concerning methods, utility and semiquantitative assessment. Histopathology **49**(4), 406–410 (2006)
37. M.M. Galloway, Texture analysis using Gray level run lengths. Comput. Graph. Image Process. **4**, 172–179 (1975)
38. A. Chu, C.M. Sega, J.F. Greenleaf, Use of gray value distribution of run lengths for texture analysis. Pattern Recogn. Lett. **11**, 415–420 (1990)
39. D. Yan, P. Wang, M. Linden, B. Knudsen, T. Randolph, Statistical methods for tissue array images-algorithmic scoring and co-training. Anal. Appl. Stat. **6**(3), 1280–1305 (2012)
40. H. Irshad, A. Veillard, L. Roux, D. Racoceanu, Methods for nuclei detection, segmentation, and classification in digital histopathology: a review—current status and future potential methodological review. IEEE Rev. Biomed. Eng. **7**, 97–114 (2014)
41. M. Sokolova, N. Japkowicz, S. Szakowicz, Beyond accuracy, F-Score and ROC: a family of discriminant measures for performance evaluation, in *AI2006: Advances in Artificial Intelligence* (2006)

Resource Management in Wireless IoT Using Gray Wolf Optimisation Framework

S. Karthick and N. Gomathi

Abstract With advancement in Wireless Internet of Things (IoT), the connectivity between the devices increases, and hence the effective resource management is considered challenging. In this paper, we use a scheduler that adopts Gray Wolf Optimization (GWO) model to learn the allocation of resources flexibly to the users. This model acts as an adequate model to allocate resources on large-scale systems and it does not avoids the setbacks like slow convergence, analytical complexity, and increase overhead due to multiple IoT inputs. The comparison between the proposed and existing resource management techniques shows that the GWO offers improved resources management efficiency than the existing methods.

Keywords Gray wolf optimization · Resource management · Wireless IoT

1 Introduction

In the light of the various applications and the advances throughout wired networking technology [1], interest for the Internet of Things (IoT) infrastructure has increased rapidly. In the Internet of Things, the word stuff is a piece of equipment that has the capacity to monitor, run, store, or cycle. Such apps have unique features, i.e., limited memory, lower battery life, and weak processing power [2]. The IoT has a strong potential to influence the course of our lives. The IoT provides various applications from home automation to healthcare systems, which allow cost-effective communication between objects and devices [3] to improve industry and society. Therefore, nearly 50 billion IoT devices are predicted to operate by 2050. Various

S. Karthick (✉)
Research Scholar, Department of Computer Science and Engineering, Vel Tech Rangarajan Dr. Sagunthala R&D Institute of Science and Technology, Avadi, Chennai, India
e-mail: karthick.usilai@gmail.com

N. Gomathi
Department of Computer Science and Engineering, Vel Tech Rangarajan Dr. Sagunthala R&D Institute of Science and Technology, Avadi, Chennai, India
e-mail: gomathin@veltechuniv.edu.in

S. S. Dash et al. (eds.), *Intelligent Computing and Applications*,
Advances in Intelligent Systems and Computing 1172,
https://doi.org/10.1007/978-981-15-5566-4_62

systems have been built to support IoT sensors or actuators leading to the proliferation of IoT networks over the last decade. Likewise, a range of OSs have been established slowly for running these tiny sensors [4].

IoT systems have minimal capacity and power and in some cases require real-time capability. In addition, heterogeneous equipment and appropriate networking and protection frameworks should be enabled. Efficient communication and maintenance of this large number of devices is one of the research community's major design priorities. In an IoT framework, the fundamental problem of science is the organized and regulated use of available resources. An IoT Resource Management Mechanism is essentially configured to effectively meet IoT system specifications [5, 6].

A variety of strategies for resource availability [7–11], such as a limited resource base, and dispersed and complex ecosystems, are being implemented to meet the challenges raised by the IoT lower end devices [12]. The main responsibility for the effective operation of the computer is an IoT operating system. Various OSs also provided various solutions to satisfy the resource requirements of low-end devices. Various systems are provided by the various OSs to ensure that the sensing nodes are working properly.

Among several OSs suggested for the use in a resource-starved network of low-end devices, low-end IoT systems typically run on a limited energy tank. It is therefore extremely important to provide an energy-efficient OS [13–15]. Such low-end machines relay sensory data using a protocol for connectivity. The messaging protocols will minimize as much energy as possible in order to be energy-efficient. Energy efficiency is required for transport layer protocols, MAC-layer, and network layer [16].

For their sensing processes, IoT devices require computing capabilities. An OS needs an efficient method and resource control system to accommodate the restricted processing capacity and memory. IoT systems are powered by batteries and are primarily used in remote areas. Therefore, energy management is very critical for an OS. The primary goal of an IoT device is to provide sensing activity for the further handling of sensed data at the base station. Efficient resources and connectivity must also be used with respect to wireless architecture, signal processing, reception of data, transmitting, and radio sleep/wake cycle. Load, index and file device data from IoT OS. Therefore, in IoT OSs, it is very desirable to provide an effective, robust, and adequate file system.

The main aim of the study is to design a scheduler that adopts GWO model to learn the allocation of resources flexibly to the users. This model acts as an adequate model to allocate resources on large-scale systems and it does not avoid the setbacks like slow convergence, analytical complexity, and increase overhead due to multiple IoT inputs.

2 Proposed Method

GWO is a system of swarm intelligence and a social intelligence with gray wolves, hunting, and leadership. There is a common social hierarchy in every group of gray wolves which control supremacy and strength. The order of gray wolves is shown in Fig. 1.

Wolves kills and consumes all the pack, is the mighty dog. Alpha, which leads the whole group to pursue, run, and eat, is the toughest dog. When the wolves are not in the pack, ill, or injured the best beta wolf follows.

Delta and omega are stronger than beta and alpha and are less dominant. As Fig. 2 demonstrates the GWO algorithm is based mainly on social understanding. The search for gray wolves is another motivation. A number of effective steps are taken by gray wolves to target, encircle, threaten, and strike. This helps you to monitor major bearings. The logical measures are as follows.

Mainly the alpha wants to eat, sleep, wake up, etc. The alpha's choices are based on the pack. Yet also there is a certain egalitarian mentality, where the other wolves are led by an alpha. During sessions, the whole pack recognizes the leader by protecting the legs. The Alpha Werewolf, as the pack will follow instructions, is also recognized as the alpha werewolf. The pack will matte only the alpha wolves. To notice, alpha is not necessarily the most important part of the package but the best part of the product administration. This showed that a kit is much more essential for organization, consistency, and power.

Beta is the second stage in the system of gray wolves. The betas are wolves who help the alpha to determine or box otherwise. The wolf of the beta can be male or female. If one of the wolves is dead or very aged, it is probably the best alpha choice. The beta wolf would adhere to alpha, but the other lower wolves should also be in charge. He is the role of an alpha counselor and a discipliner in the package. Beta enhances the entire package of alpha commands.

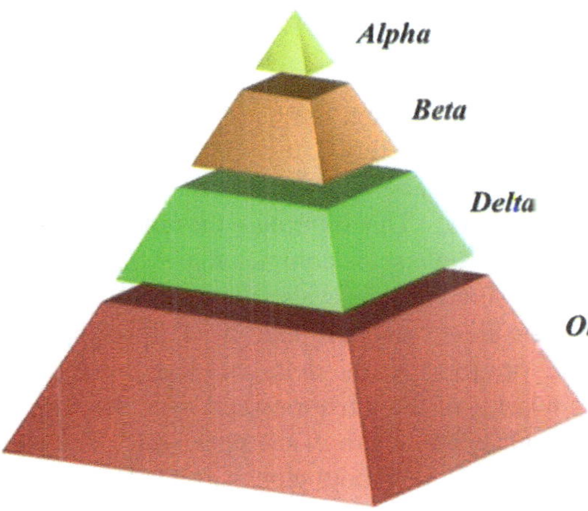

Fig. 1 Social hierarchy of wolves

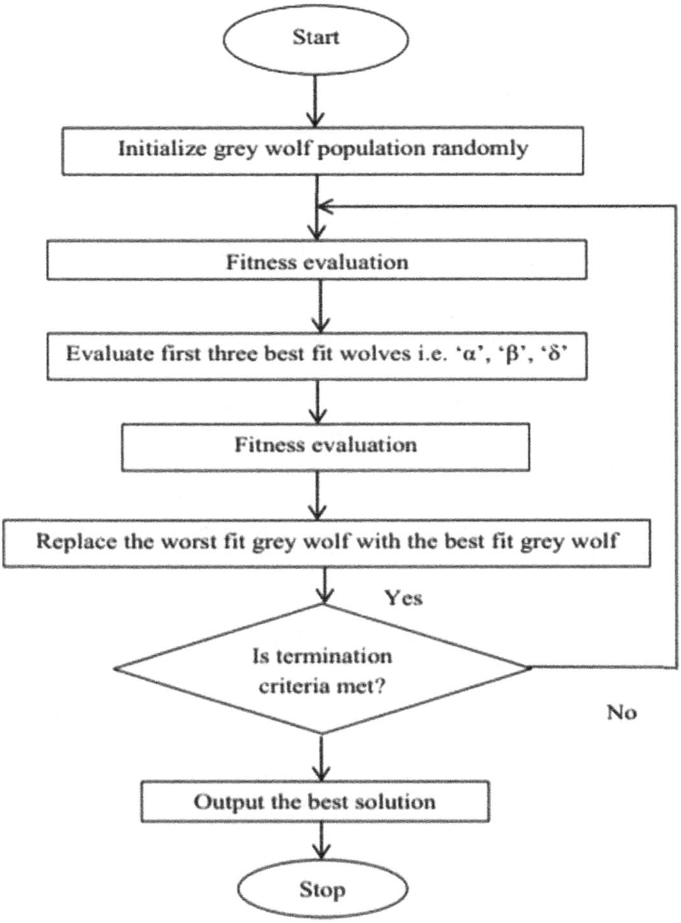

Fig. 2 Architecture of the proposed system

In the lowest ranking, the Gray Wolf is alpha. The omega is job of each wolf and always has to sell all other powerful wolves. It's the last wolf to hunt. The omega may not be an important person, but if the omega is misplaced, it experiences internal difficulties. That is because the omega of aggression and anger resides among all dogs. The whole set is completed and the domination system is maintained. The omega is in some instances also the babies in the bag.

It is considered a subordinate if an alpha, beta, or omega is not a bear. Alphas and betas are to show the delta, but they overpower the omega. Scouts will track the boundaries of the territories and alert the pack if there is a danger. It protects and guarantees the health of the packaging. The elders are the alpha or beta dogs. Hunter's tends to help both alphas and betas while catching predators and feeding the pouch. Ultimately, it is the duty of the guardians to take care of the poor, sick, and wounded wolves in the bunch.

The algorithm is given below and the flowchart is given in Fig. 2.

Step 1. The social hierarchy helps GWO to save the best solutions it has achieved so far over several iterations.
Step 2. The circular mechanism defines a circular district around solutions that can be extended as a hypersphere to higher dimensions.
Step 3. The random parameters *A* and *C* help candidate solutions to find the optimal resources for the IoT device to transmit the information with different random radius in hypersphere.
Step 4. The hunting method proposed enables candidates to find the likely position of the prey.
Step 5. The adaptive values of *a* and *A* are ensured for exploration and exploitation.
Step 6. GWO can smoothly transfer between exploration and exploitation with the adaptive values of parameters *a* and *A*.
Step 7. With decreasing *A*, half of the iterations are used for the exploration ($|A| \geq 1$) and remaining for explcitation ($|A| < 1$).
Step 8. The main parameters of GWO are adjusted finally, i.e., *a* and *C*.

3 Results and Discussions

In this section, the proposed method is compared with Radio Resource Allocation (RRA) method in [17] in terms of energy consumption and delay. The result shows that the proposed method achieves reduced energy consumption in both uplink and downlink for the data packet size of 100 and 32 bytes transmitted over the IoT network, which can be evident from Fig. 3. In Fig. 4, the result shows that the

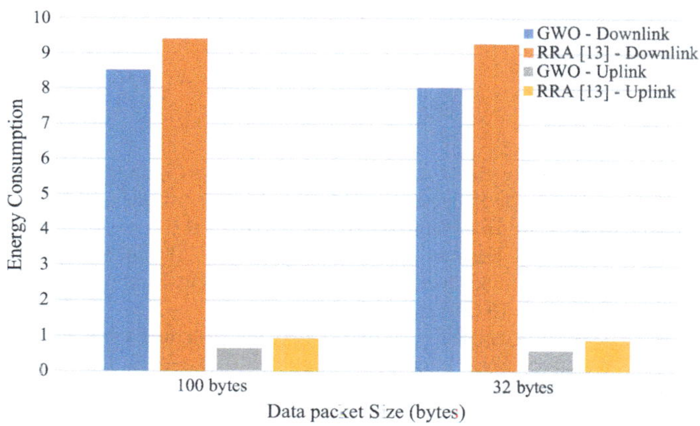

Fig. 3 Energy consumption (μJ)

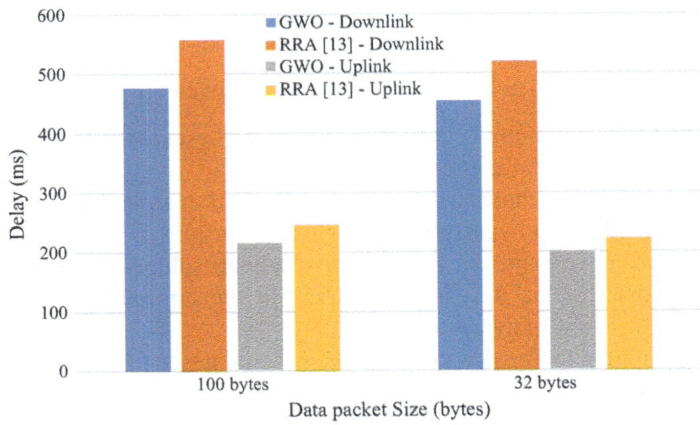

Fig. 4 Delay (ms)

proposed method achieves reduced delay in both uplink and downlink for the data packet size of 100 and 32 bytes transmitted over the IoT network.

4 Conclusions

In this paper, GWO model is adopted in WirelessIoT environment to learn the resources allocation flexibly to the IoT devices. The GWO model allocates well the required resources, avoids slow convergence, analytical complexity, and increases overhead due to multiple IoT inputs. The results show that the proposed GWO offers improved resources management efficiency in terms of increased throughput and reduced power consumption than the existing methods.

References

1. M. Weiser, The computer for the 21st century. IEEE Pervasive Comput. **1**(1), 19–25 (2002)
2. D. Miorandi, S. Sicari, F. De Pellegrini, I. Chlamtac, Internet of things: vision, applications and research challenges. Ad Hoc Netw. **10**(7), 1497–1516 (2012)
3. I. Yaqoob, E. Ahmed, I.A.T. Hashem, A.I.A. Ahmed, A. Gani, M. Imran, M. Guizani, Internet of things architecture: recent advances, taxonomy, requirements, and open challenges. IEEE Wirel. Commun. **24**(3), 10–16 (2017)
4. O. Hahm, E. Baccelli, H. Petersen, N. Tsiftes, Operating systems for low-end devices in the internet of things: a survey. IEEE Internet Things J. **3**(5), 720–734 (2015)
5. E. Baccelli, O. Hahm, M. Günes, M. Wählisch, T. Schmidt, OS for the IoT-goals, challenges, and solutions (2013)
6. N. Yuvaraj, R. Raja, C. Dhas, Analysis on improving the response time with PIDSARSA-RAL in clowdflows mining platform. EAI Endorsed Trans. Energy Web **5**(20) (2018)

7. J. Santos, T. Wauters, B. Volckaert, F. De Turck, Resource provisioning for IoT application services in smart cities, in *2017 13th International Conference on Network and Service Management (CNSM)* (IEEE, 2017), pp. 1–9
8. M. Aazam, E.N. Huh, Fog computing micro datacenter based dynamic resource estimation and pricing model for IoT, in *2015 IEEE 29th International Conference on Advanced Information Networking and Applications* (IEEE, 2015), pp. 687–694
9. A. Azari, G. Miao, Fundamental tradeoffs in resource provisioning for IoT services over cellular networks, in *2017 IEEE International Conference on Communications (ICC)* (IEEE, 2017), pp. 1–7
10. W. Li, K. Liao, Q. He, Y. Xia, Performance-aware cost-effective resource provisioning for future grid IoT-cloud system. J. Energy Eng. **145**(5), 04019016 (2019)
11. S. Nastic, S. Sehic, D.H. Le, H.L. Truong, S. Dustdar, Provisioning software-defined IoT cloud systems, in *2014 International Conference on Future Internet of Things and Cloud* (IEEE, 2014), pp. 288–295
12. M. Sivaram, A.S. Mohammed, D. Yuvaraj, V. Porkodi, V. Manikandan, N. Yuvaraj, Advanced expert system using particle swarm optimization based adaptive network based fuzzy inference system to diagnose the physical constitution of human body, in *International Conference on Emerging Technologies in Computer Engineering* (Springer, Singapore, 2019), pp. 349–362
13. M. Watfa, M. Moubarak, A. Kashani, Operating system designs in future wireless sensor networks. J. Netw. **10**(1), 1201–1214 (2010)
14. N. Yuvaraj, P. Vivekanandan, An efficient SVM based tumor classification with symmetry non-negative matrix factorization using gene expression data, in *2013 International Conference on Information Communication and Embedded Systems (ICICES)* (IEEE, 2013), pp. 761–768
15. M. Sivaram, V. Porkodi, A.S. Mohammed, V. Manikandan, N. Yuvaraj, Retransmission DBTMA protocol with fast retransmission strategy to improve the performance of MANETs. IEEE Access **7**, 85098–85109 (2019)
16. N.A. Pantazis, S.A. Nikolidakis, D.D. Vergados, Energy-efficient routing protocols in wireless sensor networks: a survey. IEEE Commun. Surv. Tutorials **15**(2), 551–591 (2012)
17. H. Malik, H. Pervaiz, M.M. Alam, Y. Le Moullec, A. Kuusik, M.A. Imran, Radio resource management scheme in NB-IoT systems. IEEE Access **6**, 15051–15064 (2018)

Less Polluted Flue Gases Obtained with Green Technology During Precious Metals Recovery from Unwanted and Discarded Electrical and Electronics Components

Rajendra Prasad Mahapatra, Satya Sai Srikant, Raghupatruni Bhima Rao, and Bijayananda Mohanty

Abstract This paper deals with less polluted flue gas recovery with well-known green technology heating process during precious and valuable metals recovery from unwanted and discarded electrical and electronic components. This unwanted and discarded electrical and electronic component is also said to be E-waste. During recovery of valuable and precious metals like gold, silver, copper, etc. after e-waste treatment with various methods, the researchers were unable to find the challenges about the treatment of released toxic and poisonous flue gases as such gases severely pollute the environment. This paper presents an innovative, clean as well as green method for flue gas treatment while recovering the valuable and precious metals from unwanted and discarded electrical and electronic components. A well known clean, green as well as eco-friendly heating process, i.e. microwave heating was applied for flue gas treatment. The results show that the treated flue gas not only minimizes toxic effect but also helps environment to get less polluted. Also the obtained minimized flue gases from the clean and green process were observed within the regulatory range of Central Pollution Control Board.

Keywords E-waste · Microwave heat treatment · Leaching · Flue gas

---

R. P. Mahapatra (✉) · B. Mohanty
National Institute of Technology Mizoram, Aizawl 796012, India
e-mail: mahapatra.rp@gmail.com

R. P. Mahapatra · S. S. Srikant
SRM Institute of Science and Technology, Modinagar 201204, India

R. B. Rao
CSIR-IMMT, Bhubaneswar 751013, India

S. S. Dash et al. (eds.), *Intelligent Computing and Applications*,
Advances in Intelligent Systems and Computing 1172,
https://doi.org/10.1007/978-981-15-5566-4_63

1 Introduction

Recently, the usage of digital materials for both business and domestic purposes has increased in an exponential way and hence problem of growing of scrapped electronics products day by day.

In India, recycling and disposal of unwanted and discarded electrical and electronic component is mostly carried out with simple methods which severely pollute the environment and causes health problems. Presently, scrapped electronics and electrical materials are being processed with disassembling, upgrading and refining [1]. It has observed with many literatures that mechanical process separates the metallic part from the scrapped components but at the same time, a huge amount of waste in form of plastics and hazardous components were produced. Hence, it is impossible to dispose the unwanted and discarded electrical and electronic components with mechanical routes.

The halogenated flame retardants (HFRs) present in the scrapped electronics and electrical components cause the poisonous and harmful dioxins formation. During the individual metals recovery from E-waste with various leaching also causes harmful dioxins and other toxic gases.

Microwave-assisted gasification systems were interfaced into a single system as shown in Fig. 1, where E-waste materials were used for analyzing the toxic gas production. However, attempts and goals were to extract the valuable metals from electronic waste. Therefore, a process was developed, which has eliminated the requirement of mechanical disassembling step. As microwave furnace generated the temperature between 1000 and 1200 °C, the harmful and poisonous gases including dioxin got decomposed. Due to this decomposition, the flue gas which produced the limited sulphur dioxide, dioxins, carbon monoxide gas, various nitrogen oxides components and others made eco-friendly process. This single-step process not only produced valuable and useful metals in alloy form but also small quantity of waste was generated. Copper has been recovered through microwave treated followed by leaching and residue treated with aqua leaching has been enriched with valuable components like gold and silver [2–6].

2 Experimental Procedures

2.1 Raw Materials

The transformers, capacitors, batteries and plastics were firstly dismantled (by sharp cutting tools) from the collected unwanted and discarded electrical and electronics PCBs samples, i.e. monitors, mobile, printer cards, motherboard of TV and computers, etc. These bared PCBs were crushed into a tiny particles having size around 230–290 μm by the crusher. An amount of 400 g of SiC powder which behaved as coupling agent [7–10] was combined with 10 kg of crushed tiny particles.

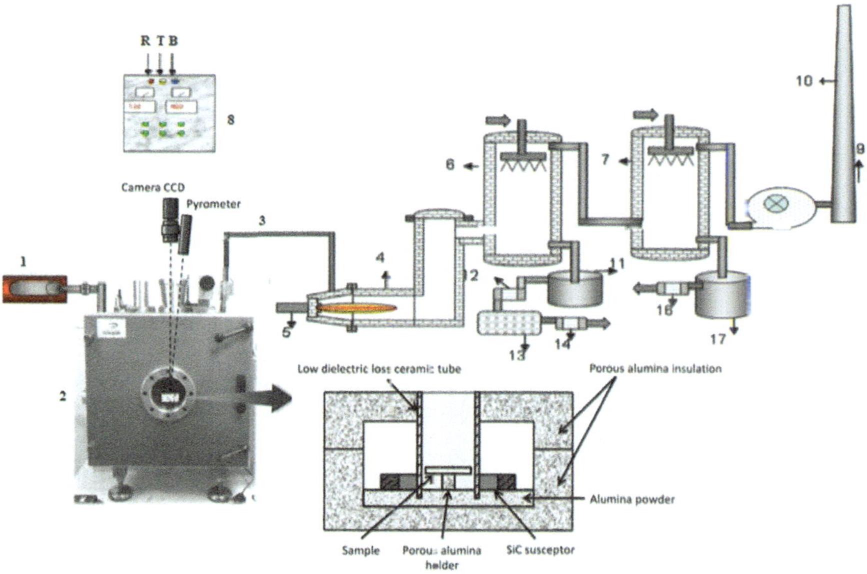

Fig. 1 Green and clean heating process arrangement of unwanted and discarded electrical and electronic components with microwave for flue gas treatment (1) Argon gas (2). Microwave Furnace (3) High temperature resisted Gas Pipe (4) Thermal oxidizer (5) Burner (6) Quench tower (7) Scrubbers (8) Power Supply (9) Cooling Fan (10) Chimney (11), (13) and (17) scrubbing fluid chamber, (12), (14) and (16) pumps to circulate and atomize scrubber solution

This mixture was then heated in the microwave furnace, as shown in Fig. 1. An inert gas (here argon) as shown in Fig. 1 was used to avoid oxidation reaction [7–9] inside the furnace. The product after microwave heating (for duration 45 min) obtained mostly consists of mixture of non-metal (ash mostly) and metal. The metallic portion separated from the mixture (after cooling) was treated by acid leaching process using Hydrochloric acid and then followed with cementation process [1].

2.2 *Flue Gas Treatment*

The investigation of flue gas has to be studied during microwave heating. The flue gas was collected in secondary chamber as shown in Fig. 1 by number 4. It consists of thermal burner. This fluid solution is precipitated with electrostatic precipitator. The gas solution was continuously scrubbed at location 7 as shown in Fig. 1. After scrubbing process, these solution has to be continuously stirred so that it can easily atomized (as obtained in location 7). The entire process represented with a single flowsheet is shown in Fig. 2 with Green and clean technology with microwave treatment process.

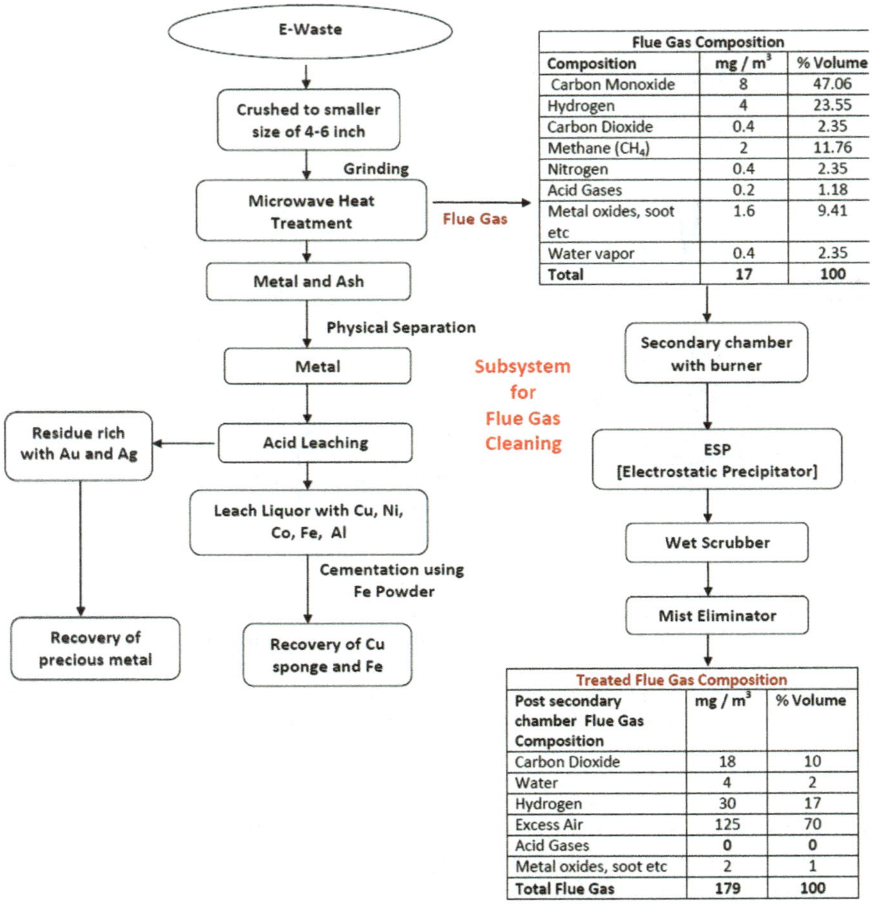

Fig. 2 Flowsheet for flue gas cleaning treatment

3 Results and Discussions

3.1 Green Technology with Microwave Heat Treatment

During microwave heating (always treated to be green and clean process) with flue gas arrangement at various time intervals, it has been analyzed that there was loss in waste powdered sample. The metallic product with 3600 g and non-metallic product with 900 g were produced for 10 kg of powdered sample after microwave treatment.

3.2 *Treatment of Flue Gas*

The typical metallic composition of the sample after microwave treatment is 12.99 % iron, 0.03 % gold, 9.79 % aluminium, 54.3 % copper, 1.13 % nickel, 0.28 % lead and 0.07 % silver. With physical separation, the weight with 5500 gm containing non-metallic part (i.e. 55 % of net loss) more of ash was easily removed. This loss was in the form of dioxins, carbon monoxide, sulphur dioxide, furans, loss of ignition (LOI), and it is known as flue gas. After scrubbing and filtering process, around 2.4 kg of metallic part was obtained. The various oxide component of nitrogen and sulphur were measured by Maritime 350 Testo gas analyzer, whereas carbon monoxide (CO) was measured by Gas chromatography (GC-2979). The dioxins and furans component was measured by SUPELCO. The hydrochloric acid (HCl) to be measured was mixed with particulate filter, acidified water and then with alkaline water. With the help of ion chromatography in other solution, the Cl^- ions were measured. The flue gas composition content was found to be in the range of NO_x: 245–250 mg per cubic metre, CO: 93–97 mg/m^3, SO_2: 190–195 mg/m^3, furans and dioxins: less than 0.15 ng/TEQ cubic metre, HCl: 80–85 mg/m^3. After scrubbing and filtering process, the total flue gas composition contents were analyzed as NO_x: 240 mg/m^3, CO: 50–55.5 mg/m^3, SO_2: 18–20 mg/m^3, Furans and dioxins: lesser than 0.12–0.15 ng, HCl with acidic materials: 60–70 mg/m^3. The assigned regulations set by Central Pollution Control Board are NO_x: 390–420 mg/m^3, SO_2: 140–200 mg/m^3, CO: 80–110 mg/m^3, Furans and dioxins: lesser than 0.15 ng/TEQ cubic metre HCl and other acidic components: 50–55 mg/m^3. It is understood except the HCl components factors (because of E-waste treatment), all other components of flue gas were found to be under within allowable standard. Hence one can say that scrubbers, gas-cyclones, absorbers, hydro-cyclones and electrostatic precipitators are very important controlling device to keep the environment clean. Moreover, the studies are still in progress to make use of flue gases in the application for alternative source for LPG and CNG and also for cooking gas.

4 Conclusion

The following conclusions were made for flue gas treatment during the recovery of some valuable metal from scrapped electronic and electrical components:

- The hazardous unwanted and discarded electrical and electronic components (e-waste sample) were treated in a microwave furnace for the useful and valuable metal recovery.
- The microwave heating is said to be clean and green process treatment as it does not pollute the environment. Also such innovative and greenery treatment generates less waste with respect to the conventional process.

- The typical metallic compositions after microwave innovative heating process are iron, copper, aluminium, lead, nickel and very few gold and silver components. With physical separation, the non-metallic was removed easily.
- With continuous process of scrubbing and filtering at various levels, the contents of flue gas were found to be well within the regulation range set by CPCB (Government of India).
- The scrubbers, gas-cyclones, absorbers, hydro-cyclones and electrostatic precipitators are very important controlling device to keep the environment clean.
- Further studies are still in progress for flue gases treatment for cooking gas and vehicle fuels applications.
- The entire innovative investigation shows system integrated with microwave furnace and flue gas treatment can be concluded as perfect eco-friendly and green process.

References

1. R.P. Mahapatra, S.S. Srikant, R.B. Rao, B. Mohanty, Recovery of basic valuable metals and alloys from E-waste using microwave heating followed by leaching and cementation process. Sadhana J. **44**(10), 209–219 (2019)
2. R.P. Mahapatra, S.S. Srikant, R.B. Rao, B. Mohanty, Microwave heating and acid leaching processes for recovery of gold and other precious metals from e-waste. Curr. Sci. **116**(3), 463–468 (2019)
3. G. Chauhan, R.J. Prashant, K.K. Pant, K.D.P. Nigam, Novel technologies and conventional processes for recovery of metals from waste electrical and electronic equipment: challenges and opportunities—a review. J. Environ. Chem. Eng. **6**, 1288–1304 (2018)
4. G.G. Wicks, D.E. Clark, R.L. Schulz, Microwave technology for waste management applications: treatment of discarded electronic circuitry (U). U.S. Department of Energy (DOE). Contract No. DE-AC09-96SR1 8500 (1995)
5. S.S. Rath, P. Nayak, P.S. Mukherjee, G.R. Chaudhury, B.K. Mishra, Treatment of electronic waste to recover metal values using thermal plasma coupled with acid leaching—a response surface modeling approach. Waste Manage. **32**, 575–583 (2012)
6. R. Khanna, R. Cayumil, P.S. Mukherjee, V. Sahajwalla, A novel recycling approach for transforming waste printed circuit boards into a material resource. Procedia Environ. Sci. **21**, 42–54 (2014)
7. S.S. Srikant, R.P. Mahapatra, T. Laxmi, Microwave energy for waste used engine oil. Inst. Eng. J. **1**, 255–261 (2015)
8. S.S. Srikant, P.S. Mukherjee, R.B. Rao, Microwave energy for waste management. Min. Proc. Technol. **3**, 948–953 (2013)
9. S.S. Srikant, P.S. Mukherjee, R.B. Rao, Prospects of microwave energy in material and mineral processing. Turk. J. Eng. Sci. Technol. **2**, 23–31 (2013)
10. C.A. Pickles, Microwaves in extractive metallurgy: part 1—a review of fundamental. Min. Eng. 1102–1111 (2009)

Secure Data Transmission in Mobile Networks Using Modified S-ACK Mechanism

P. Muthukrishnan and P. Muthu Kannan

Abstract Mobile ad hoc network is an important application in the current wifi world. There is no need of any network infrastructure. Each node acts as a transmitter that transmits the messages to the nearby nodes and receiver that receives the messages. Each node also acts as a cluster head to receive information from the virtually connected nodes. All grouped information is sent to base station. MANET is self-governing networks. The dynamic nature of MANET, where the topology is changing very rapidly and randomly. Wireless network medium used for message dispersion is prone to attack, as it is accessible by all. It is a necessity to develop a secure and efficient technique to identify cyber attackers to protect MANET from attacks. The proposed technique 'Modified S-ACK' will provide secure and efficient way of data transmission in Mobile Networks.

Keywords Wireless · Adhoc network · Nodes · Malicious attackers · MANET

1 Introduction

Mobile Ad hoc Network (MANET) comprises numerous mobile nodes that combine to form a virtually connected network without any existing network infrastructure. Each node in this network acts as a transmitter and receiver [1]. Each node receives packets and transmits to the other nodes in this network. The structure of MANET is shown in Fig. 1.

MANET is self-organizing, self-assembling wireless networks. Since the mobililty is high, the network structure of ad hoc networks is dynamic. Each node transmits and receives packets of data by random wireless channels [2]. If the distance between two nodes for data transmission is more, multihop forward technique comes

P. Muthukrishnan (✉) · P. Muthu Kannan
Saveetha School of Engineering, SIMATS, Chennai, India
e-mail: muthueng05@gmail.com

P. Muthu Kannan
e-mail: muthukannan@saveetha.com

S. S. Dash et al. (eds.), *Intelligent Computing and Applications*,
Advances in Intelligent Systems and Computing 1172,
https://doi.org/10.1007/978-981-15-5566-4_64

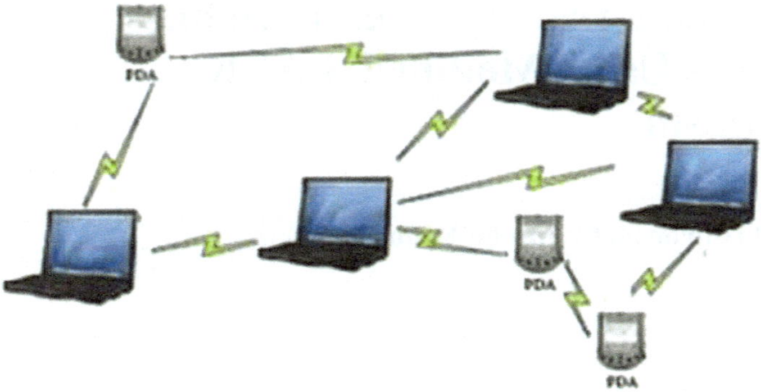

Fig. 1 Structure of MANET

into picture. Since MANETS rely on wireless transmission, sharing of message must be secured from vulnerabilities. Secured way of data transmission is needed to protect the data integrity. An insecure ad hoc network poses threat to the entire network.

In MANET, there is no central system to identify and prevent anomalies. There are many problems in MANET. One of the major problems is data security. How secured is data transmission through wireless channels? Routing table sets the routing protocols that lead nodes to transfer network topology information to build communication routes. This information is highly sensitive and may be a victim for malicious network, who attacks the network and the applications running on it [3]. There are two sources of threats to routing table. First is external attackers. An attacker could partition a network and introduce traffic congestion in the routing protocols.

Second and most severe threat is from nodes that are compromised, which might (i) misapply or share routing information to other nodes (ii) do vulnerable actions on data that induce service failures [4]. Generally, ad hoc network attacks are classified into passive attacks and active attacks. Active attacks are further categorized into internal and external attacks. Internal attacks are from internal authorized nodes inside the network [5]. It is difficult to identify those nodes that are attacking the data. External attacks are done by external nodes outside the network. They do not belong to the network and can be stopped by firewalls, decryption, and encryption techniques.

2 Characteristics of MANET

MANET comprises many independent mobile nodes that communicate with each other through radio waves. Mobile nodes that are within the range of other radio nodes can communicate easily as the transmission between the nodes will be enabled. If nodes are not within the range, then communication can be happened with the help of intermediate nodes [6]. These networks can operate from anywhere and fully dispersed and do not need an infrastructure. The characteristics of MANET are (i) Wireless communication, (ii) Nodes can be both a transmitter and a receiver, (iii) There is no central system to control, (iv) Network topology is dynamic, (v) Autonomous and self governing, as no infrastructure is needed, (vi) MANET can work from anywhere, (vii) Limited security.

The dynamic network topology due to the mobility of nodes increases the challenge to design an ad hoc network [7]. Each radio terminal or a mobile node is powered by rechargeable batteries (limited power source). There are three parts of power consumption of radio terminal. One part of power consumption for data processing inside the Radio Terminal. Second part is consumed to transmit its own information to the target. Third part of power consumption is utilized when the same Radio Terminal or the nodes are used as router. Energy consumption is the big challenge in designing ad hoc networks [8].

Generally, mobile devices have less storage and low computational capabilities. As their memory and computation is less, they are dependent on other resources for data processing [9]. A reliable network structure must be enabled through efficient and secure routing protocols.

3 Application in MANET

MANET has many applications. Some of the most common applications of ad hoc networks are listed here. MANET is used on a day to day basis for the electronic email, file transfer [10]. These days, emails are accessed from mobiles for faster communication. Due to the competitive data rates offered by the Network Service Providers, they are coming up with many new applications. Some of the applications of MANET are (i) Military zone, where the network infrastructure is unavailable. Ad hoc networks are self assembling, self governing. (ii) Disaster Recovery Management (iii) Mining (iv) Robotics.

MANET wireless systems are essential in crisis times, natural disasters like flood, earthquake, cyclone. During these times, restoring communication services is essential for people to lead their normal life. MANET does not need network infrastructure and communication is established faster than wired network [11].

There is another application used in inter-vehicular communications. Here, the secured communication happens between two vehicular nodes [12] The vehicles can communicate with each other which help to identify the real-time traffic updates, weather conditions, route identification in city environment, and highway environment. Such advanced network systems can be used for the future world.

Table 1 Types of attacks

Active attacks	Active attacks are done by the malicious nodes inside the routing protocol. This modifies the data packets or sometimes creates the false stream. Active attacks impacts the network performance
Passive attacks	Passive attacks do not modify the data packets. Instead it will aim to get the network structure, hierarchy, network information. These attacks are more complex to identify, as they do not alter the data stream. Passive attacks monitor the data transfer and collect the required network information
Wormhole attacks	Wormhole attack means attackers monitor the network and record the wireless information. They locate at different positions and create a tunnel. This tunnel between two attackers is called wormhole. Now, the data stream is forced to go through the tunnel
Denial of services	This attacks the network or any system, so that data packets are lost and does not reach the destination. DoS sends lot of void packets and creates a network congestion. By this, the services are impacted and cannot be accessed by the intended users

4 Issues in MANET

There are different issues in MANET. Some most common issues are (i) Data medium (ii) Distribution of information is needed (iii) Synchronization is needed between the sender and receiver that receives the message (iv) Some radio terminals are hidden from a sender (v) Data transmission speed should be high, otherwise there will be lag in data transfer (vi) Security. This is a critical issue. This paper revolves around this issue and discussed in detail in the next paragraph. (vii) Real time data synchronization is required for voice, video, images. (viii) Mobile storage is less. It is necessary for resource reservation for an OS. (ix) Power control is required to reduce the energy consumption at each node, routing table (x) Variation in sending and receiving the data bits leads to bigger bandwidth usage [13].

The most common security attacks in MANET systems are tabulated [14], Table 1 shows different types of attacks.

Our proposed system states how security can be provided or improved in MANET than the existing systems. In the proposed Modified S-ACK method, medium access reducing handshake technique is imposed along with node ID.

5 Existing System

A number of secure routing schemes are introduced for detecting security vulnerabilities in MANET.

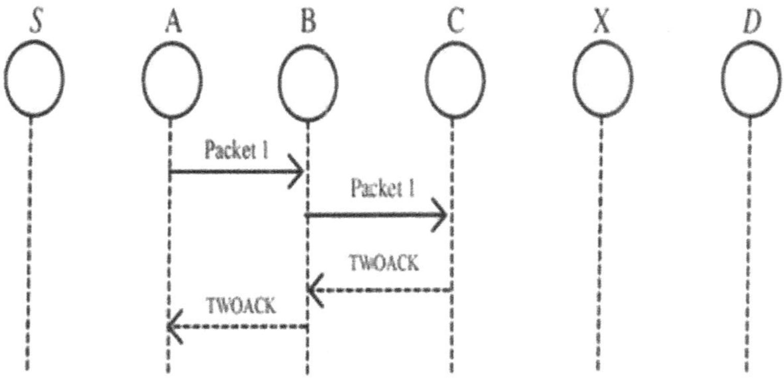

Fig. 2 Schematic representation of TWOACK

5.1 Watchdog

It is responsible for detecting malicious node behavior in the network [15]. Watchdog continuously monitors the behavior of the network and detects malicious nodes by data transmission speed of the network.

5.2 TWOACK

TWOACK is to recover the drawback of watchdog. In this, if Node A communicates with Nodes B, first Node A sends request signal to Node B. Upon receiving the request signal from Node B, it sends two ACK signals to Node A. This helps to identify the malicious node in the network. It also helps to avoid Data collision [16]. Figure 2 represents schematic representation of TWOACK.

5.3 S-ACK

In S-ACK technique, it sends data packet involving group of n number of nodes. Below example demonstrates to identify the malicious nodes in the network. Here, S-ACK involves group of three nodes, namely node A, B, and C. Node A is the source and it has to forward the packet to the destination node C. Node A sends Preq 1 to node B and node B sends Preq 1 signal to node C. Once Node C receives the signal Preq 1, it responds to node B with an acknowledgement signal Pack1. Once node B receives the acknowledgement signal Pack1, it forwards to node A. After node A receives Pack1 signal, node A is ready to forward the data packet to node C. In this network, if node A does not receive Pack1 signal, then other two nodes are

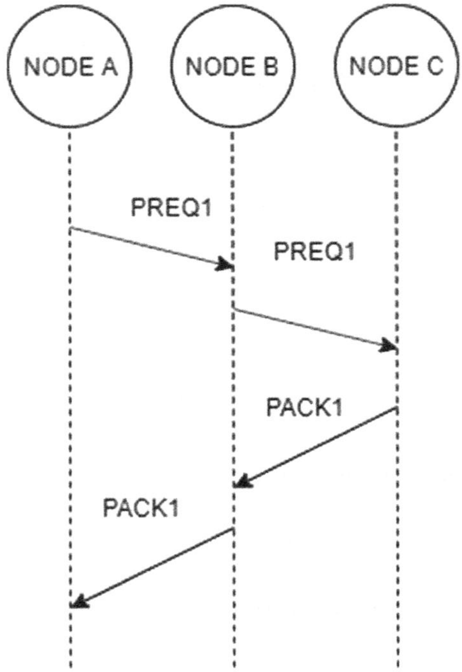

Fig. 3 Schematic representation of S-ACK

considered to be malicious. In this way, the data packets are sent from the source to the destination involving a group of three-three nodes. Figure 3 shows the schematic representation of S-ACK [17].

When comparing to Watchdog and TWOACK, this method shows more improvement in finding malicious node in the network. False identification can be avoided in S-ACK [18]. Time consumption is reduced as it does not wait for multiple acknowledgements. Figure 4 shows the flowchart of S-ACK.

6 Proposed System

In the proposed system, securing the data transmission in mobile networks using S-ACK Mechanism. In this method, data packets are sent from source to destination with the help of only two control signals Request To Send (RTS) and Clear To Send (CTS). Data is forwarded in a separate channel. Below example illustrates the operation of Modified S-ACK. Kindly refer Fig. 5 for the proposed system.

Consider there are four nodes node A, B, C, and D. Node A is the source and node D is the destination. Control signal RTS is initially sent from node A to node B. If node B is free, it sends back CTS1 signal to the neighboring nodes (node A and node C). Once node A receives CTS1 signal, it forwards the secure data to node B.

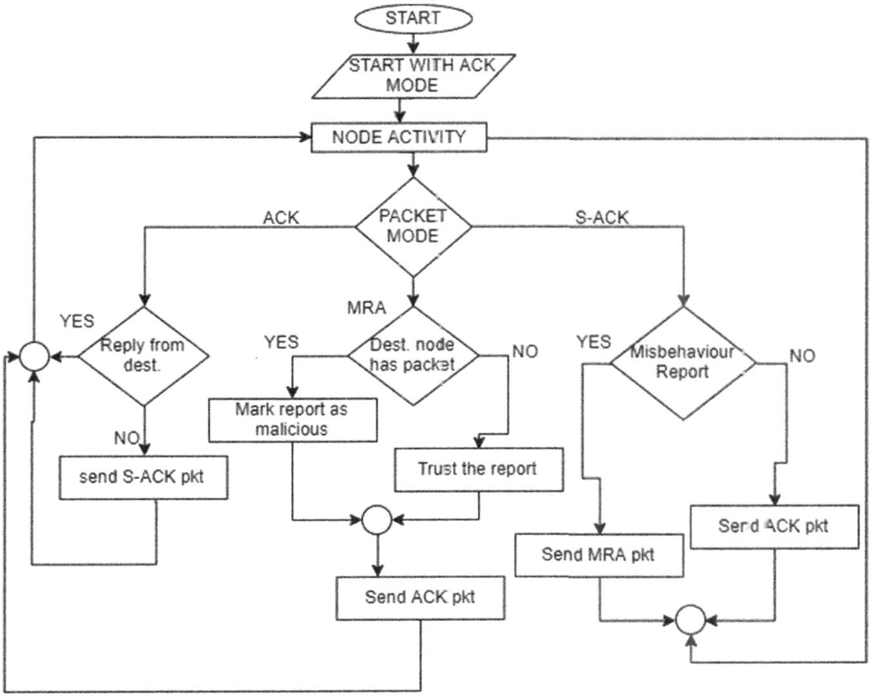

Fig. 4 Flowchart of S-ACK

At the same time, node C also receives CTS1 control signal. Once node C receives, it responds back with CTS2 signal to the neighboring nodes B and node D.

Every node has a unique identifier. Secured data is embedded with source Node Id and destination Node Id in a frame format. Frame format has source Node Id, Secure data, destination Node Id in this order. Once node B receives secure data, it checks the destination Node Id of the received data. If the Node Id does not match, it forwards to node C. The same process will continue till the data packets are received by the destination node D. Once data is sent to node D, the destination Node Id matches with the frame formatted data and receives the secure data.

Main advantages of our proposed approach are

1. Prevent data collision and detecting false misbehavioral nodes, consumes less power.
2. Data is more secured compared to contemporary approaches.
3. Acknowledgment data packets in this proposed method are true.
4. Improves the packet delivery ratio.

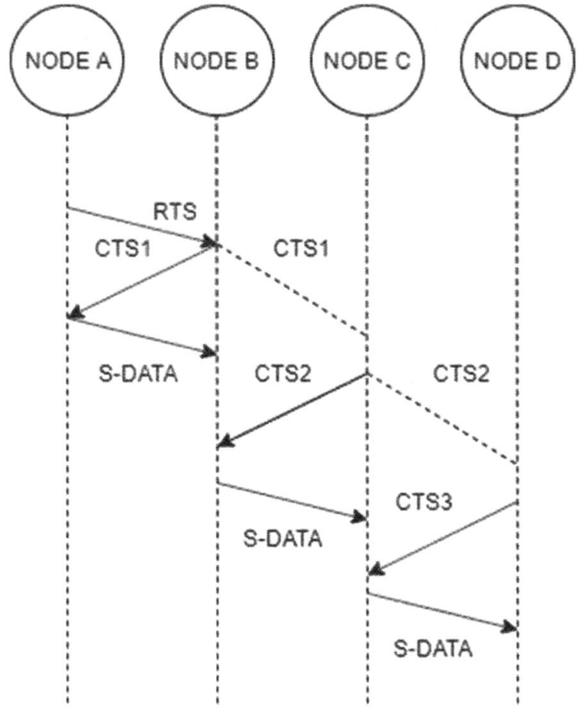

Fig. 5 Schematic diagram of modified S-ACK

7 Results and Discussion

One of the major security issues or attack in MANET is Security. Security issues like Active attack, passive attack, Wormhole attack are overcome by some existing security systems such as watchdog, TWOACK, S-ACK. TWOACK method is based on acknowledgement packets. Hence, it is important to ensure that the acknowledgment packets are true. But the disadvantage is that they are unable to identify the misbehaving nodes with the presence of false report and bad acknowledgement packets. Another drawback is the network congestion created by two acknowledgement packets. Due to the limited power consumption of MANET, such network congestion can impact the network performance. Figure 6 shows the number of attackers reduced in the proposed ratio and main the packet delivery ratio.

8 Conclusion

In the proposed modified S-ACK Mechanism, it is implemented framed data format including Source Node Id, Secure Message, destination Node Id. Only the matched Node Id receives the data. Data transmission speed is higher. It has advantages over

Fig. 6 Packet delivery ratio versus number of attackers

the other existing methods like true nature of the nodes, less power consumption, less receiver collision. All acknowledgment packets in the proposed system are true. It demonstrates higher malicious-behavior-detection rates. This technique helps to secure data in MANET by properly identifying the malicious nodes.

References

1. K. Al Agha, M.H. Berlin, T. Dang, A. Guitton, P. Minet, T. Val, J.B. Viollet, Wireless technology for industrial wireless sensor networks 56(10):4266–4278 (2009)
2. E.M. Shakshuki, N. Kang, T.R. Sheltami TR, EAACK—a secure intrusion-detection system for MANETs. IEEE Trans. Ind. Electron. **60**(3) (2013)
3. G. Jayakumar, G. Gopinath, Ad hoc mobile wireless networks routing protocol—a review. J. Comput. Sci. **3**(8), 574–582 (2007)
4. D. Johnson, D. Maltz, Dynamic source routing in ad hoc wireless networks (Chap. 5), in *Mobile Computing* (Kluwer, Norwell, 1996), pp. 153–181
5. N. Kang, E. Shakshuki, T. Sheltami, Detecting misbehaving nodes in MANETs, in *Proceedings of 12th International Conference on iiWAS, Paris, France*, 8–10 Nov 2010, pp. 215–222
6. N. Kang, E. Shakshuki, T. Sheltami, Detecting forged acknowledgements in MANETs, in *Proceedings of IEEE 25th Int. Conf. AINA, Biopolis, Singapore*, 22–25 Mar 2011, pp. 488–494

7. K. Liu, J. Deng, P.K. Varshney, K. Balakrishnan, An acknowledgment-based approach for the detection of routing misbehavior in MANETs. IEEE Trans. Mobile Comput. **6**(5), 536–550 (2007)
8. S. Marti, T.J. Giuli, K. Lai, M. Baker, Mitigating routing misbehavior in mobile ad hoc networks, in *Proceedings of 6th Annual International Conference on Mobile Computing Network, Boston, MA*, pp. 255–265 (2000)
9. N. Nasser, Y. Chen, Enhanced intrusion detection systems for discovering malicious nodes in mobile ad hoc network, in *Proceedings of IEEE International Conference on Communications, Glasgow, Scotland*, 24–28 June 2007, pp. 1154–1159
10. T. Sheltami, A. Al-Roubaiey, E. Shakshuki, A. Mahmoud, Video transmission enhancement in presence of misbehaving nodes in MANETs. Int. J. Multimed. Syst. **15**(5), 273–282 (2009)
11. T. Anantvalee, J. Wu, A survey on intrusion detection in mobile ad hoc networks, in *Wireless/Mobile Security* (Springer, New York, 2008)
12. L. Zhou, Z.J. Haas, *Network Security*. IEEE, Nov 1999, pp. 1–12
13. D.B. Johnson, A. Perrig, Y.-C. Hu, SEAD: secure efficient distance vector routing for mobile wireless ad hoc networks (Elsevier, 2003), pp. 175–192
14. J. Undercoffer, J. Parker, J. Pinkston, A. Joshi, On intrusion detection and response for mobile ad hoc networks, in *2004 IEEE International Conference on Performance, Computing, and Communications, USA*, pp. 747–752 (2004)
15. A. Bertacchini, D. Brunelli, L. Larcher, L. Benini, D. Dondi, Modeling and optimization of a solar energy harvester system for self-powered wireless sensor networks. IEEE Trans. Ind. Electron. **55**(7):2759–2766 (2008)
16. S. Vairachilai, A. Rubi, A survey on intrusion detection system in mobile adhoc networks. Int. J. Comput. Sci. Mobile Comput. **2**(12):389–393 (2013)
17. W. Liu, Y. Zhang, W. Lou, Y. Fang, Anonymous handshakes in mobile ad hoc networks, in *Military Communications Conference*, pp. 1193–1199 (2004)
18. N. Kang, T.R. Sheltami, E.M. Shakshuki, EAACK—a secure intrusion-detection system for MANETs. IEEE Trans. Ind. Electron. **60**(3), 1089–1098 (2013)

An Anonymization Approach for Dynamic Dataset with Multiple Sensitive Attributes

V. Shyamala Susan

Abstract In recent days, the personal health information is collected in different fields. When the data is shared for different reasons, it poses a major danger to the field of health care. A number of anonymization methods are implemented to maintain person's privacy. The existing methods of anonymization support only single sensitive and low-dimensional data. In our recent experiment, a method of anonymization is expected in order to anonymize high-dimensional data with multiple sensitive attributes. In line with the concept of *k*-anonymity and *l*-diversity, it combines anatomization and improved slicing strategy. The experimental findings show that it is limited to the static discharge of information only. In dynamic situations, the current technique may produce poor quality or high data loss. Hence, in this proposed approach, an anonymization model is designed in such a way to anonymize continuously growing dataset while assuring high utility.

Keywords Health care · Privacy preservation · Anonymization · Anatomization · Improved slicing · *K*-anonymity · *L*-diversity

1 Introduction

Every organization publishes data that are collected from different users. When they publish the data, the personal information of an individual may be disclosed. This violates the privacy of the person and it needs to be protected from abuse [1]. To maintain privacy of the personal information, anonymization procedures are introduced. The dataset may be classified into different categories of attributes like identifiers, quasi-identifiers, sensitive attributes, and non-sensitive attributes [2]. The *k*-anonymity generalization [3, 4] and *l-diversity* bucketization [5] are the popular privacy preservation approaches. The *k*-anonymity method in generalization [6, 7] disregards an enormous portion of information in the occasion of high-dimensional data. But these approaches are suitable only for static data release and can handle

V. Shyamala Susan (✉)
A.P.C. Mahalaxmi College for Women, Thoothukudi, India
e-mail: shyamalasusan@gmail.com

S. S. Dash et al. (eds.), *Intelligent Computing and Applications*,
Advances in Intelligent Systems and Computing 1172,
https://doi.org/10.1007/978-981-15-5566-4_65

single sensitive attribute. If all sensitive attributes are given a similar amount of privacy, they may not provide predictable outcomes on the accessible information. The real-world data contains Multiple Sensitive Attributes (MSA). To handle that situation, our recent work proposed a model to anonymize MSA. It presents a method of anonymization that integrates the assistances of anatomization and more enriched slicing according to the rule of *k*-anonymity and *l-diversity* to handle high-dimensional information with MSA. The algorithm of anatomization divides QI and SA in the dataset and publishes unmodified QI and SA in two tables: QIT and ST. The vertical partitioning stage in the enhanced slicing algorithm combines the correlated SA in ST with the correlated QI attributes in QIT, thus significantly reducing dimensionality through the use of the Modified Advanced Clustering Algorithm (m-ACA). Similarly, Metaheuristic Firefly Algorithm with Minkowsi Distance Measure (MFAMDM) approach ensures *k*-anonymity in QIT and *l-diversity* in ST in the horizontal partitioning phase. The subsequent step anonymizes the dataset by linking the two tables with the common group ID as quickly as the attributes are cut off.

The experimental findings indicate that with several SA the suggested system can efficiently maintain data privacy. The anatomization approach significantly reduces the loss of information and the slicing algorithm benefits preserve correlation and utility, thus decreasing data dimensionality and loss of information. By minimizing the time complexity, advanced clustering algorithms prove their effectiveness. Moreover, this research work tracks strictly the rule of *k*-anonymity, *l*-diversity, and thus keeps away privacy.

Now, the next objective is to anonymize a continuously growing dataset with MSA. The promising method is to anonymize the new dataset and then bring out new records periodically. These published datasets can be used for analysis by the researcher either autonomously or combine multiple datasets for better analysis. The problem in the existing method is more time is spent in anonymization as many small records are to be anonymized independently. When these small anonymized datasets are merged, they may violate the privacy of the person. Also, many independent anonymizations reduce data utility. Hence, a new method is to be designed in such a way that an entire dataset needs to be anonymized whenever new records are added to the existing dataset. In this paper, we apply an anonymization methodology for continuously growing dataset with great data utility and minimum information loss.

2 Literature Review

In order to anonymize MSA, the following researches are used. The *k*-anonymity and *l-diversity* model is introduced to safeguard the person's privacy [8]. It used data distortion to preserve privacy. A decomposition method is introduced to preserve privacy [9]. But that model cannot be extended to dynamic data release. Similarly, a model for dynamic data release is proposed [10], but that is not appropriate for high-dimensional data. Liu et al. implemented a *k*-anonymity model for MSA [11].These methods recognized characters with both type of attributes, but it is not appropriate for large dataset. An anonymization approach for multiple sensitive approaches is introduced [12], but it is not tested with a real dataset. An anatomization with slicing approach is introduced; but for Quasi-Identifiers (QI) and Sensitive Attributes (SA), the slicing algorithm must be implemented individually [13]. Hassan et al. [14] termed a model to reserve the confidentiality on multiple independent data publishing. A method for MSA using (*P*, *K*) anonymization approach, but it is not limited to low-dimensional data [1]. KC-slice methodology is suggested for dynamic data with MSA set [15]. The privacy preservation approaches that are used in ecommerce sites suffer from inference attacks [16]. Mehta and Rao [17, 18] projected a novel anonymization approach called Scalable *k*-Anonymization with minimum number of iterations in MapReduce.

2.1 Proposed Work

In this chapter, we present our current work with a static dataset, which is appropriate for MSA. Then, we extend our work to anonymize continuously growing dataset with MSA.

2.2 The Anonymization Approach for Static Dataset with MSA

Anatomization method is initially used to reduce information loss by directly discharging QI and SA. It disassociates the correlation between the two attributes and discharges double distinct tables, namely Quasi-Identifier Table (QIT) and Sensitive Table (ST). Then, enhanced slicing approach is carried out in two phases. The first phase groups the correlation between the attributes in QIT and ST and minimizes the dimensionality. The improved slicing algorithm retains privacy over partitioning both horizontally and vertically. Now, the SA is grouped by m-ACA in the vertical partitioning phase and releases many ST each with the column group ID. Similarly, QIA is grouped and released as QID. For the attributes being partitioned into columns, the m-ACA is introduced, reducing the need to reassign the data point several times in each iteration with optimum initial centroid detection. This can help to improve the clustering velocity and minimize the complexity of the computation.

Algorithm: **Modified Advanced Clustering Algorithm (m-ACA)**
Input: Input dataset $X = (x_1, ..x_n)$, the count for clusters k
Output: Vertical partitioning groups
Procedure
Begin

1. Input dataset $X = (x_1, ..x_n)$ from the dataset
2. Estimate the close centroids and allocate to close cluster

3. Identify least distance from cluster center
4. Calculate the distance between each dataset samples x_i to all the optimal centroids c_j
5. Repeat steps 1 to 5 until the number of data points extents a smaller amount of distance value
6. End for;
7. Bring the two closest clusters together in a cluster
8. Recalculate the fresh cluster center for the merged cluster until it decreases the amount of cluster to k
9. End for;

In the second phase, attributes are grouped into tuples by Metaheuristic Firefly Algorithm (MFA) ensuring *k*-anonymity in Quasi-Identifier Table (QIT) and *l-diversity* in the entire Sensitive Table (ST). In the last phase, with a shared group ID, SA is shuffled in each cluster and connected to QI in QIT. This is to guarantee that an intruder can identify a person's sensitive value with a maximum of 1/*l*. This strategy minimizes the loss of data and guarantees privacy through the slicing algorithm. The advanced algorithm of chartering reduces time and complexity.

Algorithm : Tuple-partition
Input: Dataset , the parameter ℓ
Output: Sliced Bucket
Procedure
Begin
1. Q={T}, Sliced bucket $SB = \emptyset$
2. While Q is not empty
{
2.1 Eliminate the first bucket from Q,Q=Q-{B}
2.2 Split the bucket into totally two different buckets using Firefly Algorithm(FA)
2.3 Check the tuple from Q
2.4 Fix the objective function f(X)for Q
2.5 Compute the light intensity function I_i by the objective function f(X)
2.6 Find attractiveness β by the minimum distance between the tuples
2.7 Using the intensity I_i and attractiveness β tuples are ranked
2.8 Highest ranking tuples form the bucket Q
2.9 if diversity-check(QIT, Q ∪ {B1, B2} ∪ SB, ℓ)
Q=Q U{B1,B2}
Else SB=SBU{B}
Return SB
}
3. End while
4. End

Thus, tuple partitioning guarantees ℓ-diversity in each bucket.

SLicing with Anatomization for Multiple Sensitive Attributes (SLAMSA)

The next step is to develop an anonymization group of SLicing with Anatomization for Multiple Sensitive Attributes (SLAMSA). Here, the SA is associated with a common group ID in each group with the QI characteristics in QIT. This guarantees that an individual's delicate value engaged in QIT can be extracted immediately from an opponent with a maximum likelihood of 1/*l*. A greater *l* leads to greater privacy. The method finishes when there is no longer any ungrouped SA or when it is not possible to develop fresh group. If ungrouped documents are available, they will be released as a single group. As this work follows both the principle of *k*-anonymity and *l-diversity*, it removes the overall complexity of privacy. Disclosures of membership and identity are shielded by *k*-anonymity and exposés of attributes are eliminated by *l-diversity*.

Algorithm : SLAMSA
Input: Dataset, k, l, Sensitive Attribute Table (SAT)
Output: Quasi-Identifier Table, horizontally partitioned Table;
Procedure

1. Begin
2. Divide the dataset into a set of sensitive attribute Tables and Quasi-Identifier Table w.r.t ST;
3. For each Table
   {
   3.1. For each tuple
        Assign a numerical identifier starting from 1;
4. For i=st_1 to st_m
   4.1. Form st_i w.r.t l, with no repetitive attribute values; (fitness)
   4.2. End for
5. For each group of size < k
   5.1. Select a group G of size < k randomly;
   5.2. Find the distance between G and other groups;
   5.3. Pick out the group G1 with the least distance;
   5.4. Generalize G and G1;
   5.5. end for;
6. Map generalized table and st_i (Quasi-identifier Table)
7. End

2.3 *Anonymization Approach for the Dynamic Dataset*

Assume that an anonymized table T' is released for the static dataset. Now suppose that new patient records $R = \{r_1, r_2 \ldots r_n\}$ are to be inserted into the anonymized table T'. Let this anonymized table be T''. To anonymize this table T'', the required conditions are that QI attributes in each group must be *k* anonymized and SA attributes must be *l* diverse. To ensure high data utility while anonymizing the dynamic dataset, the following conditions are suggested.

Case 1: If the new records *R* when added with T' satisfy *k*-anonymity and *l-diversity* and do not overlap with the existing anonymized table, then we can increase the records to T' as a new anonymized group.

Case 2: If the new records in *R* cannot be added directly into as an anonymized group, then these records must be inserted into an existing anonymized table T'. To preserve privacy in the new anonymized table, each record is inserted by ensuring *k*-anonymity and *l-diversity*.

Case 3: Sometimes addition or insertion of the new records into the anonymized table T' may not satisfy *k*-anonymity and *l-diversity* in each group. In such a case, we split the anonymized table into two anonymized tables for better data quality.

2.4 *How to Enforce Anonymity in All the Above Three Cases*

The add operation ensures anonymity by utilizing k-anonymity in QIT and *l-diversity* in ST in each group. The insert operation may not produce anonymity in each anonymized group when the new records are inserted in the anonymized group. It may not satisfy k-anonymity in QIT and *l-diversity* in ST. This may cause a threat to the dataset. In order to overcome such a situation, we insert the records into the waiting list. The records will be inserted from the waiting list to the anonymized group if it satisfies anonymity. If the records' waiting time is more, then they are identified to be published as a separate anonymized group. But these groups should not overlap with the existing anonymized group. The process ends when there are no ungrouped attributes in the waiting queue. The anonymized table T' can be split when it ensures k-anonymity and *l-diversity* in each group of the tables.

3 Performance Analysis

The experiment is evaluated through the dataset Cleveland Clinic Foundation and Heart Disease and Health Institute of Cardiology, and the dataset is accessible in the UCI machine learning repository. The dataset has 76 raw attributes, considering for assessment only 12 attributes for evaluation. The age, gender, and social security number are measured as QI, and further attribute that identifies the heart disease of the patient is SA. The experiment is started with 557 records for both datasets and compared with the existing KC and KCi methods. The KC slicing system utilizes the same threshold value for all sensitive data that result in high data loss, while varying threshold values are given to the SA in the KCi slicing system to decrease data loss.

3.1 *Utility*

The utilization of the proposed method is evaluated using reconstruction error. To improve the utility, the correlated attributes must be arranged in a column which is achieved by the slicing approach. The query.

Select count * from T

Where (sensitive items are present)

and $(q_1 = \text{val}_1)\, n\, (q_2 = \text{val}_2) \ldots n(q_n = \text{val}_n)$

It is measured by obtaining the distance between actual probability distribution function (Act) and estimated probability distribution function (Est) which gives KL divergence

$$\text{KL}_{\text{divergence}}\left(\text{Act}^s, \text{Est}^s\right) = \sum_{\forall \text{cell} C} \text{Act}_C^s \log \frac{\text{Act}_C^s}{\text{Est}_C^s}$$

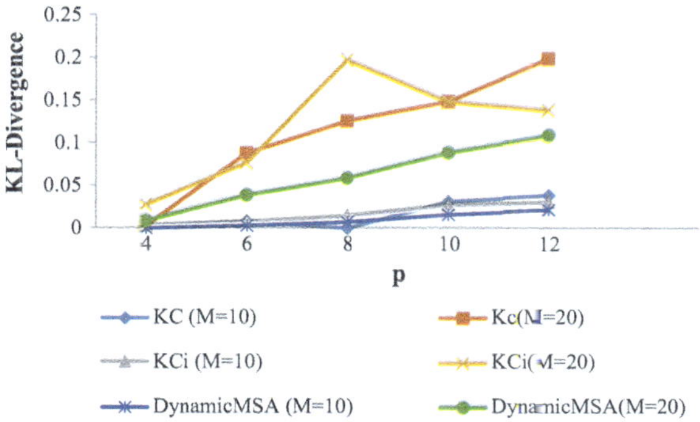

Fig. 1 Reconstruction error versus p ($r = 4$)

$$\mathrm{Act}_C^s = \frac{\text{Occurrences of } s \text{ in } C}{\text{Total Occurrences of } s \text{ in } T}.$$

The quantity of elements in G is "a," and the quantity of the same QID tuples is "b." The query's evaluated outcome converts $a \cdot b/a$. For all tuples in G with the same QID, either a or $b = 0$, which means that a minimum reconstruction mistake occurs. If Act^s is same to Est^s, $\mathrm{KL}_{\mathrm{divergence}} = 0$ which means there is a minimum reconstruction mistake.

It is necessary to evaluate the data quality from both static and dynamic datasets. Change the values of P, M, and R where P is the degree of privacy between 4 and 20 for each experiment, M is the number of SA ranging from 3 to 12, and R is the value of QID ranging from 2 to 4. Many queries are assessed by altering $q_1, q_2 \ldots q_n$ and $s_1, s_2 \ldots s_m$ randomly. The average reconstitution error is measured, and then the utility is identified.

The experimental effects are shown in the following figures.

Figure 1 shows that the data quality is lower as compared to Kc and KC_i methods. But the quality is maintained regarding to data size. Increasing the degree of privacy increases the reconstruction mistake and minimizes the utility. When the value of m is increased, there is more value matching, which limits the quality which is depicted in Fig. 2. Similarly, when R is increased, the corresponding reconstruction error is also increased due to high dimensionality but limits utility. This is depicted in Fig. 3.

The existing KC and KCi techniques show information loss as these techniques make use of suppression in each group. But this method uses advantages of anatomization that releases QID and SA directly that improves utility and limits information loss.

4 Conclusion

In this proposed work, a module is designed to anonymize a dynamic dataset with MSA. The correlation between the attributes identified using slicing algorithm and

Fig. 2 Reconstruction error versus m ($r = 4$)

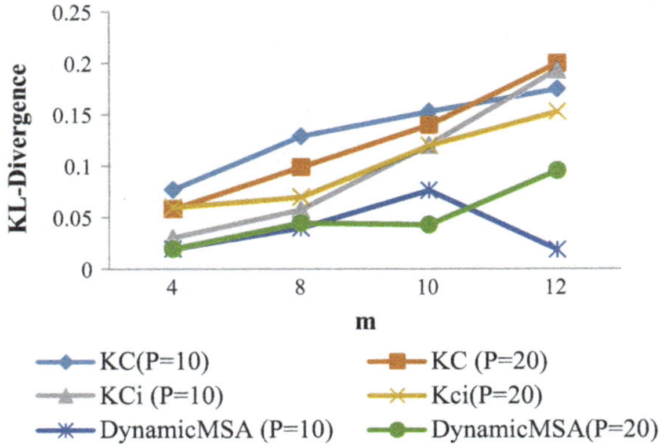

Fig. 3 Reconstruction error versus r ($m = 10$)

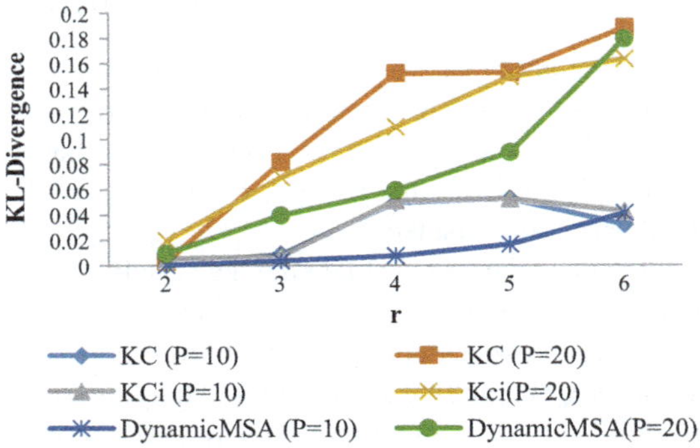

anatomization reduces the information loss. The Advanced Clustering algorithm limits time and complexity, and the design ensures *k*-anonymity and *l-diversity* to pervert privacy threats. This method can handle any number of SA in an effective way.

References

1. A. Anjum, N. Ahmad, S.U. Malik, S. Zubair, B. Shahzad. An efficient approach for publishing microdata for multiple sensitive attributes. J Supercomput. 1–29
2. S. Onashoga, B. Bamiro, A. Akinwale, J. Oguntuase, Kc-slice: a dynamic privacy-preserving data publishing technique for multisensitive attributes. Inf. Secur. J. Global Perspect. **26**(3):121–135
3. B. Gedik, L. Liu, Protecting location privacy with personalized k-anonymity: architecture and algorithms. IEEE Trans. Mob. Comput. **7**(1), 1–18 (2008)
4. K. El Emam, F.K. Dankar, Protecting privacy using k-anonymity. J. Am. Med. Inf. Assoc. **15**(5):627–637 (2008)

5. L. Ninja, L. Tiancheng, S. Venkatasubramanian, t-closeness: privacy beyond k-anonymity and l-diversity. In: *Proceedings of ICDE 2007, IEEE 23rd international conference on data engineering*, Istanbul, Turkey (2007)
6. C.C. Aggarwal, On k-anonymity and the curse of dimensionality, in *Proceedings of the 31st international conference on very large data bases, VLDB Endowment,* 30 Aug 2005
7. D. Kifer, J. Gehrke, Injecting utility into anonymized data sets, in *Proceedings of the ACM SIGMOD, International Conference on Management of Data*, 27 June 2006
8. T.S. Gal, Z. Chen, A. Gangopadhyay, A privacy protection model for patient data with multiple sensitive attributes. IGI Global **2**, 28 (2008)
9. Ye Y, Liu Y, Wang C, Lv D, Feng J (2009) Decomposition: privacy preservation for multiple sensitive attributes, in *International Conference on Database Systems for Advanced Applications* (Springer, Berlin), pp. 486–490
10. D. Das, D.K. Bhattacharyya, Decomposition + : improving l-diversity for multiple sensitive attributes, in *International Conference on Computer Science and Information Technology* (Springer, Berlin), pp. 403–412 (2012)
11. F. Liu, Y. Jia, W. Han, A new k-anonymity algorithm towards multiple sensitive attributes, in *IEEE 12th International Conference on Computer and Information Technology (CIT)* (IEEE, 2012), pp. 768–772
12. P. Usha, R. Shriram, S. Sathishkumar, Multiple sensitive attributes based privacy preserving data mining using k-anonymity. Int. J. Sci. Eng. Res. **5**(4) (2014)
13. V.S. Susan, T. Christopher., Anatomisation with slicing: a new privacy preservation approach for multiple sensitive attributes. SpringerPlus **5**(1), 964
14. A.S.M. Hasan, Q. Jiang, H. Chen, S. Wang, A new approach to privacy-preserving multiple independent data publishing. App. Sci. **8**(783), 1–22 (2018)
15. S.A. Onashoga, B.A. Bamiro, A.F. Akinwale, J.A. Oguntuase, *Privacy preserving data publishing of multiple sensitive attributes: a taxonomic review*. Presented at the International Conference on Applied Information Technology. Federal University of Agriculture, Abeokuta, Ogun State, Nigeria (2017)
16. D. Yang, Q. Bingqing, P. Cudre-Mauroux, Privacy-preserving social media data publishing for personalized ranking-based recommendation. IEEE Trans. Knowl. Data Eng. (2018). ISSN (Print):1041–4347, ISSN (Electronic):1558-2191
17. B.B. Mehta, U.P. Rao, Privacy preserving big data publishing: a scalable k-anonymization approach using MapReduce. IET Softw. **11**, 271–276 (2017). https://doi.org/10.1049/iet-sen.2016.0264
18. B.B. Mehta, U.P. Rao, Toward scalable anonymization for privacy preserving big data publishing. Recent Findings Intell. Comput. Tech. **708**, 297–304 (2018). https://doi.org/10.1007/978-981-10-8636-6. *Proceedings of the 5th ICACNI 2017*, vol. 2

Newspaper Identification in Hindi

Subhabrata Banerjee

Abstract This paper reports the creation of a newspaper identification model in Hindi. This work has two broad angles, namely, the detection of online content in Indian languages, and second, the use of world-class data. Data is a primary problem in Indic computing. This work tries to show with the use of proper data, which may remove the data-driven challenge of Indic computing. The work used annotated and unannotated gold standard Hindi data. Hand-annotated Hindi data for Named Entity Recognition (NER) is used to carry out this task. The work shows a detailed analysis of entity types present in each online newspaper. One generative and another discriminative model, namely Naïve Bayes (NB) and Support Vector Machine (SVM) based classifier, are applied here. In both the models, the success of around 90.0 in *F*-measure is received. The work is done to help the ever-growing electronic document user community. It will help a user to classify the Indic document types.

Keywords Indian language · Hindi · Gold standard data · Newspaper identification · Classifier

1 Introduction

Document means the piece of text used for teaching or giving instructions. Document is defined as "any concrete or symbolic indication, preserved or recorded, for reconstructing or for proving a phenomenon, whether physical or mental" [1]. The history of documents started in the days of Egyptian civilization. The document is divided as secret, open, or private based on the parameter of ownership. There may be various classifications of it based on content and website taxonomy. Documents come across multiple storage formats, but document means nowadays the electronic file which may contain information to be studied and stored in a retrievable form With the advent of this newer horizon of electronic use, the contents are also getting localized. In our country, there are twenty-two Indian languages. In the days of efficient

S. Banerjee (✉)
HCL Technologies, Noida, India
e-mail: subhabangalore@gmail.com

S. S. Dash et al. (eds.), *Intelligent Computing and Applications*,
Advances in Intelligent Systems and Computing 1172,
https://doi.org/10.1007/978-981-15-5566-4_66

use of electronic devices, many of the languages have a sizeable amount of online and offline data. Both governmental and non-governmental agencies are coming into practice very fast. Days are near when we would need advanced writing assistants like spell checker, grammar checker, or plagiarism analysis in these languages.

But to create such tools, we first need one sound content detection system. It will give us a highlight of the peculiarities of such a document. We need to have appropriate data to implement these systems. In our country, we did not have any useful annotated general domain data, which was built by a group of experts. The work is done in Hindi, and it is believed that the whole model right from data building to content detection be replicated in other Indian languages soon. It will help us to build a good base for the localization of software.

The problem is the identification of the source of online content. Here, the solution is built upon online newspapers. In the days of substantial electronic material, it is useful to make a tool for online detection of resources. The online content of every language is increasing, and Hindi is no less. Almost all the major newspapers and news channels have online content. Other than these, many government and business houses have sizeable online content in Hindi. It is a boon for the readers and users to have an increased number of content in their local language. But it is also an essential problem for the cybercrime detection group to detect the source of online content.

The resolution here is to work on one online content detection system for Hindi newspapers. An impressive result is reported.

The work is perhaps first of its kind in an Indian language. But there is a related group of work being worked out in natural language processing. The associated works are in the domain of author identification, sentiment analysis, and topic Modeling. Various scholars investigate each of these fields, and I am trying to give reference to only seminal contributions. References [2–4] gave pioneering work on author identification. In the Indian language, [5] has worked on the author identification of the Bengali language.

References [6–10] gave major works in sentiment analysis. References [11–13] made a significant contribution to sentiment analysis in Indian languages. References [14–17] worked sizeably well in topic modeling. References [18, 19] made an excellent contribution to the topic modeling of Indian languages.

The problem here is tackled in two angles. There is one problem of classification and the other issue of localization. The problem scope was in Hindi, so finding appropriate data was severe. A gold standard data to build the solution, with hand label entity tagging, is tried. Here, the resolution is given in three major segments. They are the statement on the problem, the resolution part, and the results and discussion part.

2 Problem

The problem in its basic form is the identification of online newspapers or content. The basis for the resolution is to exploit Indic data. The issue of this data is also addressed here, briefly.

3 Resolution

In an age of surging electronic activity of document creation, it is almost necessary to have text mining applications in each language. The recommendation here is to detect the source of the content of Hindi newspapers, in the emerging electronic behavior of the language.

In these days of good computing power, web resources, and openware, it is not a very challenging work to build a statistical classifier and solve a document classification problem. But this is possible with a resource-rich language like English or German. Finding proper resources for Indian languages is a daunting task.

To build this gap, I am working with various research teams in Kanpur and Hyderabad.

This work has two segments of resolution. The first part discusses creating the data, and the second part talks about building the classification issue.

3.1 Hindi

Hindi in Devanagari script is the lingua franca of India. It is also declared as one of the primary official languages of India. The language originated mainly from the Khariboli dialect spoken around the Delhi region. Hindi is a subgroup of Sauraseni Prakrit. It has 33 consonants and around 11 vowels. The vocabulary consists of Tatsama, Tadbhaba, Ardhtatsama, Desi, and Bideshi words. It is written left to right.

Chandrakanta by Devaki Nandan Khatri which was written in 1888 is considered as a first important prose work while it is enriched by contributions of Munshi Premchand, Swami Dayananda Saraswati, Bhartendu Harishchandra, Mahadevi Verma, Suryakant Tripathi, Ramdhari Singh Dinkar, Haribansh Rai Bacchan, Kamleshwar, Maithili Sharan Gupt, Nirmal Verma, Mohan Rakesh, Bhism Sahani, Mridula Garg, Atal Behari Vajpayee, and many others.

It is an Indo-Aryan language. Hindi documents are written primarily using Hunterian transliteration, International Alphabet of Sanskrit Transliteration (IAST), Indian languages TRANSliteration (ITRANS), ISO 15919 Transliteration of Devanagari and related Indic scripts into Latin characters (ISO 15919).

3.2 Gold Standard Data

To build gold standard data in Indian languages, few like-minded senior researchers of Natural Language Processing (NLP) domain are trying to develop hand-annotated gold standard data in various Indian languages.

> Gold standard annotated corpora are necessary resources when building and evaluating NLP systems. Manually labeled instances that are relevant to the specific NLP tasks must be created. A useful gold standard should be rich in information and include a large variety of documents and annotated instances that represent the diversity of document types and instances at stake in a specific task. This is essential to (1) either train machine-learning-based NLP systems, which need examples to learn from, or discover rules for rule-based algorithms and (2) evaluate the performance of NLP systems [20].

Gold standard data generally refers to the best data given some parameters. These parameters of data should be appropriately interpreted by an expert, which would be valid for a considerable period. The data should have detailed guidelines, annotations to be done by one expert or one editor, and data sets are collected and annotated using standard methodologies. It may be like Brown corpus is following [21], or maxims of annotation by Leech [22], or [23] guideline on annotation.

In Indic computing, many institutes and companies are working, but very few are making their data public. A recent effort is coming from Technology Development in Indian Languages (TDIL) and Jawaharlal Nehru University (JNU), but they are pretty much in their infancy. We still do not have enough data for any researcher and no proper fieldwork and analysis regarding the source of data and their taxonomies.

3.3 Our Solutions

An effort is being made to build gold standard data to be available in public domain free for anyone, primarily in Hindi and Bangla with the Indian Institute of Technology (IIT)-Kanpur and International Institute of Information Technology (IIIT)-Hyderabad. Seed data till now is collected by us after proper field study and is hand annotated. The present results are based on this.

We have the aim of creating one million hand-labeled data of Parts of Speech (POS) and NER in a few primary Indian languages. The data would be available for free in the public domain. We have created, till now, around 0.2 million words hand-annotated NER data in Hindi. Here I am using 0.1 million unbiased NER data for our use retrieved from web pages of Dainik Bhaskar, Hindusthan, and British Broadcasting Corporation (BBC) Hindi.

Figures 1 and 2 show samples of raw data and Hindi NER annotated data. A good NER of Hindi is under testing with an F-measure of 90.0+. But it is beyond the scope of this paper. We hope to publish its results soon.

After tagging the data, we stored the data in (x, y) format in Unicode Transformation Format (UTF)-8 encoded text files.

Sat 27 Jul 2019 08:25 PM IST

प्रधानमंत्री नरेंद्र मोदी, रक्षा मंत्री राजनाथ सिंह के साथ कारगिल विजय दिवस समारोह के लिए इंदिरा गांधी इंडोर स्टेडियम पहुंचे।

Fig. 1 Sample Hindi raw corpus

Sat/DATE 27/DATE Jul/DATE 2019/DATE 08:25/DATE PM/DATE IST/DATE

प्रधानमंत्री नरेंद्र/PERS मोदी/PERS, रक्षा मंत्री राजनाथ/PERS सिंह/PERS के साथ कारगिल/ORG विजय/ORG दिवस/ORG समारोह के लिए इंदिरा/LOC गांधी/LOC इंडोर/LOC स्टेडियम/LOC पहुंचे।

Fig. 2 Sample Hindi NER tagged corpus

An NB statistical classifier from Natural Language Toolkit (NLTK) and Support Vector Machine Classifier (SVC) from scikit-learn is taken.

Naïve Bayes Classifier:

In machine learning, naïve Bayes classifiers are a family of simple "probabilistic classifiers" based on applying Bayes' theorem with strong (naïve) independence assumptions between the features.

Naïve Bayes Classifier (NBC) is a probabilistic classifier. "The naïve Bayes classifier greatly simplifies learning by assuming that features are independent given class. Although independence is generally a poor assumption, in practice, naïve Bayes often competes well with more sophisticated classifiers." [24]. In this work, we have followed the NBC definition of [25].

Support Vector Machine:

"SVMs (Support Vector Machines) are a useful technique for data classification." [26]. It works out by separating hyperplane. Here, SVM, defined in [27], is used.

4 Results and Discussion

We have seen the presence of following distinct NER variation, as given in Table 1.

The results are impressive, as the papers could be detected nicely, though I am expecting better results with the help of other classifiers. Some latest neural network based models may also be tried. Though, the principle of Occam's razor and appropriate model selection acts as the best guideline.

Table 1 Representation of NER tag variation across various Hindi online resources

Resources	Person (PERS) (%)	Location (LOC) (%)	Date (DATE) (%)	Organization (ORG) (%)
BBC_HINDI	50	30	25	30
Hindusthan	40	60	30	10
Dainik Bhaskar	45	58	35	15

5 Conclusion

Here, one framework for the detection of contents is proposed in Hindi. It works with two different algorithms and on two kinds of data, one preprocessed raw data and the other a labeled data.

This kind of work with high precision gold standard data is a sizeable work of its nature. The work is evaluated with F-measure. Our group has already finished work on Hindi and Bangla NER, and as more enthusiasts are proposing to join our group, with their own mother tongue. Extending the work in Telugu and Punjabi is under active consideration. Perhaps a suite like NLTK in Indian languages is not a distant dream.

References

1. S. Briet, *Qu'est-ce que la documentation?* (Documentaires Industrielles et Techniques, Paris, 1951)
2. G. Mikros, E. Argiri, Investigating topic influence in authorship attribution, in *SIGIR 07 Workshop on Plagiarism Analysis, Authorship Identification, and Near-Duplicate Detection (PAN 07)*, ed. by B. Stein, M. Koppel, E. Stamatatos. CEUR-WS.org, July 2007
3. M. Kestemont, K. Luyckx, W. Daelemans, T. Crombez, Cross-genre authorship verification using unmasking. English Stud. **93**(3), 340–356 (2012)
4. M. Koppel, J. Schler, E. Bonchek-Dokow, Measuring differentiability: Unmasking pseudonymous authors. J. Mach. Learn. Res. **8**:1261–1276 (2007)
5. S. Phani, S. Lahiri, A. Biswas, Authorship attribution in Bengali Language. Proc. ICON **2015**, 277–282 (2015)
6. B. Liu, Sentiment analysis and subjectivity, in *Handbook of Natural Language Processing*, 2nd edn, ed. by N. Indurkhya, F.J. Damerau (2010)
7. J. Karlgren, Affect, appeal, and sentiment as factors influencing interaction with multimedia information, in *Proceedings of Theseus/Image CLEF Workshop on Visual Information Retrieval Evaluation*, pp. 8–11 (2009)
8. P.J. Stone, D.C. Dunphy, M.S. Smith, *The general inquirer: a computer approach to content analysis* (MIT Press, Cambridge, MA, 1966)
9. P. Turney, Thumbs up or thumbs down? Semantic orientation applied to unsupervised classification of reviews, in *Proceedings of the Association for Computational Linguistics*, pp. 417–424 (2002)
10. B. Pang, L. Lillian, S. Vaithyanathan, Thumbs up? Sentiment classification using machine learning techniques, in *Proceedings of the Conference on Empirical Methods in Natural Language Processing (EMNLP)* (2002), pp. 79–86

11. M.S. Akhtar, A. Kumar, A. Ekbal, P. Bhattacharyya, A hybrid deep learning architecture for sentiment analysis, in *Proceedings of COLING 2016, the 26th International Conference on Computational Linguistics: Technical Papers*, pp. 482–493, Osaka, Japan, 11–17 Dec 2016
12. N. Bansal, A. Umair, A. Mukherjee, Sentiment analysis in Hindi. Retrieved from https://cse.iitk.ac.in/users/cs671/2013/submissions/umair/project/report.pdf, IIT Kanpur Report (2013)
13. K. Starker, S. Chakraborty, A sentiment analysis system for Indian language Tweets, in *Proceedings of the Third International Conference on Mining Intelligence and Knowledge Exploration*, vol. 9468 (2015), pp. 694–702
14. D. Blei, Probabilistic topic models. Commun. ACM **55**(4), 77–84 (2012)
15. D. Newman, S. Block, Probabilistic topic decomposition of an eighteenth-century newspaper. J. Am. Soc. Inform. Sci. Technol. **57**(5), 753–767 (2006)
16. M. Steyvers, T. Griffiths," Probabilistic topic models." In T. K. Landauer, D. S. McNamara, S. Dennis, & W. Kintsch (Eds.), *Handbook of Latent Semantic Analysis*. Lawrence Erlbaum Associates Publishers, Mahwah, NJ, US (2007), pp. 427–448
17. T. Hofmann, Probabilistic latent semantic indexing, in *SIGIR '99 Proceedings of the 22nd Annual International ACM SIGIR Conference on Research and Development in Information Retrieval* (1999), , pp. 50–57
18. D. Kanojia, A. Joshi, P. Bhattacharya, M.J. Carman, Using multilingual topic models for improved alignment in English-Hindi MT. Proceedings of ICON **2015**, 245–252 (2015)
19. S. Negi, Mining bilingual topic hierarchies from unaligned text, in *Proceedings of the 5th International Joint Conference on Natural Language Processing, Chiang Mai, Thailand,* 8–13 Nov 2011, pp. 992–1000
20. L. Deleger, T. Lingren, Y. Ni, M. Kaiser, L. Stoutenborough, K. Marsolo, M. Kouril, K. Molnar, I. Solti, Preparing an annotated gold standard corpus to share with extramural investigators for de-identification research. J. Biomed. Inform. **50**, 173–183 (2014)
21. B. Santorini, Part-of-speech tagging guidelines for the Penn Treebank project. Retrieved from https://www.cis.uni-muenchen.de/~schmid/tools/TreeTagger/data/Penn-Treebank-Tagset.pdf
22. G. Leech, Corpus annotation schemes. Lit. Linguist. Comput. **8**(4), 275–281 (1993)
23. E. Hovy, J. Lavid, Towards a 'science' of corpus annotation: a new methodological challenge for corpus linguistics. Int. J. Transl. **22**(1) (2010)
24. I. Rish, An empirical study of the naive Bayes classifier, in *IJCAI 2001 Workshop on Empirical Methods in Artificial Intelligence*, vol. 3, No. 22 (2001)
25. M. Hall, A decision tree-based attribute weighting filter for Naive Bayes. Working Paper: 05/2006, Department of Computer Science, The University of Waikato, Hamilton, New Zealand (2006)
26. H. Chih-Wei, C. Chih-Chung, L. Chih-Jen, A practical guide to support vector classification. Technical Report, Department of Computer Science and Information Engineering, University of National Taiwan, Taipei (2003), pp. 1–12
27. Y. Yang, X. Liu, A re-examination of text categorization methods, in *Proceedings of the 22nd Annual International ACM SIGIR Conference on Research and Development in Information Retrieval*, Berkeley, CA (1999), pp. 42–49
28. A.R. Balamurali, A. Joshi, P. Bhattacharya, Cross-Lingual sentiment analysis for Indian languages using linked wordnets, *Proceedings of COLING 2012: Posters*, pp. 73–82, COLING 2012, Mumbai, Dec 2012

Firewall Scheduling and Routing Using pfSense

M. Muthukumar, P. Senthilkumar, and M. Jawahar

Abstract Firewalls are used to safeguard essential networks from outdoor attacks to guide network access based on the firewall access rules. The firewall system plays an important aspect that protects from rule analyst and malignant attack which provides security to all the internet users. The main purpose of this firewall system is to manage the network access to or from a secured network. Some difficulty with the process of firewall system is due to malfunction; it might be terrible to other fewer secured systems on the internal network. The detecting malignant packets are very significant in security issues. Therefore, we proposed a Similarity Index Algorithm which is to detect the malignant packets in the firewall framework. The performance of the proposed firewall system is tested using Netsim simulator environment in terms of latency and malignant packet detection rate with respective to the number of nodes or computers in network. Experimental studies were conducted in various schemes using Similarity Index Algorithm. The result shows the enhanced performance of packet delivery ratio. To achieve the above objective, three approaches have been proposed for summarization. In the first approach, firewall access rule routing and scheduling using pfSense scheme is employed. The pfSense is a software tool that provides enthusiastic support to firewall system. The pfSense can be enhanced through web system. This method adds two phases, namely rule fixture and rule matching. The evaluation of the proposed approach is done with the help of scheduling in the time interval. While comparing with existing approach, the range can be calculated with 95% of latency. In the second approach, data accessing can be processed by analyzing the real problem in packet confinement. This second approach realizes the Deep Packet Confine (DPC) and Deep Packet Assessment

M. Muthukumar (✉)
Department of Computer Science and Engineering, SRM Institute of Science and Technology
Delhi-NCR Campus, Ghaziabad, Uttar Pradesh, India
e-mail: mkumar7680@gmail.com

P. Senthilkumar · M. Jawahar
Department of Computer Science and Engineering, NREC, Secunderabad, India
e-mail: psenthilmephd@gmail.com

M. Jawahar
e-mail: mjawahar@gmail.com

S. S. Dash et al. (eds.), *Intelligent Computing and Applications*,
Advances in Intelligent Systems and Computing 1172,
https://doi.org/10.1007/978-981-15-5566-4_67

(DPA) to evaluate high traffic rate. The updated feature helps to assess disputation analyses to determine security threat. The model explains packet framing and packet filtrate. The framing model provides records to organize information and client data. The filtrate investigates entry and exit by granting communication using specified rules. The Deep packet assessment enables the client service, and summary is generated using diagnostic tools. In the third approach, experiment evaluation shows that the proposed method has the capability of perceiving a movement percentage of new attacks. It explains that the system detection can be developed by using Similarity Index Algorithm. The Similarity Index Algorithm analyzes the inward packets recognized using malignant packet detection and decides to precede packets through gateway. The implementation in firewall contributes powerful security that can be applied to all network traffic. The efficient summary is to assure data communication mechanism in the networks. The result demonstrates that the gateway operation can be turned on by investigating each packet. Finally, the result shows that the latency and malignant packet detection rate of the proposed firewall architecture are 14.74 ms and 87%, respectively.

Keywords Firewall system · Firewall access rule · pfSense

1 Introduction

In this modern world, Firewall plays an important role in network security that can be a hardware or software or combination of both. It is used to protect the network actions from the hackers and malicious nodes in network environment. In the year 1994, the idea of firewall and internet security has been introduced by Steven. M. Bellovin. A firewall is a security system that detects and manages the succeeding and departing packets based on the firewall access rules and regulation. The firewall is placed within the boundary of each computer which is connected to internet by the networks in the organization. Based on the predefined firewalls access rules, the public network and private networks can be segregated. The security of the system is dependent on the rules of the firewall configuration, or else undesired packet traffic may pass or block the desired packets. The main function of the firewall is to control the security policy and to protect the organization network from non legitimate traffic. It also provides high flexible security to online computer users. Firewalls can be achieved by testing all constrained and unconstrained network traffic according to the predefined rules [1]. Figure 1 shows the general firewall model communication with LAN, WAN, and Enterprise or Organization Networks. The firewall is situated at the junction point between the three networks such as LAN, WAN, and Enterprise or organization network along with the internet connection.

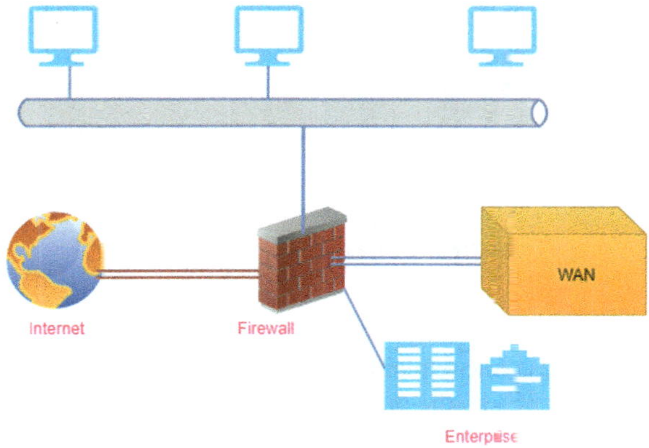

Fig. 1 General firewall model

1.1 Packet Filtrate

The packet filtrate is referred to as static packet filtering. Based on the IP address of source and destination, the incoming and outgoing packets can be controlled and monitored either to pass or halt the information.

The packet filtrate is a cost-effective security to attack against outside networks. The general packet filtrate algorithm is as follows:

Input: List of packet header data, Access rule(r), Packet (P)
Output : Allow or disallow the packet (P)
if(Access rule(r)==accept)
then *P* ← *accept*
else if (Access rule(r) =disallow)
then *P*←*Denied*
else
r is not defined
process stop
end if
//repeat or iterate the process when access rule defined
Stop

1.2 Firewall Access Rules

Firewalls can examine the firewall access rules. The access rules are network security rules that can be set by the network authority to allow traffic to their respective web hosting servers, FTP archives and daemon servers, thereby giving the computer owners immense control over the traffic that flows in and out of their systems or networks. In distributed firewalls, no two firewalls should have the same access rules and regulation [2]. Normally, all the traffic in internet can be monitored and controlled by giving the firewall installer a high level of security and protection over the network.

1.3 Firewall Access Rule Design

The access rule design is a complete group of access rules which need to be designed. This access rule explains which network traffic flows through the firewall, traffic that enters or blocks the organization network (Al-Shaer et al. 2005).

Firewall Access Rule Steps

- The system engineers must be able to connect instantly with the firewall system.
- Firewall might not be able to connect instantly with any other network devices.
- Any other device should be able to connect directly with firewall system.
- The network traffic flow should be running instantly to the specified servers.

2 Firewall Access Rule Routing

The firewall routing is used to deliver the packet from source to destination and sending it through one domain network environment to other domain environment. Routing policy allows you to manage the routing in sequence between the routing properties and the routing tables. These firewall systems support the following routing areas:

- Stateful packet filter
- Network Address Translation (NAT)
- Filtrating based on Firewall access rule
- Packet Matching
- Packet Cleaning.

3 Software Tools—pfSense

The pfSense is a freely available firewall software based on allocation of operating system. It is a physical arrangement of the computer system to make an enthusiastic support of firewall system. It can be designed and enhanced through a web-based system. The pfSense is generally set up as a boundary of firewall system, networking devices like router, repeater, and wireless access point [3].

4 Firewall Access Rule System Implementation

The firewall access rule permits to keep the malignant users out and also expand control over inherent risky users within your company. An access rule is to realize the available information and services, present inherent for spoilage and whether any security is already in place to inhibit misuse [4]. The policies are traffic rules, regulations for the network which build up the internet. Network policy allows system administrator to coordinate network elements to offer service to set of users. Each system were permitted to connect with all other neighboring systems without any limitation, then there would be no access rules for network. Network policy can be stimulated into two different ways such as fixed and energetic. A fixed policy is a set of action in a preplanned way according to a set of pre-built attributes. An energetic policy is imposed in need, and it is based on mitigating condition.

Firewall Access Rule design

- The system engineers must be able to connect instantly with the firewall system.
- Firewall might not be able to connect instantly with any other network devices.
- Any other device should be able to connect directly with firewall system.
- The network traffic flow should be running instantly to the specified servers.

Figure 2 represents the firewall access rule diagram. The firewall access rule can be done by passing the packet, matching the same access rule, blocking the unspecified access rule, and rejecting the unknown user. It can also identify the unauthorized and authorized traffic in the networks.

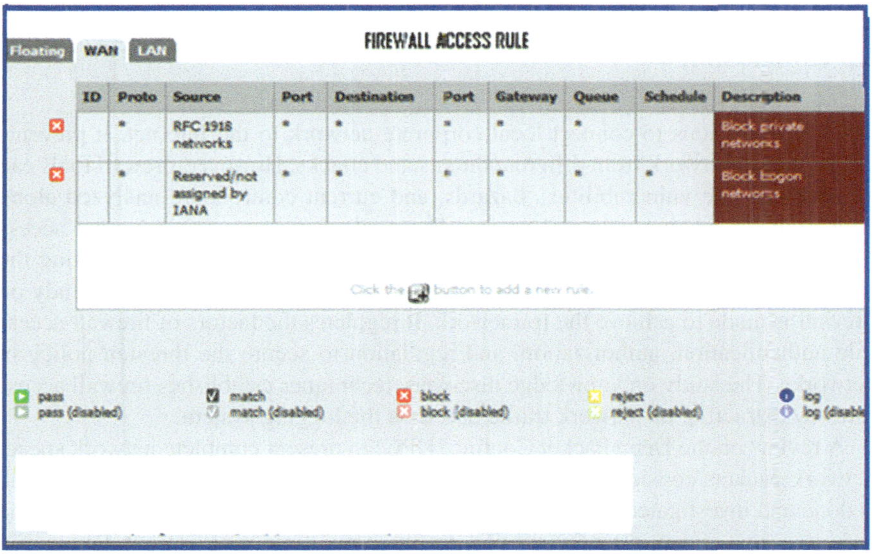

Fig. 2 Firewall access rule diagram

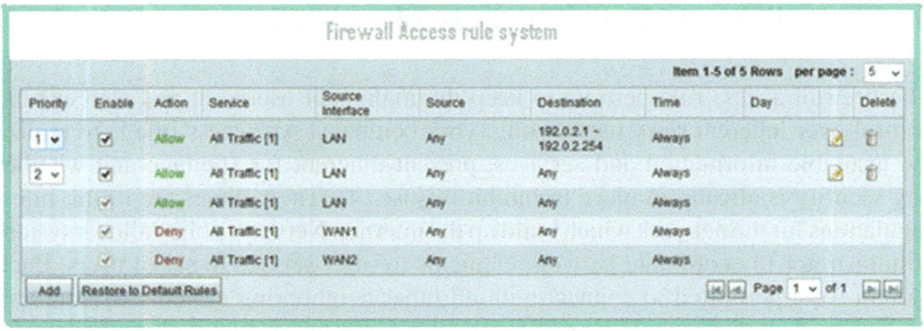

Fig. 3 Firewall access rule scheduling diagram

5 Firewall Access Rule Scheduling Implementation

The firewall access rule scheduling can be planned to be active only at transparent period of time. The scheduled access rule will proceed while they are not available, when the planned stage is not agile.

Rules for Firewall scheduling

The twisted schedule resolves only when the firewall access rule operation will be registered.

This access rule will not pertain at the edge of the schedule and will be served by pfSense, as it is not adjacent. Figure 3 represents the access rule scheduling.

6 Conclusion

Firewall is obvious to connect local corporate network to the Internet. It prevents the corporate network from different threats and attacks. However, firewall tools can be updated. The vulnerabilities, hazards, and current controls are analyzed along with establishing the collateral policy. The goal of thesis is to determine access rules for routing and scheduling using the pfSense scheme and also detecting the malignant packets. During the initial stage of investigation, an exhaustive study on firewall is made to achieve the framework. It regulates the factors of firewall access rule authentication, authorization, and regulation to secure the threaten policy in networks. The study on knowledge discovery techniques establishes firewall access rules by extracting its network traffic based on the logging system.

A review on the Deep Packet Confine (DPC) to present complete network speed, network packets consignment, and crossing a network with a high traffic flow rate is done and investigated on the Deep Packet Assessment (DPA) for determining the security threat and monitor the network traffic in real-time environment. Parametric studies on the protocol diagnostic tools and methods to confine the process of network logging traffic are successfully investigated. In the final phase, malignant packet detection algorithm is proposed in firewall architecture which efficiently detects the unauthorized packets from various network environments. The frameworks are based on the computation of the energy index of the individual port in firewall architecture. The performance of the proposed system is analyzed in terms of latency and malignant packet detection rate. This simulation resolves 14.74 ms of latency and 87% of malignant packet detection rate.

7 Limitations

- A firewall cannot guard from inner intrusion. It does anything to block internal network intruders or intrusion attack from or within the network. In organization, employee's offense or inattention cannot be inhibited by firewall system.
- A firewall cannot prevent discrete employee or internet user with device from twisting interior or exterior part of the network. This helps the user to completely pass around the firewall system.
- A firewall system provides security if it is perfectly constructed and defined the firewall access rule. A firewall administrator should design it to classify between accept and denied network traffics.
- A firewall cannot bypass your password rules or contamination of passwords.
- A firewall is ineffectual across unspecialized exemption risk.
- Firewalls cannot measure against the hybrid attacks.

8 Future Work

It is examined that the future control for this research on the corporate network security in the connection of firewall can be achieved by dealing with central research. Using proposed Knowledge Discovering Technique with log-supported database system, further inquisition can be drifted out on the impact of mesh packet filtrate. The issues measured can be recognized. It is accessible to design new mechanism that can authenticate acquaintance through proxy server. Finally, it can determine the firewall manner by the whole research to prevent malicious attacks and to acquire performance in network security.

References

1. E.S. Al-Shaer, H.H. Hamed, Modeling and management of firewall policies. IEEE Trans. Netw. Serv. Manag. (2004)
2. M.J. Chapple, J. D'Arcy, A. Striegel, An analysis of firewall rulebase (mis)management practices. ISSA J. 12–18 (2009)

3. K. Mamat, R.M. Saad, Home wireless network security using pfSense captive portal, in *Proceedings of 8th International Conference on IT in Asia 2013 (CITA'13)* (IEEE/SCOPUS/ISI). Accessed: 12 Apr 2016
4. M. Abedin, S. Nessa, L. Khan, E.S. Al-Shaer, M. Awad, Analysis of firewall policy rules using data mining techniques. Int. J. Internet Protoc. Technol. (2010)
5. C. Chao, A flexible and feasible anomaly diagnosis system for internet firewall rules, in *Proceedings of the 13th Asia-Pacific Network Operations and Management Symposium* (2011), pp. 1–8
6. Z. Chen, W. Dong, H. Li, P. Zhang, X. Chen, J. Cao, Collaborative network security in multi-tenant data center for cloud computing. Tsinghua Sci. Technol. **19**, 1 (2014)
7. K.K.R. Choo, The cyber threat landscape: challenges and future research directions. Comput. Secur. (2011)
8. T.H. Cormen, C.E. Leiserson, R.L. Rivest, C. Stein, *Introduction to Algorithms* (McGraw-Hill Higher Education, 2009)
9. A. El-Atawy, K. Ibrahim, H. Hamed, E. Al-Shaer, Policy segmentation for intelligent firewall testing, in *Proceedings of the 1st IEEE ICNP Workshop on Secure Network Protocols* (2005), pp. 67–72
10. A. El-Atawy, T. Samak, A. Wali, E.S. Al-Shaer, F. Lin, C. Pham, S. Li, An automated framework for validating firewall policy enforcement, in *IEEE International Workshop on Policies for Distributed Systems and Networks* (2007), pp. 151–160
11. M. Eslahi, M. Naseri, H. Hashim, N. Tahir, E. Saad, BYOD: current state and security challenges, in *2014 IEEE Symposium on IEEE Computer Applications and Industrial Electronics (ISCAIE)* (2014)
12. A. Gontarczyk, P. McMillan, C. Pavlovski, Cyber security zone modeling in practice, in *Proceedings of the 10th International Conference on Information Technology and Applications (ICITA)* (Sydney, Australia, 2015)
13. L. Guangcheng, Research and design of double firewall technology. Comput. CD Softw. Appl. **21**, 70–81 (2012)
14. H. Guo, T. Tang, D. Wu, The research of private network secure interconnection scheme in large-scaled enterprises, in *Genetic and Evolutionary Computing* (Springer International Publishing, 2015)
15. H. Guochao, An improved strategy for intranet security based on two firewalls. Comput. Secur. **7**, 36–38 (2012)
16. X. He, T. Chomsiri, P. Nanda, Z. Tan, Improving cloud network security using the tree-rule firewall. Future Gen. Comput. Syst. **20**, 1 (2014)
17. S. Jadhav, R. Agrawal, Fast and scalable method to resolve anomalies in firewall policies. Int. J. Adv. Innov. Res. (IJAIR) (2013)
18. H. Jang, J. Jeong, H. Kim, J. Park, A survey on interfaces to network security functions in network virtualization, in *2015 IEEE 29th International Conference on Advanced Information Networking and Applications Workshops (WAINA)* (IEEE, 2015)
19. J. Li, The research and application of multi-firewall technology in enterprise network security. Int. J. Secur. Appl. **9**(5), 153–162 (2015)
20. K. Kaur, D. Rao, Automation the process of unifying the change in the firewall performance. Int. J. Comput. Sci. Netw. Secur. (IJCSNS) (2014)
21. T. Kaur, V. Malhotra, D. Singh, Comparison of network security tools-firewall, intrusion detection system and Honeypot. Int. J. Enhanced Res. Sci. Technol. Eng. (2014)
22. S. Kikuchi, Y. Matsumoto (eds.), Performance modeling of concurrent live migration operations in cloud computing systems using prism probabilistic model checker. Cloud Comput. (CLOUD) (2011)
23. L. Kong, Research of building enterprise network security system based on cloud computing technology. Biotechnol. Indian J. (2014)
24. S. Lar, X. Liao, A. Rehman, M. Qinglu, Proactive security mechanism and design for firewall. J. Inf. Secur. **2**(3), 122–130 (2011)

25. O. Osanaiye, H. Cai, K.-K.R. Choo, A. Dehghantanha, Z. Xu, M. Dlodlo, Ensemble-based multi-filter feature selection method for DDoS detection in cloud computing. EURASIP J. Wirel. Commun. Netw. (2016)
26. O. Osanaiye, K.K.R. Choo, M. Dlodlo, Distributed denial of service (DDoS) resilience in cloud: review and conceptual cloud DDoS mitigation framework. J. Netw. Comput. Appl. (2016)
27. V. Prokhorenko, K.-K.R. Choo, H. Ashman, Web application protection techniques: a taxonomy. J. Netw. Comput. Appl. (2016)
28. S. Dandamudi, T. Eltaeib, Firewalls implementation in computer networks and their role in network security. J. Multidisc. Eng. Sci. Technol. (JMEST) **2**(3) (2015)
29. K. Salah, R. Boutaba (eds.), Estimating service response time for elastic cloud applications, in *2012 IEEE 1st International Conference on Cloud Networking* (*CLOUDNET*) (2012)
30. S. Shin, L. Xu, S. Hong, G. Gu, Enhancing network security through software defined networking (SDN), in *25th International Conference on Computer Communication and Networks* (*ICCCN*) (IEEE, 2016)
31. A. Shukhman, P. Polezhaev, Y. Ushakov, L. Legashev, V. Tarasov, N. Bakhareva, Development of network security tools for enterprise software-defined networks, in *Proceedings of the 8th International Conference on Security of Information and Networks* (ACM, 2015)
32. Y. Smirnov, V. Halimon, Double layer gateway model for connection between production network and enterprise network, in *International Review of Automatic Control* (*IREACO*) (2015)
33. A.K. Vasu, A. Sudarsan, P. Ayyappan, A. Ganesh, V. Gokul, Improving firewall performance by eliminating redundancies in access control lists. Int. J. Comput. Netw. (2014)

Undefeatable System Using Machine Learning

Anand Sharma and Uma Meena

Abstract In machine learning, computer learns from its own experience, i.e., it stores the tasks it performs and uses it as and when required. Machine learning is widely used for gaming as it consists of many algorithms which help the computer to self-learn and become an undefeatable system. Initially, the computer will have no domain knowledge except the rules of the game but gradually it will learn the game, and when all the combinations of the moves which can be played are learnt, it will become the master player, i.e., undefeatable. Games like chess, Go, Shogi, i.e., Japanese chess use reinforcement learning which comes under the domain of machine learning and maximizes the rewards [1–3]. The learning phase of the computer includes both self-learning and learning by playing with human. The paper presents a brief study of machine learning technologies which helps in making the system undefeatable. That includes Integration of Machine Learning for Behavior Modeling, Reinforcement Learning, and Goal-oriented Behavior.

Keywords Machine learning · Game playing · Reinforcement learning · Undefeatable system

1 Introduction

1.1 Puzzles

A puzzle is a game like crossword, number puzzles, word search puzzles, or logic puzzles. They are generated to be a way of entertainment. Some are easy, some are hard, and some require a little bit of head scratching. Puzzle games involve use of

A. Sharma
Department of Computer Science, SET-MUST, Lakshmangarh, India
e-mail: anand_glee@yahoo.co.in

U. Meena (✉)
Department of Computer Science, SRMIST, Delhi-NCR Campus, Ghaziabad, India
e-mail: uma.b18@gmail.com

S. S. Dash et al. (eds.), *Intelligent Computing and Applications*,
Advances in Intelligent Systems and Computing 1172,
https://doi.org/10.1007/978-981-15-5566-4_68

your brain to solve challenges that become harder over time. It is a kind of brainteaser. Envision a puzzle, where every one of the pieces appears the same, but then they all have their little contrasts all over. Such little contrasts but then you cannot build the puzzle until the point that you perceive the similitude. When you at last do, you are prepared to begin building and piece by piece the picture begins to shape. Piece by piece you match and you make. The pieces are the interminable potential outcomes you have; the decision you make to consolidate one piece with another is the decision you make which drives you to the finish of the puzzle.

1.2 Two-Player Perfect-Information Games

Two player games come under the category of multiplayer game which is played by just two players. One of the players can be a human or computer. Few games which come under this category are Chess Go, Tic Tac Toc, English Draughts, etc. [4–6]. Here, the game yields three results, i.e., either it is win, lose, or draw. This involves board game, dice game, or card game. These types of games involve timers and scoring, i.e., the player who scores maximum in the given time bound wins the game. These types of games include both indoor and outdoor games. Here, if we have one of the players as the computer, then we can train it using Reinforcement Learning, i.e., teaching the system tricks to win such that it becomes impossible for the player to defeat the system, and we get only two cases as the result that is either the system wins or it is a draw. Here, each player plays their move alternatively and is aware of the previous moves, i.e., has perfect information about the game.

1.3 Imperfect-Information and Stochastic Games

In these types of games, the player plays according to the current state and actions of the opponent player. It is a dynamic game with probabilistic transitions, i.e., not certain. In this, the process of playing according to the current state keeps on repeating which may be finite or infinite times until the end of the game is reached. These games are mostly used for analysis and modeling of discrete systems which are being operated in an unknown environment. Zero-sum games come under this category where the total gains are being added up and the losses are being separated, and therefore the total sum becomes equal to zero. Zero-sum games are mostly solved using the MinMax algorithm. The zero-sum property (if one wins, second loses) means that any result of a zero-sum situation is Pareto optimal. All the games where complete strategies are Pareto optimal are called a conflict game. Poker and Gambling are the zero-sum games since the summation of the number of wins by some players equals the number of losses of the others, and the information is also incomplete as the opponent is not sure of what the player's next move is. Markov

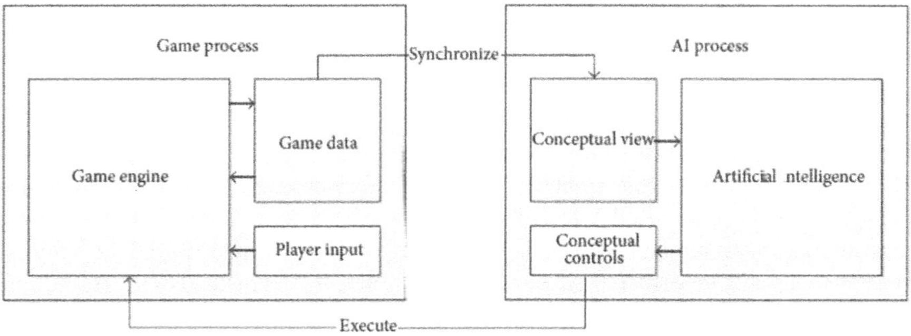

Fig. 1 Conceptual view of running artificial intelligence for games

chains are a way to statistically model random processes. They have been used in many different domains from text generation to financial modeling.

2 Machine Learning in Games

2.1 Samuel's Legacy

In 1947, Arthur L. Samuel, Professor of Electrical Engineering at the University of Illinois and pioneer of Artificial Intelligence, brought the idea of developing a checkers program. Checkers being a simpler game in comparison to chess seemed to be a perfect province for representing the power of symbolic computing with a quick programming project. He made the contribution in the tech world by making the computers learn from their experience resulting in the process of making an undefeatable system. The computer which he has trained to play the checkers game has beaten America's best checker player of that time. The steps which he used to train the computer included storing of moves and searching for the saved moves which is also called as rote learning. Then, he used reinforcement learning for evaluation functions and finally he made a program that could teach itself by playing against a copy of it. The program for checkers CHINOOK was the first program to conquest a human world championship in any game. Later, chess and Go have been evolved by using AI and Deep Learning (Fig. 1).

2.2 Book Learning

Human players do not always play with their mind instead sometimes they play with their heart. So in order to tackle such kind of moves, the computer makes use of an opening book which gives an easy way to cumulate their playing strength Opening book is pre-computed for replying a set of positions and can be straightforwardly programmed. It is also a very simple way to access the human knowledge through machine which can be originated in game playing books. The opening books can extend its libraries offline by identifying the lines that are frequently used/played and compute the best move for those positions. Negamax search with added computer evaluation is much more informative than straightforward frequency-based methods. The alternate approach used is learning from the mistakes, i.e., it gets alert when the same mistake is going to be repeated and hence prevents making the mistake again. If the information is imperfect, then we use simulation learning instead of the traditional one where the computer plays against itself and stores the information of the random moves, and using the stored statistics evaluates the quality of the current move.

2.3 Learning Search Control

This method is used to reduce the time requisite for plan generation from experience. The two approaches used in learning search control knowledge are explanation-based learning and inductive learning. The inductive learning can learn from more than one example at a single point of time and does not need the complete and tractable domain theory. One of the examples for inductive learning is Grasshopper which works in following five steps:

(i) Decision Extraction
(ii) Decision Clustering
(iii) Decision Characterization
(iv) Rule Generation
(v) Utility Optimization.

2.4 Explanation-Based Learning

The explanation-based learning takes an example and explains what it learns from it. The relevant data is taken from the example by EBL and translated in a form which can be used for problem solving. PRODIGY is a system which uses a single architecture to integrate problem solving, planning, and learning. The control rules in this can lead to the choice/rejection of the candidate and also leads to a partial ordering of the candidates. The results from the problem-solving trace generated by the central problem solver are used by the EBL module. It then constructs the explanations that describe the domain and the architecture of the problem solver using an axiomatized theory. Control rules are obtained from the results and then added to the knowledge base which later on is used to guide the search process effectively.

2.5 *Learning Playing Strategies*

It is a simple playing strategy and an interpretable procedure used to play game or subgame. Learning such a procedure is altogether harder than to figure out how to group positions as won or lost. Mesa effect occurs in the game of chess, i.e., problem of choosing move from equally good moves. So in order to resolve this problem, a simple strategy was used that checks the move which contains the shortest distance to win.

3 Proposed Undefeatable System

This section will cover the three different approaches by which we can implement undefeatable systems.

3.1 *Integrated Machine Learning for Behavior Modeling*

A decent player will learn the conduct of the adversary AI and start misusing it. The misuse of these imperfections ought to be a part of the diversion; anyway the amusement ought to likewise continue testing skilled players. To utilize a machine learning calculation, it is first important to decide the highlights [7]. We utilize supervised learning since we are going about as an instructor to the learning algorithms. We determine which cases ought to be thought of and which not. A supervised learning algorithm will part our data set into two sections: a training set and a test set. The training set serves to enable the machine learner to refresh its internal settings in view of just these cases. The test set is utilized to assess how well the inward learned capacity will perform on already imperceptible data.

3.1.1 The Problem and the Data Set

In this, the players look out to win the game, and the problem of the player is how to do it. The data set is the set of tools or inputs to be applied, following the rules of the game at the right time by analyzing the opponent's behavior to win the game. The best tool used at the best time leads to the survival of the fittest in the game. This use depends on the right analysis and the learning algorithm which includes tricks and tactics required for training of the computer.

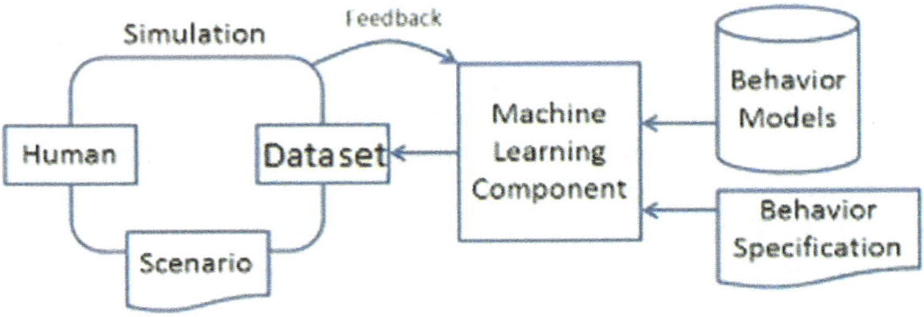

Fig. 2 Conceptual view of behavior modeling

3.1.2 Learning from the Data

The computer analyzes the moves of the opponent; in the initial stage, the performance of the system might be low, and it may even lose to the opponent. But as the number of iterations increases, the performance of the game also increases. After the complete analysis of the whole possibilities of the opponent, the learning phase of the game is completed. The most appropriate and optimized learning algorithm is used to get the best outcome. For this, the comparison of different learning algorithms should be made depending upon their execution time and their efficiency to give the desired output.

3.1.3 Ensemble Methods: Boosting

It is a set of trained classifiers. An ensemble is more precise than single classifier. The two popular methods used in this are boosting and bagging. Another term for boosting is bias. The goal is to decrease the boosting, variance, and taming the stacking alias ensemble in the supervised learning model [8]. The concept of boosting came from the question by a researcher, that is, "Can a set of weak learners create a single strong learner?" Initially, the hypothesis boosting problem was referred to as making a strong learner out of a weak learner. Boosting signifies those algorithms which attain hypothesis boosting hastily (Fig. 2).

3.2 Reinforcement Learning

Reinforcement learning comes under the domain of Machine Learning and is inspired from human behavior. In this, the computer maximizes its performance by acting correctly and attaining more points than penalties. It interacts with the environment and learns. Here, the computer is not told what action it should take instead it should decide on its own that which action will give the maximum performance. In reinforcement learning, it is not necessary to present the input and output pair and need

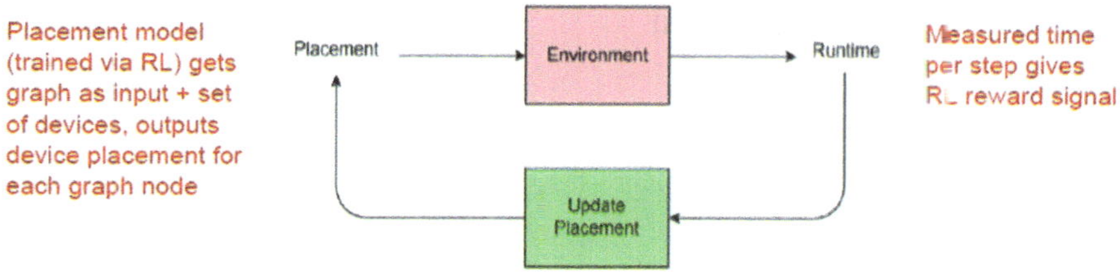

Fig. 3 Conceptual view of reinforcement learning for games

not correct the sub-optimal actions explicitly. It does not have direct input–output mapping. A good example of reinforcement learning is chess where we have thousands of moves, and it will be very difficult for a human to teach all the combinations to the computer; therefore, the computer learns from its own experiences and builds the knowledge base [9, 10]. It uses Markov Decision process, i.e., repeating the process of selecting the best action according to the environment. However, unlike to the settings in supervised learning, the agent does not receive any information regarding training from a domain expert. Instead, it may itself discover the different actions and, while doing so, will get feedback from the environment—the so-called reinforcement or reward—which can be utilized to rate the success of its own actions [11] (Fig. 3).

3.3 *Goal-Oriented Behavior*

Goal-oriented behavior control rule consists of a series of behavior in order to ample a task. Such type of control rules are delivered by reinforcement learning, planners, or human programmers. In case of exceptional cases, it becomes difficult to build a new behavior plan as done in all the working conditions. Therefore, planning reinforcement learning is partially effective instead the computer may learn subgoals and change their sequence in order to complete the task. Subgoals are the state which computer necessarily goes through for the execution of task. This will save time, be cost effective, and reduce search space. By using subgoals, the computer will be able to explore more and accelerate learning if the environment is same or similar by repeated use of subgoal. Subgoals also produce sequence of behavior for task execution. Once the computer gets fully trained using the above techniques, it will become an undefeatable system as it will know each and every trick of humans (Fig. 4).

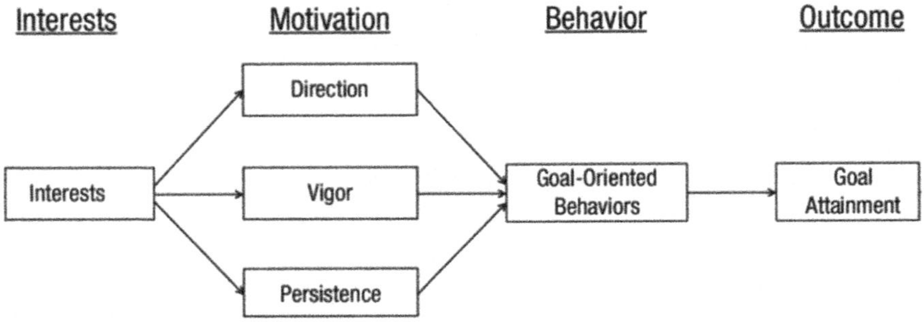

Fig. 4 Conceptual view of goal-oriented behavior for goal attainment

4 Conclusion

Earlier, the computers were not that much powerful as humans, but with the advancements in technology, they are giving a tough fight to humans. In the field of gaming, whether it is puzzle or a video game, now the computers can beat the humans with much more accuracy. In order to give computers the power to beat humans, Machine Learning is used so that the result of the game is either the computer wins or it is a draw. Machine learning is an application use of AI which provides the computer an ability to learn automatically and develop from experience without being programmed externally. It enables analysis of large amount of data at a faster speed and gives accurate results. Therefore, we use it for making the system undefeatable by using the techniques like reinforcement learning and goal-oriented behavior through which it will train itself.

References

1. Online chess games database, 365chess, 2017. https://www.365chess.com/
2. J. Schaeffer, The games computers (and people) play, in ed by M.V. Zelkowitz, *Advances in Computers*, vol. 50, pp. 189–266 (2000)
3. M. Genesereth, N. Love, B. Pell, General game playing: overview of the AAAI competition. AI Mag. **26**(2), 62–72 2005)
4. G. Tesauro, Comparison training of chess evaluation functions, in ed by J. Fürnkranz, M. Kubat, *Machines that Learn to Play Games*, Chapter 6, pp. 117–130 (2001)
5. O.E. David, N.S. Netanyahu, L. Wolf, Deepchess: End-to-end deep neural network for automatic learning in chess, in International Conference on Artificial Neural Networks (Springer, 2016), pp. 88–96
6. D. Silver, J. Schrittwieser, K. Simonyan, I. Antonoglou, A. Huang, A. Guez, T. Hubert, L. Baker, M. Lai, A. Bolton, Y. Chen, T. Lillicrap, F. Hui, L. Sifre, G. van den Driessche, T. Graepel, D. Hassabis, Mastering the game of go without human knowledge. Nature, **550**, 354– 359 (2017)
7. B. Geisler, An Empirical Study of Machine Learning Algorithms Applied to Modeling Player Behavior in a 'First Person Shooter' Video Game. M.S. Thesis, Department of Computer Sciences, University of Wisconsin Press. Madison, 2002

8. Q. Le et al., Building High-level Features Using Large Scale Unsupervised Learning, in ICML 2012: 29th International Conference on Machine Learning (2012, June)
9. V. Mnih et al., Human-level control through deep reinforcement learning. Nature **518**, 529–533 (2015)
10. M. Lai, Giraffe, Using deep reinforcement learning to play chess, Master's thesis, Imperial College London, 2015
11. S. Schiffel, M. Thielscher, Fluxplayer: a successful general game player, in *Proceedings of the AAAI Conference on Articial Intelligence* (Vancouver, 2007), pp. 1191–1196
12. S. Schiffel, M. Thielscher, Fluxplayer: a successful general game player, in *Proceedings of the AAAI Conference on Artificial Intelligence* (Vancouver, 2007), pp. 1191–1196
13. M. Bowling, Convergence and no-regret in multiagent learning, in Advances in Neural Information Processing Systems 17 (NIPS) (2005), pp. 209–216
14. C. Claus, C. Boutilier, The dynamics of reinforcement learning in cooperative multiagent systems, in *Proceedings of the National Conference on Artificial Intelligence* (Wiley, 1998), pp. 746–752
15. S. Kapetanakis, D, Kudenko, Reinforcement learning of coordination in cooperative multi-agent systems, in *Proceedings of the National Conference on Artificial Intelligence* (AAAI Press, MIT Press, Menlo Park, Cambridge, 2002), pp. 326–331

Synchronization for Nonlinear Time-Delay Chaotic Diabetes Mellitus System via State Feedback Control Strategy

Nalini Prasad Mohanty, Rajeeb Dey, Binoy Krishna Roy, and Nimai Charan Patel

Abstract Recent studies disclose that the interaction of externally injected insulin and blood glucose in some nonlinear model of diabetes mellitus results in chaotic behavior. This work deals with the design of a state feedback synchronization control technique for a nonlinear time-delay glucose–insulin regulatory system where the β-cell kinetic stakes into account in the model. A state feedback control technique is used for synchronization between master and slave system where the master system is the healthy diabetes mellitus system and the slave system is the diseased diabetes mellitus system. Here, Lipschitz condition lemma is used for approximating the system nonlinearity. A Lyapunov–Krasovskii functional approach is adopted for deriving the feedback control law which is responsible for the synchronization of chaotic behavior present in the glucose–insulin regulatory system of diabetic patients. Wirtinger inequality is utilized in the design that resulted in less conservative stability conditions in terms of linear matrix inequality. The results of simulation for the proposed synchronization technique confirm the efficacy toward healing the diabetes mellitus.

Keywords Synchronization · Nonlinear time-delay system · State feedback control · Diabetes mellitus · Wirtinger inequality

---

N. P. Mohanty (✉) · R. Dey · B. K. Roy
Electrical Engineering Department, National Institute of Technology Silchar, Silchar, Assam 788010, India
e-mail: nalini6111@gmail.com

R. Dey
e-mail: rajeeb.iitkgp@gmail.com

B. K. Roy
e-mail: bkr_nits@yahoo.co.in

N. C. Patel
Electrical Engineering Department, Government College of Engineering, Keonjhar, Odisha 758002, India
e-mail: ncpatel.iter@gmail.com

S. S. Dash et al. (eds.), *Intelligent Computing and Applications*, Advances in Intelligent Systems and Computing 1172,
https://doi.org/10.1007/978-981-15-5565-4_69

1 Introduction

The technical name of diabetes is diabetes mellitus, which refers to the change in the normal processes of human metabolism toward control mechanism disturbance for blood sugar level. The presence of primary controlling element, i.e., insulin, is either disregarded by body cells or not secreted. Generally, diabetes is categorized as gestational diabetes (GD), type 1, and type 2. The GD type of diabetes is pregnancy-based temporary condition, which affects approximately 2–4% of all pregnancies. The immune system of patient interrupts insulin-releasing cells in Type 1 diabetes. Usually 5–10% of patients are coming under this category. Whereas, the body of type 2 diabetic patient is unable to use insulin in proper way. This is the most common category (around 90–95%) in the world [1].

A linear model for glucose tolerance test is suggested by Ackerman in the year of 1964 [2]. In 1987, Bajaj introduced a basic nonlinear mathematical model using β-cell kinetics for glucose–insulin feedback system [3]. A time-delay model which incorporates two time-delays has been proposed by Sarika et al. [4]. The modified model of Sarika et al. [4] is presented by Chuedoung et al. [5].

There are many time-delay chaotic systems (TDCS) are present in nature. There are so many applications of TDCS in the field of engineering, physics, and biology [6, 7]. A system with time-delay can exhibit chaos [8, 9]. The phenomena of chaos, hyperchaos, multistability, etc., may be exhibited by systems because of the existence of time-delays in the system dynamics.

Many synchronization techniques have been reported in the literature like state feedback control [10], sliding mode control (SMC) for uncertain TDCS [11], and synchronization and anti-synchronization of TDCS using SMC which are also discussed in the literature [8, 12].

This paper has five sections. Section 2 contains the literature survey, background and motivation, and system description. Section 3 contains the design of state feedback controller (SFC) for synchronization of TDCS. The simulation result is analyzed in Sect. 4. The concluding remark is given in Sect. 5.

2 System Description

The model defined by Chuedoung et al. [5] (Fig. 1) is represented as follows in the form of delay differential equation.

$$\dot{x} = r_1 y_{\tau_1} z_{\tau_1} - r_2 x + c_1 z_{\tau_1}$$

$$\dot{y} = \frac{R_3 N}{z} - r_4 x_{\tau_2} + c_2$$

$$\dot{z} = r_5\big(y - \hat{y}\big)(T - z) + Az - r_6 z(t - z) - r_7 z \tag{1}$$

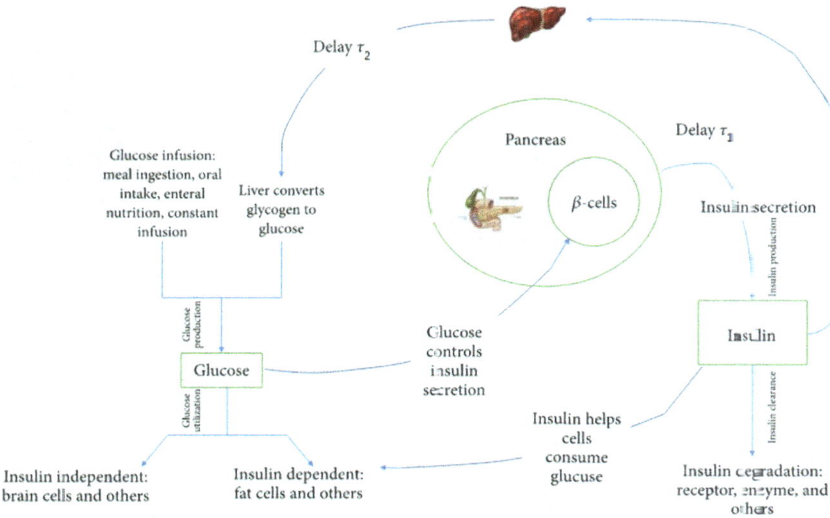

Fig. 1 Glucose–insulin system with two time-delays

where $y_{\tau_1} = y(t - \tau_1)$, $z_{\tau_1} = z(t - \tau_1)$, $z_{\tau_2} = z(t - \tau_2)$.

In Eq. (1), the insulin concentration, the glucose concentration, and the number of β-cells are represented as $x(t)$, $y(t)$, $z(t)$, respectively. The change in basal level and glucose fasting level is represented as $\hat{y}$. Due to increase in blood glucose level, the delay in insulin secretion is represented by τ_1, which is based on clinical evidence reported in [13]. Based on clinical evidence reported in [14], the delay in glucose drop is due to increase in insulin level in human body, which is represented as τ_2. The total density and normal number of β-cells are represented as T and N.

2.1 *Bifurcation Diagram*

The Bifurcation diagram is graphically represented by changing one of the system parameter with respect to other parameters that are kept fixed. The bifurcation diagram is illustrated in Fig. 2, with variation of the delay parameter $\tau_2 \in [0, 7]$, and keeping $c_1 = 0.1000$, $c_2 = 0.8000$, $r_1 = 0.4720$, $r_2 = 0.2500$, $\hat{y} = 1.4200$, $R_3 = 0.8200$, $r_4 = 0.6000$, $T = 1.5000$, $r_5 = 0.3000$, $r_6 = 0.3000$, $N = 1.2700$, $r_7 = 0.2000$, $\tau_1 = 0.0430$ with initial conditions (ICs) (3.1, 1.2, 0.4). From the bifurcation diagram, it is seen that for a range of τ_2 the system is stable, periodic, and chaotic in nature. The system is observed stable with small delays; however, it exhibits chaotic behavior with the increase in delays. The instantaneous reflection of insulin as reported by Forrest et al. [15] was 14 out of the 20 monitored Jamaican children with about 1 min response time. However, the delay for other children was 5–10 min.

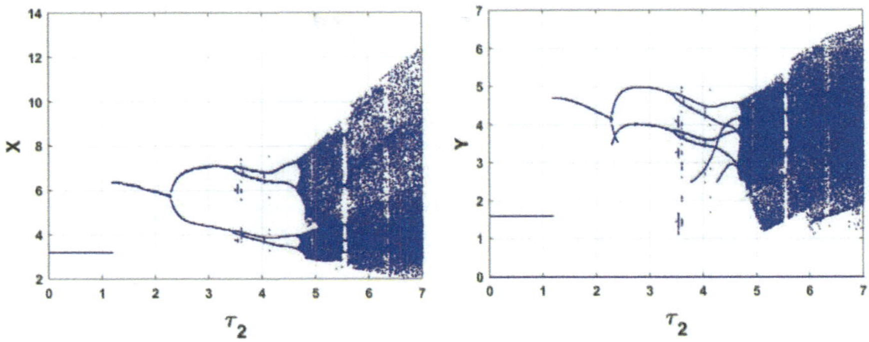

Fig. 2 Bifurcation of the system (1) with ICs (3.1, 1.2, 0.4) with respect to τ_2

3 State Feedback Control Design for Synchronization

On the basis of (1), the healthy nonlinear time-delay chaotic diabetes mellitus system (master system (MS)) can be represented in (2).

$$\begin{cases} \dot{x} = Ax + A_1 x(t-\tau_1) + A_2 x(t-\tau_2) + H\overline{\lambda}(x,t) + \overline{H}\vartheta(x(t-\tau_1),t) + C \\ x = \psi_x, t \in [-\max\{\tau_{12}, \tau_{22}\}, 0] \end{cases} \tag{2}$$

where $a_1 = R_3 N,\ a_2 = r_5 T,\ c_3 = r_5 \hat{y} T,\ A = r_5 \hat{y} + r_6 T - r_7$.

And the diseased nonlinear time-delay chaotic diabetes mellitus system (slave system (SS)) is presented in (3).

$$\begin{cases} \dot{y} = Ay + A_1 x(t-\tau_1) + A_2 x(t-\tau_2) + H\overline{\lambda}(y,t) \\ \qquad + \overline{H}\vartheta(y(t-\tau_1),t) + C + Bu(t) \\ x = \psi_x, t \in [-\max\{\tau_{12}, \tau_{22}\}, 0] \end{cases} \tag{3}$$

Let the error between the healthy and diseased uncertain system is represented as $e = x - y$. The resulting synchronization error of MS and SS is represented in (4).

$$\begin{aligned} \dot{e}(t) &= Ae(t) + A_1 e(t-\tau_1) + A_2 e(t-\tau_2) + H\overline{\lambda}(x,y,t) \\ &\quad + \overline{H}\vartheta(x(t-\tau_1), y(t-\tau_1), t) - Bu(t) \end{aligned} \tag{4}$$

For realizing the synchronization between the systems denoted in (2) and (3), the control equation can be represented as

$$u = Ke \tag{5}$$

where K is represented as gain matrix. By merging Eqs. (4) and (5), the error system becomes

$$\dot{e}(t) = Ae(t) + A_1 e(t-\tau_1) + A_2 e(t-\tau_2) + H\bar{\lambda}(x, y, t) + \overline{H}\vartheta(x(t-\tau_1), y(t-\tau_1), t) - BKe(t) \quad (6)$$

Assumption 1 Nonlinear terms $\bar{\lambda}$ and ϑ satisfy

$$\begin{aligned} &||\bar{\lambda}(x, t) - \bar{\lambda}(y, t)|| \leq ||L_{\bar{\lambda}}(x-y)||, \\ &||\vartheta(x(t-\tau_1), t) - \vartheta(y(t-\tau_1), t)|| \leq ||L_{\vartheta}(x(t-\tau_1) - y(t-\tau_1))|| \end{aligned} \quad (7)$$

where $L_{\bar{\lambda}}$ and L_{ϑ} will represent the Lipschitz constant.

Lemma 1 *For any positive definite matrix $T = T^{\mathrm{T}}$ and any differentiable function x in $[r, s] \rightarrow \mathbb{R}^n$, there exists*

$$-\int_r^s \dot{x}^T(\varphi) T \dot{x}(\varphi) \mathrm{d}\varphi \leq -\frac{1}{s-r}\Phi_0^T(r, s) T \Phi_0(r, s) - \frac{3}{s-r}\Phi_1^T(r, s) T \Phi_1(r, s), \quad (8)$$

where

$$\Phi_0(r, s) = x(s) - x(r)$$

$$\Phi_1(r, s) = x(s) + x(r) - \frac{2}{s-r}\int_r^s x(\varphi)\mathrm{d}\varphi.$$

Theorem 1 *Consider the error system (6) satisfying Assumption 1. Suppose there exist symmetric matrices P, Q_1, Q_2, Z_1, Z_2, and the scalars ε_k for $k = 1, 2$ such that the succeeding LMI is satisfied.*

$$\begin{bmatrix} \Pi & \tau_m \Xi^T Z_1 & \tau_M \Xi^T Z_2 \\ * & -Z_1 & 0 \\ * & * & -Z_2 \end{bmatrix} < 0 \quad (9)$$

where

$$\begin{aligned} \Pi = {} & e_1^T P \Xi + \Xi^T P e_1 + \Lambda - \Gamma_1^T W_1 \Gamma_1 - \Gamma_2^T W_2 \Gamma_2 + \varepsilon_1 e_1^T L_{\bar{\lambda}}^T L_{\bar{\lambda}} e_1 \\ & - \varepsilon_1 e_4^T e_4 + \varepsilon_2 e_2^T L_{\vartheta}^T L_{\vartheta} e_2 - \varepsilon_2 e_5^T e_5 \end{aligned}$$

$$\Xi = (A - BK)e_1 + A_1 e_2 + A_2 e_3 + H e_4 + \overline{H} e_5$$

$$\Gamma_1 = \mathrm{col}\{e_1 - e_2, e_1 + e_2 - 2e_6\}, \ \Gamma_2 = \mathrm{col}\{e_1 - e_3, e_1 + e_3 - 2e_7\},$$

$$W_1 = \mathrm{diag}\{Z_1, 3Z_1\}, \ W_2 = \mathrm{diag}\{Z_3, 3Z_3\}.$$

Proof An LKF candidate is constructed as

$$V(e,t) = V_1(e,t) + V_2(e,t) + V_3(e,t) \tag{10}$$

where

$$V_1(e,t) = e^T(t)Pe(t)$$

$$V_2(e,t) = \int_{t-\tau_1}^{t} e^T(\varphi)Q_1e(\varphi)\mathrm{d}\varphi + \int_{t-\tau_2}^{t} e^T(\varphi)Q_2e(\varphi)\mathrm{d}\varphi$$

$$V_3(e,t) = \int_{-\tau_1}^{0}\int_{t+s}^{t} \tau_1\dot{e}^T(\varphi)Z_1\dot{e}(\varphi)\mathrm{d}\varphi\mathrm{d}s + \int_{-\tau_2}^{0}\int_{t+s}^{t} \tau_2\dot{e}^T(\varphi)Z_2\dot{e}(\varphi)\mathrm{d}\varphi\mathrm{d}s$$

Hence, the time derivative of $V(e,t)$ is expressed by

$$\begin{aligned}\dot{V}_1(e,t) &= 2e^TP[Ae(t) + A_1e(t-\tau_1) + A_2e(t-\tau_2) + H\overline{\lambda}(x,y,t) \\ &\quad + \overline{H}\vartheta(x(t-\tau_1), y(t-\tau_1), t)] \\ &= \zeta^T\left\{e_1^TP\Xi + \Xi^TPe_1\right\}\zeta\end{aligned} \tag{11}$$

where $\Xi = (A - BK)e_1 + A_1e_2 + A_2e_3 + He_4 + \overline{H}e_5$,

$$\zeta^T = [e^T(t),\ e^T(t-\tau_1),\ e^T(t-\tau_2), \overline{\lambda}^T(x,y,t),\ \vartheta^T(x(t-\tau_1),\ y(t-\tau_1), t),\ \frac{1}{\tau_1}\int_{t-\tau_1}^{t} e^T\mathrm{d}\varphi,\ \frac{1}{\tau_2}\int_{t-\tau_2}^{t} e^T\mathrm{d}\varphi]$$

$$\begin{aligned}\dot{V}_2(e,t) &\le e^TQ_1e - e^T(t-\tau_1)Q_1e(t-\tau_1) + e^TQ_2e - e^T(t-\tau_2)Q_2e(t-\tau_2) \\ &= \zeta^T\{e_1^TQ_1e_1 - e_2^TQ_1e_2 + e_1^TQ_2e_1 - e_2^TQ_2e_2\}\zeta \\ &= \zeta^T\Lambda\zeta\end{aligned} \tag{12}$$

$$\begin{aligned}\dot{V}_3(e,t) &= \zeta^T\Xi^T\left(\tau_1^2Z_1 + \tau_2^2Z_2\right)\Xi\zeta - \int_{t-\tau_1}^{t} \tau_1\dot{e}^T(\varphi)Z_1\dot{e}(\varphi)\mathrm{d}\varphi \\ &\quad - \int_{t-\tau_2}^{t} \tau_2\dot{e}^T(\varphi)Z_2\dot{e}(\varphi)\mathrm{d}\varphi\end{aligned} \tag{13}$$

Considering Lemma 1, (13) becomes

$$-\int_{t-\tau_1}^{t} \tau_m\dot{e}^T(\varphi)Z_1\dot{e}(\varphi)\mathrm{d}\varphi - \int_{t-\tau_2}^{t} d\dot{e}^T(\varphi)Z_2\dot{e}(\varphi)\mathrm{d}\varphi \le -\zeta^T\Gamma_1^TW_1\Gamma_1\zeta - \zeta^T\Gamma_2^TW_2\Gamma_2\zeta \tag{14}$$

where

$\Gamma_1 = \mathrm{col}\{e_1 - e_2, e_1 + e_2 - 2e_6\}$, $\Gamma_2 = \mathrm{col}\{e_1 - e_3, e_1 + e_3 - 2e_7\}$,

$W_1 = \mathrm{diag}\{Z_1, 3Z_1, 5Z_1\}$, $W_2 = \mathrm{diag}\{Z_3, 3Z_3, 5Z_3\}$

So combining (11)–(14), we get

$$\dot{V}(e,t) \leq \zeta^T\{e_1^T P\Xi + \Xi^T P e_1 + \Lambda - \Gamma_1^T W_1 \Gamma_1 - \Gamma_2^T W_2 \Gamma_2 + \Xi^T\left(\tau_1^2 Z_1 + \tau_2^2 Z_2\right)\Xi\}\zeta \tag{15}$$

According to Assumption 1, we have

$$\varepsilon_1 e^T L_{\hbar}^T L_{\hbar} e - \varepsilon_1 \hbar^T(x,y,t)\hbar(x,y,t) = \zeta^T(t)\{\varepsilon_1 e_1^T L_{\hbar}^T L_{\hbar} e_1 - \varepsilon_1 e_4^T e_4\}\zeta(t) \geq 0, \tag{16}$$

$$\varepsilon_2 e^T(t-\tau_1) L_{\vartheta}^T L_{\vartheta} e(t-\tau_1) - \varepsilon_2 \vartheta^T(x(t-\tau_1), y(t-\tau_1), t)\vartheta(x(t-\tau_1), y(t-\tau_1), t) = \zeta^T(t)\{\varepsilon_2 e_2^T L_{\vartheta}^T L_{\vartheta} e_2 - \varepsilon_2 e_5^T e_5\}\zeta(t) \geq 0 \tag{17}$$

Combining (15)–(17), we get

$$\dot{V}(e,t) \leq \zeta^T\{e_1^T P\Xi + \Xi^T P e_1 + \Lambda - \Gamma_1^T W_1 \Gamma_1 - \Gamma_2^T W_2 \Gamma_2 + \Xi^T\left(\tau_1^2 Z_1 + \tau_2^2 Z_2\right)\Xi + \varepsilon_1 e_1^T L_{\hbar}^T L_{\hbar} e_1 - \varepsilon_1 e_4^T e_4 + \varepsilon_2 e_2^T L_{\vartheta}^T L_{\vartheta} e_2 - \varepsilon_2 e_5^T e_5\}\zeta$$

After some mathematical simplification and Schure complement, the inequality (9) can be produced. This proves Theorem 1.

Now an adequate condition is derived to obtain gain K.

Theorem 2 *An adequate condition for the solution in Theorem 1 is that there exist the positive definite and symmetric matrices X, $\overline{Q}_1$, $\overline{Q}_2$, $\overline{Z}_1$, $\overline{Z}_2$, matrix M of appropriate dimensions, such that the following condition is satisfied.*

$$\begin{bmatrix} \overline{\Pi} & \Psi & e_1^T X L_{\hbar}^T & e_2^T X L_{\vartheta}^T \\ * & -\Omega & 0 & 0 \\ * & * & -\overline{\varepsilon}_1 I & 0 \\ * & * & * & -\overline{\varepsilon}_2 I \end{bmatrix} < 0 \tag{18}$$

where

$\overline{\Pi} = e_1^T \overline{\Xi} X + X\overline{\Xi}^T e_1 + \overline{\Lambda} - \Gamma_1^T \overline{W}_1 \Gamma_1 - \Gamma_2^T \overline{W}_2 \Gamma_2 - \varepsilon_1 e_4^T e_4 - \varepsilon_2 e_5^T e_5$

$\overline{\Xi} = (AX - BM)e_1 + A_1 X e_2 + A_2 X e_3 + H e_4 + \overline{H} e_5$

$$\Psi = \left[\tau_1 \overline{\Xi}^T, \tau_2 \overline{\Xi}^T\right], \ \Omega = \text{diag}\left\{X\overline{Z}_1^{-1}X, X\overline{Z}_2^{-1}X\right\},$$

$$\Gamma_1 = \text{col}\{e_1 - e_2, e_1 + e_2 - 2e_6\}, \ \Gamma_2 = \text{col}\{e_1 - e_3, e_1 + e_3 - 2e_7\},$$

$$\overline{W}_1 = \text{diag}\{\overline{Z}_1, 3\overline{Z}_1\}, \ \overline{W}_2 = \text{diag}\{\overline{Z}_2, 3\overline{Z}_2\}$$

Proof Based on the Schur complements and congruence transform we can obtain (18).

4 Simulation Result

The effectiveness of the results is derived to establish the simulation for nonlinear time-delay chaotic diabetes mellitus system as follows, considering the parameters of systems (2) and (3):

$$A = \begin{bmatrix} -0.25 & 0 & 0 \\ 0 & 0 & 0 \\ 0 & 0.45 & 0.676 \end{bmatrix}, \quad A_1 = \begin{bmatrix} 0 & 0 & 0.1 \\ 0 & 0 & 0 \\ 0 & 0 & 0 \end{bmatrix}, \quad A_2 = \begin{bmatrix} 0 & 0 & 0 \\ 0.6 & 0 & 0 \\ 0 & 0 & 0 \end{bmatrix}$$

$$H = \begin{bmatrix} 0 & 0 & 0 \\ 0 & 1.0414 & 0 \\ 0 & 0 & 0.3 \end{bmatrix}, \quad \overline{H} = \begin{bmatrix} 0.4720 & 0 & 0 \\ 0 & 0 & 0 \\ 0 & 0 & 0 \end{bmatrix}, \quad B = \begin{bmatrix} 1 & 0 \\ 0 & 1 \\ 0 & 0 \end{bmatrix}$$

Attractor of the MS with ICs (3.1, 1.2, 0.4) and $\tau_1 = 0.042$, $\tau_2 = 0.07$ is shown in Fig. 3. Attractor of the SS with ICs (5.1, 1.8, 0.2) and $\boldsymbol{\tau}_1 = 3.63$, $\boldsymbol{\tau}_2 = 3.36$ is shown in Fig. 4. We can obtain the controller gain matrix:

$$K = \begin{bmatrix} 14.75 & 2.28 & 3.76 \\ 15.4 & 1.32 & 6.14 \end{bmatrix}$$

Figure 5 shows the synchronized states of MS and SS. The synchronization errors are given in Fig. 6. Figures 5 and 6 can conclude that all states of the SS are synchronized with the MS after 6.2 min.

5 Conclusion

This piece of work deals with the problem of synchronization of the nonlinear time-delay chaotic diabetes mellitus which has been addressed by utilizing a state feedback

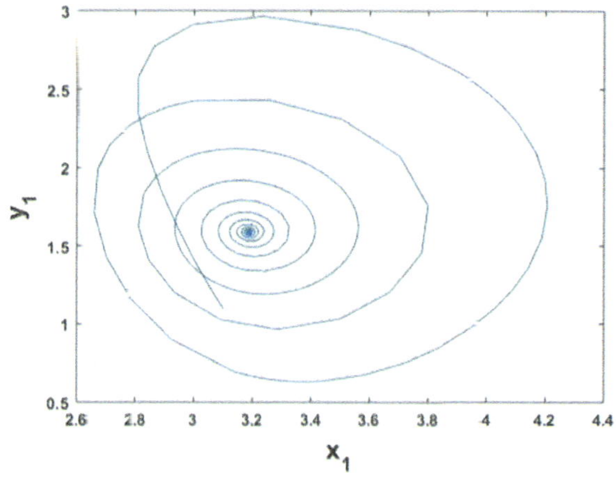

Fig. 3 Attractor of the MS with $\tau_1 = 0.042$, $\tau_2 = 0.07$, and ICs (3.1, 1.2, 0.4)

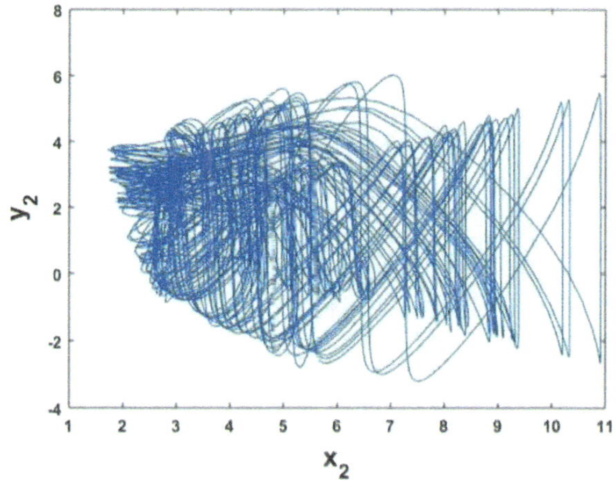

Fig. 4 Chaotic attractor of SS with $\tau_1 = 3.63$, $\tau_2 = 3.36$, and ICs (5.1, 1.8, 0.2)

controller. Here, we are adding controlled input in two states of the system. The simulation results reveal the successful achievement of the synchronization by the proposed controller. The unhealthy uncertain time-delay chaotic diabetes mellitus system follows the healthy uncertain time-delay chaotic diabetes mellitus system. The future research topic is to design observer-based synchronization control for the nonlinear time-delay chaotic diabetes mellitus system.

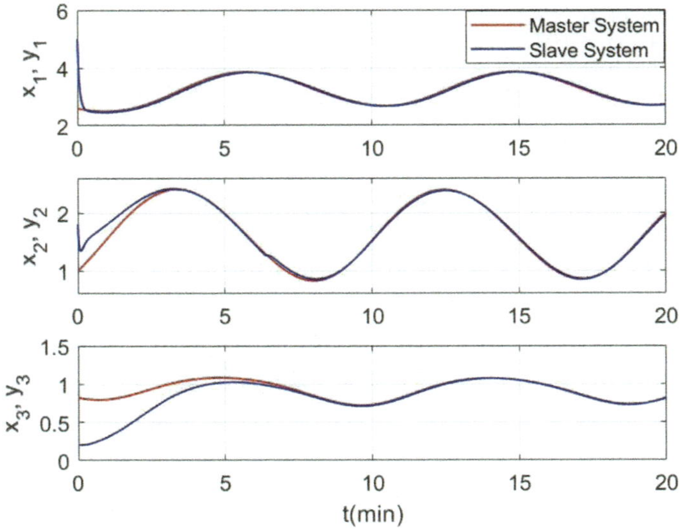

Fig. 5 State responses of SFC-based MS and SS

Fig. 6 Error responses of SFC-based MS and SS

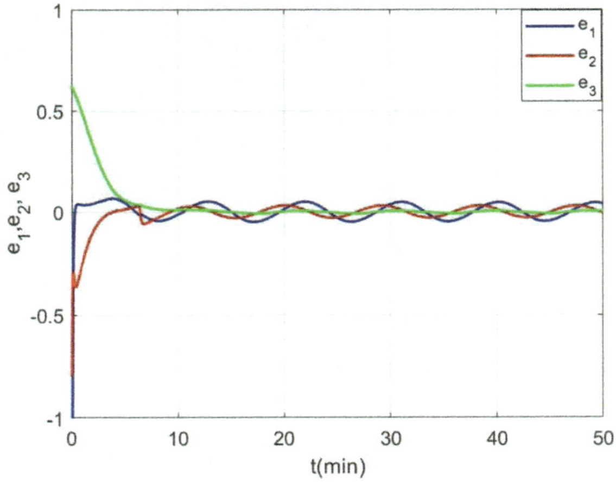

References

1. Canadian Diabetes Association, *Types of Diabetes*, Jan 2017. http://www.diabetes.ca/about-diabetes/types-of-diabetes
2. E. Ackerman, J.W. Rosevear, W.F. McGuckin, A mathematical model of the glucose-tolerance test. Phys. Med. Biol. **9**(2), 203 (1964)
3. J.S. Bajaj, G.S. Rao, J.S. Rao, R. Khardori, A mathematical model for insulin kinetics and its application to protein-deficient (malnutrition-related) diabetes mellitus (PDDM). J. Theor. Biol. **126**(4), 491–503 (1987)

4. W. Sarika, Y. Lenbury, K. Kumnungkit, W. Kunphasuruang, Modelling glucose-insulin feedback signal interchanges involving β-cells with delays. Differ. Equ. **1**(2), 1 (2008)
5. M. Chuedoung, W. Sarika, Y. Lenbury, Dynamical analysis of a nonlinear model for glucose–insulin system incorporating delays and β-cells compartment. Nonlinear Anal. Theory Methods Appl. **71**(12), e1048–e1058 (2009)
6. K. Ikeda, K. Matsumoto, High-dimensional chaotic behavior in systems with time-delayed feedback. Phys. D **29**(1–2), 223–235 (1987)
7. B. Li, Z. Zhao, R. Wang, G. Ding, Synchronization control design based on Wirtinger inequality for uncertain coronary artery time-delay system with input saturation. IEEE Access **7**, 76611–76619 (2019)
8. N.P. Mohanty, R. Dey, B.K. Roy, Switching synchronisation of a 3-D multi-state-time-delay chaotic system including externally added memristor with hidden attractors and multi-scroll via sliding mode control. Eur. Phys. J. Spec. Top. **229**, 1231–1244 (2020). https://doi.org/10.1140/epjst/e2020-900195-4 (in press)
9. N.P. Mohanty, R. Dey, B.K. Roy, A new 3-D memristive time-delay chaotic system with multi-scroll and hidden attractors. IFAC-PapersOnLine **51**(1), 580–585 (2018)
10. T. Huang, C. Li, W. Yu, G. Chen, Synchronization of delayed chaotic systems with parameter mismatches by using intermittent linear state feedback. Nonlinearity **22**(3), 569 (2009)
11. L. Li-Xiang, P. Hai-Peng, G. Bao-Zhu, X. Jin-Ming, A new sliding mode control for a class of uncertain time-delay chaotic systems. Chin. Phys. **10**(8), 708 (2001)
12. N. Vasegh, F. Khellat, Projective synchronization of chaotic time-delayed systems via sliding mode controller. Chaos Solitons Fractals **42**(2), 1054–1061 (2009)
13. P. Palumbo, S. Panunzi, A. De Gaetano, Qualitative behavior of a family of delay-differential models of the glucose-insulin system. Discrete Contin. Dyn. Syst. Ser. B **7**(2), 399 (2007)
14. R. Prager, P. Wallace, J.M. Olefsky, In vivo kinetics of insulin action on peripheral glucose disposal and hepatic glucose output in normal and obese subjects. J. Clin. Investig. **78**(2), 472–481 (1986)
15. E. Forrest, P. Robinson, M. Hazel, Insulin, growth hormone and carbohydrate tolerance in Jamaican children rehabilitated from severe malnutrition. MedCarib (1925)

Geo/G/1 System: Queues with Late and Early Arrivals

Reena Grover, Himani Chaudhary, and Geetanjali Sharma

Abstract Consider various discrete-time systems where the time axis is divided to put something into a particular space that is designed for it, known as spots. In view of the services and evacuation could be taken as spot boundaries, where their periods are integral multiple of spot span. In general, distributed service time is denoted by Geo/G/1 and single-server system with discrete time has Bernoulli arrival process. A $\text{Geo}^x/\text{G}/1$ system is the basis of all the extended models. Such particular points are in fact identical with the distribution of the system states observed at an arbitrary point on the continuous-time domain. This leads a random point which lies in the middle of a spot on the continuous-time domain with probability 1.

Keywords Discrete time · Early–late arrival · Markov chain process · Waiting time · Queue size

R. Grover (✉)
Department of Mathematics, SRM Institute of Science and Technology
Delhi-NCR Campus, Ghaziabad, UP, India
e-mail: reenadr1980@gmail.com

H. Chaudhary · G. Sharma
Department of Mathematics and Statistics, Banasthali Vidyapith,
Banasthali, Tonk, Rajasthan 304022, India
e-mail: himanichaudhary07@gmail.com

G. Sharma
e-mail: geetanjali.bu@gmail.com

S. S. Dash et al. (eds.), *Intelligent Computing and Applications*,
Advances in Intelligent Systems and Computing 1172,
https://doi.org/10.1007/978-981-15-5566-4_70

1 Introduction

1.1 Discrete-Time Systems

Let us call the entity of service a message, whose service time is always an integral multiple of a spot period. Numbers of messages that arrive during successive slots constitute a series of random variables which are independent and identical. When given messages lie in probability $\lambda(0 < \lambda < l)$ and with probability $1 - \lambda$, one can examine a mean of $1/\lambda$ spots of interarrival time I which is geometrically distributed.

$$\text{Prob}[I = l \text{ slots}] = \lambda(1 - \lambda)^{l-1}$$

where $l = 1, 2 \ldots$.

Such an arrival process is called Bernoulli process. A Geo^x/G/1 system is the basis of all the extended models.

In discrete-time queueing systems, one of the earliest investigations being made by Meisling [1], Chu and Konheim [2], and Kobayashi and Konheim [3] provide surveys of applications to computer communication networks. Bruneel and Kim [4] analyze many general systems with discrete time ones as non-independent arrivals and multiple servers.

1.2 Late Arrival Model

In late arrival model, the message arrives after a given spot just before border line of the spot. So arrived messages look at departed messages which are about to go and the departed message goes just after the message which has just reached. As message is arrived in the nth spot till the system is completed, the process and service will start as $(n + 1)$th spot starts. Hunter [5] specified this model as a delayed access with the late arrival system. Gravey and Hebuterne [6] stated late arrival model as arrival first policy.

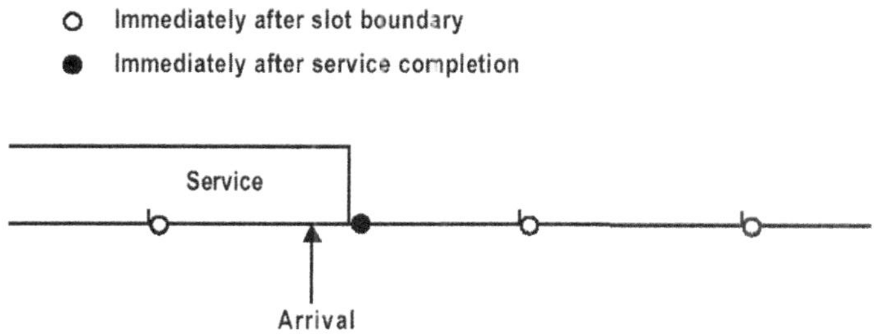

Late arrival model

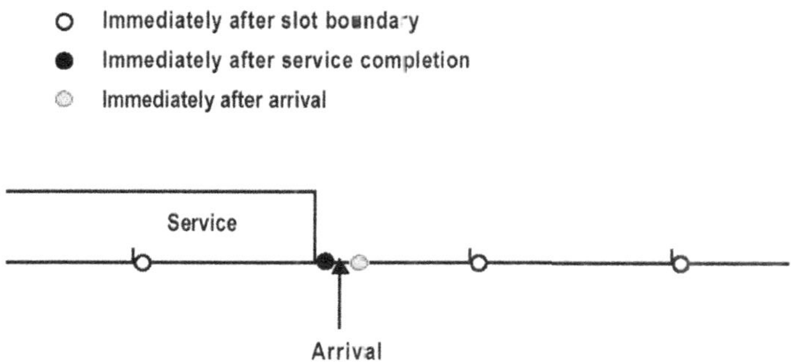

Early arrival model

Therefore, the queue size distribution of first and second moments immediately after a random spot period is

$$E[L] = \frac{\lambda^2 b^2 - \lambda\rho}{2(1-\rho)} + \rho \tag{1}$$

$$E[L^2] = \frac{\lambda^3(b - 3b^2 + 2b)}{3(1-p)} + \frac{3\lambda^2(b^2 - b)}{2(1-p)} + \frac{\lambda^4(b^2 - b)^2}{2(1-\rho)^2} + \rho \tag{2}$$

The discrete-time version of Little's theorem is

$$E[L] = \lambda E[T] \tag{3}$$

where T is the time (measured in slots) for a message that stays for arrival to service completion in the system, called response time of a message. Therefore,

$$E[T] = \frac{\lambda b^{(2)} - \rho}{2(1-\rho)} + b \tag{4}$$

which is independent of the service discipline.

If we assume the "First Come First Served (FCFS)" discipline, after service completion of messages left in system instantly is identical with messages which arrive in given duration during which the messages were in the system. Therefore, if $T(u)$ denotes the PGF for time T that a message stay in the system, then (Hunter [5]).

$$\Pi(z) = T(1-\lambda+\lambda z) \tag{5}$$

The second moment of the distribution of T is

$$E[T^2] = \frac{\lambda(2b^{(3)} - 3b^{(2)} + b)}{6(1-\rho)} + \frac{\lambda^2(b^{(2)} - b)^2}{2(1-\rho)^2} + \frac{(b^{(2)} - b)}{1-\rho} + b \tag{6}$$

$$E[W] = \frac{\lambda b^{(2)} - \rho}{2(1-\rho)} \tag{7}$$

$$E[W^2] = \frac{\lambda(2b^{(3)} - 3b^{(2)} + b)}{6(1-\rho)} + \frac{\lambda(b^{(2)} - b)^2}{2(1-\rho)^2} \tag{8}$$

$$\text{Prob}[W=0] = W(0) = \frac{1-\rho}{1-\lambda} = \frac{\text{Prob}[L=0]}{1-\lambda} > \text{Prob}[L=0] \tag{9}$$

A loose argument to justify this inequality is that an arriving message does not wait if it finds only one message in the system which is about to leave (Gravey [7]).

1.3 Early Arrival Model

In an early arrival model (Hunter [5]), the messages arrive early during a slot. If the system is empty when a message arrives, its service is started immediately, including the slot in which the message arrives when we calculate its waiting time. Gravey and Hebuterne [6] call the early arrival model the departures first policy.

From the normalization condition, we get (cf. Hunter [5], Gravey and Hebuterne [6])

$$\begin{aligned}\Pi(z) &= \frac{(1-\rho)[A(z) - z\mathrm{A}'(z)]}{(1-\lambda)[A(z) - z]} \\ &= \frac{(1-\rho)(1-z)B(1-\lambda+\lambda z)}{(1-\lambda+\lambda z)[B(1-\lambda+\lambda z) - z]}\end{aligned} \tag{10}$$

Comparing $\Pi(z)$ with early arrival model with $\Pi(z)$ with late arrival model, size of the queue observed instantly afterwards the service completion given in first model is quite less that is given for second model by the number whose PGF is given by $1-\lambda+\lambda z$, the number of arrivals (zero or one) in a slot. This is because "queue size" is observed before the possible arrival point in the early arrival model.

Since we count the slot in which a message arrives when we calculate the waiting time in the early arrival model,

$$\Pi(z) = \frac{T(1-\lambda+\lambda z)}{1-\lambda+\lambda z} \tag{11}$$

The queue size of an early arrival model is observed; it is also the queue size observed at an arbitrary point during the same slot on the continuous-time domain.

2 Output Process

A sequence of those slot boundaries at which service is completed is called an output process or a departure process. Consider the interdeparture time of a Geo/G/I system, which is defined as "time interval" between two "successive service completions." The correlation of two contiguous interdeparture times is denoted by τ and τ'. Assuming that three successive service completions occur at the end of the n_0th, $n_0+\tau$th, and $n_0+\tau+\tau'$th slots, the joint probability is obtained.

$$\Delta(u,u') \triangleq \sum_{k=1}^{\infty}\sum_{\ell=1}^{\infty} \text{Prob}[\tau=k,\tau'=1]u^k u'^{\ell} \tag{12}$$

There are five cases regarding the instantly before and after the ends of the n_0th and $n_0+\tau$th slots and the number of messages that arrive during τ slots (Table 1).

2.1 Model

Consider Geo/G/1 system where service time is geometrically distributed as

$$b(\ell) = \mu(1-\mu)^{\ell-1} \quad \ell = 1,2\ldots \tag{13}$$

where $1/mu$ is the mean service time.

Table 1 Cases before and after the ends of the n_0th and $n_0+\tau$th slots

Case	n_0-0	n_0+0	No. of arrivals during τ slots	$n_0+\tau-0$	$n_0+\tau+0$
(a)	1	0	1	1	0
(b)	1	0	≥2	≥2	≥1
(c)	2	1	0	1	0
(d)	2	1	≥1	≥2	≥1
(e)	≥3	≥2	Any	≥2	≥1

$$\Delta(u,u') = \frac{\lambda u}{1-(1-\lambda)u} \cdot \frac{\lambda u'}{1-(1-\lambda)u'} \tag{14}$$

which implies that process of departure with rate λ is a Bernoulli process.

2.2 Model

The queue size distribution $\{\pi_k^*;\ k = 0, 1, 2 \ldots\}$ of the "generating function $\pi^*(z)$" observed either at the end of a slot during with empty system or immediately after a service completion is

$$\Pi^*(z) \underline{\underline{\Delta}}\, \pi^* k^{z^k} = \frac{(1-\rho)\{B[\Lambda(z)] - z\Lambda(z)\}}{(1+\lambda-\rho)\{B[\Lambda(z)] - z\}} \tag{15}$$

The probability $\eta(l)$ that there are $l+1$ slots between an arbitrary slot boundary and the last Markov point on the condition that in the given time the system is non-empty.

$$\eta(\ell) = \frac{1}{b} \sum_{j=\ell+1}^{\infty} b(j) \quad \ell = 0, 1, 2 \ldots \tag{16}$$

In this case, the arrivals during the first l slots contribute to the system state observed at the starting of the last spot. Therefore, the probability P_k; $k \geq 1$, that there are k messages,

$$P_k = \sum_{j=1}^{k} \pi_j \sum_{\ell=0}^{\infty} \text{Prob}[k - j \text{ message arrive in } \ell \text{ slot}]\, \eta(\ell) \quad k = 1, 2 \ldots \tag{17}$$

Hence, the $P(z)$ of size distribution is $\{P_k;\ k \geq 0\}$

$$P(z) \underline{\underline{\Delta}}\, P_0 + \sum_{k=1}^{\infty} P_k z^k = \frac{(1-\rho)(1-z)B[\Lambda(z)]}{B[\Lambda(z)] - z} \tag{18}$$

This result is also given by Bruneel and Kim [4], Hunter [5], Konheim [3].

$P(z)$ in (18) observed instantly after possible spot partition is generally different from $\Pi(z)$. Instantly, the service is completed in Geo*/G/1 system. However, they are identical in the Geo/G/1 system.

Then,

$$\Pi(z) = \rho(z) \frac{1-\Lambda(z)}{\lambda(1-z)} \tag{19}$$

holds for a Geo^x/G/1 system. Bruneel and Kim [4] state that the elementary observation is true for enormous queuing models.

The event of "no service completion" includes not only a case in which the ongoing service is not completed but also a case in which the service is interrupted by the server vacation.

$$
\begin{aligned}
E[z^{L^{(n+1)}}] = \{&vE[z^{L^{(n)}+\Lambda^{(n+1)}-1}|L^{(n)} \geq 1,\ \text{service completion}] \\
&+ (1-v)E[z^{L^{(n)}+\Lambda^{(n+1)}}|L^{(n)} \geq 1,\ \text{no service completion}]\}\,\text{Prob}[L^{(n)} \geq 1] \\
&+ E[z^{\Lambda^{(n+1)}}]\,\text{Prob}[L^{(n)} = 0] \qquad (20)
\end{aligned}
$$

However, from the theorem of the total probability,

$$
\begin{aligned}
E[z^{L^{(n)}+\Lambda^{(n+1)}}|L^{(n)} \geq 1] = {}&vE[z^{L^{(n)}+\Lambda^{(n+1)}}|L^{(n)} \geq 1,\ \text{service completion}] \\
&+ (1-v)E[z^{L^{(n)}+\Lambda^{(n-1)}}|L^{(n)} \geq 1,\ \text{no service completion}] \qquad (21)
\end{aligned}
$$

3 Queue Size and Remaining Service Time

Following Dafermos and Neuts [8], Bruneel and Kim [4] and Hunter [5] recognize the "joint distribution" of "queue size" and "remaining service time" instantly at every spot boundary in the late arrival model. Let us consider the queue size $L^{(n)}$ instantly afterward the nth slot, then the remaining time of service $X_+^{(n)}$ of the message will be served at same instant, where $n = -0, 1, 2\ldots.$

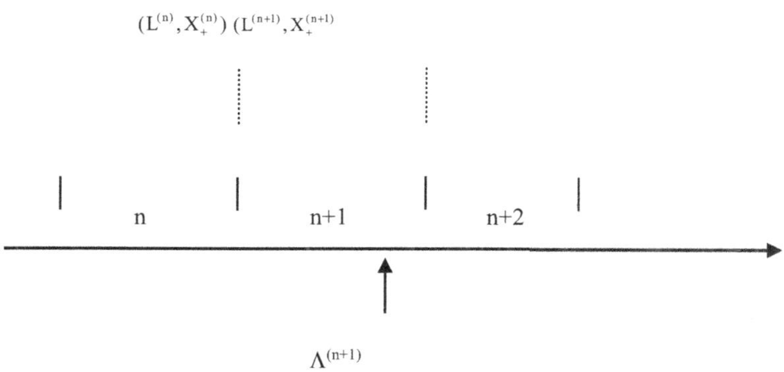

"Queue size" and "remaining service time" will be recognized instantly afterward slot boundaries. Considering $X_+^{(n)} = 0$ if $L^{(n)} = 0$. Given $L^{(n)} = 0$ if $X_+^{(n)} = 0$ because we observe the system after both arrivals, and a service completion possibly occur.

The beginning of the first slot "queue size" and "remaining service time" be $L^{(0)}$ and $X_+^{(0)}$, respectively.

The sequence $\{L^{(n)}; n=0, 1, 2....\}$ is not a "Markov chain." However, the sequence of pairs $\{L^{(n)}, X_+^{(n)}; n = 0, 1, 2....\}$ forms a "Markov chain" having (0, 0) as state space and all integer pairs (*k*, *l*) with "$k \geq 1$" and "$l \geq 1$." The number of messages which will arrive in nth slot is denoted by $\Lambda^{(n)}$, and complete service time is denoted by *X*.

$$L^{(n+1)} = X_+^{(n+1)} = 0 \quad \text{if } L^{(n)} = X_+^{(n)} = 0 \quad \text{and} \quad \Lambda^{(n+1)} = 0 \tag{22a}$$

$$\text{if } L^{(n)} = X_+^{(n)} = 1 \quad \text{and} \quad \Lambda^{(n+1)} = 0 \tag{22b}$$

$$L^{(n+1)} = \Lambda^{(n+1)} \quad \text{and} \quad X_+^{(n+1)} = X \quad \text{if } L^{(n)} = X_+^{(n)} = 0 \quad \text{or} \quad 1, \quad \text{and} \quad \Lambda^{(n+1)} \geq 1 \tag{23}$$

$$L^{(n+1)} = L^{(n)} + \Lambda^{(n+1)} \quad \text{and} \quad X_+^{(n+1)} = X_+^{(n)} - 1, \quad \text{if } L^{(n)} = 1 \quad \text{and} \quad X_+^{(n)} > 1 \tag{24}$$

$$L^{(n+1)} = L^{(n)} + \Lambda^{(n+1)} - 1 \quad \text{and} \quad X_+^{(n+1)} = X, \quad if\ L^{(n)} > 1 \quad \text{and} \quad X_+^{(n)} = 1 \tag{25}$$

A new service is therefore started at the beginning of the $n + 2$nd slot.

In order to find the steady-state distribution for $L^{(n)}, X_+^{(n)};\ n = 0, 1, 2 \ldots r$,

$$P_0 \triangleq \lim_{n\to\infty} \text{Prob}\,[L^{(n)} = X_+^{(n)} = 0] \tag{26a}$$

$$\Pi_k(\ell) \triangleq \lim_{n\to\infty} \text{Prob}\,[L^{(n)} = K, X_+^{(n)} = \ell] \quad k \geq 1, \quad \ell \geq 1 \tag{26b}$$

and

$$\Pi(z, u) \triangleq \sum_{k=1}^{\infty} \sum_{\ell=1}^{\infty} \Pi_k(\ell) z^k u^\ell \quad |z| \leq 1, \quad |u| \leq 1 \tag{26c}$$

From (22) to (25)

$$P_0 = P_0 \lambda(0) + \Pi_1(1)\,\lambda(0) \tag{27a}$$

$$\Pi_k(\ell) = P_0 \lambda(k) b(\ell) + \Pi_1(1) \lambda(k) b(\ell) + \sum_{j=1}^{k} \Pi_j(\ell + 1) \lambda(k - j)$$
$$+ \sum_{j=2}^{k+1} \Pi_j(1) \lambda(k - j + 1) b(\ell) \quad k \geq 1, \quad \ell \geq 1 \tag{27b}$$

The normalization condition is

$$P_0 + \sum_{k=1}^{\infty} \sum_{\ell=1}^{\infty} \Pi_k(\ell) = 1 \tag{27c}$$

Substituting (27b) into (26c),

$$\Pi(z, u) = [P_0 + \Pi_1(1)] \sum_{k=1}^{\infty} \lambda(k) z^k \sum_{\ell=1}^{\infty} b(\ell) u^\ell + \sum_{k=\ell}^{\infty} \sum_{\ell=1}^{\infty} z^k u^\ell \sum_{j=1}^{k} \Pi_j(\ell+1) \lambda(k-j)$$
$$+ \sum_{\ell=1}^{\infty} b(\ell) u^\ell \sum_{k=1}^{\infty} z^k \sum_{j=2}^{k+1} \Pi_j(i) \lambda(k-j+1)$$

By setting $u = \Lambda(z)$,

$$\sum_{j=1}^{\infty} \Pi_j(1) z^j = \frac{(1-\rho) z [1-\Lambda(z)] B[\Lambda(z)]}{\Lambda(z)\{B[\Lambda(z)] - z\}} \tag{28}$$

Substituting (28), then

$$\Pi(z, u) = \frac{(1-\rho) z u [1-\Lambda(z)] \{B(u) - B[\Lambda(z)]\}}{[u-\Lambda(z)]\{B[\Lambda(z)] - z\}} \tag{29}$$

From (29), the PGF $P(z)$ for the queue size instantly is denoted by

$$P(z) = P_0 + \Pi(z, 1) = \frac{(1-\rho)(1-z) B[\Lambda(z)]}{B[\Lambda(z)] - z} \tag{30a}$$

which agrees with (26); on the other hand, the PGF for the "remaining service time" instantly afterward a random slot boundary in the busy span is indicated by

$$\frac{\Pi(1, u)}{\Pi(1, 1)} = \frac{u[1 - B(u)]}{b(1-u)} \tag{30b}$$

4 Virtual Waiting Time and Unfinished Work

"Virtual waiting time" W_g (measured in slot) is in the $n + 1$st slot so that a super message could wait in analogous "Geo/G/1" system having "FCFS" discipline if it will arrive in the $n + 1$st slot. If $L^{(n)} = k \geq 1$ and $X_+^{(n)} = \ell \geq 2$, it would be $k - 1$ complete service times plus $l - 1$ slots. Finally, if $L^{(n)} = k \geq 2$ and $X_+^{(n)} = 1$, it would be exactly $k - 1$ complete service times. Hence, the PGF $W(u)$ for W is

$$W_g(u) = P_0 + \Pi_1(1) + \sum_{k=1}^{\infty}\sum_{\ell=2}^{\infty} \Pi_k(\ell)[B(u)]^{k-1}u^{\ell-1} + \sum_{k=2}^{\infty} \Pi_k(1)[B(u)]^{k-1}$$

The "unfinished work U" instantly after the n + 1st slot is the summand of messages "service time" in the waiting line, and "remaining service time" of a message will be served in that given time. Therefore, the PGF $U(u)$ for U is given by cf. Bruneel and Kim [4].

$$U(u) = W_g(u)\Lambda[B(u)] = \frac{(1-p)(1-u)\Lambda[B(u)]}{\Lambda[B(u)] - u} \tag{31a}$$

The mean of "unfinished work" is

$$E[U] = E[W_g] + \rho = \frac{\lambda b^{(2)} + \lambda^{(2)}b^{(2)} + \rho(1-2\rho)}{2(1-\rho)} \tag{31b}$$

4.1 Relationship with the Continuous-Time System

Now let us assume the "continuous-time system" by considering limit $\Delta \to \infty$.

Firstly, the arrival and process will be changed. It is given that $1/\Delta$ denoted number of slots per "time unit;" the PGF denoted by $\Lambda_c(z)$ for the number of messages that will occur during a "time unit" is

$$\Lambda_c(z) = [\Lambda(z)]^{1/\Delta} \tag{32}$$

Let b_c denote the "mean" and $b_c^{(2)}$ denote "second moment" of "service time" for new "time units."

It is obvious that b and $b^{(2)}$ is related in this manner.

$$b_c = b\Delta; \quad b_c^{(2)} = b^2\Delta^2 \tag{33a}$$

$$\rho = \lambda b = \lambda_c b_c \tag{33b}$$

is invariant, and by substitution we get

$$E[Wc] = \frac{\lambda_c^2 b_c^{(2)} - \lambda_c\rho + \lambda_c^{(2)} b_c}{2\lambda_c(1-\rho)} \tag{34}$$

where $E[Wc] = E[W]\,\Delta$ denotes "mean waiting time" expressed in new "time units."

As $\Delta \to \infty$, the Geo^x/G/1 system approaches a continuous time M^x/01/1 system. The PGF of the number of messages is represented by $G(z)$ in a bunch,

$$\Lambda_c(z) = e^{\lambda}[G(z) - 1] \tag{35}$$

if g and $g^{(2)}$ represents the "mean" and the "second factorial moment," respectively, of bunch size,

$$\lambda_c = \lambda g; \quad \lambda_c^{(2)} = \lambda_g^{(2)} + \lambda^2 g^2 \tag{36}$$

Now, substituting (36) into (34),

$$E[W_c] = \frac{\lambda g^2 b_c^2 + g^{(2)} b_c}{2g(1-\rho)} \tag{37}$$

4.2 *Early Arrival Model*

For an early arrival model, let us first consider the Markov chain $\{L_n; n = 1, 2 \ldots.\}$ of the queue size L_n instantly [5].

$$\begin{aligned} L_{(n+1)} &= A' - 1 + A'_{(n+1)} \quad L_n = 0 \\ &L_n + A'_{n+1} - 1 \quad L_n \geq 1 \end{aligned} \tag{38a}$$

where Λ' denotes the number of messages that will arrive in a slot with the condition, i.e., at least one message will arrive in that slot, and the number of messages is denoted by A'_{n+1} that will arrive between the time of service of the $n+1$st message given that $L = 0$. Number of messages A'_{n+1} will arrive between the $j-1$ last slots if the "service time" of the $n+1$st message is j slots. Therefore, PGF $A'(z)$ of A'_{n+1} is

$$A'(z) = \sum_{j=1}^{\infty} b(j)[\Lambda(z)]^{j-1} = \frac{B[\Lambda(z)]}{\Lambda(z)} \tag{38b}$$

Now,

$$\begin{aligned} \Pi(z) &= \frac{\Pi(0)[A(z) - \Lambda'(z)A'(z)]}{A(z) - z} \\ &= \frac{\lambda(0)\Pi(0)[1 - \Lambda(z)]B[\Lambda(z)]}{[1 - \lambda(0)]\Lambda(z)\{B[\Lambda(z)] - z\}} \end{aligned} \tag{39a}$$

From the normalization condition ($\Pi_{(1)} = 1$),

$$\Pi_{(0)} = \frac{[1 - \lambda(0)][1 - \rho]}{\lambda(0)\lambda} \tag{39b}$$

Hence,

$$\Pi(z) = \frac{(1-\rho)[1-\Lambda(z)]B[\Lambda(z)]}{\lambda\Lambda(z)\{B[\Lambda(z)]-z\}} \tag{40}$$

The system state in terms of "queue size" and "remaining service time" examined at random point on the "continuous-time domain" is identical, and it is examined that immediately after instantly a possible arrival point follows each "slot boundary" in the early arrival model. Queue size instantly after a possible arrival point during the nth slot is represented by $L^{(n)}$, and Message "remaining service" time which being served at the same immediately is denoted by $X^{(\mathrm{n})}$. The PGF $P(z)$ of "queue size" observed at random point on "continuous-time domain" is given by (30a).

5 Conclusion

Hence, we reach the following conclusion about the comparison of the "late and early arrival models." Distribution of the "queue size" and the remaining in the stable condition are similar in each model. The waiting time distribution of random message is also the same in both models. The size of the queue instantly after the service completion in the "early arrival model" is smaller in comparison to the "late arrival model" by the number of messages included in a possible arrival. This is because the observation point in the "early arrival model" is located before the possible "arrival point."

References

1. T. Meisling, Discrete-time queueing theory. Oper. Res. **6**(1), 96–105 (1958)
2. W.W. Chu, A.G. Konheim, On the analysis and modeling of a class of computer communication systems. IEEE Trans. Commun. **20**(3, Part 2), 645–660 (1972)
3. H. Kobayashi, A.G. Konheim, Queueing models for computer communications system analysis. IEEE Trans. Commun. **25**(1), 2–29 (1977)
4. H. Bruneel, B.G. Kim, *Discrete-Time Models for Communication Systems Including ATM*, vol. 205 (Springer Science+Business Media, 1993)
5. J.J. Hunter, *Mathematical Techniques of Applied Probability, Discrete Time Models: Techniques and Applications*, vol. 2 (Academic Press, New York, 1983)
6. A. Gravey, G. Hebuterne, Simultaneity in discrete-time single server queues with Bernoulli inputs. Perform. Eval. **14**(2), 123–131 (1992)
7. A. Gravey, On the Geo/D/1 and Geo/D/1/*n* queues. Perform. Eval. **11**(2), 117–125 (1990)
8. S.C. Dafermos, M.F. Neuts, A single server queue in discrete time. Caheirs du Centre de Recherché Operationnelle **13**, 23–40 (1971)

Intelligent Data Analysis with Classical Machine Learning

Sanjeev Kumar Punia, Manoj Kumar, and Amit Sharma

Abstract In this advance era, artificial intelligence (AI) is the basic building block for numerous innovation opportunities. The basic concept behind all artificial intelligence applications is classical machine learning (CML) technique. The intelligent data analysis (IDA) is used to analyze data in advanced reasoning systems. In this paper, our motto is to analyze the data intelligently, i.e., intelligent data analysis process using classical machine learning algorithms based on experimental data set empirical hidden regularities. It is concluded that machine learning process is the basic concept to implement any problem based on artificial intelligence, whereas intelligent data analysis uses classical machine learning concept and helps in deciding the next best available step during data analysis.

Keywords Classical machine learning (CML) · Intelligent data analysis (IDA) · Artificial intelligence (AI) · Internet of things (IoT)

1 Introduction

Presently, artificial intelligence is playing a big and active role at every step in everyone's life. Legg [1] explained that artificial intelligence is the requirement of present and future era to enhance accuracy with efficiency through different ways in every field. In parallel, artificial intelligence implements wide range of real

S. K. Punia (✉)
JIMS Engineering Management Technical Campus, Greater Noida, Uttar Pradesh, India
e-mail: drsanjeevpunia@hotmail.com

M. Kumar
School of Computer Science, University of Petroleum and Energy Studies (UPES), Dehradun, Uttrakhand, India
e-mail: wss.manojkumar@gmail.com

A. Sharma
SRMIST, NCR Campus, Modi Nagar, Ghaziabad, Uttar Pradesh, India
e-mail: Amit.krsharma123@gmail.com

S. S. Dash et al. (eds.), *Intelligent Computing and Applications*,
Advances in Intelligent Systems and Computing 1172,
https://doi.org/10.1007/978-981-15-5566-4_71

applications outside research laboratories. The advanced artificial intelligence technique concept implemented in various areas like troop control weapons, internet of things (IoT), and latest in the consulting of United States President election voting. Bogost [2] mentioned that common understanding of artificial intelligence concept is generally absent in many fields, and artificial intelligence is implemented using machine learning, neural network learning, and deep machine learning. Generally, deep machine learning is interrelated with intelligent data analysis and refers to the explosive growth of artificial intelligence.

2 Notations in Classical Machine Learning and Intelligent Data Analysis

Reference [3] explained that functional capabilities of artificial intelligence based computer systems is a replica of human natural intelligence characteristics. The cognitive reasoning is concerned with human intelligence as this also involves the problem of learning, understanding, perceiving, analysis, and classification. Finn et al. [4] explained that since origin artificial intelligence is used with all advanced computer system applications to analyze, design, develop, implement, and test all difficult problems. Any system is called intelligent if it is able to perform real reasoning in data analysis field. The intelligent data analysis is the analysis of implemented data using intelligent computer system framework. The classical machine learning technique is an important tool in artificial intelligence that uses many different mathematical models, algorithms, and methods for simulation. The classical machine learning techniques perceive relations, learn analogies, generalize result, and classify methods during difficult problem-solving process.

3 Bottleneck in Classical Machine Learning

Reference [5] demonstrates more functions and capabilities of classic intelligent data analysis by comparing four different types of machine learning problems with some machine learning tools limitations as follows.

A. **Empirical Regularities**

The different steps to solve traditional formulation machine learning problem are (i) first, characterize the training use case description (precedents) sample into different targets, (ii) second, forecast the new use case description (precedents) set into necessary targets, (iii) third, predict interpolation–extrapolation into new object properties targets, (iv) fourth, test quality construct hidden dependency into training sample, and (v) finally, the result selection for predicting new sample into new object properties targets. The solution of fixed training sample problem is to optimize the expansion of new object as the results of different extrapolation samples are different. Hence, the main realistic problem is the selection of best reliable extrapolation results. In this, the major issue is the result detection of empirical regularities problems with stable training set extension sample using classical machine learning.

B. **Data Analysis Tools Combination**

There are many classical machine learning tools to solve an interpolation–extrapolation problem, but combining the different tools is a big problem. The comparison of quality evaluation (e.g., bright and dry, green and small, etc.) is not a problem for classical machine learning compared to intelligent data analysis.

C. **Multi-model Training Sample Operation**

In a research paper, reference [6] showed that target properties have to pay more attention for default acceptance in multi-model training sample operation during classical machine learning process. The multi-model operation has multi causality problem during proximity measurement in many machine learning samples. The emergence problem predicts two different types of causation relationships (i) binary—the different value attribute set causes different values effect and (ii) ternary—the specific value attribute set with presence or absence of specific characters cause corresponding effect. In these situations, initial sample (precedent) problem has to divide into two sub-sample using special pre-processing tools. Later, these sub-samples are analyzed using specific intelligent data analysis tools as per orientation in binary or ternary causal relationship. These goal-oriented (focused) analyses of initial data sample cannot be solved using traditional machine learning techniques.

D. **Data Analysis Result Interpretation**

Reference [7] explained that many problems are related with intermediate or final interpretability results in classical machine learning techniques. These problems are more difficult to implement, especially in neural networks and decision trees. The constructive problem cannot be solved in classical machine learning using addiction technique. Many complex problems have to process the initial training samples that include poor sub-samples. Hence, choosing the right sample (precedent) from poor sub-sample also plays an important role.

4 Method Automation in Intelligent Data Analysis

Reference [8] explained automation technique for intelligent data analysis using JSM for scientific research support. Initially, main approach used in JSM automation technique implementation is discussed. Later, the proposed solutions of all four problems shown above in Sect. 3 are compared.

A. **Method Automation support in Intelligent Data Analysis**

Reference [9] proposed JSM automatic method to formalize computer-oriented inductive inference based schemes. The JSM automatic method use reasoning schemes to formalize many value-based logical reasoning. The JSM automatic method uses abductive version of constructive explanation, synthesized using classical cognitive methods. The JSM automation method implements intelligent data analysis framework using aided reasoning in (i) generation of empirical data using extended precedents stable regularity and (ii) explanation of accumulative facts using knowledge discovery implementation. Zabezhailo [10] proposed the data analysis using partially defined relationship between object and properties. The analysis of the data is based on their similar property assumption. The JSM automatic intelligent data analysis framework is divided into three samples (precedents) as (i) $£^{+}$—data set with positive properties, (ii) $£^{-}$—data set with negative properties, and (iii) $£^{*}$—data set with combined properties. Initially, separate data subset is prepared based on the three classifications explained above. The object subset properties use inference rules to implement JSM automatic intelligent data analysis framework. Polya [11] explained that automatic intelligent data analysis JSM framework value is calculated through object property as shown in Eq. 1

$$\Psi = \Psi_d \cup \Psi_d \quad (1)$$

where Ψ_d—Deductive inference rule set and Ψ_d—Plausible inference rule set.

Here, every rule of set Ψ_d is organized according to their distributive lattice pair. The first lattice is analyzed for positive empirical dependency generation. The rule of set Ψ_d is organized according to plausibility inference rule. The automatic intelligent data analysis JSM framework uses classical machine learning methods to predict the object properties of set Ψ. The different steps in prediction are (i) initially, properties of set Ψ is predicted based on sub-object Ψ_d and Ψ_d sets properties to generate all dependencies, (ii) next, the prediction result of Ψ set is added to current Ψ_d and Ψ_d sets, and (iii) finally, JSM intelligent data analysis method is repeated again and again to stabilize the Ψ set value. Reference [12] also proposed a constructive scheme (tentative test theory) to implement automatic intelligent data analysis JSM framework. The tentative test theory is formulated through language, and language is described using descriptive and argumentative function. The result of methods is tested using falsification technique based on separate discovery regularities. Josephson [13] explained that value of automatic intelligent data analysis process is evaluated based on four factors as (i) G_{Current}—goal of automatic intelligent data analysis process, (ii) T_{Result}—tentative test theory result, (iii) V_{Errors}—error correction approximate value, and (iv) G_{Next}—next goal based on new automatic intelligent data analysis process result. Finn [14] describes that JSM automatic intelligent data analysis process uses the concept of inference rules sorted by their distributive lattices pair relationship strength. The distributive lattices pair relationship strength is based on JSM automatic intelligent data process reasoning generation quality. The JSM automatic intelligent data analysis process is analog to human cognitive process steps that include induction, abduction, inference, and forecast (IAIF) as shown in Fig. 1.

Reference [15] explained that cognitive procedure scheme has significant improvement than earlier proposed schemes. The authors generate a new formalized

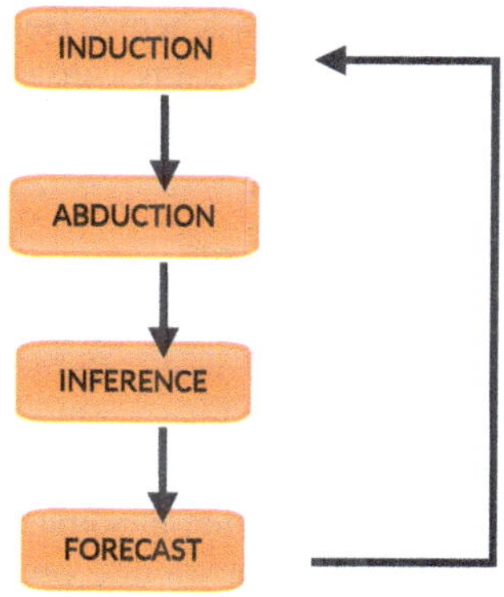

Fig. 1 JSM automatic intelligent data analysis process

algorithm by modifying the existing empirical induction technique. The JSM automatic intelligent data analysis process generates, analyzes, and corrects the empirical knowledge base. The JSM automatic intelligent data analysis processes generate and implement knowledge empirical discovery regularities hidden in expanded data sets.

B. **Bottleneck Support in Intelligent Data Analysis**

In this, we will discuss the solution of all four problems discussed above in Sect. 3 using JSM automatic intelligent data analysis process, namely (i) empirical regularities, (ii) data analysis tools combination, (iii) multi-model training sample operation, and (iv) data analysis result interpretation.

Problem Solution 1: First, empirical dependencies extension is generated through automatic intelligent data analysis process using classical machine learning. Then, interpolation–extrapolation scheme is used to search stable dependencies in extended data set. The searching of stable dependencies is based on empirical regularities. This collects a huge open data set (precedents) collection based on JSM automatic intelligent data analysis process fundamental characteristics.

Problem Solution 2: The different data analysis tools are combined and implemented based on various JSM automatic intelligent data analysis process strategies. For example, implementation of different plausible (Ψ_p) and deductive (Ψ_d) inference rules combination based on different distributive lattices. These combined automatic intelligent data analysis process reasoning tools generate and evaluate forecast dependency quality.

Problem Solution 3: The multi-model tools create many empirical dependencies problem. The selection of adequate language for special pre-processing problems uses automatic intelligent data analysis process based on classical machine learning. The selected language analyzes the homogenous sub-arrays causality based on initial empirical facts and implement for better result.

Problem Solution 4: The interpretability result is generated by causal dependencies using automatic intelligent data analysis process based on classical machine learning. The automatic intelligent data analysis process forecasts the generated result using abduction scheme. This controls the expandable element of initial training data set (precedents) using causal dependencies.

5 Conclusion

The above discussion shows that intelligent data analysis process performs better using classical machine learning method. The machine learning is the most popular and developed method used almost everywhere in all artificial intelligence based systems. However, in some areas, classical machine learning is not a best approach, e.g., (i) the knowledge presentation in the special problem based on collection theory and (ii) the reasoning formalization in cognitive procedure synthesis based on induction, abduction, analogy and reasoning.

Presently, intelligent data analysis methods implement classical machine learning which is used widely in artificial intelligence including infinite sub-areas. The classical machine learning has many implementations in different scientific and development research area. In this paper, we conclude that intelligent data analysis process became more efficient using classical machine learning process. This scenario changes the use of intelligent data analysis process in non-trivial reasoning. The advancement in machine learning provides an extended framework for better functional data set analyzation. The traditional interpolation and extrapolation scheme identifies new hidden empirical regularities in intelligent data analysis method with new rules or laws using classical machine learning.

References

1. S. Legg, M. Hutter, *A Collection of Definitions of Intelligence*. Technical Report—IDSIA, 15 June 2007. https://arxiv.org/pdf/0706.3639.pdf
2. I. Bogost, Artificial intelligence has become meaningless. *The Atlantic Monthly* (2017). https://www.theatlantic.com/technology/archive/2017/03/what-isartificialintelligence/518547/
3. S. Legg, M. Hutter, Universal intelligence: a definition of machine intelligence. Minds Mach. **17**(4):391–444 (2007). http://www.vetta.org/documents/legg-hutter-2007-universal-intelligence.pdf
4. K. Finn Victor, I. Zabezhailo Michael, Intelligent data analysis and machine learning: are they really equivalent concepts?, in *Second Russia and Pacific Conference on Computer Technology and Applications (RPC)* (2017)
5. V.K. Finn, J.S. Mill, Inductive methods in artificial intelligence systems. Sci. Techn. Inf. Process. Part I **38**(6), 385–402 (2011), Sci. Techn. Inf. Process. Part II **39**(5), 241–260 (2012)
6. O.P. Shesternikova, E.S. Pankratova, M.A. Agafonov, L.V. Vinokurova, V.K. Finn, Intelligent system for chronic pancreatitis patient's diabetes development prediction. Artif. Intell. Decis. Support **4**, 12–50 (2015). (in Russian)
7. A.Y. Volkova, Analyzing the data of different subject fields using the procedures of JSM method for automatic hypothesis generation. Autom. Doc. Math. Linguist. **45**(3), 127–139 (2011)
8. V.K. Finn, O.P. Shesternikova, On a new variant of generalized JSM method for automatic scientific researches support. Artif. Intell. Decis. Support **1**, 57–63 (2016). (in Russian)

9. M.A. Mikheenkova, A.Y. Volkova, Specification of the JSM intelligent system. Autom. Documentation Math. Linguist. **47**(4), 135–150 (2013)
10. M.I. Zabezhailo, V.K. Finn et al., Reasoning models for decision making: applications of JSM method for intelligent control systems, in *Architectures for Semiotic Modeling and Situation Analysis in Large Complex Systems—Proceedings of the Workshop of 10th IEEE Symposium on Intelligent Control* (1995), pp. 27–29
11. G. Polya, *Mathematics and Plausible Reasoning* (Princeton University Press, Princeton, NJ, 1954)
12. V.K. Finn, Distributive lattices of inductive JSM procedures. Autom. Documentation Math. Linguist. **48**(6), 65–295 (2014)
13. J.R. Josephson, Smart inductive generalizations are abductions, in *Abduction and Induction. Essays on their Relation and Integration*, ed. by P.A. Flach, A. Kakas (Kluwer Academic Publishers, Dordrecht, Boston, London, 2010), pp. 31–44
14. V.K. Finn, The synthesis of cognitive procedures and the problem of induction. Autom Documentation Math. Linguist. **43**(3), 149–195 (2010)
15. V.K. Finn, M.A. Mikheenkova, Cognitive sociology from the knowledge discovery point of view, in *Proceedings of 2nd Russia and Pacific Conference on Computer Technology and Applications* (IEEE RPC) (2017)

Author Index

S. S. Dash et al. (eds.), *Intelligent Computing and Applications*,
Advances in Intelligent Systems and Computing 1172,
https://doi.org/10.1007/978-981-15-5566-4

The manufacturer's authorised representative in the EU is Springer Nature Customer Service Centre GmbH, Europaplatz 3, 69115 Heidelberg, Germany. If you have any concerns regarding our products, please contact ProductSafety@springernature.com

Printed and bound by CPI Group (UK) Ltd, Croydon, CR0 4YY
30/09/2026
02225331-0008